Prealgebra

D0139148

FOURTH EDITION

Prealgebra

Richard N. Aufmann
Palomar College, California

Vernon C. Barker
Palomar College, California

Joanne S. Lockwood
Plymouth State University, New Hampshire

HOUGHTON MIFFLIN COMPANY
Boston   New York

Vice President and Publisher: Jack Shira
Senior Sponsoring Editor: Lynn Cox
Senior Development Editor: Dawn Nuttall
Assistant Editor: Melissa Parkin
Associate Project Editor: Kristin Penta
Editorial Assistant: Lisa C. Sullivan
Senior Production/Design Coordinator: Carol Merrigan
Manufacturing Manager: Florence Cadran
Senior Marketing Manager: Ben Rivera
Marketing Assistant: Lisa Lawler

Cover Photograph © by Don Gurewitz

Photo Credits p. 1 Spencer Grant/ PhotoEdit Inc.; p. 8 AP/ Wide World Photos; p. 16 CORBIS; p. 17 AP/ Wide World Photos; p. 20 Vince Streano/ CORBIS; p. 38 Dave Bartuff/ CORBIS p. 64 Wally McNamee/ CORBIS; p. 66 AFP/ CORBIS; p. 84 AP/ Wide World Photos; p. 87 Jean Miele/ CORBIS; p. 88 AP/ Wide World Photos; p. 99 Tannen Maury/ The Image Works; p. 100 David Young-Wolff/ PhotoEdit, Inc.; p. 115 DPA/ The ImageWorks; p. 127 AP/ Wide World Photos; p. 150 Bettmann/ CORBIS; p. 151 David Lassman/ Syracuse Newspapers/ The Image Works; p. 171 Bettmann/ CORBIS; p. 188 AP/ Wide World Photos; p. 190 Wally McNamee/ CORBIS; p. 206 Dennis MacDonald/ PhotoEdit, Inc.; p. 207 AP/ Wide World Photos; p. 208 Stephen Frank/ CORBIS; p. 232 AFP/ CORBIS; p. 237 AP/ Wide World Photos; p. 244 Bettmann/ CORBIS; p. 247 Frank Siteman/ PhotoEdit, Inc.; p. 272 AP/ Wide World Photos; p. 289 *Whirlpool and Waves at Naruto, Awa Province* by Hiroshige/ © Christie's Image/ CORBIS; p. 299 Paul Seheult/ Eye Ubiquitous/ CORBIS; p. 300 Bill Aron/ PhotoEdit, Inc.; p. 302 Stephen Ferry/ Getty Images; p. 313 AP/ Wide World Photos; p. 340 Richard Hamilton Smith/ CORBIS; p. 356 Reuters NewMedia Inc./ CORBIS; p. 363 CORBIS; p. 364 Tom Prettyman/ PhotoEdit, Inc.; p. 365 Reproduced by permission of The State Hermitage Museum, St. Petersburg, Russia/ CORBIS; p. 370 copyright PhotoDisc/ Getty Images; p. 375 David Young-Wolff/ PhotoEdit, Inc.; p. 376 Tony Freeman/ PhotoEdit, Inc.; p. 404 AP/ Wide World Photos; p. 435 Ezra Shaw/ Allsport/ Getty Images; p. 443 AP/ Wide World Photos; p. 452 AP/ Wide World Photos; p. 456 copyright 2003 PhotoDisc/ Getty Images; p. 478 Robert Brenner/ PhotoEdit, Inc.; p. 487 Helen King/ CORBIS; p. 503 Robert Brenner/ PhotoEdit, Inc.; p. 506 Reuters NewMedia Inc./ CORBIS; p. 511 Paul Conklin/ PhotoEdit Inc.; p. 518 Felicia Martinez/ PhotoEdit Inc; p. 520 David Young-Wolff/ PhotoEdit Inc.; p. 529 Yann Arthis-Bertrand/ CORBIS; p. 533 Getty Images; p. 547 AP/ Wide World Photos; p. 559 Images.com/ CORBIS; p. 562 Royalty-Free/ CORBIS; p. 567 Patrick Ward/ CORBIS; p. 607 Jerry Cooke/ CORBIS; p. 613 David Young-Wolff/ PhotoEdit, Inc; p. 623 Syracuse Newspapers/ The Image Works; p. 624 Steve Chenn/ CORBIS; p. 626 Cat Gwyn/ CORBIS; p. 634 Visuals Unlimited; p. 635 L. Clarke/ CORBIS.

Copyright © 2005 by Houghton Mifflin Company. All rights reserved.

No part of this work may be reproduced or transmitted in any form or by any means, electronic or mechanical, including photocopying and recording, or by any information storage or retrieval system without the prior written permission of Houghton Mifflin Company unless such copying is expressly permitted by federal copyright law. Address inquiries to College Permissions, Houghton Mifflin Company, 222 Berkeley Street, Boston MA 02116-3764.

Printed in the U.S.A.

Library of Congress Control Number: 20-03115486

ISBN Numbers:
Student Text: 0-618-37262-8
Instructor's Annotated Edition: 0-618-37263-6

89-WBC-07 06

Contents

Copyright © Houghton Mifflin Company. All rights reserved.

Applications

1 Whole Numbers   1

Applications

2 Integers   87

Applications

Copyright © Houghton Mifflin Company. All rights reserved.

Applications

Copyright © Houghton Mifflin Company. All rights reserved.

Copyright © Houghton Mifflin Company. All rights reserved.

Applications

Copyright © Houghton Mifflin Company. All rights reserved.

7   Measurement and Proportion   435

Copyright © Houghton Mifflin Company. All rights reserved.

Applications

10 Statistics and Probability   607

Prep Test   608

Copyright © Houghton Mifflin Company. All rights reserved.

Preface

The fourth edition of *Prealgebra* is designed to be a transition from the concrete aspects of arithmetic to the symbolic world of algebra. This text has been created to meet the needs of both the traditional college student and returning students whose mathematical proficiency may have declined during the years away from formal education.

In this new edition of *Prealgebra*, we have continued to integrate approaches suggested by AMATYC. There is an abundance of real sourced data in graphs and tables. Each chapter opens by illustrating and referencing a mathematical application within the chapter. At the end of each section there are Critical Thinking exercises and writing exercises. At the end of each chapter there is a "Focus on Problem Solving" that introduces students to various problem-solving strategies. This is followed by "Projects and Group Activities," which can be used for cooperative learning activities.

Besides the Index of Applications on the inside front cover, a chapter-by-chapter index of the variety of application problems in the text can be found in the Contents. This additional index highlights, in an easily accessible location, the importance and scope of the applications of mathematics.

One of the main challenges for students at this level is the ability to translate verbal phrases into mathematical expressions. One reason for this difficulty is that students are not exposed to verbal phrases until later in most texts. In *Prealgebra*, we introduce verbal phrases for operations as we introduce the operation. For instance, after addition concepts have been presented, we provide exercises which say "Find the sum of...." or "What is 6 more than 7?" In this way, students are constantly confronted with verbal phrases and must make a mathematical connection between the phrase and a mathematical operation.

NEW! Changes to this Edition

In response to user requests, we have expanded the coverage of subtraction of integers. The expanded explanations and additional examples in Section 2.2 will ensure greater success in student mastery of this concept.

Section 5.2, Objective A, where like terms are defined, now includes adding fractions with variables in the numerators. Also in Chapter 5, pages have been added to Section 5.3 on addition and subtraction of polynomials. The result is a more careful development of the concepts and terms associated with polynomials.

Users of the previous edition will notice that in Section 6.4, the number of exercises that require the student to translate sentences into equations has been increased.

Copyright © Houghton Mifflin Company. All rights reserved.

Section 6.5 in the previous edition has been separated into two sections for this revision. The topics presented now have expanded coverage in the exposition and additional exercises in the exercise sets. Also, new material has been added on (1) reading scatter diagrams and (2) reading graphs of equations of the form $y = mx + b$.

In this edition, calculator icons identifying calculator exercises are included in the student edition. (In the previous edition, they appeared only in the Instructor's Annotated Edition.) Additional Calculator Notes, which assist students in the operation of their calculators, have been included in the margins of the text.

Throughout the text, data problems were updated to reflect current data and trends. These application problems will demonstrate to students the variety of problems in real life that require mathematical analysis. Instructors will find that many of these problems may lead to interesting class discussions.

Another new feature of this edition is *AIM for Success*, which explains what is required of a student to be successful and how this text has been designed to foster student success. *AIM for Success* can be used as a lesson on the first day of class or as a project for students to complete to strengthen their study skills. There are suggestions for teaching this lesson in the Instructor's Resource Manual.

Related to *AIM for Success* are *Prep Tests,* which occur at the beginning of each chapter. These tests focus on the particular prerequisite skills that will be used in the upcoming chapter. The answers to these questions can be found in the Answer Appendix along with a reference (except for Chapter 1) to the objective from which the question was taken. Students who miss a question are encouraged to review the objective from which the question was taken.

The *Go Figure* problem that follows the *Prep Test* is a puzzle problem designed to engage students in problem solving.

The Chapter Summaries have been rewritten to be a more useful guide to students as they review for a test. A Chapter Summary includes Key Words and the Essential Rules and Procedures that were introduced in the chapter. Each key word and essential rule is accompanied by an example of that concept.

Copyright © Houghton Mifflin Company. All rights reserved.

Chapter Opening Features

Chapter Opener

Each chapter begins with a motivating chapter opener photo and a caption which illustrates and references a specific application from the chapter.

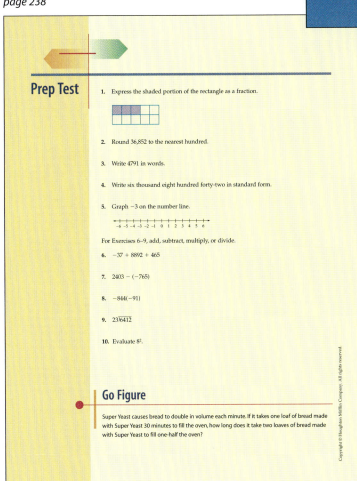

The **WEB** at the bottom of the page lets students know of additional online resources at **math.college.hmco.com/students**.

Objective-Specific Approach

Each chapter begins with a list of learning objectives which form the framework for a complete learning system. The objectives are woven throughout the text (i.e. exercises, Chapter Review Exercises, Chapter Test, Cumulative Review Exercises) as well as through the print and multimedia ancillaries. This results in a seamless learning system delivered in one consistent voice.

page 238

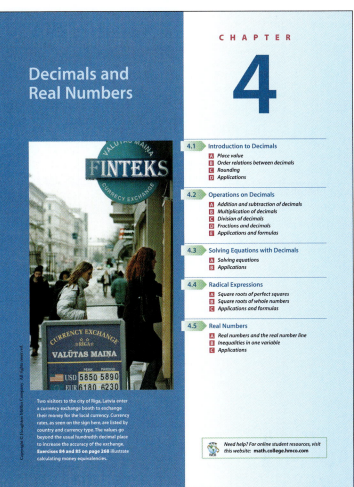

page 237

Two visitors to the city of Riga, Latvia enter a currency exchange booth to exchange their money for the local currency. Currency rates, as seen on the sign here, are listed by country and currency type. The values go beyond the usual hundredth decimal place to increase the accuracy of the exchange. **Exercises 84 and 85 on page 268** illustrate calculating money equivalencies.

Need help? For online student resources, visit this website: math.college.hmco.com

NEW! Prep Test and Go Figure

Prep Tests occur at the beginning of each chapter and test students on previously covered concepts that are required in the coming chapter. Answers are provided in the Answer Appendix. Objective references are also provided if a student needs to review specific concepts.

The **Go Figure** problem that follows the *Prep Test* is a puzzle problem designed to engage students in problem solving.

Prep Test

1. Express the shaded portion of the rectangle as a fraction.

2. Round 36,852 to the nearest hundred.

3. Write 4791 in words.

4. Write six thousand eight hundred forty-two in standard form.

5. Graph -3 on the number line.

For Exercises 6–9, add, subtract, multiply, or divide.

6. $-37 + 8892 + 465$

7. $2403 - (-765)$

8. $-844(-91)$

9. $23\overline{)6412}$

10. Evaluate 8^2.

Go Figure

Super Yeast causes bread to double in volume each minute. If it takes one loaf of bread made with Super Yeast 30 minutes to fill the oven, how long does it take two loaves of bread made with Super Yeast to fill one-half the oven?

Copyright © Houghton Mifflin Company. All rights reserved.

Copyright © Houghton Mifflin Company. All rights reserved.

XV

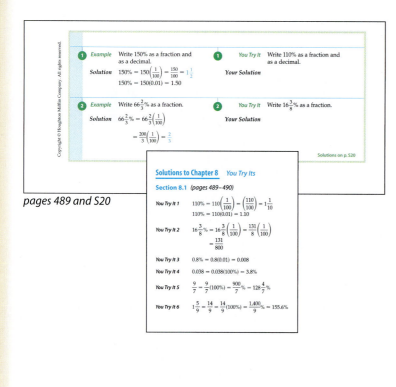

pages 489 and S20

Aufmann Interactive Method (AIM)

An Interactive Approach

Prealgebra uses an interactive style that provides a students with an opportunity to try a skill as it is presented. Each section is divided into objectives, and every objective contains one or more sets of matched-pair examples. The first example in each set is worked out; the second example, called "You Try It," is for the student to work. By solving this problem, students actively practice concepts as they are presented in the text.

There are complete worked-out solutions to these examples in an appendix. By comparing their solutions to the solutions in the appendix, students obtain immediate feedback on, and reinforcement of, the concepts.

page xxv

NEW! AIM for Success Student Preface

This new student 'how to use this book' preface explains what is required of a student to be successful and how this text has been designed to foster student success, including the Aufmann Interactive Method (AIM). *AIM for Success* can be used as a lesson on the first day of class or as a project for students to complete to strengthen their study skills. There are suggestions for teaching this lesson in the Instructor's Resource Manual.

AIM for Success

Welcome to *Prealgebra*. As you begin this course we know two important facts: (1) We want you to succeed. (2) You want to succeed. To do that requires an effort from each of us. For the next few pages, we are going to show you what is required of you to achieve that success and how you can use the features of this text to be successful.

Motivation

One of the most important keys to success is motivation. We can try to motivate you by offering interesting or important ways mathematics can benefit you. But, in the end, the motivation must come from you. On the first day of class it is easy to be motivated. Eight weeks into the term, it is harder to keep that motivation.

Take Note
Motivation alone will not lead to success. For instance, suppose a person who cannot swim is placed in a boat, taken out to the middle of a lake, and then thrown overboard. That person has a lot of motivation to swim but there is a high likelihood the person will drown without some help. Motivation gives us the desire to learn but is not the same as learning.

To stay motivated, there must be outcomes from this course that are worth your time, money, and energy. List some reasons you are taking this course. Do not make a mental list—actually write them out.

_____

_____

_____

Although we hope that one of the reasons you listed was an interest in mathematics, we know that many of you are taking this course because it is required to graduate, it is a prerequisite for a course you must take, or because it is required for your major. Although you may not agree that this course should be necessary, it is! If you are motivated to graduate or complete the requirements for your major, then use that motivation to succeed in this course. Do not become distracted from your goal to complete your education!

Commitment

To be successful, you must make a commitment to succeed. This means devoting time to math so that you achieve a better understanding of the subject.

List some activities (sports, hobbies, talents such as dance, art, or music) that you enjoy and at which you would like to become better.

ACTIVITY	TIME SPENT	TIME WISHED SPENT

xxv

Copyright © Houghton Mifflin Company. All rights reserved.

Problem Solving

Focus on Problem Solving

At the end of each chapter is a "Focus on Problem Solving" feature which introduces the student to various successful problem-solving strategies. Strategies such as drawing a diagram, applying solutions to other problems, looking for a pattern, making a table, and trial and error are some of the techniques that are demonstrated.

page 497

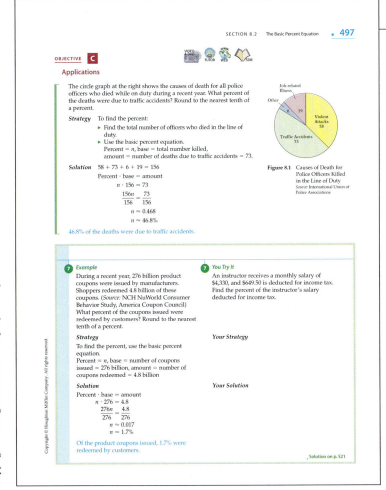

page 77

Problem-Solving Strategies

The text features a carefully developed approach to problem solving that emphasizes the importance of *strategy* when solving problems. Students are encouraged to develop their own strategies—to draw diagrams; to write out the solution steps in words—as part of their solution to a problem. In each case, model strategies are presented as guides for students to follow as they attempt the "You Try It" problem. Having students provide strategies is a natural way to incorporate writing into the math curriculum.

Copyright © Houghton Mifflin Company. All rights reserved.

Real Data and Applications

Copyright © Houghton Mifflin Company. All rights reserved.

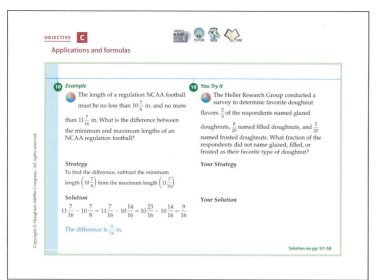

page 181

Applications

One way to motivate an interest in mathematics is through applications. Wherever appropriate, the last objective of a section presents applications that require the student to use problem-solving strategies, along with the skills covered in that section, to solve practical problems. This carefully integrated applied approach generates student awareness of the value of algebra as a real-life tool.

Applications are taken from many disciplines, including agriculture, business, carpentry, chemistry, construction, Earth science, education, manufacturing, nutrition, real estate, and sports.

page 501

Real Data

Real data examples and exercises, identified by 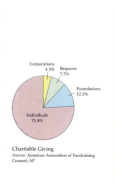, ask students to analyze and solve problems taken from actual situations. Students are often required to work with tables, graphs, and charts drawn from a variety of disciplines.

49. *Fire Science* A fire department received 24 false alarms out of a total of 200 alarms received. What percent of the alarms received were false alarms?

50. *Demographics* The table at the right shows the projected increase in population from 2000 to 2040 for each of four counties in the Central Valley of California. What percent of the 2000 population of Sacramento County is the increase in population?

County	2000 Population	Projected Increase by 2040
Sacramento	1,200,000	900,000
Kern	651,700	948,300
Fresno	794,200	705,800
San Joaquin	562,000	737,400

Source: California Department of Finance

51. *Demographics* The table at the right shows the projected increase in population from 2000 to 2040 for each of four counties in the Central Valley of California. What percent of the 2000 population of Kern County is the increase in population? Round to the nearest tenth of a percent.

52. *Business* An antiques shop owner expects to receive $16\frac{2}{3}$% of the shop's sales as profit. What is the expected profit in a month when the total sales are $24,000?

53. *Poultry* In a recent year, North Carolina produced 1,300,000 lb of turkey. This was 18.6% of the U.S. total in that year. (*Source:* U.S. Census Bureau) Calculate the U.S. total turkey production for that year. Round to the nearest million.

54. *Depreciation* A used mobile home was purchased for $43,600. This amount was 64% of the cost of the mobile home when it was new. What was the new mobile home cost?

55. *Agriculture* A farmer is given an income tax credit of 15% of the cost of some farm machinery. What tax credit would the farmer receive on farm equipment that cost $85,000?

56. *Financing* A used car is sold for $18,900. The buyer of the car makes a down payment of $3,780. What percent of the selling price is the down payment?

57. *Medicine* The active ingredient in a prescription skin cream is clobetasol propionate. It is 0.05% of the total ingredients. How many grams of clobetasol propionate are in a 30-gram tube of this cream?

58. *Charitable Giving* In a recent year, Americans gave $212 billion to charities. Use the figure at the right to determine how much of that amount came from individuals.

59. *Astronomy* The diameter of Earth is approximately 8,000 mi, and the diameter of the sun is approximately 870,000 mi. What percent of Earth's diameter is the sun's diameter?

Corporations 4.3%
Bequests 7.7%
Foundations 12.2%
Individuals 75.8%

Charitable Giving
Sources: American Association of Fundraising Counsel; AP

Copyright © Houghton Mifflin Company. All rights reserved.

Copyright © Houghton Mifflin Company. All rights reserved.

Student Pedagogy

Icons

The , TUTOR, WEB, SSM, icons at each objective head remind students of the many and varied additional resources available for each objective.

Key Words and Concepts

Key words, in bold, emphasize important terms. **Key concepts** are presented in green boxes in order to highlight these important concepts and to provide for easy reference.

Take Note

These margin notes alert students to a point requiring special attention or are used to amplify the concept under discussion.

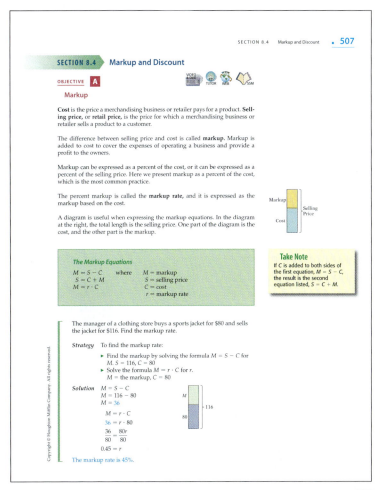

page 507

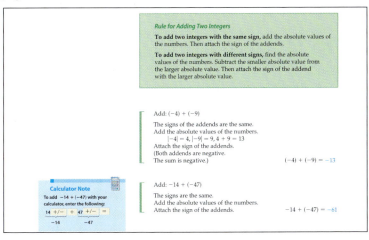

page 102

Calculator Note

These margin notes provide suggestions for using a calculator in certain situations.

Instructional Examples

Examples indicated by green brackets use explanatory comments to describe what is happening in key steps of the complete, worked-out solutions.

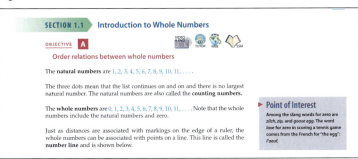

page 3

Point of Interest

These margin notes contain interesting sidelights about mathematics, its history, or its application.

Copyright © Houghton Mifflin Company. All rights reserved.

Exercises and Projects

Exercises

The exercise sets of *Prealgebra* emphasize skill building, skill maintenance, and applications. Concept-based writing or developmental exercises have been integrated with the exercise sets.

Icons identify appropriate writing ,

data analysis , and calculator exercises.

Included in each exercise set are **Critical Thinking exercises** that present extensions of topics, require analysis, or offer challenge problems. The writing exercises ask students to explain answers, write about a topic in the section, or research and report on a related topic.

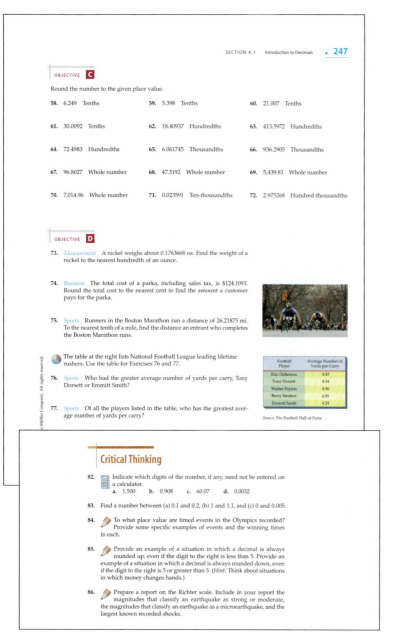

OBJECTIVE C

Round the number to the given place value.

58. 6.249 Tenths   59. 5.398 Tenths   60. 21.007 Tenths

61. 30.0092 Tenths   62. 18.40937 Hundredths   63. 413.5972 Hundredths

64. 72.4983 Hundredths   65. 6.061745 Thousandths   66. 936.2905 Thousandths

67. 96.8027 Whole number   68. 47.3192 Whole number   69. 5,439.83 Whole number

70. 7,014.96 Whole number   71. 0.023591 Ten-thousandths   72. 2.975268 Hundred-thousandths

OBJECTIVE D

73. *Measurement* A nickel weighs about 0.1763668 oz. Find the weight of a nickel to the nearest hundredth of an ounce.

74. *Business* The total cost of a parka, including sales tax, is $124.1093. Round the total cost to the nearest cent to find the amount a customer pays for the parka.

75. *Sports* Runners in the Boston Marathon run a distance of 26.21875 mi. To the nearest tenth of a mile, find the distance an entrant who completes the Boston Marathon runs.

The table at the right lists National Football League leading lifetime rushers. Use the table for Exercises 76 and 77.

76. *Sports* Who had the greater average number of yards per carry, Tony Dorsett or Emmitt Smith?

77. *Sports* Of all the players listed in the table, who has the greatest average number of yards per carry?

Football Player	Average Number of Yards per Carry
Eric Dickerson	4.43
Tony Dorsett	4.34
Walter Payton	4.36
Barry Sanders	4.99
Emmitt Smith	4.24

Source: Pro Football Hall of Fame

Critical Thinking

82. Indicate which digits of the number, if any, need not be entered on a calculator.
   a. 1.500   b. 0.908   c. 60.07   d. 0.0032

83. Find a number between (a) 0.1 and 0.2, (b) 1 and 1.1, and (c) 0 and 0.005.

84. To what place value are timed events in the Olympics recorded? Provide some specific examples of events and the winning times in each.

85. Provide an example of a situation in which a decimal is always rounded up, even if the digit to the right is less than 5. Provide an example of a situation in which a decimal is always rounded down, even if the digit to the right is 5 or greater than 5. (*Hint:* Think about situations in which money changes hands.)

86. Prepare a report on the Richter scale. Include in your report the magnitudes that classify an earthquake as strong or moderate, the magnitudes that classify an earthquake as a microearthquake, and the largest known recorded shocks.

Houghton Mifflin Company. All rights reserved.

pages 247 and 248

page 226

Projects and Group Activities

The Projects and Group Activities feature at the end of each chapter can be used as extra credit or for cooperative learning activities. The projects cover various aspects and applications of mathematics, including using patterns, earned run averages, buying a car, preparing a circle graph, and random samples.

Projects & Group Activities

Music In musical notation, notes are printed on a **staff**, which is a set of five horizontal lines and the spaces between them. The notes of a musical composition are grouped into **measures**, or **bars**. Vertical lines separate measures on a staff. The shape of a note indicates how long it should be held. The whole note has the longest time value of all notes. Each time value is divided by two in order to find the next smallest note value.

The **time signature** is a fraction that appears at the beginning of a piece of music. The numerator of the fraction indicates the number of beats in a measure. The denominator indicates what kind of note receives one beat. For example, music written in $\frac{2}{4}$ time has 2 beats to a measure, and a quarter note receives one beat. One measure in $\frac{2}{4}$ time may have 1 half note, 2 quarter notes, 4 eighth notes, or any other combination of notes totaling 2 beats. Other common time signatures include $\frac{4}{4}$, $\frac{3}{4}$, and $\frac{6}{8}$.

Copyright © Houghton Mifflin Company. All rights reserved.

End of Chapter

Chapter Summary

At the end of each chapter there is a Chapter Summary that includes Key Words and Essential Rules and Procedures that were covered in the chapter. These chapter summaries provide a single point of reference as the student prepares for a test. Each concept references the objective and the page number from the lesson where the concept is introduced.

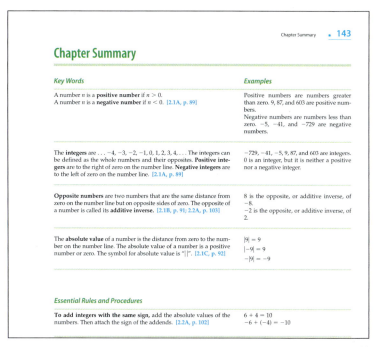

page 143

page 145

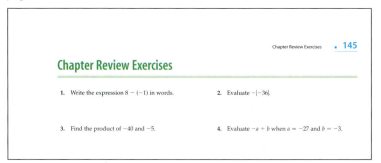

Chapter Review Exercises

Review exercises are found at the end of each chapter. These exercises are selected to help the student integrate all of the topics presented in the chapter.

page 147

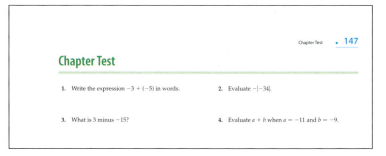

Chapter Test

The Chapter Test exercises are designed to simulate a possible test of the material in the chapter.

page 149

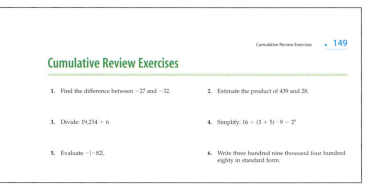

Cumulative Review Exercises

Cumulative Review Exercises, which appear at the end of each chapter (beginning with Chapter 2), help students maintain skills learned in previous chapters.

The answers to all Chapter Review Exercises, all Chapter Test exercises, and all Cumulative Review Exercises are given in the Answer Section. Along with the answer, there is a reference to the objective that pertains to each exercise.

Copyright © Houghton Mifflin Company. All rights reserved.

INSTRUCTOR RESOURCES

Prealgebra, 4e has a complete set of teaching aids for the instructor.

Instructor's Annotated Edition   This edition contains a replica of the student text and additional items just for the instructor. Answers to all exercises are provided.

Instructor's Solutions Manual with Testing   The *Instructor's Solutions Manual* contains worked-out solutions for all exercises in the text. This resource also includes ready-to-use printed Chapter Tests, cumulative tests, and final exams, as well as a print out of one example of each of the algorithmic items in *HM Testing*. A lesson plan for the *AIM for Success* and suggested course sequences are also provided in the *Instructor's Solutions Manual with Testing*.

HM ClassPrep with HM Testing CD-ROM   *HM ClassPrep* contains a multitude of text-specific resources for instructors to use to enhance the classroom experience. These resources can be easily accessed by chapter or resource type and can also link you to the text's website. *HM Testing* is our computerized test generator and contains a database of algorithmic test items as well as providing **online testing** and **gradebook** functions.

Instructor Text-specific website   The resources available on the *Class Prep CD* are also available on the instructor website at math.college.hmco.com/instructors. Appropriate items are password protected. Instructors also have access to the student part of the text's website.

STUDENT RESOURCES

Student Solutions Manual   The *Student Solutions Manual* contains complete solutions to all odd-numbered exercises in the text.

Math Study Skills Workbook by Paul D. Nolting   This workbook is designed to reinforce skills and minimize frustration for students in any math class, lab, or study skills course. It offers a wealth of study tips and sound advice on note taking, time management, and reducing math anxiety. In addition, numerous opportunities for self-assessment enable students to track their own progress.

HM Eduspace® online learning environment   *Eduspace®* is a text-specific online learning environment which combines an algorithmic tutorial program with homework capabilities. Specific content is available 24 hours a day to help you further understand your textbook.

SMARTHINKING™ live, online tutoring   Houghton Mifflin has partnered with SMARTHINKING™ to provide an easy-to-use and effective online tutorial service. **Whiteboard Simulations** and **Practice Area** promote real-time visual interaction.

Three levels of service are offered.

- **Text-specific Tutoring** provides real-time, one-on-one instruction with a specially qualified 'e-structor.'
- **Questions Any Time** allows students to submit questions to the tutor outside the scheduled hours and receive a reply within 24 hours.
- **Independent Study Resources** connect students with around-the-clock access to additional educational services, including interactive websites, diagnostic tests, and Frequently Asked Questions posed to SMARTHINKING™ e-structors.

Copyright © Houghton Mifflin Company. All rights reserved.

HM mathSpace® Student Tutorial CD-ROM   This tutorial algorithmically generates exercises and provides step-by-step solutions.

Houghton Mifflin Instructional Videos and DVDs   This text offers text-specific videos and DVDs, hosted by Dana Mosely, covering all sections of the text and providing a valuable resource for further instruction and review. Next to every objective head, the serves as a reminder that the objective is covered in a video/DVD lesson.

Student Text-specific website   Online student resources can be found at this text's website at math.college.hmco.com/students.

ACKNOWLEDGMENTS

The authors would like to thank the people who have reviewed this manuscript and provided many valuable suggestions.

Anita G. Aikman, *Collin County Community College District-Spring Creek Campus, TX*

Karen Bingham, *Clarion University of Pennsylvania, PA*

Ann Davis, *Northeastern Technical College, SC*

Dennis Donohue, *Community College of Southern Nevada, NV*

James Eckerman, *American River College, NE*

David French, *Tidewater Community College, CA*

Elizabeth Hamman, *Cypress College, CA*

Anne Haney

Sara Iaikam, *Dekalb Technical College, GA*

Kristi Korensky, *Central Community College, NE*

Angelia Reynolds, *Gulf Coast Community College, FL*

Neal L. Rogers, *Santa Ana College, CA*

Lauri Semarne

Dwight Siverson, *Colorado Northwestern Community College, CO*

Danny Whited, *Virginia Intermont, VA*

Special thanks to Christi Verity for her diligent preparation of the solutions manuals and for her contribution to the accuracy of the textbooks.

Copyright © Houghton Mifflin Company. All rights reserved.

AIM for Success

Welcome to *Prealgebra*. As you begin this course we know two important facts: (1) We want you to succeed. (2) You want to succeed. To do that requires an effort from each of us. For the next few pages, we are going to show you what is required of you to achieve that success and how you can use the features of this text to be successful.

Motivation

One of the most important keys to success is motivation. We can try to motivate you by offering interesting or important ways mathematics can benefit you. But, in the end, the motivation must come from you. On the first day of class it is easy to be motivated. Eight weeks into the term, it is harder to keep that motivation.

To stay motivated, there must be outcomes from this course that are worth your time, money, and energy. List some reasons you are taking this course. Do not make a mental list—actually write them out.

_____

_____

_____

Although we hope that one of the reasons you listed was an interest in mathematics, we know that many of you are taking this course because it is required to graduate, it is a prerequisite for a course you must take, or because it is required for your major. Although you may not agree that this course should be necessary, it is! If you are motivated to graduate or complete the requirements for your major, then use that motivation to succeed in this course. Do not become distracted from your goal to complete your education!

Take Note

Motivation alone will not lead to success. For instance, suppose a person who cannot swim is placed in a boat, taken out to the middle of a lake, and then thrown overboard. That person has a lot of motivation to swim but there is a high likelihood the person will drown without some help. Motivation gives us the desire to learn but is not the same as learning.

Commitment

To be successful, you must make a commitment to succeed. This means devoting time to math so that you achieve a better understanding of the subject.

List some activities (sports, hobbies, talents such as dance, art, or music) that you enjoy and at which you would like to become better.

ACTIVITY	TIME SPENT	TIME WISHED SPENT
_____	_____	_____
_____	_____	_____
_____	_____	_____

Copyright © Houghton Mifflin Company. All rights reserved.

Thinking about these activities, put the number of hours that you spend each week practicing these activities next to the activity. Next to that number, indicate the number of hours per week you would like to spend on these activities.

Whether you listed surfing or sailing, aerobics or restoring cars, or any other activity you enjoy, note how many hours a week you spend doing it. To succeed in math, you must be willing to commit the same amount of time. Success requires some sacrifice.

The "I Can't Do Math" Syndrome

You can do math! When you first learned the activities you listed above, you probably could not do them well. With practice, you got better. With practice, you will be better at math. Stay focused, motivated, and committed to success.

It is difficult for us to emphasize how important it is to overcome the "I Can't Do Math Syndrome." If you listen to interviews of very successful athletes after a particularly bad performance, you will note that they focus on the positive aspect of what they did, not the negative. Sports psychologists encourage athletes to always be positive—to have a "Can Do" attitude. Develop this attitude toward math.

Strategies for Success

Textbook Reconnaissance   Right now, do a 15-minute "textbook reconnaissance" of this book. Here's how:

First, read the table of contents. Do it in three minutes or less. Next, look through the entire book, page by page. Move quickly. Scan titles, look at pictures, notice diagrams.

A textbook reconnaissance shows you where a course is going. It gives you the big picture. That's useful because brains work best when going from the general to the specific. Getting the big picture before you start makes details easier to recall and understand later on.

Your textbook reconnaissance will work even better if, as you scan, you look for ideas or topics that are interesting to you. List three facts, topics, or problems that you found interesting during your textbook reconnaissance.

_____

_____

_____

The idea behind this technique is simple: It's easier to work at learning material if you know it's going to be useful to you.

Not all the topics in this book will be "interesting" to you. But that is true of any subject. Surfers find that on some days the waves are better than others, musicians find some music more appealing than other music, computer gamers find some computer games more interesting than others, car enthusiasts find some cars more exciting than others. Some car enthusiasts would prefer to have a completely restored 1957 Chevrolet than a new Ferrari.

Copyright © Houghton Mifflin Company. All rights reserved.

Know the Course Requirements   To do your best in this course, you must know exactly what your instructor requires. Course requirements may be stated in a *syllabus,* which is a printed outline of the main topics of the course, or they may be presented orally. When they are listed in a syllabus or on other printed pages, keep them in a safe place. When they are presented orally, make sure to take complete notes. In either case, it is important that you understand them completely and follow them exactly. Be sure you know the answer to each of the following questions.

1. What is your instructor's name?

2. Where is your instructor's office?

3. At what times does your instructor hold office hours?

4. Besides the textbook, what other materials does your instructor require?

5. What is your instructor's attendance policy?

6. If you must be absent from a class meeting, what should you do before returning to class? What should you do when you return to class?

7. What is the instructor's policy regarding collection or grading of homework assignments?

8. What options are available if you are having difficulty with an assignment? Is there a math tutoring center?

9. If there is a math lab at your school, where is it located? What hours is it open?

10. What is the instructor's policy if you miss a quiz?

11. What is the instructor's policy if you miss an exam?

12. Where can you get help when studying for an exam?

Remember: Your instructor wants to see you succeed. If you need help, ask! Do not fall behind.

Time Management   We know that there are demands on your time. Family, work, friends, and entertainment all compete for your time. We do not want to see you receive poor job evaluations because you are studying math. However, it is also true that we do not want to see you receive poor math test scores because you devoted too much time to work. When several competing and important tasks require your time and energy, the only way to manage the stress of being successful at both is to manage your time efficiently.

Instructors often advise students to spend twice the amount of time outside of class studying as they spend in the classroom. Time management is important if you are to accomplish this goal and succeed in school. The following activity is intended to help you structure your time more efficiently.

Take Note

Besides time management, there must be realistic ideas of how much time is available. There are very few people who can *successfully* work full-time and go to school full-time. If you work 40 hours a week, take 15 units, spend the recommended study time given at the right, and sleep 8 hours a day, you will use over 80% of the available hours in a week. That leaves less than 20% of the hours in a week for family, friends, eating, recreation, and other activities.

Copyright © Houghton Mifflin Company. All rights reserved.

Make a copy of the table below and list the name of each course you are taking this term, the number of class hours each course meets, and the number of hours you should spend studying each subject outside of class. Then fill in a weekly schedule like the one printed below. Begin by writing in the hours spent in your classes, the hours spent at work (if you have a job), and any other commitments that are not flexible with respect to the time that you do them. Then begin to write down the commitments that are more flexible, including hours spent studying. Remember to reserve time for activities such as meals and exercise. You should also schedule free time.

	Monday	Tuesday	Wednesday	Thursday	Friday	Saturday	Sunday
7 - 8 a.m.							
8 - 9 a.m.							
9 - 10 a.m.							
10 - 11 a.m.							
11 - 12 p.m.							
12 - 1 p.m.							
1 - 2 p.m.							
2 - 3 p.m.							
3 - 4 p.m.							
4 - 5 p.m.							
5 - 6 p.m.							
6 - 7 p.m.							
7 - 8 p.m.							
8 - 9 p.m.							
9 - 10 p.m.							
10 - 11 p.m.							
11 - 12 a.m.							

We know that many of you must work. If that is the case, realize that working 10 hours a week at a part-time job is equivalent to taking a three-unit class. If you must work, consider letting your education progress at a slower rate to allow you to be successful at both work and school. There is no rule that says you must finish school in a certain time frame.

Schedule Study Time   As we encouraged you to do by filling out the time management form above, schedule a certain time to study. You should think of this time like being at work or in class. Reasons for missing "study time" should be as compelling as reasons for missing work or class. "I just don't feel like it" is not a good reason to miss your scheduled study time. Although this may seem like an obvious exercise, list a few reasons you might want to study.

_____

_____

_____

Of course we have no way of knowing the reasons you listed but from our experience one reason given quite frequently is "To pass the course." There is nothing wrong with that reason. If that is the most important reason for you to study, then use it to stay focused.

One method of keeping to a study schedule is to form a *study group.* Look for people who are committed to learning, who pay attention in class, and who

Copyright © Houghton Mifflin Company. All rights reserved.

are punctual. Ask them to join your group. Choose people with similar educational goals but different methods of learning. You can gain from seeing the material from a new perspective. Limit groups to four or five people; larger groups are unwieldy.

There are many ways to conduct a *study group.* Begin with the following suggestions and see what works best for your group.

1. Test each other by asking questions. Each group member might bring two or three sample test questions to each meeting.

2. Practice teaching each other. Many of us who are teachers learned a lot about our subject when we had to explain it to someone else.

3. Compare class notes. You might ask other students about material in your notes that is difficult for you to understand.

4. Brainstorm test questions.

5. Set an agenda for each meeting. Set approximate time limits for each agenda item and determine a quitting time.

And now, probably the most important aspect of studying is that it should be done in relatively small chunks. If you can only study three hours a week for this course (probably not enough for most people), do it in blocks of one hour on three separate days, preferably after class. Three hours of studying on a Sunday is not as productive as three hours of paced study.

Text Features That Promote Success

There are 10 chapters in this text. Each chapter is divided into sections, and each section is subdivided into learning objectives. Each learning objective is labeled with a letter from A to E.

Preparing for a Chapter　Before you begin a new chapter, you should take some time to review previously learned skills. There are two ways to do this. The first is to complete the **Cumulative Review Exercises** which occurs after every chapter (except Chapter 1). For instance, turn to page 311. The questions in this review are taken from the previous chapters. The answers for all these exercises can be found on page A8. Turn to that page now and locate the answers for the Chapter 4 Cumulative Review Exercises. After the answer to the first exercise, which is 0.03879, you will see the objective reference [4.2C]. This means that this question was taken from Chapter 4, Section 2, Objective C. If you missed this question, you should return to that objective and restudy the material.

A second way of preparing for a new chapter is to complete the **Prep Test.** This test focuses on the particular skills that will be required for the new chapter. Turn to page 238 to see a Prep Test. The answers for the Prep Test are the first set of answers in the answer section for a chapter. Turn to page A6 to see the answers for the Chapter 4 Prep Test. If you answer a question incorrectly, restudy the objective from which the question was taken.

Before the class meeting in which your professor begins a new section, you should read each objective statement for that section. Next, browse through the objective material, being sure to note each word in bold type. These words indicate important concepts that you must know to learn the material. Do not worry about trying to understand all the material. Your professor is there to assist you with that endeavor. The purpose of browsing through the material is so that your brain will be prepared to accept and organize the new information when it is presented to you.

Copyright © Houghton Mifflin Company. All rights reserved.

Turn to page 3. Write down the title of the first objective in Section 1.1. Under the title of the objective, write down the words in the objective that are in bold print. It is not necessary for you to understand the meaning of these words. You are in this class to learn their meaning.

_____   _____   _____   _____

_____   _____   _____   _____

_____   _____   _____   _____

Math is Not a Spectator Sport   To learn mathematics you must be an active participant. Listening and watching your professor do mathematics is not enough. Mathematics requires that you interact with the lesson you are studying. If you filled in the blanks above, you were being interactive. There are other ways this textbook has been designed so that you can be an active learner.

Instructional Examples   Green brackets indicate an instructional example with explanatory comments for some of the steps. Using paper and pencil, you should work along as you read through the example. When you complete the example, get a clean sheet of paper. Write down the problem and then try to complete the solution without referring to your notes or the book. When you can do that, move on to the next part of the section.

page 104

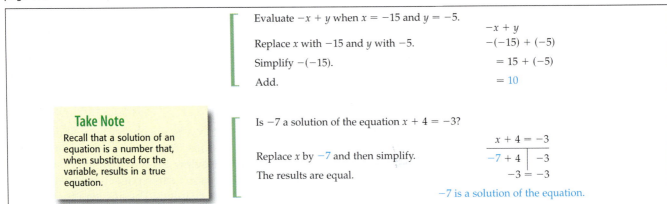

Evaluate $-x + y$ when $x = -15$ and $y = -5$.

Replace x with -15 and y with -5.

$$-x + y$$
$$-(-15) + (-5)$$

Simplify $-(-15)$.

$$= 15 + (-5)$$

Add.

$$= 10$$

Take Note

Recall that a solution of an equation is a number that, when substituted for the variable, results in a true equation.

Is -7 a solution of the equation $x + 4 = -3$?

$$x + 4 = -3$$

Replace x by -7 and then simplify.

$$-7 + 4 \mid -3$$

The results are equal.

$$-3 = -3$$

-7 is a solution of the equation.

Leaf through the book now and write down the page numbers of two other occurrences of these types of examples.

Example/You Try It Pairs   One of the key instructional features of this text is the Example/You Try It pairs.

page 180

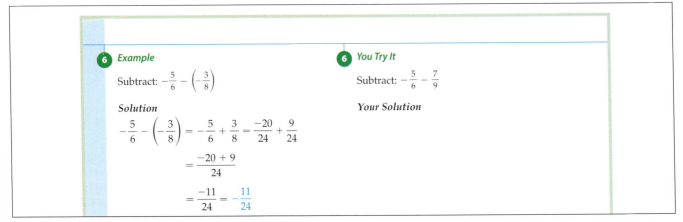

6 Example

Subtract: $-\dfrac{5}{6} - \left(-\dfrac{3}{8}\right)$

Solution

$$-\frac{5}{6} - \left(-\frac{3}{8}\right) = -\frac{5}{6} + \frac{3}{8} = \frac{-20}{24} + \frac{9}{24}$$

$$= \frac{-20 + 9}{24}$$

$$= \frac{-11}{24} = -\frac{11}{24}$$

6 You Try It

Subtract: $-\dfrac{5}{6} - \dfrac{7}{9}$

Your Solution

Copyright © Houghton Mifflin Company. All rights reserved.

Note that each Example is completely worked. Study the worked-out example carefully by working through each step. Then complete the You Try It, which is a problem similar to the Example. If you get stuck, refer to the page number following the Problem which directs you to the page on which the You Try It is solved—a complete worked-out solution is provided. Try to use the given solution to get a hint for the step you are stuck on. Then try to complete your solution.

When you have completed the solution, check your work against the solution we provided. Turn to page S7 to see the solution to You Try It 6.

Be aware that frequently there is more than one way to solve a problem. Your answer, however, should be the same as the given answer. If you have any question as to whether your method will "always work," check with your instructor or with someone in the math center.

Browse through the textbook and write down the page numbers where two other paired Example/Problem features occur.

Remember: Be an active participant in your learning process. When you are sitting in class watching and listening to an explanation, you may think that you understand. However, until you actually try to do it, you will have no confirmation of the new knowledge or skill. Most of us have had the experience of sitting in class thinking we knew how to do something only to get home and realize we didn't.

Word Problems   Word problems are difficult because we must read the problem, determine the quantity we must find, think of a method to find it, actually solve the problem, and then check the answer. In short, we must devise a *strategy* and then use that strategy to find the *solution*. Remember to check your work.

Note in the example below that solving a word problem includes stating the strategy and using the strategy to find a solution. If you have difficulty with a word problem, write down the known information. Be very specific. Write out a phrase or sentence that states what you are trying to find. Ask yourself whether there are known formulas that relate the known and unknown quantities. Do not ignore the word problems. They are an important part of mathematics.

Take Note

There is a strong connection between reading and being a successful student in math or any other subject. If you have difficulty reading, consider taking a reading course. Reading is much like other skills. There are certain things you can learn that will make you a better reader.

page 122

 Example

The daily low temperatures during one week were recorded as follows: $-10°, 2°, -1°, -9°, 1°, 0°, 3°$. Find the average daily low temperature for the week.

Strategy

To find the average daily low temperature:

▸ Add the seven temperature readings.
▸ Divide by 7.

Solution

$-10 + 2 + (-1) + (-9) + 1 + 0 + 3 = -14$

$-14 \div 7 = -2$

The average daily low temperature was $-2°$.

 You Try It

The daily high temperatures during one week were recorded as follows: $-7°, -8°, 0°, -1°, -6°, -11°, -2°$. Find the average daily high temperature for the week.

Your Strategy

Your Solution

Solution on p. S5

Copyright © Houghton Mifflin Company. All rights reserved.

Rule Boxes   Pay special attention to rules placed in boxes. These rules give you the reasons certain types of problems are solved the way they are. When you see a rule, try to rewrite the rule in your own words.

page 102

> ### *Rule for Adding Two Integers*
>
> **To add two integers with the same sign,** add the absolute values of the numbers. Then attach the sign of the addends.
>
> **To add two integers with different signs,** find the absolute values of the numbers. Subtract the smaller absolute value from the larger absolute value. Then attach the sign of the addend with the larger absolute value.

Find and write down two page numbers on which there are examples of rule boxes.

Take Note

If you are working at home and need assistance, there is online help available at math.college.hmco.com/students, at this text's website.

Chapter Exercises   When you have completed studying a section, do the exercises in the exercise set that correspond with that section. Math is a subject that needs to be learned in small sections and practiced continually in order to be mastered. Doing all of the exercises in each exercise set will help you to master the problem-solving techniques necessary for success. As you work through the exercises for a section, check your answers to the odd-numbered exercises with those in the back of the book.

Preparing for a Test   There are important features of this text that can be used to prepare for a test.

- Chapter Summary
- Chapter Review Exercises
- Chapter Test

After completing a chapter, look at the Chapter Summary. This summary is divided into two sections: Key Words and Essential Rules and Procedures. See page 143 for the Chapter Summary for Chapter 2. This summary highlights the important topics covered in the chapter. The page number following each topic refers you to the page in the text on which you can find more information about the concept.

Following the Chapter Summary are Chapter Review Exercises (see page 145) and a Chapter Test (see page 147). Doing the review exercises is an important way of testing your understanding of the chapter. The answer to each review exercise is given at the back of the book, along with its objective reference. After checking your answers, restudy any objective from which a question you missed was taken. It may be helpful to retry some exercises in that objective to reinforce your problem-solving techniques.

The Chapter Test should be used to prepare for an exam. We suggest that you try the Chapter Test a few days before your actual exam. Take the test in a quiet place and try to complete the test in the same amount of time you will

Copyright © Houghton Mifflin Company. All rights reserved.

be allowed for your exam. When taking the Chapter Test, practice the strategies of successful test takers: 1) scan the entire test to get a feel for the questions; 2) read the directions carefully; 3) work the problems that are easiest for you first; and perhaps most importantly, 4) try to stay calm.

When you have completed the Chapter Test, check your answers. If you missed a question, review the material in that objective and rework some of the exercises from that objective. This will strengthen your ability to perform the skills in that objective.

Your career goal goes here

Is it difficult to be successful? YES! Successful music groups, artists, professional athletes, chefs, and _____ have to work very hard to achieve their goals. They focus on their goals and ignore distractions. The things we ask you to do to achieve success take time and commitment. We are confident that if you follow our suggestions, you will succeed.

Copyright © Houghton Mifflin Company. All rights reserved.

Prealgebra

Whole Numbers

People looking to buy a house in a new city must first figure out what price range they can afford. Real estate prices are related to the cost of living, which varies depending on where people live. Knowing the cost of living in a new city helps people figure out what salary they need to earn there in order to maintain the standard of living they are enjoying in their present location. The **Project on page 79** shows you how to calculate the amount of money you would need to earn in another city in order to maintain your current standard of living.

Need help? For online student resources, visit this website: **math.college.hmco.com**

Copyright © Houghton Mifflin Company. All rights reserved.

Prep Test

1. Name the number of ♦s shown below.

   ♦ ♦ ♦ ♦ ♦ ♦ ♦ ♦

2. Write the numbers from 1 to 10.

   1 ___ ___ ___ ___ ___ ___ ___ ___ 10

3. Match the number with its word form.

   a. 4   **A.** five
   b. 2   **B.** one
   c. 5   **C.** zero
   d. 1   **D.** four
   e. 3   **E.** two
   f. 0   **F.** three

4. How many American flags contain the color green?

5. Write the number of states in the United States of America as a word, not a number.

Go Figure

Five adults and two children want to cross a river in a rowboat. The boat can hold one adult or two children or one child. Everyone is able to row the boat. What is the minimum number of trips that will be necessary for everyone to get to the other side?

Copyright © Houghton Mifflin Company. All rights reserved.

Copyright © Houghton Mifflin Company. All rights reserved.

SECTION 1.1   Introduction to Whole Numbers

OBJECTIVE A

Order relations between whole numbers

The **natural numbers** are 1, 2, 3, 4, 5, 6, 7, 8, 9, 10, 11, ...

The three dots mean that the list continues on and on and there is no largest natural number. The natural numbers are also called the **counting numbers.**

The **whole numbers** are 0, 1, 2, 3, 4, 5, 6, 7, 8, 9, 10, 11, ... Note that the whole numbers include the natural numbers and zero.

Just as distances are associated with markings on the edge of a ruler, the whole numbers can be associated with points on a line. This line is called the **number line** and is shown below.

The arrowhead at the right indicates that the number line continues to the right.

The **graph** of a whole number is shown by placing a heavy dot on the number line directly above the number. Shown below is the graph of 6 on the number line.

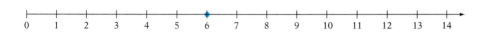

On the number line, the numbers get larger as we move from left to right. The numbers get smaller as we move from right to left. Therefore, the number line can be used to visualize the order relation between two whole numbers.

A number that appears to the right of a given number is **greater than** the given number. The symbol for *is greater than* is >.

8 is to the right of 3.
8 is greater than 3.
8 > 3

A number that appears to the left of a given number is **less than** the given number. The symbol for *is less than* is <.

5 is to the left of 12.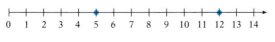
5 is less than 12.
5 < 12

An **inequality** expresses the relative order of two mathematical expressions. 8 > 3 and 5 < 12 are inequalities.

> ### Point of Interest
>
> Among the slang words for zero are *zilch, zip,* and *goose egg.* The word *love* for zero in scoring a tennis game comes from the French for "the egg": *l'oeuf.*

Take Note
An inequality symbol, < or >, points to the smaller number. The symbol opens toward the larger number.

1 *Example*   Graph 4 on the number line.

Solution

┠─┼─┼─┼─◆─┼─┼─┼─┼─┼─┼─┼─►
0   1   2   3   4   5   6   7   8   9   10   11   12

1 *You Try It*   Graph 9 on the number line.

Your Solution

┠─┼─┼─┼─┼─┼─┼─┼─┼─┼─┼─┼─►
0   1   2   3   4   5   6   7   8   9   10   11   12

---

2 *Example*   On the number line, what number is 3 units to the right of 4?

Solution

```
              3
        ────────────►
┠─┼─┼─┼─┼─┼─┼─┼─┼─┼─┼─┼─►
0  1  2  3  4  5  6  7  8  9  10 11 12
```

7 is 3 units to the right of 4.

2 *You Try It*   On the number line, what number is 4 units to the left of 11?

Your Solution

┠─┼─┼─┼─┼─┼─┼─┼─┼─┼─┼─┼─►
0   1   2   3   4   5   6   7   8   9   10   11   12

---

3 *Example*   Place the correct symbol, < or >, between the two numbers.

   a.   38   23   **b.**   0   54

Solution   **a.**   38 > 23   **b.**   0 < 54

3 *You Try It*   Place the correct symbol, < or >, between the two numbers.

   a.   47   19   **b.**   26   0

Your Solution

---

4 *Example*   Write the given numbers in order from smallest to largest.

   16, 5, 47, 0, 83, 29

Solution   0, 5, 16, 29, 47, 83

4 *You Try It*   Write the given numbers in order from smallest to largest.

   52, 17, 68, 0, 94, 3

Your Solution

Solutions on p. S1

OBJECTIVE   **B**

VIDEO & DVD   CD TUTOR   WWW WEB   SSM

Place value

When a whole number is written using the digits 0, 1, 2, 3, 4, 5, 6, 7, 8, and 9, it is said to be in **standard form**. The position of each digit in the number determines the digit's **place value**. The diagram below shows a **place-value chart** naming the first twelve place values. The number 64,273 is in standard form and has been entered in the chart.

In the number 64,273, the position of the digit 6 determines that its place value is ten-thousands.

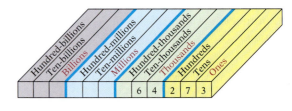

> ▶ **Point of Interest**
>
> The Romans represented numbers using M for 1,000, D for 500, C for 100, L for 50, X for 10, V for 5, and I for 1. For example, MMDCCCLXXVI represented 2,876. The Romans could represent any number up to the largest they would need for their everyday life, except zero.

When a number is written in standard form, each group of digits separated by a comma is called a **period**. The number 5,316,709,842 has four periods. The period names are shown in color in the place-value chart above.

Copyright © Houghton Mifflin Company. All rights reserved.

To write a number in words, start from the left. Name the number in each period. Then write the period name in place of the comma.

5,316,709,842 is read "five billion three hundred sixteen million seven hundred nine thousand eight hundred forty-two."

To write a whole number in standard form, write the number named in each period, and replace each period name with a comma.

Six million fifty-one thousand eight hundred seventy-four is written 6,051,874. The zero is used as a place holder for the hundred-thousands place.

The whole number 37,286 can be written in **expanded form** as

$$30,000 + 7,000 + 200 + 80 + 6$$

The place-value chart can be used to find the expanded form of a number.

▶ **Point of Interest**

George Washington used a code to communicate with his men. He had a book in which each word or phrase was represented by a three-digit number. The numbers were arbitrarily assigned to each entry. Messages appeared as a string of numbers and thus could not be decoded by the enemy.

Write the number 510,409 in expanded form.

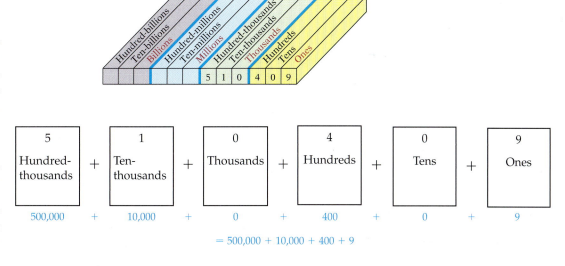

Copyright © Houghton Mifflin Company. All rights reserved.

5 *Example*

Write 82,593,071 in words.

Solution

eighty-two million five hundred ninety-three thousand seventy-one

5 *You Try It*

Write 46,032,715 in words.

Your Solution

6 *Example*

Write four hundred six thousand nine in standard form.

Solution

406,009

6 *You Try It*

Write nine hundred twenty thousand eight in standard form.

Your Solution

7 *Example*

Write 32,598 in expanded form.

Solution

30,000 + 2,000 + 500 + 90 + 8

7 *You Try It*

Write 76,245 in expanded form.

Your Solution

Solutions on p. S1

OBJECTIVE **C**

Rounding

When the distance to the sun is given as 93,000,000 mi, the number represents an approximation to the true distance. Giving an approximate value for an exact number is called **rounding.** A number is rounded to a given place value.

48 is closer to 50 than it is to 40. 48 rounded to the nearest ten is 50.

4,872 rounded to the nearest ten is 4,870.

4,872 rounded to the nearest hundred is 4,900.

A number is rounded to a given place value without using the number line by looking at the first digit to the right of the given place value.

If the digit to the right of the given place value is less than 5, replace that digit and all digits to the right of it by zeros.

Round 12,743 to the nearest hundred.

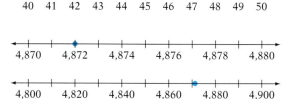

12,743 rounded to the nearest hundred is 12,700.

Copyright © Houghton Mifflin Company. All rights reserved.

If the digit to the right of the given place value is greater than or equal to 5, increase the digit in the given place value by 1, and replace all other digits to the right by zeros.

Round 46,738 to the nearest thousand.

Given place value

46,738

7 > 5

46,738 rounded to the nearest thousand is 47,000.

Round 29,873 to the nearest thousand.

Given place value

29,873

8 > 5   Round up by adding 1 to the 9 (9 + 1 = 10).
Carry the 1 to the ten-thousands place (2 + 1 = 3).

29,873 rounded to the nearest thousand is 30,000.

---

8 Example

Round 435,278 to the nearest ten-thousand.

Solution

Given place value

435,278

5 = 5

435,278 rounded to the nearest ten-thousand is 440,000.

8 You Try It

Round 529,374 to the nearest ten-thousand.

Your Solution

---

9 Example

Round 1,967 to the nearest hundred.

Solution

Given place value

1,967

6 > 5

1,967 rounded to the nearest hundred is 2,000.

9 You Try It

Round 7,985 to the nearest hundred.

Your Solution

Solutions on p. S1

Copyright © Houghton Mifflin Company. All rights reserved.

VIDEO & DVD   CD TUTOR   WEB   SSM

OBJECTIVE **D**

Applications and statistical graphs

Graphs are displays that provide a pictorial representation of data. The advantage of graphs is that they present information in a way that is easily read.

A **pictograph** uses symbols to represent information. The symbol chosen usually has a connection to the data it represents.

Figure 1.1 represents the net worth of America's richest billionaires. Each symbol represents ten billion dollars.

	Net Worth (in tens of billions of dollars)
Bill Gates	$ $ $ $ $
Warren Buffett	$ $ $ $
Paul Allen	$ $ $
Larry Ellison	$ $
Jim C. Walton	$ $

Figure 1.1 Net Worth of America's Richest Billionaires
Source: **www.Forbes.com**

Bill Gates

From the pictograph, we can see that Bill Gates has the greatest net worth. Warren Buffett's net worth is $10 billion more than Paul Allen's net worth.

A typical household in the United States has an average after-tax income of $40,550. The **circle graph** in Figure 1.2 represents how this annual income is spent. The complete circle represents the total amount, $40,550. Each sector of the circle represents the amount spent on a particular expense.

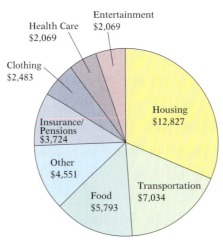

From the circle graph, we can see that the largest amount is spent on housing. We can see that the amount spent on food ($5,793) is less than the amount spent on transportation ($7,034).

Figure 1.2 Average Annual Expenses in a U.S. Household
Source: American Demographics

Copyright © Houghton Mifflin Company. All rights reserved.

The **bar graph** in Figure 1.3 shows the expected U.S. population aged 100 and over.

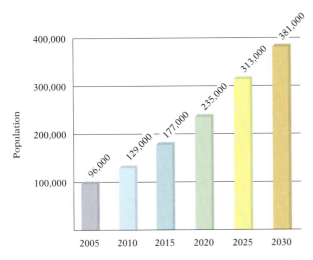

Figure 1.3 Expected U.S. Population Aged 100 and Over
Source: Census Bureau

In this bar graph, the horizontal axis is labeled with the years (2005, 2010, 2015, etc.) and the vertical axis is labeled with the numbers for the population. For each year, the height of the bar indicates the population for that year. For example, we can see that the expected population of those aged 100 and over in the year 2015 is 177,000. The graph indicates that the population of people aged 100 and over keeps increasing.

A **double-bar graph** is used to display data for the purposes of comparison.

The double-bar graph in Figure 1.4 shows the fuel efficiency of four vehicles, as rated by the Environmental Protection Agency. These are among the most fuel-efficent 2003 model-year cars for city and highway mileage.

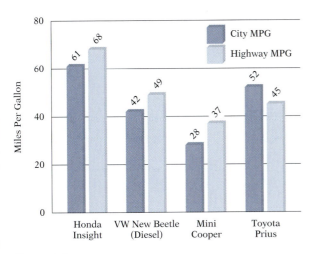

Figure 1.4

From the graph, we can see that the fuel efficiency of the Honda Insight is greater on the highway (68 mpg) than it is in city driving (61 mpg).

Copyright © Houghton Mifflin Company. All rights reserved.

The **broken-line graph** in Figure 1.5 shows the effect of inflation on the value of a $100,000 life insurance policy. (An inflation rate of 5 percent is used here.)

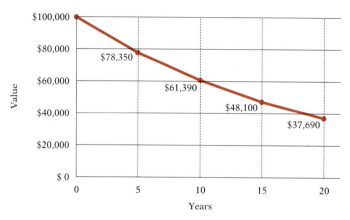

Figure 1.5 Effect of Inflation on the Value of a $100,000 Life Insurance Policy

According to the line graph, after five years the purchasing power of the $100,000 has decreased to $78,350. We can see that the value of the $100,000 keeps decreasing over the 20-year period.

Two broken-line graphs are used so that data can be compared. Figure 1.6 shows the population of California and of Texas. The figures are those of the U.S. Census for the years 1900, 1925, 1950, 1975, and 2000. The numbers are rounded to the nearest thousand.

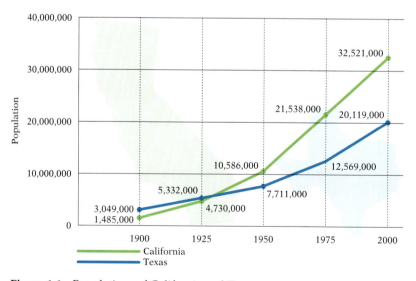

Figure 1.6 Populations of California and Texas

From the graph, we can see that the population was greater in Texas in 1900 and 1925, while the population was greater in California in 1950, 1975, and 2000.

Copyright © Houghton Mifflin Company. All rights reserved.

To solve an application problem, first read the problem carefully. The **Strategy** involves identifying the quantity to be found and planning the steps that are necessary to find that quantity. The **Solution** involves performing each operation stated in the Strategy and writing the answer.

The circle graph in Figure 1.7 shows the result of a survey of 300 people who were asked to name their favorite sport. Use this graph for Example 10 and You Try It 10.

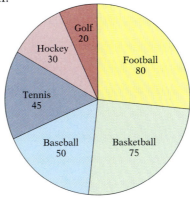

Figure 1.7   Distribution of Responses in a Survey

10 Example

According to Figure 1.7, which sport was named by the least number of people?

Strategy

To find the sport named by the least number of people, find the smallest number given in the circle graph.

Solution

The smallest number given in the graph is 20.

The sport named by the least number of people was golf.

10 You Try It

According to Figure 1.7, which sport was named by the greatest number of people?

Your Strategy

Your Solution

11 Example

 The distance between St. Louis, Missouri, and Portland, Oregon, is 2,057 mi. The distance between St. Louis, Missouri, and Seattle, Washington, is 2,135 mi. Which distance is greater, St. Louis to Portland or St. Louis to Seattle?

Strategy

To find the greater distance, compare the numbers 2,057 and 2,135.

Solution

$2{,}135 > 2{,}057$

The greater distance is from St. Louis to Seattle.

11 You Try It

The distance between Los Angeles, California, and San Jose, California, is 347 mi. The distance between Los Angeles, California, and San Francisco, California, is 387 mi. Which distance is shorter, Los Angeles to San Jose or Los Angeles to San Francisco?

Your Strategy

Your Solution

Solutions on p. S1

Copyright © Houghton Mifflin Company. All rights reserved.

The bar graph in Figure 1.8 shows the states with the most sanctioned league bowlers. Use this graph for Example 12 and You Try It 12.

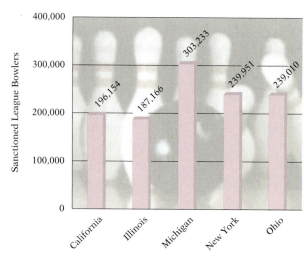

Figure 1.8 States with the Most Sanctioned League Bowlers

Sources: American Bowling Congress, Women's International Bowling Congress, Young American Bowling Alliance

---

12 **Example**

According to Figure 1.8, which state has the most sanctioned league bowlers?

Strategy

To determine which state has the most sanctioned league bowlers, locate the state that corresponds to the highest bar.

Solution

The highest bar corresponds to Michigan.

Michigan is the state with the most sanctioned league bowlers.

12 **You Try It**

According to Figure 1.8, which state has fewer sanctioned league bowlers, New York or Ohio?

Your Strategy

Your Solution

---

13 **Example**

The land area of the United States is 3,539,341 mi². What is the land area of the United States to the nearest ten-thousand square miles?

Strategy

To find the land area to the nearest ten-thousand square miles, round 3,539,341 to the nearest ten-thousand.

Solution

3,539,341 rounded to the nearest ten-thousand is 3,540,000.

To the nearest ten-thousand square miles, the land area of the United States is 3,540,000 mi².

13 **You Try It**

The land area of Canada is 3,851,809 mi². What is the land area of Canada to the nearest thousand square miles?

Your Strategy

Your Solution

Solutions on p. S1

Copyright © Houghton Mifflin Company. All rights reserved.

1.1 EXERCISES

Copyright © Houghton Mifflin Company. All rights reserved.

OBJECTIVE A

1. How do the whole numbers differ from the natural numbers?

2. Explain how to round a four-digit number to the nearest hundred.

Graph the number on the number line.

3. 2

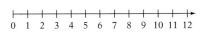

4. 7

5. 10

6. 1

7. 5

8. 11

On the number line, which number is:

9. 4 units to the left of 9

10. 5 units to the left of 8

11. 3 units to the right of 2

12. 4 units to the right of 6

13. 7 units to the left of 7

14. 8 units to the left of 11

Place the correct symbol, < or >, between the two numbers.

15. 27   39

16. 68   41

17. 0   52

18. 61   0

19. 273   194

20. 419   502

21. 2,761   3,857

22. 3,827   6,915

23. 4,610   4,061

24. 5,600   56,000

25. 8,005   8,050

26. 92,010   92,001

Write the given numbers in order from smallest to largest.

27. 21, 14, 32, 16, 11

28. 18, 60, 35, 71, 27

29. 72, 48, 84, 93, 13

30. 54, 45, 63, 28, 109

31. 26, 49, 106, 90, 77

32. 505, 496, 155, 358, 271

33. 736, 662, 204, 981, 399

34. 440, 404, 400, 444, 4,000

35. 377, 370, 307, 3,700, 3,077

OBJECTIVE **B**

Write the number in words.

36. 704

37. 508

38. 374

39. 635

40. 2,861

41. 4,790

42. 48,297

43. 53,614

44. 563,078

45. 246,053

46. 6,379,482

47. 3,842,905

Write the number in standard form.

48. seventy-five

49. four hundred ninety-six

50. two thousand eight hundred fifty-one

51. fifty-three thousand three hundred forty

52. one hundred thirty thousand two hundred twelve

53. five hundred two thousand one hundred forty

54. eight thousand seventy-three

55. nine thousand seven hundred six

Copyright © Houghton Mifflin Company. All rights reserved.

56. six hundred three thousand one hundred thirty-two

57. five million twelve thousand nine hundred seven

58. three million four thousand eight

59. eight million five thousand ten

Write the number in expanded form.

60. 6,398

61. 7,245

62. 46,182

63. 532,791

64. 328,476

65. 5,064

66. 90,834

67. 20,397

68. 400,635

69. 402,708

70. 504,603

71. 8,000,316

OBJECTIVE  C

Round the number to the given place value.

72. 3,049 Tens

73. 7,108 Tens

74. 1,638 Hundreds

75. 4,962 Hundreds

76. 17,639 Hundreds

77. 28,551 Hundreds

78. 5,326 Thousands

79. 6,809 Thousands

80. 84,608 Thousands

81. 93,825 Thousands

82. 389,702 Thousands

83. 629,513 Thousands

84. 746,898 Ten-thousands

85. 352,876 Ten-thousands

86. 36,702,599 Millions

87. 71,834,250 Millions

Copyright © Houghton Mifflin Company. All rights reserved.

OBJECTIVE D

88. *Sports* During his baseball career, Eddie Collins had a record of 743 stolen bases. Max Carey had a record of 738 stolen bases during his baseball career. Who had more stolen bases, Eddie Collins or Max Carey?

89. *Sports* During his baseball career, Ty Cobb had a record of 892 stolen bases. Billy Hamilton had a record of 937 stolen bases during his baseball career. Who had more stolen bases, Ty Cobb or Billy Hamilton?

90. *Nutrition* The figure at the right shows the annual per capita turkey consumption in different countries. (a) What is the annual per capita turkey consumption in the United States? (b) In which country is the annual per capita turkey consumption the highest?

Britain	🦃🦃🦃🦃
Canada	🦃🦃🦃🦃🦃
France	🦃🦃🦃🦃🦃🦃
Ireland	🦃🦃🦃
Israel	🦃🦃🦃🦃🦃🦃🦃🦃🦃🦃🦃
Italy	🦃🦃🦃🦃🦃
U.S.	🦃🦃🦃🦃🦃🦃🦃🦃🦃

Each 🦃 represents 2 lb.

Per Capita Turkey Consumption
Source: National Turkey Federation

91. *The Arts* The play *Hello Dolly* was performed 2,844 times on Broadway. The play *Fiddler on the Roof* was performed 3,242 times on Broadway. Which play had the greater number of performances, *Hello Dolly* or *Fiddler on the Roof*?

92. *The Arts* The play *Annie* was performed 2,377 times on Broadway. The play *My Fair Lady* was performed 2,717 times on Broadway. Which play had the greater number of performances, *Annie* or *My Fair Lady*?

93. *Nutrition* Two tablespoons of peanut butter contain 190 calories. Two tablespoons of grape jelly contain 114 calories. Which contains more calories, two tablespoons of peanut butter or two tablespoons of grape jelly?

94. *History* In 1892, the diesel engine was patented. In 1844, Samuel F. B. Morse patented the telegraph. Which was patented first, the diesel engine or the telegraph?

Samuel F. B. Morse

95. *Geography* The distance between St. Louis, Missouri, and Reno, Nevada, is 1,892 mi. The distance between St. Louis, Missouri, and San Diego, California, is 1,833 mi. Which is the shorter distance, St. Louis to Reno or St. Louis to San Diego?

96. *Consumerism* The circle graph at the right shows the result of a survey of 150 people who were asked, "What bothers you most about movie theaters?" (a) Among the respondents, what was the most often mentioned complaint? (b) What was the least often mentioned complaint?

97. *Astronomy* As measured at the equator, the diameter of the planet Uranus is 32,200 mi and the diameter of the planet Neptune is 30,800 mi. Which planet is smaller, Uranus or Neptune?

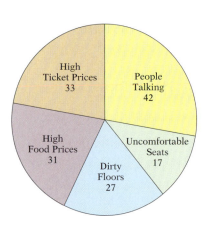

Distribution of Responses in a Survey

Copyright © Houghton Mifflin Company. All rights reserved.

98.  *Astronomy*   The diameter of Callisto, one of the moons orbiting Jupiter, is 4,890 mi. The diameter of Ganymede, another of Jupiter's moons, is 5,216 mi. Which is the larger moon, Callisto or Ganymede?

99. *Politics*   The figure below shows the length of the State of the Union Address in each of the years 1997 through 2003. (a) What was the length of the State of the Union Address in 2001? (b) In which year was the State of the Union Address the longest?

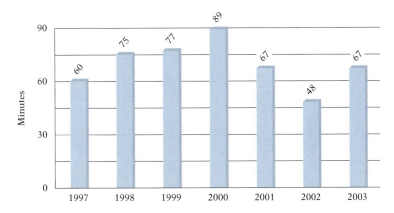

Length of the State of the Union Address
Sources: USA Today; www.whitehouse.gov; www.usconsulate.org

President Bush

100. *Geography*   The land area of Alaska is 570,833 mi^2. What is the land area of Alaska to the nearest thousand square miles?

101. *Geography*   The acreage of the Appalachian Trail is 161,546. What is the acreage of the Appalachian Trail to the nearest ten-thousand acres?

Alaska

102. *Travel*   The figure below shows the number of crashes on U.S. roadways during each of the last six months of a recent year. Also shown is the number of vehicles involved in those crashes. (a) Which was greater, the number of crashes in July or in October? (b) Were there fewer vehicles involved in crashes in July or in December?

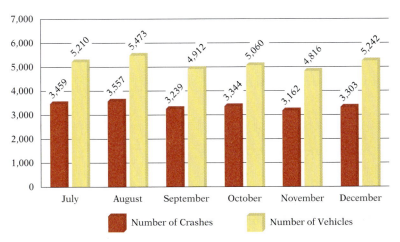

Accidents on U.S. Roadways
Source: National Highway Traffic Safety Administration

Copyright © Houghton Mifflin Company. All rights reserved.

103. *Education* Actual and projected student enrollment in elementary and secondary schools in the United States is shown in the figure at the right. Enrollment figures are for the fall of each year. The jagged line at the bottom of the vertical axis indicates that this scale is missing the tens of millions from 0 to 30,000,000. (a) During which year was enrollment the lowest? (b) Did enrollment increase or decrease between 1975 and 1980?

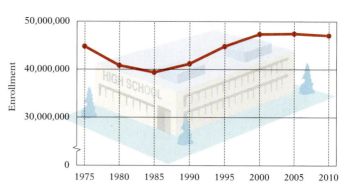

Enrollment in Elementary and Secondary Schools
Source: National Center for Education Statistics

104. *Aviation* The cruising speed of a Boeing 747 is 589 mph. What is the cruising speed of a Boeing 747 to the nearest ten miles per hour?

105. *Physics* Light travels at a speed of 299,800 km/s. What is the speed of light to the nearest thousand kilometers per second?

Critical Thinking

106. *Geography* Find the land area of the seven continents. List the continents in order from largest to smallest. List the oceans on Earth from largest to smallest.

107. *Mathematics* What is the largest three-digit number? What is the smallest five-digit number?

108. What is the total enrollment of your school? To what place value would it be reasonable to round this number? Why? To what place value is the population of your town or city rounded? Why? To what place value is the population of your state rounded? To what place value is the population of the United States rounded?

109. Look at the visual illusion pictured at the right. Describe it in words. Find other examples of visual illusions.

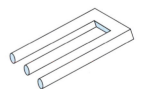

110. What is the national debt of the United States? What does this figure mean?

111. Prepare a report on the symbols used in the Egyptian numeration system.

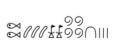

Copyright © Houghton Mifflin Company. All rights reserved.

OBJECTIVE   A

Addition of whole numbers

Addition is the process of finding the total of two or more numbers.

On Arbor Day, a community group planted 3 trees along one street and 5 trees along another street. By counting, we can see that there were a total of 8 trees planted.

$$3 \quad + \quad 5 \quad = \quad 8$$

The 3 and 5 are called **addends.** The **sum** is 8.

The basic addition facts for adding one digit to one digit should be memorized. Addition of larger numbers requires the repeated use of the basic addition facts.

To add large numbers, begin by arranging the numbers vertically, keeping the digits of the same place value in the same column.

Add: 321 + 6,472

Add the digits in each column.

$$\begin{array}{r} \;3\;2\;1 \\ +6\;4\;7\;2 \\ \hline 6\;7\;9\;3 \end{array}$$

(Thousands, Hundreds, Tens, Ones)

Find the sum of 211, 45, 23, and 410.

Remember that a *sum* is the answer to an addition problem.
Arrange the numbers vertically, keeping digits of the same place value in the same column.
Add the numbers in each column.

$$\begin{array}{r} 211 \\ 45 \\ 23 \\ +\;410 \\ \hline 689 \end{array}$$

The phrase *the sum of* was used in the example above to indicate the operation of addition. All of the phrases listed below indicate addition. An example of each is shown at the right of each phrase.

added to	6 added to 9	9 + 6
more than	3 more than 8	8 + 3
the sum of	the sum of 7 and 4	7 + 4
increased by	2 increased by 5	2 + 5
the total of	the total of 1 and 6	1 + 6
plus	8 plus 10	8 + 10

Copyright © Houghton Mifflin Company. All rights reserved.

When the sum of the numbers in a column exceeds 9, addition involves "carrying."

Add: 359 + 478

Add the ones column.
9 + 8 = 17 (1 ten + 7 ones).
Write the 7 in the ones column and carry the 1 ten to the tens column.

$$\begin{array}{ccc} \text{Hundreds} & \text{Tens} & \text{Ones} \\ & 1 & \\ 3 & 5 & 9 \\ + 4 & 7 & 8 \\ \hline & & 7 \end{array}$$

Add the tens column.
1 + 5 + 7 = 13 (1 hundred + 3 tens).
Write the 3 in the tens column and carry the 1 hundred to the hundreds column.

```
  1 1
  359
+ 478
   37
```

Add the hundreds column.
1 + 3 + 4 = 8 (8 hundreds).
Write the 8 in the hundreds column.

```
  1 1
  359
+ 478
  837
```

Calculator Note

Most scientific calculators use *algebraic logic:* the add (+), subtract (−), multiply (×), and divide (÷) keys perform the indicated operation on the number in the display and the next number keyed in. For instance, for the example at the right, enter 359 + 478 = . The display reads 837.

The bar graph in Figure 1.9 shows the seating capacity in 2003 of the five largest National Football League stadiums. What is the total seating capacity of these five stadiums? *Note:* The jagged line below 70,000 on the vertical axis indicates that this scale is missing the numbers less than 70,000.

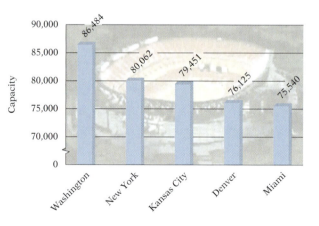

Figure 1.9 Seating Capacity of the Five Largest NFL Stadiums

Arrowhead Stadium, Kansas City

```
  86,484
  80,062
  79,451
  76,125
+ 75,540
 397,662
```

The total capacity of the five stadiums is 397,662 people.

Copyright © Houghton Mifflin Company. All rights reserved.

An important skill in mathematics is the ability to determine whether an answer to a problem is reasonable. One method of determining whether an answer is reasonable is to use estimation. An **estimate** is an approximation.

Estimation is especially valuable when using a calculator. Suppose that you are adding 1,497 and 2,568 on a calculator. You enter the number 1,497 correctly, but you inadvertently enter 256 instead of 2,568 for the second addend. The sum reads 1,753. If you quickly make an estimate of the answer, you can determine that the sum 1,753 is not reasonable and that an error has been made.

$$\begin{array}{r} 1,497 \\ +\ 2,568 \\ \hline 4,065 \end{array} \qquad \begin{array}{r} 1,497 \\ +\ \ \ 256 \\ \hline 1,753 \end{array}$$

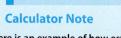

Calculator Note

Here is an example of how estimation is important when using a calculator.

To estimate the answer to a calculation, round each number to the highest place value of the number; the first digit of each number will be nonzero and all other digits will be zero. Perform the calculation using the rounded numbers.

$$\begin{array}{ccr} 1,497 & \longrightarrow & 1,000 \\ 2,568 & \longrightarrow & +\ 3,000 \\ & & \hline 4,000 \end{array}$$

As shown above, the sum 4,000 is an estimate of the sum of 1,497 and 2,568; it is very close to the actual sum, 4,065. 4,000 is not close to the incorrectly calculated sum, 1,753.

Estimate the sum of 35,498, 17,264, and 81,093.

Round each number to the nearest ten-thousand.

$$\begin{array}{rcr} 35,498 & \longrightarrow & 40,000 \\ 17,264 & \longrightarrow & 20,000 \\ 81,093 & \longrightarrow & +\ 80,000 \\ \hline & & 140,000 \end{array}$$

Add the rounded numbers.

Note that 140,000 is close to the actual sum, 133,855.

Just as the word *it* is used in language to stand for an object, a letter of the alphabet can be used in mathematics to stand for a number. Such a letter is called a **variable.**

A mathematical expression that contains one or more variables is a **variable expression.** Replacing the variables in a variable expression with numbers and then simplifying the numerical expression is called **evaluating the variable expression.**

Evaluate $a + b$ when $a = 678$ and $b = 294$.

Replace a with 678 and b with 294.

$$\begin{array}{c} a + b \\ 678 + 294 \end{array}$$

Arrange the numbers vertically.

$$\begin{array}{r} \overset{1\ 1}{678} \\ +\ 294 \\ \hline \end{array}$$

Add.

972

Copyright © Houghton Mifflin Company. All rights reserved.

Variables are often used in algebra to describe mathematical relationships. Variables are used below to describe three properties, or rules, of addition. An example of each property is shown at the right.

> **The Addition Property of Zero**
>
> $a + 0 = a$ or $0 + a = a$

$5 + 0 = 5$

The Addition Property of Zero states that the sum of a number and zero is the number. The variable a is used here to represent any whole number. It can even represent the number zero because $0 + 0 = 0$.

> **The Commutative Property of Addition**
>
> $a + b = b + a$

$5 + 7 = 7 + 5$
$12 = 12$

The Commutative Property of Addition states that two numbers can be added in either order; the sum will be the same. Here the variables a and b represent any whole numbers. Therefore, if you know that the sum of 5 and 7 is 12, then you also know that the sum of 7 and 5 is 12, because $5 + 7 = 7 + 5$.

> **The Associative Property of Addition**
>
> $(a + b) + c = a + (b + c)$

$(2 + 3) + 4 = 2 + (3 + 4)$
$5 + 4 = 2 + 7$
$9 = 9$

The Associative Property of Addition states that when adding three or more numbers, we can group the numbers in any order; the sum will be the same. Note in the example at the right above that we can add the sum of 2 and 3 to 4, or we can add 2 to the sum of 3 and 4. In either case, the sum of the three numbers is 9.

Rewrite the expression by using the Associative Property of Addition.

$(3 + x) + y$

The Associative Property of Addition states that addends can be grouped in any order.

$(3 + x) + y = 3 + (x + y)$

Copyright © Houghton Mifflin Company. All rights reserved.

An **equation** expresses the equality of two numerical or variable expressions. In the example above, $(3 + x) + y$ is an expression; it does not contain an equals sign. $(3 + x) + y = 3 + (x + y)$ is an equation; it contains an equals sign.

▶ **Point of Interest**

The equals sign (=) is generally credited to Robert Recorde. In his 1557 treatise on algebra, *The Whetstone of Whit*, he wrote, "No two things could be more equal (than two parallel lines)." His equals sign gained popular usage, even though continental mathematicians preferred a dash.

Here is another example of an equation. The **left side** of the equation is the variable expression $n + 4$. The **right side** of the equation is the number 9.

$$n + 4 = 9$$

Just as a statement in English can be true or false, an equation may be true or false. The equation shown above is *true* if the variable is replaced by 5.

$$n + 4 = 9$$
$$5 + 4 = 9 \quad \text{True}$$

The equation is *false* if the variable is replaced by 8.

$$8 + 4 = 9 \quad \text{False}$$

A **solution** of an equation is a number that, when substituted for the variable, results in a true equation. The solution of the equation $n + 4 = 9$ is 5 because replacing n by 5 results in a true equation. When 8 is substituted for n, the result is a false equation; therefore, 8 is not a solution of the equation.

10 is a solution of $x + 5 = 15$ because $10 + 5 = 15$ is a true equation.

20 is not a solution of $x + 5 = 15$ because $20 + 5 = 15$ is a false equation.

Is 9 a solution of the equation $11 = 2 + x$?

$$11 = 2 + x$$

Replace x by 9.

$$11 \mid 2 + 9$$

Simplify the right side of the equation. Compare the results. If the results are equal, the given number is a solution of the equation. If the results are not equal, the given number is not a solution.

$$11 = 11$$

Yes, 9 is a solution of the equation.

---

① Example Estimate the sum of 379, 842, 693, and 518.

Solution

$$
\begin{array}{rl}
379 & \longrightarrow \quad 400 \\
842 & \longrightarrow \quad 800 \\
693 & \longrightarrow \quad 700 \\
518 & \longrightarrow + \;\; 500 \\
\hline
& \quad\quad 2,400
\end{array}
$$

 You Try It Estimate the total of 6,285, 3,972, and 5,140.

Your Solution

---

② Example Identify the property that justifies the statement.

$$7 + 2 = 2 + 7$$

Solution The Commutative Property of Addition

 You Try It Identify the property that justifies the statement.

$$33 + 0 = 33$$

Your Solution

Solutions on p. S1

Copyright © Houghton Mifflin Company. All rights reserved.

The topic of the circle graph in Figure 1.10 is the eggs produced in the United States in a recent year. It shows where the eggs that were produced went or how they were used. Use this graph for Example 3 and You Try It 3.

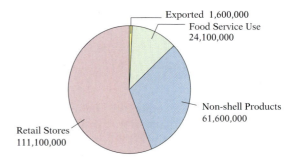

Figure 1.10 Eggs Produced in the United States (in cases)
Source: American Egg Board. *USA Today.* Copyright © November 27, 2001. Reprinted with permission.

---

3 Example Use Figure 1.10 to determine the sum of the number of cases of eggs sold by retail stores and used for non-shell products.

Solution 111,100,000 cases of eggs were sold by retail stores. 61,600,000 cases of eggs were used for non-shell products.

$$111,100,000$$
$$+61,600,000$$
$$172,700,000$$

172,700,000 cases of eggs were sold by retail stores and used for non-shell products.

3 You Try It Use Figure 1.10 to determine the total number of cases of eggs produced during the year.

Your Solution

---

4 Example Evaluate $x + y + z$ when $x = 8,427$, $y = 3,659$, and $z = 6,281$.

Solution $x + y + z$
$8,427 + 3,659 + 6,281$

$$\begin{array}{r} {}^{1}\ {}^{11} \\ 8,427 \\ 3,659 \\ +\ 6,281 \\ \hline 18,367 \end{array}$$

4 You Try It Evaluate $x + y + z$ when $x = 1,692$, $y = 4,783$, and $z = 5,046$.

Your Solution

---

5 Example Is 6 a solution of the equation $9 + y = 14$?

Solution $9 + y = 14$
$$\begin{array}{c|c} 9 + 6 & 14 \\ \hline 15 \neq 14 \end{array}$$

▶ The symbol ≠ is read "is not equal to."

No, 6 is not a solution of the equation $9 + y = 14$.

5 You Try It Is 7 a solution of the equation $13 = b + 6$?

Your Solution

Solutions on p. S1

Copyright © Houghton Mifflin Company. All rights reserved.

Copyright © Houghton Mifflin Company. All rights reserved.

OBJECTIVE **B**

Subtraction of whole numbers

Subtraction is the process of finding the difference between two numbers.

By counting, we see that the difference between $8 and $5 is $3.

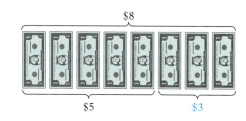

$$\$8 \quad - \quad \$5 \quad = \quad \$3$$

Minuend $-$ **Subtrahend** $=$ **Difference**

Note that addition and subtraction are related.

$$
\begin{array}{rr}
\text{Subtrahend} & 5 \\
+ \text{ Difference} & + 3 \\
\hline
= \text{ Minuend} & 8
\end{array}
$$

The fact that the sum of the subtrahend and the difference equals the minuend can be used to check subtraction.

To subtract large numbers, begin by arranging the numbers vertically, keeping the digits of the same place value in the same column. Then subtract the numbers in each column.

Find the difference between 8,955 and 2,432.

A *difference* is the answer to a subtraction problem.

Thousands | Hundreds | Tens | Ones

$$
\begin{array}{r}
8 \ 9 \ 5 \ 5 \\
- 2 \ 4 \ 3 \ 2 \\
\hline
6 \ 5 \ 2 \ 3
\end{array}
$$

Check:
$$
\begin{array}{rr}
\text{Subtrahend} & 2{,}432 \\
+ \text{ Difference} & + 6{,}523 \\
\hline
= \text{ Minuend} & 8{,}955
\end{array}
$$

In the subtraction example above, the lower digit in each place value is smaller than the upper digit. When the lower digit is larger than the upper digit, subtraction involves "borrowing."

Subtract: 692 − 378

Hundreds | Tens | Ones

$$
\begin{array}{r}
\overset{8+1}{6\ \ 9}\ 2 \\
- 3\ \ 7\ \ 8
\end{array}
$$

$8 > 2$
Borrowing is necessary.
9 tens = 8 tens + 1 ten

Hundreds | Tens | Ones

$$
\begin{array}{r}
8 + \overset{\frown}{①}10 \\
6\ \ 9\ \ 2 \\
- 3\ \ 7\ \ 8
\end{array}
$$

Borrow 1 ten from the tens column and write 10 in the ones column.

Hundreds | Tens | Ones

$$
\begin{array}{r}
8\ \ 12 \\
6\ \ 9\ \ 2 \\
- 3\ \ 7\ \ 8
\end{array}
$$

Add the borrowed 10 to 2.

Hundreds | Tens | Ones

$$
\begin{array}{r}
8\ \ 12 \\
6\ \ 9\ \ 2 \\
- 3\ \ 7\ \ 8 \\
\hline
3\ \ 1\ \ 4
\end{array}
$$

Subtract the numbers in each column.

Subtraction may involve repeated borrowing.

Subtract: $7{,}325 - 4{,}698$

$$
\begin{array}{r}
\overset{\overset{1}{}\ \overset{15}{}}{7,\ 3\ 2\ 5} \\
-\ 4,\ 6\ 9\ 8 \\
\hline
7
\end{array}
\qquad
\begin{array}{r}
\overset{\overset{2}{}\ \overset{11}{\overset{1}{}}\ \overset{15}{}}{7,\ 3\ 2\ 5} \\
-\ 4,\ 6\ 9\ 8 \\
\hline
2\ 7
\end{array}
\qquad
\begin{array}{r}
\overset{\overset{6}{7,}\ \overset{12}{\overset{2}{3}}\ \overset{11}{\overset{1}{2}}\ \overset{15}{5}}{}\\
-\ 4,\ 6\ 9\ 8 \\
\hline
2,\ 6\ 2\ 7
\end{array}
$$

Borrow 1 ten (10 ones) from the tens column and add 10 to the 5 in the ones column. Subtract $15 - 8$.

Borrow 1 hundred (10 tens) from the hundreds column and add 10 to the 1 in the tens column. Subtract $11 - 9$.

Borrow 1 thousand (10 hundreds) from the thousands column and add 10 to the 2 in the hundreds column. Subtract $12 - 6$ and $6 - 4$.

When there is a zero in the minuend, subtraction involves repeated borrowing.

Subtract: $3{,}904 - 1{,}775$

$$
\begin{array}{r}
\overset{\overset{8}{9}\ \overset{10}{0}}{3,\ 9\ 0\ 4} \\
-\ 1,\ 7\ 7\ 5 \\
\hline
\end{array}
\qquad
\begin{array}{r}
\overset{\overset{9}{}}{} \\[-6pt]
\overset{\overset{8}{9}\ \overset{10}{0}\ \overset{14}{4}}{3,\ 9\ 0\ 4} \\
-\ 1,\ 7\ 7\ 5 \\
\hline
\end{array}
\qquad
\begin{array}{r}
\overset{\overset{9}{}}{} \\[-6pt]
\overset{\overset{8}{9}\ \overset{10}{0}\ \overset{14}{4}}{3,\ 9\ 0\ 4} \\
-\ 1,\ 7\ 7\ 5 \\
\hline
2,\ 1\ 2\ 9
\end{array}
$$

There is a 0 in the tens column. Borrow 1 hundred (10 tens) from the hundreds column and write 10 in the tens column.

Borrow 1 ten from the tens column and add 10 to the 4 in the ones column.

Subtract the numbers in each column.

Note that, for the preceding example, the borrowing could be performed as shown below.

Borrow 1 from 90. ($90 - 1 = 89$. The 8 is in the hundreds column. The 9 is in the tens column.) Add 10 to the 4 in the ones column. Then subtract the numbers in each column.

$$
\begin{array}{r}
\overset{8\ \ 9\ \ 14}{3,9\,0\,4} \\
-\ 1,7\,7\,5 \\
\hline
2,1\,2\,9
\end{array}
$$

Estimate the difference between 49,601 and 35,872.

Round each number to the nearest ten-thousand.

$$
\begin{array}{rcr}
49{,}601 & \longrightarrow & 50{,}000 \\
35{,}872 & \longrightarrow & -\ 40{,}000 \\
\hline
 & & 10{,}000
\end{array}
$$

Subtract the rounded numbers.

Note that 10,000 is close to the actual difference, 13,729.

Copyright © Houghton Mifflin Company. All rights reserved.

The phrase *the difference between* was used in the example above to indicate the operation of subtraction. All of the phrases listed below indicate subtraction. An example of each is shown at the right of each phrase.

minus	10 minus 3	$10 - 3$
less	8 less 4	$8 - 4$
less than	2 less than 9	$9 - 2$
the difference between	the difference between 6 and 1	$6 - 1$
decreased by	7 decreased by 5	$7 - 5$
subtract . . . from	subtract 11 from 20	$20 - 11$

Take Note

Note the order in which the numbers are subtracted when the phrase *less than* is used. Suppose that you have $10 and I have $6 *less than* you do; then I have $6 *less than* $10, or $10 − $6 = $4.

Evaluate $c - d$ when $c = 6{,}183$ and $d = 2{,}759$.

$$c - d$$

Replace c with 6,183 and d with 2,759. $$6{,}183 - 2{,}759$$

Arrange the numbers vertically and then subtract.

$$
\begin{array}{r}
{}^{5}\;{}^{11}\,{}^{7}\,{}^{13} \\
\cancel{6},\cancel{1}\cancel{8}\cancel{3} \\
-\,2{,}759 \\
\hline
3{,}424
\end{array}
$$

Is 23 a solution of the equation $41 - n = 17$?

Replace n by 23.
Simplify the left side of the equation.
The results are not equal.

$$41 - n = 17$$
$$41 - 23 \mid 17$$
$$18 \neq 17$$

No, 23 is not a solution of the equation.

▶ **Point of Interest**

Someone who is our equal is our peer. Two make a pair. Both of the words *peer* and *pair* come from the Latin *par, paris*, meaning "equal."

---

6 Example   Subtract and check:
57,004 − 26,189

Solution

$$
\begin{array}{r}
{}^{6}\;{}^{9}\,{}^{9}\,{}^{14} \\
5\,\cancel{7},\cancel{0}\cancel{0}\,4 \\
-\,26{,}189 \\
\hline
30{,}815
\end{array}
$$

Check: 26,189
        + 30,815
        ――――――
        57,004

6 You Try It   Subtract and check:
49,002 − 31,865

Your Solution

---

7 Example   Estimate the difference between 7,261 and 4,315. Then find the exact answer.

Solution

7,261 ⟶	7,000	7,261
4,315 ⟶	− 4,000	− 4,315
	3,000	2,946

7 You Try It   Estimate the difference between 8,544 and 3,621. Then find the exact answer.

Your Solution

Solutions on p. S1

Copyright © Houghton Mifflin Company. All rights reserved.

The graph in Figure 1.11 shows the actual and projected world energy consumption in quadrillion British thermal units (Btu). Use this graph for Example 8 and You Try It 8.

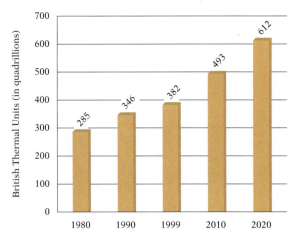

Figure 1.11   World Energy Consumption (in quadrillion British thermal units)

Sources: Energy Information Administration; Office of Energy Markets and End Use; *International Statistics Database and International Energy Annual;* World Energy Projection System

---

8 *Example*   Use Figure 1.11 to find the difference between the world energy consumption in 1980 and that projected for 2010.

Solution   2010: 493 quadrillion Btu
1980: 285 quadrillion Btu

$$\begin{array}{r} 493 \\ -\ 285 \\ \hline 208 \end{array}$$

The difference between the world energy consumption in 1980 and that projected for 2010 is 208 quadrillion Btu.

8 *You Try It*   Use Figure 1.11 to find the difference between the world energy consumption in 1990 and that projected for 2020.

Your Solution

---

9 *Example*   Evaluate $x - y$ when $x = 3{,}506$ and $y = 2{,}477$.

Solution $x - y$
$3{,}506 - 2{,}477$

$$\begin{array}{r} {\scriptstyle 4\ \ 9\ \ 16} \\ 3{,}5\cancel{0}\cancel{6} \\ -\ 2{,}4\,7\,7 \\ \hline 1{,}0\,2\,9 \end{array}$$

9 *You Try It*   Evaluate $x - y$ when $x = 7{,}061$ and $y = 3{,}229$.

Your Solution

---

10 *Example*   Is 39 a solution of the equation $24 = m - 15$?

Solution $24 = m - 15$

$$\begin{array}{c|c} 24 & 39 - 15 \\ \end{array}$$
$24 = 24$

Yes, 39 is a solution of the equation.

10 *You Try It*   Is 11 a solution of the equation $46 = 58 - p$?

Your Solution

Solutions on p. S1–S2

Copyright © Houghton Mifflin Company. All rights reserved.

Copyright © Houghton Mifflin Company. All rights reserved.

OBJECTIVE  **C**

Applications and formulas

VIDEO & DVD   CD TUTOR   WEB   SSM

One application of addition is calculating the perimeter of a figure. However, before defining perimeter, we will introduce some terms from geometry.

Two basic concepts in the study of geometry are point and line.

A **point** is symbolized by drawing a dot. A **line** is determined by two distinct points and extends indefinitely in both directions, as the arrows on the line shown at the right indicate. This line contains points *A* and *B*.

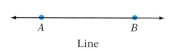
Line

A **ray** starts at a point and extends indefinitely in *one* direction. The point at which a ray starts is called the **endpoint** of the ray. Point *A* is the endpoint of the ray shown at the right.

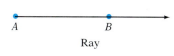
Ray

A **line segment** is part of a line and has two endpoints. The line segment shown at the right has endpoints *A* and *B*.

Line Segment

An **angle** is formed by two rays with the same endpoint. An angle is measured in **degrees**. The symbol for degrees is a small raised circle, °. A **right angle** is an angle whose measure is 90°.

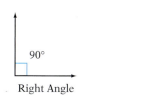
Right Angle

> **Take Note**
>
> The corner of a page of this book is a good model of a right angle.

A **plane** is a flat surface and can be pictured as a floor or a wall. Figures that lie in a plane are called **plane figures.**

Intersecting Lines

Lines in a plane can be intersecting or parallel. **Intersecting lines** cross at a point in the plane. **Parallel lines** never meet. The distance between them is always the same.

Parallel Lines

A **polygon** is a closed figure determined by three or more line segments that lie in a plane. The line segments that form the polygon are called its **sides.** The figures below are examples of polygons.

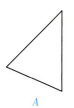
A

B

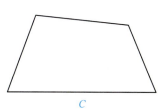
C

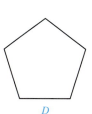
D

E

The name of a polygon is based on the number of its sides. A polygon with three sides is a **triangle**. Figure A on the previous page is a triangle. A polygon with four sides is a **quadrilateral.** Figures B and C are quadrilaterals.

Quadrilaterals are one of the most common types of polygons. Quadrilaterals are distinguished by their sides and angles. For example, a **rectangle** is a quadrilateral in which opposite sides are parallel, opposite sides are equal in length, and all four angles measure 90°.

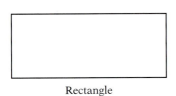

Rectangle

The **perimeter** of a plane geometric figure is a measure of the distance around the figure.

The perimeter of a triangle is the sum of the lengths of the three sides.

> ### Perimeter of a Triangle
>
> The formula for the perimeter of a triangle is $P = a + b + c$, where P is the perimeter of the triangle and a, b, and c are the lengths of the sides of the triangle.

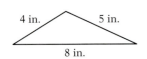

Find the perimeter of the triangle shown at the left.

Use the formula for the perimeter of a triangle. $P = a + b + c$
It does not matter which side you label a, b, or c. $P = 4 + 5 + 8$
Add. $P = 17$

The perimeter of the triangle is 17 in.

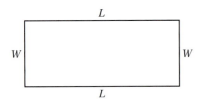

The perimeter of a quadrilateral is the sum of the lengths of its four sides.

In a rectangle, opposite sides are equal in length. Usually the length, L, of a rectangle refers to the length of one of the longer sides of the rectangle, and the width, W, refers to the length of one of the shorter sides. The perimeter can then be represented as $P = L + W + L + W$.

Use the formula $P = L + W + L + W$ to find the perimeter of the rectangle shown at the left.

Write the given formula for the perimeter of a rectangle. $P = L + W + L + W$
Substitute 32 for L and 16 for W. $P = 32 + 16 + 32 + 16$
Add. $P = 96$

The perimeter of the rectangle is 96 ft.

Copyright © Houghton Mifflin Company. All rights reserved.

In this section, some of the phrases used to indicate the operations of addition and subtraction were presented. In solving application problems, you might also look for the types of questions listed below.

Addition	*Subtraction*
How many . . . altogether?	How many more (or fewer) . . . ?
How many . . . in all?	How much is left?
How many . . . and . . . ?	How much larger (or smaller) . . . ?

The bar graph in Figure 1.12 shows the number of fatal accidents on amusement rides in the United States each year during the 1990s. Use this graph for Example 11 and You Try It 11.

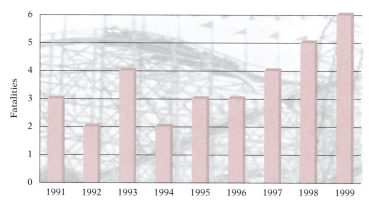

Figure 1.12   Number of Fatal Accidents on Amusement Rides
Source: USA Today, April 7, 2000

11 Example   Use Figure 1.12 to determine how many more fatal accidents occurred during the years 1995 through 1998 than occurred during the years 1991 through 1994.

Strategy   To find how many more occurred in 1995 through 1998 than occurred in 1991 through 1994:

▶ Find the total number of fatalities that occurred from 1995 to 1998 and the total number that occurred from 1991 to 1994.
▶ Subtract the smaller number from the larger.

Solution   Fatalities during 1995–1998: 15
Fatalities during 1991–1994: 11

$15 - 11 = 4$

4 more fatalities occurred from 1995 to 1998 than occurred from 1991 to 1994.

11 You Try It   Use Figure 1.12 to find the total number of fatal accidents on amusement rides during 1991 through 1999.

Your Strategy

Your Solution

Solution on p. S2

Copyright © Houghton Mifflin Company. All rights reserved.

12 Example

What is the price of a pair of skates that cost a business $109 and has a markup of $49? Use the formula $P = C + M$, where P is the price of a product to the consumer, C is the cost paid by the store for the product, and M is the markup.

Strategy

To find the price, replace C by 109 and M by 49 in the given formula and solve for P.

Solution

$P = C + M$

$P = 109 + 49$

$P = 158$

The price of the skates is $158.

12 You Try It

What is the price of a leather jacket that cost a business $148 and has a markup of $74? Use the formula $P = C + M$, where P is the price of a product to the consumer, C is the cost paid by the store for the product, and M is the markup.

Your Strategy

Your Solution

13 Example

Find the length of decorative molding needed to edge the top of the walls in a rectangular room that is 12 ft long and 8 ft wide.

Strategy

Draw a diagram.

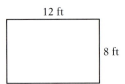

12 ft

8 ft

To find the length of molding needed, use the formula for the perimeter of a rectangle, $P = L + W + L + W$. $L = 12$ and $W = 8$.

Solution

$P = L + W + L + W$

$P = 12 + 8 + 12 + 8$

$P = 40$

40 ft of decorative molding are needed.

13 You Try It

Find the length of fencing needed to surround a rectangular corral that measures 60 ft on each side.

Your Strategy

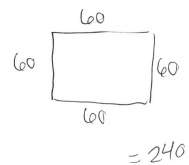

60

60   60

60

= 240

Your Solution

Solutions on p. S2

Copyright © Houghton Mifflin Company. All rights reserved.

1.2 EXERCISES

OBJECTIVE **A**

1. 🖉 Provide at least three examples of situations in which we add numbers.

2. 🖉 Explain how to estimate the sum of 3,287 and 4,916.

Add.

3. $\begin{array}{r} 732,453 \\ + 651,206 \\ \hline \end{array}$

4. $\begin{array}{r} 563,841 \\ + 726,053 \\ \hline \end{array}$

5. $\begin{array}{r} 2,879 \\ + 3,164 \\ \hline \end{array}$

6. $\begin{array}{r} 9,857 \\ + 1,264 \\ \hline \end{array}$

7. $\begin{array}{r} 45,825 \\ + 66,327 \\ \hline \end{array}$

8. $\begin{array}{r} 56,442 \\ + 71,289 \\ \hline \end{array}$

9. $\begin{array}{r} 4,037 \\ 3,342 \\ + 5,169 \\ \hline \end{array}$

10. $\begin{array}{r} 5,242 \\ 7,883 \\ + 4,165 \\ \hline \end{array}$

11. $\begin{array}{r} 67,390 \\ 42,761 \\ + 89,405 \\ \hline \end{array}$

12. $\begin{array}{r} 34,801 \\ 97,302 \\ + 68,945 \\ \hline \end{array}$

13. $\begin{array}{r} 54,097 \\ 33,432 \\ 97,126 \\ 64,508 \\ + 78,310 \\ \hline \end{array}$

14. $\begin{array}{r} 23,086 \\ 44,697 \\ 67,302 \\ 83,441 \\ + 19,843 \\ \hline \end{array}$

15. What is 88,123 increased by 80,451?

16. What is 44,765 more than 82,003?

17. What is 654 added to 7,293?

18. Find the sum of 658, 2,709, and 10,935.

19. Find the total of 216, 8,707, and 90,714.

20. Write the sum of x and y.

21. *Education*   Use the figure on the right to find the total number of undergraduates enrolled at the college in 2005.

22. *Education*   Use the figure on the right to find the total number of undergraduates enrolled at the college in 2006.

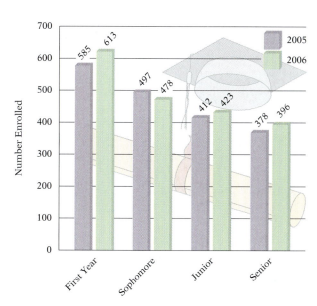

Undergraduates Enrolled in a Private College

Copyright © Houghton Mifflin Company. All rights reserved.

Estimate by rounding. Then find the exact answer.

23. $6,742 + 8,298$

24. $5,426 + 1,732$

25. $972,085 + 416,832$

26. $23,774 + 38,026$

27.
$$\begin{array}{r} 387 \\ 295 \\ 614 \\ + 702 \end{array}$$

28.
$$\begin{array}{r} 528 \\ 163 \\ 947 \\ + 275 \end{array}$$

29.
$$\begin{array}{r} 224,196 \\ 7,074 \\ + 98,531 \end{array}$$

30.
$$\begin{array}{r} 1,607 \\ 873,925 \\ + 28,744 \end{array}$$

Evaluate the variable expression $x + y$ for the given values of x and y.

31. $x = 574; y = 698$

32. $x = 359; y = 884$

33. $x = 4,752; y = 7,398$

34. $x = 6,047; y = 9,283$

35. $x = 38,229; y = 51,671$

36. $x = 74,376; y = 19,528$

Evaluate the variable expression $a + b + c$ for the given values of a, b, and c.

37. $a = 693; b = 508; c = 371$

38. $a = 177; b = 892; c = 405$

39. $a = 4,938; b = 2,615; c = 7,038$

40. $a = 6,059; b = 3,774; c = 5,136$

41. $a = 12,897; b = 36,075; c = 7,038$

42. $a = 52,847; b = 3,774; c = 5,136$

Identify the property that justifies the statement.

43. $9 + 12 = 12 + 9$

44. $8 + 0 = 8$

45. $11 + (13 + 5) = (11 + 13) + 5$

46. $0 + 16 = 16 + 0$

Copyright © Houghton Mifflin Company. All rights reserved.

47. $0 + 47 = 47$

48. $(7 + 8) + 10 = 7 + (8 + 10)$

Use the given property of addition to complete the statement.

49. The Addition Property of Zero
$28 + 0 = ?$

50. The Commutative Property of Addition
$16 + ? = 7 + 16$

51. The Associative Property of Addition
$9 + (? + 17) = (9 + 4) + 17$

52. The Addition Property of Zero
$0 + ? = 51$

53. The Commutative Property of Addition
$? + 34 = 34 + 15$

54. The Associative Property of Addition
$(6 + 18) + ? = 6 + (18 + 4)$

55. Is 38 a solution of the equation $42 = n + 4$?

56. Is 17 a solution of the equation $m + 6 = 13$?

57. Is 13 a solution of the equation $2 + h = 16$?

58. Is 41 a solution of the equation $n = 17 + 24$?

59. Is 30 a solution of the equation $32 = x + 2$?

60. Is 29 a solution of the equation $38 = 11 + z$?

OBJECTIVE **B**

61. Provide at least three examples of situations in which we subtract numbers.

62. What is the difference between an expression and an equation?

Subtract.

63.
$$\begin{array}{r} 883 \\ -\ 467 \\ \hline \end{array}$$

64.
$$\begin{array}{r} 591 \\ -\ 238 \\ \hline \end{array}$$

65.
$$\begin{array}{r} 360 \\ -\ 172 \\ \hline \end{array}$$

66.
$$\begin{array}{r} 950 \\ -\ 483 \\ \hline \end{array}$$

67.
$$\begin{array}{r} 657 \\ -\ 193 \\ \hline \end{array}$$

68.
$$\begin{array}{r} 762 \\ -\ 659 \\ \hline \end{array}$$

69.
$$\begin{array}{r} 407 \\ -\ 199 \\ \hline \end{array}$$

70.
$$\begin{array}{r} 805 \\ -\ 147 \\ \hline \end{array}$$

Copyright © Houghton Mifflin Company. All rights reserved.

| 71. | 6,814
− 3,257 | 72. | 7,361
− 4,575 | 73. | 5,000
− 2,164 | 74. | 4,000
− 1,873 |

| 75. | 3,400
− 1,963 | 76. | 7,300
− 2,562 | 77. | 30,004
− 9,856 | 78. | 70,003
− 8,246 |

79. Find the difference between 2,536 and 918.

80. What is 1,623 minus 287?

81. What is 5,426 less than 12,804?

82. Find 14,801 less 3,522.

83. Find 85,423 decreased by 67,875.

84. Write the difference between x and y.

85. *Geology* Use the figure on the right to find the difference between the maximum height to which Great Fountain erupts and the maximum height to which Valentine erupts.

86. *Geology* According to the figure on the right, how much higher is the eruption of the Giant than that of Old Faithful?

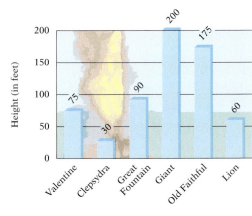

The Maximum Heights of the Eruptions of Six Geysers at Yellowstone National Park

Estimate by rounding. Then find the exact answer.

87. 7,355 − 5,219

88. 8,953 − 2,217

89. 59,126 − 20,843

90. 63,051 − 29,478

| 91. | 36,287
− 5,092 | 92. | 58,316
− 19,072 | 93. | 224,196
− 98,531 | 94. | 873,925
− 28,744 |

Copyright © Houghton Mifflin Company. All rights reserved.

Evaluate the variable expression $x - y$ for the given values of x and y.

95. $x = 50; y = 37$

96. $x = 80; y = 33$

97. $x = 914; y = 271$

98. $x = 623; y = 197$

99. $x = 740; y = 385$

100. $x = 870; y = 243$

101. $x = 8,672; y = 3,461$

102. $x = 7,814; y = 3,512$

103. $x = 1,605; y = 839$

104. $x = 1,406; y = 968$

105. $x = 23,409; y = 5,178$

106. $x = 56,397; y = 8,249$

107. Is 24 a solution of the equation $29 = 53 - y$?

108. Is 31 a solution of the equation $48 - p = 17$?

109. Is 44 a solution of the equation $t - 16 = 60$?

110. Is 25 a solution of the equation $34 = x - 9$?

111. Is 27 a solution of the equation $82 - z = 55$?

112. Is 28 a solution of the equation $72 = 100 - d$?

OBJECTIVE **C**

113. *Mathematics*   What is the sum of all the whole numbers less than 21?

114. *Mathematics*   Find the sum of all the natural numbers greater than 89 and less than 101.

115. *Mathematics*   Find the difference between the smallest four-digit number and the largest two-digit number.

116. *Demography*   The figure on the right shows the expected U.S. population aged 100 and over every two years from 2010 to 2020. (a) Which two-year period has the smallest increase in the number of people aged 100 and over? (b) Which two-year period has the greatest increase?

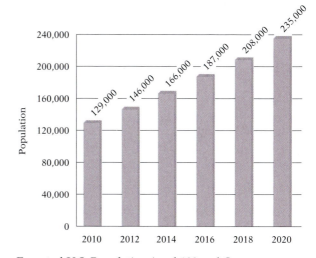

Expected U.S. Population Aged 100 and Over
Source: Census Bureau

Copyright © Houghton Mifflin Company. All rights reserved.

117. *Nutrition* You eat an apple and one cup of cornflakes with one tablespoon of sugar and one cup of milk for breakfast. Find the total number of calories consumed if one apple contains 80 calories, one cup of cornflakes has 95 calories, one tablespoon of sugar has 45 calories, and one cup of milk has 150 calories.

118. *Health* You are on a diet to lose weight and are limited to 1,500 calories per day. If your breakfast and lunch contained 950 calories, how many more calories can you consume during the rest of the day?

119. *Geometry* A rectangle has a length of 24 m and a width of 15 m. Find the perimeter of the rectangle.

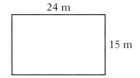

120. *Geometry* Find the perimeter of a rectangle that has a length of 18 ft and a width of 12 ft.

121. *Geometry* Find the perimeter of a triangle that has sides that measure 16 in., 12 in., and 15 in.

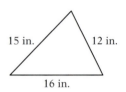

122. *Geometry* A triangle has sides of length 36 cm, 48 cm, and 60 cm. Find the perimeter of the triangle.

123. *Geometry* A rectangular playground has a length of 160 ft and a width of 120 ft. Find the length of hedge which surrounds the playground.

124. *Geometry* A rectangular vegetable garden has a length of 20 ft and a width of 14 ft. How many feet of wire fence should be purchased to surround the garden?

125. *History* The Gemini-Titan 7 space flight made 206 orbits of Earth. The Apollo-Saturn 7 space flight made 163 orbits of Earth. How many more orbits did the Gemini-Titan 7 flight make than the Apollo-Saturn 7 flight?

126. *Finances* You had $1,054 in your checking account before making a deposit of $870. Find the amount in your checking account after you made the deposit.

127. *Sports* The seating capacity of SAFECO Field in Seattle is 47,600. The seating capacity of Fenway Park in Boston is 33,871. Find the difference between the seating capacity of SAFECO Field and Fenway Park.

Fenway Park

128. *Finances* The repair bill on your car includes $358 for parts, $156 for labor, and a sales tax of $30. What is the total amount owed?

Copyright © Houghton Mifflin Company. All rights reserved.

129. *Finances*   The computer system you would like to purchase includes an operating system priced at $830, a monitor that costs $245, an extended keyboard priced at $175, and a printer that sells for $395. What is the total cost of the computer system?

130. 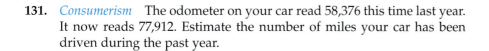 *Geography*   The area of Lake Superior is 81,000 mi^2; the area of Lake Michigan is 67,900 mi^2; the area of Lake Huron is 74,000 mi^2; the area of Lake Erie is 32,630 mi^2; and the area of Lake Ontario is 34,850 mi^2. Estimate the total area of the five Great Lakes.

131. *Consumerism*   The odometer on your car read 58,376 this time last year. It now reads 77,912. Estimate the number of miles your car has been driven during the past year.

The Great Lakes

The figure on the right shows the number of cars sold by a dealership for the first four months of 2005 and 2006. Use this graph for Exercises 132 to 134.

132. *Business*   Between which two months did car sales decrease the most in 2006? What was the amount of decrease?

133. *Business*   Between which two months did car sales increase the most in 2005? What was the amount of increase?

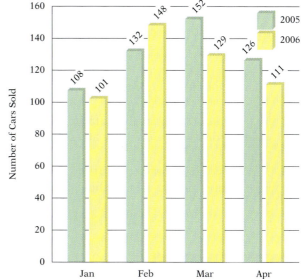

134. *Business*   In which year were more cars sold during the four months shown?

Car Sales at a Dealership

135. *Finances*   Use the formula $A = P + I$, where A is the value of an investment, P is the original investment, and I is the interest earned, to find the value of an investment that earned $775 in interest on an original investment of $12,500.

136. *Finances*   Use the formula $A = P + I$, where A is the value of an investment, P is the original investment, and I is the interest earned, to find the value of an investment that earned $484 in interest on an original investment of $8,800.

137. *Finances*   What is the mortgage loan amount on a home that sells for $290,000 with a down payment of $29,000? Use the formula $M = S - D$, where M is the mortgage loan amount, S is the selling price, and D is the down payment.

138. *Finances*   What is the mortgage loan amount on a home that sells for $236,000 with a down payment of $47,200? Use the formula $M = S - D$, where M is the mortgage loan amount, S is the selling price, and D is the down payment.

Copyright © Houghton Mifflin Company. All rights reserved.

139. *Physics*   What is the ground speed of an airplane traveling into a 25 mph head wind with an air speed of 375 mph? Use the formula $g = a - h$, where g is the ground speed, a is the air speed, and h is the speed of the head wind.

140. *Physics*   Find the ground speed of an airplane traveling into a 15 mph head wind with an air speed of 425 mph? Use the formula $g = a - h$, where g is the ground speed, a is the air speed, and h is the speed of the head wind.

In some states, the speed limit on certain sections of highway is 70 mph. To test drivers' compliance with the speed limit, the highway patrol conducted a one-week study during which they recorded the speeds of motorists on one of these sections of highway. The results are recorded in the table at the right. Use this table for Exercises 141 to 144.

Speed	Number of Cars
> 80	1,708
76 – 80	2,503
71 – 75	3,651
66 – 70	3,717
61 – 65	2,984
< 61	2,870

141. *Statistics*   (a) How many drivers were traveling at 70 mph or less? (b) How many drivers were traveling at 76 mph or more?

142. *Statistics*   Looking at the data in the table, is it possible to tell how many motorists were driving at 70 mph? Explain your answer.

143. *Statistics*   Looking at the data in the table, is it possible to tell how many motorists were driving at less than 70 mph? Explain your answer.

144. *Statistics*   Are more people driving at or below the posted speed limit, or are more people driving above the posted speed limit?

Critical Thinking

145. *Dice*   If you roll two ordinary six-sided dice and add the two numbers that appear on top, how many different sums are possible?

146. *Mathematics*   How many two-digit numbers are there? How many three-digit numbers are there?

147. Determine whether the statement is always true, sometimes true, or never true.
   a.   If a is any whole number, then $a - 0 = a$.
   b.   If a is any whole number, then $a - a = 0$.

148. Find the circulation of your local newspaper and the population of the area served by that paper. What is the difference between the area's population and the newspaper's circulation? Why would this figure be of concern to the owner of the newspaper?

149. What estimate is given for the size of the population in your state by the year 2025? What is the estimate of the size of the population in the United States by the year 2025? Estimates differ. On what basis was the estimate you recorded derived?

Copyright © Houghton Mifflin Company. All rights reserved.

Multiplication and Division of Whole Numbers

OBJECTIVE   **A**

Multiplication of whole numbers

A store manager orders six boxes of telephone answering machines. Each box contains eight answering machines. How many answering machines are ordered?

The answer can be calculated by adding 6 eights.

$$8 + 8 + 8 + 8 + 8 + 8 = 48$$

This problem involves repeated addition of the same number. The answer can be calculated by a shorter process called multiplication. **Multiplication** is the repeated addition of the same number.

There is a total of 48 dots on the 6 dominoes.

The numbers that are multiplied are called **factors.** The answer is called the **product.**

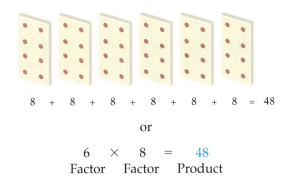

$$8 \; + \; 8 \; + \; 8 \; + \; 8 \; + \; 8 \; + \; 8 \; = \; 48$$

or

$$6 \quad \times \quad 8 \quad = \quad 48$$
Factor   Factor   Product

The times sign "×" is one symbol that is used to mean multiplication. Each of the expressions below also represents multiplication.

$$6 \cdot 8 \qquad 6(8) \qquad (6)(8) \qquad 6a \qquad 6(a) \qquad ab$$

The expression $6a$ means "6 times a." The expression ab means "a times b."

The basic facts for multiplying one-digit numbers should be memorized. Multiplication of larger numbers requires the repeated use of the basic multiplication facts.

▶ Point of Interest

The cross X was first used as a symbol for multiplication in 1631 in a book titled *The Key to Mathematics.* Also in that year, another book, *Practice of the Analytical Art,* advocated the use of a dot to indicate multiplication.

Multiply: 37(4)

Multiply 4 · 7.

4 · 7 = 28 (2 tens + 8 ones).

Write the 8 in the ones column and carry the 2 to the tens column.

$$\begin{array}{r} {\scriptstyle 2} \\ 3 \; 7 \\ \times \quad 4 \\ \hline 8 \end{array}$$

The 3 in 37 is 3 tens.

Multiply 4 · 3 tens.
Add the carry digit.

Write the 14.

$$4 \cdot 3 \text{ tens} = \quad \begin{array}{r} 12 \text{ tens} \\ + \; 2 \text{ tens} \\ \hline 14 \text{ tens} \end{array}$$

$$\begin{array}{r} {\scriptstyle 2} \\ 3 \; 7 \\ \times \quad 4 \\ \hline 14 \; 8 \end{array}$$

Copyright © Houghton Mifflin Company. All rights reserved.

In the preceding example, a number was multiplied by a one-digit number. The examples that follow illustrate multiplication by larger numbers.

Multiply: (47)(23)

Multiply by the ones digit.	Multiply by the tens digit.	Add.
$3 \cdot 47 = 141$.	$2 \cdot 47 = 94$.	

$$\begin{array}{r} 47 \\ \times\ 23 \\ \hline 141 \end{array}$$

$$\begin{array}{r} 47 \\ \times\ 23 \\ \hline 141 \\ 94 \end{array}$$

$$\begin{array}{r} 47 \\ \times\ 23 \\ \hline 141 \\ 94 \\ \hline 1{,}081 \end{array}$$

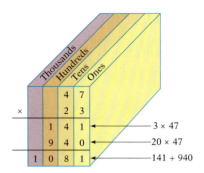

The last digit is written in the ones column.

The last digit is written in the tens column.

The place-value chart illustrates the placement of the products.

Note the placement of the products when multiplying by a factor that contains a zero.

Multiply: 439(206)

$$\begin{array}{r} 439 \\ \times\ 206 \\ \hline 2\ 634 \\ 0\ 00 \\ 87\ 8 \\ \hline 90{,}434 \end{array}$$

When working the problem, usually only one zero is written, as shown at the right. Writing this zero ensures the proper placement of the products.

$$\begin{array}{r} 439 \\ \times\ 206 \\ \hline 2\ 634 \\ 87\ 80 \\ \hline 90{,}434 \end{array}$$

Note the pattern when the following numbers are multiplied.

Multiply the nonzero part of the factors. Attach the same number of zeros in the product as the total number of zeros in the factors.

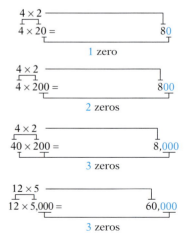

Find the product of 600 and 70.

Remember that a *product* is the answer to a multiplication problem.

$600 \cdot 70 = 42{,}000$

Copyright © Houghton Mifflin Company. All rights reserved.

Multiply: 3(20)(10)(4)

Multiply the first two numbers. $3(20)(10)4 = 60(10)(4)$

Multiply the product by the third number. $= (600)(4)$

Continue multiplying until all the numbers have been multiplied. $= 2{,}400$

Figure 1.13 shows the average weekly earnings of full-time workers in the United States. Using these figures, calculate the earnings of a female full-time worker, age 22, for working for 4 weeks.

Multiply the number of weeks (4) times the amount earned for one week ($354).

$4(354) = 1{,}416$

The average earnings of a 22-year-old, female, full-time worker, for working for 4 weeks are $1,416.

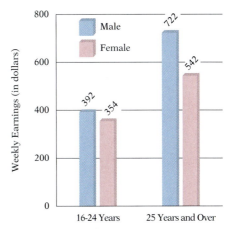

Figure 1.13   Average Weekly Earnings of Full-Time Workers
Source: Bureau of Labor Statistics

Estimate the product of 345 and 92.

Round each number to its highest place value. $345 \longrightarrow 300$
$92 \longrightarrow 90$

Multiply the rounded numbers. $300 \cdot 90 = 27{,}000$

27,000 is an estimate of the product of 345 and 92.

The phrase *the product of* was used in the example above to indicate the operation of multiplication. All of the phrases below indicate multiplication. An example of each is shown at the right of each phrase.

times	8 times 4	$8 \cdot 4$
the product of	the product of 9 and 5	$9 \cdot 5$
multiplied by	7 multiplied by 3	$3 \cdot 7$
twice	twice 6	$2 \cdot 6$

Evaluate xyz when $x = 50$, $y = 2$, and $z = 7$.

xyz means $x \cdot y \cdot z$. xyz

Replace each variable by its value. $50 \cdot 2 \cdot 7$

Multiply the first two numbers. $= 100 \cdot 7$

Multiply the product by the next number. $= 700$

Copyright © Houghton Mifflin Company. All rights reserved.

As for addition, there are properties of multiplication.

> **The Multiplication Property of Zero**
>
> $a \cdot 0 = 0$ or $0 \cdot a = 0$

$8 \cdot 0 = 0$

The Multiplication Property of Zero states that the product of a number and zero is zero. The variable a is used here to represent any whole number. It can even represent the number zero because $0 \cdot 0 = 0$.

> **The Multiplication Property of One**
>
> $a \cdot 1 = a$ or $1 \cdot a = a$

$1 \cdot 9 = 9$

The Multiplication Property of One states that the product of a number and one is the number. Multiplying a number by 1 does not change the number.

> **The Commutative Property of Multiplication**
>
> $a \cdot b = b \cdot a$

$4 \cdot 9 = 9 \cdot 4$
$36 = 36$

The Commutative Property of Multiplication states that two numbers can be multiplied in either order; the product will be the same. Here the variables a and b represent any whole numbers. Therefore, for example, if you know that the product of 4 and 9 is 36, then you also know that the product of 9 and 4 is 36 because $4 \cdot 9 = 9 \cdot 4$.

> **The Associative Property of Multiplication**
>
> $(a \cdot b) \cdot c = a \cdot (b \cdot c)$

$(2 \cdot 3) \cdot 4 = 2 \cdot (3 \cdot 4)$
$6 \cdot 4 = 2 \cdot 12$
$24 = 24$

The Associative Property of Multiplication states that when multiplying three numbers, the numbers can be grouped in any order; the product will be the same. Note in the example at the right above that we can multiply the product of 2 and 3 by 4, or we can multiply 2 by the product of 3 and 4. In either case, the product of the three numbers is 24.

Copyright © Houghton Mifflin Company. All rights reserved.

What is the solution of the equation $5x = 5$?

By the Multiplication Property of One, the product of a number and 1 is the number.

The solution is 1.

The check is shown at the right.

$$5x = 5$$

5(1)	5

$$5 = 5$$

Is 7 a solution of the equation $3m = 21$?

Replace m by 7.

Simplify the left side of the equation.

The results are equal.   Yes, 7 is a solution of the equation.

$$3m = 21$$

3(7)	21

$$21 = 21$$

 Figure 1.14 shows the average monthly savings of individuals in seven different countries. Use this graph for Example 1 and You Try It 1.

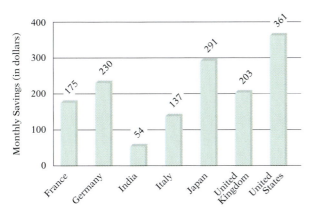

Figure 1.14   Average Monthly Savings
  Source: Taylor Nelson - Sofres for American Express

1 *Example*   Use Figure 1.14 to determine the average annual savings of individuals in Japan.

***Solution*   The average monthly savings in Japan is $291. The number of months in one year is 12.**

$$
\begin{array}{r}
291 \\
\times\ \ 12 \\
\hline
582 \\
291 \\
\hline
3{,}492
\end{array}
$$

The average annual savings of individuals in Japan is $3,492.

1 *You Try It*   According to Figure 1.14, what is the average annual savings of individuals in France?

Your Solution

Solution on p. S2

Copyright © Houghton Mifflin Company. All rights reserved.

2 *Example* Estimate the product of 2,871 and 49.

Solution

2,871 ⟶ 3,000
49 ⟶ 50

3,000 · 50 = 150,000

2 *You Try It* Estimate the product of 8,704 and 93.

Your Solution

---

3 *Example* Evaluate $3ab$ when $a = 10$ and $b = 40$.

Solution $3ab$
$3(10)(40) = 30(40)$
$= 1,200$

3 *You Try It* Evaluate $5xy$ when $x = 20$ and $y = 60$.

Your Solution

---

4 *Example* What is 800 times 300?

Solution $800 \cdot 300 = 240,000$

4 *You Try It* What is 90 multiplied by 7,000?

Your Solution

---

5 *Example* Complete the statement by using the Associative Property of Multiplication.

$(7 \cdot 8) \cdot 5 = 7 \cdot (? \cdot 5)$

Solution $(7 \cdot 8) \cdot 5 = 7 \cdot (8 \cdot 5)$

5 *You Try It* Complete the statement by using the Multiplication Property of Zero.

$? \cdot 10 = 0$

Your Solution

---

6 *Example* Is 9 a solution of the equation $82 = 9q$?

Solution

$82 = 9q$

$\dfrac{82 \,\big|\, 9(9)}{}$

$82 \neq 81$

No, 9 is not a solution of the equation.

6 *You Try It* Is 11 a solution of the equation $7a = 77$?

Your Solution

Solutions on p. S2

OBJECTIVE **B**

Exponents

Repeated multiplication of the same factor can be written in two ways:

$4 \cdot 4 \cdot 4 \cdot 4 \cdot 4$ or 4^5 ⟵ exponent
                              base

The expression 4^5 is in **exponential form.** The **exponent,** 5, indicates how many times the **base,** 4, occurs as a factor in the multiplication.

▶ Point of Interest

Lao-tzu, founder of Taoism, wrote:
Counting gave birth to Addition,
Addition gave birth to Multiplication,
Multiplication gave birth to Expo-
nentiation, Exponentiation gave
birth to all the myriad operations.

Copyright © Houghton Mifflin Company. All rights reserved.

It is important to be able to read numbers written in exponential form.

$2 = 2^1$ Read "two to the first power" or just "two."
Usually the 1 is not written.

$2 \cdot 2 = 2^2$ Read "two squared" or "two to the second power."

$2 \cdot 2 \cdot 2 = 2^3$ Read "two cubed" or "two to the third power."

$2 \cdot 2 \cdot 2 \cdot 2 = 2^4$ Read "two to the fourth power."

$2 \cdot 2 \cdot 2 \cdot 2 \cdot 2 = 2^5$ Read "two to the fifth power."

Variable expressions can contain exponents.

$x^1 = x$ x to the first power is usually written simply as x.

$x^2 = x \cdot x$ x^2 means x times x.

$x^3 = x \cdot x \cdot x$ x^3 means x occurs as a factor 3 times.

$x^4 = x \cdot x \cdot x \cdot x$ x^4 means x occurs as a factor 4 times.

Point of Interest

One billion is too large a number for most of us to comprehend. If a computer were to start counting from 1 to 1 billion, writing to the screen one number every second of every day, it would take over 31 years for the computer to complete the task.

And if a billion is a large number, consider a googol. A googol is 1 with 100 zeros after it, or 10^{100}. Edward Kasner is the mathematician credited with thinking up this number, and his nine-year-old nephew is said to have thought up the name. The two then coined the word googolplex, which is 10^{googol}.

Each place value in the place-value chart can be expressed as a power of 10.

$$\text{Ten} = 10 = 10 = 10^1$$
$$\text{Hundred} = 100 = 10 \cdot 10 = 10^2$$
$$\text{Thousand} = 1{,}000 = 10 \cdot 10 \cdot 10 = 10^3$$
$$\text{Ten-thousand} = 10{,}000 = 10 \cdot 10 \cdot 10 \cdot 10 = 10^4$$
$$\text{Hundred-thousand} = 100{,}000 = 10 \cdot 10 \cdot 10 \cdot 10 \cdot 10 = 10^5$$
$$\text{Million} = 1{,}000{,}000 = 10 \cdot 10 \cdot 10 \cdot 10 \cdot 10 \cdot 10 = 10^6$$

Note that the exponent on 10 when the number is written in exponential form is the same as the number of zeros in the number written in standard form. For example, $10^5 = 100{,}000$; the exponent on 10 is 5, and the number 100,000 has 5 zeros.

To evaluate a numerical expression containing exponents, write each factor as many times as indicated by the exponent and then multiply.

$$5^3 = 5 \cdot 5 \cdot 5 = 25 \cdot 5 = 125$$
$$2^3 \cdot 6^2 = (2 \cdot 2 \cdot 2) \cdot (6 \cdot 6) = 8 \cdot 36 = 288$$

Evaluate the variable expression c^3 when $c = 4$.

$$c^3 = c \cdot c \cdot c$$
$$4^3 = 4 \cdot 4 \cdot 4$$
$$= 16 \cdot 4 = 64$$

Replace c with 4 and then evaluate the exponential expression.

Calculator Note

A calculator can be used to evaluate an exponential expression. The y^x key (or on some calculators a x^y key or $\wedge$ key) is used to enter the exponent. For instance, for the example at the left, enter 4 y^x 3 $=$. The display reads 64.

7 Example   Write $7 \cdot 7 \cdot 7 \cdot 4 \cdot 4$ in exponential form.

Solution $7 \cdot 7 \cdot 7 \cdot 4 \cdot 4 = 7^3 \cdot 4^2$

7 You Try It   Write $2 \cdot 2 \cdot 2 \cdot 3 \cdot 3 \cdot 3 \cdot 3$ in exponential form.

Your Solution

Solution on p. S2

Copyright © Houghton Mifflin Company. All rights reserved.

8 *Example*   Evaluate 8^3.

Solution $8^3 = 8 \cdot 8 \cdot 8 = 64 \cdot 8 = 512$

8 *You Try It*   Evaluate 6^4.

Your Solution

9 *Example*   Evaluate 10^7.

Solution $10^7 = 10,000,000$

(The exponent on 10 is 7. There are 7 zeros in 10,000,000.)

9 *You Try It*   Evaluate 10^8.

Your Solution

10 *Example*   Evaluate $3^3 \cdot 5^2$.

Solution $3^3 \cdot 5^2 = (3 \cdot 3 \cdot 3) \cdot (5 \cdot 5)$
$= 27 \cdot 25 = 675$

10 *You Try It*   Evaluate $2^4 \cdot 3^2$.

Your Solution

11 *Example*   Evaluate x^2y^3 when $x = 4$ and $y = 2$.

Solution x^2y^3 (x^2y^3 means x^2 times y^3.)

$4^2 \cdot 2^3 = (4 \cdot 4) \cdot (2 \cdot 2 \cdot 2)$
$= 16 \cdot 8$
$= 128$

11 *You Try It*   Evaluate x^4y^2 when $x = 1$ and $y = 3$.

Your Solution

Solutions on p. S2

OBJECTIVE **C**

Division of whole numbers

Division is used to separate objects into equal groups.

A grocer wants to distribute 24 new products equally on 4 shelves. From the diagram, we see that the grocer would place 6 products on each shelf.

The grocer's problem could be written:

```
                                    ┌─── Number on each shelf
                                6 ←        Quotient
Number of shelves ──→ 4)24 ←──── Number of objects
          Divisor                       Dividend
```

▶ **Point of Interest**

The Chinese divided a day into 100 k'o, which was a unit equal to a little less than 15 min. Sundials were used to measure time during the daylight hours, and by A.D. 500, candles, water clocks, and incense sticks were used to measure time at night.

Note that the quotient multiplied by the divisor equals the dividend.

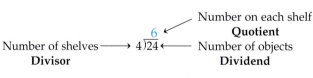

$$\frac{6}{4)24} \quad \text{because} \quad \boxed{\begin{matrix} 6 \\ \text{Quotient} \end{matrix}} \times \boxed{\begin{matrix} 4 \\ \text{Divisor} \end{matrix}} = \boxed{\begin{matrix} 24 \\ \text{Dividend} \end{matrix}}$$

Copyright © Houghton Mifflin Company. All rights reserved.

Division is also represented by the symbol ÷ or by a fraction bar. Both are read "divided by."

$$9\overline{)54}^{\;6} \qquad 54 \div 9 = 6 \qquad \frac{54}{9} = 6$$

The fact that the quotient times the divisor equals the dividend can be used to illustrate properties of division.

$0 \div 4 = 0$ because $0 \cdot 4 = 0.$

$4 \div 4 = 1$ because $1 \cdot 4 = 4.$

$4 \div 1 = 4$ because $4 \cdot 1 = 4.$

$4 \div 0 = ?$ What number can be multiplied by 0 to get 4? $? \cdot 0 = 4$
There is no number whose product with 0 is 4
because the product of a number and zero is 0.
Division by zero is undefined.

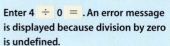
Calculator Note

Enter 4 ÷ 0 = . An error message is displayed because division by zero is undefined.

The properties of division are stated below. In these statements, the symbol ≠ is read "is not equal to."

Division Properties of Zero and One

If $a \neq 0$, $0 \div a = 0$.   Zero divided by any number other than zero is zero.

If $a \neq 0$, $a \div a = 1$.   Any number other than zero divided by itself is one.

$a \div 1 = a$ A number divided by one is the number.

$a \div 0$ is undefined.   Division by zero is undefined.

Take Note

Recall that the variable a represents any whole number. Therefore, for the first two properties, we must state that $a \neq 0$ in order to ensure that we are not dividing by zero.

The example below illustrates division of a larger whole number by a one-digit number.

Divide and check: $3{,}192 \div 4$

$$
\begin{array}{r}
7 \\
4\overline{)3{,}192} \\
-2\,8 \\
\hline
39
\end{array}
$$
Think 31 ÷ 4.
Subtract 7 × 4.
Bring down the 9.

$$
\begin{array}{r}
79 \\
4\overline{)3{,}192} \\
-2\,8 \\
\hline
39 \\
-36 \\
\hline
32
\end{array}
$$
Think 39 ÷ 4.
Subtract 9 × 4.
Bring down the 2.

$$
\begin{array}{r}
798 \\
4\overline{)3{,}192} \\
-2\,8 \\
\hline
39 \\
-36 \\
\hline
32 \\
-32 \\
\hline
0
\end{array}
$$
Think 32 ÷ 4.
Subtract 8 × 4.

Check:
$$
\begin{array}{r}
798 \\
\times\;\;\;4 \\
\hline
3{,}192
\end{array}
$$

Copyright © Houghton Mifflin Company. All rights reserved.

The place-value chart is used to show why this method works.

$$
\begin{array}{r}
7\,9\,8 \\
4\overline{)3,1\,9\,2} \\
-2\,8\,0\,0 \\
\hline
3\,9\,2 \\
-3\,6\,0 \\
\hline
3\,2 \\
-3\,2 \\
\hline
0
\end{array}
$$

7 hundreds × 4

9 tens × 4

8 ones × 4

Sometimes it is not possible to separate objects into a whole number of equal groups.

A packer at a bakery has 14 muffins to pack into 3 boxes. Each box will hold 4 muffins. From the diagram, we see that after the packer places 4 muffins in each box, there are 2 muffins left over. The 2 is called the **remainder.**

The packer's division problem could be written

Number in each box — Quotient

Number of boxes ⟶ $3\overline{)14}$ ⟵ Total number of muffins   or $4\ r2$ $3\overline{)14}$

Divisor -12 **Dividend**

                      2 ⟵ Number left over

                      Remainder

4 ⟵ Quotient

For any division problem, **(quotient · divisor) + remainder = dividend.** This result can be used to check a division problem.

Find the quotient of 389 and 24.

$$
\begin{array}{r}
16\ r5 \\
24\overline{)389} \\
-24 \\
\hline
149 \\
-144 \\
\hline
5
\end{array}
$$

Check: $(16 \cdot 24) + 5 = 384 + 5 = 389$

The phrase *the quotient of* was used in the example above to indicate the operation of division. The phrase *divided by* also indicates division.

| **the quotient of** | the quotient of 8 and 4 | $8 \div 4$ |
| **divided by** | 9 divided by 3 | $9 \div 3$ |

Copyright © Houghton Mifflin Company. All rights reserved.

Estimate the result when 56,497 is divided by 28.

Round each number to its highest place value.   56,497 ⟶ 60,000

28 ⟶ 30

Divide the rounded numbers. $60,000 \div 30 = 2,000$

2,000 is an estimate of 56,497 ÷ 28.

Evaluate $\dfrac{x}{y}$ when $x = 4,284$ and $y = 18$.

Replace x with 4,284 and y with 18. $\dfrac{x}{y}$

$\dfrac{4,284}{18}$ means 4,284 ÷ 18. $\dfrac{4,284}{18} = 238$

Is 42 a solution of the equation $\dfrac{x}{6} = 7$? $\dfrac{x}{6} = 7$

Replace x by 42.

$\dfrac{42}{6} \,\bigg|\, 7$

Simplify the left side of the equation.

The results are equal. $7 = 7$

42 is a solution of the equation.

---

12 *Example*   What is the quotient of 8,856 and 42?

Solution

$$
\begin{array}{r}
210 \text{ r}36 \\
42\overline{)8,856} \\
-\,8\,4 \\
\hline
45 \\
-\,42 \\
\hline
36 \\
-\,\,0 \\
\hline
36
\end{array}
$$

Think $42\overline{)36}$.
Subtract 0 · 42.

Check: $(210 \cdot 42) + 36$
$= 8,820 + 36 = 8,856$

12 *You Try It*   What is 7,694 divided by 24?

Your Solution

Solution on p. S2

Copyright © Houghton Mifflin Company. All rights reserved.

Figure 1.15 shows a household's annual expenses of $44,000. Use this graph for Example 13 and You Try It 13.

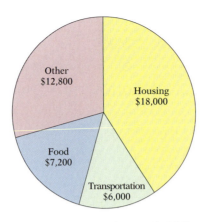

Other
$12,800

Housing
$18,000

Food
$7,200

Transportation
$6,000

Figure 1.15 Annual Household Expenses

---

13 **Example** Use Figure 1.15 to find the household's monthly expense for housing.

Solution The annual expense for housing is $18,000.

$18{,}000 \div 12 = 1{,}500$

The monthly expense is $1,500.

13 **You Try It** Use Figure 1.15 to find the household's monthly expense for food.

Your Solution

---

14 **Example** Estimate the quotient of 55,272 and 392.

Solution
$55{,}272 \longrightarrow 60{,}000$
$392 \longrightarrow 400$

$60{,}000 \div 400 = 150$

14 **You Try It** Estimate the quotient of 216,936 and 207.

Your Solution

---

15 **Example** Evaluate $\dfrac{x}{y}$ when $x = 342$ and $y = 9$.

Solution $\dfrac{x}{y}$

$\dfrac{342}{9} = 38$

15 **You Try It** Evaluate $\dfrac{x}{y}$ when $x = 672$ and $y = 8$.

Your Solution

---

16 **Example** Is 28 a solution of the equation $\dfrac{x}{7} = 4$?

Solution $\dfrac{x}{7} = 4$

$\dfrac{28}{7} \,\Big|\, 4$

$4 = 4$

Yes, 28 is a solution of the equation.

16 **You Try It** Is 12 a solution of the equation $\dfrac{60}{y} = 2$?

Your Solution

Solutions on p. S2

Copyright © Houghton Mifflin Company. All rights reserved.

OBJECTIVE **D**

Factors and prime factorization

Natural number factors of a number divide that number evenly (there is no remainder).

1, 2, 3, and 6 are natural number factors of 6 because they divide 6 evenly.

Note that both the divisor and the quotient are factors of the dividend.

$$\begin{array}{cccc} 6 & 3 & 2 & 1 \\ 1\overline{)6} & 2\overline{)6} & 3\overline{)6} & 6\overline{)6} \end{array}$$

To find the factors of a number, try dividing the number by 1, 2, 3, 4, 5, ... Those numbers that divide the number evenly are its factors. Continue this process until the factors start to repeat.

Find all the factors of 42.

$42 \div 1 = 42$	1 and 42 are factors of 42.
$42 \div 2 = 21$	2 and 21 are factors of 42.
$42 \div 3 = 14$	3 and 14 are factors of 42.
$42 \div 4$	4 will not divide 42 evenly.
$42 \div 5$	5 will not divide 42 evenly.
$42 \div 6 = 7$	6 and 7 are factors of 42.
$42 \div 7 = 6$	7 and 6 are factors of 42.

The factors are repeating.
All the factors of 42 have been found.

The factors of 42 are 1, 2, 3, 6, 7, 14, 21, and 42.

The following rules are helpful in finding the factors of a number:

2 is a factor of a number if the digit in the ones' place of the number is 0, 2, 4, 6, or 8.

436 ends in 6.
Therefore, 2 is a factor of 436
$(436 \div 2 = 218)$.

3 is a factor of a number if the sum of the digits of the number is divisible by 3.

The sum of the digits of 489 is $4 + 8 + 9 = 21$.
21 is divisible by 3.
Therefore, 3 is a factor of 489
$(489 \div 3 = 163)$.

4 is a factor of a number if the last two digits of the number are divisible by 4.

556 ends in 56.
56 is divisible by 4 $(56 \div 4 = 14)$.
Therefore, 4 is a factor of 556
$(556 \div 4 = 139)$.

5 is a factor of a number if the ones' digit of the number is 0 or 5.

520 ends in 0.
Therefore, 5 is a factor of 520
$(520 \div 5 = 104)$.

A **prime number** is a natural number greater than 1 that has exactly two natural number factors, 1 and the number itself. 7 is prime because its only factors are 1 and 7. If a number is not prime, it is a **composite** number. Because 6 has factors of 2 and 3, 6 is a composite number. The prime numbers less than 50 are

2, 3, 5, 7, 11, 13, 17, 19, 23, 29, 31, 37, 41, 43, 47

Point of Interest

Twelve is the smallest abundant number, or number whose proper divisors add up to more than the number itself. The proper divisors of a number are all of its factors except the number itself. The proper divisors of 12 are 1, 2, 3, 4, and 6, which add up to 16, which is greater than 12. There are 246 abundant numbers between 1 and 1,000.

A perfect number is one whose proper divisors add up to exactly that number. For example, the proper divisors of 6 are 1, 2, and 3, which add up to 6. There are only three perfect numbers less than 1,000: 6, 28, and 496.

Copyright © Houghton Mifflin Company. All rights reserved.

The **prime factorization** of a number is the expression of the number as a product of its prime factors. To find the prime factors of 90, begin with the smallest prime number as a trial divisor and continue with prime numbers as trial divisors until the final quotient is prime.

Find the prime factorization of 90.

$$\begin{array}{r} 45 \\ 2\overline{)90} \end{array}$$

$$\begin{array}{r} 15 \\ 3\overline{)45} \\ 2\overline{)90} \end{array}$$

$$\begin{array}{r} 5 \\ 3\overline{)15} \\ 3\overline{)45} \\ 2\overline{)90} \end{array}$$

Divide 90 by 2.

45 is not divisible by 2.
Divide 45 by 3.

Divide 15 by 3.
5 is prime.

The prime factorization of 90 is $2 \cdot 3 \cdot 3 \cdot 5$, or $2 \cdot 3^2 \cdot 5$.

Finding the prime factorization of larger numbers can be more difficult. Try each prime number as a trial divisor. Stop when the square of the trial divisor is greater than the number being factored.

Find the prime factorization of 201.

$$\begin{array}{r} 67 \\ 3\overline{)201} \end{array}$$

67 cannot be divided evenly by 2, 3, 5, 7, or 11. Prime numbers greater than 11 need not be tried because $11^2 = 121$ and $121 > 67$.

The prime factorization of 201 is $3 \cdot 67$.

---

17 *Example* Find all the factors of 40.

Solution

$40 \div 1 = 40$
$40 \div 2 = 20$
$40 \div 3$ Does not divide evenly.
$40 \div 4 = 10$
$40 \div 5 = 8$
$40 \div 6$ Does not divide evenly.
$40 \div 7$ Does not divide evenly.
$40 \div 8 = 5$ The factors are repeating.

The factors of 40 are 1, 2, 4, 5, 8, 10, 20, and 40.

17 *You Try It* Find all the factors of 30.

Your Solution

---

18 *Example* Find the prime factorization of 84.

Solution

$$\begin{array}{r} 7 \\ 3\overline{)21} \\ 2\overline{)42} \\ 2\overline{)84} \end{array}$$

$84 = 2 \cdot 2 \cdot 3 \cdot 7 = 2^2 \cdot 3 \cdot 7$

18 *You Try It* Find the prime factorization of 88.

Your Solution

Solutions on p. S3

Copyright © Houghton Mifflin Company. All rights reserved.

Example Find the prime factorization of 141.

Solution

$$\begin{array}{r} 47 \\ 3\overline{)141} \end{array}$$
▶ Try only 2, 3, 5, and 7 because $7^2 = 49$ and $49 > 47$.

$141 = 3 \cdot 47$

You Try It Find the prime factorization of 295.

Your Solution

Solution on p. S3

OBJECTIVE E

Applications and formulas

In Section 1.2, we defined perimeter as the distance around a plane figure. The perimeter of a rectangle was given as $P = L + W + L + W$. This formula is commonly written as $P = 2L + 2W$.

Perimeter of a Rectangle

The formula for the perimeter of a rectangle is $P = 2L + 2W$, where P is the perimeter of the rectangle, L is the length, and W is the width.

Take Note
Remember that $2L$ means 2 times L, and $2W$ means 2 times W.

Find the perimeter of the rectangle shown at the right.

Use the formula for the perimeter of a rectangle. $P = 2L + 2W$
Substitute 32 for L and 16 for W. $P = 2(32) + 2(16)$
Find the product of 2 and 32 and the product of $P = 64 + 32$
2 and 16.
Add. $P = 96$

The perimeter of the rectangle is 96 ft.

32 ft

16 ft

A **square** is a rectangle in which each side has the same length. Letting s represent the length of each side of a square, the perimeter of a square can be represented $P = s + s + s + s$. Note that we are adding *four* s's. We can write the addition as multiplication: $P = 4s$.

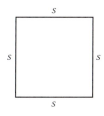

$P = s + s + s + s$
$P = 4s$

Perimeter of a Square

The formula for the perimeter of a square is $P = 4s$, where P is the perimeter and s is the length of a side of a square.

Copyright © Houghton Mifflin Company. All rights reserved.

28 km

Find the perimeter of the square shown at the left.

Use the formula for the perimeter of a square.	$P = 4s$
Substitute 28 for s.	$P = 4(28)$
Multiply.	$P = 112$

The perimeter of the square is 112 km.

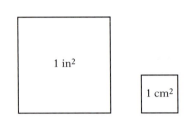

1 in²

1 cm²

Area is the amount of surface in a region. Area can be used to describe the size of a skating rink, the floor of a room, or a playground. Area is measured in square units.

A square that measures 1 inch on each side has an area of 1 square inch, which is written 1 in². A square that measures 1 centimeter on each side has an area of 1 square centimeter, which is written 1 cm².

Larger areas can be measured in square feet (ft²), square meters (m²), acres (43,560 ft²), square miles (mi²), or any other square unit.

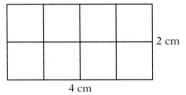

2 cm

4 cm

The area of the rectangle is 8 cm².

The area of a geometric figure is the number of squares that are necessary to cover the figure. In the figure at the left, a rectangle has been drawn and covered with squares. Eight squares, each of area 1 cm², were used to cover the rectangle. The area of the rectangle is 8 cm². Note from this figure that the area of a rectangle can be found by multiplying the length of the rectangle by its width.

> ### Area of a Rectangle
>
> The formula for the area of a rectangle is $A = LW$, where A is the area, L is the length, and W is the width of the rectangle.

10 ft

25 ft

Find the area of the rectangle shown at the left.

Use the formula for the area of a rectangle.	$A = LW$
Substitute 25 for L and 10 for W.	$A = 25(10)$
Multiply.	$A = 250$

The area of the rectangle is 250 ft².

s

$A = s \cdot s = s^2$

A square is a rectangle in which all sides are the same length. Therefore, both the length and the width of a square can be represented by s, and $A = LW = s \cdot s = s^2$.

> ### Area of a Square
>
> The formula for the area of a rectangle is $A = s^2$, where A is the area and s is the length of a side of a square.

Copyright © Houghton Mifflin Company. All rights reserved.

Find the area of the square shown at the right.

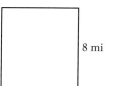

8 mi

Use the formula for the area of a square. $A = s^2$
Substitute 8 for s. $A = 8^2$
Multiply. $A = 64$

The area of the square is 64 mi².

In this section, some of the phrases used to indicate the operations of multiplication and division were presented. In solving application problems, you might also look for the following types of questions:

Multiplication	Division
per . . . How many altogether?	What is the hourly rate?
each . . . What is the total number of . . ?	Find the amount per . . .
every . . . Find the total . . .	How many does each . . . ?

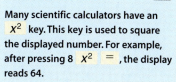

Calculator Note

Many scientific calculators have an x^2 key. This key is used to square the displayed number. For example, after pressing 8 x^2 = , the display reads 64.

Figure 1.16 shows the cost of a first-class postage stamp from the 1950s to 2003. Use this graph for Example 20 and You Try It 20.

Take Note

Each of the following indicates multiplication:

"You purchased 6 boxes of doughnuts with 12 doughnuts *per* box. *How many* doughnuts did you purchase *altogether*?"

"If *each* bottle of apple juice contains 32 oz., *what is the total number of* ounces in 8 bottles of the juice?"

"Your purchased 5 bags of oranges. *Every* bag contained 10 oranges. *Find the total number of* oranges purchased."

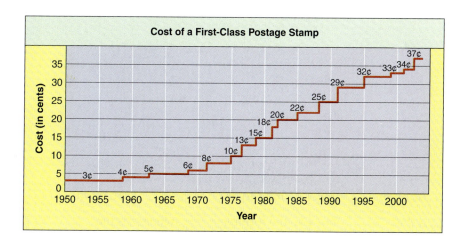

Figure 1.16 Cost of a First-Class Postage Stamp

Copyright © Houghton Mifflin Company. All rights reserved.

20 Example

How many times more expensive was a stamp in 1980 than in 1950? Use Figure 1.16.

Strategy

To find how many times more expensive a stamp was, divide the cost in 1980 (15) by the cost in 1950 (3).

Solution

15 ÷ 3 = 5

A stamp was 5 times more expensive in 1980.

20 You Try It

How many times more expensive was a stamp in 1997 than in 1960? Use Figure 1.16.

Your Strategy

Your Solution

Solution on p. S3

21 *Example*

Find the amount of sod needed to cover a football field. A football field measures 120 yd by 50 yd.

Strategy

Draw a diagram.

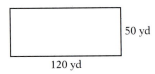

120 yd

50 yd

To find the amount of sod needed, use the formula for the area of a rectangle, $A = LW$. $L = 120$ and $W = 50$

Solution

$A = LW$

$A = 120(50)$

$A = 6,000$

6,000 ft² of sod are needed.

21 *You Try It*

A home owner wants to carpet the family room. The floor is square and measures 6 m on each side. How much carpet should be purchased?

Your Strategy

Your Solution

22 *Example*

At what rate of speed would you need to travel in order to drive a distance of 294 mi in 6 h? Use the formula $r = \dfrac{d}{t}$, where r is the average rate of speed, d is the distance, and t is the time.

Strategy

To find the rate of speed, replace d by 294 and t by 6 in the given formula and solve for r.

Solution

$r = \dfrac{d}{t}$

$r = \dfrac{294}{6} = 49$

You would need to travel at a speed of 49 mph.

22 *You Try It*

At what rate of speed would you need to travel in order to drive a distance of 486 mi in 9 h? Use the formula $r = \dfrac{d}{t}$, where r is the average rate of speed, d is the distance, and t is the time.

Your Strategy

Your Solution

Solutions on p. S3

Copyright © Houghton Mifflin Company. All rights reserved.

1.3 EXERCISES

OBJECTIVE **A**

1. 🖉 Explain how to rewrite the addition $6 + 6 + 6 + 6 + 6$ as multiplication.

2. 🖉 Provide at least three examples of situations in which we multiply numbers.

Multiply.

3. $(9)(127)$

4. $(4)(623)$

5. $(6,709)(7)$

6. $(3,608)(5)$

7. $8 \cdot 58,769$

8. $7 \cdot 60,047$

9. $\begin{array}{r} 683 \\ \times\ 71 \\ \hline \end{array}$

10. $\begin{array}{r} 591 \\ \times\ 92 \\ \hline \end{array}$

11. $\begin{array}{r} 7,053 \\ \times\ \ \ 46 \\ \hline \end{array}$

12. $\begin{array}{r} 6,704 \\ \times\ \ \ 58 \\ \hline \end{array}$

13. $\begin{array}{r} 3,285 \\ \times\ \ 976 \\ \hline \end{array}$

14. $\begin{array}{r} 5,327 \\ \times\ \ 624 \\ \hline \end{array}$

15. Find the product of 500 and 3.

16. Find 30 multiplied by 80.

17. What is 40 times 50?

18. What is twice 700?

19. What is the product of 400, 3, 20, and 0?

20. Write the product of f and g.

21. Write the product of q, r, and s.

22. 🥧 *Health*   The figure to the right shows the number of calories burned on three different exercise machines during 1 h of a light, moderate, or vigorous workout. How many calories would you burn by (a) working out vigorously on a stair climber for a total of 6 h? (b) working out moderately on a treadmill for a total of 12 h?

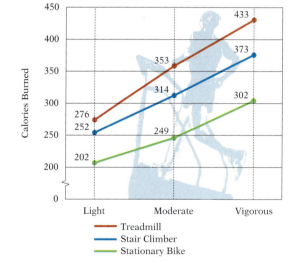

Calories Burned on Exercise Machines
Source: Journal of American Medical Association

Estimate by rounding. Then find the exact answer.

23. $3,467 \cdot 359$

24. $8,745(63)$

25. $(39,246)(29)$

26. $64,409 \cdot 67$

Copyright © Houghton Mifflin Company. All rights reserved.

27. 745(63) **28.** 432 · 91 **29.** (8,941)(726) **30.** 2,837(216)

Evaluate the expression for the given values of the variables.

31. *ab*, when *a* = 465 and *b* = 32 **32.** *cd*, when *c* = 381 and *d* = 25

33. 7*a*, when *a* = 465 **34.** 6*n*, when *n* = 382

35. *xyz*, when *x* = 5, *y* = 12, and *z* = 30 **36.** *abc*, when *a* = 4, *b* = 20, and *c* = 50

37. 2*xy*, when *x* = 67 and *y* = 23 **38.** 4*ab*, when *a* = 95 and *b* = 33

Identify the property that justifies the statement.

39. 1 · 29 = 29 **40.** (10 · 5) · 8 = 10 · (5 · 8)

41. 43 · 1 = 1 · 43 **42.** 0(76) = 0

Use the given property of multiplication to complete the statement.

43. The Commutative Property of Multiplication
19 · ? = 30 · 19

44. The Associative Property of Multiplication
(? · 6)100 = 5(6 · 100)

45. The Multiplication Property of Zero
45 · 0 = ?

46. The Multiplication Property of One
? · 77 = 77

47. Is 6 a solution of the equation 4*x* = 24? **48.** Is 0 a solution of the equation 4 = 4*n*?

49. Is 23 a solution of the equation 96 = 3*z*? **50.** Is 14 a solution of the equation 56 = 4*c*?

Copyright © Houghton Mifflin Company. All rights reserved.

51. Is 19 a solution of the equation $2y = 38$?

52. Is 11 a solution of the equation $44 = 3a$?

OBJECTIVE B

Write in exponential form.

53. $2 \cdot 2 \cdot 2 \cdot 7 \cdot 7 \cdot 7 \cdot 7 \cdot 7$

54. $3 \cdot 3 \cdot 3 \cdot 3 \cdot 3 \cdot 3 \cdot 5 \cdot 5 \cdot 5$

55. $2 \cdot 2 \cdot 3 \cdot 3 \cdot 3 \cdot 5 \cdot 5 \cdot 5 \cdot 5$

56. $7 \cdot 7 \cdot 11 \cdot 11 \cdot 11 \cdot 19 \cdot 19 \cdot 19 \cdot 19$

57. $c \cdot c$

58. $d \cdot d \cdot d$

59. $x \cdot x \cdot x \cdot y \cdot y \cdot y$

60. $a \cdot a \cdot b \cdot b \cdot b \cdot b$

Evaluate.

61. 2^5

62. 2^6

63. 10^6

64. 10^9

65. $2^3 \cdot 5^2$

66. $2^4 \cdot 3^2$

67. $3^2 \cdot 10^3$

68. $2^4 \cdot 10^2$

69. $0^2 \cdot 6^2$

70. $4^3 \cdot 0^3$

71. $2^2 \cdot 5 \cdot 3^3$

72. $5^2 \cdot 2 \cdot 3^4$

73. Find the square of 12.

74. What is the cube of 6?

75. Find the cube of 8.

76. What is the square of 11?

77. Write the fourth power of a.

78. Write the fifth power of t.

Evaluate the expression for the given values of the variables.

79. x^3y, when $x = 2$ and $y = 3$

80. x^2y, when $x = 3$ and $y = 4$

Copyright © Houghton Mifflin Company. All rights reserved.

81. ab^6, when $a = 5$ and $b = 2$

82. ab^3, when $a = 7$ and $b = 4$

83. c^2d^2, when $c = 3$ and $d = 5$

84. m^3n^3, when $m = 5$ and $n = 10$

OBJECTIVE C

85. Provide at least three examples of situations in which we divide numbers.

86. In what situation does a division problem have a remainder?

Divide.

87. $9\overline{)2,763}$

88. $4\overline{)2,160}$

89. $5\overline{)1,549}$

90. $8\overline{)1,636}$

91. $15,300 \div 6$

92. $43,500 \div 5$

93. $681 \div 32$

94. $879 \div 41$

95. $9,152 \div 62$

96. $4,161 \div 23$

97. $7,408 \div 37$

98. $5,207 \div 26$

99. $31,546 \div 78$

100. $38,976 \div 64$

101. $7,713 \div 476$

102. $8,947 \div 223$

103. Find the quotient of 7,256 and 8.

104. What is the quotient of 8,172 and 9?

105. What is 6,168 divided by 7?

106. Find 4,153 divided by 9.

107. Write the quotient of c and d.

108. *Insurance* The table at the right shows the sources of laptop insurance claims in a recent year. Claims have been rounded to the nearest ten-thousand dollars.
(a) What was the average monthly claim for theft? (b) For all sources combined, find the average claims per month.

Source	Claims (in dollars)
Accidents	560,000
Theft	300,000
Power Surge	80,000
Lightning	50,000
Transit	20,000
Water/flood	20,000
Other	110,000

Source: Safeware, The Insurance Company

Copyright © Houghton Mifflin Company. All rights reserved.

Estimate by rounding. Then find the exact answer.

109. $36{,}472 \div 47$ **110.** $62{,}176 \div 58$ **111.** $389{,}804 \div 76$ **112.** $637{,}072 \div 29$

113. $79\overline{)38{,}984}$ **114.** $53\overline{)11{,}792}$ **115.** $219\overline{)332{,}004}$ **116.** $324\overline{)632{,}124}$

Evaluate the variable expression $\dfrac{x}{y}$ for the given values of x and y.

117. $x = 48; y = 1$ **118.** $x = 56; y = 56$ **119.** $x = 79; y = 0$

120. $x = 0; y = 23$ **121.** $x = 39{,}200; y = 4$ **122.** $x = 16{,}200; y = 3$

123. Is 9 a solution of the equation $\dfrac{36}{z} = 4$?   **124.** Is 60 a solution of the equation $\dfrac{n}{12} = 5$?

125. Is 49 a solution of the equation $56 = \dfrac{x}{7}$?   **126.** Is 16 a solution of the equation $6 = \dfrac{48}{y}$?

OBJECTIVE   D

Find all the factors of the number.

127. 10   **128.** 20   **129.** 12   **130.** 9   **131.** 8

132. 16   **133.** 13   **134.** 17   **135.** 18   **136.** 24

137. 25   **138.** 36   **139.** 56   **140.** 45   **141.** 28

142. 32   **143.** 48   **144.** 64   **145.** 54   **146.** 75

Copyright © Houghton Mifflin Company. All rights reserved.

Find the prime factorization of the number.

147. 16 **148.** 24 **149.** 12 **150.** 27 **151.** 15

152. 36 **153.** 40 **154.** 50 **155.** 37 **156.** 83

157. 65 **158.** 80 **159.** 28 **160.** 49 **161.** 42

162. 81 **163.** 51 **164.** 89 **165.** 46 **166.** 120

OBJECTIVE E

Nutrition Facts	Amount/Serving	% DV*	Amount/Serving	% DV*
	Total Fat 9g	**14%**	**Total Carb.** 1g	**0%**
Serv. Size 1 oz.	Sat Fat 5g	**25%**	Fiber 0g	**0%**
Servings Per Package 12 **Calories 115**	**Cholest.** 30mg	**10%**	Sugars 0g	
Fat Cal. 80	**Sodium** 170mg	**7%**	**Protein** 7g	
*Percent Daily Values (DV) are based on a 2,000 calorie diet	Vitamin A 6% • Vitamin C 0% • Calcium 20% • Iron 0%			

167. *Nutrition* One ounce of cheddar cheese contains 115 calories. Find the number of calories in 4 oz of cheddar cheese.

168.  *Sports* During his football career, John Riggins ran the ball 2,916 times. He averaged about 4 yd per carry. About how many total yards did he gain during his career?

169. *Aviation* A plane flying from Los Angeles to Boston uses 865 gal of jet fuel each hour. How many gallons of jet fuel are used on a 5-hour flight?

170. *Geometry* Find (a) the perimeter and (b) the area of a square that measures 16 mi on each side.

16 mi

John Riggins

171. *Geometry* Find (a) the perimeter and (b) the area of a rectangle with a length of 24 m and a width of 15 m.

172. *Geometry* Find the length of fencing needed to surround a square corral that measures 55 ft on each side.

173. *Geometry* A home owner plans to fence in the area around a swimming pool in the backyard. The area to be fenced in is a square measuring 24 ft on each side. How many feet of fencing should the home owner purchase?

Copyright © Houghton Mifflin Company. All rights reserved.

174. *Geometry*   A solar panel is in the shape of a rectangle that has a width of 2 ft and a length of 3 ft. Find the area of the solar panel.

175. *Geometry*   What is the area of the floor of a two-car garage that is in the shape of a square that measures 24 ft on a side?

176. *Geometry*   A fieldstone patio is in the shape of a square that measures 9 ft on each side. What is the area of the patio?

177. *Geometry*   Find the amount of fabric needed for a rectangular flag that measures 308 cm by 192 cm.

178. *Finances*   A computer analyst doing consulting work received $5,376 for working 168 h on a project. Find the hourly rate the consultant charged.

179. *Business*   A buyer for a department store purchased 215 suits at $83 each. Estimate the total cost of the order.

180.   *Finances*   Financial advisors may predict how much money we should have saved for retirement by the ages of 35, 45, 55, and 65. One such prediction is included in the table below. (a) A couple has earnings of $100,000 per year. According to the table, by how much should their savings grow per year from age 45 to 55? (b) A couple has earnings of $50,000 per year. According to the table, by how much should their savings grow per year from age 55 to 65?

Minimum Levels of Savings Required for Married Couples to Be Prepared for Retirement				
	Savings Accumulation by Age			
Earnings	35	45	55	65
$50,000	8,000	23,000	90,000	170,000
$75,000	17,000	60,000	170,000	310,000
$100,000	34,000	110,000	280,000	480,000
$150,000	67,000	210,000	490,000	840,000

181. *Finances*   Find the total amount paid on a loan when the monthly payment is $285 and the loan is paid off in 24 months. Use the formula $A = MN$, where A is the total amount paid, M is the monthly payment, and N is the number of payments.

182. *Finances*   Find the total amount paid on a loan when the monthly payment is $187 and the loan is paid off in 36 months. Use the formula $A = MN$, where A is the total amount paid, M is the monthly payment, and N is the number of payments.

Copyright © Houghton Mifflin Company. All rights reserved.

183. *Travel* Use the formula $t = \dfrac{d}{r}$, where t is the time, d is the distance, and r is the average rate of speed, to find the time it would take to drive 513 mi at an average speed of 57 mph.

184. *Travel* Use the formula $t = \dfrac{d}{r}$, where t is the time, d is the distance, and r is the average rate of speed, to find the time it would take to drive 432 mi at an average speed of 54 mph.

185. *Investments* The current value of the stocks in a mutual fund is $10,500,000. The number of shares outstanding is 500,000. Find the value per share of the fund. Use the formula $V = \dfrac{C}{S}$, where V is the value per share, C is the current value of the stocks in the fund, and S is the number of shares outstanding.

186. *Investments* The current value of the stocks in a mutual fund is $4,500,000. The number of shares outstanding is 250,000. Find the value per share of the fund. Use the formula $V = \dfrac{C}{S}$, where V is the value per share, C is the current value of the stocks in the fund, and S is the number of shares outstanding.

New York Stock Exchange

Critical Thinking

187. *Time* There are 52 weeks in a year. Is this an exact figure or an approximation?

188. *Mathematics* 13,827 is not divisible by 4. By rearranging the digits, find the largest possible number that is divisible by 4.

189. *Mathematics* A **palindromic number** is a whole number that remains unchanged when its digits are written in reverse order. For example, 818 is a palindromic number. Find the smallest three-digit multiple of 6 that is a palindromic number.

190. Determine whether the statement is always true, sometimes true, or never true.
    a. Let a be any whole number. Then $a \cdot 0 = a$.
    b. Let a be any whole number. Then $a \cdot 1 = 1$.

191. According to the National Safety Council, in a recent year, a death resulting from an accident occurred at the rate of one every 5 min. At this rate, how many accidental deaths occurred each hour? each day? throughout the year? Explain how you arrived at your answers.

192. Prepare a monthly budget for a family of four. Explain how you arrived at the cost of each item. Annualize the budget you prepared.

Monthly Budget	
Rent	$975
Electricity	
Telephone	
Gas	
Food	

Copyright © Houghton Mifflin Company. All rights reserved.

Copyright © Houghton Mifflin Company. All rights reserved.

SECTION 1.4 ▶ # Solving Equations with Whole Numbers

OBJECTIVE A

VIDEO & DVD   CD TUTOR   WEB   SSM

Solving equations

Recall that a **solution** of an equation is a number that, when substituted for the variable, results in a true equation.

The solution of the equation $x + 5 = 11$ is 6 because when 6 is substituted for x, the result is a true equation.

$$x + 5 = 11$$
$$6 + 5 = 11$$

If 2 is subtracted from each side of the equation $x + 5 = 11$, the resulting equation is $x + 3 = 9$. Note that the solution of this equation is also 6.

$$x + 5 = 11$$
$$x + 5 - 2 = 11 - 2$$
$$x + 3 = 9 \qquad\qquad 6 + 3 = 9$$

> **Take Note**
>
> An *equation* always has an equal sign (=). An *expression* does not have an equals sign.
> $x + 5 = 11$ is an equation.
> $x + 5$ is an expression.

This illustrates the **subtraction property of equations**.

> **The same number can be subtracted from each side of an equation without changing the solution of the equation.**

The subtraction property is used to *solve* an equation. To **solve an equation** means to find a solution of the equation. That is, to solve an equation you must find a number that, when substituted for the variable, results in a true equation.

An equation such as $x = 8$ is easy to solve. The solution is 8, the number that when substituted for the variable results in the true equation $8 = 8$. In solving an equation, the goal is to get the variable alone on one side of the equation; the number on the other side of the equation is the solution.

To solve an equation in which a number is added to a variable, use the subtraction property of equations: Subtract that number from each side of the equation.

Solve: $x + 5 = 11$

Note the effect of subtracting 5 from each side of the equation and then simplifying. The variable, x, is on one side of the equation; a number, 6, is on the other side.

$$x + 5 = 11$$
$$x + 5 - 5 = 11 - 5$$
$$x + 0 = 6$$
$$x = 6$$

The solution is 6.

Check:
$$x + 5 = 11$$
$$6 + 5 \;\big|\; 11$$
$$11 = 11$$

Note that we have checked the solution. You should always check the solution of an equation.

Solve: $19 = 11 + m$

11 is added to m. Subtract 11 from each side of the equation.

$$19 = 11 + m$$
$$19 - 11 = 11 - 11 + m$$
$$8 = 0 + m$$
$$8 = m$$

The solution is 8.

Check:
$$19 = 11 + m$$
$$19 \;\big|\; 11 + 8$$
$$19 = 19$$

> **Take Note**
>
> For this equation, the variable is on the right side. The goal is to get the variable alone on the right side.

The solution of the equation $4y = 12$ is 3 because when 3 is substituted for y, the result is a true equation.

$$4y = 12$$
$$4(3) = 12$$
$$12 = 12$$

If each side of the equation $4y = 12$ is divided by 2, the resulting equation is $2y = 6$. Note that the solution of this equation is also 3.

$$4y = 12$$
$$\frac{4y}{2} = \frac{12}{2}$$
$$2y = 6 \qquad 2(3) = 6$$

This illustrates the **division property of equations.**

Each side of an equation can be divided by the same number (except zero) without changing the solution of the equation.

Solve: $30 = 5a$

a is multiplied by 5. To get a alone on the right side, divide each side of the equation by 5.

$$30 = 5a$$
$$\frac{30}{5} = \frac{5a}{5}$$
$$6 = 1a$$
$$6 = a$$

Check:
$$30 = 5a$$
$$30 \mid 5(6)$$
$$30 = 30$$

The solution is 6.

---

① Example   Solve: $9 + n = 28$

Solution
$$9 + n = 18$$
$$9 - 9 + n = 28 - 9$$
$$0 + n = 19$$
$$n = 19$$

Check:
$$9 + n = 28$$
$$9 + 19 \mid 28$$
$$28 = 28$$

The solution is 19.

① You Try It   Solve: $37 = a + 12$

Your Solution

---

② Example   Solve: $20 = 5c$

Solution
$$20 = 5c$$
$$\frac{20}{5} = \frac{5c}{5}$$
$$4 = 1c$$
$$4 = c$$

Check:
$$20 = 5c$$
$$20 \mid 5(4)$$
$$20 = 20$$

The solution is 4.

② You Try It   Solve: $3z = 36$

Your Solution

Solutions on p. S3

Copyright © Houghton Mifflin Company. All rights reserved.

Copyright © Houghton Mifflin Company. All rights reserved.

OBJECTIVE **B**

Applications and formulas

Recall that an equation states that two mathematical expressions are equal. To translate a sentence into an equation requires recognition of the words or phrases that mean "equals." Some of these phrases are

equals	is	was
is equal to	represents	is the same as

The number of scientific calculators sold by Evergreen Electronics last month is three times the number of graphing calculators the company sold this month. If it sold 225 scientific calculators last month, how many graphing calculators were sold this month?

Strategy To find the number of graphing calculators sold, write and solve an equation using x to represent the number of graphing calculators sold.

Strategy

The number of scientific calculators sold last month	is	three times the number of graphing calculators sold this month

$$225 = 3x$$

$$\frac{225}{3} = \frac{3x}{3}$$

$$75 = x$$

Evergreen Electronics sold 75 graphing calculators this month.

> **Take Note**
>
> Sentences or phrases that begin "how many…," "how much…," "find…," and "what is…" are followed by a phrase that indicates what you are looking for. (In the problem at the left, the phrase is *graphing calculators:* "How many *graphing calculators*….") Look for these phrases to determine the unknown.

---

3 *Example*

The product of seven and a number equals twenty-eight. Find the number.

Solution

The unknown number: n

The product of seven and a number	equals	twenty-eight

$$7n = 28$$

$$\frac{7n}{7} = \frac{28}{7}$$

$$n = 4$$

The number is 4.

3 *You Try It*

A number increased by four is seventeen. Find the number.

Your Solution

Solution on p. S3

4 Example

A child born in 2000 was expected to live to the age of 77. This is 29 years longer than the life expectancy of a child born in 1900. (*Sources:* U.S. Department of Health and Human Services' Administration of Aging; Census Bureau; National Center for Health Statistics) Find the life expectancy of a child born in 1900.

Strategy

To find the life expectancy of a child born in 1900, write and solve an equation using x to represent the unknown life expectancy.

Solution

Life expectancy in 2000	was	29 years longer than the life expectancy in 1900

$$77 = x + 29$$
$$77 - 29 = x + 29 - 29$$
$$48 = x$$

The life expectancy of a child born in 1900 was 48 years.

4 You Try It

In a recent year, more than 7 million people had cosmetic plastic surgery. During that year, the number of liposuctions performed was 220,159 more than the number of face lifts performed. There were 354,015 liposuctions performed. (*Source:* American Society of Plastic Surgery) How many face lifts were performed that year?

Your Strategy

Your Solution

5 Example

Use the formula $A = P + I$, where A is the value of an investment, P is the original investment, and I is the interest earned, to find the interest earned on an original investment of $12,000 that now has a value of $14,280.

Strategy

To find the interest earned, replace A by 14,280 and P by 12,000 in the given formula and solve for I.

Solution

$$A = P + I$$
$$14{,}280 = 12{,}000 + I$$
$$14{,}280 - 12{,}000 = 12{,}000 - 12{,}000 + I$$
$$2{,}280 = I$$

The interest earned on the investment is $2,280.

5 You Try It

Use the formula $A = P + I$, where A is the value of an investment, P is the original investment, and I is the interest earned, to find the interest earned on an original investment of $18,000 that now has a value of $21,060.

Your Strategy

Your Solution

Solutions on p. S3

Copyright © Houghton Mifflin Company. All rights reserved.

1.4 EXERCISES

OBJECTIVE A

Solve.

1. $x + 9 = 23$

2. $y + 17 = 42$

3. $8 + b = 33$

4. $15 + n = 54$

5. $3m = 15$

6. $8z = 32$

7. $52 = 4c$

8. $60 = 5d$

9. $16 = w + 9$

10. $72 = t + 44$

11. $28 = 19 + p$

12. $33 = 18 + x$

13. $10y = 80$

14. $12n = 60$

15. $41 = 41d$

16. $93 = 93m$

17. $b + 7 = 7$

18. $q + 23 = 23$

19. $15 + t = 91$

20. $79 + w = 88$

21. $4 + a = 25$

22. $33 = 12 + v$

23. $c + 17 = 50$

24. $100 = z + 41$

OBJECTIVE B

25. Sixteen added to a number is equal to forty. Find the number.

26. The sum of eleven and a number equals fifty-two. Find the number.

27. Five times a number is thirty. Find the number.

28. The product of ten and a number is equal to two hundred. Find the number.

29. Fifteen is three more than a number. Find the number.

30. One thousand represents three hundred fifty plus a number. Find the number.

31. A number increased by fourteen equals seventy-two. Find the number.

32. A number multiplied by twenty equals four hundred. Find the number.

Copyright © Houghton Mifflin Company. All rights reserved.

33. *Geometry* The length of a rectangle is 5 in. more than the width. The length is 17 in. Find the width of the rectangle.

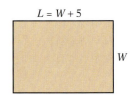

$L = W + 5$

W

34. 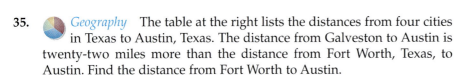 *Temperature* The average daily low temperature in Duluth, Minnesota, in June is eight times the average daily low temperature in Duluth in December. The average daily low temperature in Duluth in June is 48°. Find the average daily low temperature in Duluth in December.

35. *Geography* The table at the right lists the distances from four cities in Texas to Austin, Texas. The distance from Galveston to Austin is twenty-two miles more than the distance from Fort Worth, Texas, to Austin. Find the distance from Fort Worth to Austin.

City in Texas	Number of Miles to Austin, Texas
Corpus Christi	215
Dallas	195
Galveston	212
Houston	160

36. *Geography* The table at the right lists the distances from four cities in Texas to Austin, Texas. The distance from Houston to Austin is twice the distance from San Antonio, Texas, to Austin. Find the distance from San Antonio to Austin.

37. *Finances* Use the formula $A = MN$, where A is the total amount paid, M is the monthly payment, and N is the number of payments, to find the number of payments made on a loan for which the total amount paid is $13,968 and the monthly payment is $582.

38. *Finances* Use the formula $A = MN$, where A is the total amount paid, M is the monthly payment, and N is the number of payments, to find the number of payments made on a loan for which the total amount paid is $17,460 and the monthly payment is $485.

39. *Travel* Use the formula $d = rt$, where d is distance, r is rate of speed, and t is time, to find how long it would take to travel a distance of 1,120 mi at a speed of 140 mph.

40. *Travel* Use the formula $d = rt$, where d is distance, r is rate of speed, and t is time, to find how long it would take to travel a distance of 825 mi at a speed of 165 mph.

Critical Thinking

41. Write an equation of the form $ax = b$, where a and b are numbers and x is a variable, that (a) has 0 as the solution of the equation and (b) has 1 as the solution of the equation.

42. Write two word problems for a classmate to solve, one that is a number problem (like Exercises 25–32 on page 71) and another that involves using a formula (like Exercises 37–40 above).

Copyright © Houghton Mifflin Company. All rights reserved.

The Order of Operations Agreement

OBJECTIVE **A**

The Order of Operations Agreement

More than one operation may occur in a numerical expression. For example, the expression

$$4 + 3(5)$$

includes two arithmetic operations, addition and multiplication. The operations could be performed in different orders.

If we multiply first and then add, we have:

$$4 + 3(5)$$
$$4 + 15$$
$$19$$

If we add first and then multiply, we have:

$$4 + 3(5)$$
$$7(5)$$
$$35$$

To prevent more than one answer to the same problem, an Order of Operations Agreement is followed. By this agreement, 19 is the only correct answer.

The Order of Operations Agreement

Step 1   Do all operations inside parentheses.

Step 2   Simplify any numerical expressions containing exponents.

Step 3   Do multiplication and division as they occur from left to right.

Step 4   Do addition and subtraction as they occur from left to right.

> **Calculator Note**
>
> Many calculators use the Order of Operations Agreement shown at the left.
>
> Enter 4 ⊞ 3 ✕ 5 ⊟ into your calculator. If the answer is 19, your calculator uses the Order of Operations Agreement.

Simplify: $2(4 + 1) - 2^3 + 6 \div 2$

Perform operations in parentheses.	$2(4 + 1) - 2^3 + 6 \div 2$ $= 2(5) - 2^3 + 6 \div 2$
Simplify expressions with exponents.	$= 2(5) - 8 + 6 \div 2$
Do multiplication and division as they occur from left to right.	$= 10 - 8 + 6 \div 2$ $= 10 - 8 + 3$
Do addition and subtraction as they occur from left to right.	$= 2 + 3$ $= 5$

> **Calculator Note**
>
> Here is an example of using the parentheses keys on a calculator. To evaluate $28(103 - 78)$, enter:
>
> 28 ✕ (103 ⊟ 78) ⊟ .
>
> Note that ✕ is required on most calculators.

One or more of the above steps may not be needed to simplify an expression. In that case, proceed to the next step in the Order of Operations Agreement.

Simplify: $8 + 9 \div 3$

There are no parentheses (Step 1). There are no exponents (Step 2).	$8 + 9 \div 3$
Do the division (Step 3).	$= 8 + 3$
Do the addition (Step 4).	$= 11$

Copyright © Houghton Mifflin Company. All rights reserved.

▶ **Point of Interest**

Try this: Use the same one-digit number three times to write an expression that is equal to 30.

Evaluate $5a - (b + c)^2$ when $a = 6$, $b = 1$, and $c = 3$.

$$5a - (b + c)^2$$

Replace a with 6, b with 1, and c with 3.

$$5(6) - (1 + 3)^2$$

Use the Order of Operations Agreement to simplify the resulting numerical expression. Perform operations inside parentheses.

$$= 5(6) - (4)^2$$

Simplify expressions with exponents.

$$= 5(6) - 16$$

Do the multiplication.

$$= 30 - 16$$

Do the subtraction.

$$= 14$$

1 *Example*

Simplify: $18 \div (6 + 3) \cdot 9 - 4^2$

Solution

$$\begin{aligned}
18 \div (6 + 3) \cdot 9 - 4^2 &= 18 \div 9 \cdot 9 - 4^2 \\
&= 18 \div 9 \cdot 9 - 16 \\
&= 2 \cdot 9 - 16 \\
&= 18 - 16 \\
&= 2
\end{aligned}$$

1 *You Try It*

Simplify: $4 \cdot (8 - 3) \div 5 - 2$

Your Solution

2 *Example*

Simplify: $20 + 24(8 - 5) \div 2^2$

Solution

$$\begin{aligned}
20 + 24(8 - 5) \div 2^2 &= 20 + 24(3) \div 2^2 \\
&= 20 + 24(3) \div 4 \\
&= 20 + 72 \div 4 \\
&= 20 + 18 \\
&= 38
\end{aligned}$$

2 *You Try It*

Simplify: $16 + 3(6 - 1)^2 \div 5$

Your Solution

3 *Example*

Evaluate $(a - b)^2 + 3c$ when $a = 6$, $b = 4$, and $c = 1$.

Solution

$$(a - b)^2 + 3c$$
$$\begin{aligned}
(6 - 4)^2 + 3(1) &= (2)^2 + 3(1) \\
&= 4 + 3(1) \\
&= 4 + 3 \\
&= 7
\end{aligned}$$

3 *You Try It*

Evaluate $(a - b)^2 + 5c$ when $a = 7$, $b = 2$, and $c = 4$.

Your Solution

Solutions on p. S3–S4

Copyright © Houghton Mifflin Company. All rights reserved.

1.5 EXERCISES

OBJECTIVE **A**

1. Why do we need an Order of Operations Agreement?

2. What are the steps in the Order of Operations Agreement?

Simplify.

3. $8 \div 4 + 2$

4. $12 - 9 \div 3$

5. $6 \cdot 4 + 5$

6. $5 \cdot 7 + 3$

7. $4^2 - 3$

8. $6^2 - 14$

9. $5 \cdot (6 - 3) + 4$

10. $8 + (6 + 2) \div 4$

11. $9 + (7 + 5) \div 6$

12. $14 \cdot (3 + 2) \div 10$

13. $13 \cdot (1 + 5) \div 13$

14. $14 - 2^3 + 9$

15. $6 \cdot 3^2 + 7$

16. $18 + 5 \cdot 3^2$

17. $14 + 5 \cdot 2^3$

18. $20 + (9 - 4) \cdot 2$

19. $10 + (8 - 5) \cdot 3$

20. $3^2 + 5 \cdot (6 - 2)$

21. $2^3 + 4(10 - 6)$

22. $3^2 \cdot 2^2 + 3 \cdot 2$

23. $6(7) + 4^2 \cdot 3^2$

24. $14 - 2(6)$

25. $18 + 3(7)$

26. $2(9 - 2) + 5$

27. $6(8 - 3) - 12$

28. $15 - (7 - 1) \div 3$

29. $16 - (13 - 5) \div 4$

30. $11 + 2 - 3 \cdot 4 \div 3$

31. $17 + 1 - 8 \cdot 2 \div 4$

32. $3(5 + 3) \div 8$

Copyright © Houghton Mifflin Company. All rights reserved.

Evaluate the expression for the given values of the variables.

33. $x - 2y$, where $x = 8$ and $y = 3$

34. $x + 6y$, where $x = 5$ and $y = 4$

35. $x^2 + 3y$, where $x = 6$ and $y = 7$

36. $3x^2 + y$, where $x = 2$ and $y = 9$

37. $x^2 + y \div x$, where $x = 2$ and $y = 8$

38. $x + y^2 \div x$, where $x = 4$ and $y = 8$

39. $4x + (x - y)^2$, where $x = 8$ and $y = 2$

40. $(x + y)^2 - 2y$, where $x = 3$ and $y = 6$

41. $x^2 + 3(x - y) + z^2$, where $x = 2$, $y = 1$, and $z = 3$

42. $x^2 + 4(x - y) \div z^2$, where $x = 8$, $y = 6$, and $z = 2$

43. Use the inequality symbol $>$ to compare the expressions $11 + (8 + 4) \div 6$ and $12 + (9 - 5) \cdot 3$.

44. Use the inequality symbol $<$ to compare the expressions $3^2 + 7(4 - 2)$ and $14 - 2^3 + 20$.

Critical Thinking

45. Arrange the expressions in order from the greatest value to the least value.

$27 \div 9 + 8$ $4 + 3 \cdot 12$

$81 - 8^2$ $50 - 6(8)$

$5(10 - 2) \div 4$ $2(1 + 4)^2 \div 10$

46. What is the smallest prime number greater than $15 + (8 - 3)(2^4)$?

47. Simplify $(47 + 48 + 49 + 51 + 52 + 53) \div 100$. What do you notice that will allow you to calculate the answer mentally?

Copyright © Houghton Mifflin Company. All rights reserved.

Focus on **Problem Solving**

Copyright © Houghton Mifflin Company. All rights reserved.

Questions to Ask

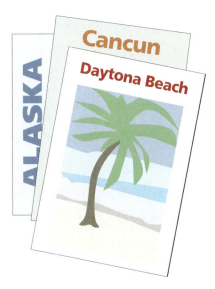

You encounter problem-solving situations every day. Some problems are easy to solve, and you may mentally solve these problems without considering the steps you are taking in order to draw a conclusion. Others may be more challenging and require more thought and consideration.

Suppose a friend suggests that you both take a trip over spring break. You'd like to go. What questions go through your mind? You might ask yourself some of the following questions:

> How much will the trip cost? What will be the cost for travel, lodging, meals, etc.?
> Are some costs going to be shared by both me and my friend?
> Can I afford it?
> How much money do I have in the bank?
> How much more money than I have now do I need?
> How much time is there to earn that much money?
> How much can I earn in that amount of time?
> How much money must I keep in the bank in order to pay the next tuition bill (or some other expense)?

These questions require different mathematical skills. Determining the cost of the trip requires *estimation;* for example, you must use your knowledge of air fares or the cost of gasoline to arrive at an estimate of these costs. If some of the costs are going to be shared, you need to *divide* those costs by 2 in order to determine your share of the expense. The question regarding how much more money you need requires *subtraction:* the amount needed minus the amount currently in the bank. To determine how much money you can earn in the given amount of time requires *multiplication*—for example, the amount you earn per week times the number of weeks to be worked. To determine whether the amount you can earn in the given amount of time is sufficient, you need to use your knowledge of *order relations* to compare the amount you can earn with the amount needed.

Facing the problem-solving situation described above may not seem difficult to you. The reason may be that you have faced similar situations before and, therefore, know how to work through this one. You may feel better prepared to deal with a circumstance such as this one because you know what questions to ask. An important aspect of learning to solve problems is learning what questions to ask. As you work through application problems in this text, try to become more conscious of the mental process you are going through. You might begin the process by asking yourself the following questions whenever you are solving an application problem:

1. Have I read the problem enough times to be able to understand the situation being described?
2. Will restating the problem in different words help me to understand the problem situation better?
3. What facts are given? (You might make a list of the information contained in the problem.)
4. What information is being asked for?
5. What relationship exists between the given facts? What relationship exists between the given facts and the solution?
6. What mathematical operations are needed in order to solve the problem?

Try to focus on the problem-solving situation, not on the computation or on getting the answer quickly. And remember, the more problems you solve, the better able you will be to solve other problems in the future, partly because you are learning what questions to ask.

Projects & Group Activities

Surveys

On page 16 is a circle graph showing the results of a survey of 150 people who were asked, "What bothers you most about movie theaters?" Note that the responses included (1) people talking in the theater, (2) high ticket prices, (3) high prices for food purchased in the theater, (4) dirty floors, (5) and uncomfortable seats.

Conduct a similar survey in your class. Ask each classmate which of the five conditions stated above is most irritating. Record the number of students who answered each one of the five possible responses. Prepare a bar graph to display the results of the survey. A model is provided below to help you get started.

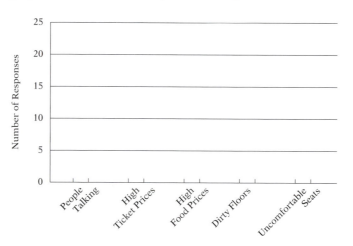

Responses to Theater-Goers Survey

Applications of Patterns in Mathematics

For the circle shown at the far left below, use a straight line to connect each dot on the circle with every other dot on the circle. How many different straight lines are there?

Follow the same procedure for each of the other circles. How many different straight lines are there in each?

Find a pattern to describe the number of dots on a circle and the corresponding number of different lines drawn. Use the pattern to determine the number

Copyright © Houghton Mifflin Company. All rights reserved.

of different lines that would be drawn in a circle with 7 dots and in a circle with 8 dots.

Now use the pattern to answer the following:

You are arranging a tennis tournament with nine players. How many singles matches will be played among the nine players if each player plays each of the other players once?

Salary Calculator

On the Internet, go to

 http://www.homefair.com/homefair/cmr/salcalc.html

This website can be used to calculate the salary you would need in order to maintain the same standard of living if you were to move to another city.

First select the state you live in now and the state you would like to move to. Then select the city closest to where you live now and the city of your choice in the state you chose to move to.

1. Is the salary greater in the city you live in now or the city you would like to move to? What is the difference between the two salaries?
2. Select a few other cities you might like to move to. Perform the same calculations. Which of the cities you selected is the most expensive to live in? the least expensive?
3. How might you determine the salaries for people in your occupation in any of the cities you selected?

Subtraction Squares

Draw a square. Write the four numbers 7, 5, 9, and 2, one at each of the four corners. Draw a second square around the first so that it goes through each of the four corners. At each corner of the second square, write the difference of the numbers at the closest corners of the smaller square: $7 - 5 = 2, 9 - 5 = 4, 9 - 2 = 7$, and $7 - 2 = 5$.

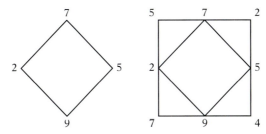

Repeat the process until you come to a pattern of four numbers that does not change. What is the pattern? Try the same procedure with any other four starting numbers. Do you end up with the same pattern? Provide an explanation for what happens.

Chapter Summary

Key Words   *Examples*

The **natural numbers** or **counting numbers** are 1, 2, 3, 4, 5, 6, 7, 8, 9, 10,.... [1.1A, p. 3]

Copyright © Houghton Mifflin Company. All rights reserved.

The **whole numbers** are 0, 1, 2, 3, 4, 5, 6, 7, 8, 9, 10,…. [1.1A, p. 3]

The symbol for "is less than" is <. The symbol for "is greater than" is >. A statement that uses the symbol < or > is an **inequality.** [1.1A, p. 3]	$3 < 7$ $9 > 2$

When a whole number is written using the digits 0, 1, 2, 3, 4, 5, 6, 7, 8, and 9, it is said to be in **standard form.** The position of each digit in the number determines that digit's **place value.** [1.1B, p. 4]	The number 598,317 is in standard form. The digit 8 is in the thousands place.

A **pictograph** represents data by using a symbol that is characteristic of the data. A **circle graph** represents data by the size of the sectors. A **bar graph** represents data by the height of the bars. A **broken-line graph** represents data by the position of the lines and shows trends or comparisons. [1.1D, pp. 8–10]

Addition is the process of finding the total of two or more numbers. The numbers being added are called **addends.** The answer is the **sum.** [1.2A, pp. 19–20]	$\begin{array}{r} {\scriptstyle 1\ 11} \\ 8,762 \\ +\ 1,359 \\ \hline 10,121 \end{array}$

Subtraction is the process of finding the difference between two numbers. The **minuend** minus the **subtrahend** equals the **difference.** [1.2B, pp. 25–26]	$\begin{array}{r} {\scriptstyle 4\ 11\ 11\ 6\ 13} \\ \cancel{52,173} \\ -34,968 \\ \hline 17,205 \end{array}$

Multiplication is the repeated addition of the same number. The numbers that are multiplied are called **factors.** The answer is the **product.** [1.3A, p. 41]	$\begin{array}{r} {\scriptstyle 4\ 5} \\ 358 \\ \times\quad 7 \\ \hline 2,506 \end{array}$

Division is used to separate objects into equal groups. The **dividend** divided by the **divisor** equals the **quotient.** For any division problem, **(quotient · divisor) + remainder = dividend.** [1.3C, pp. 48–50]	$\begin{array}{r} 93 \text{ r3} \\ 7\overline{)654} \\ -63 \\ \hline 24 \\ -21 \\ \hline 3 \end{array}$ Check: $(7 \cdot 93) + 3 = 651 + 3 = 654$

The expression 3^5 is in **exponential form.** The **exponent,** 5, indicates how many times the **base,** 3, occurs as a factor in the multiplication. [1.3B, p. 46]	$5^4 = 5 \cdot 5 \cdot 5 \cdot 5 = 625$

Natural number **factors** of a number divide that number evenly (there is no remainder). [1.3D, p. 53]	$18 \div 1 = 18$ $18 \div 2 = 9$ $18 \div 3 = 6$ $18 \div 4$ 4 does not divide 18 evenly. $18 \div 5$ 5 does not divide 18 evenly. $18 \div 6 = 3$ The factors are repeating. The factors of 18 are 1, 2, 3, 6, 9, and 18.

A number greater than 1 is a **prime** number if its only whole number factors are 1 and itself. If a number is not prime, it is a **composite number.** [1.3D, p. 53]	The prime numbers less than 20 are 2, 3, 5, 7, 11, 13, 17, and 19. The composite numbers less than 20 are 4, 6, 8, 9, 10, 12, 14, 15, 16, and 18.

Copyright © Houghton Mifflin Company. All rights reserved.

The **prime factorization** of a number is the expression of the number as a product of its prime factors. [1.3D, p. 54]

$$\begin{array}{r} 7 \\ 3\overline{)21} \\ 2\overline{)42} \end{array}$$

The prime factorization of 42 is $2 \cdot 3 \cdot 7$.

A **variable** is a letter that is used to stand for a number. A mathematical expression that contains one or more variables is a **variable expression**. Replacing the variables in a variable expression with numbers and then simplifying the numerical expression is called **evaluating the variable expression**. [1.2A, p. 21]

To evaluate the variable expression $4ab$ when $a = 3$ and $b = 2$, replace a with 3 and b with 2. Simplify the resulting expression.

$4ab$
$4(3)(2) = 12(2) = 24$

An **equation** expresses the equality of two numerical or variable expressions. An equation contains an equals sign. A **solution** of an equation is a number that, when substituted for the variable, results in a true equation. [1.2A, p. 23]

6 is a solution of the equation $5 + x = 11$ because $5 + 6 = 11$ is a true equation.

Parallel lines never meet; the distance between them is always the same. [1.2C, p. 29]

Parallel Lines

An angle is measured in **degrees.** A 90° angle is a **right angle.** [1.2C, p. 29]

90°

Right Angle

A **polygon** is a closed figure determined by three or more line segments. The line segments that form the polygon are its **sides.** A **triangle** is a three-sided polygon. A **quadrilateral** is a four-sided polygon. A **rectangle** is a quadrilateral in which opposite sides are parallel, opposite sides are equal in length, and all four angles are right angles. A **square** is a rectangle in which all sides have the same length. The **perimeter** of a plane figure is a measure of the distance around the figure, and its area is the amount of surface in the region. [1.2C, pp. 29–30; 1.3E, pp. 55–56]

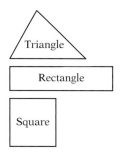

Triangle

Rectangle

Square

Essential Rules and Procedures

To round a number to a given place value: If the digit to the right of the given place value is less than 5, replace that digit and all digits to the right by zeros. If the digit to the right of the given place value is greater than or equal to 5, increase the digit in the given place value by 1, and replace all other digits to the right by zeros. [1.1C, pp. 6–7]

36,178 rounded to the nearest thousand is 36,000.

4,952 rounded to the nearest thousand is 5,000.

Copyright © Houghton Mifflin Company. All rights reserved.

To estimate the answer to a calculation: Round each number to the highest place value of that number. Perform the calculation using the rounded numbers. [1.2A, p. 21]

$$
\begin{aligned}
39{,}471 &\longrightarrow 40{,}000 \\
12{,}586 &\longrightarrow +10{,}000 \\
\hline
& 50{,}000
\end{aligned}
$$

50,000 is an estimate of the sum of 39,471 and 12,586.

Properties of Addition [1.2A, p. 22]

Addition Property of Zero $a + 0 = a$ or $0 + a = a$

$7 + 0 = 7$

Commutative Property of Addition $a + b = b + a$

$8 + 3 = 3 + 8$

Associative Property of Addition $(a + b) + c = a + (b + c)$

$(2 + 4) + 6 = 2 + (4 + 6)$

Properties of Multiplication [1.3A, p. 44]

Multiplication Property of Zero $a \cdot 0 = 0$ or $0 \cdot a = 0$

$3 \cdot 0 = 0$

Multiplication Property of One $a \cdot 1 = a$ or $1 \cdot a = a$

$6 \cdot 1 = 6$

Commutative Property of Multiplication $a \cdot b = b \cdot a$

$2 \cdot 8 = 8 \cdot 2$

Associative Property of Multiplication $(a \cdot b) \cdot c = a \cdot (b \cdot c)$

$(2 \cdot 4) \cdot 6 = 2 \cdot (4 \cdot 6)$

Division Properties of Zero and One [1.3C, p. 49]
If $a \neq 0$, $0 \div a = 0$.

$0 \div 3 = 0$

If $a \neq 0$, $a \div a = 1$.

$3 \div 3 = 1$

$a \div 1 = a$

$3 \div 1 = 3$

$a \div 0$ is undefined.

$3 \div 0$ is undefined.

Subtraction Property of Equations [1.4A, p. 67]
The same number can be subtracted from each side of an equation without changing the solution of the equation.

$$
\begin{aligned}
x + 7 &= 15 \\
x + 7 - 7 &= 15 - 7 \\
x &= 8
\end{aligned}
$$

Division Property of Equations [1.4A, p. 68]
Each side of an equation can be divided by the same number (except zero) without changing the solution of the equation.

$$
\begin{aligned}
6x &= 30 \\
\frac{6x}{6} &= \frac{30}{6} \\
x &= 5
\end{aligned}
$$

The Order of Operations Agreement [1.5A, p. 73]
Step 1 Do all operations inside parentheses.

Step 2 Simplify any numerical expressions containing exponents.

Step 3 Do multiplication and division as they occur from left to right.

Step 4 Do addition and subtraction as they occur from left to right.

$$
\begin{aligned}
5^2 - 3(2 + 4) &= 5^2 - 3(6) \\
&= 25 - 3(6) \\
&= 25 - 18 \\
&= 7
\end{aligned}
$$

Geometric Formulas [1.2C, p. 30; 1.3E, pp. 55–56]

Perimeter of a Triangle	$P = a + b + c$
Perimeter of a Rectangle	$P = 2L + 2W$
Perimeter of a Square	$P = 4s$
Area of a Rectangle	$A = LW$
Area of a Square	$A = s^2$

Find the perimeter of a triangle with sides that measure 9 m, 6 m, and 5 m.

$$
\begin{aligned}
P &= a + b + c \\
P &= 9 + 6 + 5 \\
P &= 20
\end{aligned}
$$

The perimeter of the triangle is 20 m.

Copyright © Houghton Mifflin Company. All rights reserved.

Chapter Review Exercises

1. Graph 8 on the number line.

0 1 2 3 4 5 6 7 8 9 10 11 12

2. Evaluate 10^4.

3. Find the difference between 4,207 and 1,624.

4. Write $3 \cdot 3 \cdot 5 \cdot 5 \cdot 5 \cdot 5$ in exponential notation.

5. Add: $319 + 358 + 712$

6. Round 38,729 to the nearest hundred.

7. Place the correct symbol, $<$ or $>$ between the two numbers.

247   163

8. Write thirty-two thousand five hundred nine in standard form.

9. Evaluate $2xy$ when $x = 50$ and $y = 7$.

10. Find the quotient of 15,642 and 6.

11. Subtract: $6,407 - 2,359$

12. Estimate the sum of 482, 319, 570, and 146.

13. Find all the factors of 50.

14. Is 7 a solution of the equation $24 - y = 17$?

15. Simplify: $16 + 4(7 - 5)^2 \div 8$

16. Identify the property that justifies the statement.

$10 + 33 = 33 + 10$

17. Write 4,927,036 in words.

18. Evaluate x^3y^2 when $x = 3$ and $y = 5$.

19. 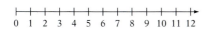 *The Film Industry* The circle graph at the right categorizes the 655 films released during a recent year by their ratings. (a) How many times more PG-13 films were released than NC-17 films? (b) How many times more R rated films were released than NC-17 films?

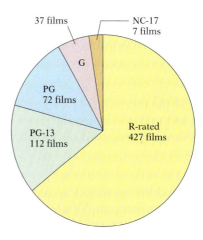

37 films   NC-17
              7 films

G

PG
72 films

PG-13
112 films

R-rated
427 films

Ratings of Films Released
Source: MPA Worldwide Market Research

Copyright © Houghton Mifflin Company. All rights reserved.

20. Divide: 6,234 ÷ 92

21. Find the product of 4 and 659.

22. Evaluate $x - y$ when $x = 270$ and $y = 133$.

23. Find the prime factorization of 90.

24. Evaluate $\dfrac{x}{y}$ when $x = 480$ and $y = 6$.

25. Complete the statement by using the Multiplication Property of One.

$? \cdot 82 = 82$

26. Solve: $36 = 4x$

27. Evaluate $x + y$ when $x = 683$ and $y = 249$.

28. Multiply: $18 \cdot 24$

29. Evaluate $(a + b)^2 - 2c$ when $a = 5$, $b = 3$, and $c = 4$.

30. *Sports* During his professional basketball career, Kareem Abdul-Jabbar had 17,440 rebounds. Elvin Hayes had 16,279 rebounds during his professional basketball career. Who had more rebounds, Abdul-Jabbar or Hayes?

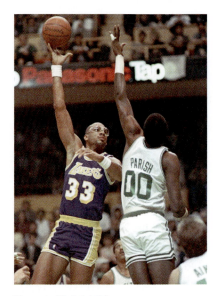

31. *Construction* A contractor quotes the cost of work on a new house, which is to have 2,800 ft² of floor space, at $65 per square foot. Find the total cost of the contractor's work on the house.

32. *Geometry* A rectangle has a length of 25 m and a width of 12 m. Find (a) the perimeter and (b) the area of the rectangle.

Kareem Abdul-Jabbar

33. *Education* The line graph at the right shows the number of students enrolled in colleges. (a) During which decade did the student population increase the most? (b) What was the amount of increase?

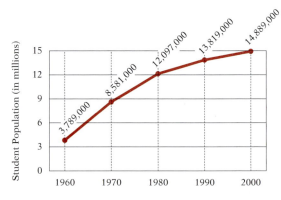

34. *Travel* Use the formula $d = rt$, where d is distance, r is rate of speed, and t is time, to find the distance traveled in 3 h by a cyclist traveling at a speed of 14 mph.

Student Enrollment in Public and Private Colleges
Source: National Center for Educational Statistics

35. *Business* Find the markup on a copy machine that cost an office supply business $1,775 and sold for $2,224. Use the formula $M = S - C$, where M is the markup on a product, S is the selling price of the product, and C is the cost of the product to the business.

Copyright © Houghton Mifflin Company. All rights reserved.

Chapter Test

1. Multiply: $3,297 \times 100$

2. Evaluate $2^4 \cdot 10^3$.

3. Find the difference between 4,902 and 873.

4. Write $x \cdot x \cdot x \cdot y \cdot y \cdot y$ in exponential notation.

5. Is 7 a solution of the equation $23 = p + 16$?

6. Round 2,961 to the nearest hundred.

7. Place the correct symbol, $<$ or $>$ between the two numbers.

7,177   7,717

8. Write eight thousand four hundred ninety in standard form.

9. Write 382,904 in words.

10. Estimate the sum of 392, 477, 519, and 648.

11. Find the product of 8 and 1,376.

12. Estimate the product of 36,479 and 58.

13. Find all the factors of 92.

14. Find the prime factorization of 240.

15. Evaluate $x - y$ when $x = 39,241$ and $y = 8,375$.

16. Identify the property that justifies the statement.

$14 + y = y + 14$

17. Evaluate $\dfrac{x}{y}$ when $x = 3,588$ and $y = 4$.

18. Simplify: $27 - (12 - 3) \div 9$

19. *Education* The table at the right shows the average annual earnings, based on level of education, for people aged 25 and older. What is the difference between average annual earnings for an individual with some college, but no degree, and for an individual with a bachelor's degree?

Educational Level	Average Annual Earnings
No high school diploma	21,400
High school diploma	28,800
Some college, no degree	32,400
Associate degree	35,400
Bachelor's degree	46,300
Master's degree	55,300

Source: Census Bureau; Bureau of Labor Statistics

Copyright © Houghton Mifflin Company. All rights reserved.

20. Solve: $68 = 17 + d$

21. Solve: $176 = 4t$

22. Evaluate $5x + (x - y)^2$ when $x = 8$ and $y = 4$.

23. Complete the statement by using the Associative Property of Addition.

$(3 + 7) + x = 3 + (? + x)$

24. *Mathematics* The sum of twelve and a number is equal to ninety. Find the number.

25. *Mathematics* What is the product of all the natural numbers less than 7?

26. *Finances* You purchase a computer system that includes an operating system priced at $850, a monitor which cost $270, an extended keyboard priced at $175, and a printer for $425. You pay for the purchase by check. You had $2,276 in your checking account before making the purchase. What was the balance in your account after making the purchase?

27. *Geometry* The length of each side of a square is 24 cm. Find (a) the perimeter and (b) the area of the square.

28. *Finances* A data processor receives a total salary of $5,690 per month. Deductions from the paycheck include $854 for taxes, $272 for retirement, and $108 for insurance. Find the data processor's monthly take home pay.

29. *Automobiles* The figure at the right shows the number of new vehicles sold with navigation systems. (a) Between which two years did the number of new vehicles sold with navigation systems increase the most? (b) What was the amount of that increase?

New Vehicles Sold with Navigation Systems
Source: J.D. Power and Associates

30. *Finances* Use the formula $C = U \cdot R$, where C is the commission earned, U is the number of units sold, and R is the rate per unit, to find the commission earned from selling 480 boxes of greeting cards when the commission rate per box is $2.

31. *Investments* The current value of the stocks in a mutual fund is $5,500,000. The number of shares outstanding is 500,000. Find the value per share of the fund. Use the formula $V = \dfrac{C}{S}$, where V is the value per share, C is the current value of the stocks in the fund, and S is the number of shares outstanding.

Copyright © Houghton Mifflin Company. All rights reserved.

2

Integers

Stock market reports involve signed numbers. Positive numbers indicate an increase in the price of a share of stock, and negative numbers indicate a decrease. Positive and negative numbers are also used to indicate whether a company has experienced a profit or loss over a specified period of time. Examples are provided in **Exercise 41 on page 124** and **Exercises 92 and 93 on page 126.**

Copyright © Houghton Mifflin Company. All rights reserved.

Need help? For online student resources, visit this website: **math.college.hmco.com**

Prep Test

1. Place the correct symbol, < or >, between the two numbers.
   54   45

2. What is the distance from 4 to 8 on the number line?

For Exercises 3–6, add, subtract, multiply, or divide.

3. $7{,}654 + 8{,}193$

4. $6{,}097 - 2{,}318$

5. 472×56

6. $144 \div 24$

7. Solve: $22 = y + 9$

8. Solve: $12b = 60$

9. What is the price of a scooter that cost a business $129 and has a markup of $43? Use the formula $P = C + M$, where P is the price of a product to a consumer, C is the cost paid by the store for the product, and M is the markup.

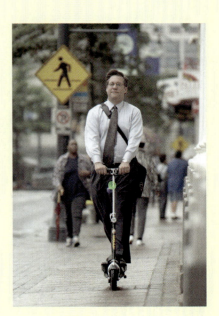

10. Simplify: $(8 - 6)^2 + 12 \div 4 \cdot 3^2$

Go Figure

If you multiply the first 20 natural numbers ($1 \cdot 2 \cdot 3 \cdot 4 \cdot 5 \cdot \cdots \cdot 17 \cdot 18 \cdot 19 \cdot 20$), how many zeros will be at the end of the number?

Copyright © Houghton Mifflin Company. All rights reserved.

Copyright © Houghton Mifflin Company. All rights reserved.

SECTION 2.1

Introduction to Integers

OBJECTIVE **A**

VIDEO & DVD   CD TUTOR   WEB   SSM

Integers and the number line

In Chapter 1, only zero and numbers greater than zero were discussed. In this chapter, numbers less than zero are introduced. Phrases such as "7 degrees below zero," "$50 in debt," and "20 feet below sea level" refer to numbers less than zero.

Numbers greater than zero are called **positive numbers.** Numbers less than zero are called **negative numbers.**

> *Positive and Negative Numbers*
>
> A number n is positive if $n > 0$.
> A number n is negative if $n < 0$.

> **▶ Point of Interest**
>
> Chinese manuscripts dating from about 250 B.C. contain the first recorded use of negative numbers. However, it was not until late in the fourteenth century that mathematicians generally accepted these numbers.

A positive number can be indicated by placing the sign + in front of the number. For example, we can write +4 instead of 4. Both +4 and 4 represent "positive 4." Usually, however, the plus sign is omitted and it is understood that the number is a positive number.

A negative number is indicated by placing a negative sign (−) in front of the number. The number −1 is read "negative one," −2 is read "negative two," and so on.

The number line can be extended to the left of zero to show negative numbers.

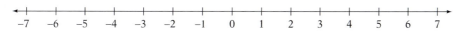

The **integers** are . . . −4, −3, −2, −1, 0, 1, 2, 3, 4, ... The integers to the right of zero are the **positive integers.** The integers to the left of zero are the **negative integers.** Zero is an integer, but it is neither positive nor negative. The point corresponding to 0 on the number line is called the **origin.**

On a number line, the numbers get larger as we move from left to right. The numbers get smaller as we move from right to left. Therefore, a number line can be used to visualize the order relation between two integers.

A number that appears to the right of a given number is greater than (>) the given number. A number that appears to the left of a given number is less than (<) the given number.

2 is to the right of −3 on the number line.
2 is greater than −3.
2 > −3

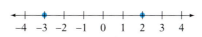

−4 is to the left of 1 on the number line.
−4 is less than 1.
−4 < 1

> **Order Relations**
>
> $a > b$ if a is to the right of b on the number line.
> $a < b$ if a is to the left of b on the number line.

1 **Example** On the number line, what number is 5 units to the right of −2?

Solution

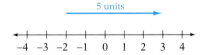

5 units

$$-4 \quad -3 \quad -2 \quad -1 \quad 0 \quad 1 \quad 2 \quad 3 \quad 4$$

3 is 5 units to the right of −2.

1 **You Try It** On the number line, what number is 4 units to the left of 1?

Your Solution

2 **Example** If G is 2 and I is 4, what numbers are B and D?

$$A \quad B \quad C \quad D \quad E \quad F \quad G \quad H \quad I$$

Solution

$$-4 \quad -3 \quad -2 \quad -1 \quad 0 \quad 1 \quad 2 \quad 3 \quad 4$$

B is −3, and D is −1.

2 **You Try It** If G is 1 and H is 2, what numbers are A and C?

$$A \quad B \quad C \quad D \quad E \quad F \quad G \quad H \quad I$$

Your Solution

3 **Example** Place the correct symbol, < or >, between the two numbers.

a. −3   −1   b. 1   −2

Solution a. −3 is to the left of −1 on the number line.

$-3 < -1$

b. 1 is to the right of −2 on the number line.

$1 > -2$

3 **You Try It** Place the correct symbol, < or >, between the two numbers.

a. 2   −5   b. −4   3

Your Solution

4 **Example** Write the given numbers in order from smallest to largest.

5, −2, 3, 0, −6

Solution −6, −2, 0, 3, 5

4 **You Try It** Write the given numbers in order from smallest to largest.

−7, 4, −1, 0, 8

Your Solution

Solutions on p. S4

Copyright © Houghton Mifflin Company. All rights reserved.

OBJECTIVE **B**

Opposites

The distance from 0 to 3 on the number line is 3 units. The distance from 0 to −3 on the number line is 3 units. 3 and −3 are the same distance from 0 on the number line, but 3 is to the right of 0 and −3 is to the left of 0.

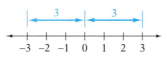

Two numbers that are the same distance from zero on the number line but on opposite sides of zero are called **opposites.**

Calculator Note

The +/− on your calculator is used to find the opposite of a number. The − is used to perform the operation of subtraction.

-3 is the opposite of 3   and   3 is the opposite of -3.

For any number n, the opposite of n is $-n$ and the opposite of $-n$ is n.

We can now define the **integers** as the whole numbers and their opposites.

A negative sign can be read as "the opposite of."

$-(3) = -3$ The opposite of positive 3 is negative 3.

$-(-3) = 3$ The opposite of negative 3 is positive 3.

Therefore, $-(a) = -a$ and $-(-a) = a$.

Note that with the introduction of negative integers and opposites, the symbols + and − can be read in different ways.

$6 + 2$	"six plus two"	+ is read "plus"
$+2$	"positive two"	+ is read "positive"
$6 - 2$	"six minus two"	− is read "minus"
-2	"negative two"	− is read "negative"
$-(-6)$	"the opposite of negative six"	− is read first as "the opposite of" and then as "negative"

When the symbols + and − indicate the operations of addition and subtraction, spaces are inserted before and after the symbol. When the symbols + and − indicate the sign of a number (positive or negative), there is no space between the symbol and the number.

5 **Example**   Find the opposite number.

a. −8   b. 15   c. a

Solution   a. 8   b. −15   c. $-a$

 5 **You Try It**   Find the opposite number.

a. 24   b. −13   c. $-b$

Your Solution

Solution on p. S4

Copyright © Houghton Mifflin Company. All rights reserved.

6 *Example* Write the expression in words.

    a. $7 - (-9)$ **b.** $-4 + 10$

Solution **a.** seven minus negative nine

    b. negative four plus ten

6 *You Try It* Write the expression in words.

    a. $-3 - 12$ **b.** $8 + (-5)$

Your Solution

7 *Example* Simplify.

    a. $-(-27)$ **b.** $-(-c)$

Solution **a.** $-(-27) = 27$

    b. $-(-c) = c$

7 *You Try It* Simplify.

    a. $-(-59)$ **b.** $-(y)$

Your Solution

Solutions on p. S4

OBJECTIVE **C**

Absolute value

The **absolute value** of a number is the distance from zero to the number on the number line. Distance is never a negative number. Therefore, the absolute value of a number is a positive number or zero. The symbol for absolute value is "| |."

The distance from 0 to 3 is 3 units. Thus $|3| = 3$ (the absolute value of 3 is 3).

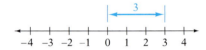

The distance from 0 to -3 is 3 units. Thus $|-3| = 3$ (the absolute value of -3 is 3).

Because the distance from 0 to 3 and the distance from 0 to -3 are the same,

$$|3| = |-3| = 3.$$

> ### Absolute Value
>
> The absolute value of a positive number is positive. $|5| = 5$
>
> The absolute value of a negative number is positive. $|-5| = 5$
>
> The absolute value of zero is zero. $|0| = 0$

Take Note

It is important to be aware that the negative sign is *in front of the absolute value.* This means $-|7| = -7$, but $|-7| = 7$.

Evaluate $-|7|$.

The negative sign is *in front of* the absolute value symbol.

Recall that a negative sign can be read as "the opposite of."

Therefore, $-|7|$ can be read "the opposite of the absolute value of 7."

$$-|7| = -7$$

Copyright © Houghton Mifflin Company. All rights reserved.

8 *Example*   Find the absolute value of **a.** 6 and **b.** −9.

  Solution   **a.** $|6| = 6$

    b. $|-9| = 9$

8 *You Try It*   Find the absolute value of **a.** −8 and **b.** 12.

  Your Solution

---

9 *Example*   Evaluate **a.** $|-27|$ and **b.** $-|-14|$.

  Solution   **a.** $|-27| = 27$

    b. $-|-14| = -14$

9 *You Try It*   Evaluate **a.** $|0|$ and **b.** $-|35|$.

  Your Solution

---

10 *Example*   Evaluate $|-x|$, where $x = -4$.

  Solution $|-x| = |-(-4)| = |4| = 4$

10 *You Try It*   Evaluate $|-y|$, where $y = 2$.

  Your Solution

---

11 *Example*   Write the given numbers in order from smallest to largest.

    $|-7|, -5, |0|, -(-4), -|-3|$

  Solution $|-7| = 7, |0| = 0,$
    $-(-4) = 4, -|-3| = -3$

    $-5, -|-3|, |0|, -(-4), |-7|$

11 *You Try It*   Write the given numbers in order from smallest to largest.

    $|6|, |-2|, -(-1), -4, -|-8|$

  Your Solution

Solutions on p. S4

OBJECTIVE   **D**

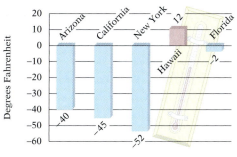

Applications

Data that are represented by negative numbers on a bar graph are shown below the horizontal axis. For instance, Figure 2.1 shows the lowest recorded temperatures, in Fahrenheit, for selected states in the United States. Hawaii's lowest recorded temperature is 12°F, which is a positive number, so the bar that represents that temperature is above the horizontal axis. The bars for the other states are below the horizontal axis and therefore represent negative numbers.

We can see from the graph that the state with the lowest recorded temperature is New York, with a temperature of −52°F.

Figure 2.1   Lowest Recorded Temperatures

Copyright © Houghton Mifflin Company. All rights reserved.

In a golf tournament, scores below par are recorded as negative numbers; scores above par are recorded as positive numbers. The winner of the tournament is the player who has the lowest score.

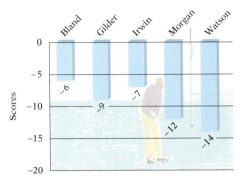

 Figure 2.2 shows the number of strokes under par for the five best finishers in the 2002 Senior Tour Championship in Oklahoma City, Oklahoma. Use this graph for Example 12 and You Try It 12.

Figure 2.2 The Top Finishers in the 2002 Senior Tour Championship

12 Example

Use Figure 2.2 to name the player who won the tournament.

Strategy

Use the bar graph and find the player with the lowest number for a score.

Solution

$-14 < -12 < -9 < -7 < -6$

The lowest number among the scores is -14.

Watson won the tournament.

12 You Try It

Use Figure 2.2 to name the player who came in third in the tournament.

Your Strategy

Your Solution

13 Example

Which is the colder temperature, $-18°F$ or $-15°F$?

Strategy

To determine which is the colder temperature, compare the numbers -18 and -15. The lower number corresponds to the colder temperature.

Solution

$-18 < -15$

The colder temperature is $-18°F$.

13 You Try It

Which is closer to blastoff, -9 s and counting or -7 s and counting?

Your Strategy

Your Solution

Solutions on p. S4

Copyright © Houghton Mifflin Company. All rights reserved.

2.1 EXERCISES

OBJECTIVE   **A**

Graph the number on the number line.

1. −5

2. −1

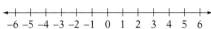

3. −6

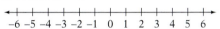

4. −2

5. x, where $x = 5$

6. x, where $x = 0$

7. x, where $x = -4$

8. x, where $x = -3$

On the number line, which number is:

9. 3 units to the right of −2?

10. 5 units to the right of −3?

11. 4 units to the left of 3?

12. 2 units to the left of −1?

13. 6 units to the right of −3?

14. 4 units to the right of −4?

For Exercises 15–18, use the following number line.

    A   B   C   D   E   F   G   H   I

15. If F is 1 and G is 2, what numbers are A and C?

16. If G is 1 and H is 2, what numbers are B and D?

17. If H is 0 and I is 1, what numbers are A and D?

18. If G is 2 and I is 4, what numbers are B and E?

Copyright © Houghton Mifflin Company. All rights reserved.

Place the correct symbol, $<$ or $>$, between the two numbers.

19. -2 -5

20. -6 -1

21. 3 -7

22. -11 -8

23. -42 27

24. 21 -34

25. 53 -46

26. -27 -39

27. -51 -20

28. -136 0

29. -131 101

30. 127 -150

Write the given numbers in order from smallest to largest.

31. $3, -7, 0, -2$

32. $-4, 8, 6, -1$

33. $-3, 1, -5, 4$

34. $-6, 2, -8, 7$

35. $9, -4, 5, 0$

36. $6, -9, -12, 8$

37. $-10, 4, 12, -5, -7$

38. $11, -8, -1, 7, -6$

39. $10, -11, -2, 5, -7$

OBJECTIVE **B**

Find the opposite of the number.

40. 22

41. 45

42. -31

43. -88

44. c

45. n

46. $-w$

47. $-d$

Write the expression in words.

48. $-(-11)$

49. $-(-13)$

50. $-(-d)$

51. $-(-p)$

52. $-2 + (-5)$

53. $5 + (-10)$

54. $6 - (-7)$

55. $-14 - (-3)$

56. $9 - 12$

57. $-13 - 8$

58. $-a - b$

59. $m + (-n)$

Copyright © Houghton Mifflin Company. All rights reserved.

Simplify.

60. $-(-5)$

61. $-(-7)$

62. $-(-38)$

63. $-(-61)$

64. $-(29)$

65. $-(46)$

66. $-(-52)$

67. $-(-73)$

68. $-(-m)$

69. $-(-z)$

70. $-(b)$

71. $-(p)$

OBJECTIVE C

Find the absolute value of the number.

72. 4

73. -4

74. -7

75. 9

76. -1

77. -11

78. 10

79. -12

Evaluate.

80. $|-15|$

81. $|-23|$

82. $-|33|$

83. $-|27|$

84. $|32|$

85. $|25|$

86. $-|-36|$

87. $-|-41|$

88. $-|-81|$

89. $-|-93|$

90. $|x|$, where $x = 7$

91. $|x|$, where $x = -10$

92. $|-x|$, where $x = 2$

93. $|-x|$, where $x = 8$

94. $|-y|$, where $y = -3$

95. $|-y|$, where $y = -6$

Place the correct symbol, $<$, $=$, or $>$, between the two numbers.

96. $|7|$ $|-9|$

97. $|-12|$ $|8|$

98. $|-5|$ $|-2|$

99. $|6|$ $|13|$

100. $|-8|$ $|3|$

101. $|-1|$ $|-17|$

102. $|-14|$ $|14|$

103. $|x|$ $|-x|$

Copyright © Houghton Mifflin Company. All rights reserved.

Write the given numbers in order from smallest to largest.

104. $|-8|, -(-3), |2|, -|-5|$

105. $-|6|, -(4), |-7|, -(-9)$

106. $-(-1), |-6|, |0|, -|3|$

107. $-|-7|, -9, -(5), |4|$

108. $-|2|, -(-8), 6, |1|, -7$

109. $-(-3), -|-8|, |5|, -|10|, -(-2)$

110. Find the values of a for which $|a| = 7$.

111. Find the values of y for which $|y| = 11$.

112. Given that x is an integer, find all values of x for which $|x| < 5$.

113. Given that c is an integer, find all values of c for which $|c| < 7$.

OBJECTIVE **D**

 The table below gives equivalent temperatures for combinations of temperature and wind speed. For example, the combination of a temperature of 15°F and a wind blowing at 10 mph has a cooling power equal to 3°F. Use this table for Exercises 114 to 119.

Wind Chill Factors														
Wind Speed (mph)	Thermometer Reading (degrees Fahrenheit)													
	25	20	15	10	5	0	−5	−10	−15	−20	−25	−30	−35	−40	−45
5	19	13	7	1	−5	−11	−16	−22	−28	−34	−40	−46	−52	−57	−63
10	15	9	3	−4	−10	−16	−22	−28	−35	−41	−47	−53	−59	−66	−72
15	13	6	0	−7	−13	−19	−26	−32	−39	−45	−51	−58	−64	−71	−77
20	11	4	−2	−9	−15	−22	−29	−35	−42	−48	−55	−61	−68	−74	−81
25	9	3	−4	−11	−17	−24	−31	−37	−44	−51	−58	−64	−71	−78	−84
30	8	1	−5	−12	−19	−26	−33	−39	−46	−53	−60	−67	−73	−80	−87
35	7	0	−7	−14	−21	−27	−34	−41	−48	−55	−62	−69	−76	−82	−89
40	6	−1	−8	−15	−22	−29	−36	−43	−50	−57	−64	−71	−78	−84	−91
45	5	−2	−9	−16	−23	−30	−37	−44	−51	−58	−65	−72	−79	−86	−93

114. *Environmental Science* Find the wind chill factor when the temperature is 5°F and the wind speed is 15 mph.

Copyright © Houghton Mifflin Company. All rights reserved.

115. *Environmental Science*   Find the wind chill factor when the temperature is 10°F and the wind speed is 20 mph.

116. *Environmental Science*   Find the cooling power of a temperature of −10°F and a 5 mph wind.

117. *Environmental Science*   Find the cooling power of a temperature of −15°F and a 10 mph wind.

118. *Environmental Science*   Which feels colder, a temperature of 0°F with a 15 mph wind or a temperature of 10°F with a 25 mph wind?

119. *Environmental Science*   Which would feel colder, a temperature of −30°F with a 5 mph wind or a temperature of −20°F with a 10 mph wind?

120. *Rocketry*   Which is closer to blastoff, −12 min and counting or −17 min and counting?

One of the measures used by a financial analyst to evaluate the financial strength of a company is *earnings per share*. This number is found by taking the total profit of the company and dividing by the number of shares of stock that the company has sold to investors. If the company has a loss instead of a profit, the earnings per share is a negative number. In a bar graph, a profit is shown by a bar extending above the horizontal axis, and a loss is shown by a bar extending below the horizontal axis. The figure on the right shows the earnings per share for Mycogen for the years 1999 through 2004. Use this graph for Exercises 121 to 124.

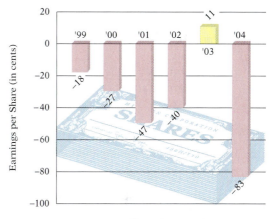

Mycogen Earnings per Share (in cents)

121. *Business*   (a) What were the earnings per share for Mycogen in 2000? (b) What were the earnings per share for Mycogen in 2002?

122. *Business*   For the years shown, in which year did Mycogen have the greatest loss?

123. *Business*   For the years shown, did Mycogen ever have a profit? If so, in what year?

124. *Business*   In which year was the Mycogen earnings per share lower, 1999 or 2001?

125. *Investments*   In the stock market, the net change in the price of a share of stock is recorded as a positive or a negative number. If the price rises, the net change is positive. If the price falls, the net change is negative. If the net change for a share of Stock A is −2 and the net change for a share of Stock B is −1, which stock showed the least net change?

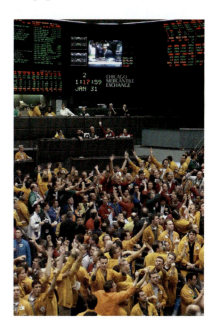

Copyright © Houghton Mifflin Company. All rights reserved.

126. *Business*   Some businesses show a profit as a positive number and a loss as a negative number. During the first quarter of this year, the loss experienced by a company was recorded as −12,575. During the second quarter of this year, the loss experienced by the company was −11,350. During which quarter was the loss greater?

127. *Business*   Some businesses show a profit as a positive number and a loss as a negative number. During the third quarter of last year, the loss experienced by a company was recorded as −26,800. During the fourth quarter of last year, the loss experienced by the company was −24,900. During which quarter was the loss greater?

Critical Thinking

128. *Mathematics* A is a point on the number line halfway between −9 and 3. B is a point halfway between A and the graph of 1 on the number line. B is the graph of what number?

129.   **a.**   Name two numbers that are 4 units from 2 on the number line.
       b.   Name two numbers that are 5 units from 3 on the number line.

130.   Determine whether the statement is always true, sometimes true, or never true.
       a.   The number $-n$ is a negative number.
       b.   A number and its opposite are different numbers.
       c. $|x| > x$
       d. $|x| > -x$
       e.   If n is a negative number, $-n$ is a positive number.
       f.   If n is a positive number, $-n$ is a negative number.

131.   The $\boxed{+/-}$ key on a calculator changes the sign of the number in the calculator's display. In other words, it changes the number in the display to its opposite. Use the $\boxed{+/-}$ key on your calculator to display each of the following numbers:
       a.   −9   **b.**   −20   **c.**   −148   **d.**   −573

132.   Given that z is an integer and $|z| < 15$, find all values of z for which $|z| > 10$.

133.   Given that x is an integer and $|x| < 10$, find all values of x for which $|x| > 6$.

134.   In your own words, describe (a) the opposite of a number, (b) the absolute value of a number, and (c) the difference between the words *negative* and *minus*.

135.   Student A, Student B, Student C, and Student D were being questioned by their teacher. The teacher knew that one of the students had left an apple on the teacher's desk but did not know which one. Student A said it was either Student B or Student D. Student D said it was neither Student B nor Student C. If both statements were false, who left the apple on the teacher's desk? Explain how you arrived at your solution.

Copyright © Houghton Mifflin Company. All rights reserved.

Copyright © Houghton Mifflin Company. All rights reserved.

SECTION 2.2   Addition and Subtraction of Integers

OBJECTIVE   **A**

Addition of integers

Not only can an integer be graphed on a number line, an integer can be represented anywhere along a number line by an arrow. A positive number is represented by an arrow pointing to the right. A negative number is represented by an arrow pointing to the left. The absolute value of the number is represented by the length of the arrow. The integers 5 and −4 are shown on the number line in the figure below.

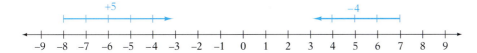

The sum of two integers can be shown on a number line. To add two integers, find the point on the number line corresponding to the first addend. At that point, draw an arrow representing the second addend. The sum is the number directly below the tip of the arrow.

$4 + 2 = 6$

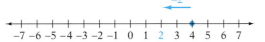

$-4 + (-2) = -6$

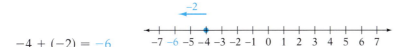

$-4 + 2 = -2$

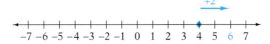

$4 + (-2) = 2$

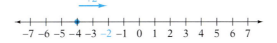

The sums shown above can be categorized by the signs of the addends.

The addends have the same sign.

$4 + 2$ positive 4 plus positive 2
$-4 + (-2)$ negative 4 plus negative 2

The addends have different signs.

$-4 + 2$ negative 4 plus positive 2
$4 + (-2)$ positive 4 plus negative 2

The rule for adding two integers depends on whether the signs of the addends are the same or different.

> ### Rule for Adding Two Integers
>
> **To add two integers with the same sign,** add the absolute values of the numbers. Then attach the sign of the addends.
>
> **To add two integers with different signs,** find the absolute values of the numbers. Subtract the smaller absolute value from the larger absolute value. Then attach the sign of the addend with the larger absolute value.

Add: $(-4) + (-9)$

The signs of the addends are the same.
Add the absolute values of the numbers.
$$|-4| = 4, |-9| = 9, 4 + 9 = 13$$
Attach the sign of the addends.
(Both addends are negative.
The sum is negative.) $(-4) + (-9) = -13$

Calculator Note

To add $-14 + (-47)$ with your calculator, enter the following:

14 [+/-] + 47 [+/-] =
 -14 -47

Add: $-14 + (-47)$

The signs are the same.
Add the absolute values of the numbers.
Attach the sign of the addends. $-14 + (-47) = -61$

Add: $6 + (-13)$

The signs of the addends are different.
Find the absolute values of the numbers.
$$|6| = 6, |-13| = 13$$
Subtract the smaller absolute value from the larger absolute value.
$$13 - 6 = 7$$
Attach the sign of the number with the larger absolute value.
$|-13| > |6|$. Attach the negative sign. $6 + (-13) = -7$

Add: $162 + (-247)$

The signs are different. Find the difference between the absolute values of the numbers.
$$247 - 162 = 85$$
Attach the sign of the number with the larger absolute value. $162 + (-247) = -85$

Add: $-8 + 8$

The signs are different. Find the difference between the absolute values of the numbers.
$$8 - 8 = 0$$ $-8 + 8 = 0$

Copyright © Houghton Mifflin Company. All rights reserved.

Note in this last example that we are adding a number and its opposite (-8 and 8), and the sum is 0. The opposite of a number is called its **additive inverse.** The opposite or additive inverse of -8 is 8, and the opposite or additive inverse of 8 is -8. **The sum of a number and its additive inverse is always zero.** This is known as the Inverse Property of Addition.

The properties of addition presented in Chapter 1 hold true for integers as well as whole numbers. These properties are repeated below, along with the Inverse Property of Addition.

The Addition Property of Zero	$a + 0 = a$ or $0 + a = a$
The Commutative Property of Addition	$a + b = b + a$
The Associative Property of Addition	$(a + b) + c = a + (b + c)$
The Inverse Property of Addition	$a + (-a) = 0$ or $-a + a = 0$

Take Note

With the Commutative Properties, the order in which the numbers appear changes. With the Associative Properties, the order in which the numbers appear remains the same.

Add: $(-4) + (-6) + (-8) + 9$

	$(-4) + (-6) + (-8) + 9$
Add the first two numbers.	$= (-10) + (-8) + 9$
Add the sum to the third number.	$= (-18) + 9$
Continue until all the numbers have been added.	$= -9$

Take Note

For the example at the left, check that the sum is the same if the numbers are added in a different order.

The price of Byplex Corporation's stock fell each trading day of the first week of June 2005. Use Figure 2.3 to find the change in the price of Byplex stock over the week's time.

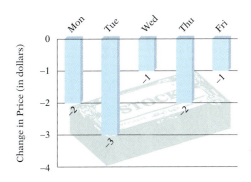

Figure 2.3   Change in Price of Byplex Corporation Stock

Add the five changes in price.

$-2 + (-3) + (-1) + (-2) + (-1)$
$= (-5) + (-1) + (-2) + (-1)$
$= -6 + (-2) + (-1)$
$= -8 + (-1) = -9$

The change in the price was -9.

This means that the price of the stock fell $9 per share.

Copyright © Houghton Mifflin Company. All rights reserved.

Evaluate $-x + y$ when $x = -15$ and $y = -5$.

$$-x + y$$

Replace x with -15 and y with -5.　　　$-(-15) + (-5)$

Simplify $-(-15)$.　　　　　　　　　　$= 15 + (-5)$

Add.　　　　　　　　　　　　　　　　$= 10$

Take Note

Recall that a solution of an equation is a number that, when substituted for the variable, results in a true equation.

Is -7 a solution of the equation $x + 4 = -3$?

$$x + 4 = -3$$

Replace x by -7 and then simplify.　　　$\dfrac{}{-7 + 4 \ \big| \ -3}$

The results are equal.　　　　　　　　$-3 = -3$

-7 is a solution of the equation.

---

① Example Add: $-97 + (-45)$

Solution $-97 + (-45) = -142$

① You Try It Add: $-38 + (-62)$

Your Solution

---

② Example Add: $81 + (-79)$

Solution $81 + (-79) = 2$

② You Try It Add: $47 + (-53)$

Your Solution

---

③ Example Add: $42 + (-12) + (-30)$

Solution $42 + (-12) + (-30)$
　　　　　$= 30 + (-30)$
　　　　　$= 0$

③ You Try It Add: $-36 + 17 + (-21)$

Your Solution

---

④ Example What is -162 increased by 98?

Solution $-162 + 98 = -64$

④ You Try It Find the sum of -154 and -37.

Your Solution

---

⑤ Example Evaluate $-x + y$ when $x = -11$ and $y = -2$.

Solution $-x + y$
　　　　　$-(-11) + (-2) = 11 + (-2)$
　　　　　　　　　　　$= 9$

⑤ You Try It Evaluate $-x + y$ when $x = -3$ and $y = -10$.

Your Solution

Solutions on p. S4

Copyright © Houghton Mifflin Company. All rights reserved.

Example Is -6 a solution of the equation $3 + y = -2$?

Solution

$$3 + y = -2$$
$$\dfrac{3 + (-6)\ \big|\ -2}{}$$
$$-3 \neq -2$$

No, -6 is not a solution of the equation.

 You Try It Is -9 a solution of the equation $2 = 11 + a$?

Your Solution

Solution on p. S4

OBJECTIVE **B**

VIDEO & DVD   CD TUTOR   WEB   SSM

Subtraction of integers

Before the rules for subtracting two integers are explained, look at the translation into words of expressions that represent the difference of two integers.

$9 - 3$	positive 9 minus positive 3
$-9 - 3$	negative 9 minus positive 3
$9 - (-3)$	positive 9 minus negative 3
$-9 - (-3)$	negative 9 minus negative 3

Note that the sign $-$ is used in two different ways. One way is as a negative sign, as in -9 (negative 9). The second way is to indicate the operation of subtraction, as in $9 - 3$ (9 minus 3).

Look at the next four expressions and decide whether the second number in each expression is a positive number or a negative number.

1. $(-10) - 8$

2. $(-10) - (-8)$

3. $10 - (-8)$

4. $10 - 8$

In expressions 1 and 4, the second number is positive 8. In expressions 2 and 3, the second number is negative 8.

Opposites are used to rewrite subtraction problems as related addition problems. Notice below that the subtraction of whole numbers is the same as the addition of the opposite number.

Subtraction		Addition of the Opposite	
$8 - 4$	$=$	$8 + (-4)$	$= 4$
$7 - 5$	$=$	$7 + (-5)$	$= 2$
$9 - 2$	$=$	$9 + (-2)$	$= 7$

Copyright © Houghton Mifflin Company. All rights reserved.

Subtraction of integers can be written as the addition of the opposite number. To subtract two integers, rewrite the subtraction expression as the first number plus the opposite of the second number. Some examples are shown below.

First number	−	second number	=	First number	+	opposite of the second number
8	−	15	=	8	+	$(-15) = -7$
8	−	(-15)	=	8	+	$15 = 23$
−8	−	15	=	−8	+	$(-15) = -23$
−8	−	(-15)	=	−8	+	$15 = 7$

> ### Rule for Subtracting Two Integers
>
> To subtract two integers, add the opposite of the second integer to the first integer.

Subtract: $(-15) - 75$

Rewrite the subtraction operation as the sum of the first number and the opposite of the second number. The opposite of 75 is −75.

$(-15) - 75$

$= (-15) + (-75)$

Add.

$= -90$

Subtract: $6 - (-20)$

Rewrite the subtraction operation as the sum of the first number and the opposite of the second number. The opposite of −20 is 20.

$6 - (-20)$

$= 6 + 20$

$= 26$

Subtract: $11 - 42$

Rewrite the subtraction operation as the sum of the first number and the opposite of the second number. The opposite of 42 is −42.

$11 - 42$

$= 11 + (-42)$

$= -31$

Take Note

$42 - 11 = 31$

$11 - 42 = -31$

$42 - 11 \neq 11 - 42$

By the Commutative Property of Addition, the order in which two numbers are added does not affect the sum; $a + b = b + a$. However, note from this last example that the order in which two numbers are subtracted *does* affect the difference. The operation of subtraction is not commutative.

Copyright © Houghton Mifflin Company. All rights reserved.

When subtraction occurs several times in an expression, rewrite each subtraction as addition of the opposite and then add.

Calculator Note

To subtract $-13 - 5 - (-8)$ with your calculator, enter the following:

$$13 \; \boxed{+/-} \quad \boxed{-} \quad 5 \quad \boxed{-} \quad 8 \; \boxed{+/-} \quad \boxed{=}$$

$\underbrace{}_{-13}$ $\underbrace{}_{-8}$

Subtract: $-13 - 5 - (-8)$

Rewrite each subtraction as addition of the opposite.

Add.

$-13 - 5 - (-8)$
$= -13 + (-5) + 8$
$= -18 + 8$
$= -10$

Simplify: $-14 + 6 - (-7)$

This problem involves both addition and subtraction. Rewrite the subtraction as addition of the opposite.

Add.

$-14 + 6 - (-7)$
$= -14 + 6 + 7$
$= -8 + 7$
$= -1$

Evaluate $a - b$ when $a = -2$ and $b = -9$.

Replace a with -2 and b with -9.

Rewrite the subtraction as addition of the opposite.

Add.

$a - b$
$-2 - (-9)$
$= -2 + 9$
$= 7$

Is -4 a solution of the equation $3 - a = 11 + a$?

Replace a by -4 and then simplify.

The results are equal.

$$3 - a = 11 + a$$

$3 - (-4)$	$11 + (-4)$
$3 + 4$	7

$$7 = 7$$

Yes, -4 is a solution of the equation.

---

7 ***Example*** Subtract: $-12 - (-17)$

Solution $-12 - (-17) = -12 + 17$
$= 5$

7 ***You Try It*** Subtract: $-35 - (-34)$

Your Solution

8 ***Example*** Subtract: $66 - (-90)$

Solution $66 - (-90) = 66 + 90$
$= 156$

8 ***You Try It*** Subtract: $83 - (-29)$

Your Solution

Solutions on p. S4

Copyright © Houghton Mifflin Company. All rights reserved.

The table below shows the boiling point and the melting point in degrees Celsius of three chemical elements. Use this table for Example 9 and You Try It 9.

Chemical Element	Boiling Point	Melting Point
Mercury	357	−39
Radon	−62	−71
Xenon	−108	−112

Radon

9 Example Use the table above to find the difference between the boiling point and the melting point of mercury.

Solution The boiling point of mercury is 357.

The melting point of mercury is −39.

$$357 − (−39) = 357 + 39$$
$$= 396$$

The difference is 396°C.

9 You Try It Use the table above to find the difference between the boiling point and the melting point of xenon.

Your Solution

10 Example What is −12 minus 8?

Solution $−12 − 8 = −12 + (−8)$
$$= −20$$

10 You Try It What is 14 less than −8?

Your Solution

11 Example Subtract 91 from 43.

Solution $43 − 91 = 43 + (−91)$
$$= −48$$

11 You Try It What is 25 decreased by 68?

Your Solution

12 Example Simplify:
$−8 − 30 − (−12) − 7 − (−14)$

Solution $−8 − 30 − (−12) − 7 − (−14)$
$$= −8 + (−30) + 12 + (−7) + 14$$
$$= −38 + 12 + (−7) + 14$$
$$= −26 + (−7) + 14$$
$$= −33 + 14$$
$$= −19$$

12 You Try It Simplify:
$−4 − (−3) + 12 − (−7) − 20$

Your Solution

Solutions on p. S4

Copyright © Houghton Mifflin Company. All rights reserved.

13 *Example*   Evaluate $-x - y$ when $x = -4$ and $y = -3$.

Solution
$$-x - y$$
$$-(-4) - (-3) = 4 - (-3)$$
$$= 4 + 3$$
$$= 7$$

13 *You Try It*   Evaluate $x - y$ when $x = -9$ and $y = 7$.

Your Solution

14 *Example*   Is 8 a solution of the equation $-2 = 6 - x$?

Solution

$-2 = 6 - x$	
-2	$6 - 8$
-2	$6 + (-8)$
$-2 =$	-2

Yes, 8 is a solution of the equation.

14 *You Try It*   Is -3 a solution of the equation $a - 5 = -8$?

Your Solution

Solutions on pp. S4–S5

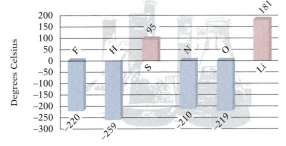

OBJECTIVE **C**

Applications and formulas

 Figure 2.4 shows the melting points in degrees Celsius of six chemical elements. The abbreviations of the elements are:

F - Fluorine   H - Hydrogen
S - Sulfur   N - Nitrogen
O - Oxygen   Li - Lithium

Use this graph for Example 15 and You Try It 15.

Figure 2.4   Melting Points of Chemical Elements

15 *Example*

Find the difference between the two lowest melting points shown in Figure 2.4.

Strategy

To find the difference, subtract the lowest melting point shown (-259) from the second lowest melting point shown (-220).

Solution

$-220 - (-259) = -220 + 259 = 39$

The difference is $39°C$.

15 *You Try It*

Find the difference between the highest and lowest melting points shown in Figure 2.4.

Your Strategy

Your Solution

Solution on p. S5

Copyright © Houghton Mifflin Company. All rights reserved.

16 *Example*

Find the temperature after an increase of 8°C from −5°C.

Strategy

To find the temperature, add the increase (8) to the previous temperature (−5).

Solution

−5 + 8 = 3

The temperature is 3°C.

16 *You Try It*

Find the temperature after an increase of 10°C from −3°C.

Your Strategy

Your Solution

17 *Example*

The average temperature on the sunlit side of the moon is approximately 215°F. The average temperature on the dark side is approximately −250°F. Find the difference between these average temperatures.

Strategy

To find the difference, subtract the average temperature on the dark side of the moon (−250) from the average temperature on the sunlit side (215).

Solution

$$215 - (-250) = 215 + 250$$
$$= 465$$

The difference is 465°F.

17 *You Try It*

The average temperature on Earth's surface is 57°F. The average temperature throughout Earth's stratosphere is −70°F. Find the difference between these average temperatures.

Your Strategy

Your Solution

18 *Example*

The distance, d, between point a and point b on the number line is given by the formula $d = |a - b|$. Use the formula to find d when $a = 7$ and $b = -8$.

Strategy

To find d, replace a by 7 and b by −8 in the given formula and solve for d.

Solution

$d = |a - b|$
$d = |7 - (-8)|$
$d = |7 + 8|$
$d = |15|$
$d = 15$

The distance between the two points is 15 units.

18 *You Try It*

The distance, d, between point a and point b on the number line is given by the formula $d = |a - b|$. Use the formula to find d when $a = -6$ and $b = 5$.

Your Strategy

Your Solution

Solutions on p. S5

Copyright © Houghton Mifflin Company. All rights reserved.

2.2 EXERCISES

OBJECTIVE **A**

1. 🖊 **a.** Explain the rule for adding two integers with the same sign.
   b. Explain the rule for adding two integers with different signs.

2. 🖊 Describe in words the Inverse Property of Addition. Provide an example of this property.

Add.

3. $-3 + (-8)$

4. $-6 + (-9)$

5. $-8 + 3$

6. $-7 + 2$

7. $-5 + 13$

8. $-4 + 11$

9. $6 + (-10)$

10. $8 + (-12)$

11. $3 + (-5)$

12. $6 + (-7)$

13. $-4 + (-5)$

14. $-12 + (-12)$

15. $-6 + 7$

16. $-9 + 8$

17. $(-5) + (-10)$

18. $(-3) + (-17)$

19. $-7 + 7$

20. $-11 + 11$

21. $(-15) + (-6)$

22. $(-18) + (-3)$

23. $0 + (-14)$

24. $-19 + 0$

25. $73 + (-54)$

26. $-89 + 62$

27. $2 + (-3) + (-4)$

28. $7 + (-2) + (-8)$

29. $-3 + (-12) + (-15)$

30. $9 + (-6) + (-16)$

31. $-17 + (-3) + 29$

32. $13 + 62 + (-38)$

33. $11 + (-22) + 4 + (-5)$

34. $-14 + (-3) + 7 + (-6)$

35. $-22 + 10 + 2 + (-18)$

36. $-6 + (-8) + 13 + (-4)$

37. $-25 + (-31) + 24 + 19$

38. $10 + (-14) + (-21) + 8$

39. What is 3 increased by -21?

40. Find 12 plus -9.

41. What is 16 more than -5?

42. What is 17 added to -7?

43. Find the total of -3, -8, and 12.

44. Find the sum of 5, -16, and -13.

Copyright © Houghton Mifflin Company. All rights reserved.

45. Write the sum of x and -7.

46. Write the total of $-a$ and b.

47. *Economics* A nation's balance of trade is the difference between its exports and imports. If the exports are greater than the imports, the result is a positive number and a *favorable balance of trade*. If the exports are less than the imports, the result is a negative number and an *unfavorable balance of trade*. The table at the right shows the unfavorable balance of trade in a recent year for the United States with four other countries. Find the total of the U.S. balance of trade with (a) Japan and Mexico, (b) Canada and Mexico, and (c) Japan and France.

U.S. Balance of Trade with Foreign Countries	
Japan	−57,931,000,000
Canada	−49,409,000,000
Mexico	−27,578,000,000
France	−10,140,000,000

Source: Bureau of Economic Analysis, U.S. Department of Commerce

Evaluate the expression for the given values of the variables.

48. $x + y$, where $x = -5$ and $y = -7$

49. $-a + b$, where $a = -8$ and $b = -3$

50. $a + b$, where $a = -8$ and $b = -3$

51. $-x + y$, where $x = -5$ and $y = -7$

52. $a + b + c$, where $a = -4$, $b = 6$, and $c = -9$

53. $a + b + c$, where $a = -10$, $b = -6$, and $c = 5$

54. $x + y + (-z)$, where $x = -3$, $y = 6$, and $z = -17$

55. $-x + (-y) + z$, where $x = -2$, $y = 8$, and $z = -11$

Identify the property that justifies the statement.

56. $-12 + 5 = 5 + (-12)$

57. $-33 + 0 = -33$

58. $-46 + 46 = 0$

59. $-7 + (3 + 2) = (-7 + 3) + 2$

Use the given property of addition to complete the statement.

60. The Associative Property of Addition
$-11 + (6 + 9) = (? + 6) + 9$

61. The Addition Property of Zero
$-13 + ? = -13$

62. The Commutative Property of Addition
$-2 + ? = -4 + (-2)$

63. The Inverse Property of Addition
$? + (-18) = 0$

Copyright © Houghton Mifflin Company. All rights reserved.

64. Is -3 a solution of the equation $x + 4 = 1$?

65. Is -8 a solution of the equation $6 = -3 + z$?

66. Is -6 a solution of the equation $6 = 12 + n$?

67. Is -8 a solution of the equation $-7 + m = -15$?

68. Is -2 a solution of the equation $3 + y = y + 3$?

69. Is -4 a solution of the equation $1 + z = z + 2$?

OBJECTIVE **B**

70. ✏️ Explain how to rewrite the subtraction $8 - (-6)$ as addition of the opposite.

71. ✏️ Explain the meaning of the words *minus* and *negative*.

Subtract.

72. $7 - 14$

73. $6 - 9$

74. $-7 - 2$

75. $-9 - 4$

76. $7 - (-2)$

77. $3 - (-4)$

78. $-6 - (-6)$

79. $-4 - (-4)$

80. $-12 - 16$

81. $-10 - 7$

82. $(-9) - (-3)$

83. $(-7) - (-4)$

84. $4 - (-14)$

85. $-4 - (-16)$

86. $(-14) - (-7)$

87. $3 - (-24)$

88. $9 - (-9)$

89. $(-41) - 65$

90. $57 - 86$

91. $-95 - (-28)$

92. How much larger is 5 than -11?

93. What is -10 decreased by -4?

94. Find -13 minus -8.

95. What is 6 less than -9?

96. Write the difference between $-y$ and 5.

97. Write $-t$ decreased by r.

Copyright © Houghton Mifflin Company. All rights reserved.

The figure at the right shows the highest and lowest temperatures ever recorded for selected regions of the world. Use this graph for Exercises 98 to 100.

98. *Temperature* What is the difference between the highest and lowest temperatures ever recorded in Africa?

99. *Temperature* What is the difference between the highest and lowest temperatures ever recorded in South America?

100. *Temperature* What is the difference between the lowest temperature recorded in Europe and the lowest temperature recorded in Asia?

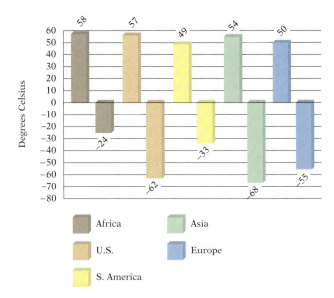

Highest and Lowest Temperatures Recorded (in degrees Celsius)

Simplify.

101. $-4 - 3 - 2$

102. $4 - 5 - 12$

103. $12 - (-7) - 8$

104. $-12 - (-3) - (-15)$

105. $4 - 12 - (-8)$

106. $-30 - (-65) - 29 - 4$

107. $-16 - 47 - 63 - 12$

108. $42 - (-30) - 65 - (-11)$

109. $12 - (-6) + 8$

110. $-7 + 9 - (-3)$

111. $-8 - (-14) + 7$

112. $-4 + 6 - 8 - 2$

113. $9 - 12 + 0 - 5$

114. $11 - (-2) - 6 + 10$

115. $5 + 4 - (-3) - 7$

116. $-1 - 8 + 6 - (-2)$

117. $-13 + 9 - (-10) - 4$

118. $6 - (-13) - 14 + 7$

Evaluate the expression for the given values of the variables.

119. $-x - y$, where $x = -3$ and $y = 9$

120. $x - (-y)$, where $x = -3$ and $y = 9$

121. $-x - (-y)$, where $x = -3$ and $y = 9$

122. $a - (-b)$, where $a = -6$ and $b = 10$

Copyright © Houghton Mifflin Company. All rights reserved.

123. $a - b - c$, where $a = 4$, $b = -2$, and $c = 9$

124. $a - b - c$, where $a = -1$, $b = 7$, and $c = -15$

125. $x - y - (-z)$, where $x = -9$, $y = 3$, and $z = 30$

126. $-x - (-y) - z$, where $x = 8$, $y = 1$, and $z = -14$

127. Is -3 a solution of the equation $x - 7 = -10$?

128. Is -4 a solution of the equation $1 = 3 - y$?

129. Is -2 a solution of the equation $-5 - w = 7$?

130. Is -8 a solution of the equation $-12 = m - 4$?

131. Is -6 a solution of the equation $-t - 5 = 7 + t$?

132. Is -7 a solution of the equation $5 + a = -9 - a$?

OBJECTIVE   C

 The elevation, or height, of places on the Earth is measured in relation to sea level, or the average level of the ocean's surface. The table below shows height above sea level as a positive number and depth below sea level as a negative number. Use the table below for Exercises 133 to 135.

Mt. Everest

Continent	Highest Elevation (in meters)		Lowest Elevation (in meters)	
Africa	Mt. Kilimanjaro	5,895	Lake Assal	−156
Asia	Mt. Everest	8,850	Dead Sea	−411
Europe	Mt. Elbrus	5,642	Caspian Sea	−28
America	Mt. Aconcagua	6,960	Death Valley	−86

133. *Geography* What is the difference in elevation (a) between Mt. Aconcagua and Death Valley and (b) between Mt. Kilimanjaro and Lake Assal?

134. *Geography* For which continent shown is the difference between the highest and lowest elevations greatest?

135. *Geography* For which continent shown is the difference between the highest and lowest elevations smallest?

136. *Temperature* Find the temperature after a rise of 9°C from −6°C.

Copyright © Houghton Mifflin Company. All rights reserved.

The table at the right shows the average temperatures at different cruising altitudes for airplanes. Use the table for Exercises 137 to 139.

Cruising Altitude	Average Temperature
12,000 ft	16°
20,000 ft	−12°
30,000 ft	−48°
40,000 ft	−70°
50,000 ft	−70°

137. *Temperature*   What is the difference between the average temperatures at 12,000 ft and at 40,000 ft?

138. *Temperature*   What is the difference between the average temperatures at 40,000 ft and at 50,000 ft?

139. *Temperature*   How much colder is the average temperature at 30,000 ft than at 20,000 ft?

140. *Sports*   Use the equation $S = N - P$, where S is a golfer's score relative to par in a tournament, N is the number of strokes made by the golfer, and P is par, to find a golfer's score relative to par when the golfer made 196 strokes and par is 208.

141. *Sports*   Use the equation $S = N - P$, where S is a golfer's score relative to par in a tournament, N is the number of strokes made by the golfer, and P is par, to find a golfer's score relative to par when the golfer made 49 strokes and par is 52.

142. *Mathematics*   The distance, d, between point a and point b on the number line is given by the formula $d = |a - b|$. Find d when $a = 6$ and $b = -15$.

143. *Mathematics*   The distance, d, between point a and point b on the number line is given by the formula $d = |a - b|$. Find d when $a = 7$ and $b = -12$.

Critical Thinking

144. *Mathematics*   Given the list of numbers at the right, find the largest difference that can be obtained by subtracting one number in the list from a different number in the list.

$5, -2, -9, 11, 14$

145. Determine whether the statement is always true, sometimes true, or never true.
   a.   The difference between a number and its additive inverse is zero.
   b.   The sum of a negative number and a negative number is a negative number.

146. The sum of two negative integers is -7. Find the integers.

147. Describe the steps involved in using a calculator to simplify $-17 - (-18) + (-5)$.

Copyright © Houghton Mifflin Company. All rights reserved.

OBJECTIVE **A**

VIDEO & DVD   CD TUTOR   WEB   SSM

Multiplication of integers

When 5 is multiplied by a sequence of decreasing integers, each product decreases by 5.

$5(3) = 15$
$5(2) = 10$
$5(1) = 5$
$5(0) = 0$

The pattern developed can be continued so that 5 is multiplied by a sequence of negative numbers. To maintain the pattern of decreasing by 5, the resulting products must be negative.

$5(-1) = -5$
$5(-2) = -10$
$5(-3) = -15$
$5(-4) = -20$

This example illustrates that the product of a positive number and a negative number is negative.

When -5 is multiplied by a sequence of decreasing integers, each product increases by 5.

$-5(3) = -15$
$-5(2) = -10$
$-5(1) = -5$
$-5(0) = 0$

The pattern developed can be continued so that -5 is multiplied by a sequence of negative numbers. To maintain the pattern of increasing by 5, the resulting products must be positive.

$-5(-1) = 5$
$-5(-2) = 10$
$-5(-3) = 15$
$-5(-4) = 20$

This example illustrates that the product of two negative numbers is positive.

The pattern for multiplication shown above is summarized in the following rule for multiplying integers.

▶ **Point of Interest**

Operations with negative numbers were not accepted until the late thirteenth century. One of the first attempts to prove that the product of two negative numbers is positive was made in the book *Ars Magna,* by Girolamo Cardan, in 1545.

> ### *Rule for Multiplying Two Integers*
>
> **To multiply two integers with the same sign,** multiply the absolute values of the factors. The product is **positive.**
>
> **To multiply two integers with different signs,** multiply the absolute values of the factors. The product is **negative.**

Multiply: $-9(12)$

The signs are different. The product is negative. $-9(12) = -108$

Multiply: $(-6)(-15)$

The signs are the same. The product is positive. $(-6)(-15) = 90$

Calculator Note

To multiply $(-6)(-15)$ with your calculator, enter the following:

6 [+/−] × 15 [+/−] =
‿   ‿
−6   −15

Copyright © Houghton Mifflin Company. All rights reserved.

Figure 2.5 shows the melting point of bromine and mercury. The melting point of helium is 7 times the melting point of mercury. Find the melting point of helium.

Multiply the melting point of mercury (−39°C) by 7.

$$-39(7) = -273$$

The melting point of helium is −273°C.

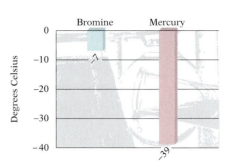

Figure 2.5 Melting Point of Chemical Elements (in degrees Celsius)

The properties of multiplication presented in Chapter 1 hold true for integers as well as whole numbers. These properties are repeated below.

The Multiplication Property of Zero	$a \cdot 0 = 0$ or $0 \cdot a = 0$
The Multiplication Property of One	$a \cdot 1 = a$ or $1 \cdot a = a$
The Commutative Property of Multiplication	$a \cdot b = b \cdot a$
The Associative Property of Multiplication	$(a \cdot b) \cdot c = a \cdot (b \cdot c)$

Take Note

For the example at the right, the product is the same if the numbers are multiplied in a different order. For instance,

$$2(-3)(-5)(-7) =$$
$$2(-3)(35) =$$
$$2(-105) =$$
$$-210$$

Multiply: $2(-3)(-5)(-7)$ $\qquad\qquad 2(-3)(-5)(-7)$

Multiply the first two numbers. $\qquad = -6(-5)(-7)$

Then multiply the product by the third number. $\qquad = 30(-7)$

Continue until all the numbers have been multiplied. $\qquad = -210$

By the Multiplication Property of One, $1 \cdot 6 = 6$ and $\mathbf{1} \cdot x = x$. Applying the rules for multiplication, we can extend this to $-1 \cdot 6 = -6$ and $\mathbf{-1} \cdot x = -x$.

Evaluate $-ab$ when $a = -2$ and $b = -9$. $\qquad\qquad -ab$

Replace a with -2 and b with -9. $\qquad\qquad -(-2)(-9)$

Simplify $-(-2)$. $\qquad\qquad = 2(-9)$

Multiply. $\qquad\qquad = -18$

Take Note

When variables are placed next to each other, it is understood that the operation is multiplication. $-ab$ means "the opposite of a times b."

Is -4 a solution of the equation $5x = -20$? $\qquad\qquad 5x = -20$

Replace x by -4 and then simplify. $\qquad\qquad 5(-4) \,|\, -20$

The results are equal. $\qquad\qquad -20 = -20$

Yes, -4 is a solution of the equation.

Copyright © Houghton Mifflin Company. All rights reserved.

1 Example Find -42 times 62.

Solution $-42 \cdot 62 = -2{,}604$

1 You Try It What is -38 multiplied by 51?

Your Solution

---

2 Example Multiply: $-5(-4)(6)(-3)$

Solution
$$-5(-4)(6)(-3) = 20(6)(-3)$$
$$= 120(-3)$$
$$= -360$$

2 You Try It Multiply: $-7(-8)(9)(-2)$

Your Solution

---

3 Example Evaluate $-5x$ when $x = -11$.

Solution
$$-5x$$
$$-5(-11) = 55$$

3 You Try It Evaluate $-9y$ when $y = 20$.

Your Solution

---

4 Example Is 5 a solution of the equation $30 = -6z$?

Solution
$$30 = -6z$$
$$\begin{array}{c|c} 30 & -6(5) \end{array}$$
$$30 \neq -30$$

No, 5 is not a solution of the equation.

4 You Try It Is -3 a solution of the equation $12 = -4a$?

Your Solution

Solutions on p. S5

OBJECTIVE **B**

VIDEO & DVD   CD TUTOR   WEB   SSM

Division of integers

For every division problem, there is a related multiplication problem.

Division: $\dfrac{8}{2} = 4$ Related multiplication: $4(2) = 8$

This fact can be used to illustrate a rule for dividing integers.

$\dfrac{12}{3} = 4$ because $4(3) = 12$ and $\dfrac{-12}{-3} = 4$ because $4(-3) = -12$.

These two division examples suggest that the quotient of two numbers with the same sign is positive. Now consider these two examples.

$\dfrac{12}{-3} = -4$ because $-4(-3) = 12$

$\dfrac{-12}{3} = -4$ because $-4(3) = -12$

These two division examples suggest that the quotient of two numbers with different signs is negative. This property is summarized next.

> **Take Note**
> Recall that the fraction bar can be read "divided by." Therefore, $\dfrac{8}{2}$ can be read "8 divided by 2."

Copyright © Houghton Mifflin Company. All rights reserved.

Rule for Dividing Two Integers

To divide two numbers with the same sign, divide the absolute values of the numbers. The quotient is **positive.**

To divide two numbers with different signs, divide the absolute values of the numbers. The quotient is **negative.**

Note from this rule that $\dfrac{12}{-3}$, $\dfrac{-12}{3}$, and $-\dfrac{12}{3}$ are all equal to -4.

If a and b are integers $(b \neq 0)$, then $\dfrac{a}{-b} = \dfrac{-a}{b} = -\dfrac{a}{b}$.

■ **Calculator Note**

To divide (-105) by (-5) with your calculator, enter the following:

$\underbrace{105\ \boxed{+/-}}_{-105}\ \boxed{\div}\ \underbrace{5\ \boxed{+/-}}_{-5}\ \boxed{=}$

Divide: $-36 \div 9$

The signs are different. The quotient is negative. $-36 \div 9 = -4$

Divide: $(-105) \div (-5)$

The signs are the same. The quotient is positive. $(-105) \div (-5) = 21$

Figure 2.6 shows the record high and low temperatures in the United States for the first four months of the year. We can read from the graph that the record low temperature for April is $-36°F$. This is four times the record low temperature for September. What is the record low temperature for September?

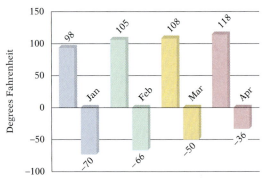

Figure 2.6   Record High and Low Temperatures, in Degrees Fahrenheit, in the United States for January, February, March, and April
Source: National Climatic Data Center, Asheville, NC, and Storm Phillips, STORMFAX, Inc.

To find the record low temperature for September, divide the record low for April (-36) by 4. $-36 \div 4 = -9$

The record low temperature in the United States for the month of September is $-9°F$.

Copyright © Houghton Mifflin Company. All rights reserved.

The division properties of zero and one, which were presented in Chapter 1, hold true for integers as well as whole numbers. These properties are repeated here.

Division Properties of Zero and One

If $a \neq 0$, $\dfrac{0}{a} = 0$.   If $a \neq 0$, $\dfrac{a}{a} = 1$.

$\dfrac{a}{1} = a$ $\dfrac{a}{0}$ is undefined.

Evaluate $a \div (-b)$ when $a = -28$ and $b = -4$.

$$a \div (-b)$$

Replace a with -28 and b with -4. $$-28 \div (-(-4))$$

Simplify $-(-4)$. $$= -28 \div (4)$$

Divide. $$= -7$$

▶ **Point of Interest**

Historical manuscripts indicate that mathematics is at least 4000 years old. Yet it was only 400 years ago that mathematicians started using variables to stand for numbers. Before that time, mathematics was written in words.

Is -4 a solution of the equation $\dfrac{-20}{x} = 5$?

$$\dfrac{-20}{x} = 5$$

Replace x by -4 and then simplify.

$$\dfrac{-20}{-4} \,\bigg|\, 5$$

The results are equal. $$5 = 5$$

Yes, -4 is a solution of the equation.

---

5 *Example* Find the quotient of -23 and -23.

Solution $-23 \div (-23) = 1$

5 *You Try It* What is 0 divided by -17?

Your Solution

---

6 *Example* Divide: $\dfrac{95}{-5}$

Solution $\dfrac{95}{-5} = -19$

6 *You Try It* Divide: $\dfrac{84}{-6}$

Your Solution

---

7 *Example* Divide: $x \div 0$

Solution Division by zero is not defined. $x \div 0$ is undefined.

7 *You Try It* Divide: $x \div 1$

Your Solution

Solutions on p. S5

Copyright © Houghton Mifflin Company. All rights reserved.

---

8 *Example* Evaluate $\dfrac{-a}{b}$ when $a = -6$ and $b = -3$.

Solution $\dfrac{-a}{b}$

$\dfrac{-(-6)}{-3} = \dfrac{6}{-3} = -2$

8 *You Try It* Evaluate $\dfrac{a}{-b}$ when $a = -14$ and $b = -7$.

Your Solution

---

9 *Example* Is -9 a solution of the equation $-3 = \dfrac{x}{3}$?

Solution $-3 = \dfrac{x}{3}$

$-3 \;\Big|\; \dfrac{-9}{3}$

$-3 = -3$

Yes, -9 is a solution of the equation.

9 *You Try It* Is -3 a solution of the equation $\dfrac{-6}{y} = -2$?

Your Solution

Solutions on p. S5

OBJECTIVE **C**

VIDEO & DVD   CD TUTOR   WEB   SSM

Applications

---

10 *Example*

The daily low temperatures during one week were recorded as follows: $-10°, 2°, -1°, -9°, 1°, 0°, 3°$. Find the average daily low temperature for the week.

Strategy

To find the average daily low temperature:

▸ Add the seven temperature readings.
▸ Divide by 7.

Solution

$-10 + 2 + (-1) + (-9) + 1 + 0 + 3 = -14$

$-14 \div 7 = -2$

The average daily low temperature was $-2°$.

10 *You Try It*

The daily high temperatures during one week were recorded as follows: $-7°, -8°, 0°, -1°, -6°, -11°, -2°$. Find the average daily high temperature for the week.

Your Strategy

Your Solution

Solution on p. S5

Copyright © Houghton Mifflin Company. All rights reserved.

2.3 EXERCISES

OBJECTIVE   **A**

1.  Give the rules for multiplying two integers.

2.  Name the operation in each expression and explain how you determined that it was that operation.
   a. $8(-7)$ **b.** $8 - 7$ **c.** $8 - (-7)$ **d.** $-xy$ **e.** $x(-y)$ **f.** $-x - y$

Multiply.

3. $-4 \cdot 6$

4. $-7 \cdot 3$

5. $-2(-3)$

6. $-5(-1)$

7. $(9)(2)$

8. $(3)(8)$

9. $5(-4)$

10. $4(-7)$

11. $-8(2)$

12. $-9(3)$

13. $(-5)(-5)$

14. $(-3)(-6)$

15. $(-7)(0)$

16. $-11(1)$

17. $14(3)$

18. $62(9)$

19. $-32(4)$

20. $-24(3)$

21. $(-8)(-26)$

22. $(-4)(-35)$

23. $9(-27)$

24. $8(-40)$

25. $-5 \cdot (23)$

26. $-6 \cdot (38)$

27. $-7(-34)$

28. $-4(-51)$

29. $4 \cdot (-8) \cdot 3$

30. $5 \cdot 7 \cdot (-2)$

31. $(-6)(5)(7)$

32. $(-9)(-9)(2)$

33. $-8(-7)(-4)$

34. $-1(4)(-9)$

35. What is twice -20?

36. Find the product of 100 and -7.

37. What is -30 multiplied by -6?

38. What is -9 times -40?

Copyright © Houghton Mifflin Company. All rights reserved.

39. Write the product of $-q$ and r.

40. Write the product of $-f$, g, and h.

41. *Business* The table at the right shows the net income for the fourth quarter of 2002 for three companies. (*Note:* Negative net income indicates a loss.) If net income were to continue throughout the year 2003 at the same level, what would be the annual net income for 2003 for (a) Midway Games, (b) Pinnacle Entertainment, and (c) Granite Broadcasting?

Company	Net Income 4th Quarter of 2002
Midway Games	−24,666,000
Pinnacle Entertainment	−6,693,000
Granite Broadcasting	−3,262,000

Source: **www.wsj.com**

Identify the property that justifies the statement.

42. $0(-7) = 0$

43. $1p = p$

44. $-8(-5) = -5(-8)$

45. $-3(9 \cdot 4) = (-3 \cdot 9)4$

Use the given property of multiplication to complete the statement.

46. The Commutative Property of Multiplication
$-3(-9) = -9(?)$

47. The Associative Property of Multiplication
$?(5 \cdot 10) = (-6 \cdot 5)10$

48. The Multiplication Property of Zero
$-81 \cdot ? = 0$

49. The Multiplication Property of One
$?(-14) = -14$

Evaluate the expression for the given values of the variables.

50. xy, when $x = -3$ and $y = -8$

51. $-xy$, when $x = -3$ and $y = -8$

52. $x(-y)$, when $x = -3$ and $y = -8$

53. $-xyz$, when $x = -6$, $y = 2$, and $z = -5$

54. $-8a$, when $a = -24$

55. $-7n$, when $n = -51$

Copyright © Houghton Mifflin Company. All rights reserved.

56. $5xy$, when $x = -9$ and $y = -2$

57. $8ab$, when $a = 7$ and $b = -1$

58. $-4cd$, when $c = 25$ and $d = -8$

59. $-5st$, when $s = -40$ and $t = -8$

60. Is -4 a solution of the equation $6m = -24$?

61. Is -3 a solution of the equation $-5x = -15$?

62. Is -6 a solution of the equation $48 = -8y$?

63. Is 0 a solution of the equation $-8 = -8a$?

64. Is 7 a solution of the equation $-3c = 21$?

65. Is 9 a solution of the equation $-27 = -3c$?

OBJECTIVE **B**

Divide.

66. $12 \div (-6)$

67. $18 \div (-3)$

68. $(-72) \div (-9)$

69. $(-64) \div (-8)$

70. $0 \div (-6)$

71. $-49 \div 1$

72. $81 \div (-9)$

73. $-40 \div (-5)$

74. $\dfrac{72}{-3}$

75. $\dfrac{44}{-4}$

76. $\dfrac{-93}{-3}$

77. $\dfrac{-98}{-7}$

78. $-114 \div (-6)$

79. $-91 \div (-7)$

80. $-53 \div 0$

81. $(-162) \div (-162)$

82. $-128 \div 4$

83. $-130 \div (-5)$

84. $(-200) \div 8$

85. $(-92) \div (-4)$

Copyright © Houghton Mifflin Company. All rights reserved.

86. Find the quotient of -700 and 70.

87. Find 550 divided by -5.

88. What is -670 divided by -10?

89. What is the quotient of -333 and -3?

90. Write the quotient of $-a$ and b.

91. Write -9 divided by x.

The figure at the right shows the net income for the fourth quarter of 2002 for three companies. (*Note:* Negative income indicates a loss. One quarter of the year is three months.) Use this figure for Exercises 92 and 93.

92. *Business* For the quarter shown, what was the average monthly net income for Fresh Choice?

93. *Business* For the quarter shown, what was the average monthly net income for Friendly Ice Cream?

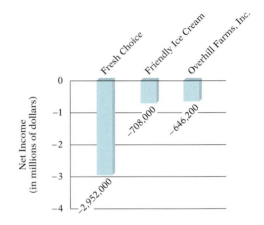

Net Income for Fourth Quarter of 2002
Source: **www.wsj.com**

Evaluate the expression for the given values of the variables.

94. $a \div b$, where $a = -36$ and $b = -4$

95. $-a \div b$, where $a = -36$ and $b = -4$

96. $a \div (-b)$, where $a = -36$ and $b = -4$

97. $(-a) \div (-b)$, where $a = -36$ and $b = -4$

98. $\dfrac{x}{y}$, where $x = -42$ and $y = -7$

99. $\dfrac{-x}{y}$, where $x = -42$ and $y = -7$

100. $\dfrac{x}{-y}$, where $x = -42$ and $y = -7$

101. $\dfrac{-x}{-y}$, where $x = -42$ and $y = -7$

Copyright © Houghton Mifflin Company. All rights reserved.

102. Is 20 a solution of the equation $\dfrac{m}{-2} = -10$?

103. Is 18 a solution of the equation $6 = \dfrac{-c}{-3}$?

104. Is 0 a solution of the equation $0 = \dfrac{a}{-4}$?

105. Is -3 a solution of the equation $\dfrac{21}{n} = 7$?

106. Is -6 a solution of the equation $\dfrac{x}{2} = \dfrac{-18}{x}$?

107. Is 8 a solution of the equation $\dfrac{m}{-4} = \dfrac{-16}{m}$?

> **OBJECTIVE** **C**

108. *Sports*   The combined scores of the top five golfers in a tournament equaled -10 (10 under par). What was the average score of the five golfers?

109. *Sports*   The combined scores of the top four golfers in a tournament equaled -12 (12 under par). What was the average score of the four golfers?

 The following figure shows the record low temperatures, in degrees Fahrenheit, in the United States for each month. Use this figure for Exercises 110 to 112.

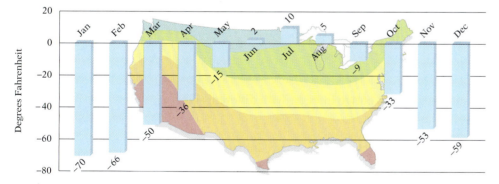

Record Low Temperatures, in Degrees Fahrenheit, in the United States
Source: National Climatic Data Center, Asheville, NC, and Storm Phillips, STORMFAX, Inc.

110. *Temperature*   What is the average record low temperature for July, August, and September?

111. *Temperature*   What is the average record low temperature for the first three months of the year?

112. *Temperature*   What is the average record low temperature for the four months with the lowest record low temperatures?

Copyright © Houghton Mifflin Company. All rights reserved.

113. *Temperature* The daily low temperatures during one week were recorded as follows: $4°, -5°, 8°, -1°, -12°, -14°, -8°$. Find the average daily low temperature for the week.

114. *Temperature* The daily high temperatures during one week were recorded as follows: $-6°, -11°, 1°, 5°, -3°, -9°, -5°$. Find the average daily high temperature for the week.

115. *Environmental Science* The wind chill factor when the temperature is $-20°F$ and the wind is blowing at 15 mph is five times the wind chill factor when the temperature is 10°F and the wind is blowing at 20 mph. If the wind chill factor at 10°F with a 20 mph wind is $-9°$, what is the wind chill factor at $-20°F$ with a 15 mph wind?

A geometric sequence is a list of numbers in which each number after the first is found by multiplying the preceding number in the list by the same number. For example, in the sequence 1, 3, 9, 27, 81, ... , each number after the first is found by multiplying the preceding number in the list by 3. To find the multiplier in a geometric sequence, divide the second number in the sequence by the first number; for the example above, $3 \div 1 = 3$.

116. *Mathematics* Find the next three numbers in the geometric sequence -5, 15, -45, ...

117. *Mathematics* Find the next three numbers in the geometric sequence 2, -4, 8, ...

118. *Mathematics* Find the next three numbers in the geometric sequence -3, -12, -48, ...

119. *Mathematics* Find the next three numbers in the geometric sequence -1, -5, -25, ...

Critical Thinking

120. Use repeated addition to show that the product of two integers with different signs is a negative number.

121. *Mathematics* (a) Find the largest possible product of two negative integers whose sum is -18. (b) Find the smallest possible sum of two negative integers whose product is 16.

122. Determine whether the statement is always true, sometimes true, or never true. (a) The product of a number and its additive inverse is a negative number. (b) The product of an odd number of negative numbers is a negative number. (c) The square of a negative number is a positive number.

123. Find all negative integers x such that $1 - 3x < 12$.

124. Describe the steps involved in using a calculator to simplify $(-2{,}491) \div (-47)$.

125. a. When is the product of two integers positive? When is it negative?
b. When is the quotient of two integers positive? When is it negative?

Copyright © Houghton Mifflin Company. All rights reserved.

Copyright © Houghton Mifflin Company. All rights reserved.

SECTION 2.4 ▶ ## Solving Equations with Integers

OBJECTIVE **A**

Solving equations

Recall that an **equation** states that two expressions are equal. Two examples of equations are shown below.

$$3x = 36 \qquad -17 = y + 9$$

In Section 1.4, we solved equations using only whole numbers. In this section, we will extend the solutions of equations to include integers.

Solving an equation requires finding a number that when substituted for the variable results in a true equation. Two important properties that are used to solve equations were discussed earlier.

> **The same number can be subtracted from each side of an equation without changing the solution of the equation.**

> **Each side of an equation can be divided by the same nonzero number without changing the solution of the equation.**

A third property of equations involves *adding* the same number to each side of an equation.

As shown at the right, the solution of the equation $x + 6 = 13$ is 7.

$$x + 6 = 13$$
$$7 + 6 = 13$$
$$13 = 13$$

If 4 is added to each side of the equation $x + 6 = 13$, the resulting equation is $x + 10 = 17$. The solution of this equation is also 7.

$$x + 6 = 13$$
$$x + 6 + 4 = 13 + 4$$
$$x + 10 = 17 \qquad 7 + 10 = 17$$

This illustrates the **addition property of equations.**

> **The same number can be added to each side of an equation without changing the solution of the equation.**

Solve: $x - 7 = 2$

7 is subtracted from the variable x. Add 7 to each side of the equation.

$$x - 7 = 2$$
$$x - 7 + 7 = 2 + 7$$

x is alone on the left side of the equation. The number on the right side is the solution.

$$x = 9$$

Check the solution.

Check:
$$\begin{array}{c|c} x - 7 = 2 \\ \hline 9 - 7 & 2 \\ 2 = 2 \end{array}$$

The solution checks.

The solution is 9.

Take Note

For this example, 13 is added to the variable. Therefore, the subtraction property is used to get the variable alone. Remember to check this solution.

Solve: $-15 = t + 13$

13 is added to the variable t. Subtract 13 from each side of the equation.

$$-15 = t + 13$$
$$-15 - 13 = t + 13 - 13$$

t is alone on the right side of the equation. The number on the left side is the solution.

$$-28 = t$$

The solution is -28.

The division property of equations is also used with integers.

Solve: $5y = -30$

The variable y is multiplied by 5. Divide each side of the equation by 5.

$$5y = -30$$
$$\frac{5y}{5} = \frac{-30}{5}$$

y is alone on the left side of the equation. The number on the right side is the solution.

$$y = -6$$

Check the solution.

Check: $$\frac{5y = -30}{5(-6) \mid -30}$$

The solution checks.

$$-30 = -30$$

The solution is -6.

Solve: $42 = -7a$

The variable a is multiplied by -7. Divide each side of the equation by -7.

$$42 = -7a$$
$$\frac{42}{-7} = \frac{-7a}{-7}$$

a is alone on the right side of the equation. The number on the left side is the solution. Remember to check the solution.

$$-6 = a$$

The solution is -6.

---

① Example   Solve: $27 = v - 13$

Solution
$$27 = v - 13$$
$$27 + 13 = v - 13 + 13$$ ▶ 13 is subtracted from v. Add 13 to each side.
$$40 = v$$

The solution is 40.

① You Try It   Solve: $-12 = x + 12$

Your Solution

② Example   Solve: $-24 = -4z$

Solution $-24 = -4z$

$$\frac{-24}{-4} = \frac{-4z}{-4}$$

$$6 = z$$

The solution is 6.

② You Try It   Solve: $14a = -28$

Your Solution

Solutions on p. S5

Copyright © Houghton Mifflin Company. All rights reserved.

Copyright © Houghton Mifflin Company. All rights reserved.

OBJECTIVE **B**

VIDEO & DVD   CD TUTOR   WWW WEB   SSM

Applications and formulas

Recall that an equation states that two mathematical expressions are equal. To translate a sentence into an equation requires that you recognize the words or phrases that mean "equals." Some of these phrases are reviewed below.

equals	is	was
is equal to	represents	is the same as

Negative fifty-six equals negative eight times a number. Find the number.

Choose a variable to represent the unknown number.

The unknown number: m

Find two verbal expressions for the same value.

Negative fifty-six	equals	negative eight times a number

Translate the expressions and then write an equation. Solve the equation.

$$-56 = -8m$$
$$\frac{-56}{-8} = \frac{-8m}{-8}$$
$$7 = m$$

The number is 7.

The high temperature today is 7°C lower than the high temperature yesterday. The high temperature today is −13°C. What was the high temperature yesterday?

Strategy   To find the high temperature yesterday, write and solve an equation using t to represent the high temperature yesterday.

Solution

The high temperature today	is	7° lower than the high temperature yesterday

$$-13 = t - 7$$
$$-13 + 7 = t - 7 + 7$$
$$-6 = t$$

The high temperature yesterday was −6°C.

A jeweler wants to make a profit of $250 on the sale of a gold bracelet that cost the jeweler $700. Use the formula $P = S - C$, where P is the profit on an item, S is the selling price, and C is the cost, to find the selling price of the bracelet.

Strategy   To find the selling price, replace P by 250 and C by 700 in the given formula and solve for S.

Solution

$$P = S - C$$
$$250 = S - 700$$
$$250 + 700 = S - 700 + 700$$
$$950 = S$$

The selling price for the gold bracelet should be $950.

3 Example

In the United States, the average household income of people age 25 to 34 is $14,886 less than the average household income of people age 45 to 54. The average household income of people age 25 to 34 is $41,414. (*Source:* Census 2000) Find the average household income of people age 45 to 54.

Strategy

To find the average income of people age 45 to 54, write and solve an equation using I to represent the average income of people age 45 to 54.

Solution

the average income of people 25 to 34	is	$14,886 less than the average income of people 45 to 54

$$41,414 = I - 14,886$$
$$41,414 + 14,886 = I - 14,886 + 14,886$$
$$56,300 = I$$

The average household income of people age 45 to 54 is $56,300.

3 You Try It

In a recent year in the United States, the number of mothers who gave birth to triplets was 112,906 less than the number of mothers who gave birth to twins. The number of mothers who gave birth to triplets was 6,742. (*Source:* National Center for Health Statistics) How many mothers gave birth to twins during that year?

Your Strategy

Your Solution

4 Example

The ground speed of an airplane flying into a wind is given by the formula $g = a - h$, where g is the ground speed, a is the air speed of the plane, and h is the speed of the head wind. Use this formula to find the air speed of a plane whose ground speed is 624 mph and for which the head wind speed is 98 mph.

Strategy

To find the air speed, replace g by 624 and h by 98 in the given formula and solve for a.

Solution

$$g = a - h$$
$$624 = a - 98$$
$$624 + 98 = a - 98 + 98$$
$$722 = a$$

The air speed of the plane is 722 mph.

4 You Try It

The ground speed of an airplane flying into a wind is given by the formula $g = a - h$, where g is the ground speed, a is the air speed of the plane, and h is the speed of the head wind. Use this formula to find the air speed of a plane whose ground speed is 250 mph and for which the head wind speed is 50 mph.

Your Strategy

Your Solution

Solutions on pp. S5–S6

Copyright © Houghton Mifflin Company. All rights reserved.

2.4 EXERCISES

OBJECTIVE A

Solve.

1. $x - 6 = 9$

2. $m - 4 = 6$

3. $8 = y - 3$

4. $12 = t - 4$

5. $x - 5 = -12$

6. $n - 7 = -21$

7. $-10 = z + 6$

8. $-21 = c + 4$

9. $x + 12 = 4$

10. $y + 7 = 2$

11. $-12 = c - 12$

12. $n - 9 = -9$

13. $6 + x = 4$

14. $12 + y = 7$

15. $12 = n - 8$

16. $19 = b - 23$

17. $3m = -15$

18. $6p = -54$

19. $-10 = 5v$

20. $-20 = 2z$

21. $-8x = -40$

22. $-4y = -28$

23. $-60 = -6v$

24. $3x = -39$

25. $5x = -100$

26. $-4n = 0$

27. $4x = 0$

28. $-15 = -15z$

29. $-2r = 16$

30. $-6p = 72$

31. $-72 = 18w$

32. $-35 = 5p$

OBJECTIVE B

33. Ten less than a number is fifteen. Find the number.

34. The difference between a number and five is twenty-two. Find the number.

35. Zero is equal to fifteen more than some number. Find the number.

36. Twenty equals the sum of a number and thirty-one. Find the number.

37. Sixteen equals negative two times a number. Find the number.

38. The product of negative six and a number is negative forty-two. Find the number.

39. Zero is equal to the product of negative six and a number. Find the number.

40. Eight times some number is negative ninety-six. Find the number.

Copyright © Houghton Mifflin Company. All rights reserved.

Use the table at the right for Exercises 41 and 42.

Year	U.S. Balance of Trade (in millions of dollars)
1960	3,508
1970	2,254
1980	−19,407
1990	−80,861
2000	−378,681

Source: Bureau of Economics Analysis, U.S. Department of Commerce

41. *Economics* The U.S. balance of trade in 1980 was $5,158 million more than the U.S. balance of trade in 1979. What was the U.S. balance of trade in 1979?

42. *Economics* The U.S. balance of trade in 1990 was $15,527 million more than the U.S. balance of trade in 1995. What was the U.S. balance of trade in 1995?

43. *Temperature* The temperature now is 5° higher than it was this morning. The temperature now is 8°C. What was the temperature this morning?

44. *Temperature* The temperature now is 9° lower than it was yesterday at this time. The temperature now is −16°C. What was the temperature yesterday at this time?

45. *Business* A car dealer wants to make a profit of $925 on the sale of a car that cost the dealer $12,600. Use the equation $P = S − C$, where P is the profit on an item, S is the selling price, and C is the cost, to find the selling price of the car.

46. *Business* An office supplier wants to make a profit of $95 on the sale of a software package that cost the supplier $385. Use the equation $P = S − C$, where P is the profit on an item, S is the selling price, and C is the cost, to find the selling price of the software.

47. *Business* The net worth of a business is given by the formula $N = A − L$, where N is the net worth, A is the assets of the business (or the amount owned), and L is the liabilities of the business (or the amount owed). Use this formula to find the assets of a business that has a net worth of $11 million and liabilities of $4 million.

48. *Business* The net worth of ABL Electronics is $43 million and it has liabilities of $14 million. Use the net worth formula $N = A − L$, where N is the net worth, A is the assets of the business (or the amount owned), and L is the liabilities of the business (or the amount owed), to find the assets of ABL Electronics.

Critical Thinking

49. For each part below, state whether the sentence is true or false. Give an example that supports your answer. (a) Zero cannot be the solution of an equation. (b) A negative number cannot be the solution of an equation. (c) If an equation contains a negative number, then the solution of the equation must be a negative number.

50. a. Find the value of $3y − 8$ given that $−3y = −36$.
   b. Find the value of $2x^2 − 18$ given that $x − 6 = −9$.

51. In your own words, explain the addition, subtraction, and division properties of equations.

Copyright © Houghton Mifflin Company. All rights reserved.

Copyright © Houghton Mifflin Company. All rights reserved.

SECTION 2.5 ▸ The Order of Operations Agreement

OBJECTIVE **A**

The Order of Operations Agreement

The Order of Operations Agreement, used in Chapter 1, is repeated here for your reference.

> *The Order of Operations Agreement*
>
> **Step 1** Do all operations inside parentheses.
>
> **Step 2** Simplify any numerical expressions containing exponents.
>
> **Step 3** Do multiplication and division as they occur from left to right.
>
> **Step 4** Do addition and subtraction as they occur from left to right.

Note how the following expressions containing exponents are simplified.

$(-3)^2 = (-3)(-3) = 9$ The (-3) is squared. Multiply -3 by -3.

$-(3)^2 = -(3 \cdot 3) = -9$ Read $-(3^2)$ as "the opposite of three squared." 3^2 is 9. The opposite of 9 is -9.

$-3^2 = -(3^2) = -9$ The expression -3^2 is the same as $-(3^2)$.

Take Note

The -3 is squared only when the negative sign is *inside* the parentheses. In $(-3)^2$, we are squaring -3; in -3^2, we are finding the opposite of 3^2.

Simplify: $8 - 4 \div (-2)$

There are no operations inside parentheses (Step 1).

There are no exponents (Step 2).

Do the division (Step 3). $8 - 4 \div (-2) = 8 - (-2)$

Do the subtraction (Step 4). $= 8 + 2 = 10$

Calculator Note

As shown above and at the left, the value of -3^2 is different from the value of $(-3)^2$. The keystrokes to evaluate each of these on your calculator are different.

To evaluate -3^2, enter

3 [x^2] [+/−]

To evaluate $(-3)^2$, enter

3 [+/−] [x^2]

Simplify: $(-3)^2 - 2(8 - 3) + (-5)$

Perform operations inside parentheses.

Simplify expressions with exponents.

Do multiplication and division as they occur from left to right.

Do addition and subtraction as they occur from left to right.

$(-3)^2 - 2(8 - 3) + (-5)$
$= (-3)^2 - 2(5) + (-5)$
$= 9 - 2(5) + (-5)$
$= 9 - 10 + (-5)$
$= 9 + (-10) + (-5)$
$= -1 + (-5)$
$= -6$

Evaluate $ab - b^2$ when $a = 2$ and $b = -6$.

$$ab - b^2$$

Replace a with 2 and each b with -6. $\qquad$ $2(-6) - (-6)^2$

Use the Order of Operations Agreement to simplify the resulting numerical expression. Simplify the exponential expression. $\qquad$ $= 2(-6) - 36$

Do the multiplication. $\qquad$ $= -12 - 36$

Do the subtraction. $\qquad$ $= -12 + (-36)$
$$= -48$$

---

① **Example** Simplify $(-4)^2$ and -4^2.

Solution $(-4)^2 = (-4)(-4) = 16$
$-4^2 = -(4 \cdot 4) = -16$

① **You Try It** Simplify $(-5)^2$ and -5^2.

Your Solution

---

② **Example** Simplify: $12 \div (-2)^2 - 5$

Solution $12 \div (-2)^2 - 5 = 12 \div 4 - 5$
$= 3 - 5$
$= 3 + (-5)$
$= -2$

② **You Try It** Simplify: $8 \div 4 \cdot 4 - (-2)^2$

Your Solution

---

③ **Example** Simplify:
$(-3)^2(5 - 7)^2 - (-9) \div 3$

Solution $(-3)^2(5 - 7)^2 - (-9) \div 3$
$= (-3)^2(-2)^2 - (-9) \div 3$
$= (9)(4) - (-9) \div 3$
$= 36 - (-9) \div 3$
$= 36 - (-3)$
$= 36 + 3$
$= 39$

③ **You Try It** Simplify:
$(-2)^2(3 - 7)^2 - (-16) \div (-4)$

Your Solution

---

④ **Example** Evaluate $6a \div (-b)$ when $a = -2$ and $b = -3$.

Solution $6a \div (-b)$
$6(-2) \div (-(-3))$
$= 6(-2) \div (3)$
$= -12 \div 3$
$= -4$

④ **You Try It** Evaluate $3a - 4b$ when $a = -2$ and $b = 5$.

Your Solution

Solutions on p. S6

Copyright © Houghton Mifflin Company. All rights reserved.

2.5 EXERCISES

OBJECTIVE **A**

Simplify.

1. $3 - 12 \div 2$

2. $-16 \div 2 + 8$

3. $2(3 - 5) - 2$

4. $2 - (8 - 10) \div 2$

5. $4 - (-3)^2$

6. $(-2)^2 - 6$

7. $4 \cdot (2 - 4) - 4$

8. $6 - 2 \cdot (1 - 3)$

9. $4 - (-2)^2 + (-3)$

10. $-3 + (-6)^2 - 1$

11. $3^3 - 4(2)$

12. $9 \div 3 - (-3)^2$

13. $3 \cdot (6 - 2) \div 6$

14. $4 \cdot (2 - 7) \div 5$

15. $2^3 - (-3)^2 + 2$

16. $6(8 - 2) \div 4$

17. $6 - 2(1 - 5)$

18. $(-2)^2 - (-3)^2 + 1$

19. $6 - (-4)(-3)^2$

20. $4 - (-5)(-2)^2$

21. $4 \cdot 2 - 3 \cdot 7$

22. $16 \div 2 - 9 \div 3$

23. $(-2)^2 - 5(3) - 1$

24. $4 - 2 \cdot 7 - 3^2$

25. $3 \cdot 2^3 + 5 \cdot (3 + 2) - 17$

26. $3 \cdot 4^2 - 16 - 4 + 3 - (1 - 2)^2$

27. $-12(6 - 8) + 1^3 \cdot 3^2 \cdot 2 - 6(2)$

28. $-3 \cdot (-2)^2 \cdot 4 \div 8 - (-12)$

29. $-27 - (-3)^2 - 2 - 7 + 6 \cdot 3$

30. $(-1) \cdot (4 - 7)^2 \div 9 + 6 - 3 - 4(2)$

31. $16 - 4 \cdot 8 + 4^2 - (-18) - (-9)$

32. $(-3)^2 \cdot (5 - 7)^2 - (-9) \div 3$

Copyright © Houghton Mifflin Company. All rights reserved.

Evaluate the variable expression given $a = -2$, $b = 4$, $c = -1$, and $d = 3$.

33. $3a + 2b$

34. $a - 2c$

35. $16 \div (ac)$

36. $6b \div (-a)$

37. $bc \div (2a)$

38. $a^2 - b^2$

39. $b^2 - c^2$

40. $2a - (c + a)^2$

41. $(b - a)^2 + 4c$

42. $\dfrac{b + c}{d}$

43. $\dfrac{d - b}{c}$

44. $\dfrac{2d + b}{-a}$

45. $\dfrac{b - d}{c - a}$

46. $\dfrac{bd}{a} \div c$

47. $(d - a)^2 \div 5$

48. $(b + c)^2 + (a + d)^2$

49. $(d - a)^2 - 3c$

50. $(b + d)^2 - 4a$

Critical Thinking

51. What is the smallest integer greater than $-2^2 - (-3)^2 + 5(4) \div 10 - (-6)$?

52. Evaluate.
    a. $1^3 + 2^3 + 3^3 + 4^3$
    b. $(-1)^3 + (-2)^3 + (-3)^3 + (-4)^3$
    c. $1^3 + 2^3 + 3^3 + 4^3 + 5^3$
    d. Based on your answers to parts (a), (b), and (c), evaluate
         $(-1)^3 + (-2)^3 + (-3)^3 + (-4)^3 + (-5)^3$.

53. **a.** Is -4 a solution of the equation $x^2 - 2x - 8 = 0$?
    b. Is -3 a solution of the equation $x^3 + 3x^2 - 5x - 15 = 0$?

54. Using the Order of Operations Agreement, explain how to evaluate Exercise 33.

55. Evaluate $a \div bc$ and $a \div (bc)$ when $a = 16$, $b = 2$, and $c = -4$. Explain why the answers are not the same.

Copyright © Houghton Mifflin Company. All rights reserved.

Focus on **Problem Solving**

Drawing Diagrams

How do you best remember something? Do you remember best what you hear? The word *aural* means "pertaining to the ear"; people with a strong aural memory remember best those things that they hear. The word *visual* means "pertaining to the sense of sight"; people with a strong visual memory remember best that which they see written down. Some people claim that their memory is in their writing hand—they remember something only if they write it down! The method by which you best remember something is probably also the method by which you can best learn something new.

In problem-solving situations, try to capitalize on your strengths. If you tend to understand the material better when you hear it spoken, read application problems aloud or have someone else read them to you. If writing helps you to organize ideas, rewrite application problems in your own words.

No matter what your main strength, visualizing a problem can be a valuable aid in problem solving. A drawing, sketch, diagram, or chart can be a useful tool in problem solving, just as calculators and computers are tools. A diagram can be helpful in gaining an understanding of the relationships inherent in a problem-solving situation. A sketch will help you to organize the given information and can lead to your being able to focus on the method by which the solution can be determined.

A tour bus drives 5 mi south, then 4 mi west, then 3 mi north, then 4 mi east. How far is the tour bus from the starting point?

Draw a diagram of the given information.

From the diagram, we can see that the solution can be determined by subtracting 3 from 5: $5 - 3 = 2$.

The bus is 2 mi from the starting point.

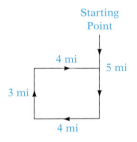

If you roll two ordinary six-sided dice and multiply the two numbers that appear on top, how many different possible products are there?

Make a chart of the possible products. In the chart below, repeated products are marked with an asterisk.

$1 \cdot 1 = 1$	$2 \cdot 1 = 2$ (*)	$3 \cdot 1 = 3$ (*)	$4 \cdot 1 = 4$ (*)	$5 \cdot 1 = 5$ (*)	$6 \cdot 1 = 6$ (*)
$1 \cdot 2 = 2$	$2 \cdot 2 = 4$ (*)	$3 \cdot 2 = 6$ (*)	$4 \cdot 2 = 8$ (*)	$5 \cdot 2 = 10$ (*)	$6 \cdot 2 = 12$ (*)
$1 \cdot 3 = 3$	$2 \cdot 3 = 6$ (*)	$3 \cdot 3 = 9$	$4 \cdot 3 = 12$ (*)	$5 \cdot 3 = 15$ (*)	$6 \cdot 3 = 18$ (*)
$1 \cdot 4 = 4$	$2 \cdot 4 = 8$	$3 \cdot 4 = 12$ (*)	$4 \cdot 4 = 16$	$5 \cdot 4 = 20$ (*)	$6 \cdot 4 = 24$ (*)
$1 \cdot 5 = 5$	$2 \cdot 5 = 10$	$3 \cdot 5 = 15$	$4 \cdot 5 = 20$	$5 \cdot 5 = 25$	$6 \cdot 5 = 30$ (*)
$1 \cdot 6 = 6$	$2 \cdot 6 = 12$	$3 \cdot 6 = 18$	$4 \cdot 6 = 24$	$5 \cdot 6 = 30$	$6 \cdot 6 = 36$

By counting the products that are not repeats, we can see that there are 18 different possible products.

Copyright © Houghton Mifflin Company. All rights reserved.

Look at Sections 1 and 2 in this chapter. You will notice that number lines are used to help you visualize the integers, as an aid in ordering integers, to help you understand the concepts of opposite and absolute value, and to illustrate addition of integers. As you begin your work with integers, you may find that sketching a number line proves helpful in coming to understand a problem or in working through a calculation that involves integers.

Projects & Group Activities

Time Zones

In 1884, a system of standard time was adopted by the International Meridian Conference. The prime meridian is a semicircle passing through Greenwich, England, and labeled 0°, as shown in the diagram at the right.

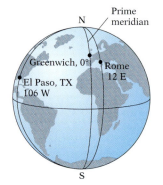

The other meridians are 15° apart, and each 15° width determines a time zone. The time zones to the east of the prime meridian are negative, and the zones to the west are positive, as shown below.

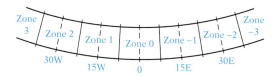

The table below gives the time zones for selected cities.

City	Time Zone	City	Time Zone
Athens	−2	London	0
Beijing	−8	Moscow	−3
Houston	6	Oslo	−1
Honolulu	11	Rio de Janeiro	3

To find the time for a city that is not in your time zone, subtract the time zone of the other city from the time zone of your city. Add that difference to the current time in your time zone. For example, say it is 2:00 P.M. in Athens. What time is it in Houston?

From the table above, the time zone in Athens is −2 and the time zone of Houston is 6.

Subtract 6 from −2. $-2 - 6 = -8$

Add that number to the current time $2:00 \text{ P.M.} + (-8) = 6:00 \text{ A.M.}$
in your time zone.

It is 6:00 A.M. in Houston.

Copyright © Houghton Mifflin Company. All rights reserved.

1. Find the time in Beijing when the time in Oslo is 11:00 P.M.

2. Find the time in Moscow when the time in Honolulu is 8:00 A.M.

3. Find the time in Rio de Janeiro when the time in Beijing is 3:00 P.M.

4. An office in Athens is open from 9:00 A.M. to 5:00 P.M. What are the times in Houston that a call to Athens can be made during Athens office hours?

5. An office in Honolulu is open from 8:00 A.M. to 4:00 P.M. What are the times in Moscow that a call to Honolulu can be made during Honolulu office hours?

Multiplication of Integers

Quadrant II   Quadrant I

Quadrant III   Quadrant IV

The grid at the left has four regions, or quadrants, numbered counterclockwise, starting at the upper right, with the Roman numerals I, II, III, IV.

1. Complete Quadrant I by multiplying each of the horizontal numbers, 1 through 5, by each of the vertical numbers, 1 through 5. The product 4(3) has been filled in for you. Complete Quadrants II, III, and IV by again multiplying each horizontal number by each vertical number.

2. What is the sign of all the products in Quadrant I? Quadrant II? Quadrant III? Quadrant IV?

3. Describe at least three patterns that you observe in the completed table.

4. How does the table show that multiplication of integers is commutative?

5. How can you use the table to find the quotient of two integers? Provide at least two examples of division of integers.

Closure

The whole numbers are said to be *closed* with respect to addition because when two whole numbers are added, the result is a whole number. The whole numbers are not closed with respect to subtraction because, for example, 4 and 7 are whole numbers but $4 - 7 = -3$, and -3 is not a whole number. Complete the table below by entering a Y if the operation is closed for those numbers and an N if it is not closed. When we discuss whether multiplication and division are closed, zero is not included because division by zero is not defined.

	Add	*Subtract*	*Multiply*	*Divide*
Whole numbers	Y	N		
Integers				

Copyright © Houghton Mifflin Company. All rights reserved.

Morse Code/Binary Code

In this chapter, the symbol "−" was described as meaning *minus, negative,* or *opposite.* You have also seen this symbol used as a hyphen or a dash.

Morse Code is a system of communication in which letters and numbers are represented using dots and dashes. At the left is the representation of the letters of the alphabet and the numbers 0 through 9 in Morse Code.

A	• —	S	• • •
B	— • • •	T	—
C	— • — •	U	• • —
D	— • •	V	• • • —
E	•	W	• — —
F	• • — •	X	— • • —
G	— — •	Y	— • — —
H	• • • •	Z	— — • •
I	• •	0	— — — — —
J	• — — —	1	• — — — —
K	— • —	2	• • — — —
L	• — • •	3	• • • — —
M	— —	4	• • • • —
N	— •	5	• • • • •
O	— — —	6	— • • • •
P	• — — •	7	— — • • •
Q	— — • —	8	— — — • •
R	• — •	9	— — — — •

1. Describe the pattern of dots and dashes used to represent the numbers 0 to 9.

2. Why might the letters E and T have been chosen to have the shortest representations (only one symbol)?

Note that Morse Code uses only two symbols. In our number system, the **decimal system,** we use 10 digits to represent any number: 0, 1, 2, 3, 4, 5, 6, 7, 8, and 9. Because 10 digits are used, the decimal system is also referred to as **base ten.**

Another number system, the **binary number system,** uses only two digits to represent any number: 0 and 1. Because two digits are used, the binary number system is also referred to as **base two.**

The first five place values of the digits in the decimal system are

Ten-thousands   Thousands   Hundreds   Tens   Ones

Each place value is a power of 10 ($10 = 10^1$, $100 = 10^2$, $1,000 = 10^3$, $10,000 = 10^4$, and so on).

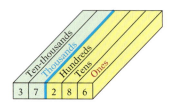

The first five place values of the digits in the binary number system are

Sixteens   Eights   Fours   Twos   Ones

Each place value is a power of 2 ($2 = 2^1$, $4 = 2^2$, $8 = 2^3$, $16 = 2^4$, and so on).

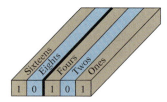

The whole numbers are represented in base two as

0
1
10 (1 two, zero ones)
11 (1 two, 1 one)
100 (1 four)
101 (1 four, 1 one)
110 (1 four, 1 two)
111 (1 four, 1 two, 1 one)
1000 (1 eight)
1001 (1 eight, 1 one), and so on.

► **Point of Interest**

One of the most important applications of the binary number system is in the area of computers. Just as information in Morse Code is relayed using only two symbols, computers operate using only two symbols, 0 and 1. Because all data, or information, in a computer is represented using only 0s and 1s, the data are said to be in binary code.

3. Write the next ten whole numbers using base two.

4. What would be the sixth and seventh place values in base two?

Copyright © Houghton Mifflin Company. All rights reserved.

Chapter Summary

Key Words	Examples
A number n is a **positive number** if $n > 0$. A number n is a **negative number** if $n < 0$. [2.1A, p. 89]	Positive numbers are numbers greater than zero. 9, 87, and 603 are positive numbers. Negative numbers are numbers less than zero. -5, -41, and -729 are negative numbers.
The **integers** are . . . $-4, -3, -2, -1, 0, 1, 2, 3, 4, \ldots$ The integers can be defined as the whole numbers and their opposites. **Positive integers** are to the right of zero on the number line. **Negative integers** are to the left of zero on the number line. [2.1A, p. 89]	-729, -41, -5, 9, 87, and 603 are integers. 0 is an integer, but it is neither a positive nor a negative integer.
Opposite numbers are two numbers that are the same distance from zero on the number line but on opposite sides of zero. The opposite of a number is called its **additive inverse.** [2.1B, p. 91; 2.2A, p. 103]	8 is the opposite, or additive inverse, of -8. -2 is the opposite, or additive inverse, of 2.
The **absolute value** of a number is the distance from zero to the number on the number line. The absolute value of a number is a positive number or zero. The symbol for absolute value is "$\mid\;\mid$". [2.1C, p. 92]	$\mid 9 \mid = 9$ $\mid -9 \mid = 9$ $-\mid 9 \mid = -9$

Essential Rules and Procedures

To add integers with the same sign, add the absolute values of the numbers. Then attach the sign of the addends. [2.2A, p. 102]	$6 + 4 = 10$ $-6 + (-4) = -10$
To add integers with different signs, find the absolute values of the numbers. Subtract the lesser absolute value from the greater absolute value. Then attach the sign of the addend with the greater absolute value. [2.2A, p. 102]	$-6 + 4 = -2$ $6 + (-4) = 2$
To subtract two integers, add the opposite of the second integer to the first integer. [2.2B, p. 106]	$6 - 4 = 6 + (-4) = 2$ $6 - (-4) = 6 + 4 = 10$ $-6 - 4 = -6 + (-4) = -10$ $-6 - (-4) = -6 + 4 = -2$
To multiply integers with the same sign, multiply the absolute values of the factors. The product is positive. [2.3A, p. 117]	$3 \cdot 5 = 15$ $-3(-5) = 15$

Copyright © Houghton Mifflin Company. All rights reserved.

To multiply integers with different signs, multiply the absolute values of the factors. The product is negative. [2.3A, p. 117]

$$-3(5) = -15$$
$$3(-5) = -15$$

To divide two numbers with the same sign, divide the absolute values of the numbers. The quotient is positive. [2.3B, p. 120]

$$15 \div 3 = 5$$
$$(-15) \div (-3) = 5$$

To divide two numbers with different signs, divide the absolute values of the numbers. The quotient is negative. [2.3B, p. 120]

$$-15 \div 3 = -5$$
$$15 \div (-3) = -5$$

Order Relations $a > b$ if a is to the right of b on the number line. $a < b$ if a is to the left of b on the number line. [2.1A, p. 90]

$$-6 > -12$$
$$-8 < 4$$

Properties of Addition [2.2A, p. 103]

Addition Property of Zero $a + 0 = a$ or $0 + a = a$

$$-6 + 0 = -6$$

Commutative Property of Addition $a + b = b + a$

$$-8 + 4 = 4 + (-8)$$

Associative Property of Addition $(a + b) + c = a + (b + c)$

$$(-5 + 4) + 6 = -5 + (4 + 6)$$

Inverse Property of Addition $a + (-a) = 0$ or $-a + a = 0$

$$7 + (-7) = 0$$

Properties of Multiplication [2.3A, p. 118]

Multiplication Property of Zero $a \cdot 0 = 0$ or $0 \cdot a = 0$

$$-9(0) = 0$$

Multiplication Property of One $a \cdot 1 = a$ or $1 \cdot a = a$

$$-3(1) = -3$$

Commutative Property of Multiplication $a \cdot b = b \cdot a$

$$-2(6) = 6(-2)$$

Associative Property of Multiplication $(a \cdot b) \cdot c = a \cdot (b \cdot c)$

$$(-2 \cdot 4) \cdot 5 = -2 \cdot (4 \cdot 5)$$

Division Properties of Zero and One [2.3B, p. 121]

If $a \neq 0, 0 \div a = 0$.

$$0 \div (-5) = 0$$

If $a \neq 0, a \div a = 1$.

$$-5 \div (-5) = 1$$

$a \div 1 = a$

$$-5 \div 1 = -5$$

$a \div 0$ is undefined.

$$-5 \div 0 \text{ is undefined.}$$

Addition Property of Equations [2.4A, p. 135]
The same number can be added to each side of an equation without changing the solution of the equation.

$$x - 4 = 12$$
$$x - 4 + 4 = 12 + 4$$
$$x = 16$$

The Order of Operations Agreement [2.5A, p. 129]

Step 1 Do all operations inside parentheses.

Step 2 Simplify any numerical expressions containing exponents.

Step 3 Do multiplication and division as they occur from left to right.

Step 4 Do addition and subtraction as they occur from left to right.

$$(-4)^2 - 3(1 - 5) = (-4)^2 - 3(-4)$$
$$= 16 - 3(-4)$$
$$= 16 - (-12)$$
$$= 16 + 12$$
$$= 28$$

Copyright © Houghton Mifflin Company. All rights reserved.

Chapter Review Exercises

1. Write the expression $8 - (-1)$ in words.

2. Evaluate $-|-36|$.

3. Find the product of -40 and -5.

4. Evaluate $-a \div b$ when $a = -27$ and $b = -3$.

5. Add: $-28 + 14$

6. Simplify: $-(-13)$

7. Graph -2 on the number line.

−6 −5 −4 −3 −2 −1 0 1 2 3 4 5 6

8. Solve: $-24 = -6y$

9. Divide: $-51 \div (-3)$

10. Find the quotient of 840 and -4.

11. Subtract: $-6 - (-7) - 15 - (-12)$

12. Evaluate $-ab$ when $a = -2$ and $b = -9$.

13. Find the sum of 18, -13, and -6.

14. Multiply: $-18(4)$

15. Simplify: $(-2)^2 - (-3)^2 \div (1 - 4)^2 \cdot 2 - 6$

16. Evaluate $-x - y$ when $x = -1$ and $y = 3$.

17. *Sports* The scores of four golfers after the final round of the 2002 Kraft Nabisco Championship are shown in the figure at the right. What is the difference between Diaz's score and Sorenstam's score?

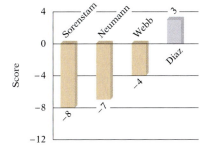

Golfers' Scores in 2002 Kraft Nabisco Championship

Copyright © Houghton Mifflin Company. All rights reserved.

18. Find the difference between -15 and -28.

19. Identify the property that justifies the statement.
$$-11(-50) = -50(-11)$$

20. Is -9 a solution of $-6 - t = 3$?

21. Simplify: $-9 + 16 - (-7)$

22. Divide: $\dfrac{0}{-17}$

23. Multiply: $-5(2)(-6)(-1)$

24. Add: $3 + (-9) + 4 + (-10)$

25. Evaluate $(a - b)^2 - 2a$ when $a = -2$ and $b = -3$.

26. Place the correct symbol, $<$ or $>$, between the two numbers.

$-8 \quad -10$

27. Complete the statement by using the Inverse Property of Addition.

$-21 + ? = 0$

28. Find the absolute value of -27.

29. Forty-eight is the product of negative six and some number. Find the number.

30. *Temperature* Which is colder, a temperature of $-4°C$ or $-12°C$?

31. *Chemistry* The figure at the right shows the boiling point in degrees Celsius of three chemical elements. The boiling point of neon is 7 times the highest boiling point shown in the table. What is the boiling point of neon?

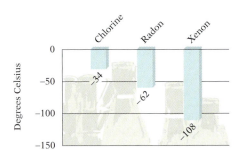

Boiling Points of Chemical Elements

32. *Temperature* Find the temperature after an increase of $5°C$ from $-8°C$.

33. *Mathematics* The distance, d, between point a and point b on the number line is given by the formula $d = |a - b|$. Find d when $a = 7$ and $b = -5$.

Copyright © Houghton Mifflin Company. All rights reserved.

Chapter Test

1. Write the expression $-3 + (-5)$ in words.

2. Evaluate $-|-34|$.

3. What is 3 minus -15?

4. Evaluate $a + b$ when $a = -11$ and $b = -9$.

5. Evaluate $(-x)(-y)$ when $x = -4$ and $y = -6$.

6. Identify the property that justifies the statement.
   $-23 + 4 = 4 + (-23)$

7. What is -360 divided by -30?

8. Find the sum of -3, -6, and 11.

9. Place the correct symbol between the two numbers.
   16   -19

10. Subtract: $7 - (-3) - 12$

11. Evaluate $a - b - c$ when $a = 6$, $b = -2$, and $c = 11$.

12. Simplify: $-(-49)$

13. Find the product of 50 and -5.

14. Write the given numbers in order from smallest to largest.
    $-|5|$, $-(-11)$, $|-9|$, $-(3)$

15. Is -9 a solution of the equation $17 - x = 8$?

16. On the number line, which number is 2 units to the right of -5?

17. ◉ *Sports* The scores of four golfers after the final round of the 2002 Masters are shown in the figure at the right. What is the difference between Woods's score and Daly's score?

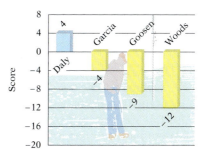

Golfers' Scores in 2002 Masters

Copyright © Houghton Mifflin Company. All rights reserved.

18. Divide: $\dfrac{0}{-16}$

19. Evaluate $2bc - (c + a)^3$ when $a = -2$, $b = 4$, and $c = -1$.

20. Find the opposite of 25.

21. Solve: $c - 11 = 5$

22. Subtract: $0 - 11$

23. Divide: $-96 \div (-4)$

24. Simplify: $16 \div 4 - 12 \div (-2)$

25. Evaluate $\dfrac{-x}{y}$ when $x = -56$ and $y = -8$.

26. Evaluate $3xy$ when $x = -2$ and $y = -10$.

27. Solve: $-11w = 121$

28. What is 14 less than 4?

29. *Temperature*   Find the temperature after an increase of 11°C from −6°C.

30. *Environmental Science*   The wind chill factor when the temperature is −25°F and the wind is blowing at 40 mph is four times the wind chill factor when the temperature is −5°F and the wind is blowing at 5 mph. If the wind chill factor at −5°F with a 5 mph wind is −16°, what is the wind chill factor at −25°F with a 40 mph wind?

31. *Temperature*   The high temperature today is 8° lower than the high temperature yesterday. The high temperature today is −13°C. What was the high temperature yesterday?

32. *Mathematics*   The distance, d, between point a and b on the number line is given by the formula $d = |a - b|$. Find d when $a = 4$ and $b = -12$.

33. *Business*   The net worth of a business is given by the formula $N = A - L$, where N is the net worth, A is the assets of the business (or the amount owned), and L is the liabilities of the business (or the amount owed). Use this formula to find the assets of a business that has a net worth of $18 million and liabilities of $6 million.

Copyright © Houghton Mifflin Company. All rights reserved.

Cumulative Review Exercises

1. Find the difference between -27 and -32.

2. Estimate the product of 439 and 28.

3. Divide: $19{,}254 \div 6$

4. Simplify: $16 \div (3 + 5) \cdot 9 - 2^4$

5. Evaluate $-|-82|$.

6. Write three hundred nine thousand four hundred eighty in standard form.

7. Evaluate $5xy$ when $x = 80$ and $y = 6$.

8. What is -294 divided by -14?

9. Subtract: $-28 - (-17)$

10. Find the sum of -24, 16, and -32.

11. Find all the factors of 44.

12. Evaluate $x^4 y^2$ when $x = 2$ and $y = 11$.

13. Round $629{,}874$ to the nearest thousand.

14. Estimate the sum of 356, 481, 294, and 117.

15. Evaluate $-a - b$ when $a = -4$ and $b = -5$.

16. Find the product of -100 and 25.

17. Find the prime factorization of 69.

18. Solve: $3x = -48$

19. Simplify: $(1 - 5)^2 \div (-6 + 4) + 8(-3)$

20. Evaluate $-c \div d$ when $c = -32$ and $d = -8$.

21. Evaluate $\dfrac{a}{b}$ when $a = 39$ and $b = -13$.

22. Place the correct symbol, $<$ or $>$, between the two numbers.

    $-62 \quad 26$

Copyright © Houghton Mifflin Company. All rights reserved.

23. What is −18 multiplied by −7?

24. Solve: 12 + p = 3

25. Write 2 · 2 · 2 · 2 · 2 · 7 · 7 in exponential notation.

26. Evaluate $4a + (a - b)^3$ when $a = 5$ and $b = 2$.

27. Add: 5,971 + 482 + 3,609

28. What is 5 less than −21?

29. Estimate the difference between 7,352 and 1,986.

30. Evaluate $3^4 \cdot 5^2$.

31. 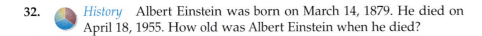 *History* The land area of the United States prior to the Louisiana Purchase was 891,364 mi². The land area of the Louisiana Purchase, which was purchased from France in 1803, was 831,321 mi². What was the land area of the United States immediately after the Louisiana Purchase?

32. *History* Albert Einstein was born on March 14, 1879. He died on April 18, 1955. How old was Albert Einstein when he died?

Albert Einstein

33. *Finances* A customer makes a down payment of $3,550 on a car costing $17,750. Find the amount that remains to be paid.

34. *Real Estate* A construction company is considering purchasing a 25-acre tract of land on which to build single-family homes. If the price is $3,690 per acre, what is the total cost of the land?

35. *Temperature* Find the temperature after an increase of 7°C from −12°C.

36. 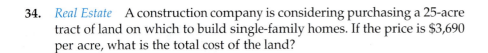 *Temperature* Record temperatures, in degrees Fahrenheit, for four states in the United States are shown at the right. (a) What is the difference between the record high and record low temperatures in Arizona? (b) For which state is the difference between the record high and record low temperatures greatest?

Record Temperatures (in degrees Fahrenheit)		
State	Lowest	Highest
Alabama	−27	112
Alaska	−80	100
Arizona	−40	128
Arkansas	−29	120

Source: The World Almanac and Book of Facts 2002

37. *Business* As a sales representative, your goal is to sell $120,000 in merchandise during the year. You sold $28,550 in merchandise during the first quarter of the year, $34,850 during the second quarter, and $31,700 during the third quarter. What must your sales for the fourth quarter be if you are to meet your goal for the year?

38. *Sports* Use the equation $S = N - P$, where S is a golfer's score relative to par in a tournament, N is the number of strokes made by the golfer, and P is par, to find a golfer's score relative to par when the golfer made 198 strokes and par is 206.

Copyright © Houghton Mifflin Company. All rights reserved.

Fractions

A New York high school band director leads his students in a rehearsal for a holiday concert. The musicians follow the movement of his baton in order to synchronize their playing. The time signature at the beginning of every piece of music tells them how many beats to play per measure. This fraction, as seen in the **Project on page 226,** serves as an essential guide for all musicians to play in time.

Copyright © Houghton Mifflin Company. All rights reserved.

Need help? For online student resources, visit this website: **math.college.hmco.com**

Prep Test

For Exercises 1 to 6, add, subtract, multiply, or divide.

1. 4×5

2. $2 \cdot 2 \cdot 2 \cdot 3 \cdot 5$

3. 9×1

4. $-6 + 4$

5. $-10 - 3$

6. $63 \div 30$

7. What is the smallest number into which both 8 and 12 divide evenly?

8. What is the greatest number that divides evenly into both 16 and 20?

9. Simplify: $8 \times 7 + 3$

10. Complete: $8 = ? + 1$

11. Place the correct symbol, $<$ or $>$, between the two numbers.
44   48

Go Figure

Maria and Pedro are siblings. Pedro has as many brothers as sisters. Maria has twice as many brothers as sisters. How many children are in the family?

Copyright © Houghton Mifflin Company. All rights reserved.

Copyright © Houghton Mifflin Company. All rights reserved.

SECTION 3.1 ▶ ## Least Common Multiple and Greatest Common Factor

OBJECTIVE   **A**

Least common multiple (LCM)

The **multiples** of a number are the products of that number and the numbers 1, 2, 3, 4, 5, ...

$4 \cdot 1 = 4$
$4 \cdot 2 = 8$
$4 \cdot 3 = 12$
$4 \cdot 4 = 16$
$4 \cdot 5 = 20$ The multiples of 4 are 4, 8, 12, 16, 20, ...
    ·
    ·
    ·

A number that is a multiple of two or more numbers is a **common multiple** of those numbers.

The multiples of 6 are 6, 12, 18, 24, 30, 36, 42, 48, 54, 60, 66, 72, ...
The multiples of 8 are 8, 16, 24, 32, 40, 48, 56, 64, 72, 80, 88, 96, ...
Some common multiples of 6 and 8 are 24, 48, and 72.

The **least common multiple (LCM)** is the smallest common multiple of two or more numbers.

The least common multiple of 6 and 8 is 24.

Listing the multiples of each number is one way to find the LCM. Another way to find the LCM uses the prime factorization of each number.

To find the LCM of 6 and 8 using prime factorization:

Write the prime factorization of each number and circle the highest power of each prime factor.

$6 = 2 \cdot ③$
$8 = ②^3$

The LCM is the product of the circled factors.

$2^3 \cdot 3 = 8 \cdot 3 = 24$

The LCM of 6 and 8 is 24.

Find the LCM of 32 and 36.

Write the prime factorization of each number and circle the highest power of each prime factor.

$32 = ②^5$
$36 = 2^2 \cdot ③^2$

The LCM is the product of the circled factors.

$2^5 \cdot 3^2 = 32 \cdot 9 = 288$

The LCM of 32 and 36 is 288.

Example Find the LCM of 12, 18, and 40.

Solution
$12 = 2^2 \cdot 3$
$18 = 2 \cdot \boxed{3^2}$
$40 = \boxed{2^3} \cdot \boxed{5}$

$LCM = 2^3 \cdot 3^2 \cdot 5$
$= 8 \cdot 9 \cdot 5 = 360$

 You Try It Find the LCM of 16, 24, and 28.

Your Solution

Solution on p. S6

OBJECTIVE B

VIDEO & DVD   CD TUTOR   WWW WEB   SSM

Greatest common factor (GCF)

Recall that a number that divides another number evenly is a **factor** of the number.

18 can be evenly divided by 1, 2, 3, 6, 9, and 18.

1, 2, 3, 6, 9, and 18 are factors of 18.

A number that is a factor of two or more numbers is a **common factor** of those numbers.

The factors of 24 are 1, 2, 3, 4, 6, 8, 12, and 24.
The factors of 36 are 1, 2, 3, 4, 6, 9, 12, 18, and 36.
The common factors of 24 and 36 are 1, 2, 3, 4, 6, and 12.

The **greatest common factor (GCF)** is the largest common factor of two or more numbers.

The greatest common factor of 24 and 36 is 12.

Take Note

12 is the GCF of 24 and 36 because 12 is the largest natural number that divides evenly into both 24 and 36.

Listing the factors of each number is one way to find the GCF. Another way to find the GCF uses the prime factorization of each number.

To find the GCF of 24 and 36 using prime factorization:

Write the prime factorization of each number and circle the lowest power of each prime factor that occurs in *both* factorizations.

$24 = 2^3 \cdot \boxed{3}$
$36 = \boxed{2^2} \cdot 3^2$

The GCF is the product of the circled factors.

$2^2 \cdot 3 = 4 \cdot 3 = 12$

Find the GCF of 12 and 30.

Write the prime factorization of each number and circle the lowest power of each prime factor that occurs in both factorizations. The prime factor 5 occurs in the prime factorization of 30 but not in the prime factorization of 12. Since 5 is not a factor in both factorizations, do not circle 5.

$12 = 2^2 \cdot \boxed{3}$
$30 = \boxed{2} \cdot 3 \cdot 5$

The GCF is the product of the circled factors.

$2 \cdot 3 = 6$

The GCF of 12 and 30 is 6.

Copyright © Houghton Mifflin Company. All rights reserved.

2 *Example*   Find the GCF of 14 and 27.

Solution $14 = 2 \cdot 7$
$27 = 3^3$

No common prime factor occurs in the factorizations.

GCF = 1

2 *You Try It*   Find the GCF of 25 and 52.

Your Solution

3 *Example*   Find the GCF of 16, 20, and 28.

Solution $16 = 2^4$
$20 = \boxed{2^2} \cdot 5$
$28 = 2^2 \cdot 7$

GCF $= 2^2 = 4$

3 *You Try It*   Find the GCF of 32, 40, and 56.

Your Solution

Solutions on p. S6

OBJECTIVE **C**

Applications

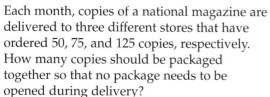

4 *Example*

Each month, copies of a national magazine are delivered to three different stores that have ordered 50, 75, and 125 copies, respectively. How many copies should be packaged together so that no package needs to be opened during delivery?

Strategy

To find the numbers of copies to be packaged together, find the GCF of 50, 75, and 125.

Solution
$50 = 2 \cdot \boxed{5^2}$
$75 = 3 \cdot 5^2$
$125 = 5^3$
GCF $= 5^2 = 25$

Each package should contain 25 copies of the magazine.

4 *You Try It*

A discount catalog offers blank CDs at reduced prices. The customer must order 20, 50, or 100 CDs. How many CDs should be packaged together so that no package needs to be opened when a clerk is filling an order?

Your Strategy

Your Solution

Solution on p. S6

Copyright © Houghton Mifflin Company. All rights reserved.

5 Example

To accommodate several activity periods and science labs after the lunch period and before the closing homeroom period, a high school wants to have both 25-minute class periods and 40-minute class periods running simultaneously in the afternoon class schedule. There is a 5-minute passing time between each class. How long a period of time must be scheduled if all students are to be in the closing homeroom period at the same time? How many 25-minute classes and 40-minute classes will be scheduled in that amount of time?

Strategy

To find the amount of time to be scheduled:

▶ Add the passing time (5 min) to the 25-minute class period and to the 40-minute class period to find the length of each period including the passing time.
▶ Find the LCM of the two time periods found in Step 1.

To find the number of 25-minute and 40-minute classes:

▶ Divide the LCM by each time period found in Step 1.

Solution

$25 + 5 = 30$
$40 + 5 = 45$

$30 = ②\cdot 3 \cdot ⑤$

$45 = ③^2 \cdot 5$

$LCM = 2 \cdot 3^2 \cdot 5 = 90$

A 90-minute time period must be scheduled.

$90 \div 30 = 3$
$90 \div 45 = 2$

There will be three 25-minute class periods and two 40-minute class periods in the 90-minute period.

5 You Try It

You and a friend are running laps at the track. You run one lap every 3 min. Your friend runs one lap every 4 min. If you start at the same time from the same place on the track, in how many minutes will both of you be at the starting point again? Will you have passed each other at some other point on the track prior to that time?

Your Strategy

Your Solution

Solution on p. S6

Copyright © Houghton Mifflin Company. All rights reserved.

3.1 EXERCISES

OBJECTIVE **A**

Find the LCM of the numbers.

1. 4 and 8

2. 3 and 9

3. 2 and 7

4. 5 and 11

5. 6 and 10

6. 8 and 12

7. 9 and 15

8. 14 and 21

9. 12 and 16

10. 8 and 14

11. 4 and 10

12. 9 and 30

13. 14 and 42

14. 16 and 48

15. 24 and 36

16. 16 and 28

17. 30 and 40

18. 45 and 60

19. 3, 5, and 10

20. 5, 10, and 20

21. 4, 8, and 12

22. 3, 12, and 18

23. 9, 36, and 45

24. 9, 36, and 72

25. 6, 9, and 15

26. 30, 40, and 60

27. 13, 26, and 39

28. 12, 48, and 72

OBJECTIVE **B**

Find the GCF of the numbers.

29. 9 and 12

30. 6 and 15

31. 18 and 30

32. 15 and 35

33. 14 and 42

34. 25 and 50

35. 16 and 80

36. 17 and 51

37. 21 and 55

38. 32 and 35

39. 8 and 36

40. 12 and 80

41. 12 and 76

42. 16 and 60

43. 24 and 30

44. 16 and 28

45. 24 and 36

46. 30 and 40

47. 45 and 75

48. 12 and 54

Copyright © Houghton Mifflin Company. All rights reserved.

49. 6, 10, and 12   **50.** 8, 12, and 20   **51.** 6, 15, and 36   **52.** 15, 20, and 30

53. 21, 63, and 84   **54.** 12, 28, and 48   **55.** 24, 36, and 60   **56.** 32, 56, and 72

OBJECTIVE **C**

57. *Business*   Two machines are filling cereal boxes. One machine, which is filling 12-ounce boxes, fills one box every 2 min. The second machine, which is filling 18-ounce boxes, fills one box every 3 min. How often are the two machines starting to fill a box at the same time?

58. *Business*   A discount catalog offers stockings at reduced prices. The customer must order 3 pairs, 6 pairs, or 12 pairs of stockings. How many pairs should be packaged together so that no package needs to be opened when a clerk is filling an order?

59. *Business*   Each week, copies of a national magazine are delivered to three different stores that have ordered 75 copies, 100 copies, and 150 copies, respectively. How many copies should be packaged together so that no package needs to be opened during delivery?

60. *Sports*   You and a friend are swimming laps at a pool. You swim one lap every 4 min. Your friend swims one lap every 5 min. If you start at the same time from the same end of the pool, in how many minutes will both of you be at the starting point again? How many times will you have passed each other in the pool prior to that time?

61. *Scheduling*   A mathematics conference is scheduling 30-minute sessions and 40-minute sessions. There will be a 10-minute break after each session. The sessions at the conference start at 9 A.M. At what time will all sessions begin at the same time once again? At what time should lunch be scheduled if all participants are to eat at the same time?

Critical Thinking

62. If x is a prime number and y is a prime number, find the LCM of x and y. Find the GCF of x and y.

63. Find the LCM of x and $2x$. Find the GCF of x and $2x$.

64. In your own words, define the least common multiple of two numbers and the greatest common factor of two numbers.

65. Explain the meaning of relatively prime factors. List three pairs of relatively prime factors.

Copyright © Houghton Mifflin Company. All rights reserved.

Copyright © Houghton Mifflin Company. All rights reserved.

SECTION 3.2 ▶ Introduction to Fractions

OBJECTIVE **A**

Proper fractions, improper fractions, and mixed numbers

A recipe calls for $\frac{1}{2}$ cup of butter; a carpenter uses a $\frac{3}{8}$-inch screw; and a stock broker might say that Sears closed down $\frac{3}{4}$. The numbers $\frac{1}{2}$, $\frac{3}{8}$, and $\frac{3}{4}$ are fractions.

A **fraction** can represent the number of equal parts of a whole. The circle at the right is divided into 8 equal parts. 3 of the 8 parts are shaded. The shaded portion of the circle is represented by the fraction $\frac{3}{8}$.

Each part of a fraction has a name.

$$\text{Fraction bar} \longrightarrow \frac{3}{8} \quad \begin{array}{l} \longleftarrow \textbf{Numerator} \\ \longleftarrow \textbf{Denominator} \end{array}$$

> ▶ **Point of Interest**
>
> The fraction bar was first used in 1050 by al-Hassar. It is also called a vinculum.

In a **proper fraction,** the numerator is smaller than the denominator. A proper fraction is less than 1.

$$\frac{1}{2} \qquad \frac{3}{8} \qquad \frac{3}{4}$$
Proper fractions

In an **improper fraction,** the numerator is greater than or equal to the denominator. An improper fraction is a number greater than or equal to 1.

$$\frac{7}{3} \qquad \frac{4}{4}$$
Improper fractions

The shaded portion of the circles at the right is represented by the improper fraction $\frac{7}{3}$.

The shaded portion of the square at the right is represented by the improper fraction $\frac{4}{4}$.

A fraction bar can be read "divided by." Therefore, the fraction $\frac{4}{4}$ can be read "4 ÷ 4." Because a number divided by itself is equal to 1, $4 \div 4 = 1$ and $\frac{4}{4} = 1$.

The shaded portion of the square above can be represented as $\frac{4}{4}$ or 1.

Since the fraction bar can be read as "divided by" and any number divided by 1 is the number, any whole number can be represented as an improper fraction. For example, $5 = \frac{5}{1}$ and $7 = \frac{7}{1}$.

Because zero divided by any number other than zero is zero, **the numerator of a fraction can be zero.**

For example, $\frac{0}{6} = 0$ because $0 \div 6 = 0$.

Recall that division by zero is not defined. Therefore, **the denominator of a fraction cannot be zero.**

For example, $\frac{9}{0}$ is not defined because $\frac{9}{0} = 9 \div 0$, and division by zero is not defined.

A **mixed number** is a number greater than 1 with a whole number part and a fractional part.

The shaded portion of the circles at the right is represented by the mixed number $2\frac{1}{2}$.

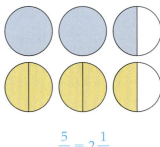

Note from the diagram at the right that the improper fraction $\frac{5}{2}$ is equal to the mixed number $2\frac{1}{2}$.

$$\frac{5}{2} = 2\frac{1}{2}$$

An improper fraction can be written as a mixed number.

To write $\frac{5}{2}$ as a mixed number, read the fraction bar as "divided by."

$\frac{5}{2}$ means $5 \div 2$.

Divide the numerator by the denominator.	To write the fractional part of the mixed number, write the remainder over the divisor.	Write the answer.
$\begin{array}{r} 2 \\ 2\overline{)5} \\ -4 \\ \hline 1 \end{array}$	$\begin{array}{r} 2\frac{1}{2} \\ 2\overline{)5} \\ -4 \\ \hline 1 \end{array}$	$\frac{5}{2} = 2\frac{1}{2}$

To write a mixed number as an improper fraction, multiply the denominator of the fractional part of the mixed number by the whole number part. The sum of this product and the numerator of the fractional part is the numerator of the improper fraction. The denominator remains the same.

Write $4\frac{5}{6}$ as an improper fraction.

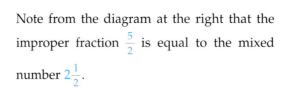

$$\begin{array}{c} + \\ 4\frac{5}{6} \\ \times \end{array} = \frac{(6 \cdot 4) + 5}{6} = \frac{24 + 5}{6} = \frac{29}{6}$$

Copyright © Houghton Mifflin Company. All rights reserved.

1 *Example*   Express the shaded portion of the circles as an improper fraction and as a mixed number.

Solution $\dfrac{19}{4}$; $4\dfrac{3}{4}$

1 *You Try It*   Express the shaded portion of the circles as an improper fraction and as a mixed number.

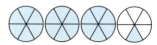

Your Solution

---

2 *Example*   Write $\dfrac{14}{5}$ as a mixed number.

Solution

$$\begin{array}{r} 2 \\ 5\overline{)14} \\ -10 \\ \hline 4 \end{array} \qquad \dfrac{14}{5} = 2\dfrac{4}{5}$$

2 *You Try It*   Write $\dfrac{26}{3}$ as a mixed number.

Your Solution

---

3 *Example*   Write $\dfrac{35}{7}$ as a whole number.

Solution

$$\begin{array}{r} 5 \\ 7\overline{)35} \\ -35 \\ \hline 0 \end{array} \qquad \dfrac{35}{7} = 5$$

▶ *Note:* The remainder is zero.

3 *You Try It*   Write $\dfrac{36}{4}$ as a whole number.

Your Solution

---

4 *Example*   Write $12\dfrac{5}{8}$ as an improper fraction.

Solution $12\dfrac{5}{8} = \dfrac{(8 \cdot 12) + 5}{8} = \dfrac{96 + 5}{8}$

$$= \dfrac{101}{8}$$

4 *You Try It*   Write $9\dfrac{4}{7}$ as an improper fraction.

Your Solution

---

5 *Example*   Write 9 as an improper fraction.

Solution $9 = \dfrac{9}{1}$

5 *You Try It*   Write 3 as an improper fraction.

Your Solution

Solutions on p. S6

Copyright © Houghton Mifflin Company. All rights reserved.

Copyright © Houghton Mifflin Company. All rights reserved.

VIDEO & DVD TUTOR CD WEB SSM

OBJECTIVE **B**

Equivalent fractions

Fractions can be graphed as points on a number line. The number lines at the right show thirds, sixths, and ninths graphed from 0 to 1.

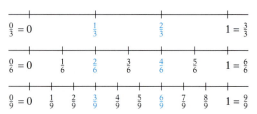

A particular point on the number line may be represented by different fractions, all of which are equal.

For example, $\dfrac{0}{3} = \dfrac{0}{6} = \dfrac{0}{9}$, $\dfrac{1}{3} = \dfrac{2}{6} = \dfrac{3}{9}$, $\dfrac{2}{3} = \dfrac{4}{6} = \dfrac{6}{9}$, and $\dfrac{3}{3} = \dfrac{6}{6} = \dfrac{9}{9}$.

Equal fractions with different denominators are called **equivalent fractions.**

$\dfrac{1}{3}$, $\dfrac{2}{6}$, and $\dfrac{3}{9}$ are equivalent fractions. $\dfrac{2}{3}$, $\dfrac{4}{6}$, and $\dfrac{6}{9}$ are equivalent fractions.

Note that we can rewrite $\dfrac{2}{3}$ as $\dfrac{4}{6}$ by multiplying both the numerator and denominator of $\dfrac{2}{3}$ by 2.

$$\dfrac{2}{3} = \dfrac{2 \cdot 2}{3 \cdot 2} = \dfrac{4}{6}$$

Also, we can rewrite $\dfrac{4}{6}$ as $\dfrac{2}{3}$ by dividing both the numerator and denominator of $\dfrac{4}{6}$ by 2.

$$\dfrac{4}{6} = \dfrac{4 \div 2}{6 \div 2} = \dfrac{2}{3}$$

This suggests the following property of fractions.

Equivalent Fractions

The numerator and denominator of a fraction can be multiplied by or divided by the same nonzero number. The resulting fraction is equivalent to the original fraction.

$$\dfrac{a}{b} = \dfrac{a \cdot c}{b \cdot c}, \quad \dfrac{a}{b} = \dfrac{a \div c}{b \div c}, \qquad \text{where} \quad b \neq 0 \quad \text{and} \quad c \neq 0$$

Write an equivalent fraction with the given denominator.

$$\dfrac{3}{8} = \dfrac{}{40}$$

Divide the larger denominator by the smaller one.

$$40 \div 8 = 5$$

Multiply the numerator and denominator of the given fraction by the quotient (5).

$$\dfrac{3}{8} = \dfrac{3 \cdot 5}{8 \cdot 5} = \dfrac{15}{40}$$

A fraction is in **simplest form** when the numerator and denominator have no common factors other than 1. The fraction $\dfrac{3}{8}$ is in simplest form because 3 and 8 have no common factors other than 1. The fraction $\dfrac{15}{50}$ is not in simplest form because the numerator and denominator have a common factor of 5.

▶ **Point of Interest**

Leonardo of Pisa, who was also called Fibonacci (c. 1175–1250), is credited with bringing the Hindu–Arabic number system to the Western world and promoting its use instead of the cumbersome Roman numeral system. He was also influential in promoting the idea of the fraction bar. His notation, however, was very different from what we have today. For instance, he wrote

$\dfrac{3}{4} \dfrac{5}{7}$ to mean $\dfrac{5}{7} + \dfrac{3}{7 \cdot 4}$.

To write a fraction in simplest form, divide the numerator and denominator of the fraction by their common factors.

Write $\frac{12}{15}$ in simplest form.

12 and 15 have a common factor of 3. Divide the numerator and denominator by 3.

$$\frac{12}{15} = \frac{12 \div 3}{15 \div 3} = \frac{4}{5}$$

Simplifying a fraction requires that you recognize the common factors of the numerator and denominator. One way to do this is to write the prime factorization of the numerator and denominator and then divide by the common prime factors.

Write $\frac{30}{42}$ in simplest form.

Write the prime factorization of the numerator and denominator. Divide by the common factors.

$$\frac{30}{42} = \frac{\overset{1}{2} \cdot \overset{1}{3} \cdot 5}{\underset{1}{2} \cdot \underset{1}{3} \cdot 7} = \frac{5}{7}$$

Write $\frac{2x}{6}$ in simplest form.

Factor the numerator and denominator. Then divide by the common factors.

$$\frac{2x}{6} = \frac{\overset{1}{2} \cdot x}{\underset{1}{2} \cdot 3} = \frac{x}{3}$$

6 Example Write an equivalent fraction with the given denominator:

$$\frac{2}{5} = \frac{}{30}.$$

Solution $30 \div 5 = 6$

$$\frac{2}{5} = \frac{2 \cdot 6}{5 \cdot 6} = \frac{12}{30}$$

$\frac{12}{30}$ is equivalent to $\frac{2}{5}$.

6 You Try It Write an equivalent fraction with the given denominator:

$$\frac{5}{8} = \frac{}{48}.$$

Your Solution

7 Example Write an equivalent fraction with the given denominator:

$$3 = \frac{}{15}.$$

Solution $3 = \frac{3}{1}$ $15 \div 1 = 15$

$$3 = \frac{3}{1} = \frac{3 \cdot 15}{1 \cdot 15} = \frac{45}{15}$$

$\frac{45}{15}$ is equivalent to 3.

7 You Try It Write an equivalent fraction with the given denominator:

$$8 = \frac{}{12}.$$

Your Solution

Solutions on p. S6

Copyright © Houghton Mifflin Company. All rights reserved.

8 **Example**   Write $\frac{18}{54}$ in simplest form.

Solution $\frac{18}{54} = \frac{\overset{1}{2} \cdot \overset{1}{3} \cdot \overset{1}{3}}{\underset{1}{2} \cdot \underset{1}{3} \cdot \underset{1}{3} \cdot 3} = \frac{1}{3}$

8 **You Try It**   Write $\frac{21}{84}$ in simplest form.

Your Solution

9 **Example**   Write $\frac{36}{20}$ in simplest form.

Solution $\frac{36}{20} = \frac{\overset{1}{2} \cdot \overset{1}{2} \cdot 3 \cdot 3}{\underset{1}{2} \cdot \underset{1}{2} \cdot 5} = \frac{9}{5}$

9 **You Try It**   Write $\frac{32}{12}$ in simplest form.

Your Solution

10 **Example**   Write $\frac{10m}{12}$ in simplest form.

Solution $\frac{10m}{12} = \frac{2 \cdot 5 \cdot m}{\underset{1}{2} \cdot 2 \cdot 3} = \frac{5m}{6}$

10 **You Try It**   Write $\frac{11t}{11}$ in simplest form.

Your Solution

Solutions on pp. S6–S7

OBJECTIVE **C**

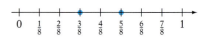

Order relations between two fractions

The number line can be used to determine the order relation between two fractions.

A fraction that appears to the left of a given fraction is less than the given fraction.

$\frac{3}{8}$ is to the left of $\frac{5}{8}$.

$$\frac{3}{8} < \frac{5}{8}$$

A fraction that appears to the right of a given fraction is greater than the given fraction.

$\frac{7}{8}$ is to the right of $\frac{3}{8}$.

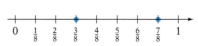

$$\frac{7}{8} > \frac{3}{8}$$

To find the order relation between two fractions with the *same* denominator, compare the numerators. The fraction with the smaller numerator is the smaller fraction. The larger fraction is the fraction with the larger numerator.

$\frac{3}{8}$ and $\frac{5}{8}$ have the same denominator. $\frac{3}{8} < \frac{5}{8}$ because $3 < 5$.

$\frac{7}{8}$ and $\frac{3}{8}$ have the same denominator. $\frac{7}{8} > \frac{3}{8}$ because $7 > 3$.

Copyright © Houghton Mifflin Company. All rights reserved.

Before comparing two fractions with *different* denominators, rewrite the fractions with a common denominator. The common denominator is the least common multiple (LCM) of the denominators of the fractions. The LCM of the denominators is sometimes called the lowest common denominator or LCD.

▶ **Point of Interest**

Archimedes (c. 287–212 B.C.) is the person who calculated that

$\pi \approx 3\frac{1}{7}$. He actually showed that

$3\frac{10}{71} < \pi < 3\frac{1}{7}$. The approximation

$3\frac{10}{71}$ is more accurate than $3\frac{1}{7}$ but more difficult to use.

Find the order relation between $\frac{5}{12}$ and $\frac{7}{18}$.

Find the LCM of the denominators.   The LCM of 12 and 18 is 36.

Write each fraction as an equivalent fraction with the LCM as the denominator.

$\frac{5}{12} = \frac{5 \cdot 3}{12 \cdot 3} = \frac{15}{36}$ ⟵ Larger numerator

$\frac{7}{18} = \frac{7 \cdot 2}{18 \cdot 2} = \frac{14}{36}$ ⟵ Smaller numerator

Compare the fractions.

$\frac{15}{36} > \frac{14}{36}$

$\frac{5}{12} > \frac{7}{18}$

---

11 *Example*   Place the correct symbol, < or >, between the two numbers.

$\frac{2}{3}$ $\quad$ $\frac{4}{7}$

Solution   The LCM of 3 and 7 is 21.

$\frac{2}{3} = \frac{14}{21}$ $\quad$ $\frac{4}{7} = \frac{12}{21}$

$\frac{14}{21} > \frac{12}{21}$

$\frac{2}{3} > \frac{4}{7}$

 11   *You Try It*   Place the correct symbol, < or >, between the two numbers.

$\frac{4}{9}$ $\quad$ $\frac{8}{21}$

Your Solution

---

12 *Example*   Place the correct symbol, < or >, between the two numbers.

$\frac{7}{12}$ $\quad$ $\frac{11}{18}$

Solution   The LCM of 12 and 18 is 36.

$\frac{7}{12} = \frac{21}{36}$ $\quad$ $\frac{11}{18} = \frac{22}{36}$

$\frac{21}{36} < \frac{22}{36}$

$\frac{7}{12} < \frac{11}{18}$

 12   *You Try It*   Place the correct symbol, < or >, between the two numbers.

$\frac{17}{24}$ $\quad$ $\frac{7}{9}$

Your Solution

Solutions on p. S7

Copyright © Houghton Mifflin Company. All rights reserved.

65 and over   Under 5

34   19

5–19 years
61

45–64 years
62

20–44 years
104

U.S. Population Distribution by Age (in millions)
Source: U.S. Bureau of the Census

VIDEO & DVD   CD TUTOR   WEB   SSM

OBJECTIVE **D**

Applications

 The graph at the left shows the U.S. population distribution by age. Use this graph for Example 13 and You Try It 13.

---

13 Example

What fraction of the total U.S. population is 20 to 44 years old?

Strategy

To find the fraction:

▶ Add the populations of all the segments to find the total U.S. population.
▶ Write a fraction with the population aged 20 to 44 in the numerator and the total population in the denominator. Write the fraction in simplest form.

Solution

$19 + 61 + 104 + 62 + 34 = 280$

$\dfrac{104}{280} = \dfrac{13}{35}$ $\dfrac{13}{35}$ of the U.S. population is aged 20 to 44.

13 You Try It

What fraction of the total U.S. population is 65 and older?

Your Strategy

Your Solution

---

14 Example

Of every dollar spent for health care, 10 cents is for prescription drugs. (*Source:* CMS, Office of the Actuary, National Health Statistics Group 2002, Blue Cross and Blue Association research, 2002) What fraction of every dollar spent for health care is for prescription drugs?

Strategy

To find the fraction, write a fraction with the amount spent for prescription drugs in the numerator and the number of cents in one dollar (100) in the denominator.

Solution

$\dfrac{10}{100} = \dfrac{1}{10}$ $\dfrac{1}{10}$ of every dollar spent for health care is for prescription drugs.

14 You Try It

Of every dollar spent for health care, 32 cents is for physicians (including clinical care). (*Source:* CMS, Office of the Actuary, National Health Statistics Group 2002, Blue Cross and Blue Association research, 2002) What fraction of every dollar spent for health care is for physicians?

Your Strategy

Your Solution

Solutions on p. S7

Copyright © Houghton Mifflin Company. All rights reserved.

3.2 EXERCISES

OBJECTIVE **A**

Express the shaded portion of the circle as a fraction.

1.

2.

3.

4.

Express the shaded portion of the circles as an improper fraction and as a mixed number.

5.

6.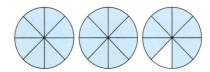

7.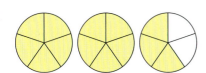

8.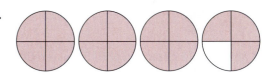

Write the improper fraction as a mixed number or a whole number.

9. $\dfrac{13}{4}$

10. $\dfrac{14}{3}$

11. $\dfrac{20}{5}$

12. $\dfrac{18}{6}$

13. $\dfrac{27}{10}$

14. $\dfrac{31}{3}$

15. $\dfrac{56}{8}$

16. $\dfrac{27}{9}$

17. $\dfrac{17}{9}$

18. $\dfrac{8}{3}$

19. $\dfrac{12}{5}$

20. $\dfrac{19}{8}$

21. $\dfrac{18}{1}$

22. $\dfrac{21}{1}$

23. $\dfrac{32}{15}$

24. $\dfrac{39}{14}$

25. $\dfrac{8}{8}$

26. $\dfrac{12}{12}$

27. $\dfrac{28}{3}$

28. $\dfrac{43}{5}$

Copyright © Houghton Mifflin Company. All rights reserved.

Write the mixed number or whole number as an improper fraction.

29. $2\frac{1}{4}$

30. $4\frac{2}{5}$

31. $5\frac{1}{2}$

32. $3\frac{2}{3}$

33. $2\frac{4}{5}$

34. $6\frac{3}{8}$

35. $7\frac{5}{6}$

36. $9\frac{1}{5}$

37. 7

38. 4

39. $8\frac{1}{4}$

40. $1\frac{7}{9}$

41. $10\frac{1}{3}$

42. $6\frac{3}{7}$

43. $4\frac{7}{12}$

44. $5\frac{4}{9}$

45. 8

46. 6

47. $12\frac{4}{5}$

48. $11\frac{5}{8}$

OBJECTIVE **B**

Write an equivalent fraction with the given denominator.

49. $\frac{1}{2} = \frac{}{12}$

50. $\frac{1}{4} = \frac{}{20}$

51. $\frac{3}{8} = \frac{}{24}$

52. $\frac{9}{11} = \frac{}{44}$

53. $\frac{2}{17} = \frac{}{51}$

54. $\frac{9}{10} = \frac{}{80}$

55. $\frac{3}{4} = \frac{}{32}$

56. $\frac{5}{8} = \frac{}{32}$

57. $6 = \frac{}{18}$

58. $5 = \frac{}{35}$

59. $\frac{1}{3} = \frac{}{90}$

60. $\frac{3}{16} = \frac{}{48}$

61. $\frac{2}{3} = \frac{}{21}$

62. $\frac{4}{9} = \frac{}{36}$

63. $\frac{6}{7} = \frac{}{49}$

64. $\frac{7}{8} = \frac{}{40}$

65. $\frac{4}{9} = \frac{}{18}$

66. $\frac{11}{12} = \frac{}{48}$

67. $7 = \frac{}{4}$

68. $9 = \frac{}{6}$

Write the fraction in simplest form.

69. $\frac{3}{12}$

70. $\frac{10}{22}$

71. $\frac{33}{44}$

72. $\frac{6}{14}$

73. $\frac{4}{24}$

Copyright © Houghton Mifflin Company. All rights reserved.

74. $\dfrac{25}{75}$ **75.** $\dfrac{8}{33}$ **76.** $\dfrac{9}{25}$ **77.** $\dfrac{0}{8}$ **78.** $\dfrac{0}{11}$

79. $\dfrac{42}{36}$ **80.** $\dfrac{30}{18}$ **81.** $\dfrac{16}{16}$ **82.** $\dfrac{24}{24}$ **83.** $\dfrac{21}{35}$

84. $\dfrac{11}{55}$ **85.** $\dfrac{16}{60}$ **86.** $\dfrac{8}{84}$ **87.** $\dfrac{12}{20}$ **88.** $\dfrac{24}{36}$

89. $\dfrac{12m}{18}$ **90.** $\dfrac{20x}{25}$ **91.** $\dfrac{4y}{8}$ **92.** $\dfrac{14z}{28}$ **93.** $\dfrac{24a}{36}$

94. $\dfrac{28z}{21}$ **95.** $\dfrac{8c}{8}$ **96.** $\dfrac{9w}{9}$ **97.** $\dfrac{18k}{3}$ **98.** $\dfrac{24t}{4}$

OBJECTIVE   **C**

Place the correct symbol, $<$ or $>$, between the two numbers.

99. $\dfrac{3}{8}$ $\dfrac{2}{5}$ **100.** $\dfrac{5}{7}$ $\dfrac{2}{3}$ **101.** $\dfrac{3}{4}$ $\dfrac{7}{9}$ **102.** $\dfrac{7}{12}$ $\dfrac{5}{8}$

103. $\dfrac{2}{3}$ $\dfrac{7}{11}$ **104.** $\dfrac{11}{14}$ $\dfrac{3}{4}$ **105.** $\dfrac{17}{24}$ $\dfrac{11}{16}$ **106.** $\dfrac{11}{12}$ $\dfrac{7}{9}$

107. $\dfrac{7}{15}$ $\dfrac{5}{12}$ **108.** $\dfrac{5}{8}$ $\dfrac{4}{7}$ **109.** $\dfrac{5}{9}$ $\dfrac{11}{21}$ **110.** $\dfrac{11}{30}$ $\dfrac{7}{24}$

111. $\dfrac{7}{12}$ $\dfrac{13}{18}$ **112.** $\dfrac{9}{11}$ $\dfrac{7}{8}$ **113.** $\dfrac{4}{5}$ $\dfrac{7}{9}$ **114.** $\dfrac{3}{4}$ $\dfrac{11}{13}$

115. $\dfrac{9}{16}$ $\dfrac{5}{9}$ **116.** $\dfrac{2}{3}$ $\dfrac{7}{10}$ **117.** $\dfrac{5}{8}$ $\dfrac{13}{20}$ **118.** $\dfrac{3}{10}$ $\dfrac{7}{25}$

Copyright © Houghton Mifflin Company. All rights reserved.

OBJECTIVE **D**

119. *Measurement* A ton is equal to 2,000 lb. What fractional part of a ton is 250 lb?

120. *Measurement* A pound is equal to 16 oz. What fractional part of a pound is 6 oz?

121. *Measurement* If a history class lasts 50 min, what fractional part of an hour is the history class?

122. *Measurement* If you sleep for 8 h one night, what fractional part of one day did you spend sleeping?

123. *Jewelry* Gold is designated by karats. Pure gold is 24 karats. What fractional part of an 18-karat gold bracelet is pure gold?

 The table at the right shows the results of a survey which asked fast-food patrons their criteria for choosing where to go for fast food. Three out of every 25 people surveyed said that the speed of the service was most important. Use this table for Exercises 124 and 125.

Fast-Food Patrons' Top Criteria for Fast-Food Restaurants	
Food Quality	$\frac{1}{4}$
Location	$\frac{13}{50}$
Menu	$\frac{4}{25}$
Price	$\frac{2}{25}$
Speed	$\frac{3}{25}$
Other	$\frac{3}{100}$

Source: Maritz Marketing Research, Inc.

124. *The Food Industry* According to the survey, do more people choose a fast-food restaurant on the basis of its location or on the basis of the quality of its food?

125. *The Food Industry* Which criterion was cited by most people?

126. *Card Games* A standard deck of playing cards consists of 52 cards. (a) What fractional part of a standard deck of cards is spades? (b) What fractional part of a standard deck of cards is aces?

127. *Education* You answer 42 questions correctly on an exam of 50 questions. Did you answer more or less than $\frac{8}{10}$ of the questions correctly?

128. *Education* To pass a real estate examination, you must answer at least $\frac{7}{10}$ of the questions correctly. If the exam has 200 questions and you answer 150 correctly, will you pass the exam?

Copyright © Houghton Mifflin Company. All rights reserved.

The table at the right shows the results of a survey in which family members in the United States were asked what concerns were serious problems for them. Nine out of every 25 people surveyed said that crime and violence were serious problems for them. Use this table for Exercises 129 to 131.

129.   *The Family*   According to the survey, which do more people consider a serious problem, the job market or the quality of their children's education?

130.   *The Family*   According to the survey, which do more people consider a serious problem, having enough time to spend with the family or staying in good health?

131.   *The Family*   Which concern did the most people cite as being a serious problem?

132.   *Sports*   Wilt Chamberlain held the record for the most field goals in a basketball game. He had 36 field goals in 63 attempts. What fraction of the number of attempts did he not have a field goal?

The table at the right shows the approximate maple syrup production during a recent year by the six states that produced the most maple syrup. Use this table for Exercises 133 and 134.

133.   *Agriculture*   What fraction of the total production by these six states was produced in New York?

134.   *Agriculture*   What fraction of the total production by these six states was produced in Wisconsin and Michigan?

The figure at the right shows how much money a data processor spends each month for various living expenses. Use this graph for Exercise 135.

135.   *Finances*   What fraction of the total monthly expenses was spent on (a) entertainment? (b) taxes?

Serious Problems for Family Members in the United States	
Crime and Violence	$\frac{9}{25}$
Divorce or separation	$\frac{13}{50}$
Drug or alcohol abuse	$\frac{21}{100}$
Enough money to pay bills	$\frac{1}{2}$
Enough time to spend with family	$\frac{23}{50}$
Job Market	$\frac{9}{25}$
Quality of children's education	$\frac{7}{20}$
Staying in good health	$\frac{2}{5}$

Source: Gallup Family Values Study Sponsored by the Church of Jesus Christ of Latter-Day Saints

Wilt Chamberlain

Production of Maple Syrup by the Top Six Producer States	
State	Maple Syrup Produced (in thousands of gallons)
Maine	175
Michigan	75
New York	200
Ohio	100
Vermont	375
Wisconsin	75

Source: National Agricultural Statistics Services

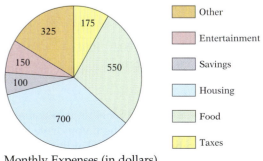

Other
Entertainment
Savings
Housing
Food
Taxes

Monthly Expenses (in dollars)

Copyright © Houghton Mifflin Company. All rights reserved.

Critical Thinking

136. Is the expression $x < \frac{4}{9}$ true when $x = \frac{3}{8}$? Is it true when $x = \frac{5}{12}$? Is it true for any negative number?

137. *Geometry* In Figure A, there are 2 rows of 5 squares, and in Figure B there are 3 rows of 7 squares. A diagonal line is drawn through each figure as shown, and the number of squares crossed by the diagonal is counted. Experiment with other arrangements of squares and develop a rule that will allow you to determine the number of squares crossed by the diagonal for m rows of n squares, where the GCF of m and n is 1.

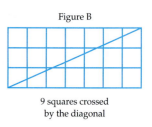

Figure A

6 squares crossed
by the diagonal

Figure B

9 squares crossed
by the diagonal

138. *Geography* What fraction of the states in the United States begin with the letter A?

139. *Business* The following circle graphs represent the fractional part of a computer company's total annual sales for different areas of the United States. (a) According to the graphs, in which year were sales in the Northeast a greater fraction of total sales, 2005 or 2006? (b) In which year were sales in the Northwest a greater fraction of total sales, 2005 or 2006?

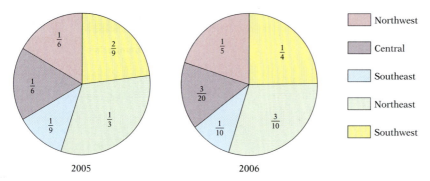

2005   2006

Computer Sales, Contribution by Region

140. **a.** On the number line, what fraction is halfway between $\frac{2}{a}$ and $\frac{4}{a}$?

    b. Find two fractions evenly spaced between $\frac{5}{b}$ and $\frac{8}{b}$.

141. Simplify $\frac{2}{1003} \cdot \frac{4}{1002} \cdot \frac{6}{1001} \cdot \cdots \cdot \frac{2002}{3} \cdot \frac{2004}{2} \cdot \frac{2006}{1}$. Write the answer in exponential form.

142. It is now 1:15 P.M. Is it possible for you to arrive in a city for a 4:00 P.M. meeting if you drive 55 mph and the city is 140 mi away? Explain how you arrived at the answer.

Copyright © Houghton Mifflin Company. All rights reserved.

Copyright © Houghton Mifflin Company. All rights reserved.

SECTION 3.3 ▶ ## Addition and Subtraction of Fractions

OBJECTIVE **A**

VIDEO & DVD CD TUTOR WEB SSM

Addition of fractions

Suppose you and a friend order a pizza. The pizza has been cut into 8 equal pieces. If you eat 3 pieces of the pizza and your friend eats 2 pieces, then together you have eaten $\frac{5}{8}$ of the pizza.

Note that in adding the fractions $\frac{3}{8}$ and $\frac{2}{8}$, the numerators are added and the denominator remains the same.

$$\frac{3}{8} + \frac{2}{8} = \frac{3+2}{8}$$

$$= \frac{5}{8}$$

Addition of Fractions

To add fractions with the same denominator, add the numerators and place the sum over the common denominator.

$$\frac{a}{b} + \frac{c}{b} = \frac{a+c}{b}, \text{ where } b \neq 0$$

Add: $\frac{5}{16} + \frac{7}{16}$

The denominators are the same. Add the numerators and place the sum over the common denominator.

$$\frac{5}{16} + \frac{7}{16} = \frac{5+7}{16}$$

Write the answer in simplest form.

$$= \frac{12}{16} = \frac{3}{4}$$

Add: $\frac{4}{x} + \frac{8}{x}$

The denominators are the same. Add the numerators and place the sum over the common denominator.

$$\frac{4}{x} + \frac{8}{x} = \frac{4+8}{x}$$

$$= \frac{12}{x}$$

Before two fractions can be added, the fractions must have the same denominator. To add fractions with different denominators, first rewrite the fractions as equivalent fractions with a common denominator. The common denominator is the least common multiple (LCM) of the denominators of the fractions. The LCM of denominators is sometimes called the least common denominator (LCD).

Copyright © Houghton Mifflin Company. All rights reserved.

Calculator Note

Some scientific calculators have a fraction key, $a^{b/c}$. It is used to perform operations on fractions. To use this key to simplify the expression at the right, enter

$\underbrace{5\ a^{b/c}\ 6}_{\frac{5}{6}}\ +\ \underbrace{3\ a^{b/c}\ 8}_{\frac{3}{8}}\ =$

Find the sum of $\dfrac{5}{6}$ and $\dfrac{3}{8}$.

The common denominator is the LCM of 6 and 8.

The LCM of 6 and 8 is 24.

Write the fractions as equivalent fractions with the common denominator.

$$\frac{5}{6} + \frac{3}{8} = \frac{20}{24} + \frac{9}{24}$$

Add the fractions.

$$= \frac{20 + 9}{24}$$

$$= \frac{29}{24} = 1\frac{5}{24}$$

During a recent year, over 42 million Americans changed homes. Figure 3.1 shows what fractions of the people moved within the same county, moved to a different county in the same state, and moved to a different state. What fractional part of those who changed homes moved outside the county they had been living in?

Add the fraction of the people who moved to a different county in the same state and the fraction who moved to a different state.

$$\frac{4}{21} + \frac{1}{7} = \frac{4}{21} + \frac{3}{21} = \frac{7}{21} = \frac{1}{3}$$

$\dfrac{1}{3}$ of the Americans who changed homes moved outside of the county they had been living in.

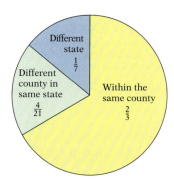

Figure 3.1 Where Americans Moved

Source: Census Bureau; *Geographical Mobility*

To add a fraction with a negative sign, rewrite the fraction with the negative sign in the numerator. Then add the numerators and place the sum over the common denominator.

Take Note

Although the sum could have been left as $\dfrac{-1}{12}$, all answers in this text are written with the negative sign in front of the fraction.

Add: $-\dfrac{5}{6} + \dfrac{3}{4}$

The common denominator is the LCM of 4 and 6.

The LCM of 4 and 6 is 12.

Rewrite with the negative sign in the numerator.

$$-\frac{5}{6} + \frac{3}{4} = \frac{-5}{6} + \frac{3}{4}$$

Rewrite each fraction in terms of the common denominator.

$$= \frac{-10}{12} + \frac{9}{12}$$

Add the fractions.

$$= \frac{-10 + 9}{12}$$

Simplify the numerator and write the negative sign in front of the fraction.

$$= \frac{-1}{12} = -\frac{1}{12}$$

Add: $-\dfrac{2}{3} + \left(-\dfrac{4}{5}\right)$

Rewrite each negative fraction with the negative sign in the numerator.

$$-\frac{2}{3} + \left(-\frac{4}{5}\right) = \frac{-2}{3} + \frac{-4}{5}$$

Rewrite each fraction as an equivalent fraction using the LCM as the denominator.

$$= \frac{-10}{15} + \frac{-12}{15}$$

Add the fractions.

$$= \frac{-10 + (-12)}{15}$$

$$= \frac{-22}{15} = -1\frac{7}{15}$$

Is $-\dfrac{2}{3}$ a solution of the equation $\dfrac{3}{4} + y = -\dfrac{1}{12}$?

$$\frac{3}{4} + y = -\frac{1}{12}$$

Replace y by $-\dfrac{2}{3}$. Then simplify.

$$\frac{3}{4} + \left(-\frac{2}{3}\right) \;\middle|\; -\frac{1}{12}$$

The common denominator is 12.

$$\frac{9}{12} + \left(\frac{-8}{12}\right) \;\middle|\; -\frac{1}{12}$$

$$\frac{9 + (-8)}{12} \;\middle|\; -\frac{1}{12}$$

The results are not equal.

$$\frac{1}{12} \ne -\frac{1}{12}$$

No, $-\dfrac{2}{3}$ is not a solution of the equation.

The mixed number $2\dfrac{1}{2}$ is the sum of 2 and $\dfrac{1}{2}$.

$$2\frac{1}{2} = 2 + \frac{1}{2}$$

Therefore, the sum of a whole number and a fraction is a mixed number.

$$2 + \frac{1}{2} = 2\frac{1}{2}$$

$$3 + \frac{4}{5} = 3\frac{4}{5}$$

$$8 + \frac{7}{9} = 8\frac{7}{9}$$

The sum of a whole number and a mixed number is a mixed number.

Add: $5 + 4\dfrac{2}{7}$

Add the whole numbers (5 and 4). $5 + 4\dfrac{2}{7} = 9\dfrac{2}{7}$

Write the fraction.

Take Note

$$5 + 4\frac{2}{7} = 5 + \left(4 + \frac{2}{7}\right)$$

$$= (5 + 4) + \frac{2}{7}$$

$$= 9 + \frac{2}{7} = 9\frac{2}{7}$$

Copyright © Houghton Mifflin Company. All rights reserved.

Copyright © Houghton Mifflin Company. All rights reserved.

To add two mixed numbers, first write the fractional parts as equivalent fractions with a common denominator. Then add the fractional parts and add the whole numbers.

Calculator Note

Use the fraction key on a calculator to enter mixed numbers. For the example at the right, enter

$\underbrace{3 \; a^{b}/c \; 5 \; a^{b}/c \; 8 \; +}_{3\frac{5}{8}}$

$\underbrace{4 \; a^{b}/c \; 7 \; a^{b}/c \; 12 \; =}_{4\frac{7}{12}}$

Add: $3\frac{5}{8} + 4\frac{7}{12}$

Write the fractions as equivalent fractions with a common denominator. The common denominator is the LCM of 8 and 12 (24).

$$3\frac{5}{8} + 4\frac{7}{12} = 3\frac{15}{24} + 4\frac{14}{24}$$

Add the fractional parts and add the whole numbers.

$$= 7\frac{29}{24}$$

Write the sum in simplest form.

$$= 7 + \frac{29}{24}$$

$$= 7 + 1\frac{5}{24}$$

$$= 8\frac{5}{24}$$

Evaluate $x + y$ when $x = 2\frac{3}{4}$ and $y = 7\frac{5}{6}$.

$$x + y$$

Replace x with $2\frac{3}{4}$ and y with $7\frac{5}{6}$.

$$2\frac{3}{4} + 7\frac{5}{6}$$

Write the fractions as equivalent fractions with a common denominator.

$$= 2\frac{9}{12} + 7\frac{10}{12}$$

Add the fractional parts and add the whole numbers.

$$= 9\frac{19}{12}$$

Write the sum in simplest form.

$$= 10\frac{7}{12}$$

 Example Add: $\frac{9}{16} + \frac{5}{12}$

Solution $\frac{9}{16} + \frac{5}{12} = \frac{27}{48} + \frac{20}{48}$

$$= \frac{27 + 20}{48} = \frac{47}{48}$$

 You Try It Add: $\frac{7}{12} + \frac{3}{8}$

Your Solution

Solution on p. S7

2 *Example*

Add: $\frac{4}{5} + \frac{3}{4} + \frac{5}{8}$

Solution

$\frac{4}{5} + \frac{3}{4} + \frac{5}{8} = \frac{32}{40} + \frac{30}{40} + \frac{25}{40} = \frac{87}{40} = 2\frac{7}{40}$

2 *You Try It*

Add: $\frac{3}{5} + \frac{2}{3} + \frac{5}{6}$

Your Solution

3 *Example*

Find the sum of $12\frac{4}{7}$ and 19.

Solution

$12\frac{4}{7} + 19 = 31\frac{4}{7}$

3 *You Try It*

What is the sum of 16 and $8\frac{5}{9}$?

Your Solution

4 *Example*

Add: $-\frac{3}{8} + \frac{3}{4} + \left(-\frac{5}{6}\right)$

Solution

$-\frac{3}{8} + \frac{3}{4} + \left(-\frac{5}{6}\right) = \frac{-3}{8} + \frac{3}{4} + \frac{-5}{6}$

$= \frac{-9}{24} + \frac{18}{24} + \frac{-20}{24}$

$= \frac{-9 + 18 + (-20)}{24}$

$= \frac{-11}{24} = -\frac{11}{24}$

4 *You Try It*

Add: $-\frac{5}{12} + \frac{5}{8} + \left(-\frac{1}{6}\right)$

Your Solution

5 *Example*

Evaluate $x + y + z$ when $x = 2\frac{1}{6}$, $y = 4\frac{3}{8}$, and $z = 7\frac{5}{9}$.

Solution

$x + y + z$

$2\frac{1}{6} + 4\frac{3}{8} + 7\frac{5}{9} = 2\frac{12}{72} + 4\frac{27}{72} + 7\frac{40}{72}$

$= 13\frac{79}{72}$

$= 14\frac{7}{72}$

5 *You Try It*

Evaluate $x + y + z$ when $x = 3\frac{5}{6}$, $y = 2\frac{1}{9}$, and $z = 5\frac{5}{12}$.

Your Solution

Solutions on p. S7

Copyright © Houghton Mifflin Company. All rights reserved.

OBJECTIVE **B**

Subtraction of fractions

► Point of Interest

The first woman mathematician for whom documented evidence exists is Hypatia (370–415). She lived in Alexandria, Egypt, and lectured at the Museum, the forerunner of our modern university. She made important contributions in mathematics, astronomy, and philosophy.

In the last objective, it was stated that in order for fractions to be added, the fractions must have the same denominator. The same is true for subtracting fractions: The two fractions must have the same denominator.

> ### Subtraction of Fractions
>
> To subtract fractions with the same denominator, subtract the numerators and place the difference over the common denominator.
>
> $$\frac{a}{b} - \frac{c}{b} = \frac{a-c}{b}, \qquad \text{where} \quad b \neq 0$$

Subtract: $\dfrac{5}{8} - \dfrac{3}{8}$

The denominators are the same. Subtract the numerators and place the difference over the common denominator.

$$\frac{5}{8} - \frac{3}{8} = \frac{5-3}{8}$$

Write the answer in simplest form.

$$= \frac{2}{8} = \frac{1}{4}$$

To subtract fractions with different denominators, first rewrite the fractions as equivalent fractions with a common denominator. The common denominator is the least common multiple (LCM) of the denominators of the fractions.

Subtract: $\dfrac{5}{12} - \dfrac{3}{8}$

The common denominator is the LCM of 12 and 8.

The LCM of 12 and 8 is 24.

Write the fractions as equivalent fractions with the common denominator.

$$\frac{5}{12} - \frac{3}{8} = \frac{10}{24} - \frac{9}{24}$$

Subtract the fractions.

$$= \frac{10-9}{24} = \frac{1}{24}$$

To subtract fractions with negative signs, first rewrite the fractions with the negative signs in the numerators.

Simplify: $-\dfrac{2}{9} - \dfrac{5}{12}$

Rewrite the negative fraction with the negative sign in the numerator.

$$-\frac{2}{9} - \frac{5}{12} = \frac{-2}{9} - \frac{5}{12}$$

Write the fractions as equivalent fractions with a common denominator.

$$= \frac{-8}{36} - \frac{15}{36}$$

Subtract the numerators and place the difference over the common denominator.

$$= \frac{-8-15}{36} = \frac{-23}{36}$$

Write the negative sign in front of the fraction.

$$= -\frac{23}{36}$$

Copyright © Houghton Mifflin Company. All rights reserved.

Subtract: $\dfrac{2}{3} - \left(-\dfrac{4}{5}\right)$

Rewrite subtraction as addition of the opposite.

$$\dfrac{2}{3} - \left(-\dfrac{4}{5}\right) = \dfrac{2}{3} + \dfrac{4}{5}$$

Write the fractions as equivalent fractions with a common denominator.

$$= \dfrac{10}{15} + \dfrac{12}{15}$$

Add the fractions.

$$= \dfrac{10 + 12}{15}$$

$$= \dfrac{22}{15} = 1\dfrac{7}{15}$$

To subtract mixed numbers when borrowing is not necessary, subtract the fractional parts and then subtract the whole numbers.

Find the difference between $5\dfrac{8}{9}$ and $2\dfrac{5}{6}$.

The LCM of 9 and 6 is 18.

Write the fractions as equivalent fractions with the LCM as the common denominator.

$$5\dfrac{8}{9} - 2\dfrac{5}{6} = 5\dfrac{16}{18} - 2\dfrac{15}{18}$$

Subtract the fractional parts and subtract the whole numbers.

$$= 3\dfrac{1}{18}$$

As in subtraction with whole numbers, subtraction of mixed numbers may involve borrowing.

Subtract: $7 - 4\dfrac{2}{3}$

Borrow 1 from 7. Write the 1 as a fraction with the same denominator as is in the fractional part of the mixed number (3).

$$7 - 4\dfrac{2}{3} = 6\dfrac{3}{3} - 4\dfrac{2}{3}$$

Note: $7 = 6 + 1 = 6 + \dfrac{3}{3} = 6\dfrac{3}{3}$

Subtract the fractional parts and subtract the whole numbers.

$$= 2\dfrac{1}{3}$$

Subtract: $9\dfrac{1}{8} - 2\dfrac{5}{6}$

Write the fractions as equivalent fractions with a common denominator.

$$9\dfrac{1}{8} - 2\dfrac{5}{6} = 9\dfrac{3}{24} - 2\dfrac{20}{24}$$

$3 < 20$. Borrow 1 from 9. Add the 1 to $\dfrac{3}{24}$.

Note: $9\dfrac{3}{24} = 9 + \dfrac{3}{24} = 8 + 1 + \dfrac{3}{24}$

$$= 8 + \dfrac{24}{24} + \dfrac{3}{24} = 8 + \dfrac{27}{24} = 8\dfrac{27}{24}$$

$$= 8\dfrac{27}{24} - 2\dfrac{20}{24}$$

Subtract.

$$= 6\dfrac{7}{24}$$

Copyright © Houghton Mifflin Company. All rights reserved.

Evaluate $x - y$ when $x = 7\dfrac{2}{9}$ and $y = 3\dfrac{5}{12}$.

$$x - y$$

Replace x with $7\dfrac{2}{9}$ and y with $3\dfrac{5}{12}$.

$$7\dfrac{2}{9} - 3\dfrac{5}{12}$$

Write the fractions as equivalent fractions with a common denominator.

$$= 7\dfrac{8}{36} - 3\dfrac{15}{36}$$

$8 < 15$. Borrow 1 from 7. Add the 1 to $\dfrac{8}{36}$.

Note: $7\dfrac{8}{36} = 6 + \dfrac{36}{36} + \dfrac{8}{36} = 6\dfrac{44}{36}$

$$= 6\dfrac{44}{36} - 3\dfrac{15}{36}$$

Subtract.

$$= 3\dfrac{29}{36}$$

---

6 **Example**

Subtract: $-\dfrac{5}{6} - \left(-\dfrac{3}{8}\right)$

Solution

$-\dfrac{5}{6} - \left(-\dfrac{3}{8}\right) = -\dfrac{5}{6} + \dfrac{3}{8} = \dfrac{-20}{24} + \dfrac{9}{24}$

$\qquad = \dfrac{-20 + 9}{24}$

$\qquad = \dfrac{-11}{24} = -\dfrac{11}{24}$

6 **You Try It**

Subtract: $-\dfrac{5}{6} - \dfrac{7}{9}$

Your Solution

---

7 **Example**

Find the difference between $8\dfrac{5}{6}$ and $2\dfrac{3}{4}$.

Solution

$8\dfrac{5}{6} - 2\dfrac{3}{4} = 8\dfrac{10}{12} - 2\dfrac{9}{12} = 6\dfrac{1}{12}$

7 **You Try It**

Find the difference between $9\dfrac{7}{8}$ and $5\dfrac{2}{3}$.

Your Solution

---

8 **Example**

Subtract: $7 - 3\dfrac{5}{13}$

Solution

$7 - 3\dfrac{5}{13} = 6\dfrac{13}{13} - 3\dfrac{5}{13} = 3\dfrac{8}{13}$

8 **You Try It**

Subtract: $6 - 4\dfrac{2}{11}$

Your Solution

Solutions on p. S7

Copyright © Houghton Mifflin Company. All rights reserved.

9 *Example*

Is $\dfrac{3}{8}$ a solution of the equation $\dfrac{2}{3} = w - \dfrac{5}{6}$?

Solution

$$\dfrac{2}{3} = w - \dfrac{5}{6}$$

$$\begin{array}{c|c} \dfrac{2}{3} & \dfrac{3}{8} - \dfrac{5}{6} \\[2mm] \dfrac{2}{3} & \dfrac{9}{24} - \dfrac{20}{24} \\[2mm] \dfrac{2}{3} & \dfrac{-11}{24} \\[2mm] \dfrac{2}{3} \neq & -\dfrac{11}{24} \end{array}$$

No, $\dfrac{3}{8}$ is not a solution of the equation.

9 *You Try It*

Is $-\dfrac{1}{4}$ a solution of the equation $\dfrac{2}{3} - v = \dfrac{11}{12}$?

Your Solution

Solution on p. S7

OBJECTIVE

VIDEO & DVD   CD TUTOR   WEB   SSM

Applications and formulas

10 *Example*

 The length of a regulation NCAA football must be no less than $10\dfrac{7}{8}$ in. and no more than $11\dfrac{7}{16}$ in. What is the difference between the minimum and maximum lengths of an NCAA regulation football?

Strategy

To find the difference, subtract the minimum length $\left(10\dfrac{7}{8}\right)$ from the maximum length $\left(11\dfrac{7}{16}\right)$.

Solution

$$11\dfrac{7}{16} - 10\dfrac{7}{8} = 11\dfrac{7}{16} - 10\dfrac{14}{16} = 10\dfrac{23}{16} - 10\dfrac{14}{16} = \dfrac{9}{16}$$

The difference is $\dfrac{9}{16}$ in.

10 *You Try It*

 The Heller Research Group conducted a survey to determine favorite doughnut flavors. $\dfrac{2}{5}$ of the respondents named glazed doughnuts, $\dfrac{8}{25}$ named filled doughnuts, and $\dfrac{3}{20}$ named frosted doughnuts. What fraction of the respondents did not name glazed, filled, or frosted as their favorite type of doughnut?

Your Strategy

Your Solution

Solution on pp. S7–S8

Copyright © Houghton Mifflin Company. All rights reserved.

11 *Example*

A chef has $1\frac{1}{2}$ c of granulated sugar and wants to make both of the recipes below. Does the chef have enough granulated sugar for both recipes?

Crème Brulée

$\frac{2}{3}$ c cream

$\frac{2}{3}$ c milk

$\frac{2}{3}$ c granulated sugar

1 vanilla bean

6 egg yokes

Chocolate Chip Cookies

$2\frac{1}{4}$ c flour

1 tsp salt

1 tsp baking soda

$\frac{3}{4}$ c granulated sugar

$\frac{3}{4}$ c brown sugar

8 oz chocolate chips

$\frac{1}{4}$ tsp vanilla

$\frac{1}{2}$ lb butter

2 eggs

Strategy

To determine if the chef has enough sugar:

▸ Add the number of cups of sugar from the crème brulée recipe $\left(\frac{2}{3}\right)$ to the number of cups of sugar from the chocolate chip cookies recipe $\left(\frac{3}{4}\right)$.

▸ Compare the sum to the number of cups of sugar the chef has $\left(1\frac{1}{2}\right)$.

Solution

$$\frac{2}{3} + \frac{3}{4} = \frac{8}{12} + \frac{9}{12} = \frac{17}{12} = 1\frac{5}{12}$$

$$1\frac{1}{2} = 1\frac{6}{12}$$

$$1\frac{6}{12} > 1\frac{5}{12}$$

The chef has enough sugar for both recipes.

11 *You Try It*

The dimensions of finished lumber are $\frac{1}{4}$ in. less in thickness and $\frac{1}{2}$ in. less in width than the given dimensions. For instance, a finished 2 by 4 is actually $1\frac{3}{4}$ in. thick and $3\frac{1}{2}$ in. wide. Nail sizes are measured in *pennys*. A 2-penny nail is 1 in. long. The length of the nail increases by $\frac{1}{4}$ in. for each 1-penny increase in size. For instance, a 3-penny nail is $1\frac{1}{4}$ in. long; a 4-penny nail is $1\frac{1}{2}$ in. long. What size penny nail is needed to nail four 1 by 6 pieces of lumber together so that the nail extends $\frac{1}{2}$ in. into the fourth board?

Your Strategy

Your Solution

Solution on p. S8

Copyright © Houghton Mifflin Company. All rights reserved.

3.3 EXERCISES

Copyright © Houghton Mifflin Company. All rights reserved.

OBJECTIVE **A**

Add.

1. $\dfrac{4}{11} + \dfrac{5}{11}$

2. $\dfrac{3}{7} + \dfrac{2}{7}$

3. $\dfrac{2}{3} + \dfrac{1}{3}$

4. $\dfrac{1}{2} + \dfrac{1}{2}$

5. $\dfrac{5}{6} + \dfrac{5}{6}$

6. $\dfrac{3}{8} + \dfrac{7}{8}$

7. $\dfrac{7}{18} + \dfrac{13}{18} + \dfrac{1}{18}$

8. $\dfrac{8}{15} + \dfrac{2}{15} + \dfrac{11}{15}$

9. $\dfrac{7}{b} + \dfrac{9}{b}$

10. $\dfrac{3}{y} + \dfrac{6}{y}$

11. $\dfrac{5}{c} + \dfrac{4}{c}$

12. $\dfrac{2}{a} + \dfrac{8}{a}$

13. $\dfrac{1}{x} + \dfrac{4}{x} + \dfrac{6}{x}$

14. $\dfrac{8}{n} + \dfrac{5}{n} + \dfrac{3}{n}$

15. $\dfrac{1}{4} + \dfrac{2}{3}$

16. $\dfrac{2}{3} + \dfrac{1}{2}$

17. $\dfrac{7}{15} + \dfrac{9}{20}$

18. $\dfrac{4}{9} + \dfrac{1}{6}$

19. $\dfrac{2}{3} + \dfrac{1}{12} + \dfrac{5}{6}$

20. $\dfrac{3}{8} + \dfrac{1}{2} + \dfrac{5}{12}$

21. $\dfrac{7}{12} + \dfrac{3}{4} + \dfrac{4}{5}$

22. $\dfrac{7}{11} + \dfrac{1}{2} + \dfrac{5}{6}$

23. $-\dfrac{3}{4} + \dfrac{2}{3}$

24. $-\dfrac{7}{12} + \dfrac{5}{8}$

25. $\dfrac{2}{5} + \left(-\dfrac{11}{15}\right)$

26. $\dfrac{1}{4} + \left(-\dfrac{1}{7}\right)$

27. $\dfrac{3}{8} + \left(-\dfrac{1}{2}\right) + \dfrac{7}{12}$

28. $-\dfrac{7}{12} + \dfrac{2}{3} + \left(-\dfrac{4}{5}\right)$

29. $\dfrac{2}{3} + \left(-\dfrac{5}{6}\right) + \dfrac{1}{4}$

30. $-\dfrac{5}{8} + \dfrac{3}{4} + \dfrac{1}{2}$

31. $8 + 7\dfrac{2}{3}$

32. $6 + 9\dfrac{3}{5}$

33. $2\dfrac{1}{6} + 3\dfrac{1}{2}$

34. $1\dfrac{3}{10} + 4\dfrac{3}{5}$

35. $8\dfrac{3}{5} + 6\dfrac{9}{20}$

36. $7\dfrac{5}{12} + 3\dfrac{7}{9}$

37. $5\dfrac{5}{12} + 4\dfrac{7}{9}$

38. $2\dfrac{11}{12} + 3\dfrac{7}{15}$

39. $2\dfrac{1}{4} + 3\dfrac{1}{2} + 1\dfrac{2}{3}$

40. $1\dfrac{2}{3} + 2\dfrac{5}{6} + 4\dfrac{7}{9}$

Solve.

41. What is $-\dfrac{5}{6}$ added to $\dfrac{4}{9}$?

42. What is $\dfrac{7}{12}$ added to $-\dfrac{11}{16}$?

43. Find the total of $\dfrac{2}{7}$, $\dfrac{3}{14}$, and $\dfrac{1}{4}$.

44. Find the total of $\dfrac{1}{3}$, $\dfrac{5}{18}$, and $\dfrac{2}{9}$.

45. What is $-\dfrac{2}{3}$ more than $-\dfrac{5}{6}$?

46. What is $-\dfrac{7}{12}$ more than $-\dfrac{5}{9}$?

47. Find $3\dfrac{7}{12}$ plus $2\dfrac{5}{8}$.

48. Find $5\dfrac{4}{9}$ plus $6\dfrac{5}{6}$.

49. Find $\dfrac{7}{8}$ increased by $1\dfrac{1}{3}$.

50. Find the sum of $7\dfrac{11}{15}$, $2\dfrac{7}{10}$, and $5\dfrac{2}{5}$.

Evaluate the variable expression $x + y$ for the given values of x and y.

51. $x = \dfrac{3}{5}, y = \dfrac{4}{5}$

52. $x = \dfrac{5}{8}, y = \dfrac{3}{8}$

53. $x = \dfrac{2}{3}, y = -\dfrac{3}{4}$

54. $x = -\dfrac{3}{8}, y = \dfrac{2}{9}$

55. $x = \dfrac{5}{6}, y = \dfrac{8}{9}$

56. $x = \dfrac{3}{10}, y = -\dfrac{7}{15}$

57. $x = -\dfrac{5}{8}, y = -\dfrac{1}{6}$

58. $x = -\dfrac{3}{8}, y = -\dfrac{5}{6}$

Copyright © Houghton Mifflin Company. All rights reserved.

Evaluate the variable expression $x + y + z$ for the given values of x, y, and z.

59. $x = \dfrac{3}{8}, y = \dfrac{1}{4}, z = \dfrac{7}{12}$

60. $x = \dfrac{5}{6}, y = \dfrac{2}{3}, z = \dfrac{7}{24}$

61. $x = 1\dfrac{1}{2}, y = 3\dfrac{3}{4}, z = 6\dfrac{5}{12}$

62. $x = 7\dfrac{2}{3}, y = 2\dfrac{5}{6}, z = 5\dfrac{4}{9}$

63. $x = 4\dfrac{3}{5}, y = 8\dfrac{7}{10}, z = 1\dfrac{9}{20}$

64. $x = 2\dfrac{3}{14}, y = 5\dfrac{5}{7}, z = 3\dfrac{1}{2}$

65. Is $-\dfrac{3}{5}$ a solution of the equation $z + \dfrac{1}{4} = -\dfrac{7}{20}$?

66. Is $\dfrac{3}{8}$ a solution of the equation $\dfrac{3}{4} = t + \dfrac{3}{8}$?

67. Is $-\dfrac{5}{6}$ a solution of the equation $\dfrac{1}{4} + x = -\dfrac{7}{12}$?

68. Is $-\dfrac{4}{5}$ a solution of the equation $0 = q + \dfrac{4}{5}$?

 The figure at the right shows how the money borrowed on home equity loans is spent. Use this graph for Exercises 69 and 70.

69. *Loans*   What fractional part of the money borrowed on home equity loans is spent on debt consolidation and home improvement?

70. *Loans*   What fractional part of the money borrowed on home equity loans is spent on home improvement, cars, and tuition?

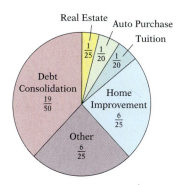

How Money Borrowed on Home Equity Loans Is Spent

Source: Consumer Bankers Association

OBJECTIVE   **B**

Subtract.

71. $\dfrac{7}{12} - \dfrac{5}{12}$

72. $\dfrac{17}{20} - \dfrac{9}{20}$

73. $\dfrac{11}{24} - \dfrac{7}{24}$

74. $\dfrac{39}{48} - \dfrac{23}{48}$

75. $\dfrac{8}{d} - \dfrac{3}{d}$

76. $\dfrac{12}{y} - \dfrac{7}{y}$

77. $\dfrac{5}{n} - \dfrac{10}{n}$

78. $\dfrac{6}{c} - \dfrac{13}{c}$

Copyright © Houghton Mifflin Company. All rights reserved.

79. $\dfrac{3}{7} - \dfrac{5}{14}$

80. $\dfrac{7}{8} - \dfrac{5}{16}$

81. $\dfrac{2}{3} - \dfrac{1}{6}$

82. $\dfrac{5}{21} - \dfrac{1}{6}$

83. $\dfrac{11}{12} - \dfrac{2}{3}$

84. $\dfrac{9}{20} - \dfrac{1}{30}$

85. $-\dfrac{1}{2} - \dfrac{3}{8}$

86. $-\dfrac{5}{6} - \dfrac{1}{9}$

87. $-\dfrac{3}{10} - \dfrac{4}{5}$

88. $-\dfrac{7}{15} - \dfrac{3}{10}$

89. $-\dfrac{5}{12} - \left(-\dfrac{2}{3}\right)$

90. $-\dfrac{3}{10} - \left(-\dfrac{5}{6}\right)$

91. $-\dfrac{5}{9} - \left(-\dfrac{11}{12}\right)$

92. $-\dfrac{5}{8} - \left(-\dfrac{7}{12}\right)$

93. $4\dfrac{11}{18} - 2\dfrac{5}{18}$

94. $3\dfrac{7}{12} - 1\dfrac{1}{12}$

95. $8\dfrac{3}{4} - 2$

96. $6\dfrac{5}{9} - 4$

97. $8\dfrac{5}{6} - 7\dfrac{3}{4}$

98. $5\dfrac{7}{8} - 3\dfrac{2}{3}$

99. $7 - 3\dfrac{5}{8}$

100. $6 - 2\dfrac{4}{5}$

101. $10 - 4\dfrac{8}{9}$

102. $5 - 2\dfrac{7}{18}$

103. $7\dfrac{3}{8} - 4\dfrac{5}{8}$

104. $11\dfrac{1}{6} - 8\dfrac{5}{6}$

105. $12\dfrac{5}{12} - 10\dfrac{17}{24}$

106. $16\dfrac{1}{3} - 11\dfrac{5}{12}$

107. $6\dfrac{2}{3} - 1\dfrac{7}{8}$

108. $7\dfrac{7}{12} - 2\dfrac{5}{6}$

109. $10\dfrac{2}{5} - 8\dfrac{7}{10}$

110. $5\dfrac{5}{6} - 4\dfrac{7}{8}$

Copyright © Houghton Mifflin Company. All rights reserved.

Solve.

111. What is $-\dfrac{7}{12}$ minus $\dfrac{7}{9}$?

112. What is $\dfrac{3}{5}$ decreased by $-\dfrac{7}{10}$?

113. What is $-\dfrac{2}{3}$ less than $-\dfrac{7}{8}$?

114. Find the difference between $-\dfrac{1}{6}$ and $-\dfrac{8}{9}$.

115. Find 8 less $1\dfrac{7}{12}$.

116. Find 9 minus $5\dfrac{3}{20}$.

Evaluate the variable expression $x - y$ for the given values of x and y.

117. $x = \dfrac{8}{9}, y = \dfrac{5}{9}$

118. $x = \dfrac{5}{6}, y = \dfrac{1}{6}$

119. $x = -\dfrac{11}{12}, y = \dfrac{5}{12}$

120. $x = -\dfrac{15}{16}, y = \dfrac{5}{16}$

121. $x = -\dfrac{2}{3}, y = -\dfrac{3}{4}$

122. $x = -\dfrac{5}{12}, y = -\dfrac{5}{9}$

123. $x = -\dfrac{3}{10}, y = -\dfrac{7}{15}$

124. $x = -\dfrac{5}{6}, y = -\dfrac{2}{15}$

125. $x = 5\dfrac{7}{9}, y = 4\dfrac{2}{3}$

126. $x = 9\dfrac{5}{8}, y = 2\dfrac{3}{16}$

127. $x = 7\dfrac{9}{10}, y = 3\dfrac{1}{2}$

128. $x = 6\dfrac{4}{9}, y = 1\dfrac{1}{6}$

129. $x = 5, y = 2\dfrac{7}{9}$

130. $x = 8, y = 4\dfrac{5}{6}$

131. $x = 10\dfrac{1}{2}, y = 5\dfrac{7}{12}$

132. $x = 9\dfrac{2}{15}, y = 6\dfrac{11}{15}$

133. Is $-\dfrac{3}{4}$ a solution of the equation $\dfrac{4}{5} = \dfrac{31}{20} - y$?

134. Is $\dfrac{5}{8}$ a solution of the equation $-\dfrac{1}{4} = x - \dfrac{7}{8}$?

135. Is $-\dfrac{3}{5}$ a solution of the equation $x - \dfrac{1}{4} = -\dfrac{17}{20}$?

136. Is $-\dfrac{2}{3}$ a solution of the equation $\dfrac{2}{3} - x = 0$?

Copyright © Houghton Mifflin Company. All rights reserved.

The figure at the right shows the sources of federal income. The category "Social Security, Medicare" includes unemployment and other retirement taxes. The category "Miscellaneous Taxes" includes excise, customs, estate, and gift taxes. Use this graph for Exercises 137 and 138.

137. *Taxes*   What is the difference between the fraction of federal income from personal income taxes and the fraction from corporate income taxes?

138. *Taxes*   What is the difference between the fraction of federal income from the Social Security category and the fraction from miscellaneous taxes?

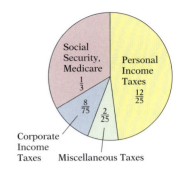

Sources of Federal Income
Source: Department of the Treasury

OBJECTIVE **C**

139. *Real Estate*   You purchased $3\frac{1}{4}$ acres of land and then sold $1\frac{1}{2}$ acres of the property. How many acres of the property do you own now?

140. *Carpentry*   A $2\frac{3}{4}$-foot piece is cut from a 6-foot board. Find the length of the remaining piece of board.

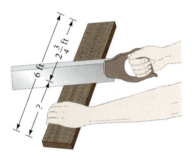

141. *Community Service*   You are required to contribute 20 h of community service to the town in which your college is located. After you have contributed $12\frac{1}{4}$ h, how many more hours of community service are still required of you?

142. *Horse Racing*   The 3-year-olds in the Kentucky Derby run $1\frac{1}{4}$ mi. The horses in the Belmont Stakes run $1\frac{1}{2}$ mi, and they run $1\frac{3}{16}$ mi in the Preakness Stakes. How much farther do the horses run in the Kentucky Derby than in the Preakness Stakes? How much farther do they run in the Belmont Stakes than in the Preakness Stakes?

143. *Sports*   A boxer is put on a diet to gain 15 lb in four weeks. The boxer gains $4\frac{1}{2}$ lb the first week and $3\frac{3}{4}$ lb the second week. How much weight must the boxer gain during the third and fourth weeks in order to gain a total of 15 lb?

144. *Construction*   A roofer and an apprentice are roofing a newly constructed house. In one day, the roofer completes $\frac{1}{3}$ of the job and the apprentice completes $\frac{1}{4}$ of the job. How much of the job remains to be done? Working at the same rate, can the roofer and the apprentice complete the job in one more day?

Copyright © Houghton Mifflin Company. All rights reserved.

145. *Sociology* The table at the right shows the results of a survey in which adults in the United States were asked how many evening meals they cook at home during an average week. (a) Which response was given most frequently? (b) What fraction of the adult population cooks two or fewer dinners at home per week? (c) What fraction of the adult population cooks five or more dinners at home per week? Is this less than half or more than half of the people?

Responses to the question, "How many evening meals do you cook at home each week?"	
0	$\frac{2}{25}$
1	$\frac{1}{20}$
2	$\frac{1}{10}$
3	$\frac{13}{100}$
4	$\frac{3}{20}$
5	$\frac{21}{100}$
6	$\frac{9}{100}$
7	$\frac{19}{100}$

Source: Millward Brown for Whirlpool

146. *Wages* A student worked $4\frac{1}{3}$ h, 5 h, and $3\frac{2}{3}$ h this week at a part-time job. The student is paid $9 an hour. How much did the student earn this week?

During the second half of the 1900s, greenskeepers mowed the grass on golf putting surfaces progressively lower. The table at the right shows the average grass height by decade. Use this table for Exercises 147 and 148.

Average Height of Grass on Golf Putting Surfaces	
Decade	Height (in inches)
1950s	$\frac{1}{4}$
1960s	$\frac{7}{32}$
1970s	$\frac{3}{16}$
1980s	$\frac{5}{32}$
1990s	$\frac{1}{8}$

Source: Golf Course Superintendents Association of America

147. *Sports* What was the difference between the average height of the grass in the 1980s and the 1950s?

148. *Sports* Calculate the difference between the average grass height in the 1970s and the 1960s.

149. *Geometry* You want to fence in the triangular plot of land shown at the right. How many feet of fencing do you need? Use the formula $P = a + b + c$.

150. *Geometry* The course of a yachting race is in the shape of a triangle with sides that measure $4\frac{3}{10}$, $3\frac{7}{10}$, and $2\frac{1}{2}$ mi. Find the total length of the course. Use the formula $P = a + b + c$.

151. *Geometry* A flower garden in the yard of an historical home is in the shape of a triangle as shown at the right. The wooden beams lining the edge of the garden need to be replaced. Find the total length of wood beams which must be purchased in order to replace the old beams. Use the formula $P = a + b + c$.

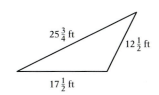

Copyright © Houghton Mifflin Company. All rights reserved.

The table at the right shows three U.S. Olympic gold medalists in the high jump during the 1900s. Use this table for Exercises 152 and 153.

Olympic Gold Medalists in the High Jump		
Year	Athlete	Height of Jump (in feet)
1920	Richard Landon	$6\frac{1}{3}$
1924	Harold Osborn	$6\frac{1}{2}$
1996	Charles Austin	$7\frac{5}{6}$

Source: The World Almanac and Book of Facts

152. *Sports* Find the difference between the height of Charles Austin's jump and the height of Richard Landon's jump.

153. *Sports* How much higher was Charles Austin's jump than Harold Osborn's jump?

154. *Demographics* Three-twentieths of the men in the United States are left-handed. (*Source:* Scripps Survey Research Center Poll) What fraction of the men in the United States are not left-handed?

Charles Austin

Critical Thinking

155. The figure at the right is divided into 5 parts. Is each part of the figure $\frac{1}{5}$ of the figure? Why or why not?

156. Draw a diagram that illustrates the addition of two fractions with the same denominator.

157. Use the diagram at the right to illustrate the sum of $\frac{1}{8}$ and $\frac{5}{6}$. Why does the figure contain 24 squares? Would it be possible to illustrate the sum of $\frac{1}{8}$ and $\frac{5}{6}$ if there were 48 squares in the figure? What if there were 16 squares? Make a list of the possible number of squares that could be used to illustrate the sum of $\frac{1}{8}$ and $\frac{5}{6}$.

158. A researcher completed a study of the ages of students at a college. The fraction of the total number of students enrolled in the college in various age groups was then recorded in the figure at the right. Are the results that are displayed in the circle graph possible? Explain your answer.

159. A local humane society reported that $\frac{3}{5}$ of the households in the city owned some type of pet. The report went on to say that $\frac{1}{6}$ of the households had a bird, $\frac{2}{5}$ had a dog, $\frac{3}{10}$ had a cat, and $\frac{1}{20}$ of the households had a different animal as a pet. The sum of $\frac{1}{6}$, $\frac{2}{5}$, $\frac{3}{10}$, and $\frac{1}{20}$ is $\frac{11}{12}$, which is more than $\frac{3}{5}$. Is this possible? Explain your answer.

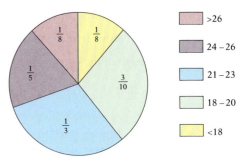

Age Distribution of Enrolled College Students

Copyright © Houghton Mifflin Company. All rights reserved.

Copyright © Houghton Mifflin Company. All rights reserved.

SECTION 3.4 ▶ ## Multiplication and Division of Fractions

OBJECTIVE **A**

Multiplication of fractions

To multiply two fractions, multiply the numerators and multiply the denominators.

> **Multiplication of Fractions**
>
> The product of two fractions is the product of the numerators over the product of the denominators.
>
> $$\frac{a}{b} \cdot \frac{c}{d} = \frac{ac}{bd},\qquad \text{where}\quad b \neq 0\quad \text{and}\quad d \neq 0$$

Note that fractions do not need to have the same denominator in order to be multiplied.

After multiplying two fractions, write the product in simplest form.

Multiply: $\frac{2}{5} \cdot \frac{1}{3}$

Multiply the numerators.
Multiply the denominators.
$$\frac{2}{5} \cdot \frac{1}{3} = \frac{2 \cdot 1}{5 \cdot 3} = \frac{2}{15}$$

The product $\frac{2}{5} \cdot \frac{1}{3}$ can be read "$\frac{2}{5}$ times $\frac{1}{3}$" or "$\frac{2}{5}$ of $\frac{1}{3}$."

Reading the times sign as "of" is useful in diagraming the product of two fractions.

$\frac{1}{3}$ of the bar at the right is shaded.

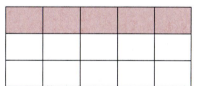

Shade $\frac{2}{5}$ of the $\frac{1}{3}$ already shaded.

$\frac{2}{15}$ of the bar is now shaded.

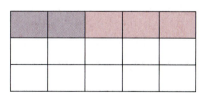

$$\frac{2}{5} \text{ of } \frac{1}{3} = \frac{2}{5} \cdot \frac{1}{3} = \frac{2}{15}$$

If a is a natural number, then $\frac{1}{a}$ is called the **reciprocal** or **multiplicative inverse** of a. Note that $a \cdot \frac{1}{a} = \frac{a}{1} \cdot \frac{1}{a} = \frac{a}{a} = 1$.

The product of a number and its multiplicative inverse is 1.

$$\frac{1}{8} \cdot 8 = 8 \cdot \frac{1}{8} = 1$$

Multiply: $\dfrac{3}{8} \cdot \dfrac{4}{9}$

Multiply the numerators.
Multiply the denominators.

$$\dfrac{3}{8} \cdot \dfrac{4}{9} = \dfrac{3 \cdot 4}{8 \cdot 9}$$

Express the fraction in simplest form by first writing the prime factorization of each number.

$$= \dfrac{3 \cdot 2 \cdot 2}{2 \cdot 2 \cdot 2 \cdot 3 \cdot 3}$$

Divide by the common factors and write the product in simplest form.

$$= \dfrac{1}{6}$$

The sign rules for multiplying positive and negative fractions are the same rules used to multiply integers.

The product of two numbers with the same sign is positive.
The product of two numbers with different signs is negative.

Multiply: $-\dfrac{3}{4} \cdot \dfrac{8}{15}$

The signs are different.
The product is negative.

$$-\dfrac{3}{4} \cdot \dfrac{8}{15} = -\left(\dfrac{3}{4} \cdot \dfrac{8}{15} \right)$$

Multiply the numerators.
Multiply the denominators.

$$= -\dfrac{3 \cdot 8}{4 \cdot 15}$$

Write the product in simplest form.

$$= -\dfrac{3 \cdot 2 \cdot 2 \cdot 2}{2 \cdot 2 \cdot 3 \cdot 5}$$

$$= -\dfrac{2}{5}$$

Multiply: $-\dfrac{3}{8}\left(-\dfrac{2}{5}\right)\left(-\dfrac{10}{21}\right)$

$$-\dfrac{3}{8}\left(-\dfrac{2}{5}\right)\left(-\dfrac{10}{21}\right)$$

Use the Order of Operations Agreement.
Multiply the first two fractions. The product is positive.

$$= \left(\dfrac{3}{8} \cdot \dfrac{2}{5} \right)\left(-\dfrac{10}{21} \right)$$

The product of the first two fractions and the third fraction is negative.

$$= -\left(\dfrac{3}{8} \cdot \dfrac{2}{5} \cdot \dfrac{10}{21} \right)$$

Multiply the numerators.
Multiply the denominators.

$$= -\dfrac{3 \cdot 2 \cdot 10}{8 \cdot 5 \cdot 21}$$

Write the product in simplest form.

$$= -\dfrac{3 \cdot 2 \cdot 2 \cdot 5}{2 \cdot 2 \cdot 2 \cdot 5 \cdot 3 \cdot 7}$$

$$= -\dfrac{1}{14}$$

Copyright © Houghton Mifflin Company. All rights reserved.

Thus, the product of three negative fractions is negative. We can modify the rule for multiplying positive and negative fractions to say that **the product of an odd number of negative fractions is negative and the product of an even number of negative fractions is positive.**

▶ **Point of Interest**

Try this: What is the result if you take one-third of a half-dozen and add to it one-fourth of the product of the result and 8?

To multiply a whole number by a fraction or a mixed number, first write the whole number as a fraction with a denominator of 1.

Multiply: $3 \cdot \dfrac{5}{8}$

Write the whole number 3 as the fraction $\dfrac{3}{1}$.

$$3 \cdot \dfrac{5}{8} = \dfrac{3}{1} \cdot \dfrac{5}{8}$$

Multiply the fractions. There are no common factors in the numerator and denominator.

$$= \dfrac{3 \cdot 5}{1 \cdot 8}$$

Write the improper fraction as a mixed number.

$$= \dfrac{15}{8} = 1\dfrac{7}{8}$$

Multiply: $\dfrac{x}{7} \cdot \dfrac{y}{5}$

Multiply the numerators.
Multiply the denominators.

$$\dfrac{x}{7} \cdot \dfrac{y}{5} = \dfrac{x \cdot y}{7 \cdot 5}$$

Write the product in simplest form.

$$= \dfrac{xy}{35}$$

When a factor is a mixed number, first write the mixed number as an improper fraction. Then multiply.

Find the product of $-4\dfrac{1}{6}$ and $2\dfrac{7}{10}$.

The signs are different.
The product is negative.

$$-4\dfrac{1}{6} \cdot 2\dfrac{7}{10} = -\left(4\dfrac{1}{6} \cdot 2\dfrac{7}{10}\right)$$

Write each mixed number as an improper fraction.

$$= -\left(\dfrac{25}{6} \cdot \dfrac{27}{10}\right)$$

Multiply the fractions.

$$= -\dfrac{25 \cdot 27}{6 \cdot 10}$$

$$= -\dfrac{5 \cdot 5 \cdot 3 \cdot 3 \cdot 3}{2 \cdot 3 \cdot 2 \cdot 5}$$

Write the product in simplest form.

$$= -\dfrac{45}{4} = -11\dfrac{1}{4}$$

Copyright © Houghton Mifflin Company. All rights reserved.

Is $-\frac{2}{3}$ a solution of the equation $\frac{3}{4}x = -\frac{1}{2}$?

Replace x by $-\frac{2}{3}$ and then simplify.

$$
\begin{array}{c|c}
\frac{3}{4}x = -\frac{1}{2} \\
\hline
\frac{3}{4}\left(-\frac{2}{3}\right) & -\frac{1}{2} \\
-\dfrac{3 \cdot 2}{4 \cdot 3} & -\dfrac{1}{2} \\
-\dfrac{3 \cdot 2}{2 \cdot 2 \cdot 3} & -\dfrac{1}{2}
\end{array}
$$

The results are equal.

$$-\frac{1}{2} = -\frac{1}{2}$$

Yes, $-\frac{2}{3}$ is a solution of the equation.

---

❶ Example Multiply: $\frac{7}{9} \cdot \frac{3}{14} \cdot \frac{2}{5}$

Solution $\dfrac{7}{9} \cdot \dfrac{3}{14} \cdot \dfrac{2}{5} = \dfrac{7 \cdot 3 \cdot 2}{9 \cdot 14 \cdot 5}$

$$= \frac{7 \cdot 3 \cdot 2}{3 \cdot 3 \cdot 2 \cdot 7 \cdot 5} = \frac{1}{15}$$

❶ You Try It Multiply: $\frac{5}{12} \cdot \frac{9}{35} \cdot \frac{7}{8}$

Your Solution

---

❷ Example Multiply: $\frac{6}{x} \cdot \frac{8}{y}$

Solution $\dfrac{6}{x} \cdot \dfrac{8}{y} = \dfrac{6 \cdot 8}{x \cdot y}$

$$= \frac{48}{xy}$$

❷ You Try It Multiply: $\frac{y}{10} \cdot \frac{z}{7}$

Your Solution

---

❸ Example Multiply: $-\frac{3}{4}\left(\frac{1}{2}\right)\left(-\frac{8}{9}\right)$

Solution $-\dfrac{3}{4}\left(\dfrac{1}{2}\right)\left(-\dfrac{8}{9}\right)$

$$= \frac{3}{4} \cdot \frac{1}{2} \cdot \frac{8}{9}$$

▶ The product of two negative fractions is positive.

$$= \frac{3 \cdot 1 \cdot 8}{4 \cdot 2 \cdot 9}$$

$$= \frac{3 \cdot 1 \cdot 2 \cdot 2 \cdot 2}{2 \cdot 2 \cdot 2 \cdot 3 \cdot 3} = \frac{1}{3}$$

❸ You Try It Multiply: $-\frac{1}{3}\left(-\frac{5}{12}\right)\left(\frac{8}{15}\right)$

Your Solution

Solutions on p. S8

Copyright © Houghton Mifflin Company. All rights reserved.

4 *Example*   What is the product of $\frac{7}{12}$ and 4?

4   *You Try It*   Find the product of $\frac{8}{9}$ and 6.

Solution

$$\frac{7}{12} \cdot 4 = \frac{7}{12} \cdot \frac{4}{1}$$

$$= \frac{7 \cdot 4}{12 \cdot 1}$$

$$= \frac{7 \cdot 2 \cdot 2}{2 \cdot 2 \cdot 3 \cdot 1}$$

$$= \frac{7}{3}$$

$$= 2\frac{1}{3}$$

Your Solution

5 *Example*   Multiply: $-7\frac{1}{2} \cdot 4\frac{2}{5}$

5   *You Try It*   Multiply: $3\frac{6}{7} \cdot 2\frac{4}{9}$

Solution

$$-7\frac{1}{2} \cdot 4\frac{2}{5} = -\left(\frac{15}{2} \cdot \frac{22}{5}\right)$$

$$= -\frac{15 \cdot 22}{2 \cdot 5}$$

$$= -\frac{3 \cdot 5 \cdot 2 \cdot 11}{2 \cdot 5}$$

$$= -\frac{33}{1} = -33$$

Your Solution

6 *Example*   Evaluate the variable expression xy when $x = 1\frac{4}{5}$ and $y = -\frac{5}{6}$.

6   *You Try It*   Evaluate the variable expression xy when $x = 5\frac{1}{8}$ and $y = \frac{2}{3}$.

Solution xy

$$1\frac{4}{5}\left(-\frac{5}{6}\right) = -\left(\frac{9}{5} \cdot \frac{5}{6}\right)$$

$$= -\frac{9 \cdot 5}{5 \cdot 6}$$

$$= -\frac{3 \cdot 3 \cdot 5}{5 \cdot 2 \cdot 3}$$

$$= -\frac{3}{2} = -1\frac{1}{2}$$

Your Solution

Solutions on p. S8

Copyright © Houghton Mifflin Company. All rights reserved.

Copyright © Houghton Mifflin Company. All rights reserved.

OBJECTIVE **B**

Division of fractions

The **reciprocal** of a fraction is that fraction with the numerator and denominator interchanged.

The reciprocal of $\dfrac{3}{4}$ is $\dfrac{4}{3}$.

The reciprocal of $\dfrac{a}{b}$ is $\dfrac{b}{a}$.

The process of interchanging the numerator and denominator of a fraction is called **inverting** the fraction.

To find the reciprocal of a whole number, first rewrite the whole number as a fraction with a denominator of 1. Then invert the fraction.

$$6 = \frac{6}{1}$$

The reciprocal of 6 is $\dfrac{1}{6}$.

Reciprocals are used to rewrite division problems as related multiplication problems. Look at the following two problems:

$$6 \div 2 = 3 \qquad\qquad 6 \cdot \frac{1}{2} = 3$$

6 divided by 2 equals 3.   6 times the reciprocal of 2 equals 3.

Division is defined as multiplication by the reciprocal. Therefore, "divided by 2" is the same as "times $\dfrac{1}{2}$." Fractions are divided by making this substitution.

Division of Fractions

To divide two fractions, multiply by the reciprocal of the divisor.

$$\frac{a}{b} \div \frac{c}{d} = \frac{a}{b} \cdot \frac{d}{c}, \qquad \text{where} \quad b \neq 0, \quad c \neq 0, \quad \text{and} \quad d \neq 0$$

Divide: $\dfrac{2}{5} \div \dfrac{3}{4}$

Rewrite the division as multiplication by the reciprocal.

$$\frac{2}{5} \div \frac{3}{4} = \frac{2}{5} \cdot \frac{4}{3}$$

Multiply the fractions.

$$= \frac{2 \cdot 4}{5 \cdot 3}$$

$$= \frac{2 \cdot 2 \cdot 2}{5 \cdot 3} = \frac{8}{15}$$

The sign rules for dividing positive and negative fractions are the same rules used to divide integers.

> **The quotient of two numbers with the same sign is positive.**
> **The quotient of two numbers with different signs is negative.**

▶ **Point of Interest**

Try this: What number when multiplied by its reciprocal is equal to 1?

Simplify: $-\dfrac{7}{10} \div \left(-\dfrac{14}{15}\right)$

The signs are the same.
The quotient is positive.

$$-\frac{7}{10} \div \left(-\frac{14}{15}\right) = \frac{7}{10} \div \frac{14}{15}$$

Rewrite the division as multiplication by the reciprocal.

$$= \frac{7}{10} \cdot \frac{15}{14}$$

Multiply the fractions.

$$= \frac{7 \cdot 15}{10 \cdot 14}$$

$$= \frac{7 \cdot 3 \cdot 5}{2 \cdot 5 \cdot 2 \cdot 7}$$

$$= \frac{3}{4}$$

To divide a fraction and a whole number, first write the whole number as a fraction with a denominator of 1.

Find the quotient of $\dfrac{3}{4}$ and 6.

Write the whole number 6 as the fraction $\dfrac{6}{1}$.

$$\frac{3}{4} \div 6 = \frac{3}{4} \div \frac{6}{1}$$

Rewrite the division as multiplication by the reciprocal.

$$= \frac{3}{4} \cdot \frac{1}{6}$$

Multiply the fractions.

$$= \frac{3 \cdot 1}{4 \cdot 6}$$

$$= \frac{3 \cdot 1}{2 \cdot 2 \cdot 2 \cdot 3}$$

$$= \frac{1}{8}$$

Take Note

$\dfrac{3}{4} \div 6 = \dfrac{1}{8}$ means that if $\dfrac{3}{4}$ is divided into 6 equal parts, each equal part is $\dfrac{1}{8}$. For example, if 6 people share $\dfrac{3}{4}$ of a pizza, each person eats $\dfrac{1}{8}$ of the pizza.

When a number in a quotient is a mixed number, first write the mixed number as an improper fraction. Then divide the fractions.

Divide: $\dfrac{2}{3} \div 1\dfrac{1}{4}$

Write the mixed number $1\dfrac{1}{4}$ as an improper fraction.

$$\frac{2}{3} \div 1\frac{1}{4} = \frac{2}{3} \div \frac{5}{4}$$

Rewrite the division as multiplication by the reciprocal.

$$= \frac{2}{3} \cdot \frac{4}{5}$$

Multiply the fractions.

$$= \frac{2 \cdot 4}{3 \cdot 5} = \frac{8}{15}$$

Copyright © Houghton Mifflin Company. All rights reserved.

7 *Example*   Divide: $\dfrac{4}{5} \div \dfrac{8}{15}$

Solution $\dfrac{4}{5} \div \dfrac{8}{15} = \dfrac{4}{5} \cdot \dfrac{15}{8}$

$= \dfrac{4 \cdot 15}{5 \cdot 8}$

$= \dfrac{2 \cdot 2 \cdot 3 \cdot 5}{5 \cdot 2 \cdot 2 \cdot 2}$

$= \dfrac{3}{2} = 1\dfrac{1}{2}$

7 *You Try It*   Divide: $\dfrac{5}{6} \div \dfrac{10}{27}$

Your Solution

8 *Example*   Divide: $\dfrac{x}{2} \div \dfrac{y}{4}$

Solution $\dfrac{x}{2} \div \dfrac{y}{4} = \dfrac{x}{2} \cdot \dfrac{4}{y}$

$= \dfrac{x \cdot 4}{2 \cdot y}$

$= \dfrac{x \cdot 2 \cdot 2}{2 \cdot y} = \dfrac{2x}{y}$

8 *You Try It*   Divide: $\dfrac{x}{8} \div \dfrac{y}{6}$

Your Solution

9 *Example*   What is the quotient of 6 and $-\dfrac{3}{5}$?

Solution $6 \div \left(-\dfrac{3}{5}\right) = -\left(\dfrac{6}{1} \div \dfrac{3}{5}\right)$

$= -\left(\dfrac{6}{1} \cdot \dfrac{5}{3}\right)$

$= -\dfrac{6 \cdot 5}{1 \cdot 3}$

$= -\dfrac{2 \cdot 3 \cdot 5}{1 \cdot 3}$

$= -\dfrac{10}{1} = -10$

9 *You Try It*   Find the quotient of 4 and $-\dfrac{6}{7}$.

Your Solution

Solutions on pp. S8–S9

Copyright © Houghton Mifflin Company. All rights reserved.

10 *Example*   Divide: $3\frac{4}{15} \div 2\frac{1}{10}$

Solution $3\frac{4}{15} \div 2\frac{1}{10} = \frac{49}{15} \div \frac{21}{10}$

$$= \frac{49}{15} \cdot \frac{10}{21}$$

$$= \frac{49 \cdot 10}{15 \cdot 21}$$

$$= \frac{7 \cdot 7 \cdot 2 \cdot 5}{3 \cdot 5 \cdot 3 \cdot 7}$$

$$= \frac{14}{9} = 1\frac{5}{9}$$

10 *You Try It*   Divide: $4\frac{3}{8} \div 3\frac{1}{2}$

Your Solution

11 *Example*   Evaluate $x \div y$ when $x = 3\frac{1}{8}$ and $y = 5$.

Solution $x \div y$

$$3\frac{1}{8} \div 5 = \frac{25}{8} \div \frac{5}{1}$$

$$= \frac{25}{8} \cdot \frac{1}{5}$$

$$= \frac{25 \cdot 1}{8 \cdot 5}$$

$$= \frac{5 \cdot 5 \cdot 1}{2 \cdot 2 \cdot 2 \cdot 5} = \frac{5}{8}$$

11 *You Try It*   Evaluate $x \div y$ when $x = 2\frac{1}{4}$ and $y = 9$.

Your Solution

Solutions on p. S9

OBJECTIVE **C**

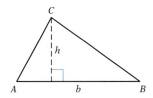

Applications and formulas

Figure *ABC* is a triangle. *AB* is the **base,** *b*, of the triangle. The line segment from *C* that forms a right angle with the base is the **height,** *h*, of the triangle. The formula for the area of a triangle is given below. Use this formula for Example 12 and You Try It 12.

Area of Triangle

The formula for the area of a triangle is $A = \dfrac{1}{2}bh$, where A is the area of the triangle, b is the base, and h is the height.

Copyright © Houghton Mifflin Company. All rights reserved.

12 Example

A riveter uses metal plates that are in the shape of a triangle and have a base of 12 cm and a height of 6 cm. Find the area of one metal plate.

Strategy

To find the area, use the formula for the area of a triangle, $A = \frac{1}{2}bh$. $b = 12$ and $h = 6$.

Solution

$A = \dfrac{1}{2}bh$

$A = \dfrac{1}{2}(12)(6)$

$A = 36$

The area is 36 cm².

6 cm

12 cm

12 You Try It

Find the amount of felt needed to make a banner that is in the shape of a triangle with a base of 18 in. and a height of 9 in.

Your Strategy

Your Solution

13 Example

A 12-foot board is cut into pieces $2\frac{1}{2}$ ft long for use as bookshelves. What is the length of the remaining piece after as many shelves as possible are cut?

Strategy

To find the length of the remaining piece:

▶ Divide the total length (12) by the length of each shelf $\left(2\frac{1}{2}\right)$. The quotient is the number of shelves cut, with a certain fraction of a shelf left over.
▶ Multiply the fraction left over by the length of a shelf.

Solution

$12 \div 2\dfrac{1}{2} = \dfrac{12}{1} \div \dfrac{5}{2} = \dfrac{12}{1} \cdot \dfrac{2}{5} = \dfrac{12 \cdot 2}{1 \cdot 5} = \dfrac{24}{5} = 4\dfrac{4}{5}$

4 shelves, each $2\frac{1}{2}$ ft long, can be cut from the board. The piece remaining is $\frac{4}{5}$ of $2\frac{1}{2}$ ft long.

$\dfrac{4}{5} \cdot 2\dfrac{1}{2} = \dfrac{4}{5} \cdot \dfrac{5}{2} = \dfrac{4 \cdot 5}{5 \cdot 2} = 2$

The length of the remaining piece is 2 ft.

13 You Try It

The Booster Club is making 22 sashes for the high school band members. Each sash requires $1\frac{3}{8}$ yd of material at a cost of $12 per yard. Find the total cost of the material.

Your Strategy

Your Solution

Copyright © Houghton Mifflin Company. All rights reserved.

3.4 EXERCISES

OBJECTIVE   **A**

1. 🖊 Explain why you need a common denominator when adding or subtracting two fractions and why you don't need a common denominator when multiplying or dividing two fractions.

2. 🖊 The product of 1 and a number is $\dfrac{3}{8}$. Find the number. Explain how you arrived at the answer.

Multiply.

3. $\dfrac{2}{3} \cdot \dfrac{9}{10}$

4. $\dfrac{3}{8} \cdot \dfrac{4}{5}$

5. $-\dfrac{6}{7} \cdot \dfrac{11}{12}$

6. $\dfrac{5}{6} \cdot \left(-\dfrac{2}{5}\right)$

7. $\dfrac{14}{15} \cdot \dfrac{6}{7}$

8. $\dfrac{15}{16} \cdot \dfrac{4}{9}$

9. $-\dfrac{6}{7} \cdot \dfrac{0}{10}$

10. $\dfrac{5}{12} \cdot \dfrac{3}{0}$

11. $\left(-\dfrac{4}{15}\right) \cdot \left(-\dfrac{3}{8}\right)$

12. $\left(-\dfrac{3}{4}\right) \cdot \left(-\dfrac{2}{9}\right)$

13. $-\dfrac{3}{4} \cdot \dfrac{1}{2}$

14. $-\dfrac{8}{15} \cdot \dfrac{5}{12}$

15. $\dfrac{9}{x} \cdot \dfrac{7}{y}$

16. $\dfrac{4}{c} \cdot \dfrac{8}{d}$

17. $-\dfrac{y}{5} \cdot \dfrac{z}{6}$

18. $-\dfrac{a}{10} \cdot \left(-\dfrac{b}{6}\right)$

19. $\dfrac{2}{3} \cdot \dfrac{3}{8} \cdot \dfrac{4}{9}$

20. $\dfrac{5}{7} \cdot \dfrac{1}{6} \cdot \dfrac{14}{15}$

21. $-\dfrac{7}{12} \cdot \dfrac{5}{8} \cdot \dfrac{16}{25}$

22. $\dfrac{5}{12} \cdot \left(-\dfrac{1}{3}\right) \cdot \left(-\dfrac{8}{15}\right)$

23. $\left(-\dfrac{3}{5}\right) \cdot \dfrac{1}{2} \cdot \left(-\dfrac{5}{8}\right)$

24. $\dfrac{5}{6} \cdot \left(-\dfrac{2}{3}\right) \cdot \dfrac{3}{25}$

25. $6 \cdot \dfrac{1}{6}$

26. $\dfrac{1}{10} \cdot 10$

27. $\dfrac{3}{4} \cdot 8$

28. $\dfrac{5}{7} \cdot 14$

29. $12 \cdot \left(-\dfrac{5}{8}\right)$

30. $24 \cdot \left(-\dfrac{3}{8}\right)$

Copyright © Houghton Mifflin Company. All rights reserved.

31. $-16 \cdot \dfrac{7}{30}$

32. $-9 \cdot \dfrac{7}{15}$

33. $\dfrac{6}{7} \cdot 0$

34. $0 \cdot \dfrac{9}{11}$

35. $\dfrac{5}{22} \cdot 2\dfrac{1}{5}$

36. $\dfrac{4}{15} \cdot 1\dfrac{7}{8}$

37. $3\dfrac{1}{2} \cdot 5\dfrac{3}{7}$

38. $2\dfrac{1}{4} \cdot 1\dfrac{1}{3}$

39. $3\dfrac{1}{3} \cdot \left(-\dfrac{7}{10}\right)$

40. $2\dfrac{1}{4} \cdot \left(-\dfrac{7}{9}\right)$

41. $-1\dfrac{2}{3} \cdot \left(-\dfrac{3}{5}\right)$

42. $-2\dfrac{1}{8} \cdot \left(-\dfrac{4}{17}\right)$

43. $3\dfrac{1}{3} \cdot 2\dfrac{1}{3}$

44. $3\dfrac{1}{4} \cdot 2\dfrac{2}{3}$

45. $3\dfrac{1}{3} \cdot (-9)$

46. $-2\dfrac{1}{2} \cdot 4$

47. $8 \cdot 5\dfrac{1}{4}$

48. $3 \cdot 2\dfrac{1}{9}$

49. $3\dfrac{1}{2} \cdot 1\dfrac{5}{7} \cdot \dfrac{11}{12}$

50. $2\dfrac{2}{3} \cdot \dfrac{8}{9} \cdot 1\dfrac{5}{16}$

51. Find the product of $\dfrac{3}{4}$ and $\dfrac{14}{15}$.

52. Find the product of $\dfrac{12}{25}$ and $\dfrac{5}{16}$.

53. Find $-\dfrac{9}{16}$ multiplied by $\dfrac{4}{27}$.

54. Find $\dfrac{3}{7}$ multiplied by $-\dfrac{14}{15}$.

55. What is the product of $-\dfrac{7}{24}, \dfrac{8}{21}$, and $\dfrac{3}{7}$?

56. What is the product of $-\dfrac{5}{13}, -\dfrac{26}{75}$, and $\dfrac{5}{8}$?

57. What is $4\dfrac{4}{5}$ times $\dfrac{3}{8}$?

58. What is $5\dfrac{1}{3}$ times $\dfrac{3}{16}$?

59. Find the product of $-2\dfrac{2}{3}$ and $-1\dfrac{11}{16}$.

60. Find the product of $1\dfrac{3}{11}$ and $5\dfrac{1}{2}$.

Copyright © Houghton Mifflin Company. All rights reserved.

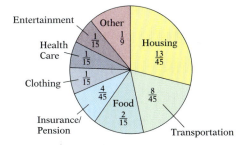

A typical household in the United States has an average after-tax income of $45,000. The graph at the right represents how this annual income is spent. Use this graph for Exercises 61 and 62.

How a Typical U.S. Household Spends Its Annual Income
Source: Based on data from American Demographics

61. *Cost of Living*   Find the amount of money a typical household in the United States spends on housing per year.

62. *Cost of Living*   How much money does a typical household in the United States spend annually on food?

Evaluate the variable expression xy for the given values of x and y.

63. $x = -\dfrac{5}{16}, y = \dfrac{7}{15}$ **64.** $x = -\dfrac{2}{5}, y = -\dfrac{5}{6}$ **65.** $x = \dfrac{4}{7}, y = 6\dfrac{1}{8}$ **66.** $x = 6\dfrac{3}{5}, y = 3\dfrac{1}{3}$

67. $x = -49, y = \dfrac{5}{14}$ **68.** $x = -\dfrac{3}{10}, y = -35$ **69.** $x = 1\dfrac{3}{13}, y = -6\dfrac{1}{2}$ **70.** $x = -3\dfrac{1}{2}, y = -2\dfrac{2}{7}$

Evaluate the variable expression xyz for the given values of x, y, and z.

71. $x = \dfrac{3}{8}, y = \dfrac{2}{3}, z = \dfrac{4}{5}$ **72.** $x = 4, y = \dfrac{0}{8}, z = 1\dfrac{5}{9}$ **73.** $x = 2\dfrac{3}{8}, y = -\dfrac{3}{19}, z = -\dfrac{4}{9}$

74. $x = \dfrac{4}{5}, y = -15, z = \dfrac{7}{8}$ **75.** $x = \dfrac{5}{6}, y = -3, z = 1\dfrac{7}{15}$ **76.** $x = 4\dfrac{1}{2}, y = 3\dfrac{5}{9}, z = 1\dfrac{7}{8}$

77. Is $-\dfrac{1}{3}$ a solution of the equation $\dfrac{3}{4}y = -\dfrac{1}{4}$?   **78.** Is $\dfrac{2}{5}$ a solution of the equation $-\dfrac{5}{6}z = \dfrac{1}{3}$?

79. Is $\dfrac{3}{4}$ a solution of the equation $\dfrac{4}{5}x = \dfrac{5}{3}$?   **80.** Is $\dfrac{1}{2}$ a solution of the equation $\dfrac{3}{4}p = \dfrac{3}{2}$?

81. Is $-\dfrac{1}{6}$ a solution of the equation $6x = 1$?   **82.** Is $-\dfrac{4}{5}$ a solution of the equation $\dfrac{5}{4}n = -1$?

Copyright © Houghton Mifflin Company. All rights reserved.

OBJECTIVE **B**

Divide.

83. $\dfrac{5}{7} \div \dfrac{2}{5}$

84. $\dfrac{3}{8} \div \dfrac{2}{3}$

85. $\dfrac{4}{7} \div \left(-\dfrac{4}{7}\right)$

86. $-\dfrac{5}{7} \div \left(-\dfrac{5}{6}\right)$

87. $0 \div \dfrac{7}{9}$

88. $0 \div \dfrac{4}{5}$

89. $\left(-\dfrac{1}{3}\right) \div \dfrac{1}{2}$

90. $\left(-\dfrac{3}{8}\right) \div \dfrac{7}{8}$

91. $-\dfrac{5}{16} \div \left(-\dfrac{3}{8}\right)$

92. $\left(-\dfrac{3}{4}\right) \div \left(-\dfrac{5}{6}\right)$

93. $\dfrac{0}{1} \div \dfrac{1}{9}$

94. $\dfrac{1}{2} \div \left(-\dfrac{8}{0}\right)$

95. $6 \div \dfrac{3}{4}$

96. $8 \div \dfrac{2}{3}$

97. $\dfrac{3}{4} \div (-6)$

98. $-\dfrac{2}{3} \div 8$

99. $\dfrac{9}{10} \div 0$

100. $\dfrac{2}{11} \div 0$

101. $\dfrac{5}{12} \div \left(-\dfrac{15}{32}\right)$

102. $\dfrac{3}{8} \div \left(-\dfrac{5}{12}\right)$

103. $\left(-\dfrac{2}{3}\right) \div (-4)$

104. $\left(-\dfrac{4}{9}\right) \div (-6)$

105. $\dfrac{8}{x} \div \left(-\dfrac{y}{4}\right)$

106. $-\dfrac{9}{m} \div \dfrac{n}{7}$

107. $\dfrac{b}{6} \div \dfrac{5}{d}$

108. $\dfrac{y}{10} \div \dfrac{4}{z}$

109. $3\dfrac{1}{3} \div \dfrac{5}{8}$

110. $5\dfrac{1}{2} \div \dfrac{1}{4}$

111. $5\dfrac{3}{5} \div \left(-\dfrac{7}{10}\right)$

112. $6\dfrac{8}{9} \div \left(-\dfrac{31}{36}\right)$

113. $-1\dfrac{1}{2} \div 1\dfrac{3}{4}$

114. $-1\dfrac{3}{5} \div 3\dfrac{1}{10}$

Copyright © Houghton Mifflin Company. All rights reserved.

115. $5\dfrac{1}{2} \div 11$

116. $4\dfrac{2}{3} \div 7$

117. $5\dfrac{2}{7} \div 1$

118. $9\dfrac{5}{6} \div 1$

119. $-16 \div 1\dfrac{1}{3}$

120. $-9 \div \left(-3\dfrac{3}{5}\right)$

121. $2\dfrac{4}{13} \div 1\dfrac{5}{26}$

122. $3\dfrac{3}{8} \div 2\dfrac{7}{16}$

123. Find the quotient of $\dfrac{9}{10}$ and $\dfrac{3}{4}$.

124. Find the quotient of $\dfrac{3}{5}$ and $\dfrac{12}{25}$.

125. What is $-\dfrac{15}{24}$ divided by $\dfrac{3}{5}$?

126. What is $\dfrac{5}{6}$ divided by $-\dfrac{10}{21}$?

127. Find $\dfrac{7}{8}$ divided by $3\dfrac{1}{4}$.

128. Find $-\dfrac{3}{8}$ divided by $2\dfrac{1}{4}$.

129. What is the quotient of $-3\dfrac{5}{11}$ and $3\dfrac{4}{5}$?

130. What is the quotient of $-10\dfrac{1}{5}$ and $-1\dfrac{7}{10}$?

Evaluate the variable expression $x \div y$ for the given values of x and y.

131. $x = -\dfrac{5}{8}, y = -\dfrac{15}{2}$

132. $x = -\dfrac{14}{3}, y = -\dfrac{7}{9}$

133. $x = \dfrac{1}{7}, y = 0$

134. $x = \dfrac{4}{0}, y = 12$

135. $x = -18, y = \dfrac{3}{8}$

136. $x = 20, y = -\dfrac{5}{6}$

137. $x = -\dfrac{1}{2}, y = -3\dfrac{5}{8}$

138. $x = 4\dfrac{3}{8}, y = 7$

139. $x = 6\dfrac{2}{5}, y = -4$

140. $x = -2\dfrac{5}{8}, y = 1\dfrac{3}{4}$

141. $x = -3\dfrac{2}{5}, y = -1\dfrac{7}{10}$

142. $x = -5\dfrac{2}{5}, y = -9$

Copyright © Houghton Mifflin Company. All rights reserved.

The table at the right shows the net weight of four different boxes of cereal. Use this table for Exercises 143 and 144.

Cereal	Net Weight
Kellogg Honey Crunch Corn Flakes	24 oz
Nabisco Instant Cream of Wheat	28 oz
Post Shredded Wheat	18 oz
Quaker Oats	41 oz

143. *The Food Industry*   Find the number of $\frac{3}{4}$-ounce servings in a box of the Kellogg Honey Crunch Corn Flakes.

144. *The Food Industry*   Find the number of $1\frac{1}{4}$-ounce servings in a box of Shredded Wheat.

OBJECTIVE   **C**

Solve.

145. *Sports*   A chukker is one period of play in a polo match. A chukker lasts $7\frac{1}{2}$ min. Find the length of time in four chukkers.

146. *History*   The Assyrian calendar was based on the phases of the moon. One lunation was $29\frac{1}{2}$ days long. There were 12 lunations in one year. Find the number of days in one year in the Assyrian calendar.

147. *Measurement*   One rod is equal to $5\frac{1}{2}$ yd. How many feet are in one rod? How many inches are in one rod?

148. *Travel*   A car used $12\frac{1}{2}$ gal of gasoline on a 275-mile trip. How many miles can this car travel on 1 gal of gasoline?

149. *Housework*   According to a national survey, the average couple spends $4\frac{1}{2}$ h cleaning house each week. How many hours does the average couple spend cleaning house each year?

150. *Business*   A factory worker can assemble a product in $7\frac{1}{2}$ min. How many products can the worker assemble in one hour?

151. *Real Estate*   A developer purchases $25\frac{1}{2}$ acres of land and plans to set aside 3 acres for an entranceway to a housing development to be built on the property. Each house will be built on a $\frac{3}{4}$-acre plot of land. How many houses does the developer plan to build on the property?

Copyright © Houghton Mifflin Company. All rights reserved.

152. *Consumerism*   You are planning a barbecue for 25 people. You want to serve $\frac{1}{4}$-pound hamburger patties to your guests and you estimate each person will eat two hamburgers. How much hamburger meat should you buy for the barbecue?

153. *Board Games*   A wooden travel game board has hinges which allow the board to be folded in half. If the dimensions of the open board are 14 in. by 14 in. by $\frac{7}{8}$ in., what are the dimensions of the board when it is closed?

154. *Carpentry*   A 16-foot board is cut into pieces $2\frac{1}{2}$ ft long for use as bookshelves. What is the length of the remaining piece after as many shelves as possible are cut?

155. *Nutrition*   According to the Center for Science in the Public Interest, the average teenage boy drinks $3\frac{1}{3}$ cans of soda per day. The average teenage girl drinks $2\frac{1}{3}$ cans of soda per day.
   a.   The average teenage boy drinks how many cans of soda per week?
   b.   If a can of soda contains 150 calories, how many calories does the average teenage boy consume each week in soda?
   c.   How many more cans of soda per week does the average teenage boy drink than the average teenage girl?

156. *Wages*   Find the total wages of an employee who worked $26\frac{1}{2}$ h this week and who earns an hourly wage of $12.

157. *Geometry*   Find the area of a rectangle that has a length of $8\frac{1}{2}$ yd and a width of 5 yd.

158. *Geometry*   What is the area of a rectangular recreational area that has a length of $3\frac{1}{4}$ mi and a width of $1\frac{1}{2}$ mi?

159. *Geometry*   A sail is in the shape of a triangle with a base of 12 m and a height of 16 m. How much canvas was needed to make the body of the sail?

160. *Geometry*   A vegetable garden is in the shape of a triangle with a base of 21 ft and a height of 13 ft. Find the area of the vegetable garden.

161. *Geometry*   A city plans to plant grass seed in a public playground that has the shape of a triangle with a height of 24 m and a base of 20 m. Each bag of grass seed will seed 120 m². How many bags of seed should be purchased?

Copyright © Houghton Mifflin Company. All rights reserved.

162. *Oceanography* The pressure on a submerged object is given by $P = 15 + \frac{1}{2}D$, where D is the depth in feet and P is the pressure measured in pounds per square inch. Find the pressure on a diver who is at a depth of $12\frac{1}{2}$ ft.

163. *Sports* Find the rate of a hiker who walked $4\frac{2}{3}$ mi in $1\frac{1}{3}$ h. Use the equation $r = \frac{d}{t}$, where r is the rate in miles per hour, d is the distance, and t is the time.

164. *Physics* Find the amount of force necessary to push a 75-pound crate across a floor where the coefficient of friction is $\frac{3}{8}$. Use the equation $F = \mu N$, where F is the force, μ is the coefficient of friction, and N is the weight of the crate. Force is measured in pounds.

Critical Thinking

165. *Cartography* On a map, two cities are $3\frac{1}{8}$ in. apart. If $\frac{1}{8}$ in. on the map represents 50 mi, what is the number of miles in the distance between the two cities?

166. Determine whether the statement is always true, sometimes true, or never true.
    a. Let n be an even number. Then $\frac{1}{2}n$ is a whole number.

    b. Let n be an odd number. Then $\frac{1}{2}n$ is an improper fraction.

167. Show by example that each of the following properties of multiplication of fractions is *not* satisfied by division of fractions: (a) Commutative Property, (b) Associative Property, (c) Inverse Property.

168. 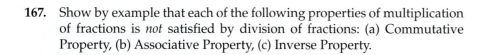 A box of stationery is to contain pieces of notepaper that measure $5\frac{3}{4}$ in. by $7\frac{3}{4}$ in. The notepaper is to be folded in half for mailing. What size envelopes would you design for this stationery? Explain how you arrived at your decision.

169. 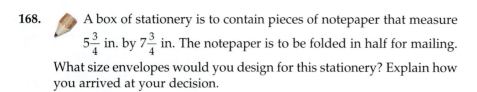 On page 206, Exercise 146 describes the Assyrian calendar. Our calendar is based on the solar year. One solar year is $365\frac{1}{4}$ days. Use this fact to explain leap years.

170. Draw a floor plan of your home or apartment.

Copyright © Houghton Mifflin Company. All rights reserved.

Copyright © Houghton Mifflin Company. All rights reserved.

SECTION 3.5 ▶ **Solving Equations with Fractions**

OBJECTIVE   **A**

Solving equations

Earlier in the text, you solved equations using the subtraction, addition, and division properties of equations. These properties are reviewed below.

> **The same number can be subtracted from each side of an equation without changing the solution of the equation.**

> **The same number can be added to each side of an equation without changing the solution of the equation.**

> **Each side of an equation can be divided by the same nonzero number without changing the solution of the equation.**

A fourth property of equations involves *multiplying* each side of an equation by the same nonzero number.

As shown at the right, the solution of the equation $3x = 12$ is 4.

$$3x = 12$$
$$3 \cdot 4 = 12$$
$$12 = 12$$

If each side of the equation $3x = 12$ is multiplied by 2, the resulting equation is $6x = 24$. The solution of this equation is also 4.

$$3x = 12$$
$$2 \cdot 3x = 2 \cdot 12$$
$$6x = 24 \qquad 6 \cdot 4 = 24$$

This illustrates the **multiplication property of equations.**

> **Each side of an equation can be multiplied by the same nonzero number without changing the solution of the equation.**

Solve: $\dfrac{x}{5} = 2$

The variable is divided by 5. Multiply each side of the equation by 5.

$$\frac{x}{5} = 2$$

Note that $5 \cdot \dfrac{x}{5} = \dfrac{5}{1} \cdot \dfrac{x}{5} = \dfrac{5x}{5} = x$.

$$5 \cdot \frac{x}{5} = 5 \cdot 2$$

The variable x is alone on the left side of the equation. The number on the right side is the solution.

$$x = 10$$

Check the solution.

Check: $\dfrac{x}{5} = 2$

$$\frac{10}{5} \;\Big|\; 2$$

The solution checks.

$$2 = 2$$

The solution is 10.

Recall that the product of a number and its reciprocal is 1. For instance,

$$\frac{3}{4} \cdot \frac{4}{3} = 1 \quad \text{and} \quad \left(-\frac{5}{2}\right)\left(-\frac{2}{5}\right) = 1$$

Multiplying each side of an equation by the reciprocal of a number is useful when solving equations in which the variable is multiplied by a fraction.

Solve: $\frac{3}{4}a = 12$

$\frac{3}{4}$ multiplies the variable a. Note the effect of multiplying each side of the equation by $\frac{4}{3}$, the reciprocal of $\frac{3}{4}$.

Note: $\frac{4}{3} \cdot \frac{3}{4} \cdot a = 1 \cdot a = a$.

The result is an equation with the variable alone on the left side of the equation. The number on the right side is the solution.

$$\frac{3}{4}a = 12$$

$$\frac{4}{3} \cdot \frac{3}{4}a = \frac{4}{3} \cdot 12$$

$$a = 16$$

The solution is 16.

Take Note

Remember to check your solution.

Check: $\dfrac{3}{4}a = 12$

$$\begin{array}{c|c} \dfrac{3}{4}(16) & 12 \\ \hline 12 & = 12 \end{array}$$

Solve: $6 = -\frac{3c}{5}$

$$-\frac{3c}{5} = -\left(\frac{3}{5} \cdot \frac{c}{1}\right) = -\frac{3}{5} \cdot c$$

$-\frac{3}{5}$ multiplies the variable c. Multiply each side of the equation by $-\frac{5}{3}$, the reciprocal of $-\frac{3}{5}$.

c is alone on the right side of the equation. The number on the left side is the solution.

Check your solution.

$$6 = -\frac{3c}{5}$$

$$6 = -\frac{3}{5} \cdot c$$

$$-\frac{5}{3} \cdot 6 = -\frac{5}{3}\left(-\frac{3}{5} \cdot c\right)$$

$$-10 = c$$

Check: $6 = -\dfrac{3c}{5}$

$$\begin{array}{c|c} 6 & \dfrac{-3(-10)}{5} \\[2mm] 6 & \dfrac{30}{5} \\[2mm] & 6 = 6 \end{array}$$

The solution is -10.

As shown in the following Example 1 and You Try It 1, the addition and subtraction properties of equations can be used to solve equations that contain fractions.

Copyright © Houghton Mifflin Company. All rights reserved.

1 **Example**   Solve: $y + \dfrac{2}{3} = \dfrac{3}{4}$

Solution

$$y + \frac{2}{3} = \frac{3}{4}$$

$$y + \frac{2}{3} - \frac{2}{3} = \frac{3}{4} - \frac{2}{3} \qquad \blacktriangleright \frac{2}{3} \text{ is added to } y.$$

$$y = \frac{9}{12} - \frac{8}{12} \qquad \text{Subtract } \frac{2}{3}.$$

$$y = \frac{1}{12}$$

The solution is $\dfrac{1}{12}$.

1 **You Try It**   Solve: $-\dfrac{1}{5} = z - \dfrac{5}{6}$

Your Solution

---

2 **Example**   Solve: $-\dfrac{3}{5} = \dfrac{6}{7}c$

Solution

$$-\frac{3}{5} = \frac{6}{7}c$$

$$\frac{7}{6}\left(-\frac{3}{5}\right) = \frac{7}{6} \cdot \frac{6}{7}c$$

$$-\frac{7}{10} = c$$

The solution is $-\dfrac{7}{10}$.

2 **You Try It**   Solve: $26 = 4x$

Your Solution

Solutions on p. S9

OBJECTIVE **B**

Applications

3 **Example**

Three-eighths times a number is equal to negative one-fourth. Find the number.

Solution

The unknown number: y

Three-eighths times a number	is equal to	negative one-fourth

$$\frac{3}{8}y = -\frac{1}{4}$$

$$\frac{8}{3} \cdot \frac{3}{8}y = \frac{8}{3}\left(-\frac{1}{4}\right)$$

$$y = -\frac{2}{3}$$

The number is $-\dfrac{2}{3}$.

3 **You Try It**

Negative five-sixths is equal to ten-thirds of a number. Find the number.

Your Solution

Solution on p. S9

Copyright © Houghton Mifflin Company. All rights reserved.

4 Example

One-third of all of the sugar produced by Sucor, Inc. is brown sugar. This year Sucor produced 250,000 lb of brown sugar. How many pounds of sugar were produced by Sucor?

Strategy

To find the number of pounds of sugar produced, write and solve an equation using x to represent the number of pounds of sugar produced.

Solution

One-third of the sugar produced	is	brown sugar

$$\frac{1}{3}x = 250{,}000$$

$$3 \cdot \frac{1}{3}x = 3 \cdot 250{,}000$$

$$x = 750{,}000$$

Sucor produced 750,000 lb of sugar.

4 You Try It

The number of computer software games sold by BAL Software in January was three-fifths of all the software products sold by the company. BAL Software sold 450 computer software games in January. Find the total number of software products sold in January.

Your Strategy

Your Solution

5 Example

The average score on exams taken during a semester is given by $A = \frac{T}{N}$, where A is the average score, T is the total number of points scored on all tests, and N is the number of tests. Find the total number of points scored by a student whose average score for 6 tests was 84.

Strategy

To find the total number of points scored, replace A with 84 and N with 6 in the given formula and solve for T.

Solution

$$A = \frac{T}{N}$$

$$84 = \frac{T}{6}$$

$$6 \cdot 84 = 6 \cdot \frac{T}{6}$$

$$504 = T$$

The total number of points scored was 504.

5 You Try It

The average score on exams taken during a semester is given by $A = \frac{T}{N}$, where A is the average score, T is the total number of points scored on all tests, and N is the number of tests. Find the total number of points scored by a student whose average score for 5 tests was 73.

Your Strategy

Your Solution

Solutions on pp. S9–S10

Copyright © Houghton Mifflin Company. All rights reserved.

3.5 EXERCISES

OBJECTIVE **A**

Solve.

1. $\dfrac{x}{4} = 9$

2. $8 = \dfrac{y}{2}$

3. $-3 = \dfrac{m}{4}$

4. $\dfrac{n}{5} = -2$

5. $\dfrac{2}{5}x = 10$

6. $\dfrac{3}{4}z = 12$

7. $-\dfrac{5}{6}w = 10$

8. $-\dfrac{1}{2}x = 3$

9. $\dfrac{1}{4} + y = \dfrac{3}{4}$

10. $\dfrac{5}{9} = t - \dfrac{1}{9}$

11. $x + \dfrac{1}{4} = \dfrac{5}{6}$

12. $\dfrac{7}{8} = y - \dfrac{1}{6}$

13. $-\dfrac{2x}{3} = -\dfrac{1}{2}$

14. $-\dfrac{4a}{5} = \dfrac{2}{3}$

15. $\dfrac{5n}{6} = -\dfrac{2}{3}$

16. $\dfrac{7z}{8} = -\dfrac{5}{16}$

17. $-\dfrac{3}{8}t = -\dfrac{1}{4}$

18. $-\dfrac{3}{4}t = -\dfrac{7}{8}$

19. $4a = 6$

20. $6z = 10$

21. $-9c = 12$

22. $-10z = 28$

23. $-2x = \dfrac{8}{9}$

24. $-5y = -\dfrac{15}{16}$

OBJECTIVE **B**

25. A number minus one-third equals one-half. Find the number.

26. The sum of a number and one-fourth is one-sixth. Find the number.

27. Three-fifths times a number is nine-tenths. Find the number.

28. The product of negative two-thirds and a number is five-sixths. Find the number.

29. The quotient of a number and negative four is three-fourths. Find the number.

30. A number divided by negative two equals two-fifths. Find the number.

31. Negative three-fourths of a number is equal to one-sixth. Find the number.

32. Negative three-eighths equals the product of two-thirds and some number. Find the number.

Copyright © Houghton Mifflin Company. All rights reserved.

33. *Education* A female's average salary prior to earning a master's in business administration is one-half her salary after earning the degree. Her average salary prior to earning the master's degree is $40,000. (*Source:* The Graduate Management Admission Council) What is her salary after earning the master's degree?

34. *Education* During the 2002–2003 academic year, the average cost for tuition and fees at a four-year public college was $\frac{2}{9}$ the average cost for tuition and fees at a four-year private college. The average cost for tuition and fees at a four-year public college was $4000. (*Source:* College Board) What was the average cost for tuition and fees at a four-year private college?

35. *Catering* The number of quarts of orange juice in a fruit punch recipe is three-fifths of the total number of quarts in the punch. The number of quarts of orange juice in the punch is 15. Find the total number of quarts in the punch.

36. *The Electorate* The number of people who voted in an election for mayor of a city was two-thirds of the total number of eligible voters. There were 24,416 people who voted in the election. Find the number of eligible voters.

37. *Cost of Living* The amount of rent paid by a mechanic is $\frac{2}{5}$ of the mechanic's monthly income. Using the figure at the right, determine the mechanic's monthly income.

38. *Travel* The average number of miles per gallon for a car is calculated using the formula $a = \frac{m}{g}$, where a is the average number of miles per gallon and m is the number of miles traveled on g gallons of gas. Use this formula to find the number of miles a car can travel on 16 gal of gas if the car averages 26 mi per gallon.

39. *Travel* The average number of miles per gallon for a truck is calculated using the formula $a = \frac{m}{g}$, where a is the average number of miles per gallon and m is the number of miles traveled on g gallons of gas. Use this formula to find the number of miles a truck can travel on 38 gal of diesel fuel if the truck averages 14 mi per gallon.

Critical Thinking

40. If $\frac{3}{8}x = -\frac{1}{4}$, is $6x$ greater than -1 or less than -1?

41. Given $-\frac{x}{2} = \frac{2}{3}$, select the best answer from the choices below.
    a. $-9x > 10$ b. $-6x < 8$ c. $-9x > 10$ and $-6x < 8$

42. Explain why dividing each side of $3x = 6$ by 3 is the same as multiplying each side of the equation by $\frac{1}{3}$.

Copyright © Houghton Mifflin Company. All rights reserved.

Exponents, Complex Fractions, and the Order of Operations Agreement

OBJECTIVE **A**

Exponents

Copyright © Houghton Mifflin Company. All rights reserved.

Recall that an exponent indicates the repeated multiplication of the same factor. For example,

$$3^5 = 3 \cdot 3 \cdot 3 \cdot 3 \cdot 3$$

The exponent, 5, indicates how many times the base, 3, occurs as a factor in the multiplication.

The base of an exponential expression can be a fraction, for example, $\left(\frac{2}{3}\right)^4$. To evaluate this expression, write the factor as many times as indicated by the exponent and then multiply.

$$\left(\frac{2}{3}\right)^4 = \frac{2}{3} \cdot \frac{2}{3} \cdot \frac{2}{3} \cdot \frac{2}{3} = \frac{2 \cdot 2 \cdot 2 \cdot 2}{3 \cdot 3 \cdot 3 \cdot 3} = \frac{16}{81}$$

> ▶ **Point of Interest**
>
> René Descartes (1596–1650) was the first mathematician to extensively use exponential notation as it is used today. However, for some unknown reason, he always used *xx* for *x*².

Evaluate $\left(-\frac{3}{5}\right)^2 \cdot \left(\frac{5}{6}\right)^3$.

$$\left(-\frac{3}{5}\right)^2 \cdot \left(\frac{5}{6}\right)^3$$

Write each factor as many times as indicated by the exponent.

$$= \left(-\frac{3}{5}\right) \cdot \left(-\frac{3}{5}\right) \cdot \frac{5}{6} \cdot \frac{5}{6} \cdot \frac{5}{6}$$

Multiply. The product of two negative numbers is positive.

$$= \frac{3}{5} \cdot \frac{3}{5} \cdot \frac{5}{6} \cdot \frac{5}{6} \cdot \frac{5}{6}$$

$$= \frac{3 \cdot 3 \cdot 5 \cdot 5 \cdot 5}{5 \cdot 5 \cdot 6 \cdot 6 \cdot 6}$$

Write the product in simplest form.

$$= \frac{5}{24}$$

Evaluate x^3 when $x = 2\frac{1}{2}$.

$$x^3$$

Replace x with $2\frac{1}{2}$.

$$\left(2\frac{1}{2}\right)^3$$

Write the mixed number as an improper fraction.

$$= \left(\frac{5}{2}\right)^3$$

Write the base as many times as indicated by the exponent.

$$= \frac{5}{2} \cdot \frac{5}{2} \cdot \frac{5}{2}$$

Multiply.

$$= \frac{125}{8}$$

Write the improper fraction as a mixed number.

$$= 15\frac{5}{8}$$

① *Example* Evaluate $\left(-\dfrac{3}{4}\right)^3 \cdot 8^2$.

Solution $\left(-\dfrac{3}{4}\right)^3 \cdot 8^2$

$= \left(-\dfrac{3}{4}\right)\left(-\dfrac{3}{4}\right)\left(-\dfrac{3}{4}\right) \cdot 8 \cdot 8$

$= -\left(\dfrac{3}{4} \cdot \dfrac{3}{4} \cdot \dfrac{3}{4} \cdot \dfrac{8}{1} \cdot \dfrac{8}{1}\right)$

$= -\dfrac{3 \cdot 3 \cdot 3 \cdot 8 \cdot 8}{4 \cdot 4 \cdot 4 \cdot 1 \cdot 1} = -27$

① *You Try It* Evaluate $\left(\dfrac{2}{9}\right)^2 \cdot (-3)^4$.

Your Solution

---

② *Example* Evaluate $x^2 y^2$ when $x = 1\dfrac{1}{2}$ and $y = \dfrac{2}{3}$.

Solution $x^2 y^2$

$\left(1\dfrac{1}{2}\right)^2 \cdot \left(\dfrac{2}{3}\right)^2 = \left(\dfrac{3}{2}\right)^2 \cdot \left(\dfrac{2}{3}\right)^2$

$= \dfrac{3}{2} \cdot \dfrac{3}{2} \cdot \dfrac{2}{3} \cdot \dfrac{2}{3}$

$= \dfrac{3 \cdot 3 \cdot 2 \cdot 2}{2 \cdot 2 \cdot 3 \cdot 3} = 1$

② *You Try It* Evaluate $x^4 y^3$ when $x = 2\dfrac{1}{3}$ and $y = \dfrac{3}{7}$.

Your Solution

Solutions on p. S10

OBJECTIVE **B**

Complex fractions

A **complex fraction** is a fraction whose numerator or denominator contains one or more fractions. Examples of complex fractions are shown below.

Main fraction bar $\longrightarrow$ $\dfrac{\dfrac{3}{4}}{\dfrac{7}{8}}$ $\dfrac{4}{3 - \dfrac{1}{2}}$ $\dfrac{\dfrac{9}{10} + \dfrac{3}{5}}{\dfrac{5}{6}}$ $\dfrac{3\dfrac{1}{2} \cdot 2\dfrac{5}{8}}{\left(4\dfrac{2}{3}\right) \div \left(3\dfrac{1}{5}\right)}$

Look at the first example given above and recall that the fraction bar can be read "divided by."

Therefore, $\dfrac{\dfrac{3}{4}}{\dfrac{7}{8}}$ can be read "$\dfrac{3}{4}$ divided by $\dfrac{7}{8}$" and can be written $\dfrac{3}{4} \div \dfrac{7}{8}$. This

is the division of two fractions and can be simplified by multiplying by the reciprocal, as shown at the top of the next page.

Copyright © Houghton Mifflin Company. All rights reserved.

$$\frac{\dfrac{3}{4}}{\dfrac{7}{8}} = \frac{3}{4} \div \frac{7}{8} = \frac{3}{4} \cdot \frac{8}{7} = \frac{3 \cdot 8}{4 \cdot 7} = \frac{6}{7}$$

To simplify a complex fraction, first simplify the expression above the main fraction bar and the expression below the main fraction bar; the result is one number in the numerator and one number in the denominator. Then rewrite the complex fraction as a division problem by reading the main fraction bar as "divided by."

Simplify: $\dfrac{4}{3 - \dfrac{1}{2}}$

The numerator (4) is already simplified. Simplify the expression in the denominator.

Note: $3 - \dfrac{1}{2} = \dfrac{6}{2} - \dfrac{1}{2} = \dfrac{5}{2}$

$$\frac{4}{3 - \dfrac{1}{2}} = \frac{4}{\dfrac{5}{2}}$$

Rewrite the complex fraction as division.

$$= 4 \div \frac{5}{2}$$

Divide.

$$= \frac{4}{1} \div \frac{5}{2}$$

$$= \frac{4}{1} \cdot \frac{2}{5}$$

Write the answer in simplest form.

$$= \frac{8}{5} = 1\frac{3}{5}$$

Simplify: $\dfrac{-\dfrac{9}{10} + \dfrac{3}{5}}{1\dfrac{1}{4}}$

Simplify the expression in the numerator.

Note: $-\dfrac{9}{10} + \dfrac{3}{5} = \dfrac{-9}{10} + \dfrac{6}{10} = \dfrac{-3}{10} = -\dfrac{3}{10}$

$$\frac{-\dfrac{9}{10} + \dfrac{3}{5}}{1\dfrac{1}{4}} = \frac{-\dfrac{3}{10}}{\dfrac{5}{4}}$$

Write the mixed number in the denominator as an improper fraction.

Rewrite the complex fraction as division. The quotient will be negative.

$$= -\left(\frac{3}{10} \div \frac{5}{4} \right)$$

Divide by multiplying by the reciprocal.

$$= -\left(\frac{3}{10} \cdot \frac{4}{5} \right)$$

$$= -\frac{6}{25}$$

Copyright © Houghton Mifflin Company. All rights reserved.

Evaluate $\dfrac{wx}{yz}$ when $w = 1\frac{1}{3}$, $x = 2\frac{5}{8}$, $y = 4\frac{1}{2}$, and $z = 3\frac{1}{3}$.

$$\dfrac{wx}{yz}$$

Replace each variable with its given value.

$$\dfrac{1\frac{1}{3} \cdot 2\frac{5}{8}}{4\frac{1}{2} \cdot 3\frac{1}{3}}$$

Simplify the numerator.

Note: $1\frac{1}{3} \cdot 2\frac{5}{8} = \frac{4}{3} \cdot \frac{21}{8} = \frac{7}{2}$

Simplify the denominator.

Note: $4\frac{1}{2} \cdot 3\frac{1}{3} = \frac{9}{2} \cdot \frac{10}{3} = 15$

$$= \dfrac{\frac{7}{2}}{15}$$

Rewrite the complex fraction as division.

$$= \frac{7}{2} \div 15$$

Divide by multiplying by the reciprocal.

Note: $15 = \frac{15}{1}$; the reciprocal of $\frac{15}{1}$ is $\frac{1}{15}$.

$$= \frac{7}{2} \cdot \frac{1}{15} = \frac{7}{30}$$

3

Example   Is $\frac{2}{3}$ a solution of $\dfrac{x + \frac{1}{2}}{x} = \frac{7}{4}$?

Solution

$$\dfrac{x + \frac{1}{2}}{x} = \frac{7}{4}$$

$$\begin{array}{c|c} \dfrac{\frac{2}{3} + \frac{1}{2}}{\frac{2}{3}} & \dfrac{7}{4} \\[3ex] \dfrac{\frac{7}{6}}{\frac{2}{3}} & \dfrac{7}{4} \\[3ex] \frac{7}{6} \div \frac{2}{3} & \frac{7}{4} \\[2ex] \frac{7}{6} \cdot \frac{3}{2} & \frac{7}{4} \\[2ex] \frac{7}{4} = & \frac{7}{4} \end{array}$$

Yes, $\frac{2}{3}$ is a solution of the equation.

3

You Try It   Is $-\frac{1}{2}$ a solution of $\dfrac{2y - 3}{y} = -2$?

Your Solution

Solution on p. S10

Copyright © Houghton Mifflin Company. All rights reserved.

Copyright © Houghton Mifflin Company. All rights reserved.

 Example

Evaluate the variable expression $\dfrac{x - y}{z}$ when

$x = 4\dfrac{1}{8}$, $y = 2\dfrac{5}{8}$, and $z = \dfrac{3}{4}$.

Solution

$\dfrac{x - y}{z}$

$\dfrac{4\dfrac{1}{8} - 2\dfrac{5}{8}}{\dfrac{3}{4}} = \dfrac{\dfrac{3}{2}}{\dfrac{3}{4}} = \dfrac{3}{2} \div \dfrac{3}{4} = \dfrac{3}{2} \cdot \dfrac{4}{3} = 2$

 You Try It

Evaluate the variable expression $\dfrac{x}{y - z}$ when

$x = 2\dfrac{4}{9}$, $y = 3$, and $z = 1\dfrac{1}{3}$.

Your Solution

Solution on p. S10

OBJECTIVE **C**

VIDEO & DVD   CD TUTOR   WWW WEB   SSM

The Order of Operations Agreement

The Order of Operations Agreement applies in simplifying expressions containing fractions.

> ### The Order of Operations Agreement
>
> **Step 1** Do all operations inside parentheses.
>
> **Step 2** Simplify any numerical expressions containing exponents.
>
> **Step 3** Do multiplication and division as they occur from left to right.
>
> **Step 4** Do addition and subtraction as they occur from left to right.

Simplify: $\left(\dfrac{1}{2}\right)^2 + \left(\dfrac{2}{3} \div \dfrac{5}{9}\right) \cdot \dfrac{5}{6}$

$$\left(\dfrac{1}{2}\right)^2 + \left(\dfrac{2}{3} \div \dfrac{5}{9}\right) \cdot \dfrac{5}{6}$$

Do the operation inside the parentheses (Step 1).

$$= \left(\dfrac{1}{2}\right)^2 + \left(\dfrac{6}{5}\right) \cdot \dfrac{5}{6}$$

Simplify the exponential expression (Step 2).

$$= \dfrac{1}{4} + \left(\dfrac{6}{5}\right) \cdot \dfrac{5}{6}$$

Do the multiplication (Step 3).

$$= \dfrac{1}{4} + 1$$

Do the addition (Step 4).

$$= 1\dfrac{1}{4}$$

A fraction bar acts like parentheses. Therefore, simplify the numerator and denominator of a fraction as part of Step 1 in the Order of Operations Agreement.

Simplify: $6 - \dfrac{2+1}{15-8} \div \dfrac{3}{14}$

$$6 - \dfrac{2+1}{15-8} \div \dfrac{3}{14}$$

Perform operations above and below the fraction bar.

$$= 6 - \dfrac{3}{7} \div \dfrac{3}{14}$$

Do the division.

$$= 6 - \left(\dfrac{3}{7} \cdot \dfrac{14}{3}\right)$$

$$= 6 - 2$$

Do the subtraction.

$$= 4$$

Evaluate $\dfrac{w+x}{y} - z$ when $w = \dfrac{3}{4}$, $x = \dfrac{1}{4}$, $y = 2$, and $z = \dfrac{1}{3}$.

$$\dfrac{w+x}{y} - z$$

Replace each variable with its given value.

$$\dfrac{\dfrac{3}{4} + \dfrac{1}{4}}{2} - \dfrac{1}{3}$$

Simplify the numerator of the complex fraction.

$$= \dfrac{1}{2} - \dfrac{1}{3}$$

Do the subtraction.

$$= \dfrac{1}{6}$$

 Example Simplify: $\left(-\dfrac{2}{3}\right)^2 \div \dfrac{7-2}{13-4} - \dfrac{1}{3}$

Solution

$$\left(-\dfrac{2}{3}\right)^2 \div \dfrac{7-2}{13-4} - \dfrac{1}{3}$$

$$= \left(-\dfrac{2}{3}\right)^2 \div \dfrac{5}{9} - \dfrac{1}{3}$$

$$= \dfrac{4}{9} \div \dfrac{5}{9} - \dfrac{1}{3}$$

$$= \dfrac{4}{9} \cdot \dfrac{9}{5} - \dfrac{1}{3}$$

$$= \dfrac{4}{5} - \dfrac{1}{3} = \dfrac{7}{15}$$

 You Try It Simplify: $\left(-\dfrac{1}{2}\right)^3 \cdot \dfrac{7-3}{4-9} + \dfrac{4}{5}$

Your Solution

Solution on p. S10

Copyright © Houghton Mifflin Company. All rights reserved.

3.6 EXERCISES

Copyright © Houghton Mifflin Company. All rights reserved.

OBJECTIVE **A**

Evaluate.

1. $\left(\dfrac{3}{4}\right)^2$ **2.** $\left(\dfrac{5}{8}\right)^2$ **3.** $\left(-\dfrac{1}{6}\right)^3$ **4.** $\left(-\dfrac{2}{7}\right)^3$

5. $\left(2\dfrac{1}{4}\right)^2$ **6.** $\left(3\dfrac{1}{2}\right)^2$ **7.** $\left(\dfrac{5}{8}\right)^3 \cdot \left(\dfrac{2}{5}\right)^2$ **8.** $\left(\dfrac{3}{5}\right)^3 \cdot \left(\dfrac{1}{3}\right)^2$

9. $\left(\dfrac{18}{25}\right)^2 \cdot \left(\dfrac{5}{9}\right)^3$ **10.** $\left(\dfrac{2}{3}\right)^3 \cdot \left(\dfrac{5}{6}\right)^2$ **11.** $\left(\dfrac{4}{5}\right)^4 \cdot \left(-\dfrac{5}{8}\right)^3$ **12.** $\left(-\dfrac{9}{11}\right)^2 \cdot \left(\dfrac{1}{3}\right)^4$

13. $7^2 \cdot \left(\dfrac{2}{7}\right)^3$ **14.** $4^3 \cdot \left(\dfrac{5}{12}\right)^2$ **15.** $4 \cdot \left(\dfrac{4}{7}\right)^2 \cdot \left(-\dfrac{3}{4}\right)^3$ **16.** $3 \cdot \left(\dfrac{2}{5}\right)^2 \cdot \left(-\dfrac{1}{6}\right)^2$

Evaluate the variable expression for the given values of x and y.

17. x^4, when $x = \dfrac{2}{3}$ **18.** y^3, when $y = -\dfrac{3}{4}$

19. $x^4 y^2$, when $x = \dfrac{5}{6}$ and $y = -\dfrac{3}{5}$ **20.** $x^5 y^3$, when $x = -\dfrac{5}{8}$ and $y = \dfrac{4}{5}$

21. $x^3 y^2$, when $x = \dfrac{2}{3}$ and $y = 1\dfrac{1}{2}$ **22.** $x^2 y^4$, when $x = 2\dfrac{1}{3}$ and $y = \dfrac{3}{7}$

OBJECTIVE **B**

Simplify.

23. $\dfrac{\frac{9}{16}}{\frac{3}{4}}$ **24.** $\dfrac{\frac{7}{24}}{\frac{3}{8}}$ **25.** $\dfrac{-\frac{5}{6}}{\frac{15}{16}}$ **26.** $\dfrac{\frac{7}{12}}{-\frac{5}{18}}$

27. $\dfrac{\frac{2}{3} + \frac{1}{2}}{7}$

28. $\dfrac{-5}{\frac{3}{8} - \frac{1}{4}}$

29. $\dfrac{2 + \frac{1}{4}}{\frac{3}{8}}$

30. $\dfrac{1 - \frac{3}{4}}{\frac{5}{12}}$

31. $\dfrac{\frac{9}{25}}{\frac{4}{5} - \frac{1}{10}}$

32. $\dfrac{-\frac{5}{7}}{\frac{4}{7} - \frac{3}{14}}$

33. $\dfrac{\frac{1}{3} - \frac{3}{4}}{\frac{1}{6} + \frac{2}{3}}$

34. $\dfrac{\frac{9}{14} - \frac{1}{7}}{\frac{9}{14} + \frac{1}{7}}$

35. $\dfrac{3 + 2\frac{1}{3}}{5\frac{1}{6} - 1}$

36. $\dfrac{4 - 3\frac{5}{8}}{2\frac{1}{2} - \frac{3}{4}}$

37. $\dfrac{5\frac{2}{3} - 1\frac{1}{6}}{3\frac{5}{8} - 2\frac{1}{4}}$

38. $\dfrac{3\frac{1}{4} - 2\frac{1}{2}}{4\frac{3}{4} + 1\frac{1}{2}}$

Evaluate the expression for the given values of the variables.

39. $\dfrac{x + y}{z}$, when $x = \dfrac{2}{3}$, $y = \dfrac{3}{4}$, and $z = \dfrac{1}{12}$

40. $\dfrac{x}{y + z}$, when $x = \dfrac{8}{15}$, $y = \dfrac{3}{5}$, and $z = \dfrac{2}{3}$

41. $\dfrac{xy}{z}$, when $x = \dfrac{3}{4}$, $y = -\dfrac{2}{3}$, and $z = \dfrac{5}{8}$

42. $\dfrac{x}{yz}$, when $x = -\dfrac{5}{12}$, $y = \dfrac{8}{9}$, and $z = -\dfrac{3}{4}$

43. $\dfrac{x - y}{z}$, when $x = 2\dfrac{5}{8}$, $y = 1\dfrac{1}{4}$, and $z = 1\dfrac{3}{8}$

44. $\dfrac{x}{y - z}$, when $x = 2\dfrac{3}{10}$, $y = 3\dfrac{2}{5}$, and $z = 1\dfrac{4}{5}$

45. Is $-\dfrac{3}{4}$ a solution of the equation $\dfrac{4x}{x + 5} = -\dfrac{4}{3}$?

46. Is $-\dfrac{4}{5}$ a solution of the equation

$$\dfrac{15y}{\frac{3}{10} + y} = -24?$$

Copyright © Houghton Mifflin Company. All rights reserved.

OBJECTIVE **C**

Simplify.

47. $\dfrac{3}{7} \cdot \dfrac{14}{15} + \dfrac{4}{5}$

48. $\dfrac{3}{5} \div \dfrac{6}{7} + \dfrac{4}{5}$

49. $\left(\dfrac{5}{6}\right)^2 - \dfrac{5}{9}$

50. $\left(\dfrac{3}{5}\right)^2 - \dfrac{3}{10}$

51. $\dfrac{3}{4} \cdot \left(\dfrac{11}{12} - \dfrac{7}{8}\right) + \dfrac{5}{16}$

52. $\dfrac{7}{18} + \dfrac{5}{6} \cdot \left(\dfrac{2}{3} - \dfrac{1}{6}\right)$

53. $\dfrac{11}{16} - \left(\dfrac{3}{4}\right)^2 + \dfrac{7}{8}$

54. $\left(-\dfrac{2}{3}\right)^2 - \dfrac{7}{18} + \dfrac{5}{6}$

55. $\left(1\dfrac{1}{3} - \dfrac{5}{6}\right) + \dfrac{7}{8} \div \left(-\dfrac{1}{2}\right)^2$

56. $\left(\dfrac{1}{4}\right)^2 \div \left(2\dfrac{1}{2} - \dfrac{3}{4}\right) + \dfrac{5}{7}$

57. $\left(\dfrac{2}{3}\right)^2 + \dfrac{8-7}{3-9} \div \dfrac{3}{8}$

58. $\left(\dfrac{1}{3}\right)^2 \cdot \dfrac{14-5}{6-10} + \dfrac{3}{4}$

59. $\dfrac{1}{2} + \dfrac{\dfrac{13}{25}}{4 - \dfrac{3}{4}} \div \dfrac{1}{5}$

60. $\dfrac{4}{5} + \dfrac{3 - \dfrac{7}{9}}{\dfrac{5}{6}} \cdot \dfrac{3}{8}$

61. $\left(\dfrac{2}{3}\right)^2 + \dfrac{\dfrac{5}{8} - \dfrac{1}{4}}{\dfrac{2}{3} - \dfrac{1}{6}} \cdot \dfrac{8}{9}$

Evaluate the expression for the given values of the variables.

62. $x^2 + \dfrac{y}{z}$, when $x = -\dfrac{2}{3}$, $y = \dfrac{5}{8}$, and $z = \dfrac{3}{4}$

63. $\dfrac{x}{y} - z^2$, when $x = \dfrac{5}{6}$, $y = \dfrac{1}{3}$, and $z = -\dfrac{3}{4}$

64. $x - y^3 z$, when $x = \dfrac{5}{6}$, $y = \dfrac{1}{2}$, and $z = \dfrac{8}{9}$

65. $xy^3 + z$, when $x = \dfrac{9}{10}$, $y = \dfrac{1}{3}$, and $z = \dfrac{7}{15}$

Copyright © Houghton Mifflin Company. All rights reserved.

66. $\dfrac{wx}{y} + z$, when $w = \dfrac{4}{5}$, $x = \dfrac{5}{8}$, $y = \dfrac{3}{4}$, and $z = \dfrac{2}{3}$

67. $\dfrac{w}{xy} - z$, when $w = 2\dfrac{1}{2}$, $x = 4$, $y = \dfrac{3}{8}$, and $z = \dfrac{2}{3}$

68. Is $-\dfrac{1}{2}$ a solution of the equation $\dfrac{-8z}{z + \dfrac{5}{6}} - 4z = -14$?

69. Is $-\dfrac{1}{3}$ a solution of the equation $\dfrac{12w}{\dfrac{1}{6} - w} = -7$?

Critical Thinking

70. *Computers* A computer can perform 600,000 operations in one second. To the nearest minute, how many minutes will it take for the computer to perform 10^8 operations?

Place the correct symbol, $<$ or $>$, between the two numbers.

71. $\left(\dfrac{9}{10}\right)^3$ 1^5

72. $(-3)^3$ $(-2)^5$

73. $\left(-1\dfrac{1}{10}\right)^2$ $(0.9)^2$

74. Simplify: $\dfrac{\dfrac{3}{x} + \dfrac{2}{x}}{\dfrac{5}{6}}$

75. Given x is a whole number, for what value of x will the expression $\left(\dfrac{3}{4}\right)^2 + x^5 \div \dfrac{7}{8}$ have a minimum value? What is the minimum value?

76. Which of the variables u, v, w, x, and y can be doubled so that $\dfrac{u + \dfrac{v}{w}}{\dfrac{x}{y}}$ is

(a) halved or (b) doubled?

77. A farmer died and left 17 horses to be divided among 3 children. The first child was to receive one-half of the horses, the second child one-third of the horses, and the third child one-ninth of the horses. The executor for the family's estate realized that 17 horses could not be divided by halves, thirds, or ninths and so added a neighbor's horse to the farmer's. With 18 horses, the executor gave 9 horses to the first child, 6 horses to the second child, and 2 horses to the third child. This accounted for the 17 horses, so the executor returned the borrowed horse to the neighbor. Explain why this worked.

Copyright © Houghton Mifflin Company. All rights reserved.

Focus on Problem Solving

Common Knowledge

An application problem may not provide all the information that is needed to solve the problem. Sometimes, however, the necessary information is common knowledge.

> You are traveling by bus from Boston to New York. The trip is 4 hours long. If the bus leaves Boston at 10 A.M., what time should you arrive in New York?
>
> What other information do you need to solve this problem?
>
> You need to know that, using a 12-hour clock, the hours run
>
> 10 A.M.
> 11 A.M.
> 12 P.M.
> 1 P.M.
> 2 P.M.
>
> Four hours after 10 A.M. is 2 P.M.
>
> You should arrive in New York at 2 P.M.

> You purchase a 37¢ stamp at the Post Office and hand the clerk a one-dollar bill. How much change do you receive?
>
> What information do you need to solve this problem?
>
> You need to know that there are 100¢ in one dollar.
>
> Your change is 100¢ − 37¢.
>
> $100 - 37 = 63$
>
> You receive 63¢ in change.

What information do you need to know to solve each of the following problems?

1. You sell a dozen tickets to a fundraiser. Each ticket costs $10. How much money do you collect?

2. The weekly lab period for your science course is one hour and twenty minutes long. Find the length of the science lab period in minutes.

3. An employee's monthly salary is $3750. Find the employee's annual salary.

4. A survey revealed that eighth graders spend an average of 3 hours each day watching television. Find the total time an eighth grader spends watching TV each week.

5. You want to buy a carpet for a room that is 15 feet wide and 18 feet long. Find the amount of carpet that you need.

Copyright © Houghton Mifflin Company. All rights reserved.

Projects & Group Activities

Music

In musical notation, notes are printed on a **staff,** which is a set of five horizontal lines and the spaces between them. The notes of a musical composition are grouped into **measures,** or **bars.** Vertical lines separate measures on a staff. The shape of a note indicates how long it should be held. The whole note has the longest time value of all notes. Each time value is divided by two in order to find the next smallest note value.

Notes

Whole $\frac{1}{2}$ $\frac{1}{4}$ $\frac{1}{8}$ $\frac{1}{16}$ $\frac{1}{32}$ $\frac{1}{64}$

The **time signature** is a fraction that appears at the beginning of a piece of music. The numerator of the fraction indicates the number of beats in a measure. The denominator indicates what kind of note receives one beat. For example, music written in $\frac{2}{4}$ time has 2 beats to a measure, and a quarter note receives one beat. One measure in $\frac{2}{4}$ time may have 1 half note, 2 quarter notes, 4 eighth notes, or any other combination of notes totaling 2 beats. Other common time signatures include $\frac{4}{4}$, $\frac{3}{4}$, and $\frac{6}{8}$.

1. Explain the meaning of the 6 and the 8 in the time signature $\frac{6}{8}$.

2. Give some possible combinations of notes in one measure of a piece of music written in $\frac{4}{4}$ time.

3. What does a dot at the right of a note indicate? What is the effect of a dot at the right of a half note? at the right of a quarter note? at the right of an eighth note?

4. Symbols called rests are used to indicate periods of silence in a piece of music. What symbols are used to indicate the different time values of rests?

5. Find some examples of musical compositions written in different time signatures. Use a few measures from each to show that the sum of the time values of the notes and rests in each measure equals the numerator of the time signature.

Copyright © Houghton Mifflin Company. All rights reserved.

Construction

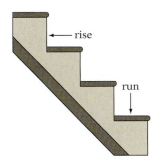

Suppose you are involved in building your own home. Design a stairway from the first floor of the house to the second floor. Here are some of the questions you will need to answer:

What is the distance from the floor of the first story to the floor of the second story?

Typically, what is the number of steps in a stairway?

What is a reasonable length for the run of each step?

What width wood is being used to build the staircase?

In designing the stairway, remember that each riser should be the same height and each run should be the same length. And the width of the wood used for the steps will have to be incorporated into the calculation.

Fractions of Diagrams

The diagram below has been broken up into nine areas separated by heavy lines. Eight of the areas have been labeled **A** through **H**. The ninth area is shaded. Determine which lettered areas would have to be shaded so that half of the entire diagram is shaded and half is not shaded. Write down the strategy that you or your group use to arrive at the solution. Compare your strategy with that of other individual students or groups.

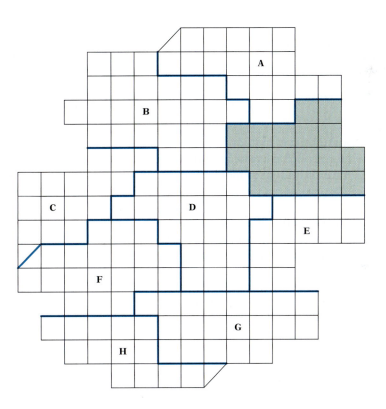

Puzzle from the Middle Ages

From the Middle Ages through the nineteenth century, it was quite common for the aristocracy to support court mathematicians and challenge other courts to contests between the mathematicians. In one such contest, Fibonacci (1170–1230), also known as Leonardo of Pisa, was given the following problem.

Find a fraction that is a square $\left[\text{for instance, } \frac{16}{25} \text{ because } \frac{16}{25} = \left(\frac{4}{5} \right)^2 \right]$ and has the following property: If 5 is added to the fraction, the new fraction is still a square, and if 5 is subtracted from the original fraction, the new fraction is still a square. *Hint:* One of the fractions is $\frac{2{,}401}{144} = \left(\frac{49}{12} \right)^2$.

Copyright © Houghton Mifflin Company. All rights reserved.

Using Patterns in Experimentation

Show how to cut a pie into the greatest number of pieces with only five straight cuts of a knife. An illustration showing how five cuts can produce 13 pieces is at the right. The correct answer, however, has more than 13 pieces.

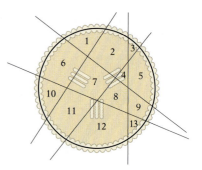

A reasonable question to ask is "How do I know when I have the maximum number of pieces?" To determine the answer, we suggest that you start with one cut, then two cuts, then three cuts, ... Try to discover a pattern for the greatest number of pieces that each number of cuts can produce.

Chapter Summary

Key Words	Examples
A number that is a multiple of two or more numbers is a **common multiple** of those numbers. The **least common multiple** *(LCM)* is the smallest common multiple of two or more numbers. [3.1A, p. 153]	12, 24, 36, 48, . . . are common multiples of 4 and 6. The LCM of 4 and 6 is 12.
A number that is a factor of two or more numbers is a **common factor** of those numbers. The **greatest common factor** *(GCF)* is the largest common factor of two or more numbers. [3.1B, p. 154]	The common factors of 12 and 16 are 1, 2, and 4. The GCF of 12 and 16 is 4.
A **fraction** can represent the number of equal parts of a whole. In a fraction, the **fraction bar** separates the **numerator** and the **denominator.** [3.2A, p. 159]	In the fraction $\frac{3}{4}$, the numerator is 3 and the denominator is 4.
In a **proper fraction,** the numerator is smaller than the denominator; a proper fraction is a number less than 1. In an **improper fraction,** the numerator is greater than or equal to the denominator; an improper fraction is a number greater than or equal to 1. A **mixed number** is a number greater than 1 with a whole number part and a fractional part. [3.2A, pp. 159–160]	$\frac{2}{5}$ is a proper fraction. $\frac{7}{6}$ is an improper fraction. $4\frac{1}{10}$ is a mixed number; 4 is the whole number part and $\frac{1}{10}$ is the fractional part.
Equal fractions with different denominators are called **equivalent fractions.** [3.2B, p. 162]	$\frac{3}{4}$ and $\frac{6}{8}$ are equivalent fractions.

Copyright © Houghton Mifflin Company. All rights reserved.

A fraction is in **simplest form** when the numerator and denominator have no common factors other than 1. [3.2B, p. 162]

The fraction $\frac{11}{12}$ is in simplest form.

The **reciprocal** of a fraction is that fraction with the numerator and denominator interchanged. [3.4B, p. 196]

The reciprocal of $\frac{3}{8}$ is $\frac{8}{3}$.

The reciprocal of 5 is $\frac{1}{5}$.

A **complex fraction** is a fraction whose numerator or denominator contains one or more fractions. [3.6B, p. 216]

$\dfrac{\frac{2}{3} - \frac{5}{8}}{\frac{1}{9}}$ is a complex fraction.

Essential Rules and Procedures

To find the LCM of two or more numbers, write the prime factorization of each number and circle the highest power of each prime factor. The LCM is the product of the circled factors. [3.1A, p. 153]

$12 = \textcircled{$2^2$} \cdot 3$
$18 = 2 \cdot \textcircled{3^2}$
The LCM of 12 and 18 is $2^2 \cdot 3^2 = 36$.

To find the GCF of two or more numbers, write the prime factorization of each number and circle the lowest power of each prime factor that occurs in each factorization. The GCF is the product of the circled factors. [3.1B, p. 154]

$12 = 2^2 \cdot \textcircled{3}$
$18 = \textcircled{2} \cdot 3^2$
The GCF of 12 and 18 is $2 \cdot 3 = 6$.

To write an improper fraction as a mixed number, divide the numerator by the denominator. [3.2A, p. 160]

$\dfrac{29}{6} = 29 \div 6 = 4\dfrac{5}{6}$

To write a mixed number as an improper fraction, multiply the denominator of the fractional part of the mixed number by the whole number part. Add this product and the numerator of the fractional part. The sum is the numerator of the improper fraction. The denominator remains the same. [3.2A, p. 160]

$3\dfrac{2}{5} = \dfrac{5 \times 3 + 2}{5} = \dfrac{17}{5}$

To write a fraction in simplest form, divide the numerator and denominator of the fraction by their common factors. [3.2B, p. 163]

$\dfrac{30}{45} = \dfrac{2 \cdot \overset{1}{\cancel{3}} \cdot \overset{1}{\cancel{5}}}{\underset{1}{\cancel{3}} \cdot 3 \cdot \underset{1}{\cancel{5}}} = \dfrac{2}{3}$

To add fractions with the same denominators, add the numerators and place the sum over the common denominator.

$\dfrac{5}{12} + \dfrac{11}{12} = \dfrac{16}{12} = 1\dfrac{1}{3}$

$\dfrac{a}{b} + \dfrac{c}{b} = \dfrac{a + c}{b}$, where $b \neq 0$ [3.3A, p. 173]

Copyright © Houghton Mifflin Company. All rights reserved.

To subtract fractions with the same denominators, subtract the numerators and place the difference over the common denominator. $\dfrac{a}{b} - \dfrac{c}{b} = \dfrac{a-c}{b}$, where $b \neq 0$ [3.3B, p. 178]

$$\frac{9}{16} - \frac{5}{16} = \frac{4}{16} = \frac{1}{4}$$

To add or subtract fractions with different denominators, first rewrite the fractions as equivalent fractions with a common denominator. The common denominator is the least common multiple (LCM) of the denominators of the fractions. Then add or subtract the fractions. [3.3A/3.3B, pp. 173, 178]

$$\frac{7}{8} + \frac{5}{6} = \frac{21}{24} + \frac{20}{24} = \frac{41}{24} = 1\frac{17}{24}$$

$$\frac{2}{3} - \frac{7}{16} = \frac{32}{48} - \frac{21}{48} = \frac{11}{48}$$

To multiply two fractions, multiply the numerators; this is the numerator of the product. Multiply the denominators; this is the denominator of the product. $\dfrac{a}{b} \cdot \dfrac{c}{d} = \dfrac{ac}{bd}$, where $b \neq 0$ and $d \neq 0$ [3.4A, p. 191]

$$\frac{3}{4} \cdot \frac{2}{9} = \frac{3 \cdot 2}{4 \cdot 9} = \frac{3 \cdot 2}{2 \cdot 2 \cdot 3 \cdot 3} = \frac{1}{6}$$

To divide two fractions, multiply the first fraction by the reciprocal of the second fraction. $\dfrac{a}{b} \div \dfrac{c}{d} = \dfrac{a}{b} \cdot \dfrac{d}{c}$, where $b \neq 0$, $c \neq 0$, and $d \neq 0$ [3.4B, p. 196]

$$\frac{8}{15} \div \frac{4}{5} = \frac{8}{15} \cdot \frac{5}{4} = \frac{8 \cdot 5}{15 \cdot 4}$$
$$= \frac{2 \cdot 2 \cdot 2 \cdot 5}{3 \cdot 5 \cdot 2 \cdot 2} = \frac{2}{3}$$

The formula for the area of a triangle is $A = \dfrac{1}{2}bh$. [3.4C, p. 199]

Find the area of a triangle with a base measuring 6 ft and a height of 3 ft.

$$A = \frac{1}{2}bh = \frac{1}{2}(6)(3) = 9$$

The area is 9 ft².

Multiplication Property of Equations Each side of an equation can be multiplied by the same number (except zero) without changing the solution of the equation. [3.5A, p. 209]

$$\frac{5}{6}x = 10$$
$$\frac{6}{5} \cdot \frac{5}{6}x = \frac{6}{5} \cdot 10$$
$$x = 12$$

To simplify a complex fraction, simplify the expression above the main fraction bar and simplify the expression below the main fraction bar. Then rewrite the complex fraction as a division problem by reading the main fraction bar as "divided by." [3.6B, p. 217]

$$\frac{-\dfrac{8}{9} + \dfrac{2}{3}}{1\dfrac{1}{5}} = \frac{-\dfrac{8}{9} + \dfrac{6}{9}}{\dfrac{6}{5}} = \frac{-\dfrac{2}{9}}{\dfrac{6}{5}}$$

$$= -\frac{2}{9} \div \frac{6}{5} = -\frac{2}{9} \cdot \frac{5}{6} = -\frac{5}{27}$$

The Order of Operations Agreement [3.6C, p. 219]
Step 1 Do all operations inside parentheses.
Step 2 Simplify any numerical expressions containing exponents.
Step 3 Do multiplication and division as they occur from left to right.
Step 4 Do addition and subtraction as they occur from left to right.

$$\left(\frac{1}{3}\right)^2 + \left(\frac{7}{12} - \frac{5}{6}\right) \cdot (-4)$$
$$= \left(\frac{1}{3}\right)^2 + \left(-\frac{1}{4}\right) \cdot (-4)$$
$$= \frac{1}{9} + \left(-\frac{1}{4}\right) \cdot (-4) = \frac{1}{9} + 1 = 1\frac{1}{9}$$

Copyright © Houghton Mifflin Company. All rights reserved.

Chapter Review Exercises

1. Write $\frac{19}{2}$ as a mixed number.

2. Subtract: $6\frac{2}{9} - 3\frac{7}{18}$

3. Evaluate $x \div y$ when $x = 2\frac{5}{8}$ and $y = 1\frac{3}{4}$.

4. Multiply: $\left(-2\frac{1}{3}\right) \cdot \frac{3}{7}$

5. Divide: $3\frac{3}{4} \div 1\frac{7}{8}$

6. Find the product of 3 and $\frac{8}{9}$.

7. Evaluate $\frac{x}{y + z}$ when $x = \frac{7}{8}$, $y = \frac{4}{5}$, and $z = -\frac{1}{2}$.

8. Place the correct symbol, $<$ or $>$, between the two numbers.

   $\frac{3}{5}$ $\frac{7}{15}$

9. Find the LCM of 50 and 75.

10. Add: $6\frac{11}{15} + 4\frac{7}{10}$

11. Evaluate xy when $x = 8$ and $y = \frac{5}{12}$.

12. Express the shaded portion of the circles as an improper fraction and as a mixed number.

13. Place the correct symbol, $<$ or $>$, between the two numbers.

    $\frac{7}{8}$ $\frac{17}{20}$

14. Simplify: $\dfrac{\frac{5}{8} - \frac{1}{4}}{\frac{1}{2} + \frac{1}{8}}$

15. Write a fraction that is equivalent to $\frac{4}{9}$ and has a denominator of 72.

16. Evaluate x^2y^3 when $x = \frac{8}{9}$ and $y = -\frac{3}{4}$.

17. Evaluate $ab^2 - c$ when $a = 4$, $b = \frac{1}{2}$, and $c = \frac{5}{7}$.

18. Find the GCF of 42 and 63.

19. Write $2\frac{5}{14}$ as an improper fraction.

20. Evaluate $x + y + z$ when $x = \frac{5}{8}$, $y = -\frac{3}{4}$, and $z = \frac{1}{2}$.

21. Find the quotient of $\frac{5}{9}$ and $-\frac{2}{3}$.

22. Simplify: $\frac{2}{5} \div \frac{4}{7} + \frac{3}{8}$

Copyright © Houghton Mifflin Company. All rights reserved.

23. Multiply: $5\frac{1}{4} \cdot \frac{8}{9} \cdot (-3)$

24. Find the difference between $\frac{2}{3}$ and $\frac{11}{18}$.

25. Subtract: $\frac{7}{8} - \left(-\frac{5}{6}\right)$

26. Evaluate $\left(-\frac{3}{8}\right)^2 \cdot 4^2$.

27. Find the sum of $3\frac{7}{12}$ and $5\frac{1}{2}$.

28. Write $\frac{30}{105}$ in the simplest form.

29. Evaluate $a - b$ when $a = 7$ and $b = 2\frac{3}{10}$.

30. Solve: $-\frac{5}{9} = \frac{1}{6} + p$

31. *Measurement* What fractional part of an hour is 40 min?

32. *Geometry* An exercise course has stations set up along a path that is in the shape of a triangle that has sides that measure $12\frac{1}{12}$ yd, $29\frac{1}{3}$ yd, and $26\frac{3}{4}$ yd. What is the entire length of the exercise course? Use the formula $P = a + b + c$.

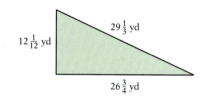

33. *Health* A wrestler is put on a diet to gain 12 lb in four weeks. The wrestler gains $3\frac{1}{2}$ lb the first week and $2\frac{1}{4}$ lb the second week. How much weight must the wrestler gain during the third and fourth weeks in order to gain a total of 12 lb?

34. *Business* An employee hired for piecework can assemble a unit in $2\frac{1}{2}$ min. How many units can this employee assemble during an 8-hour day?

35. *Wages* Find the overtime pay due an employee who worked $6\frac{1}{4}$ h of overtime this week. The employee's overtime rate is $24 an hour.

36. *Physics* What is the final velocity, in feet per second, of an object dropped from a plane with a starting velocity of 0 ft/s and a fall of $15\frac{1}{2}$ s? Use the formula $V = S + 32t$, where V is the final velocity of a falling object, S is its starting velocity, and t is the time of the fall.

Copyright © Houghton Mifflin Company. All rights reserved.

Chapter Test

1. Write $\frac{18}{7}$ as a mixed number.

2. Subtact: $7\frac{3}{4} - 3\frac{5}{6}$

3. Evaluate xy when $x = 6\frac{3}{7}$ and $y = 3\frac{1}{2}$.

4. Find the product of $-\frac{2}{3}$ and $-\frac{7}{8}$.

5. Find the LCM of 30 and 45.

6. Add: $\frac{11}{12} + \left(-\frac{3}{8}\right)$

7. Evaluate x^3y^2 when $x = 1\frac{1}{2}$ and $y = \frac{5}{6}$.

8. Write $3\frac{4}{5}$ as an improper fraction.

9. What is $-\frac{7}{12}$ divided by $-\frac{3}{4}$?

10. Simplify: $\frac{2}{7} \div \frac{3}{14} + \frac{2}{3}$

11. Evaluate $\frac{x}{yz}$ when $x = \frac{7}{20}$, $y = \frac{2}{15}$, and $z = \frac{3}{8}$.

12. Find the GCF of 18 and 54.

13. How much larger is $\frac{13}{14}$ than $\frac{16}{21}$?

14. Write $\frac{60}{75}$ in simplest form.

15. Evaluate $x + y + z$ when $x = 1\frac{3}{8}$, $y = \frac{1}{2}$, and $z = \frac{5}{6}$.

16. Place the correct symbol, $<$ or $>$, between the two numbers.

    $\dfrac{5}{6}$ $\dfrac{11}{15}$

17. Evaluate $a^2b - c^2$ when $a = \frac{2}{3}$, $b = 9$, and $c = \frac{3}{5}$.

18. Simplify: $\dfrac{\frac{3}{4} - \frac{1}{3}}{\frac{1}{6} + \frac{1}{3}}$

Copyright © Houghton Mifflin Company. All rights reserved.

19. Evaluate $\frac{x-y}{z^3}$ when $x = \frac{4}{9}$, $y = \frac{10}{27}$, and $z = \frac{2}{3}$.

20. Evaluate $x \div y$ when $x = -\frac{8}{9}$ and $y = \frac{16}{27}$.

21. Solve: $\frac{3x}{5} = -\frac{3}{10}$

22. Is $-\frac{3}{4}$ a solution of the equation $z + \frac{1}{5} = \frac{11}{20}$?

23. Multiply: $2\frac{7}{8} \cdot \frac{2}{11} \cdot 4$

24. Solve: $x + \frac{1}{3} = \frac{5}{6}$

25. Write a fraction that is equivalent to $\frac{3}{7}$ and has a denominator of 28.

26. A number minus one-half is equal to one-third. Find the number.

27. *Measurement* What fractional part of a day is 10 h?

28. *Health* A patient is put on a diet to lose 30 lb in three months. The patient loses $11\frac{1}{6}$ lb during the first month and $8\frac{5}{8}$ lb during the second month. Find the amount of weight the patient must lose during the third month to achieve the goal.

29. *Consumerism* You are planning a barbecue for 35 people. You want to serve $\frac{1}{4}$-pound hamburger patties to your guests and you estimate each person will eat two hamburgers. How much hamburger meat should you buy for the barbecue?

30. *Geometry* Find the amount of felt needed to make a pennant that is in the shape of a triangle with a base of 20 in. and a height of 12 in. Use the formula $A = \frac{1}{2}bh$.

31. *Community Service* You are required to contribute 20 h of community service to the town in which your college is located. On one occasion you work $7\frac{1}{4}$ h, and on another occasion you work $2\frac{3}{4}$ h. How many more hours of community service are still required of you?

32. *Business* An employee hired for piecework can assemble a unit in $4\frac{1}{2}$ min. How many units can this employee assemble in 6 h?

33. *Investments* Use the equation $C = SN$, where C is the cost of the shares of stock in a stock purchase, S is the cost per share, and N is the number of shares purchased, to find the cost of purchasing 400 shares of stock selling for $12\frac{3}{4}$ per share.

Copyright © Houghton Mifflin Company. All rights reserved.

Cumulative Review Exercises

1. Evaluate $3a + (a - b)^3$ when $a = 4$ and $b = 1$.

2. Find the product of 4 and $\frac{7}{8}$.

3. Add: $4\frac{7}{9} + 3\frac{5}{6}$

4. Subtract: $-42 - (-27)$

5. Find the GCF of 72 and 108.

6. Multiply: $3\frac{1}{13} \cdot 5\frac{1}{5}$

7. Find the quotient of $\frac{8}{9}$ and $-\frac{4}{5}$.

8. Subtract: $-\frac{2}{3} - \left(-\frac{2}{5}\right)$

9. Simplify: $\dfrac{\frac{1}{5} + \frac{1}{4}}{\frac{1}{4} - \frac{1}{5}}$

10. Place the correct symbol, $<$ or $>$, between the two numbers.

$$\frac{7}{11} \qquad \frac{4}{5}$$

11. Divide: $-2\frac{1}{3} \div 1\frac{2}{7}$

12. Multiply: $-\frac{3}{8} \cdot \frac{2}{5} \cdot \left(-\frac{4}{9}\right)$

13. Evaluate abc when $a = \frac{4}{7}$, $b = 1\frac{1}{6}$, and $c = 3$.

14. Subtract: $8\frac{3}{4} - 1\frac{5}{7}$

15. Subtract $-\frac{3}{8}$ from $\frac{7}{12}$.

16. Simplify: $\frac{2}{5} \div \frac{9 - 6}{3 + 7} + \left(-\frac{1}{2}\right)^2$

17. Evaluate $a - b$ when $a = \frac{3}{4}$ and $b = -\frac{7}{8}$.

18. Find the sum of $1\frac{9}{16}$ and $4\frac{5}{8}$.

19. Solve: $28 = -7y$

20. Write $\frac{41}{9}$ as a mixed number.

21. Find the difference between $\frac{5}{14}$ and $\frac{9}{42}$.

22. Evaluate $x^3 y^4$ when $x = \frac{7}{12}$ and $y = \frac{6}{7}$.

Copyright © Houghton Mifflin Company. All rights reserved.

23. Evaluate $2a - (b - a)^2$ when $a = 2$ and $b = -3$.

24. Add: $6,847 + 3,501 + 924$

25. Evaluate $(x - y)^3 + 5x$ when $x = 8$ and $y = 6$.

26. Solve: $x + \frac{4}{5} = \frac{1}{4}$

27. Estimate the difference between 89,357 and 66,042.

28. Simplify: $-8 - (-12) - (-15) - 32$

29. Write $7\frac{3}{4}$ as an improper fraction.

30. Find the prime factorization of 140.

31. *Health* The chart at the right shows the calories burned per hour as a result of different aerobic activities. Suppose you weigh 150 lb. According to the chart, how many more calories would you burn by bicycling at 12 mph for 4 h than by walking at a rate of 3 mph for 5 h?

Activity	100 lb	150 lb
Bicycling, 6 mph	160	240
Bicycling, 12 mph	270	410
Jogging, 5 1/2 mph	440	660
Jogging, 7 mph	610	920
Jumping rope	500	750
Tennis, singles	265	400
Walking, 2 mph	160	240
Walking, 3 mph	210	320
Walking, 4 1/2 mph	295	440

32. *Demographics* The Census Bureau projects that the population of New England will increase to 15,321,000 in 2020 from 13,581,000 in 2000. Find the projected increase in the population of New England during the 20-year period.

33. *Currency* The average life span of the $1 bill is one-sixth the average life span of the $100 bill. The average life span of the $1 bill is $1\frac{1}{2}$ years. (*Source:* Federal Reserve System; Bureau of Engraving and Printing) What is the average life span of the $100 bill?

34. *Geometry* Find the length of fencing needed to surround a square dog pen that measures $16\frac{1}{2}$ ft on each side. Use the formula $P = 4s$.

35. *Travel* A bicyclist rode for $\frac{3}{4}$ h at a rate of $5\frac{1}{2}$ mph. Use the equation $d = rt$, where d is the distance traveled, r is the rate of travel, and t is the time, to find the distance traveled by the bicyclist.

36. *Oceanography* The pressure on a submerged object is given by $P = 15 + \frac{1}{2}D$, where D is the depth in feet and P is the pressure measured in pounds per square inch. Find the pressure on a diver who is at a depth of $14\frac{3}{4}$ ft.

Copyright © Houghton Mifflin Company. All rights reserved.

Decimals and Real Numbers

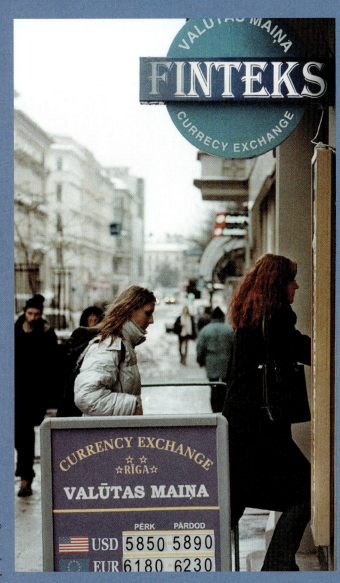

Copyright © Houghton Mifflin Company. All rights reserved.

Two visitors to the city of Riga, Latvia enter a currency exchange booth to exchange their money for the local currency. Currency rates, as seen on the sign here, are listed by country and currency type. The values go beyond the usual hundredth decimal place to increase the accuracy of the exchange. **Exercises 84 and 85 on page 268** illustrate calculating money equivalencies.

Need help? For online student resources, visit this website: **math.college.hmco.com**

Prep Test

1. Express the shaded portion of the rectangle as a fraction.

2. Round 36,852 to the nearest hundred.

3. Write 4791 in words.

4. Write six thousand eight hundred forty-two in standard form.

5. Graph -3 on the number line.

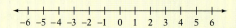

For Exercises 6–9, add, subtract, multiply, or divide.

6. $-37 + 8892 + 465$

7. $2403 - (-765)$

8. $-844(-91)$

9. $23\overline{)6412}$

10. Evaluate 8^2.

Go Figure

Super Yeast causes bread to double in volume each minute. If it takes one loaf of bread made with Super Yeast 30 minutes to fill the oven, how long does it take two loaves of bread made with Super Yeast to fill one-half the oven?

Copyright © Houghton Mifflin Company. All rights reserved.

Copyright © Houghton Mifflin Company. All rights reserved.

SECTION 4.1 ▶ **Introduction to Decimals**

OBJECTIVE **A**

VIDEO & DVD | CD TUTOR | WWW WEB | SSM

Place value

The price tag on a sweater reads $61.88. The number 61.88 is in **decimal notation.** A number written in decimal notation is often called simply a **decimal.**

A number written in decimal notation has three parts.

61	.	88
Whole number part	Decimal point	Decimal part

The decimal part of the number represents a number less than one. For example, $.88 is less than one dollar. The decimal point (.) separates the whole number part from the decimal part.

The position of a digit in a decimal determines the digit's place value. The place-value chart is extended to the right to show the place value of digits to the right of a decimal point.

In the decimal 458.302719, the position of the digit 7 determines that its place value is ten-thousandths.

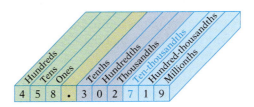

Note the relationship between fractions and numbers written in decimal notation.

seven tenths	seven hundredths	seven thousandths
$\frac{7}{10} = 0.7$	$\frac{7}{100} = 0.07$	$\frac{7}{1,000} = 0.007$
1 zero in 10	2 zeros in 100	3 zeros in 1,000
1 decimal place in 0.7	2 decimal places in 0.07	3 decimal places in 0.007

To write a decimal in words, write the decimal part of the number as though it were a whole number, and then name the place value of the last digit.

0.9684   nine thousand six hundred eighty-four ten-thousandths

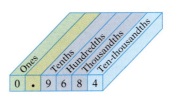

The decimal point in a decimal is read as "and."

372.516   three hundred seventy-two and five hundred sixteen thousandths

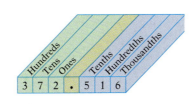

▶ **Point of Interest**

The idea that all fractions should be represented in tenths, hundredths, and thousandths was presented in 1585 in Simon Stevin's publication *De Thiende* and its French translation, *La Disme,* which was well read and accepted by the French. This may help to explain why the French accepted the metric system so easily two hundred years later.

In *De Thiende,* Stevin argued in favor of his notation by including examples for astronomers, tapestry makers, surveyors, tailors, and the like. He stated that using decimals would enable calculations to be "performed …with as much ease as counter-reckoning."

To write a decimal in standard form when it is written in words, write the whole number part, replace the word *and* with a decimal point, and write the decimal part so that the last digit is in the given place-value position.

four and twenty-three <u>hundredths</u>

3 is in the hundredths place.   4.2<u>3</u>

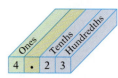

When writing a decimal in standard form, you may need to insert zeros after the decimal point so that the last digit is in the given place-value position.

ninety-one and eight <u>thousandths</u>

8 is in the thousandths place.   91.00<u>8</u>
Insert two zeros so that the 8 is in
the thousandths place.

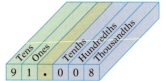

sixty-five <u>ten-thousandths</u>

5 is in the ten-thousandths place.   0.006<u>5</u>
Insert two zeros so that the 5 is in
the ten-thousandths place.

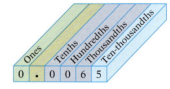

---

1 **Example** Name the place value of the digit 8 in the number 45.687.

Solution The digit 8 is in the hundredths place.

1 **You Try It** Name the place value of the digit 4 in the number 907.1342.

Your Solution

---

2 **Example** Write $\frac{43}{100}$ as a decimal.

Solution $\frac{43}{100} = 0.43$ ▶ forty-three hundredths

2 **You Try It** Write $\frac{501}{1,000}$ as a decimal.

Your Solution

---

3 **Example** Write 0.289 as a fraction.

Solution $0.289 = \frac{289}{1,000}$ ▶ 289 thousandths

3 **You Try It** Write 0.67 as a fraction.

Your Solution

---

4 **Example** Write 293.50816 in words.

Solution two hundred ninety-three and fifty thousand eight hundred sixteen hundred-thousandths

4 **You Try It** Write 55.6083 in words.

Your Solution

Solutions on p. S10

Copyright © Houghton Mifflin Company. All rights reserved.

⑤ *Example*   Write twenty-three and two hundred forty-seven millionths in standard form.

⑤ *You Try It*   Write eight hundred six and four hundred ninety-one hundred-thousandths in standard form.

Solution   23.000247

Your Solution

Solution on p. S10

Copyright © Houghton Mifflin Company. All rights reserved.

OBJECTIVE **B**

Order relations between decimals

A whole number can be written as a decimal by writing a decimal point to the right of the last digit. For example,

$62 = 62.$ $497 = 497.$

You know that \$62 and \$62.00 both represent sixty-two dollars. Any number of zeros may be written to the right of the decimal point in a whole number without changing the value of the number.

$62 = 62.00 = 62.0000$ $497 = 497.0 = 497.000$

Also, any number of zeros may be written to the right of the last digit in a decimal without changing the value of the number.

$0.8 = 0.80 = 0.800$ $1.35 = 1.350 = 1.3500 = 1.35000 = 1.350000$

This fact is used to find the order relation between two decimals.

To compare two decimals, write the decimal part of each number so that each has the same number of decimal places. Then compare the two numbers.

Place the correct symbol, $<$ or $>$, between the two numbers 0.693 and 0.71.

0.693 has 3 decimal places.
0.71 has 2 decimal places.
Write 0.71 with 3 decimal places. $0.71 = 0.710$

Compare 0.693 and 0.710.
693 thousandths $<$ 710 thousandths $0.693 < 0.710$

Remove the zero written in 0.710. $0.693 < 0.71$

Place the correct symbol, $<$ or $>$, between the two numbers 5.8 and 5.493.

Write 5.8 with 3 decimal places. $5.8 = 5.800$

Compare 5.800 and 5.493.
The whole number part (5) is the same.
800 thousandths $>$ 493 thousandths $5.800 > 5.493$

Remove the extra zeros written in 5.800. $5.8 > 5.493$

> ▶ **Point of Interest**
>
> The decimal point did not make its appearance until the early 1600s. Stevin's notation used subscripts with circles around them after each digit: 0 for ones, 1 for tenths (which he called "primes"), 2 for hundredths (called "seconds"), 3 for thousandths ("thirds"), and so on. For example, 1.375 would have been written
>
> 1   3   7   5
> ⓪   ①   ②   ③

6 *Example* Place the correct symbol, < or >, between the two numbers.

0.039   0.1001

Solution 0.039 = 0.0390

0.0390 < 0.1001

0.039 < 0.1001

6 *You Try It* Place the correct symbol, < or >, between the two numbers.

0.065   0.0802

Your Solution

7 *Example* Write the given numbers in order from smallest to largest.

1.01, 1.2, 1.002, 1.1, 1.12

Solution 1.010, 1.200, 1.002, 1.100, 1.120
1.002, 1.010, 1.100, 1.120, 1.200

1.002, 1.01, 1.1, 1.12, 1.2

7 *You Try It* Write the given numbers in order from smallest to largest.

3.03, 0.33, 0.3, 3.3, 0.03

Your Solution

Solutions on p. S10

OBJECTIVE C

Rounding

In general, rounding decimals is similar to rounding whole numbers except that the digits to the right of the given place value are dropped instead of being replaced by zeros.

If the digit to the right of the given place value is less than 5, that digit and all digits to the right are dropped.

Round 6.9237 to the nearest hundredth.

Given place value (hundredths)

6.9237

3 < 5   Drop the digits 3 and 7.

6.9237 rounded to the nearest hundredth is 6.92.

If the digit to the right of the given place value is greater than or equal to 5, increase the digit in the given place value by 1, and drop all digits to its right.

Round 12.385 to the nearest tenth.

Given place value (tenths)

12.385

8 > 5   Increase 3 by 1 and drop all digits to the right of 3.

12.385 rounded to the nearest tenth is 12.4.

Copyright © Houghton Mifflin Company. All rights reserved.

Round 0.46972 to the nearest thousandth.

Given place value (thousandths)

0.46972

7 > 5   Round up by adding 1 to the 9 (9 + 1 = 10).
Carry the 1 to the hundredths' place
(6 + 1 = 7).

0.46972 rounded to the nearest thousandth is 0.470.

Note that in this example, the zero in the given place value is not dropped. This indicates that the number is rounded to the nearest thousandth. If we dropped the zero and wrote 0.47, it would indicate that the number was rounded to the nearest hundredth.

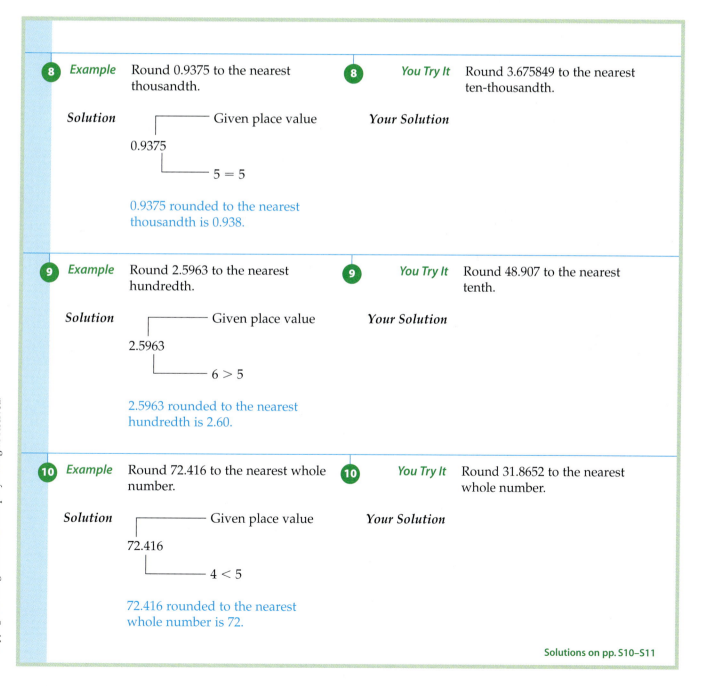

8 *Example*   Round 0.9375 to the nearest
thousandth.

Solution

Given place value

0.9375

5 = 5

0.9375 rounded to the nearest
thousandth is 0.938.

8   *You Try It*   Round 3.675849 to the nearest
ten-thousandth.

Your Solution

9 *Example*   Round 2.5963 to the nearest
hundredth.

Solution

Given place value

2.5963

6 > 5

2.5963 rounded to the nearest
hundredth is 2.60.

9   *You Try It*   Round 48.907 to the nearest
tenth.

Your Solution

10 *Example*   Round 72.416 to the nearest whole
number.

Solution

Given place value

72.416

4 < 5

72.416 rounded to the nearest
whole number is 72.

10   *You Try It*   Round 31.8652 to the nearest
whole number.

Your Solution

Solutions on pp. S10–S11

Copyright © Houghton Mifflin Company. All rights reserved.

Babe Ruth

VIDEO & DVD / CD TUTOR / WEB / SSM

OBJECTIVE D

Applications

 The table below shows the number of home runs hit, for every 100 times at bat, by four Major League baseball players. Use this table for Example 11 and You Try It 11.

Home Runs Hit for Every 100 At-Bats	
Harmon Killebrew	7.03
Ralph Kiner	7.09
Babe Ruth	8.05
Ted Williams	6.76

Source: Major League Baseball

11 *Example*

According to the table above, who had more home runs for every 100 times at bat, Ted Williams or Babe Ruth?

Strategy

To determine who had more home runs for every 100 times at bat, compare the numbers 6.76 and 8.05.

Solution

8.05 > 6.76

Babe Ruth had more home runs for every 100 at-bats.

11 *You Try It*

According to the table above, who had more home runs for every 100 times at bat, Harmon Killebrew or Ralph Kiner?

Your Strategy

Your Solution

12 *Example*

 On average, an American goes to the movies 4.56 times per year. To the nearest whole number, how many times per year does an American go to the movies?

Strategy

To find the number, round 4.56 to the nearest whole number.

Solution

4.56 rounded to the nearest whole number is 5.

An American goes to the movies about 5 times per year.

12 *You Try It*

 One of the driest cities in the Southwest is Yuma, Arizona, with an average annual precipitation of 2.65 in. To the nearest inch, what is the average annual precipitation in Yuma?

Your Strategy

Your Solution

Solutions on p. S11

Copyright © Houghton Mifflin Company. All rights reserved.

4.1 EXERCISES

Name the place value of the digit 5.

1. 76.31587

2. 291.508

3. 432.09157

4. 0.0006512

5. 38.2591

6. 0.0000853

Write the fraction as a decimal.

7. $\dfrac{3}{10}$

8. $\dfrac{9}{10}$

9. $\dfrac{21}{100}$

10. $\dfrac{87}{100}$

11. $\dfrac{461}{1,000}$

12. $\dfrac{853}{1,000}$

13. $\dfrac{93}{1,000}$

14. $\dfrac{61}{1,000}$

Write the decimal as a fraction.

15. 0.1

16. 0.3

17. 0.47

18. 0.59

19. 0.289

20. 0.601

21. 0.09

22. 0.013

Write the number in words.

23. 0.37

24. 25.6

25. 9.4

26. 1.004

27. 0.0053

28. 41.108

29. 0.045

30. 3.157

31. 26.04

Copyright © Houghton Mifflin Company. All rights reserved.

Write the number in standard form.

32. six hundred seventy-two thousandths

33. three and eight hundred six ten-thousandths

34. nine and four hundred seven ten-thousandths

35. four hundred seven and three hundredths

36. six hundred twelve and seven hundred four thousandths

37. two hundred forty-six and twenty-four thousandths

38. two thousand sixty-seven and nine thousand two ten-thousandths

39. seventy-three and two thousand six hundred eighty-four hundred-thousandths

OBJECTIVE **B**

Place the correct symbol, $<$ or $>$, between the two numbers.

40. 0.16   0.6

41. 0.7   0.56

42. 5.54   5.45

43. 3.605   3.065

44. 0.047   0.407

45. 9.004   9.04

46. 1.0008   1.008

47. 9.31   9.031

48. 7.6005   7.605

49. 4.6   40.6

50. 0.31502   0.3152

51. 0.07046   0.07036

Write the given numbers in order from smallest to largest.

52. 0.39, 0.309, 0.399

53. 0.66, 0.699, 0.696, 0.609

54. 0.24, 0.024, 0.204, 0.0024

55. 1.327, 1.237, 1.732, 1.372

56. 0.06, 0.059, 0.061, 0.0061

57. 21.87, 21.875, 21.805, 21.78

Copyright © Houghton Mifflin Company. All rights reserved.

OBJECTIVE   C

Round the number to the given place value.

58. 6.249   Tenths

59. 5.398   Tenths

60. 21.007   Tenths

61. 30.0092   Tenths

62. 18.40937   Hundredths

63. 413.5972   Hundredths

64. 72.4983   Hundredths

65. 6.061745   Thousandths

66. 936.2905   Thousandths

67. 96.8027   Whole number

68. 47.3192   Whole number

69. 5,439.83   Whole number

70. 7,014.96   Whole number

71. 0.023591   Ten-thousandths

72. 2.975268   Hundred-thousandths

OBJECTIVE   D

73. *Measurement*   A nickel weighs about 0.1763668 oz. Find the weight of a nickel to the nearest hundredth of an ounce.

74. *Business*   The total cost of a parka, including sales tax, is $124.1093. Round the total cost to the nearest cent to find the amount a customer pays for the parka.

75. *Sports*   Runners in the Boston Marathon run a distance of 26.21875 mi. To the nearest tenth of a mile, find the distance an entrant who completes the Boston Marathon runs.

The table at the right lists National Football League leading lifetime rushers. Use the table for Exercises 76 and 77.

76. *Sports*   Who had the greater average number of yards per carry, Tony Dorsett or Emmitt Smith?

77. *Sports*   Of all the players listed in the table, who has the greatest average number of yards per carry?

Football Player	Average Number of Yards per Carry
Eric Dickerson	4.43
Tony Dorsett	4.34
Walter Payton	4.36
Barry Sanders	4.99
Emmitt Smith	4.24

Source: Pro Football Hall of Fame

78.   *Health*   The average life expectancy in Great Britain is 75.3 years. The average life expectancy in Italy is 75.5 years. (*Source:* U.S. Centers for Disease Control) In which country is the average life expectancy higher, Great Britain or Italy?

Copyright © Houghton Mifflin Company. All rights reserved.

79. *The Olympics* The length of the marathon footrace in the Olympics is 42.195 km. What is the length of this race to the nearest tenth of a kilometer?

80. *Consumerism* Charge accounts generally require a minimum payment on the balance in the account each month. Use the minimum payment schedule shown below to determine the minimum payment due on the given account balances.

a. $187.93
b. $342.55
c. $261.48
d. $16.99
e. $310.00
f. $158.32
g. $200.10

If the New Balance Is:	The Minimum Required Payment Is:
Up to $20.00	The new balance
$20.01 to $200.00	$20.00
$200.01 to $250.00	$25.00
$250.01 to $300.00	$30.00
$300.01 to $350.00	$35.00
$350.01 to $400.00	$40.00

81. *Consumerism* Shipping and handling charges when ordering online generally are based on the dollar amount of the order. Use the table shown below to determine the cost of shipping each order.

a. $12.42
b. $23.56
c. $47.80
d. $66.91
e. $35.75
f. $20.00
g. $18.25

If the Amount Ordered Is:	The Shipping and Handling Charge Is:
$10.00 and under	$1.60
$10.01 to $20.00	$2.40
$20.01 to $30.00	$3.60
$30.01 to $40.00	$4.70
$40.01 to $50.00	$6.00
$50.01 and up	$7.00

Critical Thinking

82. Indicate which digits of the number, if any, need not be entered on a calculator.
   a. 1.500   **b.** 0.908   **c.** 60.07   **d.** 0.0032

83. Find a number between (a) 0.1 and 0.2, (b) 1 and 1.1, and (c) 0 and 0.005.

84. To what place value are timed events in the Olympics recorded? Provide some specific examples of events and the winning times in each.

85. Provide an example of a situation in which a decimal is always rounded up, even if the digit to the right is less than 5. Provide an example of a situation in which a decimal is always rounded down, even if the digit to the right is 5 or greater than 5. (*Hint:* Think about situations in which money changes hands.)

86. Prepare a report on the Richter scale. Include in your report the magnitudes that classify an earthquake as strong or moderate, the magnitudes that classify an earthquake as a microearthquake, and the largest known recorded shocks.

Copyright © Houghton Mifflin Company. All rights reserved.

Copyright © Houghton Mifflin Company. All rights reserved.

SECTION 4.2 ▸   **Operations on Decimals**

OBJECTIVE **A**

VIDEO & DVD   CD TUTOR   WEB   SSM

Addition and subtraction of decimals

To add decimals, write the numbers so that the decimal points are on a vertical line. Add as you would with whole numbers. Then write the decimal point in the sum directly below the decimal points in the addends.

Add: 0.326 + 4.8 + 57.23

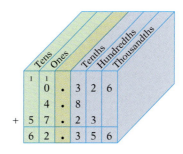

Note that placing the decimal points on a vertical line ensures that digits of the same place value are added.

> **Point of Interest**
>
> Try this: Six different numbers are added together and their sum is 11. Four of the six numbers are 4, 3, 2, and 1. Find the other two numbers.

Find the sum of 0.64, 8.731, 12, and 5.9.

Arrange the numbers vertically, placing the decimal points on a vertical line.

Add the numbers in each column.

Write the decimal point in the sum directly below the decimal points in the addends.

$$
\begin{array}{r}
\scriptstyle 1\ 2 \\
0.64 \\
8.731 \\
12. \\
+\ 5.9 \\
\hline
27.271
\end{array}
$$

To subtract decimals, write the numbers so that the decimal points are on a vertical line. Subtract as you would with whole numbers. Then write the decimal point in the difference directly below the decimal point in the subtrahend.

Subtract and check: 31.642 − 8.759

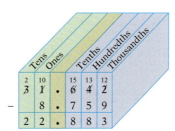

Note that placing the decimal points on a vertical line ensures that digits of the same place value are subtracted.

Check: Subtrahend   8.759
        + Difference   + 22.883
        ──────────────── ────────
        = Minuend   31.642

Subtract and check: $5.4 - 1.6832$

Insert zeros in the minuend so that it has the same number of decimal places as the subtrahend.

$$\begin{array}{r} 5.4000 \\ -\ 1.6832 \end{array}$$

Subtract and then check.

$$\begin{array}{r} {}^{4\ 13\ 9\ 9\ 10} \\ \cancel{5.4000} \\ -\ 1.6832 \\ \hline 3.7168 \end{array} \quad \text{Check:} \quad \begin{array}{r} 1.6832 \\ +\ 3.7168 \\ \hline 5.4000 \end{array}$$

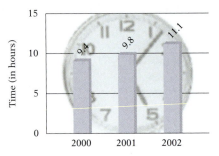 Figure 4.1 shows the average number of hours Internet users spent online each week in 2000, 2001, and 2002. On average, how many more hours per week did Internet users spend online in 2002 than in 2000?

Figure 4.1 Average Time Internet Users Spent Online Each Week
Source: UCLA Center for Communication Policy

Subtract the number of hours given for 2000 from the number given for 2002.

$11.1 - 9.4 = 1.7$

On average, Internet users spent 1.7 more hours online per week in 2002 than in 2000.

The sign rules for adding and subtracting decimals are the same rules used to add and subtract integers.

Simplify: $-36.087 + 54.29$

The signs of the addends are different. Subtract the smaller absolute value from the larger absolute value.

$54.29 - 36.087 = 18.203$

Attach the sign of the number with the larger absolute value.

$|54.29| > |-36.087|$

The sum is positive.

$$-36.087 + 54.29 = 18.203$$

> **Take Note**
>
> Recall that the absolute value of a number is the distance from zero to the number on the number line. The absolute value of a number is a positive number or zero.
>
> $|54.29| = 54.29$
>
> $|-36.087| = 36.087$

Recall that the opposite or additive inverse of n is $-n$ and the opposite of $-n$ is n. To find the opposite of a number, change the sign of the number.

Simplify: $-2.86 - 10.3$

Rewrite subtraction as addition of the opposite. The opposite of 10.3 is -10.3.

$$-2.86 - 10.3$$
$$= -2.86 + (-10.3)$$

The signs of the addends are the same. Add the absolute values of the numbers. Attach the sign of the addends.

$$= -13.16$$

Copyright © Houghton Mifflin Company. All rights reserved.

Evaluate $c - d$ when $c = 6.731$ and $d = -2.48$.

	$c - d$
Replace c with 6.731 and d with -2.48.	$6.731 - (-2.48)$
Rewrite subtraction as addition of the opposite.	$= 6.731 + 2.48$
Add.	$= 9.211$

> ▶ **Point of Interest**
>
> **Try this brain teaser. You have two U.S. coins that add up to $.55. One is not a nickel. What are these two coins?**

Recall that to estimate the answer to a calculation, round each number to the highest place value of the number; the first digit of each number will be nonzero and all other digits will be zero. Perform the calculation using the rounded numbers.

Estimate the sum of 23.037 and 16.7892.

Round each number to the nearest ten.

$$23.037 \longrightarrow 20$$
$$16.7892 \longrightarrow +\ 20$$
$$\overline{\hphantom{16.7892 \longrightarrow +\ 2}40}$$

Add the rounded numbers.

40 is an estimate of the sum of 23.037 and 16.7892. Note that 40 is very close to the actual sum of 39.8262.

$$23.037$$
$$+\ 16.7892$$
$$\overline{39.8262}$$

When a number in an estimation is a decimal less than 1, round the decimal so that there is one nonzero digit.

Estimate the difference between 4.895 and 0.6193.

Round 4.895 to the nearest one.
Round 0.6193 to the nearest tenth.
Subtract the rounded numbers.

$$4.895 \longrightarrow 5.0$$
$$0.6193 \longrightarrow -\ 0.6$$
$$\overline{\hphantom{0.6193 \longrightarrow -\ 0}4.4}$$

4.4 is an estimate of the difference between 4.895 and 0.6193.
It is close to the actual difference of 4.2757.

$$4.8950$$
$$-\ 0.6193$$
$$\overline{4.2757}$$

---

1 **Example**   Add: $35.8 + 182.406 + 71.0934$

Solution
$$\overset{1\quad 1}{}$$
$$35.8$$
$$182.406$$
$$+\ 71.0934$$
$$\overline{289.2994}$$

1 **You Try It**   Add: $8.64 + 52.7 + 0.39105$

Your Solution

---

2 **Example**   What is -251.49 more than -638.7?

Solution $-638.7 + (-251.49) = -890.19$

2 **You Try It**   What is 4.002 minus 9.378?

Your Solution

Solutions on p. S11

Copyright © Houghton Mifflin Company. All rights reserved.

③ Example Subtract and check: $73 - 8.16$

Solution

$$\begin{array}{r} {\scriptstyle 6\ 12\ \ 9\ 10} \\ 7\cancel{3}.\cancel{0}\cancel{0} \\ -\ \ 8.16 \\ \hline 64.84 \end{array}$$

Check:
$$\begin{array}{r} 8.16 \\ +\ 64.84 \\ \hline 73.00 \end{array}$$

③ You Try It Subtract and check: $25 - 4.91$

Your Solution

---

④ Example Estimate the sum of 0.3927, 0.4856, and 0.2104.

Solution

$$\begin{array}{rcr} 0.3927 & \longrightarrow & 0.4 \\ 0.4856 & \longrightarrow & 0.5 \\ 0.2104 & \longrightarrow & +\ 0.2 \\ & & \hline \\ & & 1.1 \end{array}$$

④ You Try It Estimate the sum of 6.514, 8.903, and 2.275.

Your Solution

---

⑤ Example Evaluate $x + y + z$ when $x = -1.6$, $y = 7.9$, and $z = -4.8$.

Solution

$x + y + z$
$-1.6 + 7.9 + (-4.8) = 6.3 + (-4.8)$
$\qquad\qquad\qquad\qquad = 1.5$

⑤ You Try It Evaluate $x + y + z$ when $x = -7.84$, $y = -3.05$, and $z = 2.19$.

Your Solution

---

⑥ Example Is -4.3 a solution of the equation $9.7 - b = 5.4$?

Solution

$$\begin{array}{r|l} 9.7 - b = 5.4 \\ \hline 9.7 - (-4.3) & 5.4 \\ 9.7 + 4.3 & 5.4 \\ 14.0 & \neq 5.4 \end{array}$$

No, -4.3 is not a solution of the equation.

⑥ You Try It Is -23.8 a solution of the equation $-m + 16.9 = 40.7$?

Your Solution

Solutions on p. S11

OBJECTIVE B

Multiplication of decimals

Decimals are multiplied as though they were whole numbers; then the decimal point is placed in the product. Writing the decimals as fractions shows where to write the decimal point in the product.

$$0.4 \cdot 2 = \frac{4}{10} \cdot \frac{2}{1} = \frac{8}{10} = 0.8$$

1 decimal place in 0.4   1 decimal place in 0.8

Copyright © Houghton Mifflin Company. All rights reserved.

$$0.4 \cdot 0.2 = \frac{4}{10} \cdot \frac{2}{10} = \frac{8}{100} = 0.08$$

1 decimal place in 0.4   2 decimal places in 0.08
1 decimal place in 0.2

$$0.4 \cdot 0.02 = \frac{4}{10} \cdot \frac{2}{100} = \frac{8}{1,000} = 0.008$$

1 decimal place in 0.4   3 decimal places in 0.008
2 decimal places in 0.02

To multiply decimals, multiply the numbers as you would whole numbers. Then write the decimal point in the product so that the number of decimal places in the product is the sum of the numbers of decimal places in the factors.

Multiply: (32.41)(7.6)

```
   32.41     2 decimal places
 ×   7.6     1 decimal place
 ───────
   19446
  22687
 ───────
 246.316     3 decimal places
```

Calculator Note

Scientific calculators have a floating decimal point. This means that the decimal point is automatically placed in the answer. For example, for the product at the left, enter

32 . 41 × 7 . 6 =

The display reads 246.316, with the decimal point in the correct position.

Estimating the product of 32.41 and 7.6 shows that the decimal point has been correctly placed.

Round 32.41 to the nearest ten.   32.41 ⟶   30
Round 7.6 to the nearest one.   7.6 ⟶   × 8
Multiply the two numbers.   ─────
                                                        240

240 is an estimate of (32.41)(7.6). It is close to the actual product 246.316.

Multiply: 0.061(0.08)

```
   0.061      3 decimal places
 × 0.08       2 decimal places
 ───────
 0.00488      5 decimal places
```

Insert two zeros between the 4 and the decimal point so that there are 5 decimal places in the product.

Figure 4.2 shows U.S. Postal Service rates for express mail. How much would it cost a company to mail 25 Express Mail packages, each weighing 0.75 oz, post office to addressee?

Multiply the rate per package (17.85) by the number of packages (25).

17.85(25) = 446.25

The cost would be $446.25.

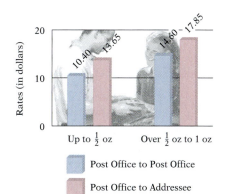

Figure 4.2   U.S. Postal Service Rates for Express Mail

Copyright © Houghton Mifflin Company. All rights reserved.

To multiply a decimal by a power of 10 (10, 100, 1,000, . . .), move the decimal point to the right the same number of places as there are zeros in the power of 10.

$$2.7935 \cdot \underline{10} \qquad = 27.935$$

1 zero   1 decimal place

$$2.7935 \cdot \underline{100} \qquad = 279.35$$

2 zeros   2 decimal places

$$2.7935 \cdot 1,\underline{000} \qquad = 2,793.5$$

3 zeros   3 decimal places

$$2.7935 \cdot \underline{10,000} \qquad = 27,935.$$

4 zeros   4 decimal places

$$2.7935 \cdot \underline{100,000} \qquad = 279,350.$$

A zero must be inserted before the decimal point.

5 zeros   5 decimal places

Note that if the power of 10 is written in exponential notation, the exponent indicates how many places to move the decimal point.

$$2.7935 \cdot 10^1 = 27.935$$

1 decimal place

$$2.7935 \cdot 10^2 = 279.35$$

2 decimal places

$$2.7935 \cdot 10^3 = 2,793.5$$

3 decimal places

$$2.7935 \cdot 10^4 = 27,935.$$

4 decimal places

$$2.7935 \cdot 10^5 = 279,350.$$

5 decimal places

Find the product of 64.18 and 10^3.

The exponent on 10 is 3. Move the decimal point in 64.18 three places to the right.

$$64.18 \cdot 10^3 = 64,180$$

Evaluate $100x$ when $x = 5.714$.

$$100x$$

Replace x with 5.714.

$$100(5.714)$$

Multiply. There are two zeros in 100. Move the decimal point in 5.714 two places to the right.

$$= 571.4$$

Copyright © Houghton Mifflin Company. All rights reserved.

The sign rules for multiplying decimals are the same rules used to multiply integers.

The product of two numbers with the same sign is positive.
The product of two numbers with different signs is negative.

Multiply: $(-3.2)(-0.008)$

The signs are the same.
The product is positive.
Multiply the absolute values of
the numbers. $(-3.2)(-0.008) = 0.0256$

Is -0.6 a solution of the equation $4.3a = -2.58$?

$$4.3a = -2.58$$

Replace a by -0.6 and then simplify. $4.3(-0.6) \mid -2.58$
The results are equal. $-2.58 = -2.58$

Yes, -0.6 is a solution of the equation.

---

7 *Example*   Multiply: $0.00073(0.052)$

Solution
$$\begin{array}{r} 0.00073 \\ \times\quad 0.052 \\ \hline 146 \\ 365 \\ \hline 0.00003796 \end{array}$$

7 *You Try It*   Multiply: $0.000081(0.025)$

Your Solution

---

8 *Example*   Estimate the product of 0.7639 and 0.2188.

Solution
$$\begin{array}{r} 0.7639 \longrightarrow \quad 0.8 \\ 0.2188 \longrightarrow \times\ 0.2 \\ \hline 0.16 \end{array}$$

8 *You Try It*   Estimate the product of 6.407 and 0.959.

Your Solution

---

9 *Example*   What is 835.294 multiplied by 1,000?

Solution $835.294 \cdot 1{,}000 = 835{,}294$

9 *You Try It*   Find the product of 1.756 and 10^4.

Your Solution

---

10 *Example*   Multiply: $-3.42(6.1)$

Solution $-3.42(6.1) = -20.862$

10 *You Try It*   Multiply: $(-0.7)(-5.8)$

Your Solution

Solutions on p. S11

Copyright © Houghton Mifflin Company. All rights reserved.

Example Evaluate $50ab$ when $a = -0.9$ and $b = -0.2$.

Solution $50ab$
$$50(-0.9)(-0.2) = -45(-0.2)$$
$$= 9$$

 You Try It Evaluate $25xy$ when $x = -0.8$ and $y = 0.6$.

Your Solution

Solution on p. S11

OBJECTIVE **C**

Division of decimals

Point of Interest

Benjamin Banneker (1731–1806) was the first African American to earn distinction as a mathematician and a scientist. He was on the survey team that determined the boundaries of Washington, D.C. The mathematics of surveying requires extensive use of decimals.

To divide decimals, move the decimal point in the divisor to the right so that the divisor is a whole number. Move the decimal point in the dividend the same number of places to the right. Place the decimal point in the quotient directly above the decimal point in the dividend. Then divide as you would with whole numbers.

Divide: $29.585 \div 4.85$

$$4.85\overline{)29.58.5}$$

Move the decimal point 2 places to the right in the divisor. Move the decimal point 2 places to the right in the dividend. Place the decimal point in the quotient. Then divide as shown at the right.

$$
\begin{array}{r}
6.1 \\
485\overline{)2958.5} \\
-2910 \\
\hline
48\ 5 \\
-48\ 5 \\
\hline
0
\end{array}
$$

Moving the decimal point the same number of places in the divisor and the dividend does not change the quotient because the process is the same as multiplying the numerator and denominator of a fraction by the same number. For the last example,

$$4.85\overline{)29.585} = \frac{29.585}{4.85} = \frac{29.585 \cdot 100}{4.85 \cdot 100} = \frac{2958.5}{485} = 485\overline{)2958.5}$$

In division of decimals, rather than writing the quotient with a remainder, we usually round the quotient to a specified place value. The symbol $\approx$, which is read "is approximately equal to," is used to indicate that the quotient is an approximate value after being rounded.

Divide and round to the nearest tenth: $0.86 \div 0.7$

$$
\begin{array}{r}
1.22 \approx 1.2 \\
0.7\overline{)0.8.60} \\
-7 \\
\hline
1\ 6 \\
-1\ 4 \\
\hline
20 \\
-14 \\
\hline
6
\end{array}
$$

To round the quotient to the nearest tenth, the division must be carried to the hundredths place. Therefore, zeros must be inserted in the dividend so that the quotient has a digit in the hundredths place.

Copyright © Houghton Mifflin Company. All rights reserved.

Figure 4.3 shows average hourly earnings in the United States. How many times greater were the average hourly earnings in 2000 than in 1970? Round to the nearest whole number.

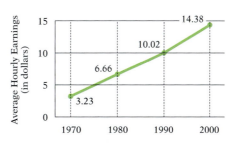

Figure 4.3   Average Hourly Earnings
Source: Statistical Abstract of the United States; Bureau of Labor Statistics

Divide the 2000 average hourly earnings (14.38) by those in 1970 (3.23).

$14.38 \div 3.23 \approx 4$

The average hourly earnings in 2000 were about 4 times greater than in 1970.

To divide a decimal by a power of 10 (10, 100, 1,000, 10,000, . . .), move the decimal point to the left the same number of places as there are zeros in the power of 10.

$462.81 \div 1\underline{0} \qquad = 46.281$

  1 zero   1 decimal place

$462.81 \div 1\underline{00} \qquad = 4.6281$

  2 zeros   2 decimal places

$462.81 \div 1,\underline{000} \quad = 0.46281$

  3 zeros   3 decimal places

$462.81 \div 1\underline{0,000} \quad = 0.046281$

  4 zeros   4 decimal places

A zero must be inserted between the decimal point and the 4.

$462.81 \div 1\underline{00,000} = 0.0046281$

  5 zeros   5 decimal places

Two zeros must be inserted between the decimal point and the 4.

If the power of 10 is written in exponential notation, the exponent indicates how many places to move the decimal point.

$462.81 \div 10^1 = 46.281$

    1 decimal place

$462.81 \div 10^2 = 4.6281$

    2 decimal places

$462.81 \div 10^3 = 0.46281$

    3 decimal places

$462.81 \div 10^4 = 0.046281$

    4 decimal places

$462.81 \div 10^5 = 0.0046281$

    5 decimal places

Copyright © Houghton Mifflin Company. All rights reserved.

Find the quotient of 3.59 and 100.

There are two zeros in 100. Move the
decimal point in 3.59 two places to the left. $3.59 \div 100 = 0.0359$

What is the quotient of 64.79 and 10^4?

The exponent on 10 is 4. Move the decimal
point in 64.79 four places to the left. $64.79 \div 10^4 = 0.006479$

The sign rules for dividing integers are the same rules used to divide decimals.

The quotient of two numbers with the same sign is positive.
The quotient of two numbers with different signs is negative.

Divide: $-1.16 \div 2.9$

The signs are different.
The quotient is negative.
Divide the absolute values of the numbers. $-1.16 \div 2.9 = -0.4$

Evaluate $c \div d$ when $c = -8.64$ and $d = -0.4$.

Replace c with -8.64 and d with -0.4. $c \div d$
$(-8.64) \div (-0.4)$

The signs are the same. The quotient is positive.
Divide the absolute values of the numbers. $= 21.6$

---

12 *Example*   Divide: $431.97 \div 7.26$

Solution
$$
\begin{array}{r}
5\,9.5 \\
7.26.\overline{)4\,3\,1.9\,7.0} \\
-3\,6\,3\,0 \\
\hline
6\,8\,9\,7 \\
-6\,5\,3\,4 \\
\hline
3\,6\,3\,0 \\
-3\,6\,3\,0 \\
\hline
0
\end{array}
$$

12 *You Try It*   Divide: $314.746 \div 6.53$

Your Solution

---

13 *Example*   Estimate the quotient of 8.37 and
0.219.

Solution $8.37 \longrightarrow 8$
$0.219 \longrightarrow 0.2$

$8 \div 0.2 = 40$

13 *You Try It*   Estimate the quotient of 62.7
and 3.45.

Your Solution

Solutions on p. S11

Copyright © Houghton Mifflin Company. All rights reserved.

14 *Example*   Divide and round to the nearest hundredth: $448.2 \div 53$

14 *You Try It*   Divide and round to the nearest thousandth: $519.37 \div 86$

Solution

$$
\begin{array}{r}
8.4\,5\,6 \approx 8.46 \\
53\overline{)4\,4\,8.2\,0\,0} \\
-4\,2\,4 \\
\hline
2\,4\,2 \\
-2\,1\,2 \\
\hline
3\,0\,0 \\
-2\,6\,5 \\
\hline
3\,5\,0 \\
-3\,1\,8 \\
\hline
3\,2
\end{array}
$$

Your Solution

---

15 *Example*   Find the quotient of 592.4 and 10^4.

15 *You Try It*   What is 63.7 divided by 100?

Solution $592.4 \div 10^4 = 0.05924$

Your Solution

---

16 *Example*   Divide and round to the nearest tenth: $-6.94 \div -1.5$

16 *You Try It*   Divide and round to the nearest tenth: $-25.7 \div 0.31$

Solution   The quotient is positive.

$-6.94 \div (-1.5) \approx 4.6$

Your Solution

---

17 *Example*   Evaluate $\dfrac{x}{y}$ when $x = -76.8$ and $y = 0.8$.

17 *You Try It*   Evaluate $\dfrac{x}{y}$ when $x = -40.6$ and $y = -0.7$.

Solution $\dfrac{x}{y}$

$\dfrac{-76.8}{0.8} = -96$

Your Solution

---

18 *Example*   Is -0.4 a solution of the equation $\dfrac{8}{x} = -20$?

18 *You Try It*   Is -1.2 a solution of the equation $-2 = \dfrac{d}{-0.6}$?

Solution $\dfrac{8}{x} = -20$

$$
\begin{array}{c|c}
\dfrac{8}{-0.4} & -20 \\
\hline
-20 = -20
\end{array}
$$

Yes, -0.4 is a solution of the equation.

Your Solution

Solutions on p. S11

Copyright © Houghton Mifflin Company. All rights reserved.

OBJECTIVE **D**

VIDEO & DVD   CD TUTOR   WEB   SSM

Fractions and decimals

Because the fraction bar can be read "divided by," any fraction can be written as a decimal. To write a fraction as a decimal, divide the numerator of the fraction by the denominator.

Take Note

The fraction bar can be read "divided by."

$$\frac{3}{4} = 3 \div 4$$

Dividing the numerator by the denominator results in a remainder of 0. The decimal 0.75 is a terminating decimal.

Convert $\frac{3}{4}$ to a decimal.

$$
\begin{array}{r}
0.75 \\
4\overline{)3.00} \\
-2\,8 \\
\hline
20 \\
-20 \\
\hline
0
\end{array}
$$

0.75 ⟵ This is a **terminating decimal.**

0 ⟵ The remainder is zero.

$$\frac{3}{4} = 0.75$$

Convert $\frac{5}{11}$ to a decimal.

$$
\begin{array}{r}
0.4545 \\
11\overline{)5.0000} \\
-4\,4 \\
\hline
60 \\
-55 \\
\hline
50 \\
-44 \\
\hline
60 \\
-55 \\
\hline
5
\end{array}
$$

0.4545 ⟵ This is a **repeating decimal.**

5 ⟵ The remainder is never zero.

Take Note

No matter how far we carry out the division, the remainder is never zero. The decimal $0.\overline{45}$ is a repeating decimal.

$$\frac{5}{11} = 0.\overline{45}$$ The bar over the digits 45 is used to show that these digits repeat.

Convert $2\frac{4}{9}$ to a decimal.

Write the fractional part of the mixed number as a decimal. Divide the numerator by the denominator.

$$
\begin{array}{r}
0.444 = 0.\overline{4} \\
9\overline{)4.000}
\end{array}
$$

The whole number part of the mixed number is the whole number part of the decimal.

$$2\frac{4}{9} = 2.\overline{4}$$

To convert a decimal to a fraction, remove the decimal point and place the decimal part over a denominator equal to the place value of the last digit in the decimal.

↱ hundredths   ↱ hundredths   ↱ tenths

$$0.57 = \frac{57}{100} \qquad\qquad 7.65 = 7\frac{65}{100} = 7\frac{13}{20} \qquad 8.6 = 8\frac{6}{10} = 8\frac{3}{5}$$

Copyright © Houghton Mifflin Company. All rights reserved.

Convert 4.375 to a fraction.

The 5 in 4.375 is in the thousandths place. Write 0.375 as a fraction with a denominator of 1,000.

$$4.375 = 4\frac{375}{1,000}$$

Simplify the fraction.

$$= 4\frac{3}{8}$$

Calculator Note

Some calculators *truncate* a decimal number that exceeds the calculator display. This means that the digits beyond the calculator's display are not shown. For this type of calculator, $\frac{2}{3}$ would be shown as 0.66666666. Other calculators *round* a decimal number when the calculator display is exceeded. For this type of calculator, $\frac{2}{3}$ would be shown as 0.66666667.

To find the order relation between a fraction and a decimal, first rewrite the fraction as a decimal. Then compare the two decimals.

Find the order relation between $\frac{6}{7}$ and 0.855.

Write the fraction as a decimal. Round to one more place value than the given decimal. (0.855 has 3 decimal places; round to 4 decimal places.)

$$\frac{6}{7} \approx 0.8571$$

Compare the two decimals.

$$0.8571 > 0.8550$$

Replace the decimal approximation of $\frac{6}{7}$ with $\frac{6}{7}$.

$$\frac{6}{7} > 0.855$$

---

19 *Example*   Convert $\frac{5}{8}$ to a decimal.

Solution $\begin{array}{r} 0.625 \\ 8\overline{)5.000} \end{array}$ $\frac{5}{8} = 0.625$

19   *You Try It*   Convert $\frac{4}{5}$ to a decimal.

Your Solution

---

20 *Example*   Convert $3\frac{1}{3}$ to a decimal.

Solution   Write $\frac{1}{3}$ as a decimal.

$$\begin{array}{r} 0.333 \\ 3\overline{)1.000} \end{array} = 0.\overline{3}$$

$$3\frac{1}{3} = 3.\overline{3}$$

20   *You Try It*   Convert $1\frac{5}{6}$ to a decimal.

Your Solution

---

21 *Example*   Convert 7.25 to a fraction.

Solution $7.25 = 7\frac{25}{100} = 7\frac{1}{4}$

21   *You Try It*   Convert 6.2 to a fraction.

Your Solution

Solutions on pp. S11–S12

Copyright © Houghton Mifflin Company. All rights reserved.

22 *Example* Place the correct symbol, < or >, between the two numbers.

$$0.845 \qquad \frac{5}{6}$$

Solution $\dfrac{5}{6} \approx 0.8333$

$0.8450 > 0.8333$

$0.845 > \dfrac{5}{6}$

22 *You Try It* Place the correct symbol, < or >, between the two numbers.

$$0.588 \qquad \frac{7}{12}$$

Your Solution

Solution on p. S12

OBJECTIVE **E**

Applications and Formulas

23 *Example*

A one-year subscription to a monthly magazine costs $93. The price of each issue at the newsstand is $9.80. How much would you save per issue by buying a year's subscription rather than buying each issue at the newsstand?

Strategy

To find the amount saved:

▶ Find the subscription price per issue by dividing the cost of the subscription (93) by the number of issues (12).

▶ Subtract the subscription price per issue from the newsstand price (9.80).

Solution

```
      7.75        9.80
12)93.00        -7.75
  -84           2.05
   90
  -84
   60
  -60
    0
```

The savings would be $2.05 per issue.

23 *You Try It*

You hand a postal clerk a ten-dollar bill to pay for the purchase of twelve 37¢ stamps. How much change do you receive?

Your Strategy

Your Solution

Solution on p. S12

Copyright © Houghton Mifflin Company. All rights reserved.

Figure 4.4 shows the breakdown by age group of Americans who are hearing impaired. Use this graph for Example 24 and You Try It 24.

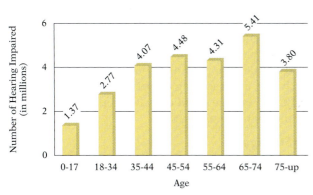

Figure 4.4   Breakdown by Age Group of Hearing-Impaired Americans
Source: American Speech-Language-Hearing Association

24 *Example*

Use Figure 4.4 to determine whether the number of hearing-impaired individuals under the age of 45 is more or less than the number of hearing impaired who are over the age of 64.

Strategy

To make the comparison:

▶ Find the number of hearing-impaired individuals under the age of 45 by adding the numbers who are aged 0–17 (1.37 million), aged 18–34 (2.77 million), and aged 35–44 (4.07 million).

▶ Find the number of hearing-impaired individuals over the age of 64 by adding the numbers who are aged 65–74 (5.41 million) and aged 75 and older (3.80 million).

▶ Compare the two sums.

Solution

1.37 + 2.77 + 4.07 = 8.21

5.41 + 3.80 = 9.21

8.21 < 9.21

The number of hearing-impaired individuals under the age of 45 is less than the number of hearing impaired who are over the age of 64.

24 *You Try It*

Is the number of hearing-impaired individuals who are aged 65–74 more or less than 4 times the number of hearing-impaired individuals who are aged 0–17? Use Figure 4.4.

Your Strategy

Your Solution

Copyright © Houghton Mifflin Company. All rights reserved.

Solution on p. S12

25 Example

An overseas flight charges $12.80 for each kilogram or part of a kilogram over 50 kg of luggage weight. How much extra must be paid for three pieces of luggage weighing 21.4, 19.3, and 16.8 kg?

Strategy

To find the extra charge:

▶ Add the three weights (21.4, 19.3, and 16.8) to find the total weight of the luggage.
▶ Subtract 50 kg from the total weight of the luggage to find the excess weight.
▶ Round the difference up to the nearest whole number.
▶ Multiply the charge per kilogram of excess weight (12.80) by the excess weight.

Solution

$21.4 + 19.3 + 16.8 = 57.5$

$57.5 - 50 = 7.5$

7.5 rounded up to the nearest whole number is 8.

$12.80(8) = 102.40$

The extra charge for the luggage is $102.40.

25 You Try It

A health food store buys nuts in 100-pound containers and repackages the nuts in cellophane bags for resale. Each cellophane bag costs $.06, two pounds of nuts are placed in each bag, and each bag of nuts is then sold for $12.50. Find the profit on a 100-pound container of nuts costing $475.

Your Strategy

Your Solution

26 Example

Use the formula $P = BF$, where P is the insurance premium, B is the base rate, and F is the rating factor, to find the insurance premium due on an insurance policy with a base rate of $342.50 and a rating factor of 2.2.

Strategy

To find the insurance premium due, replace B by 342.50 and F by 2.2 in the given formula and solve for P.

Solution

$P = BF$
$P = 342.50(2.2)$
$P = 753.50$

The insurance premium due is $753.50.

26 You Try It

Use the formula $P = BF$, where P is the insurance premium, B is the base rate, and F is the rating factor, to find the insurance premium due on an insurance policy with a base rate of $276.25 and a rating factor of 1.8.

Your Strategy

Your Solution

Solutions on p. S12

Copyright © Houghton Mifflin Company. All rights reserved.

4.2 EXERCISES

OBJECTIVE **A**

Add or subtract.

1. $1.864 + 39 + 25.0781$

2. $2.04 + 35.6 + 4.918$

3. $35.9 + 8.217 + 146.74$

4. $12 + 73.59 + 6.482$

5. $36.47 - 15.21$

6. $85.69 - 2.13$

7. $28 - 6.74$

8. $5 - 1.386$

9. $6.02 - 3.252$

10. $0.92 - 0.0037$

11. $-42.1 - 8.6$

12. $-6.57 - 8.933$

13. $5.73 - 9.042$

14. $-31.894 + 7.5$

15. $-9.37 + 3.465$

16. $1.09 - (-8.3)$

17. $-19 - (-2.65)$

18. $3.18 - 5.72 - 6.4$

19. $-12.3 - 4.07 + 6.82$

20. $-8.9 + 7.36 - 14.2$

21. $-5.6 - (-3.82) - 17.409$

22. Find the sum of 2.536, 14.97, 8.014, and 21.67.

23. Find the total of 6.24, 8.573, 19.06, and 22.488.

24. What is 6.9217 decreased by 3.4501?

25. What is 8.9 less than 62.57?

26. How much greater is 5 than 1.63?

27. What is the sum of -65.47 and -32.91?

28. Find 382.9 more than -430.6.

29. Find -138.72 minus 510.64.

30. What is 4.793 less than -6.82?

31. How much greater is -31 than -62.09?

Copyright © Houghton Mifflin Company. All rights reserved.

Estimate by rounding. Then find the exact answer.

32. $45.06 + 80.71$

33. $6.408 + 5.917$

34. $0.24 + 0.38 + 0.96$

35. $56.87 - 23.24$

36. $6.272 - 1.848$

37. $0.931 - 0.628$

38. $5.37 + 26.49$

39. $87.65 - 49.032$

40. $387.6 - 54.92$

41. *Education* The graph at the right shows where U.S. children in grades K–12 are being educated. Figures are in millions of children.
a. Find the total number of children in grades K–12.
b. How many more children are being educated in public school than in private school?

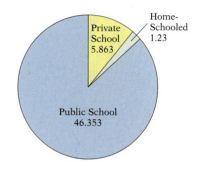

Where Children in Grades K–12 are Being Educated in the United States
Source: U.S. Department of Education; Home School Legal Defense Association

Evaluate the variable expression $x + y$ for the given values of x and y.

42. $x = 62.97; y = -43.85$

43. $x = 5.904; y = -7.063$

44. $x = -125.41; y = 361.55$

45. $x = -6.175; y = -19.49$

Evaluate the variable expression $x + y + z$ for the given values of x, y, and z.

46. $x = 41.33; y = -26.095; z = 70.08$

47. $x = -6.059; y = 3.884; z = 15.71$

48. $x = 81.72; y = 36.067; z = -48.93$

49. $x = -16.219; y = 47; z = -2.3885$

Copyright © Houghton Mifflin Company. All rights reserved.

Evaluate the variable expression $x - y$ for the given values of x and y.

50. $x = 43.29; y = 18.76$

51. $x = 6.029; y = -4.708$

52. $x = -16.329; y = 4.54$

53. $x = -21.073; y = 6.48$

54. $x = -3.69; y = -1.527$

55. $x = -8.21; y = -6.798$

56. Is -1.2 a solution of the equation $6.4 = 5.2 + a$?

57. Is -2.8 a solution of the equation $0.8 - p = 3.6$?

58. Is -0.5 a solution of the equation $x - 0.5 = 1$?

59. Is 36.8 a solution of the equation $27.4 = y - 9.4$?

OBJECTIVE   B

Multiply.

60. $0.9(0.3)$

61. $(3.4)(0.5)$

62. $(0.72)(3.7)$

63. $8.29(0.004)$

64. $-5.2(0.8)$

65. $(-6.3)(-2.4)$

66. $(1.9)(-3.7)$

67. $-1.3(4.2)$

68. $-8.1(-7.5)$

69. $1.31(-0.006)$

70. $-10(0.59)$

71. $(-100)(4.73)$

72. What is the product of 5.92 and 100?

73. What is 1,000 times 4.25?

74. Find 0.82 times 10^2.

75. Find the product of 6.71 and 10^4.

76. Find the product of 2.7, -16, and 3.04.

77. What is the product of 0.06, -0.4, and -1.5?

Copyright © Houghton Mifflin Company. All rights reserved.

Estimate by rounding. Then find the exact answer.

78. 86.4(4.2)

79. (9.81)(0.77)

80. 0.238(8.2)

81. (6.88)(9.97)

82. (8.432)(0.043)

83. 28.45(1.13)

The table at the right shows currency exchange rates for several foreign countries. To determine how many Swiss francs would be exchanged for 1,000 U.S. dollars, multiply the number of francs exchanged for 1 U.S. dollar (1.339) by 1,000: 1,000(1.339) = 1,339. Use this table for Exercises 84 and 85.

Country and Monetary Unit	Number of Units Exchanged for 1 U.S. Dollar
Britain (Pound)	0.6336
Canada (Dollar)	1.485
European Union (Euro)	0.9188
Japan (Yen)	117.6
Mexico (Peso)	11.104
Switzerland (Franc)	1.339

84. *Exchange Rates* How many Mexican pesos would be exchanged for 5,000 U.S. dollars?

85. *Exchange Rates* How many British pounds would be exchanged for 20,000 U.S. dollars?

Evaluate the expression for the given values of the variables.

86. xy, when $x = 5.68$ and $y = 0.2$

87. ab, when $a = 6.27$ and $b = 8$

88. $40c$, when $c = 2.5$

89. $10t$, when $t = -4.8$

90. xy, when $x = -3.71$ and $y = 2.9$

91. ab, when $a = 0.379$ and $b = -0.22$

92. ab, when $a = 452$ and $b = -0.86$

93. cd, when $c = -2.537$ and $d = -9.1$

94. cd, when $c = -4.259$ and $d = -6.3$

95. Is -8 a solution of the equation $1.6 = -0.2z$?

96. Is -1 a solution of the equation $-7.9c = -7.9$?

97. Is -10 a solution of the equation $-83.25r = 8.325$?

98. Is -3.6 a solution of the equation $32.4 = -9w$?

Copyright © Houghton Mifflin Company. All rights reserved.

OBJECTIVE **C**

Divide.

99. $16.15 \div 0.5$

100. $7.02 \div 3.6$

101. $27.08 \div (-0.4)$

102. $-8.919 \div 0.9$

103. $(-3.312) \div (-0.8)$

104. $84.66 \div (-1.7)$

105. $-2.501 \div 0.41$

106. $1.003 \div (-0.59)$

Divide. Round to the nearest tenth.

107. $55.63 \div 8.8$

108. $1.873 \div 1.4$

109. $(-52.8) \div (-9.1)$

110. $-6.824 \div 0.053$

Divide. Round to the nearest hundredth.

111. $6.457 \div 8$

112. $19.07 \div 0.54$

113. $0.0416 \div (-0.53)$

114. $(-31.792) \div (-0.86)$

115. Find the quotient of 52.78 and 10.

116. What is 37,942 divided by 1,000?

117. What is the quotient of 48.05 and 10^2?

118. Find 9.407 divided by 10^3.

In Exercises 119 to 122, round answers to the nearest tenth.

119. Find the quotient of -19.04 and 0.75.

120. What is the quotient of -21.892 and -0.96?

121. Find 27.735 divided by -60.3.

122. What is -13.97 divided by 28.4?

Copyright © Houghton Mifflin Company. All rights reserved.

Estimate by rounding. Then divide and round to the nearest hundredth.

123. $42.43 \div 3.8$

124. $678 \div 0.71$

125. $6.398 \div 5.5$

126. $0.994 \div 0.456$

127. $1.237 \div 0.021$

128. $421.093 \div 4.087$

129. $33.14 \div 4.6$

130. $129.38 \div 4.47$

131. *The Internet* The graph at the right shows the expected growth of digital subscriber lines (DSL). Figures given are in millions. How many times greater was the DSL market in 2003 than in 2001? Round to the nearest tenth.

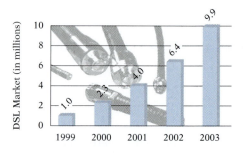

Digital Subscriber Lines (DSL) Market
Source: Copyright © 1999 *USA Today.* Reprinted with permission.

Evaluate the variable expression $\dfrac{x}{y}$ for the given values of x and y.

132. $x = 52.8; y = 0.4$

133. $x = 3.542; y = 0.7$

134. $x = -2.436; y = 0.6$

135. $x = 0.648; y = -2.7$

136. $x = 26.22; y = -6.9$

137. $x = -8.034; y = -3.9$

138. $x = -64.05; y = -6.1$

139. $x = -2.501; y = 0.41$

140. $x = 1.003; y = -0.59$

141. Is 24.8 a solution of the equation $\dfrac{q}{-8} = -3.1$?

142. Is 0.48 a solution of the equation $\dfrac{-6}{z} = -12.5$?

143. Is -8.4 a solution of the equation $21 = \dfrac{t}{0.4}$?

144. Is -0.9 a solution of the equation $\dfrac{-2.7}{a} = \dfrac{a}{-0.3}$?

OBJECTIVE D

Convert the fraction to a decimal. Place a bar over repeating digits of a repeating decimal.

145. $\dfrac{3}{8}$

146. $\dfrac{7}{15}$

147. $\dfrac{8}{11}$

148. $\dfrac{9}{16}$

149. $\dfrac{7}{12}$

Copyright © Houghton Mifflin Company. All rights reserved.

150. $\dfrac{5}{3}$ **151.** $\dfrac{7}{4}$ **152.** $2\dfrac{3}{4}$ **153.** $1\dfrac{1}{2}$ **154.** $3\dfrac{2}{9}$

155. $4\dfrac{1}{6}$ **156.** $\dfrac{3}{25}$ **157.** $2\dfrac{1}{4}$ **158.** $6\dfrac{3}{5}$ **159.** $3\dfrac{8}{9}$

Convert the decimal to a fraction.

160. 0.6   **161.** 0.2   **162.** 0.25   **163.** 0.75   **164.** 0.48

165. 0.125   **166.** 0.325   **167.** 2.5   **168.** 3.4   **169.** 4.55

170. 9.95   **171.** 1.72   **172.** 5.68   **173.** 0.045   **174.** 0.085

Place the correct symbol, $<$ or $>$, between the two numbers.

175. $\dfrac{9}{10}$ 0.89   **176.** $\dfrac{7}{20}$ 0.34   **177.** $\dfrac{4}{5}$ 0.803   **178.** $\dfrac{3}{4}$ 0.706

179. 0.444 $\dfrac{4}{9}$ **180.** 0.72 $\dfrac{5}{7}$ **181.** 0.13 $\dfrac{3}{25}$ **182.** 0.25 $\dfrac{13}{50}$

183. $\dfrac{5}{16}$ 0.312   **184.** $\dfrac{7}{18}$ 0.39   **185.** $\dfrac{10}{11}$ 0.909   **186.** $\dfrac{8}{15}$ 0.543

OBJECTIVE E

187. *Finances* If you earn an annual salary of $47,619, what is your monthly salary?

188. *Finances* You pay $947.60 a year in car insurance. The insurance is paid in four equal payments. Find the amount of each payment.

Copyright © Houghton Mifflin Company. All rights reserved.

189. *Temperature* On January 22, 1943, in Spearfish, South Dakota, the temperature fell from 12.22°C at 9:00 A.M. to −20°C at 9:27 A.M. How many degrees did the temperature fall during the 27-minute period?

190. *Temperature* On January 10, 1911, in Rapid City, South Dakota, the temperature fell from 12.78°C at 7:00 A.M. to −13.33°C at 7:15 A.M. How many degrees did the temperature fall during the 15-minute period?

191. *Consumerism* A case of diet cola costs $8.89. If there are 24 cans in a case, find the cost per can. Round to the nearest cent.

192. *Travel* You travel 295 mi on 12.5 gal of gasoline. How many miles can you travel on one gallon of gasoline?

193. *Consumerism* It costs $.038 an hour to operate an electric motor. How much does it cost to operate the motor for 90 h?

194. *Travel* When the Massachusetts Turnpike opened, the toll for a passenger car that traveled the entire 136 mi of it was $5.60. Find the cost per mile. Round to the nearest cent.

195. *Taxes* For tax purposes, the standard deduction on tax returns for the business use of a car in 2002 was 36.5¢ per mile. Find the amount deductible on a 2002 tax return for driving a business car 11,842 mi during the year.

The graph at the right shows that the worldwide consumption of cigarettes has been increasing. Use this table for Exercises 196 and 197.

196. *Health* **a.** Find the increase in cigarette consumption from 1950 to 1990. **b.** How many times greater was the cigarette consumption in 2000 than in 1960?

197. *Health* **a.** During which 10-year period was the increase in cigarette consumption greatest? **b.** During which 10-year period was the increase in cigarette consumption the least?

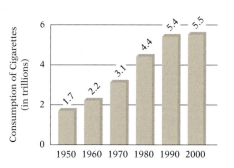

Consumption of Cigarettes (in trillions)

1950: 1.7 1960: 2.2 1970: 3.1 1980: 4.4 1990: 5.4 2000: 5.5

Worldwide Consumption of Cigarettes
Source: The Tobacco Atlas; U.S. Department of Agriculture

198. *Consumerism* You make a down payment of $125 on a camcorder and agree to make payments of $34.17 a month for 9 months. Find the total cost of the camcorder.

Copyright © Houghton Mifflin Company. All rights reserved.

199. *Finances*   You have a monthly budget of $1620. This month you have already spent $62.78 for the telephone bill, $164.93 for food, $35.50 for gasoline, $560 for your share of the rent, and $291.62 for a loan repayment. How much money do you have left in the budget for the remainder of the month?

200. *Finances*   You had a balance of $347.08 in your checking account. You then made a deposit of $189.53 and wrote a check for $62.89. Find the new balance in your checking account.

201. *Finances*   A bookkeeper earns a salary of $660 for a 40-hour week. This week the bookkeeper worked 6 h of overtime at a rate of $24.75 for each hour of overtime worked. Find the bookkeeper's total income for the week.

202. *Computers*   The list below shows the average number of hours per week that students use a computer. On average, how many more hours per year does a 2nd grade student use a computer than a 5th grade student?

Grade Level	Average Number of Hours of Computer Use Per Week
Pre Kindergarten – Kindergarten	3.9
1st – 3rd	4.9
4th – 6th	4.2
7th – 8th	6.9
9th – 12th	6.7

Source: Find/SVP American Learning Household Survey

203. *Consumerism*   Using the menu shown below, estimate the bill for the following order: 1 soup, 1 cheese sticks, 1 blackened swordfish, 1 chicken divan, and 1 carrot cake.

Appetizers
Soup of the Day $5.75
Cheese Sticks $6.25
Potato Skins $6.50

Entrees
Roast Prime Rib $28.95
Blackened Swordfish $26.95
Chicken Divan $24.95

Desserts
Carrot Cake $7.25
Ice Cream Pie $8.50
Cheese Cake $9.75

204. *Consumerism*   Using the menu shown above, estimate the bill for the following order: 1 potato skins, 1 cheese sticks, 1 roast prime rib, 1 chicken divan, 1 ice cream pie, and 1 cheese cake.

Copyright © Houghton Mifflin Company. All rights reserved.

205. *Life Expectancy* The graph below shows the life expectancy at birth for males and females.

    a. Has life expectancy increased for both males and females with every 10-year period shown in the graph?

    b. Did males or females have a longer life expectancy in 2000? How much longer?

    c. During which year shown in the graph was the difference between male life expectancy and female life expectancy greatest?

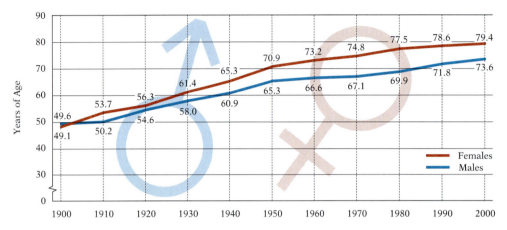

Life Expectancies of Males and Females in the United States

206. *Business* For $135, a druggist purchases 5 L of cough syrup and repackages it in 250-milliliter bottles. Each bottle costs the druggist $.55. Each bottle of cough syrup is sold for $11.89. Find the profit on the 5 L of cough syrup. (*Hint:* There are 1000 milliliters in 1 liter.)

207. *Business* A confectioner ships holiday packs of candy and nuts anywhere in the United States. At the right is a price list for nuts and candy, and below that is a table of shipping charges to zones in the United States. Find the cost of sending the following orders to the given mail zones. For any fraction of a pound, use the next higher weight. Sixteen ounces is equal to one pound.

Code	Description	Price
112	Almonds 16 oz	$4.75
116	Cashews 8 oz	$2.90
117	Cashews 16 oz	$5.50
130	Macadamias 7 oz	$5.25
131	Macadamias 16 oz	$9.95
149	Pecan halves 8 oz	$6.25
155	Mixed nuts 8 oz	$4.80
160	Cashew brittle 8 oz	$1.95
182	Pecan roll 8 oz	$3.70
199	Chocolate peanuts 8 oz	$1.90

a. Code	Quantity	b. Code	Quantity	c. Code	Quantity
116	2	112	1	117	3
130	1	117	4	131	1
149	3	131	2	155	2
182	4	160	3	160	4
Mail to Zone 4.		182	5	182	1
		Mail to Zone 3.		199	3
				Mail to Zone 2.	

Pounds	Zone 1	Zone 2	Zone 3	Zone 4
1 – 3	$6.55	$6.85	$7.25	$7.75
4 – 6	$7.10	$7.40	$7.80	$8.30
7 – 9	$7.50	$7.80	$8.20	$8.70
10 – 12	$7.90	$8.20	$8.60	$9.10

208. *Transportation* A taxi costs $2.50 plus $.30 for each $\frac{1}{8}$ mi driven. Find the cost of hiring a taxi to get from the airport to your hotel, a distance of 4.5 mi.

Copyright © Houghton Mifflin Company. All rights reserved.

209. *Geometry*   Find the perimeter of a rectangle that measures 4.5 in. by 3.25 in. Use the formula $P = 2L + 2W$.

210. *Geometry*   Find the perimeter of a rectangle that measures 2.8 m by 6.4 m. Use the formula $P = 2L + 2W$.

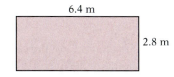

6.4 m

2.8 m

211. *Geometry*   Find the area of a rectangle that measures 4.5 in. by 3.25 in. Use the formula $A = LW$.

212. *Geometry*   Find the area of a rectangle that has a length of 7.8 cm and a width of 4.6 cm. Use the formula $A = LW$.

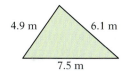

4.6 cm

7.8 cm

213. *Geometry*   Find the perimeter of a triangle with sides that measure 2.8 m, 4.75 m, and 6.4 m. Use the formula $P = a + b + c$.

214. *Geometry*   The lengths of three sides of a triangle are 7.5 m, 6.1 m, and 4.9 m. Find the perimeter of the triangle. Use the formula $P = a + b + c$.

4.9 m   6.1 m

7.5 m

215. *Geometry*   The length of each side of a square is 3.5 ft. Find the perimeter of the square. Use the formula $P = 4s$.

3.5 ft

3.5 ft

216. *Business*   Use the formula $M = S - C$, where M is the markup on a consumer product, S is the selling price, and C is the cost of the product to the business, to find the markup on a product that cost a business $1,653.19 and has a selling price of $2,231.81.

217. *Accounting*   The amount of an employee's earnings that is subject to federal withholding is called federal earnings. Find the federal earnings for an employee who earns $694.89 and has a withholding allowance of $132.69. Use the formula $F = E - W$, where F is the federal earnings, E is the employee's earnings, and W is the withholding allowance.

218. *Consumerism*   Use the formula $M = \dfrac{C}{N}$, where M is the cost per mile for a rental car, C is the total cost, and N is the number of miles driven, to find the cost per mile when the total cost of renting a car is $260.16 and you drive the car 542 mi.

219. *Physics*   Find the force exerted on a falling object that has a mass of 4.25 kg. Use the formula $F = ma$, where F is the force exerted by gravity on a falling object, m is the mass of the object, and a is the acceleration of gravity. The acceleration of gravity is -9.80 m/s² (meters per second squared). The force is measured in newtons.

Copyright © Houghton Mifflin Company. All rights reserved.

220. *Utilities*   Find the cost of operating a 1800-watt TV set for 5 h at a cost of $.06 per kilowatt-hour. Use the formula $c = 0.001wtk$, where c is the cost of operating an appliance, w is the number of watts, t is the time in hours, and k is the cost per kilowatt-hour.

221. *Finances*   Find the equity on a home that is valued at $225,000 when the homeowner has $167,853.25 in loans on the property. Use the formula $E = V - L$, where E is the equity, V is the value of the home, and L is the loan amount on the property.

Critical Thinking

222. Find the product of 1.0035 and 1.00079 without using a calculator. Then find the product using a calculator and compare the two numbers. If your calculator has an eight-digit display, what number did the calculator display? Some calculators **truncate** the product, which means that the digits that cannot be displayed are discarded. Other calculators round the answer to the rightmost place value in the calculator's display. Determine which method your calculator uses to handle approximate answers. If the decimal places in a negative number are truncated, is the resulting number greater than, less than, or equal to the original number?

223. *Business*   A ballpoint pen priced at 50¢ was not selling. When the price was reduced to a different whole number of cents, the entire stock sold for $31.93. How many cents were charged per pen when the price was reduced?

224. Determine whether the statement is always true, sometimes true, or never true.
  a. The product of an even number of negative factors is a negative number.
  b. The sum of an odd number of negative addends is a negative number.
  c. If $a \geq 0$, then $|a| = a$.
  d. If $a \leq 0$, then $|a| = -a$.

225. Using the method, presented in this section, of estimating the sum of two decimals, what is the largest amount by which the estimate of the sum of two decimals with tenths, hundredths, and thousandths places could differ from the exact sum? Assume that the number in the thousandths place is not zero.

226. Convert $\frac{1}{9}, \frac{2}{9}, \frac{3}{9},$ and $\frac{4}{9}$ to decimals. Describe the pattern. Use the pattern to convert $\frac{5}{9}, \frac{7}{9},$ and $\frac{8}{9}$ to decimals.

227. Explain how baseball batting averages are determined.

228. Prepare a report on the Kelvin scale. The report should include a definition of absolute zero and an explanation of how to convert from Kelvin to Celsius and from Celsius to Kelvin.

Copyright © Houghton Mifflin Company. All rights reserved.

SECTION 4.3 ▶ Solving Equations with Decimals

VIDEO & DVD   CD TUTOR   WEB   SSM

OBJECTIVE **A**

Solving equations

The properties of equations discussed earlier are restated here.

> **The same number can be added to each side of an equation without changing the solution of the equation.**
>
> **The same number can be subtracted from each side of an equation without changing the solution of the equation.**
>
> **Each side of an equation can be multiplied by the same nonzero number without changing the solution of the equation.**
>
> **Each side of an equation can be divided by the same nonzero number without changing the solution of the equation.**

Solve: $3.4 = a - 3.56$

3.56 is subtracted from the variable a. Add 3.56 to each side of the equation.

a is alone on the right side of the equation. The number on the left side is the solution.

$$3.4 = a - 3.56$$
$$3.4 + 3.56 = a - 3.56 + 3.56$$
$$6.96 = a$$

The solution is 6.96.

Take Note

Remember to check the solutions for all equations.

Check:
$$3.4 = a - 3.56$$
$$\begin{array}{c|c} 3.4 & 6.96 - 3.56 \end{array}$$
$$3.4 = 3.4$$

Solve: $-1.25y = 3.875$

The variable is multiplied by -1.25. Divide each side of the equation by -1.25.

y is alone on the left side of the equation. The number on the right side is the solution.

$$-1.25y = 3.875$$
$$\frac{-1.25y}{-1.25} = \frac{3.875}{-1.25}$$
$$y = -3.1$$

The solution is -3.1.

---

1 *Example*   Solve: $4.56 = 9.87 + z$

Solution
$$4.56 = 9.87 + z$$
$$4.56 - 9.87 = 9.87 - 9.87 + z$$
$$-5.31 = z$$

The solution is -5.31.

1 *You Try It*   Solve: $a - 1.23 = -6$

Your Solution

---

2 *Example*   Solve: $\dfrac{x}{2.45} = -0.3$

Solution
$$\frac{x}{2.45} = -0.3$$
$$2.45 \cdot \frac{x}{2.45} = 2.45(-0.3)$$
$$x = -0.735$$

The solution is -0.735.

2 *You Try It*   Solve: $-2.13 = -0.71c$

Your Solution

Solutions on p. S12

Copyright © Houghton Mifflin Company. All rights reserved.

OBJECTIVE **B**

Applications

3 *Example*

The cost of operating an electrical appliance is given by the formula $c = 0.001wtk$, where c is the cost of operating the appliance, w is the number of watts, t is the number of hours, and k is the cost per kilowatt-hour. Find the cost per kilowatt-hour if it costs $.60 to operate a 2,000-watt television for 5 h.

Strategy

To find the cost per kilowatt-hour, replace c by 0.60, w by 2,000, and t by 5 in the given formula and solve for k.

Solution

$$c = 0.001wtk$$
$$0.60 = 0.001(2,000)(5)k$$
$$0.60 = 10k$$
$$\frac{0.60}{10} = \frac{10k}{10}$$
$$0.06 = k$$

It costs $.06 per kilowatt-hour.

3 *You Try It*

The net worth of a business is given by the equation $N = A - L$, where N is the net worth, A is the assets of the business (the amount owned) and L is the liabilities of the business (the amount owed). Use the net worth equation to find the assets of a business that has a net worth of $24.3 billion and liabilities of $17.9 billion.

Your Strategy

Your Solution

4 *Example*

The total of the monthly payments for an installment loan is the product of the number of months of the loan and the monthly payment. The total of the monthly payments for a 48-month, new-car loan is $20,433.12. What is the monthly payment?

Strategy

To find the monthly payment, write and solve an equation using m to represent the amount of the monthly payment.

Solution

The total of the monthly payments	is	the product of the number of months of the loan and the monthly payment

$$20,433.12 = 48m$$
$$425.69 = m \quad \blacktriangleright \text{ Divide each side by 48.}$$

The monthly payment is $425.69.

4 *You Try It*

The selling price of a product is the sum of the amount paid by the store for the product and the amount of the markup. The selling price of a titanium golf club is $295.50, and the amount paid by the store for the golf club is $223.75. Find the markup.

Your Strategy

Your Solution

Solutions on p. S12

Copyright © Houghton Mifflin Company. All rights reserved.

4.3 EXERCISES

OBJECTIVE A

Solve. Write the answer as a decimal.

1. $y + 3.96 = 8.45$

2. $x - 2.8 = 1.34$

3. $-9.3 = c - 15$

4. $-28 = x - 3.27$

5. $7.3 = -\dfrac{n}{1.1}$

6. $-5.1 = \dfrac{y}{3.2}$

7. $-7x = 8.4$

8. $1.44 = -0.12t$

9. $y - 0.234 = -0.09$

10. $9 = z + 0.98$

11. $6.21r = -1.863$

12. $-78.1a = 85.91$

13. $-0.001 = x + 0.009$

14. $5 = 43.5 + c$

15. $\dfrac{x}{2} = -0.93$

16. $-1.03 = -\dfrac{z}{3}$

17. $-9.85y = 2.0685$

18. $7w = -0.014$

19. $-6v = 15$

20. $-55 = -40x$

21. $0.908 = 2.913 + x$

22. $-76.51 = y - 43.9$

23. $\dfrac{t}{-2.1} = -7.8$

24. $\dfrac{w}{0.02} = -9.64$

OBJECTIVE B

25. *Business*   Xlint Office Supplies wants to make a profit of $8.50 on the sale of each desk calendar that costs the store $15.23. Use the equation $P = S - C$, where P is the profit on an item, S is the selling price, and C is the cost, to find the selling price of the calendar.

26. *Cost of Living*   The average cost per mile to operate a car is given by the equation $M = \dfrac{C}{N}$, where M is the average cost per mile, C is the total cost of operating the car, and N is the number of miles the car is driven. Use this formula to find the total cost of operating a car for 25,000 mi when the average cost per mile is $.34.

Copyright © Houghton Mifflin Company. All rights reserved.

27. *Physics* The average acceleration of an object is given by $a = \frac{v}{t}$, where a is the average acceleration, v is the velocity, and t is the time. Find the velocity after 6.3 s of an object whose acceleration is 16 ft/s² (feet per second squared).

28. *Accounting* The fundamental accounting equation is $A = L + S$, where A is the assets of a company, L is the liabilities of the company, and S is the stockholders' equity. Find the stockholders' equity in a company whose assets are $34.8 million and whose liabilities are $29.9 million.

29. *Cost of Living* The cost of operating an electrical appliance is given by the formula $c = 0.001wtk$, where c is the cost of operating the appliance, w is the number of watts, t is the number of hours, and k is the cost per kilowatt-hour. Find the cost per kilowatt-hour if it costs $.01 to operate a 1,000-watt microwave for 4 h.

30. *Business* The markup on an item in a store equals the difference between the selling price of the item and the cost of the item. Find the selling price of a package of golf balls for which the cost is $6.54 and the markup is $3.46.

31. *Consumerism* The total of the monthly payments for a car lease is the product of the number of months of the lease and the monthly lease payment. The total of the monthly payments for a 60-month car lease is $15,387. Find the monthly lease payment.

32. *Geometry* The area of a rectangle is 210 in². If the width of the rectangle is 10.5 in., what is the length? Use the formula $A = LW$.

$A = 210$ in²

33. *Geometry* The length of a rectangle is 18 ft. If the area is 225 ft², what is the width of the rectangle? Use the formula $A = LW$.

34. *Geometry* The perimeter of a square picture frame is 33 in. Find the length of each side of the frame. Use the formula $P = 4s$.

35. *Geometry* A square rug has a perimeter of 30 ft. Find the length of each edge of the rug. Use the formula $P = 4s$.

Critical Thinking

36. Solve: $0.\overline{33}x = 7$

37. a. Make up an equation of the form $x - b = c$ for which $x = -0.96$.
b. Make up an equation of the form $ax = b$ for which $x = 2.1$.

38. For the equation $0.375x = 0.6$, a student offered the solution shown at the right. Is this a correct method of solving the equation? Explain your answer.

39. Consider the equation $12 = \frac{x}{a}$, where a is any positive number. Explain how increasing values of a affect the solution, x, of the equation.

$0.375x = 0.6$

$\frac{375}{1,000}x = \frac{6}{10}$

$\frac{3}{8}x = \frac{3}{5}$

$\frac{8}{3} \cdot \frac{3}{8}x = \frac{8}{3} \cdot \frac{3}{5}$

$x = \frac{8}{5} = 1.6$

Copyright © Houghton Mifflin Company. All rights reserved.

SECTION 4.4 ▶ Radical Expressions

Copyright © Houghton Mifflin Company. All rights reserved.

OBJECTIVE **A**

Square roots of perfect squares

Recall that the square of a number is equal to the number multiplied times itself.

$$3^2 = 3 \cdot 3 = 9$$

The square of an integer is called a **perfect square.**

9 is a perfect square because 9 is the square of 3: $3^2 = 9$.

The numbers 1, 4, 9, 16, 25, 36, 49, 64, 81, and 100 are perfect squares.

$$1^2 = 1$$
$$2^2 = 4$$
$$3^2 = 9$$
$$4^2 = 16$$
$$5^2 = 25$$
$$6^2 = 36$$
$$7^2 = 49$$
$$8^2 = 64$$
$$9^2 = 81$$
$$10^2 = 100$$

Larger perfect squares can be found by squaring 11, squaring 12, squaring 13, and so on.

Note that squaring the negative integers results in the same list of numbers.

$$(-1)^2 = 1$$
$$(-2)^2 = 4$$
$$(-3)^2 = 9$$
$$(-4)^2 = 16, \text{ and so on.}$$

Perfect squares are used in simplifying square roots. The symbol for square root is $\sqrt{}$.

> ### *Square Root*
>
> A square root of a positive number x is a number whose square is x.
>
> If $a^2 = x$, then $\sqrt{x} = a$.

The expression $\sqrt{9}$, read "the square root of 9," is equal to the number that when squared is equal to 9.

Since $3^2 = 9$, $\sqrt{9} = 3$.

Every positive number has two square roots, one a positive number and one a negative number. The symbol $\sqrt{}$ is used to indicate the positive square root of a number. When the negative square root of a number is to be found, a negative sign is placed in front of the square root symbol. For example,

$$\sqrt{9} = 3 \quad \text{and} \quad -\sqrt{9} = -3$$

► **Point of Interest**

The radical symbol was first used in 1525, when it was written as √. Some historians suggest that the radical symbol also developed into the symbols for "less than" and "greater than." Because typesetters of that time did not want to make additional symbols, the radical was rotated to the position > and used as a "greater than" symbol and rotated to ← and used for the "less than" symbol. Other evidence, however, suggests that the "less than" and "greater than" symbols were developed independently of the radical symbol.

The square root symbol, $\sqrt{}$, is also called a **radical.** The number under the radical is called the **radicand.** In the radical expression $\sqrt{9}$, 9 is the radicand.

Simplify: $\sqrt{49}$

$\sqrt{49}$ is equal to the number that when squared equals 49. $7^2 = 49$.

$$\sqrt{49} = 7$$

Simplify: $-\sqrt{49}$

The negative sign in front of the square root symbol indicates the negative square root of 49. $(-7)^2 = 49$.

$$-\sqrt{49} = -7$$

Simplify: $\sqrt{25} + \sqrt{81}$

Simplify each radical expression.

Since $5^2 = 25$, $\sqrt{25} = 5$.

Since $9^2 = 81$, $\sqrt{81} = 9$.

Add.

$$\sqrt{25} + \sqrt{81} = 5 + 9$$
$$= 14$$

Simplify: $5\sqrt{64}$

The expression $5\sqrt{64}$ means 5 times $\sqrt{64}$.

Simplify $\sqrt{64}$.

Multiply.

$$5\sqrt{64} = 5 \cdot 8$$
$$= 40$$

Simplify: $6 + 4\sqrt{9}$

Simplify $\sqrt{9}$.

Use the Order of Operations Agreement.

$$6 + 4\sqrt{9} = 6 + 4 \cdot 3$$
$$= 6 + 12$$
$$= 18$$

Simplify: $\sqrt{\dfrac{1}{9}}$

$\sqrt{\dfrac{1}{9}}$ is equal to the number that when squared equals $\dfrac{1}{9}$. $\left(\dfrac{1}{3}\right)^2 = \dfrac{1}{9}$.

$$\sqrt{\dfrac{1}{9}} = \dfrac{1}{3}$$

Note that the square root of $\dfrac{1}{9}$ is equal to the square root of the numerator $\left(\sqrt{1} = 1\right)$ over the square root of the denominator $\left(\sqrt{9} = 3\right)$.

Copyright © Houghton Mifflin Company. All rights reserved.

Evaluate $\sqrt{xy}$ when $x = 5$ and $y = 20$.

$$\sqrt{xy}$$

Replace x with 5 and y with 20. $\sqrt{5 \cdot 20}$

Simplify under the radical. $= \sqrt{100}$

Take the square root of 100. $10^2 = 100$. $= 10$

Take Note

The radical is a grouping symbol. Therefore, when simplifying numerical expressions, simplify the radicand as part of step one of the Order of Operations Agreement.

---

1 *Example*   Simplify: $\sqrt{121}$

Solution   Since $11^2 = 121$, $\sqrt{121} = 11$.

1 *You Try It*   Simplify: $-\sqrt{144}$

Your Solution

---

2 *Example*   Simplify: $\sqrt{\dfrac{4}{25}}$

Solution   Since $\left(\dfrac{2}{5}\right)^2 = \dfrac{4}{25}$, $\sqrt{\dfrac{4}{25}} = \dfrac{2}{5}$.

2 *You Try It*   Simplify: $\sqrt{\dfrac{81}{100}}$

Your Solution

---

3 *Example*   Simplify: $\sqrt{36} - 9\sqrt{4}$

Solution
$$\begin{aligned}
\sqrt{36} - 9\sqrt{4} &= 6 - 9 \cdot 2 \\
&= 6 - 18 \\
&= 6 + (-18) \\
&= -12
\end{aligned}$$

3 *You Try It*   Simplify: $4\sqrt{16} - \sqrt{9}$

Your Solution

---

4 *Example*   Evaluate $6\sqrt{ab}$ when $a = 2$ and $b = 8$.

Solution
$$6\sqrt{ab}$$
$$\begin{aligned}
6\sqrt{2 \cdot 8} &= 6\sqrt{16} \\
&= 6(4) \\
&= 24
\end{aligned}$$

4 *You Try It*   Evaluate $5\sqrt{a + b}$ when $a = 17$ and $b = 19$.

Your Solution

Solutions on p. S12

Copyright © Houghton Mifflin Company. All rights reserved.

VIDEO & DVD CD TUTOR WEB SSM

OBJECTIVE B

Square roots of whole numbers

In the last objective, the radicand in each radical expression was a perfect square. Since the square root of a perfect square is an integer, the exact value of each radical expression could be found.

If the radicand is not a perfect square, the square root can only be approximated. For example, the radicand in the radical expression $\sqrt{2}$ is 2, and 2 is not a perfect square. The square root of 2 can be approximated to any desired place value.

To the nearest tenth:	$\sqrt{2} \approx 1.4$	$(1.4)^2 = 1.96$
To the nearest hundredth:	$\sqrt{2} \approx 1.41$	$(1.41)^2 = 1.9881$
To the nearest thousandth:	$\sqrt{2} \approx 1.414$	$(1.414)^2 = 1.999396$
To the nearest ten-thousandth:	$\sqrt{2} \approx 1.4142$	$(1.4142)^2 = 1.99996164$

The square of each approximation gets closer and closer to 2 as the number of place values in the decimal approximation increases. But no matter how many place values are used to approximate $\sqrt{2}$, the digits never terminate or repeat. In general, the square root of any number that is not a perfect square can only be approximated.

Calculator Note

The way in which you evaluate the square root of a number depends on the type of calculator you have. Here are two possible keystrokes to find $\sqrt{27}$:

27 [√] [=]

or

[√] 27 [ENTER]

The first method is used on many scientific calculators. The second method is used on many graphing calculators.

Approximate $\sqrt{11}$ to the nearest ten-thousandth.

11 is not a perfect square.

Use a calculator to approximate $\sqrt{11}$.

$\sqrt{11} \approx 3.3166$

Calculator Note

To evaluate $3\sqrt{5}$ on a calculator, enter either

3 [×] 5 [√] [=]

or

3 [√] 5 [ENTER]

Round the number in the display to the desired place value.

Approximate $3\sqrt{5}$ to the nearest ten-thousandth.

$3\sqrt{5}$ means 3 times $\sqrt{5}$.

$3\sqrt{5} \approx 6.7082$

Between what two whole numbers is the value of $\sqrt{41}$?

Since the number 41 is between the perfect squares 36 and 49, the value of $\sqrt{41}$ is between $\sqrt{36}$ and $\sqrt{49}$.

$\sqrt{36} = 6$ and $\sqrt{49} = 7$,

so the value of $\sqrt{41}$ is between the whole numbers 6 and 7.

This can be written using inequality symbols as $6 < \sqrt{41} < 7$, which is read

"the square root of 41 is greater than 6 and less than 7."

Use a calculator to verify that $\sqrt{41} \approx 6.4$, which is between 6 and 7.

Copyright © Houghton Mifflin Company. All rights reserved.

Sometimes we are not interested in an approximation of the square root of a number, but rather the exact value in simplest form.

A radical expression is in simplest form when the radicand contains no factor, other than 1, that is a perfect square. The Product Property of Square Roots is used to simplify radical expressions.

> ### *Product Property of Square Roots*
> If a and b are positive numbers, then $\sqrt{a \cdot b} = \sqrt{a} \cdot \sqrt{b}$.

The Product Property of Square Roots states that the square root of a product is equal to the product of the square roots. For example,

$$\sqrt{4 \cdot 9} = \sqrt{4} \cdot \sqrt{9}$$

Note that $\sqrt{4 \cdot 9} = \sqrt{36} = 6$ and $\sqrt{4} \cdot \sqrt{9} = 2 \cdot 3 = 6$.

Simplify: $\sqrt{50}$

Think: What perfect square is a factor of 50?

Begin with a perfect square that is larger than 50.

Then test each successively smaller perfect square.

$8^2 = 64$; 64 is too big.
$7^2 = 49$; 49 is not a factor of 50.
$6^2 = 36$; 36 is not a factor of 50.
$5^2 = 25$; 25 is a factor of 50. $(50 = 25 \cdot 2)$

Write $\sqrt{50}$ as $\sqrt{25 \cdot 2}$. $\sqrt{50} = \sqrt{25 \cdot 2}$

Use the Product Property of Square Roots. $= \sqrt{25} \cdot \sqrt{2}$

Simplify $\sqrt{25}$. $= 5 \cdot \sqrt{2}$

The radicand 2 contains no factor other than 1 that $= 5\sqrt{2}$
is a perfect square. The radical expression $5\sqrt{2}$ is in
simplest form.

Remember that $5\sqrt{2}$ means 5 times $\sqrt{2}$. Using a calculator,
$5\sqrt{2} \approx 5(1.4142) = 7.071$ and $\sqrt{50} \approx 7.071$.

> **Calculator Note**
>
> The keystrokes to evaluate $5\sqrt{2}$ on a calculator are either
>
> 5 × 2 √ =
>
> or
>
> 5 √ 2 ENTER
>
> Round the number in the display to the desired place value.

---

5   *Example*   Approximate $4\sqrt{17}$ to the nearest ten-thousandth.

5   *You Try It*   Approximate $5\sqrt{23}$ to the nearest ten-thousandth.

Solution $4\sqrt{17} \approx 16.4924$

Your Solution

Solution on p. S13

Copyright © Houghton Mifflin Company. All rights reserved.

6 Example Between what two whole numbers is the value of $\sqrt{79}$?

Solution 79 is between the perfect squares 64 and 81.

$\sqrt{64} = 8$ and $\sqrt{81} = 9$.

$8 < \sqrt{79} < 9$

6 You Try It Between what two whole numbers is the value of $\sqrt{57}$?

Your Solution

7 Example Simplify: $\sqrt{32}$

Solution $6^2 = 36$; 36 is too big.
$5^2 = 25$; 25 is not a factor of 32.
$4^2 = 16$; 16 is a factor of 32.
$\sqrt{32} = \sqrt{16 \cdot 2}$
$= \sqrt{16} \cdot \sqrt{2}$
$= 4 \cdot \sqrt{2}$
$= 4\sqrt{2}$

7 You Try It Simplify: $\sqrt{80}$

Your Solution

Solutions on p. S13

OBJECTIVE **C**

Applications and formulas

8 Example

Find the range of a submarine periscope that is 8 ft above the surface of the water. Use the formula $R = 1.4\sqrt{h}$, where R is the range in miles and h is the height in feet of the periscope above the surface of the water. Round to the nearest hundredth.

Strategy

To find the range, replace h by 8 in the given formula and solve for R.

Solution

$R = 1.4\sqrt{h}$
$R = 1.4\sqrt{8}$
$R \approx 3.96$

The range of the periscope is 3.96 mi.

8 You Try It

Find the range of a submarine periscope that is 6 ft above the surface of the water. Use the formula $R = 1.4\sqrt{h}$, where R is the range in miles and h is the height in feet of the periscope above the surface of the water. Round to the nearest hundredth.

Your Strategy

Your Solution

Solution on p. S13

Copyright © Houghton Mifflin Company. All rights reserved.

4.4 EXERCISES

OBJECTIVE **A**

1. ✎ Describe (a) how to find the square root of a perfect square and (b) how to simplify the square root of a number that is not a perfect square.

2. ✎ Explain why $2\sqrt{2}$ is in simplest form and $\sqrt{8}$ is not in simplest form.

Simplify.

3. $\sqrt{36}$

4. $\sqrt{1}$

5. $-\sqrt{9}$

6. $-\sqrt{1}$

7. $\sqrt{169}$

8. $\sqrt{196}$

9. $\sqrt{225}$

10. $\sqrt{81}$

11. $-\sqrt{25}$

12. $-\sqrt{64}$

13. $-\sqrt{100}$

14. $-\sqrt{4}$

15. $\sqrt{8+17}$

16. $\sqrt{40+24}$

17. $\sqrt{49}+\sqrt{9}$

18. $\sqrt{100}+\sqrt{16}$

19. $\sqrt{121}-\sqrt{4}$

20. $\sqrt{144}-\sqrt{25}$

21. $3\sqrt{81}$

22. $8\sqrt{36}$

23. $-2\sqrt{49}$

24. $-6\sqrt{121}$

25. $5\sqrt{16}-4$

26. $7\sqrt{64}+9$

27. $3+10\sqrt{1}$

28. $14-3\sqrt{144}$

29. $\sqrt{4}-2\sqrt{16}$

30. $\sqrt{144}+3\sqrt{9}$

31. $5\sqrt{25}+\sqrt{49}$

32. $20\sqrt{1}-\sqrt{36}$

33. $\sqrt{\dfrac{1}{100}}$

34. $\sqrt{\dfrac{1}{81}}$

35. $\sqrt{\dfrac{9}{16}}$

36. $\sqrt{\dfrac{25}{49}}$

37. $\sqrt{\dfrac{1}{4}}+\sqrt{\dfrac{1}{64}}$

38. $\sqrt{\dfrac{1}{36}}-\sqrt{\dfrac{1}{144}}$

Copyright © Houghton Mifflin Company. All rights reserved.

Evaluate the expression for the given values of the variables.

39. $-4\sqrt{xy}$, where $x = 3$ and $y = 12$

40. $-3\sqrt{xy}$, where $x = 20$ and $y = 5$

41. $8\sqrt{x + y}$, where $x = 19$ and $y = 6$

42. $7\sqrt{x + y}$, where $x = 34$ and $y = 15$

43. $5 + 2\sqrt{ab}$, where $a = 27$ and $b = 3$

44. $6\sqrt{ab} - 9$, where $a = 2$ and $b = 32$

45. $\sqrt{a^2 + b^2}$, where $a = 3$ and $b = 4$

46. $\sqrt{c^2 - a^2}$, where $a = 6$ and $c = 10$

47. $\sqrt{c^2 - b^2}$, where $b = 12$ and $c = 13$

48. $\sqrt{b^2 - 4ac}$, where $a = 1$, $b = -4$, and $c = -5$

49. What is the sum of five and the square root of nine?

50. Find eight more than the square root of four.

51. Find the difference between six and the square root of twenty-five.

52. What is seven decreased by the square root of sixteen?

53. What is negative four times the square root of eighty-one?

54. Find the product of negative three and the square root of forty-nine.

OBJECTIVE **B**

 Approximate to the nearest ten-thousandth.

55. $\sqrt{3}$

56. $\sqrt{7}$

57. $\sqrt{10}$

58. $\sqrt{19}$

59. $2\sqrt{6}$

60. $10\sqrt{21}$

61. $3\sqrt{14}$

62. $6\sqrt{15}$

63. $-4\sqrt{2}$

64. $-5\sqrt{13}$

65. $-8\sqrt{30}$

66. $-12\sqrt{53}$

Copyright © Houghton Mifflin Company. All rights reserved.

Between what two whole numbers is the value of the radical expression?

67. $\sqrt{23}$ **68.** $\sqrt{47}$ **69.** $\sqrt{29}$ **70.** $\sqrt{71}$

71. $\sqrt{62}$ **72.** $\sqrt{103}$ **73.** $\sqrt{130}$ **74.** $\sqrt{95}$

Simplify.

75. $\sqrt{8}$ **76.** $\sqrt{12}$ **77.** $\sqrt{45}$ **78.** $\sqrt{18}$ **79.** $\sqrt{20}$

80. $\sqrt{44}$ **81.** $\sqrt{27}$ **82.** $\sqrt{56}$ **83.** $\sqrt{48}$ **84.** $\sqrt{28}$

85. $\sqrt{75}$ **86.** $\sqrt{96}$ **87.** $\sqrt{63}$ **88.** $\sqrt{72}$ **89.** $\sqrt{98}$

90. $\sqrt{108}$ **91.** $\sqrt{112}$ **92.** $\sqrt{200}$ **93.** $\sqrt{175}$ **94.** $\sqrt{180}$

OBJECTIVE   **C**

95. *Earth Science*   A tsunami is a great sea wave produced by underwater earthquakes or volcanic eruption. Find the velocity of a tsunami when the depth of the water is 100 ft. Use the formula $v = 3\sqrt{d}$, where v is the velocity in feet per second of a tsunami as it approaches land and d is the depth in feet of the water.

96. *Earth Science*   A tsunami is a great sea wave produced by underwater earthquakes or volcanic eruption. Find the velocity of a tsunami when the depth of the water is 144 ft. Use the formula $v = 3\sqrt{d}$, where v is the velocity in feet per second of a tsunami as it approaches land and d is the depth in feet of the water.

97. *Physics*   If an object is dropped from a plane, how long will it take for the object to fall 144 ft? Use the formula $t = \sqrt{\dfrac{d}{16}}$, where t is the time in seconds that the object falls and d is the distance in feet that the object falls.

Copyright © Houghton Mifflin Company. All rights reserved.

98. *Physics* If an object is dropped from a bridge, how long will it take for the object to fall 64 ft? Use the formula $t = \sqrt{\dfrac{d}{16}}$, where t is the time in seconds that the object falls and d is the distance in feet that the object falls.

99. *Astronautics* The weight of an object is related to the distance the object is above the surface of Earth. A formula for this relationship is $d = 4{,}000\sqrt{\dfrac{E}{S}} - 4{,}000$, where E is the object's weight on the surface of Earth and S is the object's weight at a distance of d miles above Earth's surface. A space explorer, who weighs 144 lb on the surface of Earth, weighs 36 lb in space. How far above Earth's surface is the space explorer?

100. *Astronautics* The weight of an object is related to the distance the object is above the surface of Earth. A formula for this relationship is $d = 4{,}000\sqrt{\dfrac{E}{S}} - 4{,}000$, where E is the object's weight on the surface of Earth and S is the object's weight at a distance of d miles above Earth's surface. A space explorer, who weighs 189 lb on the surface of Earth, weighs 21 lb in space. How far above Earth's surface is the space explorer?

Critical Thinking

101. List the whole numbers between $\sqrt{4}$ and $\sqrt{100}$.

102. Simplify. **a.** $\sqrt{0.81}$ **b.** $-\sqrt{0.64}$ **c.** $\sqrt{2\dfrac{7}{9}}$ **d.** $-\sqrt{3\dfrac{1}{16}}$

103. List the expressions $\sqrt{\dfrac{1}{4} + \dfrac{1}{8}}$, $\sqrt{\dfrac{1}{3} + \dfrac{1}{9}}$, and $\sqrt{\dfrac{1}{5} + \dfrac{1}{6}}$ in order from smallest to largest.

104. **a.** Use the expressions $\sqrt{16 + 9}$ and $\sqrt{16} + \sqrt{9}$ to show that $\sqrt{a + b} \neq \sqrt{a} + \sqrt{b}$.
 b. Use the expressions $\sqrt{16 - 9}$ and $\sqrt{16} - \sqrt{9}$ to show that $\sqrt{a - b} \neq \sqrt{a} - \sqrt{b}$.

105. **a.** Find the two-digit perfect square that has exactly nine factors.
 b. Find two whole numbers such that their difference is 10, the smaller number is a perfect square, and the larger number is two less than a perfect square.

106. Find a perfect square that is between 350 and 400. Explain the strategy you used to find it.

107. Explain what the expression $3\sqrt{16}$ means.

Copyright © Houghton Mifflin Company. All rights reserved.

OBJECTIVE **A**

Real numbers and the real number line

A **rational number** is the quotient of two integers.

> *Rational Numbers*
>
> A rational number is a number that can be written in the form $\dfrac{a}{b}$, where a and b are integers and $b \neq 0$.

Each of the three numbers shown at the right is a rational number.

$$\frac{3}{4} \qquad \frac{-2}{9} \qquad \frac{13}{-5}$$

An integer can be written as the quotient of the integer and 1. Therefore, every integer is a rational number.

$$6 = \frac{6}{1} \qquad\qquad -8 = \frac{-8}{1}$$

A mixed number can be written as the quotient of two integers. Therefore, every mixed number is a rational number.

$$1\frac{4}{7} = \frac{11}{7} \qquad\qquad 3\frac{2}{5} = \frac{17}{5}$$

Recall from Section 4.2 that a fraction can be written as a decimal by dividing the numerator of the fraction by the denominator. The result is either a terminating decimal or a repeating decimal.

To convert $\dfrac{3}{8}$ to a decimal, read the fraction bar as "divided by."

$\dfrac{3}{8} = 3 \div 8 = 0.375$. This is an example of a terminating decimal.

To convert $\dfrac{6}{11}$ to a decimal, divide 6 by 11.

$\dfrac{6}{11} = 6 \div 11 = 0.\overline{54}$. This is an example of a repeating decimal.

Every rational number can be written either as a terminating decimal or as a repeating decimal. All terminating and repeating decimals are rational numbers.

Some numbers have decimal representations that never terminate or repeat, for example,

$$0.12122122212222\ldots$$

The pattern in this number is one more 2 following each successive 1 in the number. There is no repeating block of digits. This number is an **irrational number.** Other examples of irrational numbers include π (which is presented in Chapter 9) and square roots of integers that are not perfect squares.

> **Take Note**
>
> Rational numbers are fractions such as $-\dfrac{4}{5}$ or $\dfrac{10}{7}$, where the numerator and denominator are integers. Rational numbers are also represented by repeating decimals such as 0.2626262 . . . and by terminating decimals such as 1.83. An irrational number is neither a repeating decimal nor a terminating decimal. For instance, 1.45445444544445 . . . is an irrational number.

Copyright © Houghton Mifflin Company. All rights reserved.

> ## Irrational Numbers
>
> An **irrational number** is a number whose decimal representation never terminates or repeats.

The rational numbers and the irrational numbers taken together are called the **real numbers.**

> ## Real Numbers
>
> The **real numbers** are all the rational numbers together with all the irrational numbers.

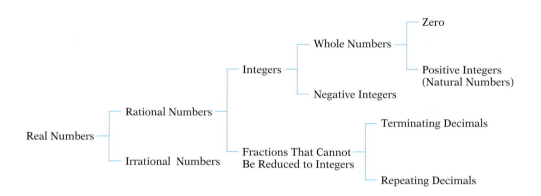

The number line is also called the **real number line.** Every real number corresponds to a point on the real number line, and every point on the real number line corresponds to a real number.

Graph $3\frac{1}{2}$ on the real number line.

$3\frac{1}{2}$ is a positive number and is therefore to the right of zero on the number line. Draw a solid dot three and one-half units to the right of zero on the number line.

Graph -2.5 on the real number line.

-2.5 is a negative number and is therefore to the left of zero on the number line. Draw a solid dot two and one-half units to the left of zero on the number line.

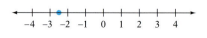

Copyright © Houghton Mifflin Company. All rights reserved.

Graph the real numbers greater than 2.

To graph the real numbers greater than 2 would mean that a solid dot should be placed above every number to the right of 2 on the number line. It is not possible to list all the real numbers greater than 2. It is not even possible to list all the real numbers between 2 and 3, or even to give the smallest real number greater than 2. The number 2.0000000001 is greater than 2 and is certainly very close to 2, but even smaller numbers greater than 2 can be written by inserting more and more zeros after the decimal point. Therefore, the graph of the real numbers greater than 2 is shown by drawing a heavy line to the right of 2.

The arrow indicates that the heavy line continues without end. The real numbers greater than 2 do not include the number 2. The parenthesis on the graph indicates that 2 is not included in the graph.

Graph the real numbers between 1 and 3.

The real numbers between 1 and 3 do not include the number 1 or the number 3; thus parentheses are drawn at 1 and 3. Draw a heavy line between 1 and 3 to indicate all the real numbers between these two numbers.

1   *Example*   Graph 0.5 on the real number line.

Solution   Draw a solid dot one-half unit to the right of zero on the number line.

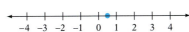

1   *You Try It*   Graph $-1\frac{1}{2}$ on the real number line.

Your Solution

2   *Example*   Graph the real numbers less than −1.

Solution   The real numbers less than −1 are to the left of −1 on the number line. Draw a parenthesis at −1. Draw a heavy line to the left of −1. Draw an arrow at the left of the line.

2   *You Try It*   Graph the real numbers greater than −2.

Your Solution

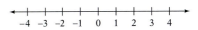

Solutions on p. S13

Copyright © Houghton Mifflin Company. All rights reserved.

3 *Example* Graph the real numbers between −3 and 0.

Solution Draw a parenthesis at −3 and a parenthesis at 0. Draw a heavy line between −3 and 0.

3 *You Try It* Graph the real numbers between −1 and 4.

Your Solution

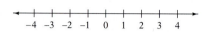

Solution on p. S13

OBJECTIVE **B**

Inequalities in one variable

Recall that the symbol for "is greater than" is >, and the symbol for "is less than" is <. The symbol ≥ means "is greater than or equal to." The symbol ≤ means "is less than or equal to."

The statement 5 < 5 is a false statement because 5 is not less than 5.

5 < 5 False

The statement 5 ≤ 5 is a true statement because 5 is "less than <u>or</u> equal to" 5; 5 is equal to 5.

5 ≤ 5 True

An **inequality** contains the symbol >, <, ≥, or ≤, and expresses the relative order of two mathematical expressions.

$$4 > -3$$
$$-9.7 < 0$$
$$6 + 2 \ge 1$$
$$x \le 5$$

Inequalities

The inequality $x \le 5$ is read "x is less than or equal to 5."

For the inequality $x > -3$, which values of the variable listed below make the inequality true?

a. −6 **b.** −3.9 **c.** 0 **d.** $\sqrt{7}$

Replace x in $x > -3$ with each number, and determine whether each inequality is true.

a.	**b.**	**c.**	**d.**
$x > -3$	$x > -3$	$x > -3$	$x > -3$
$-6 > -3$	$-3.9 > -3$	$0 > -3$	$\sqrt{7} > -3$
False	False	True	True

The numbers 0 and $\sqrt{7}$ make the inequality true.

There are many values of the variable x that will make the inequality $x > -3$ true; any number greater than −3 makes the inequality true. Replacing x with any number less than −3 will result in a false statement.

What values of the variable x make the inequality $x \le 4$ true?

All real numbers less than or equal to 4 make the inequality true.

Copyright © Houghton Mifflin Company. All rights reserved.

The numbers that make an inequality true can be graphed on the real number line.

> Graph $x > 1$.
>
> The numbers that, when substituted for x, make this inequality true are all the real numbers greater than 1. The numbers greater than 1 are all the numbers to the right of 1 on the number line. The parenthesis on the graph indicates that 1 is not included in the numbers greater than 1.

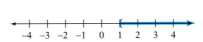

> Graph $x \geq 1$.
>
> The numbers that make this inequality true are all the real numbers greater than or equal to 1. The bracket at 1 indicates that 1 is included in the numbers greater than or equal to 1.

Note: **For $<$ or $>$, draw a parenthesis on the graph. For $\leq$ or $\geq$, draw a bracket.**

4 **Example**   For the inequality $x \leq -6$, which values of the variable that are listed below make the inequality true?

a. -12 b. -6 c. 0 d. $\sqrt{5}$

Solution   a. $x \leq -6$
                    $-12 \leq -6$ True

b. $x \leq -6$
     $-6 \leq -6$ True

c. $x \leq -6$
     $0 \leq -6$ False

d. $x \leq -6$
     $\sqrt{5} \leq -6$ False

The numbers -12 and -6 make the inequality true.

4 **You Try It**   For the inequality $x \geq 4$, which values of the variable that are listed below make the inequality true?

a. -1 b. 0 c. 4 d. $\sqrt{26}$

Your Solution

5 **Example**   What values of the variable x make the inequality $x < 8$ true?

Solution   All real numbers less than 8 make the inequality true.

5 **You Try It**   What values of the variable x make the inequality $x > -7$ true?

Your Solution

Solutions on p. S13

Copyright © Houghton Mifflin Company. All rights reserved.

6 *Example 6*   Graph $x \le 3$.

Solution   Draw a bracket at 3.
Draw an arrow to the left of 3.

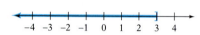

6 *You Try It 6*   Graph $x \ge -4$.

Your Solution

Solution on p. S13

VIDEO & DVD
CD TUTOR
WWW WEB
SSM

OBJECTIVE   C

Applications

Solving application problems requires recognition of the verbal phrases that translate into mathematical symbols. Below is a partial list of the phrases used to indicate each of the four inequality symbols.

$<$ is less than	$>$ is greater than
	is more than
	exceeds

$\le$ is less than or equal to	$\ge$ is greater than or equal to
maximum	minimum
at most	at least
or less	or more

---

7 *Example*

The minimum wage at the company you work for is $9.25 an hour. Write an inequality for the wages at the company. Is it possible for an employee to earn $9.15 an hour?

Strategy

▶ To write the inequality, let w represent the wages. Since $9.25 is a minimum wage, all wages are greater than or equal to $9.25.

▶ To determine whether a wage of $9.15 is possible, replace w in the inequality by 9.15. If the inequality is true, it is possible. If the inequality is false, it is not possible.

Solution

$w \ge 9.25$

$9.15 \ge 9.25$ ▶ False

It is not possible for an employee to earn $9.15 an hour.

7 *You Try It*

On the highway near your home, motorists who exceed a speed of 55 mph are ticketed. Write an inequality for the speeds at which a motorist is ticketed. Will a motorist traveling at 58 mph be ticketed?

Your Strategy

Your Solution

Solution on p. S13

Copyright © Houghton Mifflin Company. All rights reserved.

4.5 EXERCISES

OBJECTIVE A

Graph the number on the real number line.

1. $2\frac{1}{2}$

$\leftarrow \mid \; + \; + \; + \; + \; + \; + \; + \; + \; + \; + \; + \; + \mid \rightarrow$
$\quad -6 \; -5 \; -4 \; -3 \; -2 \; -1 \;\; 0 \;\; 1 \;\; 2 \;\; 3 \;\; 4 \;\; 5 \;\; 6$

2. $-2\frac{1}{2}$

$\leftarrow \mid \; + \; + \; + \; + \; + \; + \; + \; + \; + \; + \; + \; + \mid \rightarrow$
$\quad -6 \; -5 \; -4 \; -3 \; -2 \; -1 \;\; 0 \;\; 1 \;\; 2 \;\; 3 \;\; 4 \;\; 5 \;\; 6$

3. -3.5

$\leftarrow \mid \; + \; + \; + \; + \; + \; + \; + \; + \; + \; + \; + \; + \mid \rightarrow$
$\quad -6 \; -5 \; -4 \; -3 \; -2 \; -1 \;\; 0 \;\; 1 \;\; 2 \;\; 3 \;\; 4 \;\; 5 \;\; 6$

4. -0.5

$\leftarrow \mid \; + \; + \; + \; + \; + \; + \; + \; + \; + \; + \; + \; + \mid \rightarrow$
$\quad -6 \; -5 \; -4 \; -3 \; -2 \; -1 \;\; 0 \;\; 1 \;\; 2 \;\; 3 \;\; 4 \;\; 5 \;\; 6$

5. $-4\frac{1}{2}$

$\leftarrow \mid \; + \; + \; + \; + \; + \; + \; + \; + \; + \; + \; + \; + \mid \rightarrow$
$\quad -6 \; -5 \; -4 \; -3 \; -2 \; -1 \;\; 0 \;\; 1 \;\; 2 \;\; 3 \;\; 4 \;\; 5 \;\; 6$

6. $\frac{1}{2}$

$\leftarrow \mid \; + \; + \; + \; + \; + \; + \; + \; + \; + \; + \; + \; + \mid \rightarrow$
$\quad -6 \; -5 \; -4 \; -3 \; -2 \; -1 \;\; 0 \;\; 1 \;\; 2 \;\; 3 \;\; 4 \;\; 5 \;\; 6$

7. 1.5

$\leftarrow \mid \; + \; + \; + \; + \; + \; + \; + \; + \; + \; + \; + \; + \mid \rightarrow$
$\quad -6 \; -5 \; -4 \; -3 \; -2 \; -1 \;\; 0 \;\; 1 \;\; 2 \;\; 3 \;\; 4 \;\; 5 \;\; 6$

8. 5.5

$\leftarrow \mid \; + \; + \; + \; + \; + \; + \; + \; + \; + \; + \; + \; + \mid \rightarrow$
$\quad -6 \; -5 \; -4 \; -3 \; -2 \; -1 \;\; 0 \;\; 1 \;\; 2 \;\; 3 \;\; 4 \;\; 5 \;\; 6$

Graph.

9. the real numbers greater than 6

$\leftarrow \mid \; + \; + \; + \; + \; + \; + \; + \; + \; + \; + \; + \; + \mid \rightarrow$
$\quad -6 \; -5 \; -4 \; -3 \; -2 \; -1 \;\; 0 \;\; 1 \;\; 2 \;\; 3 \;\; 4 \;\; 5 \;\; 6$

10. the real numbers greater than 1

$\leftarrow \mid \; + \; + \; + \; + \; + \; + \; + \; + \; + \; + \; + \; + \mid \rightarrow$
$\quad -6 \; -5 \; -4 \; -3 \; -2 \; -1 \;\; 0 \;\; 1 \;\; 2 \;\; 3 \;\; 4 \;\; 5 \;\; 6$

11. the real numbers less than 0

$\leftarrow \mid \; + \; + \; + \; + \; + \; + \; + \; + \; + \; + \; + \; + \mid \rightarrow$
$\quad -6 \; -5 \; -4 \; -3 \; -2 \; -1 \;\; 0 \;\; 1 \;\; 2 \;\; 3 \;\; 4 \;\; 5 \;\; 6$

12. the real numbers less than 2

$\leftarrow \mid \; + \; + \; + \; + \; + \; + \; + \; + \; + \; + \; + \; + \mid \rightarrow$
$\quad -6 \; -5 \; -4 \; -3 \; -2 \; -1 \;\; 0 \;\; 1 \;\; 2 \;\; 3 \;\; 4 \;\; 5 \;\; 6$

13. the real numbers greater than -1

$\leftarrow \mid \; + \; + \; + \; + \; + \; + \; + \; + \; + \; + \; + \; + \mid \rightarrow$
$\quad -6 \; -5 \; -4 \; -3 \; -2 \; -1 \;\; 0 \;\; 1 \;\; 2 \;\; 3 \;\; 4 \;\; 5 \;\; 6$

14. the real numbers greater than -4

$\leftarrow \mid \; + \; + \; + \; + \; + \; + \; + \; + \; + \; + \; + \; + \mid \rightarrow$
$\quad -6 \; -5 \; -4 \; -3 \; -2 \; -1 \;\; 0 \;\; 1 \;\; 2 \;\; 3 \;\; 4 \;\; 5 \;\; 6$

Copyright © Houghton Mifflin Company. All rights reserved.

15. the real numbers less than -5

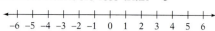

16. the real numbers less than -3

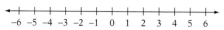

17. the real numbers between 2 and 5

18. the real numbers between 4 and 6

19. the real numbers between -4 and 0

20. the real numbers between 0 and 3

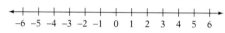

21. the real numbers between -2 and 6

22. the real numbers between -1 and 5

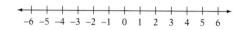

23. the real numbers between -6 and 1

24. the real numbers between -5 and 0

OBJECTIVE **B**

25. For the inequality $x > 9$, which numbers listed below make the inequality true?

   a. -3.8 **b.** 0 **c.** 9 **d.** $\sqrt{101}$

26. For the inequality $x \le 5$, which numbers listed below make the inequality true?

   a. $-\sqrt{11}$ **b.** 0 **c.** 5 **d.** 5.01

27. For the inequality $x \ge -2$, which numbers listed below make the inequality true?

   a. -6 **b.** -2 **c.** 0.4 **d.** $\sqrt{17}$

28. For the inequality $x \le -7$, which numbers listed below make the inequality true?

   a. -14 **b.** -7 **c.** -1.3 **d.** $-\sqrt{2}$

What values of the variable x make the inequality true?

29. $x < 3$ **30.** $x > -6$ **31.** $x \ge -1$ **32.** $x \le 5$

Copyright © Houghton Mifflin Company. All rights reserved.

Graph the inequality on the real number line.

33. $x < -2$

34. $x > 4$

35. $x \geq 0$

36. $x \leq -3$

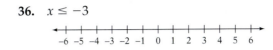

37. $x > -5$

38. $x < -1$

39. $x \leq 2$

40. $x \geq 6$

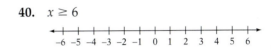

OBJECTIVE C

41. *Business* Each sales representative for a company must sell at least 50,000 units per year. Write an inequality for the number of units a sales representative must sell. Has a representative who sold 49,000 units this past year met the sales goal?

42. *Health* A health official recommends a cholesterol level of less than 220 units. Write an inequality for the acceptable cholesterol levels. Is a cholesterol level of 238 within the recommended levels?

43. *Education* A part-time student can take a maximum of 9 credit hours per semester. Write an inequality for the number of credit hours a part-time student can take. Does a student taking 8.5 credit hours fulfill the requirement for being a part-time student?

44. *Community Service* A service organization will receive a bonus of $200 for collecting more than 1,750 lb of aluminum cans during a collection drive. Write an inequality for the number of cans that must be collected in order to earn the bonus. If 1,705.5 lb of aluminum cans are collected, will the organization receive the bonus?

45. *Finances* Your monthly budget allows you to spend at most $2,400 per month. Write an inequality for the amount of money you can spend per month. Have you kept within your budget during a month in which you spent $2,380.50?

Copyright © Houghton Mifflin Company. All rights reserved.

46. *Education* In order to get a B in a history course, you must earn more than 80 points on the final exam. Write an inequality for the number of points you need to score on the final exam. Will a score of $80\frac{1}{2}$ earn you a B in the course?

47. *Produce* Eggs should not be stored at temperatures greater than 85°F. Write an inequality for the temperatures at which eggs should not be stored. Is it safe to store eggs at a temperature of 86.5°F?

48. *Sports* According to NCAA rules, the diameter of the ring on a basketball hoop is to be $\frac{5}{8}$ in. or less. Write an inequality for the diameter of the ring on a basketball hoop. Does a ring with a diameter of $\frac{9}{16}$ in. meet the NCAA regulations?

Critical Thinking

49. Classify each number as a whole number, an integer, a positive integer, a negative integer, a rational number, an irrational number, and/or a real number.

a. -2 b. 18 c. $-\frac{9}{37}$ d. -6.606 e. $4.5\overline{6}$ f. $3.050050005\dots$

50. Using the variable x, write an inequality to represent the graph.

a.

b.

51. For the given inequality, which of the numbers in parentheses make the inequality true?

a. $|x| < 9$ $(-2.5, 0, 9, 15.8)$ b. $|x| > -3$ $(-6.3, -3, 0, 6.7)$

c. $|x| \geq 4$ $(-1.5, 0, 4, 13.6)$ d. $|x| \leq 5$ $(-4.9, 0, 2.1, 5)$

52. Given that a, b, c, and d are real numbers, which will ensure that $a + c < b + d$?

a. $a < b$ and $c < d$ b. $a > b$ and $c > d$

c. $a < b$ and $c > d$ d. $a > b$ and $c < d$

53. Determine whether the statement is always true, sometimes true, or never true.

a. Given that $a > 0$ and $b > 0$, then $ab > 0$.

b. Given that $a < 0$, then $a^2 > 0$.

c. Given that $a > 0$ and $b > 0$, then $a^2 > b$.

54. Enter -4 on your calculator and then press the square root key. What is in the calculator's display? Explain why.

55. In your own words, define (a) a rational number, (b) an irrational number, and (c) a real number.

Copyright © Houghton Mifflin Company. All rights reserved.

Focus on **Problem Solving**

From Concrete to Abstract

As you progress in your study of algebra, you will find that the problems become less concrete and more abstract. Problems that are concrete provide information pertaining to a specific instance. Abstract problems are theoretical; they are stated without reference to a specific instance. Let's look at an example of an abstract problem.

How many cents are in d dollars?

How can you solve this problem? Are you able to solve the same problem if the information given is concrete?

How many cents are in 5 dollars?

You know that there are 100 cents in 1 dollar. To find the number of cents in 5 dollars, multiply 5 by 100.

$$100 \cdot 5 = 500 \qquad \text{There are 500 cents in 5 dollars.}$$

Use the same procedure to find the number of cents in d dollars: multiply d by 100.

$$100 \cdot d = 100d \qquad \text{There are } 100d \text{ cents in } d \text{ dollars.}$$

This problem might be taken a step further:

If one pen costs c cents, how many pens can be purchased with d dollars?

Consider the same problem using numbers in place of the variables.

If one pen costs 25 cents, how many pens can be purchased with 2 dollars?

To solve this problem, you need to calculate the number of cents in 2 dollars (multiply 2 by 100) and divide the result by the cost per pen (25 cents).

$$\frac{100 \cdot 2}{25} = \frac{200}{25} = 8 \qquad \text{If one pen costs 25 cents,} \\ \text{8 pens can be purchased with 2 dollars.}$$

Use the same procedure to solve the related abstract problem. Calculate the number of cents in d dollars (multiply d by 100), and divide the result by the cost per pen (c cents).

$$\frac{100 \cdot d}{c} = \frac{100d}{c} \qquad \text{If one pen costs } c \text{ cents,} \\ \frac{100d}{c} \text{ pens can be purchased} \\ \text{with } d \text{ dollars.}$$

At the heart of the study of algebra is the use of variables. It is the variables in the problems above that make them abstract. But it is variables that allow us to generalize situations and state rules about mathematics.

Copyright © Houghton Mifflin Company. All rights reserved.

Try the following problems.

1. How many nickels are in *d* dollars?

2. How long can you talk on a pay phone if you have only *d* dollars and the call costs *c* cents per minute?

3. If you travel *m* miles on one gallon of gasoline, how far can you travel on *g* gallons of gasoline?

4. If you walk a mile in *x* minutes, how far can you walk in *h* hours?

5. If one photocopy costs *n* nickels, how many photocopies can you make for *q* quarters?

Projects & Group Activities

Averages We often discuss temperature in terms of average high or average low temperature. Temperatures collected over a period of time are analyzed to determine, for example, the average high temperature for a given month in your city or state. The following activity is planned to help you better understand the concept of "average."

1. Choose two cities in the United States. We will refer to them as City X and City Y. Over an eight-day period, record the daily high temperature each day in each city.

2. Determine the average high temperature for City X for the eight-day period. (Add the eight numbers, and then divide the sum by 8.) Do not round your answer.

3. Subtract the average high temperature for City X from each of the eight daily high temperatures for City X. You should have a list of eight numbers; the list should include positive numbers, negative numbers, and possibly zero.

4. Find the sum of the list of eight differences recorded in Step 3.

5. Repeat Steps 2 through 4 for City Y.

6. Compare the two sums found in Step 5 for City X and City Y.

7. If you were to conduct this activity again, what would you expect the outcome to be? Use the results to explain what an average high temperature means. In your own words, explain what "average" means.

Copyright © Houghton Mifflin Company. All rights reserved.

Sequences

Suppose you are offered a 30-day job that pays $.01 the first day, $.02 the second day, $.04 the third day, and so on. Each day you work, your earnings are twice your earnings for the previous day. Would you accept this job over a 30-day job that pays $50,000 per day?

Day 1

For the job in which earnings double each day, make a guess as to your earnings on the 30th day of work, and your total earnings over the 30-day period. Then calculate these figures. You may be surprised at the results!

Day 2

The list of numbers that indicates your earnings each day is an ordered list of numbers, called a **sequence**.

Day 3

0.01, 0.02, 0.04, 0.08, 0.16, 0.32, 0.64, 1.28, 2.56, 5.12, . . .

This list is ordered because the position of a number in the list indicates the day on which that amount was earned. For example, the 8th term of the sequence is 1.28, and $1.28 is earned on the 8th day.

Each of the numbers of a sequence is called a **term** of the sequence. A formula can be used to find a specific term of the sequence given above.

$$\text{term } t = (\text{first term})(2^{t-1})$$

For example,

$$\text{term } 5 = (0.01)(2^{5-1}) = 0.01(2^4) = 0.01(16) = 0.16$$

The amount earned on day 5 is $.16.

1. Use the formula to find the amount earned on day 15 and on day 20.

2. What amount is earned on the 30th day?

A formula can be used to find the total amount earned after any given day.

$$\text{day } d = \frac{(\text{first term})(1 - 2^d)}{1 - 2}$$

For example,

$$\text{day } 5 = \frac{(0.01)(1 - 2^5)}{1 - 2} = \frac{(0.01)(1 - 32)}{1 - 2}$$

$$= \frac{(0.01)(-31)}{-1} = 0.31$$

The total amount earned during the first 5 days is $.31.

3. Use the formula to find the total amount earned by the end of day 15 and by the end of day 20.

4. What is the total amount earned after 30 days? How does this compare with being paid $50,000 per day?

Take Note

Note that the Order of Operations Agreement is used to simplify the expression at the right.

Copyright © Houghton Mifflin Company. All rights reserved.

Customer Billing Chris works at B & W Garage as an auto mechanic and has just completed an engine overhaul for a customer. To determine the cost of the repair job, Chris keeps a list of times worked and parts used. A parts list and a list of the times worked are shown below. Use these tables, and the fact that the charge for labor is $46.75 per hour, to determine the total cost for parts and labor.

Parts Used		Time Spent	
Item	Quantity	Day	Hours
Gasket set	1	Monday	7.0
Ring set	1	Tuesday	7.5
Valves	8	Wednesday	6.5
Wrist pins	8	Thursday	8.5
Valve springs	16	Friday	9.0
Rod bearings	8		
Main bearings	5		
Valve seals	16		
Timing chain	1		

Price List		
Item Number	Description	Unit Price
27345	Valve spring	$9.25
41257	Main bearing	$17.49
54678	Valve	$16.99
29753	Ring set	$169.99
45837	Gasket set	$174.90
23751	Timing chain	$50.49
23765	Fuel pump	$429.99
28632	Wrist pin	$13.55
34922	Rod bearing	$4.69
2871	Valve seal	$1.69

Chapter Summary

Key Words

Examples

Key Words	Examples
A number written in **decimal notation** has three parts: a whole number part, a decimal point, and a decimal part. The **decimal part** of a number represents a number less than one. A number written in decimal notation is often simply called a **decimal.** [4.1A, p. 239]	For the decimal 31.25, 31 is the whole number part and 25 is the decimal part.
The square of an integer is called a **perfect square.** [4.4A, p. 281]	$1^2 = 1, 2^2 = 4, 3^2 = 9, 4^2 = 16, 5^2 = 25, \ldots,$ so $1, 4, 9, 16, 25, \ldots$ are perfect squares.
A **square root** of a positive number x is a number whose square is x. The symbol for square root is $\sqrt{}$, which is called a **radical sign.** The number under the radical is called the **radicand.** [4.4B, pp. 281–282]	$\sqrt{25} = 5$ because $5^2 = 25$. In the expression $\sqrt{25}$, 25 is the radicand.
A radical expression is in **simplest form** when the radicand contains no factor, other than 1, that is a perfect square. [4.4B, p. 285]	$\sqrt{18}$ is not in simplest form because the radicand, 18, contains the factor 9, and 9 is a perfect square.

Copyright © Houghton Mifflin Company. All rights reserved.

A **rational number** is a number that can be written in the form $\frac{a}{b}$, where a and b are integers and $b \neq 0$. Every rational number can be written either as a terminating decimal or as a repeating decimal. All terminating and repeating decimals are rational numbers. [4.5A, p. 291]

$\frac{7}{16} = 0.4375$, a terminating decimal.

$\frac{4}{15} = 0.2\overline{6}$, a repeating decimal.

An **irrational number** is a number whose decimal representation never terminates or repeats. [4.5A, p. 292]

π, $\sqrt{3}$, and $0.23233233323333\ldots$ are irrational numbers.

The **real numbers** are all the rational numbers together with all the irrational numbers. [4.5A, p. 292]

An **inequality** contains the symbol $>$, $<$, $\geq$, or $\leq$ and expresses the relative order of two mathematical expressions. [4.5B, p. 294]

$3.9 \leq 5$

$5 \leq 5$

$8.3 \geq 8$

$8 \geq 8$

$x < 7$

Essential Rules and Procedures

To write a decimal in words, write the decimal part as though it were a whole number. Then name the place value of the last digit. The decimal point is read as "and." [4.1A, p. 239]

The decimal 12.875 is written in words as twelve and eight hundred seventy-five thousandths.

To write a decimal in standard form when it is written in words, write the whole number part, replace the word *and* with a decimal point, and write the decimal part so that the last digit is in the given place-value position. [4.1A, p. 240]

The decimal forty-nine and sixty-three thousandths is written in standard form as 49.063.

To compare two decimals, write the decimal part of each number so that each has the same number of decimal places. Then compare the two numbers. [4.1B, p. 241]

$1.790 > 1.789$

$0.8130 < 0.8315$

To round a decimal, use the same rules used with whole numbers, except drop the digits to the right of the given place value instead of replacing them with zeros. [4.1C, p. 242]

2.7134 rounded to the nearest tenth is 2.7.

0.4687 rounded to the nearest hundredth is 0.47.

To add or subtract decimals, write the decimals so that the decimal points are on a vertical line. Add or subtract as you would with whole numbers. Then write the decimal point in the answer directly below the decimal points in the given numbers. [4.2A, p. 249]

$$\begin{array}{r} \overset{1\ 1}{} \\ 1.35 \\ 20.8 \\ +\ 0.76 \\ \hline 22.91 \end{array}$$

$$\begin{array}{r} \overset{2\ 15\ \ \ 6\ 10}{3\cancel{3}.8\cancel{7}\cancel{0}} \\ -\ 9.641 \\ \hline 26.229 \end{array}$$

Copyright © Houghton Mifflin Company. All rights reserved.

To estimate the answer to a calculation, round each number to the highest place value of the number; the first digit of each number will be nonzero, and all other digits will be zero. If a number is a decimal less than one, round the decimal so that there is one nonzero digit. Perform the calculation using the rounded numbers. [4.2A, p. 251]

$$
\begin{array}{rcr}
35.87 & \longrightarrow & 40 \\
61.09 & \longrightarrow & + 60 \\
\hline
& & 100
\end{array}
$$

$$
\begin{array}{rcr}
0.3876 & \longrightarrow & 0.4 \\
0.5472 & \longrightarrow & + 0.5 \\
\hline
& & 0.9
\end{array}
$$

To multiply decimals, multiply the numbers as you would whole numbers. Then write the decimal point in the product so that the number of decimal places in the product is the sum of the decimal places in the factors. [4.2A, p. 253]

$$
\begin{array}{rl}
26.83 & \text{2 decimal places} \\
\times \ 0.45 & \text{2 decimal places} \\
\hline
13415 & \\
10732 \ \ & \\
\hline
12.0735 & \text{4 decimal places}
\end{array}
$$

To multiply a decimal by a power of 10, move the decimal point to the right the same number of places as there are zeros in the power of 10. If the power of 10 is written in exponential notation, the exponent indicates how many places to move the decimal point. [4.2B, p. 254]

$3.97 \cdot 10,000 = 39,700$

$0.641 \cdot 10^5 = 64,100$

To divide decimals, move the decimal point in the divisor to the right so that the divisor is a whole number. Move the decimal point in the dividend the same number of places to the right. Place the decimal point in the quotient directly above the decimal point in the dividend. Then divide as you would with whole numbers. [4.2C, p. 256]

$$
\begin{array}{r}
6.2 \\
0.39\overline{)2.41.8} \\
\underline{-2\,3\,4} \\
7\,8 \\
\underline{-7\,8} \\
0
\end{array}
$$

To divide a decimal by a power of 10, move the decimal point to the left the same number of places as there are zeros in the power of 10. If the power of 10 is written in exponential notation, the exponent indicates how many places to move the decimal point. [4.2C, p. 257]

$972.8 \div 1,000 = 0.9728$

$61.305 \div 10^4 = 0.0061305$

To write a fraction as a decimal, divide the numerator of the fraction by the denominator. [4.2D, p. 260]

$\dfrac{7}{8} = 7 \div 8 = 0.875$

To convert a decimal to a fraction, remove the decimal point and place the decimal part over a denominator equal to the place value of the last digit in the decimal. [4.2D, p. 260]

0.85 is eighty-five <u>hundredths</u>.

$0.85 = \dfrac{85}{100} = \dfrac{17}{20}$

To find the order relation between a decimal and a fraction, first rewrite the fraction as a decimal. Then compare the two decimals. [4.2D, p. 261]

Because $\dfrac{3}{11} \approx 0.273$, and $0.273 > 0.26$, $\dfrac{3}{11} > 0.26$.

Square Root
For $x > 0$, if $a^2 = x$, then $\sqrt{x} = a$. [4.4A, p. 281]

Because $6^2 = 36$, $\sqrt{36} = 6$.

Product Property of Square Roots
If a and b are positive numbers, then $\sqrt{a \cdot b} = \sqrt{a} \cdot \sqrt{b}$. [4.4B, p. 285]

$\sqrt{4 \cdot 25} = \sqrt{4} \cdot \sqrt{25}$

Copyright © Houghton Mifflin Company. All rights reserved.

Chapter Review Exercises

1. Approximate $3\sqrt{47}$ to the nearest ten-thousandth.

2. Find the product of 0.918 and 10^5.

3. Simplify: $-\sqrt{121}$

4. Subtract: $-3.981 - 4.32$

5. Evaluate $a + b + c$ when $a = 80.59$, $b = -3.647$, and $c = 12.3$.

6. Write five and thirty-four thousandths in standard form.

7. Simplify: $\sqrt{100} - 2\sqrt{49}$

8. Find the quotient of 14.2 and 10^3.

9. Solve: $4.2z = -1.428$

10. Place the correct symbol, $<$ or $>$, between the two numbers.

    8.039   8.31

11. Evaluate $\dfrac{x}{y}$ when $x = 0.396$ and $y = 3.6$.

12. Multiply: $(9.47)(0.26)$

13. For the inequality $x \geq -1$, what numbers listed below make the inequality true?
    a. -6 **b.** -1 **c.** -0.5 **d.** $\sqrt{10}$

14. Place the correct symbol, $<$ or $>$, between the two numbers.

    $\dfrac{3}{7}$ 0.429

15. Convert 0.28 to a fraction.

16. Divide and round to the nearest tenth: $-6.8 \div 47.92$

17. *Labor Force* The graph at the right shows that the number of older workers is expected to increase during the first decade of the twenty-first century. Find the projected increase in the number of workers over the age of 65.

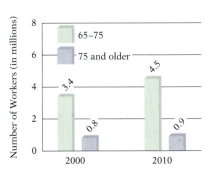

Older Workers in the Labor Force
Source: Bureau of Labor Statistics

Copyright © Houghton Mifflin Company. All rights reserved.

18. Graph all the real numbers between −6 and −2.

-6 −5 −4 −3 −2 −1 0 1 2 3 4 5 6

19. Graph $x \geq -3$.

-6 −5 −4 −3 −2 −1 0 1 2 3 4 5 6

20. Find the sum of −247.8 and −193.4.

21. Find the quotient of 614.3 and 100.

22. Evaluate $a - b$ when $a = 80.32$ and $b = 29.577$.

23. Simplify: $\sqrt{90}$

24. Evaluate $60st$ when $s = 5$ and $t = -3.7$.

25. Estimate the difference between 506.81 and 64.1.

26. *Education*   A student must have a grade point average of at least 3.5 to qualify for a certain scholarship. Write an inequality for the grade point average a student must have in order to qualify for the scholarship. Does a student who has a grade point average of 3.48 qualify for the scholarship?

27. *Chemistry*   The boiling point of bromine is 58.8°C. The melting point of bromine is −7.2°C. Find the difference between the boiling point and the melting point of bromine.

28. *History*   The figure at the right shows the monetary cost of four wars. (a) What is the difference between the monetary costs of the two World Wars? (b) How many times greater was the monetary cost of the Vietnam War than World War I?

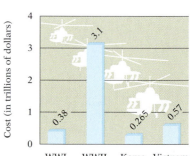

Monetary Cost of War
Source: Congressional Research Service Using Numbers from the *Statistical Abstract of the United States*

29. *Consumerism*   A 7-ounce jar of instant coffee costs $11.78. Find the cost per ounce. Round to the nearest cent.

30. *Consumerism*   The total of the monthly payments for a car lease is the product of the number of months of the lease and the monthly lease payment. The total of the monthly payments for a 24-month car lease is $9,977.76. Find the monthly lease payment.

31. *Business*   Use the formula $P = C + M$, where P is the price of a product to a customer, C is the cost paid by a store for the product, and M is the markup, to find the price of a treadmill that costs a business $369.99 and has a markup of $129.50.

32. *Physics*   The velocity of a falling object is given by the formula $v = \sqrt{64d}$, where v is the velocity in feet per second and d is the distance the object has fallen. Find the velocity of an object that has fallen a distance of 25 ft.

Copyright © Houghton Mifflin Company. All rights reserved.

Chapter Test

1. Write nine and thirty-three thousandths in standard form.

2. Place the correct symbol, $<$ or $>$, between the two numbers.

   4.003　4.009

3. Round 6.051367 to the nearest thousandth.

4. Find the difference between -30 and -7.247.

5. Evaluate $x - y$ when $x = 6.379$ and $y = -8.28$.

6. Estimate the difference between 92.34 and 17.95.

7. Find the total of 4.58, -3.9, and 6.017.

8. What is the product of -2.5 and 7.36?

9. Evaluate $-20cd$ when $c = 0.5$ and $d = -6.4$.

10. Solve: $5.488 = -3.92p$

11. Simplify: $\sqrt{256} - 2\sqrt{121}$

12. Find the quotient of 84.96 and 100.

13. Evaluate $\dfrac{x}{y}$ when $x = 52.7$ and $y = -6.2$.

14. Place the correct symbol, $<$ or $>$, between the two numbers.

    0.22　$\dfrac{2}{9}$

15. Approximate $2\sqrt{46}$ to the nearest ten-thousandth.

16. Simplify: $\sqrt{68}$

17. 🥧 *The Film Industry*　The table at the right shows six James Bond films released between 1960 and 1970 and their gross box office income, in millions of dollars, in the United States. How much greater was the gross from *Thunderball* than the gross from *On Her Majesty's Secret Service*?

Film	U.S. Box Office Gross
Dr. No	$16.1
On Her Majesty's Secret Service	$22.8
From Russia with Love	$24.8
You Only Live Twice	$43.1
Goldfinger	$51.1
Thunderball	$63.6

Source: **www.worldwideboxoffice.com**

Copyright © Houghton Mifflin Company. All rights reserved.

18. Is -2.5 a solution of the equation $8.4 = 5.9 + a$?

19. Multiply: $8.973 \cdot 10^4$

20. Graph the real numbers between -2 and 2.

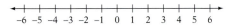

21. Graph $x \geq 3$.

22. Evaluate $x + y$ when $x = -233.81$ and $y = 71.3$.

23. Solve: $-8v = 26$

24. 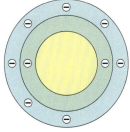 *Chemistry* The boiling point of fluorine is $-188.14°C$. The melting point of fluorine is $-219.62°C$. Find the difference between the boiling point and the melting point of fluorine.

Flourine

25. *Physics* The velocity of a falling object is given by the formula $v = \sqrt{64d}$, where v is the velocity in feet per second and d is the distance the object has fallen. Find the velocity of an object that has fallen a distance of 16 ft.

26. *Accounting* The fundamental accounting equation is $A = L + S$, where A is the assets of the company, L is the liabilities of the company, and S is the stockholders' equity. Find the stockholders' equity in a company whose assets are \$48.2 million and whose liabilities are \$27.6 million.

27. *Geometry* The lengths of the three sides of a triangle are 8.75 m, 5.25 m, and 4.5 m. Find the perimeter of the triangle. Use the formula $P = a + b + c$.

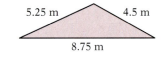

28. *Business* Each sales representative for a company must sell at least 65,000 units per year. Write an inequality for the number of units a sales representative must sell. Has a representative who sold 57,000 units this year met the sales goal?

29. *Physics* Find the force exerted on a falling object that has a mass of 5.75 kg. Use the formula $F = ma$, where F is the force exerted by gravity on a falling object, m is the mass of the object, and a is the acceleration of gravity. The acceleration of gravity is -9.80 m/s². The force is measured in newtons.

30. *Temperature* On January 19, 1892, the temperature in Fort Assiniboine, Montana, rose to 2.78°C from $-20.56°C$ in a period of only 15 min. Find the difference between these two temperatures.

Copyright © Houghton Mifflin Company. All rights reserved.

Cumulative Review Exercises

1. Find the quotient of 387.9 and 10^4.

2. Evaluate $(x + y)^2 - 2z$ when $x = -3$, $y = 2$, and $z = -5$.

3. Solve: $-9.8 = -0.49c$

4. Write eight million seventy-two thousand ninety-two in standard form.

5. Graph all the real numbers between -4 and 1.

6. Graph $x \le -2$.

7. Find the difference between -23 and -19.

8. Estimate the sum of 372, 541, 608, and 429.

9. Simplify: $\sqrt{192}$

10. Evaluate $x \div y$ when $x = 3\frac{2}{3}$ and $y = 2\frac{4}{9}$.

11. What is -36.92 increased by 18.5?

12. Simplify: $\left(\frac{5}{9}\right)\left(-\frac{3}{10}\right)\left(-\frac{6}{7}\right)$

13. Evaluate $x^4 y^2$ when $x = 2$ and $y = 10$.

14. Find the prime factorization of 260.

15. Convert $\frac{19}{25}$ to a decimal.

16. Approximate $10\sqrt{91}$ to the nearest ten-thousandth.

17. *Labor* The figure at the right shows the number of vacation days per year that are legally mandated in several countries.
    a. Which country mandates more vacation days, Ireland or Sweden?
    b. How many times more vacation days does Austria mandate than Switzerland?

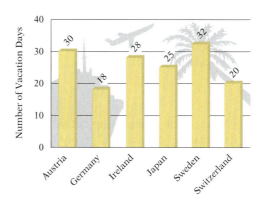

Number of Legally Mandated Vacation Days
Source: Economic Policy Institute; World Almanac

Copyright © Houghton Mifflin Company. All rights reserved.

18. Divide: $\dfrac{-8}{0}$

19. Simplify: $-\dfrac{5}{7} + \dfrac{4}{21}$

20. Simplify: $4\sqrt{25} - \sqrt{81}$

21. Estimate the product of 62.8 and 0.47.

22. Simplify: $5(3 - 7) \div (-4) + 6(2)$

23. Evaluate $\dfrac{a}{b + c}$ when $a = \dfrac{3}{8}$, $b = \dfrac{1}{2}$, and $c = \dfrac{3}{4}$.

24. Evaluate $x - y + z$ when $x = \dfrac{5}{12}$, $y = -\dfrac{3}{8}$, and $z = -\dfrac{3}{4}$.

25. Divide and round to the nearest tenth: $2.617 \div 0.93$

26. *Consumerism*   Your cellular phone company charges $39.99 per month, which includes 50 min of free air time, and $.75 for each additional minute after the first 50. What is your cellular phone service bill for a month in which you had 87 min of calls?

27. *Temperature*   On December 24, 1924, in Fairfield, Montana, the temperature fell from 17.22°C at noon to −29.4°C at midnight. How many degrees did the temperature fall in the 12-hour period?

28. *Consumerism*   Use the formula $C = \dfrac{M}{N}$, where C is the cost per visit at a health club, M is the membership fee, and N is the number of visits to the club, to find the cost per visit when your annual membership fee at a health club is $390 and you visit the club 125 times during the year.

29. *Business*   The figure at the right shows how the average salesperson spends the workweek. (a) On average, how many hours per week does a salesperson work? (b) Does the average salesperson spend more time face-to-face selling or doing both administrative work and placing service calls?

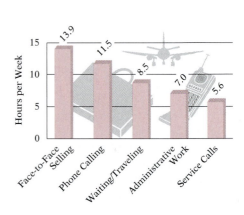

Average Salesperson's Workweek
Source: Dartnell's 28th Survey of Sales Force Compensation

30. *Physics*   The relationship between the velocity of a car and its braking distance is given by the formula $v = \sqrt{20d}$, where v is the velocity in miles per hour and d is its braking distance in feet. How fast is a car going when its braking distance is 45 ft?

Copyright © Houghton Mifflin Company. All rights reserved.

Variable Expressions

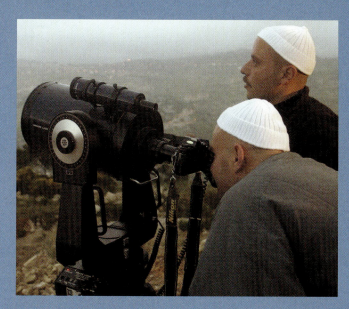

Telescopes allow scientists to look at objects that are extremely large but far away. They also allow people to view exciting events in space, such as the partial eclipse of the sun that these two men in Aley, Lebanon are observing. Astronomers use scientific notation to describe distances in space. Scientific notation replaces very large and very small numbers with more concise expressions, making these numbers easier to read and write. **Exercises 55–58 on page 356** provide examples of situations in which scientific notation is used.

Need help? For online student resources, visit this website: **math.college.hmco.com**

Copyright © Houghton Mifflin Company. All rights reserved.

Prep Test

1. Place the correct symbol, $<$ or $>$, between the two numbers.
   54   45

For Exercises 2–6, add, subtract, multiply, or divide.

2. $-19 + 8$

3. $26 - 38$

4. $-2(44)$

5. $-\dfrac{3}{4}(-8)$

6. $3.97 \cdot 10^4$

7. Simplify: $(-3)^2$

8. Simplify: $(8 - 6)^2 + 12 \div 4 \cdot 3^2$

Go Figure

Luis, Kim, Reggie, and Dave are standing in line. Dave is not first. Kim is between Luis and Reggie. Luis is between Dave and Kim. Give the order in which the men are standing.

Copyright © Houghton Mifflin Company. All rights reserved.

Copyright © Houghton Mifflin Company. All rights reserved.

SECTION 5.1   **Properties of Real Numbers**

OBJECTIVE   **A**

Application of the Properties of Real Numbers

The Properties of Real Numbers describe the way operations on numbers can be performed. These properties have been stated in previous chapters but are restated here for review. The properties are used to rewrite variable expressions.

PROPERTIES OF REAL NUMBERS

The Commutative Property of Addition If a and b are real numbers, then $a + b = b + a$.	$7 + 12 = 12 + 7$ $19 = 19$

The Commutative Property of Addition states that when we add two numbers, the numbers can be added in either order; the sum is the same.

The Commutative Property of Multiplication If a and b are real numbers, then $a \cdot b = b \cdot a$.	$7 \cdot (-2) = (-2) \cdot 7$ $-14 = -14$

The Commutative Property of Multiplication states that when we multiply two numbers, the numbers can be multiplied in either order; the product is the same.

The Associative Property of Addition If a, b, and c are real numbers, then $(a + b) + c = a + (b + c)$.	$(7 + 3) + 8 = 7 + (3 + 8)$ $10 + 8 = 7 + 11$ $18 = 18$

The Associative Property of Addition states that when we add three or more numbers, the numbers can be grouped in any order; the sum is the same.

The Associative Property of Multiplication If a, b, and c are real numbers, then $(a \cdot b) \cdot c = a \cdot (b \cdot c)$.	$(4 \cdot 5) \cdot 3 = 4 \cdot (5 \cdot 3)$ $20 \cdot 3 = 4 \cdot 15$ $60 = 60$

The Associative Property of Multiplication states that when we multiply three or more factors, the factors can be grouped in any order; the product is the same.

> ### The Addition Property of Zero
>
> If a is a real number, then
> $a + 0 = 0 + a = a$.

$$(-7) + 0 = 0 + (-7) = -7$$

The Addition Property of Zero states that the sum of a number and zero is the number.

> ### The Multiplication Property of Zero
>
> If a is a real number, then
> $a \cdot 0 = 0 \cdot a = 0$.

$$5 \cdot 0 = 0 \cdot 5 = 0$$

The Multiplication Property of Zero states that the product of a number and zero is zero.

> ### The Multiplication Property of One
>
> If a is a real number, then
> $a \cdot 1 = 1 \cdot a = a$.

$$9 \cdot 1 = 1 \cdot 9 = 9$$

The Multiplication Property of One states that the product of a number and 1 is the number.

> ### The Inverse Property of Addition
>
> If a is a real number, then
> $a + (-a) = (-a) + a = 0$.

$$2 + (-2) = (-2) + 2 = 0$$

The sum of a number and its opposite is zero.
$-a$ is the opposite of a. $-a$ is also called the **additive inverse** of a.
a is the opposite of $-a$, or a is the *additive inverse* of $-a$.
The sum of a number and its additive inverse is zero.

> ### The Inverse Property of Multiplication
>
> If a is a real number and $a \neq 0$, then
> $a \cdot \dfrac{1}{a} = \dfrac{1}{a} \cdot a = 1$.

$$4 \cdot \frac{1}{4} = \frac{1}{4} \cdot 4 = 1$$

The product of a nonzero number and its reciprocal is one.
$\frac{1}{a}$ is the reciprocal of a. $\frac{1}{a}$ is also called the **multiplicative inverse** of a.
a is the reciprocal of $\frac{1}{a}$, or a is the *multiplicative inverse* of $\frac{1}{a}$.
The product of a nonzero number and its multiplicative inverse is one.

Copyright © Houghton Mifflin Company. All rights reserved.

The Properties of Real Numbers can be used to rewrite a variable expression in a simpler form. This process is referred to as *simplifying* the variable expression.

Simplify: $5 \cdot (4x)$

| Use the Associative Property of Multiplication. | $5 \cdot (4x) = (5 \cdot 4)x$ |
| Multiply 5 times 4. | $= 20x$ |

Simplify: $(6x) \cdot 2$

Use the Commutative Property of Multiplication.	$(6x) \cdot 2 = 2 \cdot (6x)$
Use the Associative Property of Multiplication.	$= (2 \cdot 6)x$
Multiply 2 times 6.	$= 12x$

Simplify: $(5y)(3y)$

Use the Commutative and Associative Properties of Multiplication.	$(5y)(3y) = 5 \cdot y \cdot 3 \cdot y$
	$= 5 \cdot 3 \cdot y \cdot y$
	$= (5 \cdot 3)(y \cdot y)$
Write $y \cdot y$ in exponential form. Multiply 5 times 3.	$= 15y^2$

By the Multiplication Property of One, the product of 1 and x is x.

$$1 \cdot x = x$$
$$1x = x$$

Just as the product of 1 and x is written x, the product of -1 and x is written $-x$.

$$-1 \cdot x = -x$$
$$-1x = -x$$

Simplify: $(-2)(-x)$

Write $-x$ as $-1x$.	$(-2)(-x) = (-2)(-1x)$
Use the Associative Property of Multiplication.	$= [(-2)(-1)]x$
Multiply -2 times -1.	$= 2x$

> **Take Note**
>
> Brackets, [], are used as a grouping symbol to group the factors -2 and -1 because parentheses have already been used in the expression to show that -2 and -1 are being multiplied. The expression $[(-2)(-1)]$ is considered easier to read than $((-2)(-1))$.

Simplify: $-4t + 9 + 4t$

Use the Commutative Property of Addition.	$-4t + 9 + 4t = -4t + 4t + 9$
Use the Associative Property of Addition.	$= (-4t + 4t) + 9$
Use the Inverse Property of Addition.	$= 0 + 9$
Use the Addition Property of Zero.	$= 9$

Copyright © Houghton Mifflin Company. All rights reserved.

1 *Example* Simplify: $-5(7b)$

Solution $-5(7b) = (-5 \cdot 7)b$
$\qquad\qquad\quad = -35b$

1 *You Try It* Simplify: $-6(-3p)$

Your Solution

2 *Example* Simplify: $(-4r)(-9t)$

Solution $(-4r)(-9t) = [(-4)(-9)](r \cdot t)$
$\qquad\qquad\qquad\quad = 36rt$

2 *You Try It* Simplify: $(-2m)(-8n)$

Your Solution

3 *Example* Simplify: $(-8)(-z)$

Solution $(-8)(-z) = (-8)(-1z)$
$\qquad\qquad\quad = [(-8)(-1)]z$
$\qquad\qquad\quad = 8z$

3 *You Try It* Simplify: $(-12)(-d)$

Your Solution

4 *Example* Simplify: $-5y + 5y + 7$

Solution $-5y + 5y + 7 = 0 + 7$
$\qquad\qquad\qquad\quad = 7$

4 *You Try It* Simplify: $6n + 9 + (-6n)$

Your Solution

Solutions on p. S13

OBJECTIVE **B**

VIDEO & DVD   CD TUTOR   WEB   SSM

The Distributive Property

Consider the numerical expression $6 \cdot (7 + 9)$.

This expression can be evaluated by applying the Order of Operations Agreement.

Simplify the expression inside the parentheses.
Multiply.
$\qquad 6 \cdot (7 + 9) = 6 \cdot 16$
$\qquad\qquad\qquad\quad = 96$

There is an alternative method of evaluating this expression.

Multiply each number inside the parentheses by 6 and add the products.
$\qquad 6 \cdot (7 + 9) = 6 \cdot 7 + 6 \cdot 9$
$\qquad\qquad\qquad\quad = 42 + 54$
$\qquad\qquad\qquad\quad = 96$

Each method produced the same result. The second method uses the **Distributive Property**, which is another of the Properties of Real Numbers.

The Distributive Property

If a, b, and c are real numbers, then $a(b + c) = ab + ac$.

Copyright © Houghton Mifflin Company. All rights reserved.

The Distributive Property is used to remove parentheses from a variable expression.

> Simplify $3(5a + 4)$ by using the Distributive Property.
>
> Use the Distributive Property. $\qquad 3(5a + 4) = 3(5a) + 3(4)$
>
> Simplify. $\qquad\qquad\qquad\qquad\qquad = 15a + 12$

> Simplify $-4(2a + 3)$ by using the Distributive Property.
>
> Use the Distributive Property. $\qquad -4(2a + 3) = -4(2a) + (-4)(3)$
>
> Simplify. $\qquad\qquad\qquad\qquad\qquad = -8a + (-12)$
>
> Rewrite addition of the opposite as subtraction. $\qquad\qquad\qquad\qquad = -8a - 12$

The Distributive Property can also be stated in terms of subtraction.

$$a(b - c) = ab - ac$$

> Simplify $5(2x - 4y)$ by using the Distributive Property.
>
> Use the Distributive Property. $\qquad 5(2x - 4y) = 5(2x) - 5(4y)$
>
> Simplify. $\qquad\qquad\qquad\qquad\qquad = 10x - 20y$

> Simplify $-3(2x - 8)$ by using the Distributive Property.
>
> Use the Distributive Property. $\qquad -3(2x - 8) = -3(2x) - (-3)(8)$
>
> Simplify. $\qquad\qquad\qquad\qquad\qquad = -6x - (-24)$
>
> Rewrite the subtraction as addition of the opposite. $\qquad\qquad\qquad\qquad = -6x + 24$

The Distributive Property can be extended to more than two addends inside the parentheses. For example,

$$4(2a + 3b - 5c) = 4(2a) + 4(3b) - 4(5c)$$

$$= 8a + 12b - 20c$$

The Distributive Property is used to remove the parentheses from an expression that has a negative sign in front of the parentheses. Just as $-x = -1 \cdot x$, the expression $-(x + y) = -1(x + y)$. Therefore,

$$-(x + y) = -1(x + y) = -1x - 1y = -x - y$$

When a negative sign precedes parentheses, remove the parentheses and change the sign of *each* term inside the parentheses.

> Rewrite the expression $-(4a - 3b + 7)$ without parentheses.
>
> Remove the parentheses and change the sign of each term inside the parentheses. $\qquad -(4a - 3b + 7) = -4a + 3b - 7$

Copyright © Houghton Mifflin Company. All rights reserved.

5 *Example*

Simplify by using the Distributive Property:
$6(5c - 12)$

Solution

$6(5c - 12) = 6(5c) - 6(12)$
$\qquad\qquad = 30c - 72$

5 *You Try It*

Simplify by using the Distributive Property:
$-7(2k - 5)$

Your Solution

6 *Example*

Simplify by using the Distributive Property:
$-4(-2a - b)$

Solution

$-4(-2a - b) = -4(-2a) - (-4)(b)$
$\qquad\qquad = 8a + 4b$

6 *You Try It*

Simplify by using the Distributive Property:
$-4(x - 2y)$

Your Solution

7 *Example*

Simplify by using the Distributive Property:
$-2(3m - 8n + 5)$

Solution

$-2(3m - 8n + 5)$
$= -2(3m) - (-2)(8n) + (-2)(5)$
$= -6m + 16n - 10$

7 *You Try It*

Simplify by using the Distributive Property:
$3(-2v + 3w - 7)$

Your Solution

8 *Example*

Simplify by using the Distributive Property:
$3(2a + 6b - 5c)$

Solution

$3(2a + 6b - 5c) = 3(2a) + 3(6b) - 3(5c)$
$\qquad\qquad\qquad = 6a + 18b - 15c$

8 *You Try It*

Simplify by using the Distributive Property:
$-4(2x - 7y - z)$

Your Solution

9 *Example*

Rewrite $-(5x + 3y - 2z)$ without parentheses.

Solution

$-(5x + 3y - 2z) = -5x - 3y + 2z$

9 *You Try It*

Rewrite $-(c - 9d + 1)$ without parentheses.

Your Solution

Solutions on p. S13

Copyright © Houghton Mifflin Company. All rights reserved.

5.1 EXERCISES

OBJECTIVE   A

Identify the Property of Real Numbers that justifies the statement.

1. $3 \cdot (4 \cdot 7) = (3 \cdot 4) \cdot 7$

2. $a + 0 = a$

3. $x + 7 = 7 + x$

4. $12 \cdot a = a \cdot 12$

5. $4r + (-4r) = 0$

6. $5 + (a + 7) = (a + 7) + 5$

7. $\dfrac{2}{3} \cdot \dfrac{3}{2} = 1$

8. $-\dfrac{2}{3} + \dfrac{2}{3} = 0$

9. $a(bc) = (bc)a$

10. $1 \cdot x = x$

11. $\dfrac{1}{2}(2x) = \left(\dfrac{1}{2} \cdot 2\right)x$ **a.** _____
$\quad\quad = 1 \cdot x$ **b.** _____
$\quad\quad = x$ **c.** _____

12. $(5x + 6) + (-6) = 5x + [6 + (-6)]$ **a.** _____
$\quad\quad\quad\quad\quad\quad = 5x + 0$ **b.** _____
$\quad\quad\quad\quad\quad\quad = 5x$ **c.** _____

Use the given Property of Real Numbers to complete the statement.

13. The Associative Property of Addition
$x + (4 + y) = ?$

14. The Commutative Property of Multiplication
$v \cdot w = ?$

15. The Inverse Property of Multiplication
$5 \cdot ? = 1$

16. The Inverse Property of Multiplication
$? \cdot \dfrac{3}{4} = 1$

17. The Multiplication Property of Zero
$a \cdot ? = 0$

18. The Inverse Property of Multiplication
For $a \neq 0, a \cdot \dfrac{1}{a} = ?$

19. The Inverse Property of Addition
$-7y + ? = 0$

20. The Inverse Property of Addition
$\dfrac{2}{3}x + ? = 0$

21. The multiplicative inverse of $-\dfrac{2}{3}$ is ? .

22. For $a \neq 0$, the multiplicative inverse of $-\dfrac{2}{a}$ is ? .

Copyright © Houghton Mifflin Company. All rights reserved.

Simplify the variable expression.

23. $6(2x)$

24. $3(4y)$

25. $-5(3x)$

26. $-3(6z)$

27. $(3t) \cdot 7$

28. $(9r) \cdot 5$

29. $(-3p) \cdot 7$

30. $(-4w) \cdot 6$

31. $(-2)(-6q)$

32. $(-3)(-5m)$

33. $\dfrac{1}{2}(4x)$

34. $\dfrac{2}{3}(6n)$

35. $-\dfrac{5}{3}(9w)$

36. $-\dfrac{2}{5}(10v)$

37. $-\dfrac{1}{2}(-2x)$

38. $-\dfrac{1}{3}(-3x)$

39. $(2x)(3x)$

40. $(4k)(6k)$

41. $(-3x)(9x)$

42. $(4b)(-12b)$

43. $\left(\dfrac{1}{2}x\right)(2x)$

44. $\left(\dfrac{1}{3}h\right)(3h)$

45. $\left(-\dfrac{2}{3}\right)(x)\left(-\dfrac{3}{2}\right)$

46. $\left(-\dfrac{4}{3}\right)(z)\left(-\dfrac{3}{4}\right)$

47. $6\left(\dfrac{1}{6}c\right)$

48. $9\left(\dfrac{1}{9}v\right)$

49. $-5\left(-\dfrac{1}{5}a\right)$

50. $-9\left(-\dfrac{1}{9}s\right)$

51. $\dfrac{4}{5}w \cdot 15$

52. $\dfrac{7}{5}y \cdot 30$

53. $2v \cdot 8w$

54. $3m \cdot 7n$

55. $(-4b)(7c)$

56. $(-3k)(-6m)$

57. $3x + (-3x)$

58. $7xy + (-7xy)$

59. $-12h + 12h$

60. $5 + 8y + (-8y)$

61. $9 + 2m + (-2m)$

62. $12 - 3m + 3m$

Copyright © Houghton Mifflin Company. All rights reserved.

63. $8x + 7 + (-8x)$ **64.** $13v + 12 + (-13v)$ **65.** $6t - 15 + (-6t)$ **66.** $10z - 4 + (-10z)$

67. $8 + (-8) - 5y$ **68.** $12 + (-12) - 7b$ **69.** $(-4) + 4 + 13b$ **70.** $-7 + 7 - 15t$

OBJECTIVE **B**

Simplify by using the Distributive Property.

71. $2(5z + 2)$ **72.** $3(4n + 5)$ **73.** $6(2y + 5z)$

74. $4(7a + 2b)$ **75.** $3(7x - 9)$ **76.** $9(3w - 7)$

77. $-(2x - 7)$ **78.** $-(3x + 4)$ **79.** $-(-4x - 9)$

80. $-(-5y - 12)$ **81.** $-5(y + 3)$ **82.** $-4(x + 5)$

83. $-6(2x - 3)$ **84.** $-3(7y - 4)$ **85.** $-5(4n - 8)$

86. $-4(3c - 2)$ **87.** $-8(-6z + 3)$ **88.** $-2(-3k + 9)$

89. $-6(-4p - 7)$ **90.** $-5(-8c - 5)$ **91.** $5(2a + 3b + 1)$

92. $5(3x + 9y + 8)$ **93.** $4(3x - y - 1)$ **94.** $3(2x - 3y + 7)$

Copyright © Houghton Mifflin Company. All rights reserved.

95. $9(4m - n + 2)$

96. $-4(3x + 2y - 5)$

97. $-6(-2v + 3w + 7)$

98. $-7(-2b - 4)$

99. $-4(-5x - 1)$

100. $-9(3x - 6y)$

101. $5(4a - 5b + c)$

102. $-4(-2m - n + 3)$

103. $-6(3p - 2r - 9)$

Rewrite without parentheses.

104. $-(4x + 6y - 8z)$

105. $-(5a - 9b + 7)$

106. $-(-6m + 3n + 1)$

107. $-(11p - 2q - r)$

Critical Thinking

108. Determine whether the statement is true or false. If the statement is false, give an example that illustrates that it is false.
   a. Division is a commutative operation.
   b. Division is an associative operation.
   c. Subtraction is an associative operation.
   d. Subtraction is a commutative operation.

109. Is the statement "any number divided by itself is one" a true statement? If not, for what number or numbers is the statement not true?

110. Does every real number have an additive inverse? If not, which real numbers do not have an additive inverse?

111. Does every real number have a multiplicative inverse? If not, which real numbers do not have a multiplicative inverse?

112. ✏ In your own words, explain the Distributive Property.

113. ✏ Explain why division by zero is not allowed.

114. ✏ Give examples of two operations that occur in everyday experience that are not commutative (for example, putting on socks and then shoes).

Copyright © Houghton Mifflin Company. All rights reserved.

Copyright © Houghton Mifflin Company. All rights reserved.

SECTION 5.2 ▶ **Variable Expressions in Simplest Form**

OBJECTIVE **A**

Addition of like terms

A variable expression is shown at the right. The expression can be rewritten by writing subtraction as addition of the opposite. A **term** of a variable expression is one of the addends of the expression.

$$4y^3 - 3xy + x - 9$$

$$4y^3 + (-3xy) + x + (-9)$$

The variable expression has 4 terms: $4y^3$, $-3xy$, x, and -9.

The term -9 is a **constant term**, or simply a **constant**. The terms $4y^3$, $-3xy$, and x are **variable terms**.

Each variable term consists of a **numerical coefficient** and a **variable part**. The table at the right gives the numerical coefficient and the variable part of each variable term.

Term	Numerical Coefficient	Variable Part
$4y^3$	4	y^3
$-3xy$	-3	xy
x	1	x

For an expression such as x, the numerical coefficient is 1 ($x = 1x$). The numerical coefficient for $-x$ is -1 ($-x = -1x$). The numerical coefficient of $-xy$ is -1 ($-xy = -1xy$). Usually the 1 is not written.

For the variable expression at the right, state:

$$9x^2 - x - 7yz^2 + 8$$

a. The number of terms
b. The coefficient of the second term
c. The variable part of the third term
d. The constant term

   a. There are 4 terms: $9x^2$, $-x$, $-7yz^2$, and 8.
   b. The coefficient of the second term is -1.
   c. The variable part of the third term is yz^2.
   d. The constant term is 8.

Like terms of a variable expression have the same variable part. Constant terms are also like terms.

   For the expression $13ab + 4 - 2ab - 10$, the terms $13ab$ and $-2ab$ are like variable terms, and 4 and -10 are like constant terms.

For the expression at the right, note that $5y^2$ and $-3y$ are not like terms because $y^2 = y \cdot y$, and $y \cdot y \neq y$. However, $6xy$ and $9yx$ are like variable terms because $xy = yx$ by the Commutative Property of Multiplication.

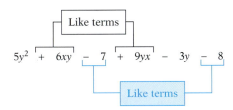

For the variable expression $7 - 9x^2 - 8x - 9 + 4x$, state which terms are like terms.

The terms $-8x$ and $4x$ are like variable terms.

The terms 7 and -9 are like constant terms.

Variable expressions containing like terms are simplified by using an alternate form of the Distributive Property.

> ### *Alternative Form of the Distributive Property*
>
> If a, b, and c are real numbers, then $ac + bc = (a + b)c$.

Simplify: $6c + 7c$

$6c$ and $7c$ are like terms.
Use the Alternative Form of
the Distributive Property.
Then simplify.

$$6c + 7c = (6 + 7)c$$
$$= 13c$$

This example shows that to simplify a variable expression with like terms, add the coefficients of the like terms. Adding or subtracting the like terms of a variable expression is called **combining like terms.**

Simplify: $6a + 7 - 9a + 3$

Use the Commutative Property
of Addition to rearrange terms
so that like terms are together.

$$6a + 7 - 9a + 3$$
$$= 6a + 7 + (-9a) + 3$$
$$= 6a + (-9a) + 7 + 3$$

Use the Alternative Form of the
Distributive Property to add like
variable terms. Add the like
constant terms.

$$= [6 + (-9)]a + (7 + 3)$$
$$= -3a + 10$$

Simplify: $4x^2 - 7x + x^2 - 12x$

Use the Commutative Property
of Addition to rearrange terms
so that like terms are together.

$$4x^2 - 7x + x^2 - 12x$$
$$= 4x^2 + (-7x) + x^2 + (-12x)$$
$$= 4x^2 + x^2 + (-7x) + (-12x)$$

Use the Alternative Form of the
Distributive Property to add like
terms.

$$= (4 + 1)x^2 + [-7 + (-12)]x$$
$$= 5x^2 + (-19)x$$
$$= 5x^2 - 19x$$

---

1 *Example*   Simplify: $\dfrac{3x}{7} + \dfrac{2x}{7}$

Solution $\dfrac{3x}{7} + \dfrac{2x}{7} = \dfrac{3x + 2x}{7}$

$\quad = \dfrac{(3 + 2)x}{7} = \dfrac{5x}{7}$

1 *You Try It*   Simplify: $\dfrac{x}{5} + \dfrac{2x}{5}$

Your Solution

Solution on p. S13

Copyright © Houghton Mifflin Company. All rights reserved.

2 *Example*   Simplify:
$9y - 3z - 12y + 3z + 2$

Solution $9y - 3z - 12y + 3z + 2$
$= 9y - 12y - 3z + 3z + 2$
$= -3y + 0z + 2$
$= -3y + 2$

2 *You Try It*   Simplify:
$12a^2 - 8a + 3 - 16a^2 + 8a$

Your Solution

---

3 *Example*   Simplify:
$6b^2 - 9ab + 3b^2 - ab$

Solution $6b^2 - 9ab + 3b^2 - ab$
$= 6b^2 + 3b^2 - 9ab - ab$
$= 9b^2 - 10ab$

3 *You Try It*   Simplify:
$-7x^2 + 4xy + 8x^2 - 12xy$

Your Solution

---

4 *Example*   Simplify:
$6u + 7v - 8 + 9u - 12v + 14$

Solution $6u + 7v - 8 + 9u - 12v + 14$
$= 6u + 9u + 7v - 12v - 8 + 14$
$= 15u - 5v + 6$

4 *You Try It*   Simplify:
$-2r + 7s - 12 - 8r + s + 8$

Your Solution

---

5 *Example*   Simplify:
$5r^2t - 6rt^2 + 8rt^2 - 9r^2t$

Solution $5r^2t - 6rt^2 + 8rt^2 - 9r^2t$
$= 5r^2t - 9r^2t - 6rt^2 + 8rt^2$
$= -4r^2t + 2rt^2$

5 *You Try It*   Simplify:
$8x^2y - 15xy^2 + 12xy^2 - 7x^2y$

Your Solution

Solutions on p. S13

OBJECTIVE **B**

General variable expressions

General variable expressions are simplified by repeated use of the Properties of the Real Numbers.

Simplify: $7(2a - 4b) - 3(4a - 2b)$

Use the Distributive Property to remove parentheses.

$7(2a - 4b) - 3(4a - 2b)$
$= 14a - 28b - 12a + 6b$

Use the Commutative Property of Addition to rearrange terms.

$= 14a - 12a - 28b + 6b$

Use the Alternative Form of the Distributive Property to combine like terms.

$= 2a - 22b$

Copyright © Houghton Mifflin Company. All rights reserved.

To simplify variable expressions that contain grouping symbols within other grouping symbols, simplify inside the inner grouping symbols first.

Simplify: $2x - 4[3 - 2(6x + 5)]$

Use the Distributive Property to remove the parentheses.	$2x - 4[3 - 2(6x + 5)]$ $= 2x - 4[3 - 12x - 10]$
Combine like terms inside the brackets.	$= 2x - 4[-12x - 7]$
Use the Distributive Property to remove the brackets.	$= 2x + 48x + 28$
Combine like terms.	$= 50x + 28$

Simplify: $2a^2 + 3[4(2a^2 - 5) - 4(3a - 1)]$

Use the Distributive Property to remove both sets of parentheses.	$2a^2 + 3[4(2a^2 - 5) - 4(3a - 1)]$ $= 2a^2 + 3[8a^2 - 20 - 12a + 4]$
Combine like terms inside the brackets.	$= 2a^2 + 3[8a^2 - 12a - 16]$
Use the Distributive Property to remove the brackets.	$= 2a^2 + 24a^2 - 36a - 48$
Combine like terms.	$= 26a^2 - 36a - 48$

---

⑥ Example Simplify:
$4 - 3(2a - b) + 4(3a + 2b)$

Solution $4 - 3(2a - b) + 4(3a + 2b)$
$= 4 - 6a + 3b + 12a + 8b$
$= 6a + 11b + 4$

⑥ You Try It Simplify:
$6 - 4(2x - y) + 3(x - 4y)$

Your Solution

---

⑦ Example Simplify:
$7y - 4(2y - 3z) - (6y - 4z)$

Solution $7y - 4(2y - 3z) - (6y - 4z)$
$= 7y - 8y + 12z - 6y + 4z$
$= -7y + 16z$

⑦ You Try It Simplify:
$8c - 4(3c - 8) - 5(c + 4)$

Your Solution

---

⑧ Example Simplify:
$9v - 4[2(1 - 3v) - 5(2v + 4)]$

Solution $9v - 4[2(1 - 3v) - 5(2v + 4)]$
$= 9v - 4[2 - 6v - 10v - 20]$
$= 9v - 4[-16v - 18]$
$= 9v + 64v + 72$
$= 73v + 72$

⑧ You Try It Simplify:
$6p + 5[3(2 - 3p) - 2(5 - 4p)]$

Your Solution

Solutions on pp. S13–S14

Copyright © Houghton Mifflin Company. All rights reserved.

5.2 EXERCISES

OBJECTIVE   **A**

List the terms of the variable expression. Then underline the constant term.

1. $3x^2 + 4x - 9$ **2.** $-7y^2 - 2y + 6$ **3.** $b + 5$ **4.** $8n^2 - 1$

List the variable terms of the expression. Then underline the variable part of each term.

5. $9a^2 - 12a + 4b^2$ **6.** $6x^2y + 7xy^2 + 11$ **7.** $3x^2 + 16$ **8.** $-2n^2 + 5n - 8$

State the coefficients of the variable terms.

9. $x^2 - 6x - 7$ **10.** $-x + 15$ **11.** $12a^2 + 4ab - 1$ **12.** $x^2y - x + y$

Simplify by combining like terms.

13. $7a + 9a$ **14.** $8c + 15c$ **15.** $12x + 15x$ **16.** $9b + 24b$

17. $9z - 6z$ **18.** $12h - 4h$ **19.** $9x - x$ **20.** $12y - y$

21. $8z - 15z$ **22.** $2p - 13p$ **23.** $w - 7w$ **24.** $y - 9y$

25. $12v - 12v$ **26.** $11c - 11c$ **27.** $9s - 8s$ **28.** $6n - 5n$

29. $\dfrac{n}{5} + \dfrac{3n}{5}$ **30.** $\dfrac{2n}{9} + \dfrac{5n}{9}$ **31.** $\dfrac{x}{4} + \dfrac{x}{4}$ **32.** $\dfrac{5x}{8} + \dfrac{3x}{8}$

33. $\dfrac{8y}{7} - \dfrac{4y}{7}$ **34.** $\dfrac{5y}{3} - \dfrac{y}{3}$ **35.** $\dfrac{5c}{6} - \dfrac{c}{6}$ **36.** $\dfrac{9d}{10} - \dfrac{7d}{10}$

Copyright © Houghton Mifflin Company. All rights reserved.

37. $4x - 3y + 2x$

38. $3m - 6n + 4m$

39. $4r + 8p - 2r + 5p$

40. $-12t - 6s + 9t + 4s$

41. $9w - 5v - 12w + 7v$

42. $3c - 8 + 7c - 9$

43. $-4p + 9 - 5p + 2$

44. $-6y - 17 + 4y + 9$

45. $8p + 7 - 6p - 7$

46. $9m - 12 + 2m + 12$

47. $7h + 15 - 7h - 9$

48. $7v^2 - 9v + v^2 - 8v$

49. $9y^2 - 8 + 4y^2 + 9$

50. $r^2 + 4r - 8r - 5r^2$

51. $3w^2 - 7 - 9 + 9w^2$

52. $4c - 7c^2 + 8c - 8c^2$

53. $9w^2 - 15w + w - 9w^2$

54. $12v^2 + 15v - 14v - 12v^2$

55. $7a^2b + 5ab^2 - 2a^2b + 3ab^2$

56. $3xy^2 + 2x^2y - 7xy^2 - 4x^2y$

57. $8a - 9b + 2 - 8a + 9b + 3$

58. $10v + 12w - 9 - v - 12w + 9$

59. $6x^2 - 7x + 1 + 5x^2 + 5x - 1$

60. $4y^2 + 7y + 1 + y^2 - 10y + 9$

61. $-3b^2 + 6b + 1 + 11b^2 - 8b - 1$

62. $-4z^2 - 6z + 1 - z^2 + 7z + 8$

OBJECTIVE **B**

Simplify.

63. $5x + 2(x + 1)$

64. $6y + 2(2y + 3)$

65. $9n - 3(2n - 1)$

Copyright © Houghton Mifflin Company. All rights reserved.

66. $12x - 2(4x - 6)$

67. $7a - (3a - 4)$

68. $9m - 4(2m - 3)$

69. $7 + 2(2a - 3)$

70. $5 + 3(2y - 8)$

71. $6 + 4(2x + 9)$

72. $4 + 3(7d + 7)$

73. $8 - 4(3x - 5)$

74. $13 - 7(4y + 3)$

75. $2 - 9(2m + 6)$

76. $4 - 7(6w - 9)$

77. $3(6c + 5) + 2(c + 4)$

78. $7(2k - 5) + 3(4k - 3)$

79. $2(a - 2b) + 3(2a + 3b)$

80. $4(3x - 6y) + 5(2x - 3y)$

81. $6(7z - 5) - 3(9z - 6)$

82. $8(2t + 4) - 4(3t - 1)$

83. $-2(6y + 2) + 3(4y - 5)$

84. $-3(2a - 5) - 2(4a + 3)$

85. $-5(x - 2y) - 4(2x + 3y)$

86. $-6(-x - 3y) - 2(-3x + 9y)$

87. $2 - 3(2v - 1) + 2(2v + 4)$

88. $5 - 2(3x + 5) - 3(4x - 1)$

89. $2c - 3(c + 4) - 2(2c - 3)$

90. $5m - 2(3m + 2) - 4(m - 1)$

91. $8a + 3(2a - 1) + 6(4 - 2a)$

92. $9z - 2(2z - 7) + 4(3 - 5z)$

93. $3n - 2[5 - 2(2n - 4)]$

94. $6w + 4[3 - 5(6w - 2)]$

95. $9x - 3[8 - 2(5 - 3x)]$

Copyright © Houghton Mifflin Company. All rights reserved.

96. $11y - 7[2(2y - 5) + 3(7 - 5y)]$

97. $-3v - 6[2(3 - 2v) - 5(3v - 7)]$

98. $8b - 3[2(3 - 5b) - 4(3b - 4)]$

99. $21r - 4[3(4 - 5r) - 3(2 - 7r)]$

100. $7y^2 - 2[3(2y - 4) + 3(2y^2)]$

101. $9z^2 - 3[4(2z + 3) - 3(2z^2 - 6)]$

Critical Thinking

102. The square and the rectangle at the right can be used to illustrate algebraic expressions. Note at the right the expression for $2x + 1$. The expression below is $3(x + 2)$.

x	1	1	x	1	1	x	1	1

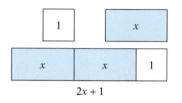

$2x + 1$

Rearrange these rectangles so that the x's are together and the 1's are together. Write a mathematical expression for the rearranged figure. Using similar squares and rectangles, draw figures that represent the expressions $2 + 3x$, $5x$, $2(2x + 3)$, $4x + 3$, and $4x + 6$. Does the figure for $2(2x + 3)$ equal the figure for $4x + 6$? How does this relate to the Distributive Property? Does the figure for $2 + 3x$ equal the figure for $5x$? How does this relate to combining like terms?

103. The procedure for multiplying whole numbers with more than one digit is based on the Distributive Property. Examine the following.

$$7 \cdot 435 = 7(400 + 30 + 5) = 7 \cdot 400 + 7 \cdot 30 + 7 \cdot 5$$
$$= 2,800 + 210 + 35 = 3,045$$

Use the Distributive Property to multiply 527 by 6.

$$\begin{array}{r} {}^{2\ 3} \\ 4\,3\,5 \\ \times\quad 7 \\ \hline 3{,}0\,4\,5 \end{array}$$

104. Explain why the simplification of the expression $2 + 3(2x + 4)$ shown at the right is incorrect. What is the correct simplification?

Why is this incorrect?

$$2 + 3(2x + 4) = 5(2x + 4)$$
$$= 10x + 20$$

105. It was stated in this section that the variable terms x^2 and x are not like terms. Use measurements of distance and area to show that these terms would not be combined as measurements.

Copyright © Houghton Mifflin Company. All rights reserved.

Copyright © Houghton Mifflin Company. All rights reserved.

SECTION 5.3 ▷ ## Addition and Subtraction of Polynomials

OBJECTIVE A

Addition of polynomials

A **monomial** is a number, a variable, or a product of numbers and variables. The expressions below are all monomials.

7	b	$\dfrac{2}{3}a$	$12xy^2$
A number	A variable	A product of a number and a variable	A product of a number and variables

The expression $3\sqrt{x}$ is not a monomial because $\sqrt{x}$ cannot be written as a product of variables.

The expression $\dfrac{2x}{y^2}$ is not a monomial because it is a quotient of variables.

A **polynomial** is a variable expression in which the terms are monomials.

A polynomial of *one* term is a **monomial.** $-7x^2$ is a monomial.

A polynomial of *two* terms is a **binomial.** $4x + 2$ is a binomial.

A polynomial of *three* terms is a **trinomial.** $7x^2 + 5x - 7$ is a trinomial.

Polynomials with more than three terms do not have special names.

> **Take Note**
>
> The expression $x + y + z$ has 3 terms; it is a trinomial. The expression xyz has 1 term; it is a monomial.

The terms of a polynomial in one variable are usually arranged so that the exponents of the variable decrease from left to right. This is called **descending order.**

$5x^3 - 4x^2 + 6x - 1$

$7z^4 + 4z^3 + z - 6$

$2y^4 + y^3 - 2y^2 + 4y - 5$

To add polynomials, add the coefficients of the like terms. Either a horizontal format or a vertical format can be used.

Use a horizontal format to add $(6x^3 + 4x^2 - 7) + (-12x^2 + 4x - 8)$.

Use the Commutative and Associative Properties of Addition to rearrange the terms so that like terms are grouped together. Then combine like terms.

$(6x^3 + 4x^2 - 7) + (-12x^2 + 4x - 8) = 6x^3 + (4x^2 - 12x^2) + 4x + (-7 - 8)$
$$= 6x^3 - 8x^2 + 4x - 15$$

Use a vertical format to add $(-4x^2 + 6x - 9) + (12 - 8x + 2x^3)$.

Arrange the terms of each polynomial in descending order with like terms in the same column.

$$\begin{array}{r} -4x^2 + 6x - 9 \\ 2x^3 \quad\quad\ - 8x + 12 \\ \hline \end{array}$$

Combine the terms in each column. $2x^3 - 4x^2 - 2x + 3$

1 *Example*

Use a horizontal format to add

$(8x^2 - 4x - 9) + (2x^2 + 9x - 9)$.

Solution $(8x^2 - 4x - 9) + (2x^2 + 9x - 9)$
  $= (8x^2 + 2x^2) + (-4x + 9x) + (-9 - 9)$
  $= 10x^2 + 5x - 18$

1 *You Try It*

Use a horizontal format to add

$(-4x^3 + 2x^2 - 8) + (4x^3 + 6x^2 - 7x + 5)$.

Your Solution

2 *Example*

Use a vertical format to add

$(-5x^3 + 4x^2 - 7x + 9) + (5x - 11 + 2x^3)$.

Solution Arrange the terms of each polynomial in descending order with like terms in the same column.

$$\begin{array}{r} -5x^3 + 4x^2 - 7x + 9 \\ \underline{2x^3 + 5x - 11} \\ -3x^3 + 4x^2 - 2x - 2 \end{array}$$

2 *You Try It*

Use a vertical format to add

$(6x^3 + 2x + 8) + (2x^2 - 12x - 8 - 9x^3)$.

Your Solution

3 *Example*

Find the total of $8y^2 + 3y - 5$ and $4y^2 + 9$.

Solution $(8y^2 + 3y - 5) + (4y^2 + 9)$
  $= (8y^2 + 4y^2) + 3y + (-5 + 9)$
  $= 12y^2 + 3y + 4$

3 *You Try It*

What is the sum of $6a^4 - 5a^2 + 7$ and $8a^4 + 3a^2 - 1$?

Your Solution

Solutions on p. S14

OBJECTIVE B

VIDEO & DVD   CD TUTOR   WEB   SSM

Subtraction of polynomials

The **opposite** of the polynomial $(3x^2 - 7x + 8)$ is $-(3x^2 - 7x + 8)$.

To find the opposite of a polynomial, change the sign of each term of the polynomial.

$-(3x^2 - 7x + 8) = -3x^2 + 7x - 8$

As another example, the opposite of $4x^2 + 5x - 9$ is $-4x^2 - 5x + 9$.

To subtract two polynomials, add the opposite of the second polynomial to the first. Polynomials are subtracted by using either a horizontal or a vertical format.

Copyright © Houghton Mifflin Company. All rights reserved.

Use a horizontal format to subtract $(5a^2 - a + 2) - (-2a^3 + 3a - 3)$.

$$(5a^2 - a + 2) - (-2a^3 + 3a - 3)$$

Rewrite subtraction as addition of the opposite polynomial. The opposite of $-2a^2 + 3a - 3$ is $2a^3 - 3a + 3$.

$$= (5a^2 - a + 2) + (2a^3 - 3a + 3)$$

Combine like terms.

$$= 2a^3 + 5a^2 - 4a + 5$$

Use a vertical format to subtract $(3y^3 + 4y + 9) - (2y^2 + 4y - 21)$.

The opposite of $(2y^2 + 4y - 21)$ is $(-2y^2 - 4y + 21)$.

$$
\begin{array}{r}
3y^3 + 4y + 9 \\
\underline{- 2y^2 - 4y + 21} \\
3y^3 - 2y^2 + 30
\end{array}
$$

Add the opposite of $2y^2 + 4y - 21$ to the first polynomial.

---

④ Example

Use a horizontal format to subtract $(7c^2 - 9c - 12) - (9c^2 + 5c - 8)$.

Solution   The opposite of $9c^2 + 5c - 8$ is $-9c^2 - 5c + 8$.

Add the opposite of $9c^2 + 5c - 8$ to the first polynomial.

$$(7c^2 - 9c - 12) - (9c^2 + 5c - 8)$$
$$= (7c^2 - 9c - 12) + (-9c^2 - 5c + 8)$$
$$= -2c^2 - 14c - 4$$

④ You Try It

Use a horizontal format to subtract $(-4w^3 + 8w - 8) - (3w^3 - 4w^2 + 2w - 1)$.

Your Solution

---

⑤ Example

Use a vertical format to subtract $(3k^2 - 4k + 1) - (k^3 + 3k^2 - 6k - 8)$.

Solution   The opposite of $k^3 + 3k^2 - 6k - 8$ is $-k^3 - 3k^2 + 6k + 8$.

Add the opposite of $k^3 + 3k^2 - 6k - 8$ to the first polynomial.

$$
\begin{array}{r}
3k^2 - 4k + 1 \\
\underline{-k^3 - 3k^2 + 6k + 8} \\
-k^3 + 2k + 9
\end{array}
$$

⑤ You Try It

Use a vertical format to subtract $(13y^3 - 6y - 7) - (4y^2 - 6y - 9)$.

Your Solution

Solutions on p. S14

Copyright © Houghton Mifflin Company. All rights reserved.

Example Find the difference between $3z^2 - 4z + 1$ and $5z^2 - 8$.

Solution $(3z^2 - 4z + 1) - (5z^2 - 8)$
$= (3z^2 - 4z + 1) + (-5z^2 + 8)$
$= -2z^2 - 4z + 9$

 You Try It What is the difference between $-6n^4 + 5n^2 - 10$ and $4n^2 + 2$?

Your Solution

Solution on p. S14

OBJECTIVE

Applications

A company's **revenue** is the money the company earns by selling its products. A company's **cost** is the money it spends to manufacture and sell its products. A company's **profit** is the difference between its revenue and cost. This relationship is expressed by the formula $P = R - C$, where P is the profit, R is the revenue, and C is the cost. This formula is used in the example below.

A company manufactures and sells wood stoves. The total monthly cost, in dollars, to produce n wood stoves is $30n + 2000$. The company's monthly revenue, in dollars, obtained from selling all n wood stoves is $-0.4n^2 + 150n$. Express in terms of n the company's monthly profit.

$R = -0.4n^2 + 150n$,
$C = 30n + 2000$
Rewrite subtraction as addition of the opposite.

$P = R - C$
$P = (-0.4n^2 + 150n) - (30n + 2000)$

$P = (-0.4n^2 + 150n) + (-30n - 2000)$

$P = -0.4n^2 + (150n - 30n) - 2000$
$P = -0.4n^2 + 120n - 2000$

The company's monthly profit is $(-0.4n^2 + 120n - 2000)$ dollars.

 Example The distance from Acton to Boyd is $(y^2 + y + 7)$ miles. The distance from Boyd to Carlyle is $(y^2 - 3)$ miles. Find the distance from Acton to Carlyle.

Acton   Boyd   Carlyle

Solution $(y^2 + y + 7) + (y^2 - 3)$
$= (y^2 + y^2) + y + (7 - 3)$
$= 2y^2 + y + 4$

The distance from Acton to Carlyle is $(2y^2 + y + 4)$ miles.

 You Try It The distance from Dover to Engel is $(5y^2 - y)$ miles. The distance from Engel to Farley is $(7y^2 + 4)$ miles. Find the distance from Dover to Farley.

Your Solution

Dover   Engel   Farley

Solution on p. S14

Copyright © Houghton Mifflin Company. All rights reserved.

5.3 EXERCISES

OBJECTIVE **A**

State whether or not the expression is a monomial.

1. 17

2. $3x^4$

3. $\dfrac{17}{x}$

4. $\sqrt{6x}$

5. $\dfrac{2}{3}y$

6. $\dfrac{2}{3y}$

7. $\dfrac{\sqrt{y}}{3}$

8. $\dfrac{y}{3}$

State whether or not the expression is a polynomial.

9. $\dfrac{1}{5}x^3 + \dfrac{1}{2}x$

10. $\dfrac{1}{5x^2} + \dfrac{1}{2x}$

11. $\sqrt{x} + 5$

12. $x + \sqrt{5}$

How many terms does the polynomial have?

13. $3x^2 - 8x + 7$

14. $5y^3 + 6$

15. $9x^2y^3z^5$

16. $n^4 + 2n^3 - n^2 + 3n + 6$

State whether the polynomial is a monomial, a binomial, or a trinomial.

17. $8x^4 - 6x^2$

18. $4a^2b^2 + 9ab + 10$

19. $7a^3bc^5$

20. $y + 1$

Write the polynomial in descending order.

21. $8x^2 - 2x + 3x^3 - 6$

22. $7y - 8 + 2y^2 + 4y^3$

23. $2a - 3a^2 + 5a^3 + 1$

24. $b - 3b^2 + b^4 - 2b^3$

25. $4 - b^2$

26. $1 - y^4$

Add. Use a horizontal format.

27. $(5y^2 + 3y - 7) + (6y^2 - 7y + 9)$

28. $(7m^2 - 9m - 8) + (5m^2 + 10m + 4)$

29. $(-4b^2 + 9b + 11) + (7b^2 - 12b - 13)$

30. $(-8x^2 - 11x - 15) + (4x^2 - 12x + 13)$

Copyright © Houghton Mifflin Company. All rights reserved.

31. $(3w^3 + 8w^2 - 2w) + (5w^2 - 6w - 5)$

32. $(11p^3 - 9p^2 - 6p) + (10p^2 - 8p + 4)$

33. $(3a^2 - 7 + 2a - 9a^3) + (7a^3 - 12a^2 - 10a + 8)$

34. $(9x - 8x^2 - 12 + 7x^3) + (-3x^3 - 7x^2 + 5x - 9)$

35. $(7t^3 - 8t - 15) + (8t - 20 + 7t^2)$

36. $(8y^2 - 3y - 1) + (3y - 1 - 6y^3 - 8y^2)$

37. Find the sum of $6t^2 - 8t - 15$ and $7t^2 + 8t - 20$.

38. What is $8y^2 - 3y - 1$ plus $-6y^2 + 3y - 1$?

Add. Use a vertical format.

39. $(5k^2 - 7k - 8) + (6k^2 + 9k - 10)$

40. $(8v^2 - 9v + 12) + (12v^2 - 11v - 2)$

41. $(8x^3 - 9x + 2) + (9x^3 + 9x - 7)$

42. $(13z^3 - 7z^2 + 4z) + (10z^2 + 5z - 9)$

43. $(12b^3 + 9b^2 + 5b - 10) + (4b^3 + 5b^2 - 5b + 11)$

44. $(5a^3 - a^2 + 4a - 19) + (-a^3 + a^2 - 7a + 19)$

45. $(8p^3 - 7p) + (9p^2 - 7 + p)$

46. $(12c^3 + 9c) + (-7c^2 - 8 - c)$

47. $(7a^2 - 7 - 6a) + (-6a^3 - 7a^2 + 6a - 10)$

48. $(12x^2 + 8 + 7x) + (3x^3 - 12x^2 - 7x - 11)$

49. Find the total of $9d^4 - 7d^2 + 5$ and $-6d^4 - 3d^2 - 8$.

50. What is the sum of $8z^3 + 5z^2 - 4z + 7$ and $-3z^3 - z^2 + 6z - 2$?

OBJECTIVE **B**

Write the opposite of the polynomial.

51. $8x^3 + 5x^2 - 3x - 6$

52. $7y^4 - 4y^2 + 10$

53. $-9a^3 + a^2 - 2a + 9$

Subtract. Use a horizontal format.

54. $(3x^2 - 2x - 5) - (x^2 + 7x - 3)$

55. $(7y^2 - 8y - 10) - (3y^2 + 2y - 9)$

Copyright © Houghton Mifflin Company. All rights reserved.

56. $(11b^3 - 2b^2 + 1) - (6b^2 - 12b - 13)$

57. $(13w^3 + 3w^2 - 9) - (7w^3 - 9w + 10)$

58. $(8z^3 - 9z^2 + 4z + 12) - (10z^3 - z^2 + 4z - 9)$

59. $(15t^3 - 9t^2 + 8t + 11) - (17t^3 - 9t^2 - 8t + 6)$

60. $(9y^3 + 8y) - (-17y^2 + 5)$

61. $(8p^3 + 14p) - (9p^2 - 12)$

62. $(-6r^3 + 9r + 19) - (6r^3 + 19 - 16r)$

63. $(-4v^2 + 8v - 2) - (6v^3 + 7v + 1 - 13v^2)$

64. Find the difference between $10b^2 - 7b + 4$ and $8b^2 + 5b - 14$.

65. What is $7m^2 - 3m - 6$ minus $2m^2 - m + 5$?

Subtract. Use a vertical format.

66. $(4a^2 + 9a - 11) - (2a^2 - 3a - 9)$

67. $(8b^2 - 7b - 6) - (5b^2 + 8b + 12)$

68. $(6z^3 + 4z^2 + 1) - (3z^3 - 8z - 9)$

69. $(10y^3 - 8y - 13) - (6y^2 + 2y + 7)$

70. $(8y^2 - 9y - 16) - (3y^3 - 4y^2 + 2y + 5)$

71. $(4a^2 + 8a + 12) - (3a^3 + 4a^2 + 7a - 12)$

72. $(10b^3 - 7b) - (8b^2 + 14)$

73. $(7m - 6) - (2m^3 - m^2)$

74. $(5n^3 - 4n - 9 + 8n^2) - (2n^3 + 8n^2 + 4n - 9)$

75. $(4q^3 + 7q^2 + 8q - 9) - (-8q - 9 + 7q^2 + 14q^3)$

76. What is $8x^3 - 5x^2 + 6x$ less than $x^2 - 4x + 7$?

77. What is the difference between $7x^4 + 3x^2 - 11$ and $-5x^4 - 8x^2 + 6$?

OBJECTIVE **C**

78. *Distance* The distance from Ashley to Wyle is $(4x^2 + 3x - 5)$ kilometers. The distance from Wyle to Erie is $(6x^2 - x + 7)$ kilometers. Find the distance from Ashley to Erie.

Copyright © Houghton Mifflin Company. All rights reserved.

79. *Distance* The distance from Haley to Lincoln is $(2y^2 + y - 4)$ kilometers. The distance from Lincoln to Bedford is $(5y^2 - y + 3)$ kilometers. Find the distance from Haley to Bedford.

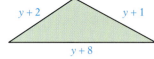

80. *Geometry* Find the perimeter of the triangle shown at the right. The dimensions given are in feet. Use the formula $P = a + b + c$.

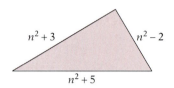

81. *Geometry* Find the perimeter of the triangle shown at the right. The dimensions given are in meters. Use the formula $P = a + b + c$.

For Exercises 82 to 85, use the formula $P = R - C$, where P is the profit, R is the revenue, and C is the cost.

82. *Business* A company's total monthly cost, in dollars, for manufacturing and selling n videotapes per month is $35n + 2000$. The company's monthly revenue, in dollars, from selling all n videotapes is $-0.2n^2 + 175n$. Express in terms of n the company's monthly profit.

83. *Business* A company manufactures and sells snowmobiles. The total monthly cost, in dollars, to produce n snowmobiles is $50n + 4000$. The company's revenue, in dollars, obtained from selling all n snowmobiles is $-0.6n^2 + 250n$. Express in terms of n the company's monthly profit.

84. *Business* A company's total monthly cost, in dollars, for manufacturing and selling n portable CD players per month is $75n + 6000$. The company's revenue, in dollars, from selling all n portable CD players is $-0.4n^2 + 800n$. Express in terms of n the company's monthly profit.

85. *Business* A company's total monthly cost, in dollars, for manufacturing and selling n pairs of in-line skates per month is $100n + 1500$. The company's revenue, in dollars, from selling all n pairs is $-n^2 + 800n$. Express in terms of n the company's monthly profit.

Critical Thinking

86. What polynomial must be added to $3x^2 - 6x + 9$ so that the sum is $4x^2 + 3x - 2$?

87. What polynomial must be subtracted from $2x^2 - x - 2$ so that the difference is $5x^2 + 3x + 1$?

88. In your own words, explain the meaning of monomial, binomial, trinomial, and polynomial. Give an example of each. Give an example of an expression that is not a polynomial.

Copyright © Houghton Mifflin Company. All rights reserved.

Copyright © Houghton Mifflin Company. All rights reserved.

SECTION 5.4   ## Multiplication of Monomials

OBJECTIVE   **A**

Multiplication of monomials

Recall that in the exponential expression 3^4, 3 is the base and 4 is the exponent. The exponential expression 3^4 means to multiply 3, the base, 4 times. Therefore, $3^4 = 3 \cdot 3 \cdot 3 \cdot 3 = 81$.

For the variable exponential expression x^6, x is the base and 6 is the exponent. The exponent indicates the number of times the base occurs as a factor. Therefore,

$$\overbrace{\text{Multiply } x \text{ 6 times.}}$$
$$x^6 = x \cdot x \cdot x \cdot x \cdot x \cdot x$$

The product of exponential expressions with the *same* base can be simplified by writing each expression in factored form and writing the result with an exponent.

$$x^3 \cdot x^2 = \overbrace{(x \cdot x \cdot x)}^{\text{3 factors}} \cdot \overbrace{(x \cdot x)}^{\text{2 factors}}$$
$$= \underbrace{x \cdot x \cdot x \cdot x \cdot x}_{\text{5 factors}}$$
$$= x^5$$

Note that adding the exponents results in the same product.

$$x^3 \cdot x^2 = x^{3+2} = x^5$$

This suggests the following rule for multiplying exponential expressions.

Rule for Multiplying Exponential Expressions

If m and n are positive integers, then $x^m \cdot x^n = x^{m+n}$.

Simplify: $a^4 \cdot a^5$

The bases are the same.
Add the exponents.

$$a^4 \cdot a^5 = a^{4+5}$$
$$= a^9$$

Simplify: $c^3 \cdot c^4 \cdot c$

The bases are the same.
Add the exponents. Note that $c = c^1$.

$$c^3 \cdot c^4 \cdot c = c^{3+4+1}$$
$$= c^8$$

Simplify: $x^5 y^3$

The bases are *not* the same. The exponential expression is in simplest form.

$x^5 y^3$ is in simplest form.

Simplify: $(4x^3)(2x^2)$

Use the Commutative and Associative Properties of Multiplication to rearrange and group like factors.

$(4x^3)(2x^2) = (4 \cdot 2)(x^3 \cdot x^2)$

Multiply the coefficients. Multiply variables with the same base by adding the exponents.

$= 8x^{3+2}$
$= 8x^5$

Simplify: $(a^3b^2)(a^4)$

Multiply variables with the same base by adding the exponents.

$(a^3b^2)(a^4) = a^{3+4}b^2$
$= a^7b^2$

Simplify: $(-2v^3z^5)(5v^2z^6)$

Multiply the coefficients of the monomials. Multiply variables with the same base by adding the exponents.

$(-2v^3z^5)(5v^2z^6) = [(-2)5](v^{3+2})(z^{5+6})$
$= -10v^5z^{11}$

1 Example Simplify: $(-6c^5)(7c^8)$

Solution $(-6c^5)(7c^8) = [(-6)7](c^{5+8})$
$= -42c^{13}$

1 You Try It Simplify: $(-7a^4)(4a^2)$

Your Solution

2 Example Simplify: $(-5ab^3)(4a^5)$

Solution $(-5ab^3)(4a^5) = (-5 \cdot 4)(a \cdot a^5)b^3$
$= -20a^{1+5}b^3$
$= -20a^6b^3$

2 You Try It Simplify: $(8m^3n)(-3n^5)$

Your Solution

3 Example Simplify: $(6x^3y^2)(4x^4y^5)$

Solution $(6x^3y^2)(4x^4y^5)$
$= (6 \cdot 4)(x^3 \cdot x^4)(y^2 \cdot y^5)$
$= 24x^{3+4}y^{2+5}$
$= 24x^7y^7$

3 You Try It Simplify: $(12p^4q^3)(-3p^5q^2)$

Your Solution

Solutions on p. S14

Copyright © Houghton Mifflin Company. All rights reserved.

Copyright © Houghton Mifflin Company. All rights reserved.

OBJECTIVE **B**

Powers of monomials

The expression $(x^4)^3$ is an example of a *power of a monomial;* the monomial x^4 is raised to the third (3) power.

The power of a monomial can be simplified by writing the power in factored form and then using the Rule for Multiplying Exponential Expressions.

$$(x^4)^3 = x^4 \cdot x^4 \cdot x^4$$
$$= x^{4+4+4} = x^{12}$$

Note that multiplying the exponent inside the parentheses by the exponent outside the parentheses results in the same product.

$$(x^4)^3 = x^{4\cdot3} = x^{12}$$

This suggests the following rule for simplifying powers of monomials.

Rule for Simplifying the Power of an Exponential Expression

If m and n are positive integers, then $(x^m)^n = x^{m \cdot n}$.

Simplify: $(z^2)^5$

Use the Rule for Simplifying the Power of an Exponential Expression.

$$(z^2)^5 = z^{2\cdot5} = z^{10}$$

The expression $(a^2b^3)^2$ is the *power of the product* of the two exponential expressions a^2 and b^3. The power of a product of exponential expressions can be simplified by writing the product in factored form and then using the Rule for Multiplying Exponential Expressions.

Write the power of the product of the monomial in factored form.
Use the Rule for Multiplying Exponential Expressions.

$$(a^2b^3)^2 = (a^2b^3)(a^2b^3)$$
$$= a^{2+2}b^{3+3}$$
$$= a^4b^6$$

Note that multiplying each exponent inside the parentheses by the exponent outside the parentheses results in the same product.

$$(a^2b^3)^2 = a^{2\cdot2}b^{3\cdot2}$$
$$= a^4b^6$$

Rule for Simplifying Powers of Products

If m, n, and p are positive integers, then $(x^m y^n)^p = x^{m\cdot p}y^{n\cdot p}$.

Simplify: $(x^4y)^6$

Multiply each exponent inside the parentheses by the exponent outside the parentheses. Remember that $y = y^1$.

$(x^4y)^6 = x^{4 \cdot 6}y^{1 \cdot 6}$

$= x^{24}y^6$

Simplify: $(5x^2)^3$

Multiply each exponent inside the parentheses by the exponent outside the parentheses. Note that $5 = 5^1$.

$(5x^2)^3 = 5^{1 \cdot 3}x^{2 \cdot 3}$

$= 5^3x^6$

Evaluate 5^3.

$= 125x^6$

Simplify: $(-a^5)^4$

Multiply each exponent inside the parentheses by the exponent outside the parentheses. Note that $-a^5 = (-1)a^5 = (-1)^1a^5$.

$(-a^5)^4 = (-1)^{1 \cdot 4}a^{5 \cdot 4}$

$= (-1)^4a^{20}$

$= 1a^{20} = a^{20}$

Simplify: $(3m^5p^2)^4$

Multiply each exponent inside the parentheses by the exponent outside the parentheses.

$(3m^5p^2)^4 = 3^{1 \cdot 4}m^{5 \cdot 4}p^{2 \cdot 4}$

$= 3^4m^{20}p^8$

Evaluate 3^4.

$= 81m^{20}p^8$

---

④ ***Example***   Simplify: $(-2x^4)^3$

Solution $(-2x^4)^3 = (-2)^{1 \cdot 3}x^{4 \cdot 3}$

$= (-2)^3x^{12}$

$= -8x^{12}$

④   ***You Try It***   Simplify: $(-y^4)^5$

Your Solution

---

⑤ ***Example***   Simplify: $(-2p^3r)^4$

Solution $(-2p^3r)^4 = (-2)^{1 \cdot 4}p^{3 \cdot 4}r^{1 \cdot 4}$

$= (-2)^4p^{12}r^4$

$= 16p^{12}r^4$

⑤   ***You Try It***   Simplify: $(-3a^4bc^2)^3$

Your Solution

Solutions on p. S14

Copyright © Houghton Mifflin Company. All rights reserved.

5.4 EXERCISES

OBJECTIVE **A**

Multiply.

1. $a^4 \cdot a^5$ **2.** $y^5 \cdot y^8$ **3.** $x^9 \cdot x^7$ **4.** $d^6 \cdot d$

5. $n^4 \cdot n^2$ **6.** $p^7 \cdot p^3$ **7.** $z^3 \cdot z \cdot z^4$ **8.** $b \cdot b^2 \cdot b^6$

9. $(a^3b^2)(a^5b)$ **10.** $(xy^5)(x^3y^7)$ **11.** $(-m^3n)(m^6n^2)$ **12.** $(-r^4t^3)(r^2t^9)$

13. $(2x^3)(5x^4)$ **14.** $(6x^3)(9x)$ **15.** $(8x^2y)(xy^5)$ **16.** $(4a^3b^4)(3ab^5)$

17. $(-4m^3)(3m^4)$ **18.** $(6r^2)(-4r)$ **19.** $(7v^3)(-2w)$ **20.** $(-9a^3)(4b^2)$

21. $(ab^2c^3)(-2b^3c^2)$ **22.** $(4x^2y^3)(-5x^5)$ **23.** $(4b^4c^2)(6a^3b)$ **24.** $(3xy^5)(5y^2z)$

25. $(-8r^2t^3)(-5rt^4v)$ **26.** $(-4ab^3c^2)(b^3c)$ **27.** $(9mn^4p)(-3mp^2)$ **28.** $(-3v^2wz)(-4vz^4)$

29. $(2x)(3x^2)(4x^4)$ **30.** $(5a^2)(4a)(3a^5)$ **31.** $(3ab)(2a^2b^3)(a^3b)$

32. $(4x^2y)(3xy^5)(2x^2y^2)$ **33.** $(-xy^5)(3x^2)(5y^3)$ **34.** $(-6m^3n)(-mn^2)(m)$

35. $(8rt^3)(-2r^3v^2)(-3t^5v^2)$ **36.** $(-y^5z)(-2x^3z)(-3xy^4)$ **37.** $(-5ac^3)(-4b^3c)(-3a^2b^2)$

38. Find the product of $7x^2y^3z^5$ and $3xy^4$.   **39.** What is $2ab^6$ times $-4a^5b^4$?

Copyright © Houghton Mifflin Company. All rights reserved.

OBJECTIVE **B**

Simplify.

40. $(x^3)^5$

41. $(b^2)^4$

42. $(z^6)^3$

43. $(p^4)^7$

44. $(y^{10})^2$

45. $(c^7)^4$

46. $(d^9)^2$

47. $(3x)^2$

48. $(2y)^3$

49. $(x^2y^3)^6$

50. $(m^4n^2)^3$

51. $(r^3t)^4$

52. $(a^2b)^5$

53. $(-y^2)^2$

54. $(-z^3)^2$

55. $(2x^4)^3$

56. $(3n^3)^3$

57. $(-2a^2)^3$

58. $(-3b^3)^2$

59. $(3x^2y)^2$

60. $(4a^4b^5)^3$

61. $(2a^3bc^2)^3$

62. $(4xy^3z^2)^2$

63. $(-mn^5p^3)^4$

Critical Thinking

64. *Geometry*   Find the area of the rectangle shown at the right. The dimensions given are in meters. Use the formula $A = LW$.

65. *Geometry*   Find the area of the rectangle shown at the right. The dimensions given are in feet. Use the formula $A = LW$.

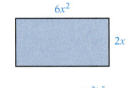

66. *Geometry*   Find the area of the square shown at the right. The dimensions given are in centimeters. Use the formula $A = s^2$.

67. Evaluate $(2^3)^2$ and $2^{(3^2)}$. Are the results the same? If not, which expression has the larger value?

68. ✏ If n is a positive integer and $x^n = y^n$, does $x = y$? Explain your answer.

Copyright © Houghton Mifflin Company. All rights reserved.

Copyright © Houghton Mifflin Company. All rights reserved.

SECTION 5.5 ## Multiplication of Polynomials

OBJECTIVE **A**

Multiplication of a polynomial by a monomial

Recall that the Distributive Property states that if a, b, and c are real numbers, then $a(b + c) = ab + ac$. The Distributive Property is used to multiply a polynomial by a monomial. Each term of the polynomial is multiplied by the monomial.

Multiply: $y^2(4y^2 + 3y - 7)$

Use the Distributive Property. Multiply each term of the polynomial by y^2.	$y^2(4y^2 + 3y - 7)$
	$= y^2(4y^2) + y^2(3y) - y^2(7)$
Use the Rule for Multiplying Exponential Expressions.	$= 4y^4 + 3y^3 - 7y^2$

Multiply: $3x^3(4x^4 - 2x + 5)$

Use the Distributive Property. Multiply each term of the polynomial by $3x^3$.	$3x^3(4x^4 - 2x + 5)$
	$= 3x^3(4x^4) - 3x^3(2x) + 3x^3(5)$
Use the Rule for Multiplying Exponential Expressions.	$= 12x^7 - 6x^4 + 15x^3$

Multiply: $-3a(6a^4 - 3a^2)$

Use the Distributive Property. Multiply each term of the polynomial by $-3a$.	$-3a(6a^4 - 3a^2)$
	$= -3a(6a^4) - (-3a)(3a^2)$
Use the Rule for Multiplying Exponential Expressions.	$= -18a^5 - (-9a^3)$
Rewrite $-(-9a^3)$ as $+ 9a^3$.	$= -18a^5 + 9a^3$

---

1 *Example*   Multiply: $-2x(7x - 4y)$

Solution $-2x(7x - 4y)$
$= -2x(7x) - (-2x)(4y)$
$= -14x^2 + 8xy$

1 *You Try It*   Multiply: $-3a(-6a + 5b)$

Your Solution

---

2 *Example*   Multiply:
$2xy(3x^2 - xy + 2y^2)$

Solution $2xy(3x^2 - xy + 2y^2)$
$= 2xy(3x^2) - 2xy(xy) + 2xy(2y^2)$
$= 6x^3y - 2x^2y^2 + 4xy^3$

2 *You Try It*   Multiply:
$3mn^2(2m^2 - 3mn - 1)$

Your Solution

Solutions on p. S14

OBJECTIVE **B**

Multiplication of two binomials

In the previous objective, a monomial and a polynomial were multiplied. Using the Distributive Property, we multiplied each term of the polynomial by the monomial. Two binomials are also multiplied by using the Distributive Property. Each term of one binomial is multiplied by the other binomial.

Multiply: $(x + 2)(x + 6)$

Use the Distributive Property. Multiply each term of $(x + 6)$ by $(x + 2)$.	$(x + 2)(x + 6)$ $= (x + 2)x + (x + 2)6$
Use the Distributive Property again to multiply $(x + 2)x$ and $(x + 2)6$.	$= x(x) + 2(x) + x(6) + 2(6)$ $= x^2 + 2x + 6x + 12$
Simplify by combining like terms.	$= x^2 + 8x + 12$

> **Take Note**
>
> FOIL is just a method of remembering to multiply each term of one binomial by the other binomial. It is based on the Distributive Property.
>
> $(2x + 3)(3x + 4)$
> $= 2x(3x + 4) + 3(3x + 4)$
> F O I L
> $= 6x^2 + 8x + 9x + 12$
> $= 6x^2 + 17x + 12$

Because it is frequently necessary to multiply two binomials, the terms of the binomials are labeled as shown in the diagram below and the product is computed by using a method called **FOIL.** The letters of FOIL stand for **F**irst, **O**uter, **I**nner, and **L**ast. The FOIL method is based on the Distributive Property and involves adding the product of the first terms, the outer terms, the inner terms, and the last terms.

The product $(2x + 3)(3x + 4)$ is shown below using FOIL.

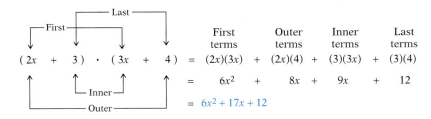

Multiply $(4x - 3)(2x + 3)$ using the FOIL method.

$(4x - 3)(2x + 3) = (4x)(2x) + (4x)(3) + (-3)(2x) + (-3)(3)$

$ = 8x^2 + 12x - 6x - 9$

$ = 8x^2 + 6x - 9$

 Example **3**

Multiply: $(2x - 3)(x + 2)$

Solution

$(2x - 3)(x + 2)$

$ = (2x)(x) + (2x)(2) + (-3)(x) + (-3)(2)$

$ = 2x^2 + 4x - 3x - 6$

$ = 2x^2 + x - 6$

 You Try It **3**

Multiply: $(3c + 7)(3c - 7)$

Your Solution

Solution on p. S14

Copyright © Houghton Mifflin Company. All rights reserved.

5.5 EXERCISES

Multiply.

1. $x(x^2 - 3x - 4)$

2. $y(3y^2 + 4y - 8)$

3. $4a(2a^2 + 3a - 6)$

4. $3b(6b^2 - 5b - 7)$

5. $-2a(3a^2 + 9a - 7)$

6. $-4x(x^2 - 3x - 7)$

7. $m^3(4m - 9)$

8. $r^2(2r^2 + 7)$

9. $2x^3(5x^2 - 6xy + 2y^2)$

10. $4b^4(3a^2 + 4ab - b^2)$

11. $-6r^5(r^2 - 2r - 6)$

12. $-5y^4(3y^2 - 6y^3 + 7)$

13. $4a^2(3a^2 + 6a - 7)$

14. $5b^3(2b^2 - 4b - 9)$

15. $-2n^2(3 - 4n^3 - 5n^5)$

16. $-4x^3(6 - 4x^2 - 5x^4)$

17. $ab^2(3a^2 - 4ab + b^2)$

18. $x^2y^3(5y^3 - 6xy - x^3)$

19. $-x^2y^3(4x^5y^2 - 5x^3y - 7x)$

20. $-a^2b^4(3a^6b^4 + 6a^3b^2 - 5a)$

21. $6r^2t^3(1 - rt - r^3t^3)$

22. What is $3p$ times $4p^2 + 5p - 8$?

23. What is the product of $-4q$ and $-9q + 7$?

Multiply. Use the FOIL method.

24. $(x + 4)(x + 6)$

25. $(y + 9)(y + 3)$

26. $(a - 6)(a - 7)$

Copyright © Houghton Mifflin Company. All rights reserved.

27. $(x + 6)(x + 5)$

28. $(y + 4)(y + 3)$

29. $(a - 3)(a - 8)$

30. $(3c + 4)(2c + 3)$

31. $(5z + 2)(2z + 1)$

32. $(3v - 7)(4v + 3)$

33. $(8c - 7)(5c + 3)$

34. $(8x - 3)(5x - 4)$

35. $(5v - 3)(2v - 1)$

36. $(4n - 9)(4n - 5)$

37. $(7t - 2)(5t + 4)$

38. $(3y - 4)(4y + 7)$

39. $(8x + 5)(3x - 2)$

40. $(4a - 5)(4a + 5)$

41. $(5r + 2)(5r - 2)$

42. What is $2b + 3$ multiplied by $3b + 8$?

43. Find the product of $7y + 5$ and $3y - 8$.

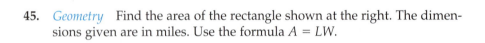

Critical Thinking

44. *Geometry* Find the area of the rectangle shown at the right. The dimensions given are in inches. Use the formula $A = LW$.

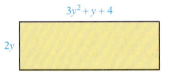
$3y^2 + y + 4$
$2y$

45. *Geometry* Find the area of the rectangle shown at the right. The dimensions given are in miles. Use the formula $A = LW$.

$2x + 3$
$x - 6$

46. *Geometry* Find the area of the square shown at the right. The dimensions given are in meters. Use the formula $A = s^2$.

$2x + 1$

47. State whether the statement is true or false. If the statement is false, change the statement to make it true.
  a. $(5 + x)^2 = 25 + x^2$ **b.** $(5x)^2 = 25x^2$ **c.** $(a - 4)^2 = a^2 - 16$

48. Simplify: $(a - b)^2 - (a + b)^2$

49. 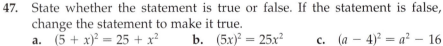 Explain the similarities and differences between multiplying two binomials and multiplying a monomial and a binomial.

Copyright © Houghton Mifflin Company. All rights reserved.

Copyright © Houghton Mifflin Company. All rights reserved.

SECTION 5.6 ▸ Division of Monomials

OBJECTIVE A

Division of monomials

The quotient of two exponential expressions with the *same* base can be simplified by writing each expression in factored form, dividing by the common factors, and then writing the result with an exponent.

$$\frac{x^6}{x^2} = \frac{\overset{1}{\cancel{x}} \cdot \overset{1}{\cancel{x}} \cdot x \cdot x \cdot x \cdot x}{\underset{1}{\cancel{x}} \cdot \underset{1}{\cancel{x}}} = x^4$$

Note that subtracting the exponents results in the same quotient.

$$\frac{x^6}{x^2} = x^{6-2} = x^4$$

This example suggests that to divide monomials with like bases, subtract the exponents.

> **Rule for Dividing Exponential Expressions**
>
> If m and n are positive integers and $x \neq 0$, then $\frac{x^m}{x^n} = x^{m-n}$.

Simplify: $\frac{c^8}{c^5}$

Use the Rule for Dividing Exponential Expressions.

$$\frac{c^8}{c^5} = c^{8-5} = c^3$$

Simplify: $\frac{x^5 y^7}{x^4 y^2}$

Use the Rule for Dividing Exponential Expressions by subtracting the exponents of the like bases. Note that $x^{5-4} = x^1$ but that the exponent 1 is not written.

$$\frac{x^5 y^7}{x^4 y^2} = x^{5-4} y^{7-2} = xy^5$$

The expression at the right has been simplified in two ways: dividing by common factors, and using the Rule for Dividing Exponential Expressions.

$$\frac{x^3}{x^3} = \frac{\overset{1}{\cancel{x}} \cdot \overset{1}{\cancel{x}} \cdot \overset{1}{\cancel{x}}}{\underset{1}{\cancel{x}} \cdot \underset{1}{\cancel{x}} \cdot \underset{1}{\cancel{x}}} = 1$$

$$\frac{x^3}{x^3} = x^{3-3} = x^0$$

Because $\frac{x^3}{x^3} = 1$ and $\frac{x^3}{x^3} = x^0$, 1 must equal x^0. Therefore, the following definition of zero as an exponent is used.

> **Zero as an Exponent**
>
> If $x \neq 0$, then $x^0 = 1$. The expression 0^0 is not defined.

Simplify: 15^0

Any nonzero expression to the zero power is 1.

$15^0 = 1$

Simplify: $(4t^3)^0$, $t \neq 0$

Any nonzero expression to the zero power is 1.

$(4t^3)^0 = 1$

Simplify: $-(2r)^0$, $r \neq 0$

Any nonzero expression to the zero power is 1. Because the negative sign is in front of the parentheses, the answer is -1.

$-(2r)^0 = -1$

The expression at the right has been simplified in two ways: dividing by common factors, and using the Rule for Dividing Exponential Expressions.

$$\frac{x^3}{x^5} = \frac{\overset{1}{\cancel{x}} \cdot \overset{1}{\cancel{x}} \cdot \overset{1}{\cancel{x}}}{\underset{1}{\cancel{x}} \cdot \underset{1}{\cancel{x}} \cdot \underset{1}{\cancel{x}} \cdot x \cdot x} = \frac{1}{x^2}$$

$$\frac{x^3}{x^5} = x^{3-5} = x^{-2}$$

Because $\frac{x^3}{x^5} = \frac{1}{x^2}$ and $\frac{x^3}{x^5} = x^{-2}$, $\frac{1}{x^2}$ must equal x^{-2}. Therefore, the following definition of a negative exponent is used.

Definition of Negative Exponents

If n is a positive integer and $x \neq 0$, then $x^{-n} = \frac{1}{x^n}$ and $\frac{1}{x^{-n}} = x^n$.

An exponential expression is in simplest form when there are no negative exponents in the expression.

Simplify: y^{-7}

Use the Definition of Negative Exponents to rewrite the expression with a positive exponent.

$y^{-7} = \frac{1}{y^7}$

Simplify: $\frac{1}{c^{-4}}$

Use the Definition of Negative Exponents to rewrite the expression with a positive exponent.

$\frac{1}{c^{-4}} = c^4$

Copyright © Houghton Mifflin Company. All rights reserved.

A numerical expression with a negative exponent can be evaluated by first rewriting the expression with a positive exponent.

Evaluate: 2^{-3}

Use the Definition of Negative Exponents to write the expression with a positive exponent. Then simplify.

$$2^{-3} = \frac{1}{2^3} = \frac{1}{8}$$

Take Note

Note from the example at the left that 2^{-3} is a *positive* number. A negative exponent does not indicate a negative number.

Sometimes applying the Rule for Dividing Exponential Expressions results in a quotient that contains a negative exponent. If this happens, use the Definition of Negative Exponents to rewrite the expression with a positive exponent.

Simplify: $\dfrac{p^4}{p^7}$

Use the Rule for Dividing Exponential Expressions.

$$\frac{p^4}{p^7} = p^{4-7}$$

Use the Definition of Negative Exponents to rewrite the expression with a positive exponent.

$$= p^{-3}$$

$$= \frac{1}{p^3}$$

---

1 *Example* Simplify: $\dfrac{1}{a^{-8}}$

Solution $\dfrac{1}{a^{-8}} = a^8$

1 *You Try It* Simplify: $\dfrac{1}{d^{-6}}$

Your Solution

---

2 *Example* Simplify: 3^{-4}

Solution $3^{-4} = \dfrac{1}{3^4} = \dfrac{1}{81}$

2 *You Try It* Simplify: 4^{-2}

Your Solution

---

3 *Example* Simplify: $\dfrac{b^2}{b^9}$

Solution $\dfrac{b^2}{b^9} = b^{2-9} = b^{-7} = \dfrac{1}{b^7}$

3 *You Try It* Simplify: $\dfrac{n^6}{n^{11}}$

Your Solution

Solutions on p. S14

Copyright © Houghton Mifflin Company. All rights reserved.

VIDEO & DVD · CD TUTOR · WWW WEB · SSM

OBJECTIVE B

Scientific notation

Very large and very small numbers are encountered in the natural sciences. For example, the mass of an electron is 0.00000000000000000000000000000911 kg. Numbers such as this are difficult to read, so a more convenient system called **scientific notation** is used. In scientific notation, a number is expressed as the product of two factors, one a number between 1 and 10, and the other a power of ten.

> **Take Note**
> There are two steps in writing a number in scientific notation:
> (1) determine the number between 1 and 10, and
> (2) determine the exponent on 10.

To express a number in scientific notation, write it in the form $a \times 10^n$, where a is a number between 1 and 10 and n is an integer.

For numbers greater than 10, move the decimal point to the right of the first digit. The exponent n is positive and equal to the number of places the decimal point has been moved.

$$240,000 = 2.4 \times 10^5$$
$$93,000,000 = 9.3 \times 10^7$$

For numbers less than 1, move the decimal point to the right of the first nonzero digit. The exponent n is negative. The absolute value of the exponent is equal to the number of places the decimal point has been moved.

$$0.0003 = 3.0 \times 10^{-4}$$
$$0.0000832 = 8.32 \times 10^{-5}$$

Changing a number written in scientific notation to decimal notation also requires moving the decimal point.

When the exponent is positive, move the decimal point to the right the same number of places as the exponent.

$$3.45 \times 10^6 = 3,450,000$$
$$2.3 \times 10^8 = 230,000,000$$

When the exponent is negative, move the decimal point to the left the same number of places as the absolute value of the exponent.

$$8.1 \times 10^{-3} = 0.0081$$
$$6.34 \times 10^{-7} = 0.000000634$$

---

4 Example Write 824,300,000,000 in scientific notation.

Solution The number is greater than 10. Move the decimal point 11 places to the left. The exponent on 10 is 11.

$$824,300,000,000 = 8.243 \times 10^{11}$$

4 You Try It Write 0.000000961 in scientific notation.

Your Solution

---

5 Example Write 6.8×10^{-10} in decimal notation.

Solution The exponent on 10 is negative. Move the decimal point 10 places to the left.

$$6.8 \times 10^{-10} = 0.00000000068$$

5 You Try It Write 7.329×10^6 in decimal notation.

Your Solution

Solutions on p. S14

Copyright © Houghton Mifflin Company. All rights reserved.

5.6 EXERCISES

OBJECTIVE A

Simplify.

1. 27^0

2. $(3x)^0$

3. $-(17)^0$

4. $-(2a)^0$

5. 3^{-2}

6. 4^{-3}

7. 2^{-3}

8. 5^{-2}

9. x^{-5}

10. v^{-3}

11. w^{-8}

12. m^{-9}

13. y^{-1}

14. d^{-4}

15. $\dfrac{1}{a^{-5}}$

16. $\dfrac{1}{c^{-6}}$

17. $\dfrac{1}{b^{-3}}$

18. $\dfrac{1}{y^{-7}}$

19. $\dfrac{a^8}{a^2}$

20. $\dfrac{c^{12}}{c^5}$

21. $\dfrac{q^5}{q}$

22. $\dfrac{r^{10}}{r}$

23. $\dfrac{m^4 n^7}{m^3 n^5}$

24. $\dfrac{a^5 b^6}{a^3 b^2}$

25. $\dfrac{t^4 u^8}{t^2 u^5}$

26. $\dfrac{b^{11} c^4}{b^4 c}$

27. $\dfrac{x^4}{x^9}$

28. $\dfrac{r^2}{r^5}$

29. $\dfrac{b}{b^5}$

30. $\dfrac{m^5}{m^8}$

OBJECTIVE B

Write the number in scientific notation.

31. 2,370,000

32. 75,000

33. 0.00045

34. 0.000076

35. 309,000

36. 819,000,000

37. 0.000000601

38. 0.00000000096

39. 57,000,000,000

40. 934,800,000,000

41. 0.000000017

42. 0.0000009217

Copyright © Houghton Mifflin Company. All rights reserved.

Write the number in decimal notation.

43. 7.1×10^5

44. 2.3×10^7

45. 4.3×10^{-5}

46. 9.21×10^{-7}

47. 6.71×10^8

48. 5.75×10^9

49. 7.13×10^{-6}

50. 3.54×10^{-8}

51. 5×10^{12}

52. 1.0987×10^{11}

53. 8.01×10^{-3}

54. 4.0162×10^{-9}

55. *Physics*   Light travels approximately 16,000,000,000 mi in one day. Write this number in scientific notation.

56. *Earth Science*   Write the mass of Earth, which is approximately 5,980,000,000,000,000,000,000,000 kg, in scientific notation.

57. *Physics*   The electric charge on an electron is 0.00000000000000000016 coulomb. Write this number in scientific notation.

58. *Physics*   The length of an infrared light wave is approximately 0.0000037 m. Write this number in scientific notation.

59. *Computers*   One unit used to measure the speed of a computer is the picosecond. One picosecond is 0.000000000001 of a second. Write this number in scientific notation.

60. *Economics*   What was the U.S. trade deficit in 2002? Use the graph at the right. Write the answer in scientific notation.

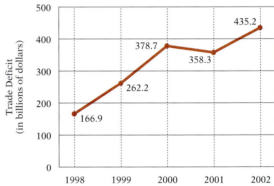

The U.S. Trade Deficit

Source: U.S. Department of Commerce, Bureau of Economic Analysis

Critical Thinking

61. Place the correct symbol, $<$ or $>$, between the two numbers.
   a. $3.45 \times 10^{-14} \ ? \ 6.45 \times 10^{-15}$ **b.** $5.23 \times 10^{18} \ ? \ 5.23 \times 10^{17}$
   c. $3.12 \times 10^{12} \ ? \ 4.23 \times 10^{11}$ **d.** $-6.81 \times 10^{-24} \ ? \ -9.37 \times 10^{-25}$

62. **a.** Evaluate 3^{-x} when $x = -2, -1, 0, 1$, and 2.
   b. Evaluate 2^{-x} when $x = -2, -1, 0, 1$, and 2.

63. In your own words, explain how to divide exponential expressions.

Copyright © Houghton Mifflin Company. All rights reserved.

Copyright © Houghton Mifflin Company. All rights reserved.

SECTION 5.7   ## Verbal Expressions and Variable Expressions

OBJECTIVE **A**

VIDEO & DVD   CD TUTOR   WEB   SSM

Translation of verbal expressions into variable expressions

One of the major skills required in applied mathematics is translating a verbal expression into a mathematical expression. Doing so requires recognizing the verbal phrases that translate into mathematical operations. Following is a partial list of the verbal phrases used to indicate the different mathematical operations.

Addition	more than	8 more than w	$w + 8$
	the sum of	the sum of z and 9	$z + 9$
	the total of	the total of r and s	$r + s$
	increased by	x increased by 7	$x + 7$
Subtraction	less than	12 less than b	$b - 12$
	the difference between	the difference between x and 1	$x - 1$
	decreased by	17 decreased by a	$17 - a$
Multiplication	times	negative 2 times c	$-2c$
	the product of	the product of x and y	xy
	of	three-fourths of m	$\dfrac{3}{4}m$
	twice	twice d	$2d$
Division	divided by	v divided by 15	$\dfrac{v}{15}$
	the quotient of	the quotient of y and 3	$\dfrac{y}{3}$
Power	the square of or the second power of	the square of x	x^2
	the cube of or the third power of	the cube of r	r^3
	the fifth power of	the fifth power of a	a^5

▶ **Point of Interest**

The way in which expressions are symbolized has changed over time. Here are some expressions as they may have appeared in the early sixteenth century.

R p. 9 for $x + 9$. The symbol R was used for a variable to the first power. The symbol p. was used for plus.

R m. 3 for $x - 3$. The symbol R is again the variable. The symbol m. is used for minus.

The square of a variable was designated by Q, and the cube was designated by C. The expression $x^3 + x^2$ was written C p. Q.

Translating a phrase that contains the word *sum, difference, product,* or *quotient* can sometimes cause a problem. In the examples at the right, note where the operation symbol is placed.

the *sum* of x and y $x + y$

the *difference* between x and y $x - y$

the *product* of x and y $x \cdot y$

the *quotient* of x and y $\dfrac{x}{y}$

Take Note

The expression $3(c + 5)$ must have parentheses. If we write $3 \cdot c + 5$, then by the Order of Operations Agreement, only the c is multiplied by 3, but we want the 3 multiplied by the *sum* of c and 5.

Translate "three times the sum of c and five" into a variable expression.

Identify words that indicate the mathematical operations.

3 <u>times</u> the <u>sum of</u> c and 5

Use the identified words to write the variable expression. Note that the phrase <u>times the sum of</u> requires parentheses.

$3(c + 5)$

The sum of two numbers is thirty-seven. If x represents the smaller number, translate "twice the larger number" into a variable expression.

Write an expression for the larger number by subtracting the smaller number, x, from the sum.

larger number: $37 - x$

Identify the words that indicate the mathematical operations on the larger number.

<u>twice</u> the larger number

Use the identified words to write a variable expression.

$2(37 - x)$

 Example

Translate "the quotient of r and the sum of r and four" into a variable expression.

Solution

the <u>quotient of</u> r and the <u>sum of</u> r and four

$\dfrac{r}{r + 4}$

 You Try It

Translate "twice x divided by the difference between x and seven" into a variable expression.

Your Solution

 Example

Translate "the sum of the square of y and six" into a variable expression.

Solution

the <u>sum of</u> the <u>square</u> of y and six

$y^2 + 6$

 You Try It

Translate "the product of negative three and the square of d" into a variable expression.

Your Solution

Solutions on p. S14

Copyright © Houghton Mifflin Company. All rights reserved.

OBJECTIVE **B**

Translation and simplification of verbal expressions

After a verbal expression is translated into a variable expression, it may be possible to simplify the variable expression.

Translate "a number plus five less than the product of eight and the number" into a variable expression. Then simplify.

The letter x is chosen for the unknown number. Any letter could be used.	the unknown number: x
Identify words that indicate the mathematical operations.	$\underline{x}$ $\underline{\text{plus}}$ 5 $\underline{\text{less than}}$ the $\underline{\text{product}}$ $\underline{\text{of}}$ 8 and x
Use the identified words to write the variable expression.	$x + (8x - 5)$
Simplify the expression by adding like terms.	$x + 8x - 5$ $9x - 5$

Translate "five less than twice the difference between a number and seven" into a variable expression. Then simplify.

	the unknown number: x
Identify words that indicate the mathematical operations.	5 $\underline{\text{less than}}$ $\underline{\text{twice}}$ the $\underline{\text{difference}}$ $\underline{\text{between}}$ x and 7
Use the identified words to write the variable expression.	$2(x - 7) - 5$
Simplify the expression.	$2x - 14 - 5$ $2x - 19$

---

③ Example

The sum of two numbers is twenty-eight. Using x to represent the smaller number, translate "the sum of the smaller number and three times the larger number" into a variable expression. Then simplify.

Solution

The smaller number is x.
The larger number is $28 - x$.
the $\underline{\text{sum of}}$ the smaller number and three $\underline{\text{times}}$ the larger number

$x + 3(28 - x)$ ▶ This is the variable expression.
$x + 84 - 3x$ ▶ Simplify.
$-2x + 84$

③ You Try It

The sum of two numbers is sixteen. Using x to represent the smaller number, translate "the difference between the larger number and twice the smaller number" into a variable expression. Then simplify.

Your Solution

Solution on p. S14

Copyright © Houghton Mifflin Company. All rights reserved.

Example 4

Translate "eight more than the product of four and the total of a number and twelve" into a variable expression. Then simplify.

Solution

Let the unknown number be x.

8 more than the product of 4 and the total of x and 12

$4(x + 12) + 8$ ▶ This is the variable expression.
$4x + 48 + 8$ ▶ Now simplify.
$4x + 56$

You Try It 4

Translate "the difference between fourteen and the sum of a number and seven" into a variable expression. Then simplify.

Your Solution

Solution on p. S15

OBJECTIVE **C**

Applications

Many applications of mathematics require that you identify the unknown quantity, assign a variable to that quantity, and then attempt to express other unknowns in terms of that quantity.

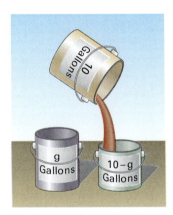

Ten gallons of paint were poured into two containers of different sizes. Express the amount of paint poured into the smaller container in terms of the amount poured into the larger container.

Assign a variable to the amount of paint poured into the larger container. (Any variable can be used.)

gallons of paint poured into the larger container: g

Express the amount of paint in the smaller container in terms of g. (g gallons of paint were poured into the larger container.)

The number of gallons of paint in the smaller container is $10 - g$.

Example 5

A cyclist is riding at twice the speed of a runner. Express the speed of the cyclist in terms of the speed of the runner.

Solution

the speed of the runner: r
the speed of the cyclist is twice r: $2r$

You Try It 5

A mixture of candy contains three pounds more of milk chocolate than of caramel. Express the amount of milk chocolate in the mixture in terms of the amount of caramel in the mixture.

Your Solution

Solution on p. S15

Copyright © Houghton Mifflin Company. All rights reserved.

5.7 EXERCISES

Translate into a variable expression.

1. three more than t

2. the total of twice q and five

3. five less than the product of six and m

4. seven subtracted from the product of eight and d

5. the difference between three times b and seven

6. the difference between six times c and twelve

7. the product of n and seven

8. the quotient of nine times k and seven

9. twice the sum of three and w

10. six times the difference between y and eight

11. four times the difference between twice r and five

12. seven times the total of p and ten

13. the quotient of v and the difference between v and four

14. x divided by the sum of x and one

15. four times the square of t

16. six times the cube of q

17. the sum of the square of m and the cube of the m

18. the difference between the square of d and d

19. The sum of two numbers is thirty-one. Using s to represent the smaller number, translate "five more than the larger number" into a variable expression.

20. The sum of two numbers is seventy-four. Using L to represent the larger number, translate "the quotient of the larger number and the smaller number" into a variable expression.

Translate into a variable expression. Then simplify.

21. a number decreased by the total of the number and twelve

22. a number decreased by the difference between six and the number

Copyright © Houghton Mifflin Company. All rights reserved.

23. the difference between two thirds of a number and three eighths of the number

24. two more than the total of a number and five

25. twice the sum of seven times a number and six

26. five times the product of seven and a number

27. the sum of eleven times a number and the product of three and the number

28. a number plus the product of the number and ten

29. nine times the sum of a number and seven

30. a number added to the product of four and the number

31. seven more than the sum of a number and five

32. a number minus the sum of the number and six

33. the product of seven and the difference between a number and four

34. six times the difference between a number and three

35. the difference between ten times a number and the product of three and the number

36. fifteen more than the difference between a number and seven

37. the sum of a number and twice the difference between the number and four

38. the difference between a number and the total of three times the number and five

39. seven times the difference between a number and fourteen

40. the product of three and the sum of a number and twelve

41. the product of eight and the sum of a number and ten

42. the difference between the square of a number and the total of twelve and the square of the number

43. a number increased by the difference between seven times the number and eight

44. the product of ten and the total of a number and one

45. five increased by twice the sum of a number and fifteen

46. eleven less than the difference between a number and eight

Copyright © Houghton Mifflin Company. All rights reserved.

47. fourteen decreased by the sum of a number and thirteen

48. eleven minus the sum of a number and six

49. the product of eight times a number and two

50. eleven more than a number added to the difference between the number and seventeen

51. a number plus nine added to the difference between four times the number and three

52. the sum of a number and ten added to the difference between the number and eleven

53. The sum of two numbers is nine. Using y to represent the smaller number, translate "five times the larger number" into a variable expression. Then simplify.

54. The sum of two numbers is fourteen. Using p to represent the smaller number, translate "eight less than the larger number" into a variable expression. Then simplify.

55. The sum of two numbers is seventeen. Using m to represent the larger number, translate "nine less than three times the smaller number" into a variable expression. Then simplify.

56. The sum of two numbers is nineteen. Using k to represent the larger number, translate "the difference between twice the smaller number and ten" into a variable expression. Then simplify.

OBJECTIVE C

57. *Astronomy*   The distance from Earth to the sun is approximately 390 times the distance from Earth to the moon. Express the distance from Earth to the sun in terms of the distance from Earth to the moon.

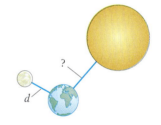

58. *Physics*   The length of an infrared ray is twice the length of an ultraviolet ray. Express the length of the infrared ray in terms of the ultraviolet ray.

59. *Genetics*   The human genome contains 11,000 more genes than does the roundworm genome. (*Source:* Celera, USA TODAY research) Express the number of genes in the human genome in terms of the number of genes in the roundworm genome.

60. *Astronomy*   The planet Saturn has 11 more moons than Jupiter. (*Source:* NASA) Express the number of moons Saturn has in terms of the number of moons Jupiter has.

61. *Food Mixtures*   A mixture contains three times as many peanuts as cashews. Express the amount of peanuts in the mixture in terms of the amount of cashews in the mixture.

Copyright © Houghton Mifflin Company. All rights reserved.

62. 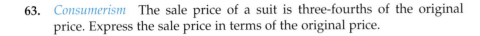 *Taxes* According to the Internal Revenue Service, it takes five times as long to fill out Schedule A (itemized deductions) than Schedule B (interest and dividends). Express the amount of time it takes to fill out Schedule A in terms of the time it takes to fill out Schedule B.

63. *Consumerism* The sale price of a suit is three-fourths of the original price. Express the sale price in terms of the original price.

64. *Travel* One cyclist drives six miles per hour faster than another cyclist. Express the speed of the faster cyclist in terms of the speed of the slower cyclist.

65. *Sports* A fishing line three feet long is cut into two pieces, one shorter than the other. Express the length of the shorter piece in terms of the length of the longer piece.

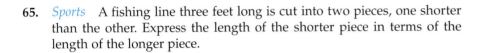

66. *Investments* The dividend paid on a company's stock is one twentieth of the price of the stock. Express the dividend paid on the stock in terms of the price of the stock.

67. *Carpentry* A twelve-foot board is cut into two pieces of different lengths. Express the length of the longer piece in terms of the length of the shorter piece.

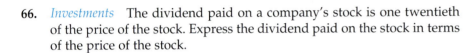

Critical Thinking

68. *Geometry* A wire whose length is given as x inches is bent into a square. Express the length of a side of the square in terms of x.

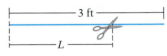

69. *Chemistry* The chemical formula for water is H_2O. This formula means that there are two hydrogen atoms and one oxygen atom in each molecule of water. If x represents the number of atoms of oxygen in a glass of pure water, express the number of hydrogen atoms in the glass of water.

70. *Mechanics* A block-and-tackle system is designed so that pulling five feet on one end of a rope will move a weight on the other end three feet. If x represents the distance the rope is pulled, express the distance the weight will move in terms of x.

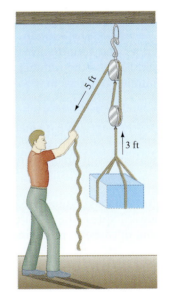

71. *Mechanics* A mechanical gear is designed so that a larger wheel makes four turns as a smaller wheel makes seven turns. Express the number of turns made by the larger wheel in terms of the number made by the smaller wheel.

72. Translate the expressions $3x + 4$ and $3(x + 4)$ into phrases.

73. Explain the similarities and differences between the expressions "the difference between x and 5" and "5 less than x."

Copyright © Houghton Mifflin Company. All rights reserved.

Focus on Problem Solving

Look for a Pattern

A very useful problem-solving strategy is to look for a pattern. We illustrate this strategy below using a fairly old problem.

A legend says that a peasant invented the game of chess and gave it to a very rich king as a present. The king so enjoyed the game that he gave the peasant the choice of anything in the kingdom. The peasant's request was simple. "Place one grain of wheat on the first square, 2 grains on the second square, 4 grains on the third square, 8 on the fourth square, and continue doubling the number of grains until the last square of the chessboard is reached." How many grains of wheat must the king give the peasant?

A chessboard consists of 64 squares. To find the total number of grains of wheat on the 64 squares, we begin by looking at the amount of wheat on the first few squares.

Square 1	Square 2	Square 3	Square 4	Square 5	Square 6	Square 7	Square 8
1	2	4	8	16	32	64	128
1	3	7	15	31	63	127	255

The bottom row of numbers represents the sum of the number of grains of wheat up to and including that square. For instance, the number of grains of wheat on the first 7 squares is

$$1 + 2 + 4 + 8 + 16 + 32 + 64 = 127$$

One pattern to observe is that the number of grains of wheat on a square can be expressed by a power of 2.

Number of grains on square $n = 2^{n-1}$.

For example, the number of grains on square $7 = 2^{7-1} = 2^6 = 64$.

A second pattern of interest is that the number *below* a square (the total number of grains up to and including that square) is one less than the number of grains of wheat *on* the next square. For example, the number *below* square 7 is one less than the number *on* square 8 ($128 - 1 = 127$). From this observation, the number of grains of wheat on the first eight squares is the number on square 8 (128) plus one less than the number on square 8 (127); the total number of grains of wheat on the first eight squares is $128 + 127 = 255$.

From this observation,

$$\begin{array}{rcl} \text{Number of grains of} & = & \text{number of grains} \quad + \quad \text{one less than the number} \\ \text{wheat on the chessboard} & & \text{on square 64} \qquad\qquad \text{of grains on square 64} \\ & = & 2^{64-1} + 2^{64-1} - 1 \\ & = & 2^{63} + 2^{63} - 1 \approx 18{,}000{,}000{,}000{,}000{,}000{,}000 \end{array}$$

To give you an idea of the magnitude of this number, this is more wheat than has been produced in the world since chess was invented.

Suppose that the same king decided to have a banquet in the long dining room of the palace. The king had 50 square tables and each table could seat only one person on each side. The king pushed the tables together to form one long banquet table. How many people can sit at this table? *Hint:* Try constructing a pattern by using 2 tables, 3 tables, and 4 tables.

Copyright © Houghton Mifflin Company. All rights reserved.

Projects & Group Activities

Multiplication of Polynomials

Section 5.5 introduced multiplying a polynomial by a monomial and multiplying two binomials. Multiplying a binomial times a polynomial of three or more terms requires the repeated application of the Distributive Property. Each term of one polynomial is multiplied by the other polynomial. For the product $(2y - 3)(y^2 + 2y + 5)$ shown below, note that the Distributive Property is used twice. The final result is simplified by combining like terms.

$$(2y - 3)(y^2 + 2y + 5) = (2y - 3)y^2 + (2y - 3)2y + (2y - 3)5$$
$$= 2y(y^2) - 3(y^2) + 2y(2y) - 3(2y) + 2y(5) - 3(5)$$
$$= 2y^3 - 3y^2 + 4y^2 - 6y + 10y - 15$$
$$= 2y^3 + y^2 + 4y - 15$$

In Exercises 1–6, find the product of the polynomials.

1. $(y + 6)(y^2 - 3y + 4)$

2. $(2b^2 + 4b - 5)(2b + 3)$

3. $(2a^2 + 4a - 5)(3a + 1)$

4. $(3z^2 - 5z + 7)(z - 3)$

5. $(x - 3)(x^3 + 2x^2 - 4x - 5)$

6. $(c^3 + 3c^2 - 4c + 5)(2c - 3)$

7. **a.** Multiply: $(x + 1)(x - 1)$
   b. Multiply: $(x + 1)(-x^2 + x - 1)$
   c. Multiply: $(x + 1)(x^3 - x^2 + x - 1)$
   d. Multiply: $(x + 1)(-x^4 + x^3 - x^2 + x - 1)$
   e. Use the pattern of the answers to parts a–d to multiply $(x + 1)(x^5 - x^4 + x^3 - x^2 + x - 1)$.
   f. Use the pattern of the answers to parts a–e to multiply $(x + 1)(-x^6 + x^5 - x^4 + x^3 - x^2 + x - 1)$.

Chapter Summary

Key Words	*Examples*
The **additive inverse** of a number a is $-a$. The additive inverse of a number is also called the **opposite** number. [5.1A, p. 316]	The opposite of 15 is -15. The opposite of -24 is 24.
The **multiplicative inverse** of a nonzero number a is $\frac{1}{a}$. The multiplicative inverse of a number is also called the **reciprocal** of the number. [5.1A, p. 316]	The reciprocal of -3 is $-\frac{1}{3}$. The reciprocal of $\frac{6}{7}$ is $\frac{7}{6}$.

Copyright © Houghton Mifflin Company. All rights reserved.

A **term** of a variable expression is one of the addends of the expression. A **variable term** consists of a **numerical coefficient** and a **variable part**. A **constant term** has no variable part. [5.2A, p. 325]

The variable expression $-3x^2 + 2x - 5$ has three terms, $-3x^2$, $2x$, and -5.

$-3x^2$ and $2x$ are variable terms.

-5 is a constant term.

For the term $-3x^2$, the coefficient is -3 and the variable part is x^2.

Like terms of a variable expression have the same variable part. Constant terms are also like terms. [5.2A, p. 325]

$-6a^3b^2$ and $4a^3b^2$ are like terms.

A **monomial** is a number, a variable, or a product of numbers and variables. [5.3A, p. 333]

5 is a number, y is a variable, $8a^2b^2$ is a product of numbers and variables.

5, y, and $8a^2b^2$ are monomials.

A **polynomial** is a variable expression in which the terms are monomials. A polynomial of one term is a **monomial.** A polynomial of two terms is a **binomial.** A polynomial of three terms is a **trinomial.** [5.3A, p. 333]

5, y, and $8a^2b^2$ are monomials.

$x + 9$, $y^2 - 3$, and $6a + 7b$ are binomials.

$x^2 + 2x - 1$ is a trinomial.

The terms of a polynomial in one variable are usually arranged so that the exponents of the variable decrease from left to right. This is called **descending order.** [5.3A, p. 333]

$7y^4 + 5y^3 - y^2 + 6y - 8$ is written in descending order.

Essential Rules and Procedures

Properties of Addition [5.1A, pp. 315–316]

Addition Property of Zero $a + 0 = a$ or $0 + a = a$

$-16 + 0 = -16$

Commutative Property of Addition $a + b = b + a$

$-9 + 5 = 5 + (-9)$

Associative Property of Addition $(a + b) + c = a + (b + c)$

$(-6 + 4) + 2 = -6 + (4 + 2)$

Inverse Property of Addition $a + (-a) = 0$ **or** $-a + a = 0$

$8 + (-8) = 0$

Properties of Multiplication [5.1A, pp. 315–316]

Multiplication Property of Zero $a \cdot 0 = 0$ or $0 \cdot a = 0$

$-3(0) = 0$

Multiplication Property of One $a \cdot 1 = a$ or $1 \cdot a = a$

$-7(1) = -7$

Commutative Property of Multiplication $a \cdot b = b \cdot a$

$-5(10) = 10(-5)$

Associative Property of Multiplication $(a \cdot b) \cdot c = a \cdot (b \cdot c)$

$(-3 \cdot 4) \cdot 6 = -3 \cdot (4 \cdot 6)$

Inverse Property of Multiplication For $a \neq 0$, $a \cdot \dfrac{1}{a} = \dfrac{1}{a} \cdot a = 1$.

$8 \cdot \dfrac{1}{8} = 1$

Distributive Property [5.1B, p. 318]

$a(b + c) = ab + ac$

Alternative Form of the Distributive Property [5.2A, p. 326]

$ac + bc = (a + b)c$

$5(4x - 3) = 5(4x) - 5(3) = 20x - 15$

$8a + 7b = (8 + 7)b = 15b$

Copyright © Houghton Mifflin Company. All rights reserved.

To add polynomials, combine like terms, which means to add the coefficients of the like terms. [5.3A, p. 333]

$$(8x^2 + 2x - 9) + (-3x^2 + 5x - 7)$$
$$= (8x^2 - 3x^2) + (2x + 5x) + (-9 - 7)$$
$$= 5x^2 + 7x - 16$$

To subtract two polynomials, add the opposite of the second polynomial to the first polynomial. [5.3B, pp. 334–335]

$$(3y^2 - 8y + 6) - (-y^2 + 4y - 5)$$
$$= (3y^2 - 8y + 6) + (y^2 - 4y + 5)$$
$$= 4y^2 - 12y + 11$$

Rule for Multiplying Exponential Expressions [5.4A, pp. 341]
$$x^m \cdot x^n = x^{m+n}$$

$$b^5 \cdot b^4 = b^{5+4} = b^9$$

Rule for Simplifying the Power of an Exponential Expression [5.4B, p. 343]
$$(x^m)^n = x^{m \cdot n}$$

$$(y^3)^7 = y^{3(7)} = y^{21}$$

Rule for Simplifying Powers of Products [5.4B, p. 343]
$$(x^m y^n)^p = x^{m \cdot p} y^{n \cdot p}$$

$$(x^6 y^4 z^5)^2 = x^{6(2)} y^{4(2)} z^{5(2)} = x^{12} y^8 z^{10}$$

The FOIL Method [5.5B, p. 348]
To multiply two binomials, add the products of the First terms, the Outer terms, the Inner terms, and the Last terms.

$$(4x + 3)(2x - 5)$$
$$= (4x)(2x) + (4x)(-5) + (3)(2x) + (3)(-5)$$
$$= 8x^2 - 20x + 6x - 15$$
$$= 8x^2 - 14x - 15$$

Rule for Dividing Exponential Expressions [5.6A, p. 351]
For $x \neq 0$, $\dfrac{x^m}{x^n} = x^{m-n}$.

$$\frac{y^8}{y^3} = y^{8-3} = y^5$$

Zero as an Exponent [5.6A, p. 351]
For $x \neq 0$, $x^0 = 1$.
The expression 0^0 is not defined.

$$17^0 = 1$$
$$(5y)^0 = 1, \, y \neq 0$$

Definition of Negative Exponents [5.6A, p. 352]
For $x \neq 0$, $x^{-n} = \dfrac{1}{x^n}$ and $\dfrac{1}{x^{-n}} = x^n$.

$$x^{-6} = \frac{1}{x^6} \text{ and } \frac{1}{x^{-6}} = x^6$$

Scientific Notation [5.6B, p. 354]
To express a number in scientific notation, write it in the form $a \times 10^n$, where a is a number between 1 and 10 and n is an integer.

If the number is greater than 10, the exponent on 10 will be positive and equal to the number of decimal places the decimal point is moved.

$$367{,}000{,}000 = 3.67 \times 10^8$$

If the number is less than 1, the exponent on 10 will be negative. The absolute value of the exponent equals the number of places the decimal point has been moved.

$$0.0000059 = 5.9 \times 10^{-6}$$

To change a number written in scientific notation to decimal notation, move the decimal point to the right if the exponent on 10 is positive and to the left if the exponent on 10 is negative. Move the decimal point the same number of places as the absolute value of the exponent on 10.

$$2.418 \times 10^7 = 24{,}180{,}000$$
$$9.06 \times 10^{-5} = 0.0000906$$

Copyright © Houghton Mifflin Company. All rights reserved.

Chapter Review Exercises

1. Simplify: $4z^2 + 3z - 9z + 2z^2$

2. Multiply: $-2(9z + 1)$

3. Add: $(3z^2 + 4z - 7) + (7z^2 - 5z - 8)$

4. Multiply: $(2m^3n)(-4m^2n)$

5. Evaluate: 3^{-5}

6. Write the additive inverse of $\dfrac{3}{7}$.

7. Multiply: $\dfrac{2}{3}\left(\dfrac{3}{2}x\right)$

8. Simplify: $-5(2s - 5t) + 6(3t + s)$

9. Multiply: $(-5xy^4)(-3x^2y^3)$

10. Multiply: $(7a + 6)(3a - 4)$

11. Subtract:
    $(6b^3 - 7b^2 + 5b - 9) - (9b^3 - 7b^2 + b + 9)$

12. Simplify: $(2z^4)^5$

13. Multiply: $-\dfrac{3}{4}(-8w)$

14. Multiply: $5xyz^2(-3x^2z + 6yz^2 - x^3y^4)$

15. Write the multiplicative inverse of $-\dfrac{9}{4}$.

16. Multiply: $-4(3c - 8)$

17. Simplify: $2m - 6n + 7 - 4m + 6n + 9$

18. Multiply: $(4a^3b^8)(-3a^2b^7)$

19. Identify the property that justifies the statement.
    $a(b + c) = ab + ac$

20. Simplify: $(p^2q^3)^3$

21. Simplify: $\dfrac{a^4}{a^{11}}$

22. Write 0.0000397 in scientific notation.

Copyright © Houghton Mifflin Company. All rights reserved.

23. Identify the property that justifies the statement.
$a + b = b + a$

24. Add: $(9y^3 + 8y^2 - 10) + (-6y^3 + 8y - 9)$

25. Simplify: $8(2c - 3d) - 4(c - 5d)$

26. Multiply: $7(2m - 6)$

27. Simplify: $\dfrac{x^3 y^5}{xy}$

28. Simplify: $7a^2 + 9 - 12a^2 + 3a$

29. Multiply: $(3p - 9)(4p + 7)$

30. Multiply: $-2a^2b(4a^3 - 5ab^2 + 3b^4)$

31. Simplify: $-12x + 7y + 15x - 11y$

32. Simplify: $-7(3a - 4b) - 5(3b - 4a)$

33. Simplify: c^{-5}

34. Subtract:
$(12x^3 + 9x^2 - 5x - 1) - (6x^3 + 9x^2 + 5x - 1)$

35. Write 2.4×10^5 in decimal notation.

36. *Geometry* Find the perimeter of the triangle shown at the right. The dimensions given are in feet. Use the formula $P = a + b + c$.

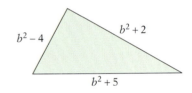

$b^2 - 4$ $b^2 + 2$ $b^2 + 5$

37. Translate "nine less than the quotient of four times a number and seven" into a variable expression.

38. Translate "the sum of three times a number and twice the difference between the number and seven" into a variable expression. Then simplify.

39. *Chemistry* Avogadro's number is used in chemistry, and its value is approximately 602,300,000,000,000,000,000,000. Express this number in scientific notation.

40. *Food Mixtures* Thirty pounds of a blend of coffee beans uses only mocha java and expresso beans. Express the number of pounds of expresso beans in the blend in terms of the number of pounds of mocha java beans in the blend.

Copyright © Houghton Mifflin Company. All rights reserved.

Chapter Test

1. Simplify: $\frac{2}{3}\left(-\frac{3}{2}r\right)$

2. Simplify: $-3(5y - 7)$

3. Simplify: $7y - 3 - 4y + 6$

4. Simplify: $4x^2 - 2z + 7z - 8x^2$

5. Simplify: $2a - 4b + 12 - 5a - 2b + 6$

6. Write the multiplicative inverse of $\frac{5}{4}$.

7. Simplify: $-2(3x - 4y) + 5(2x + y)$

8. Simplify: $9 - 2(4b - a) + 3(3b - 4a)$

9. Write 0.00000079 in scientific notation.

10. Write 4.9×10^6 in decimal notation.

11. Add: $(4x^2 - 2x - 2) + (2x^2 - 3x + 7)$

12. Simplify: $(v^2w^5)^4$

13. Simplify: $(3m^2n^3)^3$

14. Multiply: $(-5v^2z)(2v^3z^2)$

15. Multiply: $(3p - 8)(2p + 5)$

16. Multiply: $(2m^2n^2)(-4mn^3 + 2m^3 - 3n^4)$

17. Complete the statement by using the Commutative Property of Addition.
$3z + ? = 4w + 3z$

18. Simplify: $\dfrac{x^2y^5}{xy^2}$

Copyright © Houghton Mifflin Company. All rights reserved.

19. Simplify: a^{-5}

20. Identify the property that justifies the statement.
$2 \cdot (c \cdot d) = (2 \cdot c) \cdot d$

21. Subtract:
$(5a^3 - 6a^2 + 4a - 8) - (8a^3 - 7a^2 + 4a + 2)$

22. Simplify: $\dfrac{1}{c^{-6}}$

23. Identify the property that justifies the statement.
$6(s + t) = 6s + 6t$

24. Complete the statement by using the Multiplication Property of Zero.
$6w \cdot 0 = ?$

25. Multiply: $(3x - 7y)(3x + 7y)$

26. Write the additive inverse of $-\dfrac{4}{7}$.

27. Write 720,000,000 in scientific notation.

28. Multiply: $(3a - 6)(4a + 2)$

29. Simplify: $2(4a - 3b) + 3(5a - 2b)$

30. Simplify: $\dfrac{m^4 n^2}{m^2 n^5}$

31. Translate "five more than three times a number" into a variable expression.

32. Translate "the sum of a number and four times the difference between the number and seven" into a variable expression and then simplify.

33. *Food Mixtures* A muffin batter contains 3 c more flour than sugar. Express the amount of flour in the batter in terms of the amount of sugar in the batter.

Copyright © Houghton Mifflin Company. All rights reserved.

Cumulative Review Exercises

1. Find the quotient of 4.712 and -0.38.

2. Simplify: $9v - 10 + 5v + 8$

3. Multiply: $(3x - 5)(2x + 4)$

4. Evaluate $-a - b$ when $a = \dfrac{11}{24}$ and $b = -\dfrac{5}{6}$.

5. Simplify: $\sqrt{81} + 3\sqrt{25}$

6. Graph the real numbers greater than -3.

7. Simplify: $\dfrac{1}{x^{-7}}$

8. Solve: $-4t = 36$

9. Write 0.00000084 in scientific notation.

10. Add: $(5x^2 - 3x + 2) + (4x^2 + x - 6)$

11. Evaluate $-5\sqrt{x + y}$ when $x = 18$ and $y = 31$.

12. Simplify: $\dfrac{\dfrac{5}{8} + \dfrac{3}{4}}{3 - \dfrac{1}{2}}$

13. Multiply: $(-3a^2b)(4a^5b^8)$

14. Simplify: $\dfrac{x^3}{x^5}$

15. Evaluate x^3y^2 when $x = \dfrac{2}{5}$ and $y = 2\dfrac{1}{2}$.

16. Simplify: $-8p(6)$

17. Estimate the difference between 829.43 and 567.109.

18. Multiply: $-3ab^2(4a^2b + 5ab - 2ab^2)$

19. Simplify: $6(5x - 4y) - 12(x - 2y)$

20. Evaluate $\dfrac{a}{-b}$ when $a = -56$ and $b = -8$.

Copyright © Houghton Mifflin Company. All rights reserved.

21. Convert 0.5625 to a fraction.

22. Simplify: $6 \cdot (-2)^3 \div 12 - (-8)$

23. Simplify: $\sqrt{300}$

24. Subtract: $(8y^2 - 7y + 4) - (3y^2 - 5y + 9)$

25. Evaluate $-6cd$ when $c = -\frac{2}{9}$ and $d = \frac{3}{4}$.

26. Simplify: $-(3a^2)^0$

27. Simplify: $(2a^4b^3)^5$

28. Evaluate $(a - b)^2 + 5c$ when $a = -4$, $b = 6$, and $c = -2$.

29. Find the product of $2\frac{4}{5}$ and $\frac{6}{7}$.

30. Write 6.23×10^{-5} in decimal notation.

31. Translate "the quotient of ten and the difference between a number and nine" into a variable expression.

32. Translate "two less than twice the sum of a number and four" into a variable expression. Then simplify.

33. *Meteorology* The average annual precipitation in Seattle, Washington, is 38.6 in. The average annual precipitation in El Paso, Texas, is 7.82 in. Find the difference between the average annual precipitation in Seattle and the average annual precipitation in El Paso.

34. *Environmental Science* The graph at the right shows the amount of trash produced per person per day in the United States. On average, how much more trash did a person in the United States throw away during 2000 than during 1960?

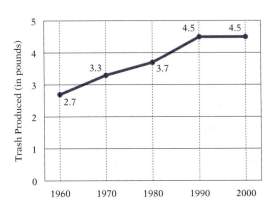

Trash Production per Person per Day in the United States
Source: U.S. Environmental Protection Agency

35. *Astronomy* The distance from Neptune to the sun is approximately 30 times the distance from Earth to the sun. Express the distance from Neptune to the sun in terms of the distance from Earth to the sun.

36. *Investments* The cost, C, of the shares of stock in a stock purchase is equal to the cost per share, S, times the number of shares purchased, N. Use the equation $C = SN$ to find the cost of purchasing 200 shares of stock selling for $\$15\frac{3}{8}$ per share.

Copyright © Houghton Mifflin Company. All rights reserved.

First-Degree Equations

Renting a video is a simple process when each item has a unique bar code and UPC Identification Number. A laser scans the product and interprets the coded numbers to identify things such as the product being purchased and its price. Each product has a different numerical pattern. The **Project on page 426** shows how modular equations are used to form these specific identifications.

Copyright © Houghton Mifflin Company. All rights reserved.

Need help? For online student resources, visit this website: **math.college.hmco.com**

Prep Test

1. Subtract: $8 - 12$

2. Multiply: $-\dfrac{3}{4}\left(-\dfrac{4}{3}\right)$

3. Multiply: $-\dfrac{5}{8}(16)$

4. Simplify: $\dfrac{-3}{-3}$

5. Simplify: $-16 + 7y + 16$

6. Simplify: $8x - 9 - 8x$

7. Evaluate $2x + 3$ when $x = -4$.

8. Given $y = -4x + 5$, find the value of y when $x = -2$.

Go Figure

How can you cut a donut into 8 equal pieces with three cuts of the knife?

Copyright © Houghton Mifflin Company. All rights reserved.

SECTION 6.1 ▶ **Equations of the Form $x + a = b$ and $ax = b$**

OBJECTIVE   **A**

Equations of the form $x + a = b$

Recall that an **equation** expresses the equality of two mathematical expressions. The display at the right shows some examples of equations.

$$3x - 7 = 4x + 9$$
$$y = 3x - 6$$
$$2z^2 - 5z + 10 = 0$$
$$\frac{3}{x} + 7 = 9$$

The first equation in the display above is a **first-degree equation in one variable.** The equation has one variable, x, and each instance of the variable is the first power (the exponent on x is 1). First-degree equations in one variable are the topic of Sections 1 through 4 of this chapter. The second equation is a first-degree equation in two variables. These are discussed in Section 5. The remaining equations are not first-degree equations and will not be discussed in this text.

Which of the equations shown at the right are first-degree equations in one variable?

1. $5x + 4 = 9 - 3(2x + 1)$

2. $\sqrt{x} + 9 = 10$

3. $p = -14$

4. $2x - 5 = x^2 - 9$

Equation 1 is a first-degree equation in one variable.

Equation 2 is not a first-degree equation in one variable. First-degree equations do not contain square roots of variable expressions.

Equation 3 is a first-degree equation in one variable.

Equation 4 is not a first-degree equation in one variable. First-degree equations in one variable do not have exponents greater than 1 on the variable.

Recall that a **solution** of an equation is a number that, when substituted for the variable, results in a true equation.

15 is a solution of the equation $x - 5 = 10$ because $15 - 5 = 10$ is a true equation.

20 is not a solution of $x - 5 = 10$ because $20 - 5 = 10$ is a false equation.

To **solve an equation** means to determine the solutions of the equation. The simplest equation to solve is an equation of the form **variable = constant**. The constant is the solution.

Consider the equation $x = 7$, which is in the form *variable = constant*. The solution is 7 because $7 = 7$ is a true equation.

Copyright © Houghton Mifflin Company. All rights reserved.

Find the solution of the equation $y = 3 + 7$.

Simplify the right side of the equation.

$$y = 3 + 7$$
$$y = 10$$

The solution is 10.

Note that replacing x in $x + 8 = 12$ by 4 results in a true equation. The solution of the equation $x + 8 = 12$ is 4.

$$x + 8 = 12$$
$$4 + 8 = 12$$
$$12 = 12$$

If 5 is added to each side of $x + 8 = 12$, the solution is still 4.

$$x + 8 = 12$$
$$x + 8 + 5 = 12 + 5$$
$$x + 13 = 17$$

Check:
$$x + 13 = 17$$
$$\overline{4 + 13 \ \vert \ 17}$$
$$17 = 17$$

If -3 is added to each side of $x + 8 = 12$, the solution is still 4.

$$x + 8 = 12$$
$$x + 8 + (-3) = 12 + (-3)$$
$$x + 5 = 9$$

Check:
$$x + 5 = 9$$
$$\overline{4 + 5 \ \vert \ 9}$$
$$9 = 9$$

These examples suggest that adding the same number to each side of an equation does not change the solution of the equation. This is called the Addition Property of Equations.

Addition Property of Equations

The same number or variable expression can be added to each side of an equation without changing the solution of the equation.

This property is used in solving equations. Note the effect of adding, to each side of the equation $x + 8 = 12$, the opposite of the constant term 8. After simplifying, the equation is in the form *variable = constant*. The solution is the constant, 4.

$$x + 8 = 12$$
$$x + 8 + (-8) = 12 + (-8)$$
$$x + 0 = 4$$
$$x = 4$$

Check the solution.

Check:
$$x + 8 = 12$$
$$\overline{4 + 8 \ \vert \ 12}$$
$$12 = 12$$

The solution checks.

The solution is 4.

The goal in solving an equation is to rewrite it in the form **variable = constant**. **The Addition Property of Equations is used to remove a term from one side of an equation by adding the opposite of that term to each side of the equation.** The resulting equation has the same solution as the original equation.

Copyright © Houghton Mifflin Company. All rights reserved.

Solve: $m - 9 = 2$

Remove the constant term -9 from the left side of the equation by adding 9, the opposite of -9, to each side of the equation. Then simplify.

$$m - 9 = 2$$
$$m - 9 + 9 = 2 + 9$$
$$m + 0 = 11$$
$$m = 11$$

You should check the solution.   The solution is 11.

In each of the equations above, the variable appeared on the left side of the equation, and the equation was rewritten in the form *variable = constant*. For some equations, it may be more practical to work toward the goal of *constant = variable*, as shown in the example below.

Solve: $12 = n - 8$

The variable is on the right side of the equation. The goal is to rewrite the equation in the form *constant = variable*.

Remove the constant term from the right side of the equation by adding 8 to each side of the equation. Then simplify.

$$12 = n - 8$$
$$12 + 8 = n - 8 + 8$$
$$20 = n + 0$$
$$20 = n$$

You should check the solution.   The solution is 20.

Because subtraction is defined in terms of addition, the Addition Property of Equations allows the same number to be subtracted from each side of an equation without changing the solution of the equation.

Solve: $z + 9 = 6$

The goal is to rewrite the equation in the form *variable = constant*.

Add the opposite of 9 to each side of the equation. This is equivalent to subtracting 9 from each side of the equation. Then simplify.

$$z + 9 = 6$$
$$z + 9 - 9 = 6 - 9$$
$$z + 0 = -3$$
$$z = -3$$

The solution checks.   The solution is -3.

> **Take Note**
>
> Remember to check the solution.
>
> Check: $z + 9 = 6$
> $-3 + 9 \,|\, 6$
> $6 = 6$

Solve: $5 + x - 9 = -10$

Simplify the left side of the equation by combining the constant terms.

$$5 + x - 9 = -10$$
$$x - 4 = -10$$

Add 4 to each side of the equation.

Simplify.

$$x - 4 + 4 = -10 + 4$$
$$x + 0 = -6$$
$$x = -6$$

-6 checks as a solution.   The solution is -6.

Copyright © Houghton Mifflin Company. All rights reserved.

① *Example*   Solve: $6 + x = 4$

Solution $6 + x = 4$
$6 - 6 + x = 4 - 6$
$x = -2$

The solution is -2.

① *You Try It*   Solve: $7 + y = 12$

Your Solution

② *Example*   Solve: $a + \dfrac{3}{4} = \dfrac{1}{4}$

Solution $a + \dfrac{3}{4} = \dfrac{1}{4}$

$a + \dfrac{3}{4} - \dfrac{3}{4} = \dfrac{1}{4} - \dfrac{3}{4}$

$a = -\dfrac{2}{4} = -\dfrac{1}{2}$

The solution is $-\dfrac{1}{2}$.

② *You Try It*   Solve: $b - \dfrac{3}{8} = \dfrac{1}{2}$

Your Solution

③ *Example*   Solve: $7x - 4 - 6x = 3$

Solution $7x - 4 - 6x = 3$
    $x - 4 = 3$ ▶ Combine like terms.

$x - 4 + 4 = 3 + 4$
    $x = 7$

The solution is 7.

③ *You Try It*   Solve: $-5r + 3 + 6r = 1$

Your Solution

Solutions on p. S15

OBJECTIVE **B**

Equations of the form $ax = b$

Note that replacing x by 3 in $4x = 12$ results in a true equation. The solution of the equation is 3.

$4x = 12$
$4(3) = 12$
$12 = 12$

If each side of the equation $4x = 12$ is multiplied by 2, the solution is still 3.

$4x = 12$
$2(4x) = 2(12)$
$8x = 24$

Check:
$$\begin{array}{c|c} 8x = 24 \\ \hline 8(3) & 24 \\ 24 = 24 \end{array}$$

If each side of the equation $4x = 12$ is multiplied by -3, the solution is still 3.

$4x = 12$
$-3(4x) = -3(12)$
$-12x = -36$

Check:
$$\begin{array}{c|c} -12x = -36 \\ \hline -12(3) & -36 \\ -36 = -36 \end{array}$$

Copyright © Houghton Mifflin Company. All rights reserved.

These examples suggest that multiplying each side of an equation by the same nonzero number does not change the solution of the equation. This is called the Multiplication Property of Equations.

> ## Multiplication Property of Equations
>
> Each side of an equation can be multiplied by the same nonzero number without changing the solution of the equation.

This property is used in solving equations. Note the effect of multiplying each side of the equation $4x = 12$ by $\frac{1}{4}$, the reciprocal of the coefficient 4. After simplifying, the equation is in the form *variable = constant*.

$$4x = 12$$
$$\frac{1}{4} \cdot 4x = \frac{1}{4} \cdot 12$$
$$1 \cdot x = 3$$
$$x = 3$$

The solution is 3.

The Multiplication Property of Equations is used to remove a coefficient from a variable term of an equation by multiplying each side of the equation by the reciprocal of the coefficient. The resulting equation will have the same solution as the original equation.

Solve: $\frac{3}{4}x = -9$

The goal is to rewrite the equation in the form *variable = constant*.

Multiply each side of the equation by $\frac{4}{3}$, the reciprocal of $\frac{3}{4}$. After simplifying, the equation is in the form *variable = constant*.

$$\frac{3}{4}x = -9$$
$$\frac{4}{3} \cdot \frac{3}{4}x = \frac{4}{3} \cdot (-9)$$
$$1 \cdot x = -12$$
$$x = -12$$

You should check this solution.   The solution is -12.

Because division is defined in terms of multiplication, the Multiplication Property of Equations allows each side of an equation to be divided by the same nonzero number without changing the solution of the equation.

Solve: $-2x = 8$

Multiply each side of the equation by the reciprocal of -2. This is equivalent to dividing each side of the equation by -2.

$$-2x = 8$$
$$\frac{-2x}{-2} = \frac{8}{-2}$$
$$1 \cdot x = -4$$
$$x = -4$$

Check the solution.

Check:
$$-2x = 8$$
$$-2(-4) \mid 8$$
$$8 = 8$$

The solution checks.   The solution is -4.

Copyright © Houghton Mifflin Company. All rights reserved.

When using the Multiplication Property of Equations, multiply each side of the equation by the reciprocal of the coefficient when the coefficient is a fraction. Divide each side of the equation by the coefficient when the coefficient is an integer or a decimal.

4 Example

Solve: $48 = -12y$

Solution

$48 = -12y$

$\dfrac{48}{-12} = \dfrac{-12y}{-12}$

$-4 = y$

The solution is -4.

4 You Try It

Solve: $-60 = 5d$

Your Solution

5 Example

Solve: $\dfrac{2x}{3} = 12$

Solution

$\dfrac{2x}{3} = 12$

$\dfrac{3}{2}\left(\dfrac{2}{3}x\right) = \dfrac{3}{2}(12)$ ▸ $\dfrac{2x}{3} = \dfrac{2}{3}x$

$x = 18$

The solution is 18.

5 You Try It

Solve: $10 = \dfrac{-2x}{5}$

Your Solution

6 Example

Solve and check: $3y - 7y = 8$

Solution

$3y - 7y = 8$

$-4y = 8$ ▸ Combine like terms.

$\dfrac{-4y}{-4} = \dfrac{8}{-4}$

$y = -2$

Check: $\begin{array}{r|l} 3y - 7y = 8 \\ \hline 3(-2) - 7(-2) & 8 \\ -6 - (-14) & 8 \\ -6 + 14 & 8 \\ 8 = 8 \end{array}$

-2 checks as the solution.

The solution is -2.

6 You Try It

Solve and check: $\dfrac{1}{3}x - \dfrac{5}{6}x = 4$

Your Solution

Solutions on p. S15

Copyright © Houghton Mifflin Company. All rights reserved.

6.1 EXERCISES

OBJECTIVE **A**

1. Classify each equation as one of the form $x + a = b$ or $ax = b$. Explain your reasoning.

   a. $7 + p = -23$ **b.** $-16 = -2s$ **c.** $-\dfrac{7}{8}g = 49$ **d.** $2.8 = q - 9$

2. What is the solution of the equation $x = 9$? Use your answer to explain why the goal in solving the equations in this section is to get the variable alone on one side of the equation.

Solve.

3. $x + 3 = 9$ **4.** $y + 6 = 8$ **5.** $4 + x = 13$ **6.** $9 + y = 14$

7. $m - 12 = 5$ **8.** $n - 9 = 3$ **9.** $x - 3 = -2$ **10.** $y - 6 = -1$

11. $a + 5 = -2$ **12.** $b + 3 = -3$ **13.** $3 + m = -6$ **14.** $5 + n = -2$

15. $8 = x + 3$ **16.** $7 = y + 5$ **17.** $3 = w - 6$ **18.** $4 = y - 3$

19. $-7 = -7 + m$ **20.** $-9 = -9 + n$ **21.** $-3 = v + 5$ **22.** $-1 = w + 2$

23. $-5 = 1 + x$ **24.** $-3 = 4 + y$ **25.** $3 = -9 + m$ **26.** $4 = -5 + n$

27. $4 + x - 7 = 3$ **28.** $12 + y - 4 = 8$ **29.** $8t + 6 - 7t = -6$ **30.** $-5z + 5 + 6z = 12$

31. $y + \dfrac{4}{7} = \dfrac{6}{7}$ **32.** $z + \dfrac{3}{5} = \dfrac{4}{5}$ **33.** $x - \dfrac{3}{8} = \dfrac{1}{8}$ **34.** $a - \dfrac{1}{6} = \dfrac{5}{6}$

35. $c + \dfrac{2}{3} = \dfrac{3}{4}$ **36.** $n + \dfrac{1}{3} = \dfrac{2}{5}$ **37.** $w - \dfrac{1}{4} = \dfrac{3}{8}$ **38.** $t - \dfrac{1}{3} = \dfrac{1}{2}$

OBJECTIVE **B**

Solve.

39. $3x = 9$ **40.** $8a = 16$ **41.** $4c = -12$ **42.** $5z = -25$

Copyright © Houghton Mifflin Company. All rights reserved.

43. $-2r = 16$

44. $-6p = 72$

45. $-4m = -28$

46. $-12x = -36$

47. $-3y = 0$

48. $-7a = 0$

49. $12 = 2c$

50. $28 = 7x$

51. $-72 = 18v$

52. $35 = -5p$

53. $-68 = -17t$

54. $-60 = -15y$

55. $12x = 30$

56. $9v = 15$

57. $-6a = 21$

58. $-8c = 20$

59. $28 = -12y$

60. $36 = -16z$

61. $-52 = -18a$

62. $-40 = -30w$

63. $\frac{2}{3}x = 4$

64. $\frac{3}{4}y = 9$

65. $\frac{1}{3}a = -12$

66. $\frac{3y}{5} = -15$

67. $-\frac{4c}{7} = 16$

68. $-\frac{5n}{8} = 20$

69. $-\frac{z}{4} = -3$

70. $-\frac{3x}{8} = -15$

71. $8 = \frac{4}{5}y$

72. $10 = -\frac{5}{6}c$

73. $\frac{5y}{6} = \frac{7}{12}$

74. $\frac{-3v}{4} = -\frac{7}{8}$

75. $7y - 9y = 10$

76. $8w - 5w = 9$

77. $m - 4m = 21$

78. $2a - 6a = 10$

Critical Thinking

79. (a) Solve the equation $x + a = b$ for x. Is the solution you have written valid for all real numbers a and b? (b) Solve the equation $ax = b$ for x. Is the solution you have written valid for all real numbers a and b?

80. Solve: **a.** $\dfrac{2}{\frac{1}{x}} = 8$ **b.** $\dfrac{3}{\frac{2}{x}} = 6$

81. ✏ Write out the steps for solving the equation $\frac{2}{3}x = 6$. Identify each Property of Real Numbers or Property of Equations as you use it.

Copyright © Houghton Mifflin Company. All rights reserved.

Copyright © Houghton Mifflin Company. All rights reserved.

SECTION 6.2 **Equations of the Form $ax + b = c$**

OBJECTIVE **A**

Equations of the form $ax + b = c$

To solve an equation such as $3w - 5 = 16$, both the Addition and Multiplication Properties of Equations are used.

$$3w - 5 = 16$$

First add the opposite of the constant term -5 to each side of the equation.

$$3w - 5 + 5 = 16 + 5$$
$$3w = 21$$

Divide each side of the equation by the coefficient of w.

$$\frac{3w}{3} = \frac{21}{3}$$

The equation is in the form *variable = constant*.

$$w = 7$$

Check the solution.

Check: $\begin{array}{c|c} 3w - 5 = 16 \\ \hline 3(7) - 5 & 16 \\ 21 - 5 & 16 \\ & 16 = 16 \end{array}$

> **Take Note**
> Note that the Order of Operations Agreement applies to evaluating the expression $3(7) - 5$.

7 checks as the solution.

The solution is 7.

Solve: $8 = 4 - \dfrac{2}{3}x$

The variable is on the right side of the equation. Work toward the goal of *constant = variable*.

$$8 = 4 - \frac{2}{3}x$$

Subtract 4 from each side of the equation.

$$8 - 4 = 4 - 4 - \frac{2}{3}x$$
$$4 = -\frac{2}{3}x$$

Multiply each side of the equation by $-\dfrac{3}{2}$.

$$-\frac{3}{2} \cdot 4 = \left(-\frac{3}{2}\right)\left(-\frac{2}{3}x\right)$$

The equation is in the form *constant = variable*.

$$-6 = x$$

You should check the solution.

The solution is -6.

> **Take Note**
> Always check the solution.
>
> Check: $8 = 4 - \dfrac{2}{3}x$
>
> $\begin{array}{c|c} 8 & 4 - \dfrac{2}{3}(-6) \\ 8 & 4 + 4 \\ 8 = 8 \end{array}$

---

1 *Example*

Solve: $4m - 7 + m = 8$

Solution

$$4m - 7 + m = 8$$
$$5m - 7 = 8 \qquad \blacktriangleright \text{ Combine like terms.}$$
$$5m - 7 + 7 = 8 + 7$$
$$5m = 15$$
$$m = 3 \qquad \blacktriangleright \text{ Divide each side by 5.}$$

The solution is 3.

1 *You Try It*

Solve: $5v + 3 - 9v = 9$

Your Solution

Solution on p. S15

VIDEO & DVD   CD TUTOR   WWW WEB   SSM

OBJECTIVE B

Applications

Some application problems can be solved by using a known formula. Here is an example.

You can afford a maximum monthly car payment of $250. Find the maximum loan amount you can afford. Use the formula $P = 0.02076L$, where P is the amount of a car payment on a 60-month loan at a 9% interest rate and L is the amount of the loan.

Strategy   To find the maximum loan amount, replace the variable P in the formula by its value (250) and solve for L.

Solution

$$P = 0.02076L$$
$$250 = 0.02076L \qquad \text{Replace } P \text{ by 250.}$$
$$\frac{250}{0.02076} = \frac{0.02076L}{0.02076} \qquad \text{Divide each side of the equation by 0.02076.}$$
$$12{,}042.39 \approx L$$

The maximum loan amount you can afford is $12,042.39.

> **Calculator Note**
>
> To solve for L, use your calculator: 250 ÷ 0.02076. Then round the answer to the nearest cent.

---

2 *Example*

An accountant uses the straight-line depreciation equation $V = C - 4{,}500t$ to determine the value V, after t years, of a computer that originally cost C dollars. Use this formula to determine in how many years the value of a computer that originally cost $39,000 will be worth $25,500.

Strategy

To find the number of years, replace each of the variables by their value and solve for t.
$V = 25{,}500$, $C = 39{,}000$.

Solution

$$V = C - 4{,}500t$$
$$25{,}500 = 39{,}000 - 4{,}500t$$
$$25{,}500 - 39{,}000 = 39{,}000 - 39{,}000 - 4{,}500t$$
$$-13{,}500 = -4{,}500t$$
$$\frac{-13{,}500}{-4{,}500} = \frac{-4{,}500t}{-4{,}500}$$
$$3 = t$$

In 3 years, the computer will have a value of $25,500.

2 *You Try It*

The pressure P, in pounds per square inch, at a certain depth in the ocean is approximated by the equation $P = 15 + \frac{1}{2}D$, where D is the depth in feet. Use this formula to find the depth when the pressure is 45 pounds per square inch.

Your Strategy

Your Solution

Solution on p. S15

Copyright © Houghton Mifflin Company. All rights reserved.

6.2 EXERCISES

OBJECTIVE **A**

1. ✎ In your own words, state the Addition Property of Equations. Explain when this property is used.

2. ✎ In your own words, state the Multiplication Property of Equations. Explain when this property is used.

Solve.

3. $5y + 1 = 11$

4. $3x + 5 = 26$

5. $2z - 9 = 11$

6. $7p - 2 = 26$

7. $12 = 2 + 5a$

8. $29 = 1 + 7v$

9. $-5y + 8 = 13$

10. $-7p + 6 = -8$

11. $-12a - 1 = 23$

12. $-15y - 7 = 38$

13. $10 - c = 14$

14. $3 - x = 1$

15. $4 - 3x = -5$

16. $8 - 5x = -12$

17. $-33 = 3 - 4z$

18. $-41 = 7 - 8v$

19. $-4t + 16 = 0$

20. $-6p - 72 = 0$

21. $5a + 9 = 12$

22. $7c + 5 = 20$

23. $2t - 5 = 2$

24. $3v - 1 = 4$

25. $8x + 1 = 7$

26. $6y + 5 = 8$

Copyright © Houghton Mifflin Company. All rights reserved.

388 ∎ CHAPTER 6   First-Degree Equations

27. $4z - 5 = 1$

28. $8 = 5 + 6p$

29. $25 = 11 + 8v$

30. $-4 = 11 + 6z$

31. $-3 = 7 + 4y$

32. $9w - 4 = 17$

33. $8a - 5 = 31$

34. $5 - 8x = 5$

35. $7 - 12y = 7$

36. $-3 - 8z = 11$

37. $-9 - 12y = 5$

38. $5n - \dfrac{2}{9} = \dfrac{43}{9}$

39 $6z - \dfrac{1}{3} = \dfrac{5}{3}$

40. $7y - \dfrac{2}{5} = \dfrac{12}{5}$

41. $3p - \dfrac{5}{8} = \dfrac{19}{8}$

42. $\dfrac{3}{4}x - 1 = 2$

43. $\dfrac{4}{5}y + 3 = 11$

44. $\dfrac{5t}{6} + 4 = -1$

45. $\dfrac{3v}{7} - 2 = 10$

46. $\dfrac{2a}{5} - 5 = 7$

47. $\dfrac{4z}{9} + 23 = 3$

48. $\dfrac{x}{3} + 6 = 1$

49. $\dfrac{y}{4} + 5 = 2$

50. $17 = 20 + \dfrac{3}{4}x$

51. $\dfrac{2}{5}y - 3 = 1$

52. $\dfrac{7}{3}v + 2 = 8$

53. $5 - \dfrac{7}{8}y = 2$

54. $3 - \dfrac{5}{2}z = 6$

55. $\dfrac{3}{5}y + \dfrac{1}{4} = \dfrac{3}{4}$

56. $\dfrac{5}{6}x - \dfrac{2}{3} = \dfrac{5}{3}$

Copyright © Houghton Mifflin Company. All rights reserved.

57. $\dfrac{3}{5} = \dfrac{2}{7}t + \dfrac{1}{5}$

58. $\dfrac{10}{3} = \dfrac{9}{5}w - \dfrac{2}{3}$

59. $\dfrac{z}{3} - \dfrac{1}{2} = \dfrac{1}{4}$

60. $\dfrac{a}{6} + \dfrac{1}{4} = \dfrac{3}{8}$

61. $5.6t - 5.1 = 1.06$

62. $7.2 + 5.2z = 8.76$

63. $6.2 - 3.3t = -12.94$

64. $2.4 - 4.8v = 13.92$

65. $6c - 2 - 3c = 10$

66. $12t + 6 + 3t = 16$

67. $4y + 5 - 12y = -3$

68. $7m - 15 - 10m = 6$

69. $17 = 12p - 5 - 6p$

70. $29 = 4x + 5 - 9x$

71. $3 = 6n + 23 - 10n$

OBJECTIVE **B**

To determine the depreciated value of an X-ray machine, an accountant uses the formula $V = C - 5{,}500t$, where V is the depreciated value of the machine in t years and C is the original cost. Use this formula for Exercises 72 and 73.

72. *Accounting* An X-ray machine originally cost $70,000. In how many years will the depreciated value be $48,000?

73. *Accounting* An X-ray machine originally cost $63,000. In how many years will the depreciated value be $47,500? Round to the nearest tenth.

The formula for the monthly car payment for a 60-month car loan at a 9% interest rate is $P = 0.02076L$, where P is the monthly car payment and L is the amount of the loan. Use this formula for Exercises 74 and 75.

74. *Consumerism* If you can afford a maximum monthly car payment of $300, what is the maximum loan amount you can afford? Round to the nearest cent.

Copyright © Houghton Mifflin Company. All rights reserved.

75. *Consumerism* If you can afford a maximum of $325 for a monthly car payment, what is the largest loan amount you can afford? Round to the nearest cent.

The world record time for a 1-mile race can be approximated by the formula $t = 17.08 - 0.0067y$, where y is the year of the race and between 1900 and 2000, and t is the time, in minutes, of the race. Use this formula for Exercises 76 and 77.

76. *Sports* Approximate the year in which the first 4-minute mile was run. The actual year was 1954.

77. *Sports* In 1985, the world record for a 1-mile race was 3.77 min. For what year does the equation predict this record time?

Black ice is an ice covering on roads that is especially difficult to see and therefore extremely dangerous for motorists. The distance a car traveling 30 mph will slide after its brakes are applied is related to the outside air temperature by the formula $C = \frac{1}{4}D - 45$, where C is the Celsius temperature and D is the distance in feet the car will slide. Use this formula for Exercises 78 and 79.

78. *Physics* Determine the distance a car will slide on black ice when the outside air temperature is $-3°C$.

79. *Physics* Determine the distance a car will slide on black ice when the outside air temperature is $-11°C$.

Critical Thinking

80. Solve: $x \div 28 = 1{,}481$ remainder 25

81. Make up an equation of the form $ax + b = c$ that has -3 as its solution.

82. If a, b, and c are real numbers, is it always possible to solve the equation $ax + b = c$? If not, what values of a, b, or c must be excluded?

83. Does the sentence "Solve $2x - 3(4x + 1)$" make sense? Why or why not?

84. Explain in your own words the steps you would take to solve the equation $\frac{2}{3}x + 4 = 10$. State the Property of Real Numbers or the Property of Equations that is used at each step.

Copyright © Houghton Mifflin Company. All rights reserved.

SECTION 6.3 General First-Degree Equations

OBJECTIVE A

Equations of the form $ax + b = cx + d$

An equation that contains variable terms on both the left and the right side is solved by repeated application of the Addition Property of Equations. The Multiplication Property of Equations is then used to remove the coefficient of the variable and write the equation in the form *variable = constant*.

Solve: $5z - 4 = 8z + 5$

The goal is to rewrite the equation in the form *variable = constant*.

Use the Addition Property of Equations to remove $8z$ from the right side by subtracting $8z$ from each side of the equation. After simplifying, there is only one variable term in the equation.

$$5z - 4 = 8z + 5$$

$$5z - 8z - 4 = 8z - 8z + 5$$

$$-3z - 4 = 5$$

Solve this equation by following the procedure developed in the last section. Using the Addition Property of Equations, add 4 to each side of the equation.

$$-3z - 4 + 4 = 5 + 4$$

$$-3z = 9$$

Divide each side of the equation by -3. After simplifying, the equation is in the form *variable = constant*.

$$\frac{-3z}{-3} = \frac{9}{-3}$$

$$z = -3$$

Check the solution.

$$\begin{array}{c|c} \multicolumn{2}{c}{5z - 4 = 8z + 5} \\ \hline 5(-3) - 4 & 8(-3) + 5 \\ -15 - 4 & -24 + 5 \\ \multicolumn{2}{c}{-19 = -19} \end{array}$$

-3 checks as a solution.

The solution is -3.

Point of Interest

Evariste Galois (1812–1832), even though he was killed in a duel at the age of 21, made significant contributions to solving equations. There is a branch of mathematics, called Galois Theory, that shows what kinds of equations can and cannot be solved. In fact, Galois, fearing he would be killed the next morning, stayed up all night before the duel, frantically writing notes pertaining to this new branch of mathematics.

---

1 Example

Solve: $2c + 5 = 8c + 2$

Solution

$$2c + 5 = 8c + 2$$
$$2c - 8c + 5 = 8c - 8c + 2$$
$$-6c + 5 = 2$$
$$-6c + 5 - 5 = 2 - 5$$
$$-6c = -3$$
$$\frac{-6c}{-6} = \frac{-3}{-6}$$
$$c = \frac{1}{2}$$

The solution is $\frac{1}{2}$.

1 You Try It

Solve: $r - 7 = 5 - 3r$

Your Solution

Solution on p. S15

Copyright © Houghton Mifflin Company. All rights reserved.

Example

Solve: $6a + 3 - 9a = 3a + 7$

Solution

$$6a + 3 - 9a = 3a + 7$$
$$-3a + 3 = 3a + 7$$
$$-3a - 3a + 3 = 3a - 3a + 7$$
$$-6a + 3 = 7$$
$$-6a + 3 - 3 = 7 - 3$$
$$-6a = 4$$
$$\frac{-6a}{-6} = \frac{4}{-6}$$
$$a = -\frac{2}{3}$$

The solution is $-\frac{2}{3}$.

▶ Combine like terms.

▶ $\frac{4}{-6} = -\frac{2}{3}$

▶ Remember to check the solution.

 You Try It

Solve: $4a - 2 + 5a = 2a - 2 + 3a$

Your Solution

Solution on p. S16

OBJECTIVE **B**

Equations with parentheses

When an equation contains parentheses, one of the steps in solving the equation requires the use of the Distributive Property. The Distributive Property is used to remove parentheses from a variable expression.

Solve: $6 - 2(3x - 1) = 3(3 - x) + 5$

The goal is to rewrite the equation in the form *variable = constant*.

Use the Distributive Property to remove parentheses. Then combine like terms on each side of the equation.

$$6 - 2(3x - 1) = 3(3 - x) + 5$$
$$6 - 6x + 2 = 9 - 3x + 5$$
$$8 - 6x = 14 - 3x$$

Using the Addition Property of Equations, add $3x$ to each side of the equation. After simplifying, there is only one variable term in the equation.

$$8 - 6x + 3x = 14 - 3x + 3x$$
$$8 - 3x = 14$$

Subtract 8 from each side of the equation. After simplifying, there is only one constant term in the equation.

$$8 - 8 - 3x = 14 - 8$$
$$-3x = 6$$

Divide each side of the equation by -3, the coefficient of x. The equation is in the form *variable = constant*.

$$\frac{-3x}{-3} = \frac{6}{-3}$$
$$x = -2$$

-2 checks as a solution.

The solution is -2.

Calculator Note

A calculator can be used to check the solution. First evaluate the left side of the equation for $x = -2$. Enter

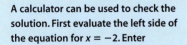

The display reads 20. Then evaluate the right side of the equation for $x = -2$. Enter

3 (3 − 2 +/−) + 5

The display reads 20, the same value as the left side of the equation. The solution checks.

The solution to this last equation illustrates the steps involved in solving first-degree equations.

Copyright © Houghton Mifflin Company. All rights reserved.

Steps in Solving General First-Degree Equations

1. Use the Distributive Property to remove parentheses.

2. Combine like terms on each side of the equation.

3. Rewrite the equation with only one variable term.

4. Rewrite the equation with only one constant term.

5. Rewrite the equation so that the coefficient of the variable is 1.

3 *Example*

Solve: $4 - 3(2t + 1) = 15$

Solution

$$4 - 3(2t + 1) = 15$$
$$4 - 6t - 3 = 15$$
$$-6t + 1 = 15$$
$$-6t + 1 - 1 = 15 - 1$$
$$-6t = 14$$
$$\frac{-6t}{-6} = \frac{14}{-6} \qquad \blacktriangleright \frac{14}{-6} = -\frac{7}{3}$$
$$t = -\frac{7}{3} \qquad \blacktriangleright \text{This solution checks.}$$

The solution is $-\dfrac{7}{3}$.

3 *You Try It*

Solve: $6 - 5(3y + 2) = 26$

Your Solution

4 *Example*

Solve: $5x - 3(2x - 3) = 4(x - 2)$

Solution

$$5x - 3(2x - 3) = 4(x - 2)$$
$$5x - 6x + 9 = 4x - 8$$
$$-x + 9 = 4x - 8$$
$$-x - 4x + 9 = 4x - 4x - 8$$
$$-5x + 9 = -8$$
$$-5x + 9 - 9 = -8 - 9$$
$$-5x = -17$$
$$\frac{-5x}{-5} = \frac{-17}{-5}$$
$$x = \frac{17}{5} \qquad \blacktriangleright \text{This solution checks.}$$

The solution is $\dfrac{17}{5}$.

4 *You Try It*

Solve: $2w - 7(3w + 1) = 5(5 - 3w)$

Your Solution

Solutions on p. S16

Copyright © Houghton Mifflin Company. All rights reserved.

OBJECTIVE C

Applications

Take Note

Pat and Chris are on a 10-foot seesaw, with Pat 4 ft from the fulcrum. Pat weighs 60 lb. Chris weighs 40 lb. We know the seesaw balances because

$$F_1 x = F_2(d - x)$$
$$60(4) = 40(10 - 4)$$
$$60(4) = 40(6)$$
$$240 = 240$$

A lever system is shown at the right. It consists of a lever, or bar; a fulcrum; and two forces, F_1 and F_2. The distance d represents the length of the lever, x represents the distance from F_1 to the fulcrum, and $d - x$ represents the distance from F_2 to the fulcrum.

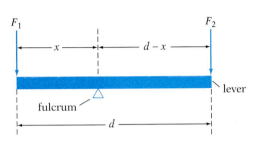

When a lever system balances, $F_1 x = F_2(d - x)$. This is known as Archimedes' Principle of Levers.

---

5 *Example*

A lever 10 ft long is used to move a 100-pound rock. The fulcrum is placed 2 ft from the rock. What minimum force must be applied to the other end of the lever to move the rock?

Strategy

To find the minimum force needed, replace the variables F_1, d, and x by the given values, and solve for F_2.
$F_1 = 100, d = 10, x = 2$

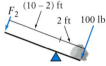

Solution

$$F_1 x = F_2(d - x)$$
$$100 \cdot 2 = F_2(10 - 2)$$
$$200 = 8F_2$$
$$\frac{200}{8} = \frac{8F_2}{8}$$
$$25 = F_2$$

Check:

$$\begin{array}{c|c} F_1 x = F_2(d - x) \\ \hline 100 \cdot 2 & 25(10 - 2) \\ 200 & 25(8) \\ 200 = 200 \end{array}$$

25 checks as the solution.

The minimum force required is 25 lb.

5 *You Try It*

A lever is 25 ft long. A force of 45 lb is applied to one end of the lever, and a force of 80 lb is applied to the other end. What is the location of the fulcrum when the system balances?

Your Strategy

Your Solution

Solution on p. S16

Copyright © Houghton Mifflin Company. All rights reserved.

6.3 EXERCISES

OBJECTIVE **A**

Solve.

1. $4x + 3 = 2x + 9$

2. $6z + 5 = 3z + 20$

3. $7y - 6 = 3y + 6$

4. $8w - 5 = 5w + 10$

5. $12m + 11 = 5m + 4$

6. $8a + 9 = 2a - 9$

7. $7c - 5 = 2c - 25$

8. $7r - 1 = 5r - 13$

9. $2n - 3 = 5n - 18$

10. $4t - 7 = 10t - 25$

11. $3z + 5 = 19 - 4z$

12. $2m + 3 = 23 - 8m$

13. $5v - 3 = 4 - 2v$

14. $3r - 8 = 2 - 2r$

15. $7 - 4a = 2a$

16. $5 - 3x = 5x$

17. $12 - 5y = 3y - 12$

18. $8 - 3m = 8m - 14$

19. $7r = 8 + 2r$

20. $-2w = 4 - 5w$

21. $5a + 3 = 3a + 10$

22. $7y + 3 = 5y + 12$

Copyright © Houghton Mifflin Company. All rights reserved.

23. $9w - 2 = 5w + 4$

24. $7n - 3 = 3n + 6$

25. $x - 7 = 5x - 21$

26. $3y - 4 = 9y - 24$

27. $5n - 1 + 2n = 4n + 8$

28. $3y + 1 + y = 2y + 11$

29. $3z - 2 - 7z = 4z + 6$

30. $2a + 3 - 9a = 3a + 33$

31. $4t - 8 + 12t = 3 - 4t - 11$

32. $6x - 5 + 9x = 7 - 4x - 12$

OBJECTIVE **B**

Solve.

33. $3(4y + 5) = 25$

34. $5(3z - 2) = 8$

35. $-2(4x + 1) = 22$

36. $-3(2x - 5) = 30$

37. $5(2k + 1) - 7 = 28$

38. $7(3t - 4) + 8 = -6$

39. $3(3v - 4) + 2v = 10$

40. $4(3x + 1) - 5x = 25$

41. $3y + 2(y + 1) = 12$

42. $7x + 3(x + 2) = 33$

Copyright © Houghton Mifflin Company. All rights reserved.

43. $7v - 3(v - 4) = 20$

44. $15m - 4(2m - 5) = 34$

45. $6 + 3(3x - 3) = 24$

46. $9 + 2(4p - 3) = 24$

47. $9 - 3(4a - 2) = 9$

48. $17 - 8(x - 3) = 1$

49. $3(2z - 5) = 4z + 1$

50. $4(3z - 1) = 5z + 17$

51. $2 - 3(5x + 2) = 2(3 - 5x)$

52. $5 - 2(3y + 1) = 3(2 - 3y)$

53. $4r + 11 = 5 - 2(3r + 3)$

54. $3v + 6 = 9 - 4(2v - 2)$

55. $7n - 2 = 5 - (9 - n)$

56. $8x - 5 = 7 - 2(5 - x)$

OBJECTIVE   C

Use the lever system equation $F_1x = F_2(d - x)$, for Exercises 57 to 60.

57. *Physics* Two people are sitting 15 ft apart on a seesaw. One person weighs 180 lb; the second person weighs 120 lb. How far from the 180-pound person should the fulcrum be placed so that the seesaw balances?

58. *Physics* Two children are sitting on a seesaw that is 10 ft long. One child weighs 60 lb; the second child weighs 90 lb. How far from the 90-pound child should the fulcrum be placed so that the seesaw balances?

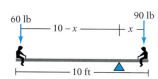

59. *Physics* A metal bar 8 ft long is used to move a 150-pound rock. The fulcrum is placed 1.5 ft from the rock. What minimum force must be applied to the other end of the bar to move the rock? Round to the nearest tenth.

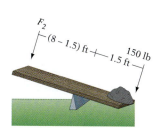

Copyright © Houghton Mifflin Company. All rights reserved.

60. *Physics* A screwdriver 9 in. long is used as a lever to open a can of paint. The tip of the screwdriver is placed under the lip of the lid with the fulcrum 0.15 in. from the lip. A force of 30 lb is applied to the other end of the screwdriver. Find the force on the lip of the top of the can.

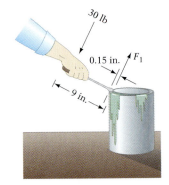

To determine the break-even point, or the number of units that must be sold so that no profit or loss occurs, an economist uses the formula $Px = Cx + F$, where P is the selling price per unit, x is the number of units that must be sold to break even, C is the cost to make each unit, and F is the fixed cost. Use this formula for Exercises 61 to 64.

61. *Business* A business analyst has determined that the selling price per unit for a laser printer is $1,600. The cost to make the laser printer is $950, and the fixed cost is $211,250. Find the break-even point.

62. *Business* An economist has determined that the selling price per unit for a gas barbecue is $325. The cost to make one gas barbecue is $175, and the fixed cost is $39,000. Find the break-even point.

63. *Business* A manufacturer of thermostats determines that the cost per unit for a programmable thermostat is $38 and that the fixed cost is $24,400. The selling price for the thermostat is $99. Find the break-even point.

64. *Business* A manufacturing engineer determines that the cost per unit for a computer mouse is $12 and that the fixed cost is $19,240. The selling price for the computer mouse is $49. Find the break-even point.

Critical Thinking

65. If $5a - 4 = 3a + 2$, what is the value of $4a^3$?

66. If $3 + 2(4a - 3) = 5$ and $4 - 3(2 - 3b) = 11$, which is larger, a or b?

67. Explain the problem with the demonstration shown at the right that suggests $2 = 3$.

$$2x + 5 = 3x + 5$$
$$2x + 5 - 5 = 3x + 5 - 5 \quad \blacktriangleright \text{ Subtract 5 from each side of the equation.}$$
$$2x = 3x$$
$$\frac{2x}{x} = \frac{3x}{x} \quad \blacktriangleright \text{ Divide each side of the equation by } x.$$
$$2 = 3$$

68. The equation $x = x + 1$ has no solution, whereas the solution of the equation $2x + 3 = 3$ is zero. Is there a difference between no solution and a solution of zero? Explain your answer.

69. Archimedes supposedly said, "Give me a long enough lever and I can move the world." Explain what Archimedes meant by that statement.

Copyright © Houghton Mifflin Company. All rights reserved.

Copyright © Houghton Mifflin Company. All rights reserved.

SECTION 6.4 ▶ **Translating Sentences into Equations**

OBJECTIVE   **A**

Translate a sentence into an equation and solve

An equation states that two mathematical expressions are equal. Therefore, to translate a sentence into an equation requires recognition of the words or phrases that mean "equals." Some of these words and phrases are listed below.

equals	is	represents
amounts to	totals	is the same as

Translate "five less than four times a number is four more than the number" into an equation and solve.

Assign a variable to the unknown number.	the unknown number: n

Find two verbal expressions for the same value.

Five less than four times a number	is	four more than the number

Write an equation.

Solve the equation.

$$4n - 5 = n + 4$$

Subtract n from each side.

$$3n - 5 = 4$$

Add 5 to each side.

$$3n = 9$$

Divide each side by 3.

$$n = 3$$

The solution checks.

The number is 3.

> **Take Note**
>
> You can check the solution to a translation problem.
> Check:
>
> | 5 less than 4 times 3 | 4 more than 3 |
> |---|---|
> | $4 \cdot 3 - 5$ | $3 + 4$ |
> | $12 - 5$ | 7 |
> | $7 = 7$ | |

---

1 *Example*

Translate "eight less than three times a number equals five times the number" into an equation and solve.

Solution

the unknown number: x

Eight less than three times a number	equals	five times the number

$$3x - 8 = 5x$$
$$3x - 3x - 8 = 5x - 3x$$
$$-8 = 2x$$
$$\frac{-8}{2} = \frac{2x}{2}$$
$$-4 = x$$

-4 checks as the solution.

The number is -4.

1 *You Try It*

Translate "six more than one-half a number is the total of the number and nine" into an equation and solve.

Your Solution

Solution on p. S16

2 *Example*

Translate "four more than five times a number is six less than three times the number" into an equation and solve.

Solution

the unknown number: m

Four more than five times a number	is	six less than three times the number

$$5m + 4 = 3m - 6$$
$$5m - 3m + 4 = 3m - 3m - 6$$
$$2m + 4 = -6$$
$$2m + 4 - 4 = -6 - 4$$
$$2m = -10$$
$$\frac{2m}{2} = \frac{-10}{2}$$
$$m = -5$$

-5 checks as the solution.

The number is -5.

2 *You Try It*

Translate "seven less than a number is equal to five more than three times the number" into an equation and solve.

Your Solution

3 *Example*

The sum of two numbers is 9. Eight times the smaller number is five less than three times the larger number. Find the numbers.

Solution

the smaller number: p
the larger number: $9 - p$

Eight times the smaller number	is	five less than three times the larger number

$$8p = 3(9 - p) - 5$$
$$8p = 27 - 3p - 5$$
$$8p = 22 - 3p$$
$$8p + 3p = 22 - 3p + 3p$$
$$11p = 22$$
$$\frac{11p}{11} = \frac{22}{11}$$
$$p = 2$$

$9 - p = 9 - 2 = 7$

These numbers check as solutions.

The smaller number is 2.
The larger number is 7.

3 *You Try It*

The sum of two numbers is 14. One more than three times the smaller number equals the sum of the larger number and three. Find the two numbers.

Your Solution

Solutions on p. S16

Copyright © Houghton Mifflin Company. All rights reserved.

OBJECTIVE **B**

Applications

4 *Example*

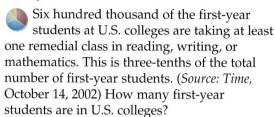 Six hundred thousand of the first-year students at U.S. colleges are taking at least one remedial class in reading, writing, or mathematics. This is three-tenths of the total number of first-year students. (*Source: Time, October 14, 2002*) How many first-year students are in U.S. colleges?

Strategy

To find the number of first-year students, write and solve an equation using n to represent the number of first-year students.

Solution

| 600,000 | is | $\frac{3}{10}$ of the first-year students |

$$600,000 = \frac{3}{10}n$$
$$\frac{10}{3} \cdot 600,000 = \frac{10}{3} \cdot \frac{3}{10}n$$
$$2,000,000 = n$$

The total number of first-year students at U.S. colleges is 2,000,000.

4 *You Try It*

In 2000, Americans put an average of 11,988 miles on their cars. This was 3,175 miles more than they put on their cars in 1980. (*Source:* Energy Information Administration, *Monthly Energy Review,* October 2002) On average, how many miles did Americans put on their cars in 1980?

Your Strategy

Your Solution

---

5 *Example*

A wallpaper hanger charges a fee of $50 plus $28 for each roll of wallpaper used in a room. If the total charge for hanging wallpaper is $218, how many rolls of wallpaper were used?

Strategy

To find the number of rolls of wallpaper used, write and solve an equation using n to represent the number of rolls of wallpaper.

Solution

| $50 plus $28 for each roll of wallpaper | is | $218 |

$$50 + 28n = 218$$
$$50 - 50 + 28n = 218 - 50$$
$$28n = 168$$
$$\frac{28n}{28} = \frac{168}{28}$$
$$n = 6$$

The wallpaper hanger used 6 rolls.

5 *You Try It*

The fee charged by a ticketing agency for a concert is $9.50 plus $57.50 for each ticket purchased. If your total charge for tickets is $527, how many tickets are you purchasing?

Your Strategy

Your Solution

Solutions on p. S16–S17

Copyright © Houghton Mifflin Company. All rights reserved.

6 Example

A bank charges a checking account customer a fee of $12 per month plus $1.50 for each use of an ATM. For the month of July, the customer was charged $24. How many times did this customer use an ATM during the month of July?

Strategy

To find the number of times an ATM was used, write and solve an equation using n for the number of times an ATM was used.

Solution

| $12.00 plus $1.50 per ATM use | is | $24 |

$$12 + 1.50n = 24$$
$$12 - 12 + 1.50n = 24 - 12$$
$$1.50n = 12$$
$$\frac{1.50n}{1.50} = \frac{12}{1.50}$$
$$n = 8$$

The customer used the ATM 8 times in July.

6 You Try It

An auction web site charges a customer a fee of $10 to place an ad plus $5.50 for each day the ad is posted on the web site. A customer is charged $43 for advertising a used trumpet on this web site. For how many days did the customer have the ad for the trumpet posted on the web site?

Your Strategy

Your Solution

7 Example

A guitar wire 22 in. long is cut into two pieces. The longer piece is 4 in. more than twice the shorter piece. Find the length of the shorter piece.

Strategy

To find the length, write and solve an equation using x to represent the length of the shorter piece and $22 - x$ to represent the length of the longer piece.

Solution

| The longer piece | is | 4 in. more than twice the shorter piece |

$$22 - x = 2x + 4$$
$$22 - x - 2x = 2x - 2x + 4$$
$$22 - 3x = 4$$
$$22 - 22 - 3x = 4 - 22$$
$$-3x = -18$$
$$\frac{-3x}{-3} = \frac{-18}{-3}$$
$$x = 6$$

The shorter piece is 6 in.

7 You Try It

A board 18 ft long is cut into two pieces. One foot more than twice the shorter piece is 2 ft less than the longer piece. Find the length of each piece.

Your Strategy

Your Solution

Solutions on p. S17

Copyright © Houghton Mifflin Company. All rights reserved.

6.4 EXERCISES

Translate into an equation and solve.

1. The sum of a number and twelve is twenty. Find the number.

2. The difference between nine and a number is seven. Find the number.

3. Three-fifths of a number is negative thirty. Find the number.

4. The quotient of a number and six is twelve. Find the number.

5. Four more than three times a number is thirteen. Find the number.

6. The sum of twice a number and five is fifteen. Find the number.

7. The difference between nine times a number and six is twelve. Find the number.

8. Six less than four times a number is twenty-two. Find the number.

9. The sum of a number and twice the number is nine. Find the number.

10. Eleven more than negative four times a number is three. Find the number.

11. Seventeen less than the product of five and a number is three. Find the number.

12. Eight less than the product of eleven and a number is negative nineteen. Find the number.

13. Seven more than the product of six and a number is eight less than the product of three and the number. Find the number.

14. Fifteen less than the product of four and a number is the product of five and the number decreased by ten. Find the number.

15. Forty equals nine less than the product of seven and a number. Find the number.

16. Twenty-three equals the difference between eight and the product of five and a number. Find the number.

17. Twice the difference between a number and twenty-five is three times the number. Find the number.

18. Four times a number is three times the difference between thirty-five and the number. Find the number.

19. The product of four and the number minus three equals eight less than the product of six and the number. Find the number.

20. Five less than the product of four and a number is the product of three and the sum of the number and seven. Find the number.

Copyright © Houghton Mifflin Company. All rights reserved.

21. Six more than twice the sum of three times a number and eight is negative two. Find the number.

22. Three times the difference between four times a number and seven is fifteen. Find the number.

23. The sum of two numbers is twenty. Three times the smaller is equal to two times the larger. Find the two numbers.

24. The sum of two numbers is fifteen. One less than three times the smaller is equal to the larger. Find the two numbers.

25. The sum of two numbers is twenty-one. Twice the smaller number is three more than the larger number. Find the two numbers.

26. The sum of two numbers is thirty. Three times the smaller number is twice the larger number. Find the two numbers.

27. The sum of two numbers is twenty-three. The larger number is five more than twice the smaller number. Find the two numbers.

28. The sum of two numbers is twenty-five. The larger number is five less than four times the smaller number. Find the two numbers.

OBJECTIVE **B**

Write an equation and solve.

29. *Consumerism* As a consequence of depreciation, the value of a car one year after it was purchased is $19,900. This is four-fifths of its original value. Find the original value of the car.

30. *The Arts* An oil painting was purchased at an auction this year for $250,000. This is two and one-half times the value of the painting five years ago. What was the value of the painting five years ago?

31. *The Millennium* Five million people greeted the new millennium on New Year's Eve, December 31, 1999, in Rio de Janeiro. This is two and one-half times the number of people who were in Times Square in New York City that evening. How many people greeted the new millennium in Times Square in New York City?

32. *Computers* The vertical resolution of the 15-inch Flat Panel Apple Studio Display is 854 pixels. This is two-thirds of its horizontal resolution. Find the horizontal resolution of the Apple Studio Display.

Copyright © Houghton Mifflin Company. All rights reserved.

33. *Lotteries* Of every dollar spent on a Powerball ticket, 50 cents is used for prizes. This is 44 cents more than the amount spent for advertising and administrative costs. Of every dollar spent on a Powerball ticket, how much is used for advertising and administrative costs?

34. *Education*   In 1950, the vocabulary of the average 14-year-old in the United States was 25,000 words. This is 5,000 words more than twice the number of words in the vocabulary of the average 14-year-old in the United States in 2000. (*Source: Time,* February 14, 2000) Find the number of words in the vocabulary of the average 14-year-old in the United States in 2000.

35. *Recycling*   Americans recycle 18 million tons of paper each year. This is 2 million more than four times the number of tons of paper that is thrown away by American office workers each year. Find the number of tons of paper that is thrown away by American office workers each year.

36. *Consumerism*   A technical information hotline charges a customer $18 plus $1.50 per minute to answer questions about software. For how many minutes did a customer who received a bill for $34.50 use this service?

37. *Consumerism*   The total cost to paint the inside of a house was $2,692. This cost included $250 for materials and $66 per hour for labor. How many hours of labor were required?

38. *Consumerism*   The cellular phone service for a business executive is $30 per month plus $.80 per minute of phone use. In a month when the executive's cellular phone bill was $63.60, how many minutes did the executive use the phone?

39. *Carpentry*   A 12-foot board is cut into two pieces. Twice the length of the shorter piece is three feet less than the longer piece. Find the length of each piece.

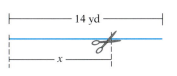

40. *Sports*   A 14-yard fishing line is cut into two pieces. Three times the length of the longer piece is four times the length of the shorter piece. Find the length of each piece.

Copyright © Houghton Mifflin Company. All rights reserved.

41. *Financial Aid* Seven thousand dollars is divided into two scholarships. Twice the amount of the smaller scholarship is $1,000 less than the larger scholarship. What is the amount of the larger scholarship?

42. *Investments* An investment of $10,000 is divided into two accounts, one for stocks and one for mutual funds. The value of the stock account is $2,000 less than twice the value of the mutual fund account. Find the amount in each account.

43. *Food Mixtures* A 10-pound blend of coffee contains Colombian coffee, French Roast, and Java. There is 1 lb more of French Roast than of Colombian and 2 lb more of Java than of French Roast. How many pounds of each are in the mixture?

44. *Agriculture* A 60-pound soil supplement contains nitrogen, iron, and potassium. There is twice as much potassium as iron and three times as much nitrogen as iron. How many pounds of each are in the soil supplement?

Critical Thinking

An equation that is never true is called a **contradiction.** For example, the equation $x = x + 1$ is a contradiction. There is no value of x that will make the equation true. An equation that is true for all real numbers is called an **identity.** The equation $x + x = 2x$ is an identity. This equation is true for any real number. A **conditional equation** is one that is true for some real numbers and false for some real numbers. The equation $2x = 4$ is a conditional equation. This equation is true when x is 2 and false for any other real number. Determine whether each equation below is a contradiction, identity, or conditional equation. If it is a conditional equation, find the solution.

45. $6x + 2 = 5 + 3(2x - 1)$ **46.** $3 - 2(4x + 1) = 5 + 8(1 - x)$

47. $6 + 4(2y + 1) = 5 - 8y$ **48.** $3t - 5(t + 1) = 2(2 - t) - 9$

49. $3v - 2 = 5v - 2(2 + v)$ **50.** $9z = 15z$

51. It is always important to check the answer to an application problem to be sure the answer makes sense. Consider the following problem. A 4-quart mixture of fruit juices is made from apple juice and cranberry juice. There are 6 more quarts of apple juice than of cranberry juice. Write and solve an equation for the number of quarts of each juice used. Does the answer to this question make sense? Explain.

52. A formula is an equation that relates variables in a known way. Find two examples of formulas that are used in your college major. Explain what each of the variables represents.

Copyright © Houghton Mifflin Company. All rights reserved.

SECTION 6.5 ▷ The Rectangular Coordinate System

OBJECTIVE **A**

The rectangular coordinate system

Before the fifteenth century, geometry and algebra were considered separate branches of mathematics. That all changed when René Descartes, a French mathematician who lived from 1596 to 1650, founded analytic geometry. In this geometry, a **coordinate system** is used to study relationships between variables.

A **rectangular coordinate system** is formed by two number lines, one horizontal and one vertical, that intersect at the zero point of each line. The point of intersection is called the **origin.** The two lines are called **coordinate axes,** or simply **axes.**

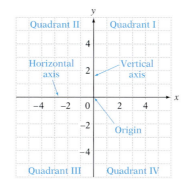

The axes determine a **plane,** which can be thought of as a large, flat sheet of paper. The two axes divide the plane into four regions called **quadrants,** which are numbered counterclockwise from I to IV.

> ▶ **Point of Interest**
>
> Although Descartes is given credit for introducing analytic geometry, others, notably Pierre Fermat, were working on the same concept. Nowhere in Descartes's work is there a coordinate system as we draw it with two axes. Descartes did not use the word *coordinate* in his work. This word was introduced by Gottfried Leibnitz, who also first used the words *abscissa* and *ordinate*.

Each point in the plane can be identified by a pair of numbers called an **ordered pair.** The first number of the pair measures a horizontal distance and is called the **abscissa,** or *x***-coordinate.** The second number of the pair measures a vertical distance and is called the **ordinate,** or *y***-coordinate.** The ordered pair (x, y) associated with a point is also called the **coordinates** of the point.

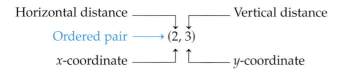

To **graph,** or **plot,** a point in the plane, place a dot at the location given by the ordered pair. The **graph of an ordered pair** is the dot drawn at the coordinates of the point in the plane. The points whose coordinates are $(3, 4)$ and $(-2.5, -3)$ are graphed in the figures below.

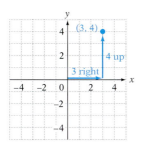 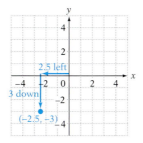

Copyright © Houghton Mifflin Company. All rights reserved.

The points whose coordinates are (3, −1) and (−1, 3) are shown graphed at the right. Note that the graphed points are in different locations. **The order of the coordinates of an ordered pair is important.**

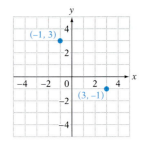

Each point in the plane is associated with an ordered pair, and each ordered pair is associated with a point in the plane. Although only the labels for integers are given on a coordinate grid, the graph of any ordered pair can be approximated. For example, the points whose coordinates are (−2.3, 4.1) and ($\sqrt{2}$, −$\sqrt{3}$) are shown in the graph at the right.

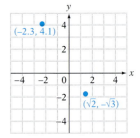

① **Example** Graph the ordered pairs (−2, −3), (3, −2), (1, 3), and (4, 1).

Solution

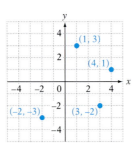

① **You Try It** Graph the ordered pairs (−1, 3), (1, 4), (−4, 0), and (−2, −1).

Your Solution

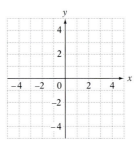

② **Example** Find the coordinates of each point.

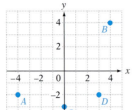

Solution $A(-4, -2)$ $C(0, -3)$
  $B(4, 4)$ $D(3, -2)$

② **You Try It** Find the coordinates of each point.

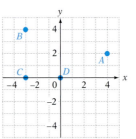

Your Solution

Solutions on p. S17

Copyright © Houghton Mifflin Company. All rights reserved.

Copyright © Houghton Mifflin Company. All rights reserved.

OBJECTIVE **B**

VIDEO & DVD   CD TUTOR   WEB   SSM

Scatter diagrams

Discovering a relationship between two variables is an important task in the study of mathematics. These relationships occur in many forms and in a wide variety of applications. Here are some examples:

A botanist wants to know the relationship between the number of bushels of wheat yielded per acre and the amount of watering per acre.

An environmental scientist wants to know the relationship between the incidence of skin cancer and the amount of ozone in the atmosphere.

A business analyst wants to know the relationship between the price of a product and the number of products that are sold at that price.

A researcher may investigate the relationship between two variables by means of *regression analysis*, which is a branch of statistics. The study of the relationship between two variables may begin with a **scatter diagram,** which is a graph of the ordered pairs of the known data.

The following table gives data collected by a university registrar comparing the grade point averages of graduating high school seniors and their scores on a national test.

GPA, x	3.50	3.50	3.25	3.00	3.00	2.75	2.50	2.50	2.00	2.00	1.50
Test, y	1,500	1,100	1,200	1,200	1,000	1,000	1,000	900	800	900	700

The scatter diagram for these data is shown below.

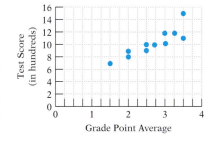

Each ordered pair represents the GPA and test score for a student. For example, the ordered pair (2.75, 1,000) indicates a student with a GPA of 2.75 who had a test score of 1,000.

The dot on the scatter diagram at (3, 1,200) represents the student with a GPA of 3.00 and a test score of 1,200.

3 *Example*

A nutritionist collected data on the number of grams of sugar and grams of fiber in 1-ounce servings of six brands of cereal. The data are recorded in the following table. Graph the scatter diagram for the data.

Sugar, x	6	8	6	5	7	5
Fiber, y	2	1	4	4	2	3

Solution

Graph the ordered pairs on the rectangular coordinate system. The horizontal axis represents the grams of sugar. The vertical axis represents the grams of fiber.

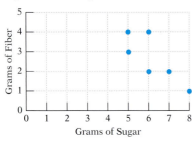

3 *You Try It*

A sports statistician collected data on the total number of yards gained by a college football team and the number of points scored by the team. The data are recorded in the following table. Graph the scatter diagram for the data.

Yards, x	300	400	350	400	300	450
Points, y	18	24	14	21	21	30

Your Solution

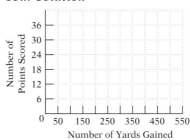

4 *Example*

To test a heart medicine, a doctor measured the heart rates, in beats per minute, of five patients before and after they took the medication. The results are recorded in the scatter diagram. One patient's heart rate before taking the medication was 75 beats per minute. What was this patient's heart rate after taking the medication?

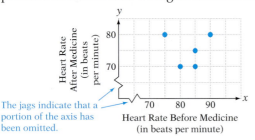

The jags indicate that a portion of the axis has been omitted.

Solution

Locate 75 beats per minute on the x-axis. Follow the vertical line from 75 to a point plotted in the diagram. Follow a horizontal line from that point to the y-axis. Read the number where that line intersects the y-axis.

The ordered pair is (75, 80), which indicates that the patient's heart rate before taking the medication was 75 and the heart rate after taking the medication was 80.

4 *You Try It*

A study by the FAA showed that narrow, over-the-wing emergency exit rows slow passenger evacuation. The scatter diagram below shows the space between seats, in inches, and the evacuation time, in seconds, for a group of 35 passengers. What was the evacuation time when the space between seats was 20 in.?

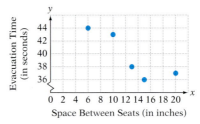

Your Solution

Solutions on p. S17

Copyright © Houghton Mifflin Company. All rights reserved.

6.5 EXERCISES

OBJECTIVE   A

1.   Explain how to locate the point $(-4, 3)$ in a rectangular coordinate system.

2.   Explain how to locate the point $(2, -5)$ in a rectangular coordinate system.

In which quadrant does the given point lie?

3. $(5, 4)$

4. $(3, -2)$

5. $(-8, 1)$

6. $(-7, -6)$

7.   Describe the signs of the coordinates of a point plotted in (a) Quadrant I and (b) Quadrant III.

8.   Describe the signs of the coordinates of a point plotted in (a) Quadrant II and (b) Quadrant IV.

9. Graph the ordered pairs $(5, 2)$, $(3, -5)$, $(-2, 1)$, and $(0, 3)$.

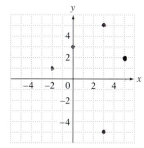

10. Graph the ordered pairs $(-3, -3)$, $(5, -1)$, $(-2, 4)$, and $(0, -5)$.

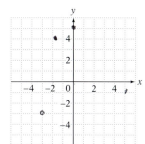

11. Graph the ordered pairs $(-2, -3)$, $(1, -1)$, $(-4, 5)$, and $(-1, 0)$.

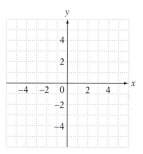

12. Graph the ordered pairs $(2, 5)$, $(0, 0)$, $(3, -4)$, and $(-1, 4)$.

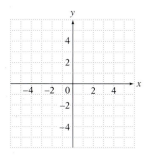

13. Graph the ordered pairs $(2, -5)$, $(-4, -1)$, $(-3, 1)$, and $(0, 2)$.

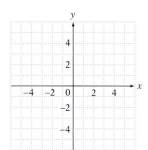

14. Graph the ordered pairs $(3, 1)$, $(4, -3)$, $(-2, 5)$, and $(-4, -2)$.

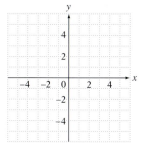

Copyright © Houghton Mifflin Company. All rights reserved.

15. Graph the ordered pairs $(-1, -5)$, $(-2, 3)$, $(4, -1)$, and $(-3, 0)$.

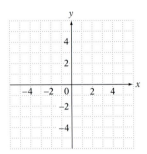

16. Graph the ordered pairs $(4, -5)$, $(-3, 2)$, $(5, 0)$, and $(-5, -1)$.

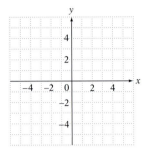

17. Graph the ordered pairs $(4, 1)$, $(-3, 5)$, $(4, 0)$, and $(-1, -2)$.

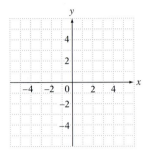

18. Find the coordinates of each point.

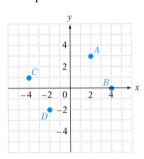

19. Find the coordinates of each point.

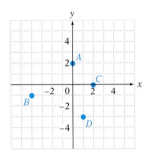

20. Find the coordinates of each point.

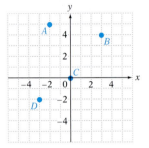

21. Find the coordinates of each point.

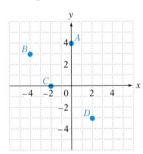

22. Find the coordinates of each point.

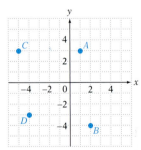

23. Find the coordinates of each point.

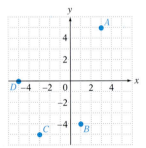

24. Find the coordinates of each point.

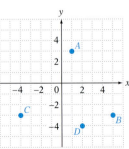

25. Find the coordinates of each point.

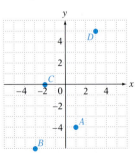

26. Find the coordinates of each point.

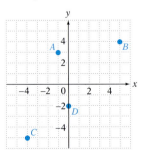

Copyright © Houghton Mifflin Company. All rights reserved.

27. **a.** Name the abscissas of points A and C.
b. Name the ordinates of points B and D.

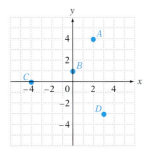

28. **a.** Name the abscissas of points A and C.
b. Name the ordinates of points B and D.

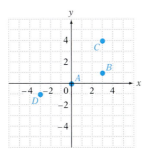

29. **a.** Name the abscissas of points A and C.
b. Name the ordinates of points B and D.

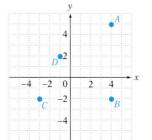

OBJECTIVE B

30. *Business* The number of miles, in thousands, a rental car is driven and the cost to service that vehicle were recorded by the manager of the rental agency. The data are recorded in the following table. Graph the scatter diagram for the data.

Miles (in thousands), x	10	10	5	20	15	5
Cost of service, y	100	250	250	500	300	150

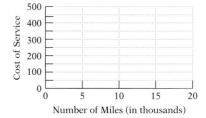

31. *Employment* The number of years of previous work experience and the monthly salary of a person who completes a bachelor's degree in marketing are recorded in the following table. Graph the scatter diagram of these data.

Years experience, x	2	0	5	2	3	1
Salary (in hundreds), y	30	25	45	35	30	35

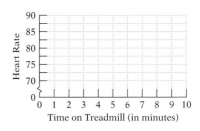

32. *Physiology* An exercise physiologist measured the time, in minutes, a person spent on a treadmill at a fast walk and the heart rate of that person. The results are recorded in the following table. Draw a scatter diagram of these data.

Time on treadmill, x	2	10	5	8	6	5
Heart rate, y	75	90	80	90	85	85

33. *Criminology* Sherlock Holmes solved a crime by recognizing a relationship between the length, in inches, of a person's stride and the height of that person. The data for six people are recorded in the table below. Graph the scatter diagram of these data.

Length of stride, x	15	25	20	25	15	30
Height, y	60	70	65	65	65	75

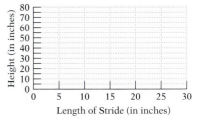

Copyright © Houghton Mifflin Company. All rights reserved.

34. 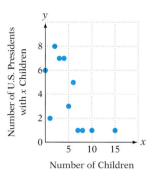 *U.S. Presidents* The scatter diagram at the right pairs number of children with the number of U.S. presidents who had that number of children. How many presidents had 5 children?

35. *Sports* The scatter diagram at the right shows the record times for races of different lengths at a junior high school track meet. What was the record time for the 800-meter race?

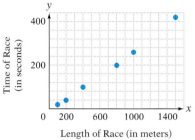

36. 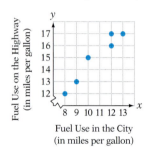 *Fuel Efficiency* The American Council for an Energy-Efficient Economy releases rankings of environmentally friendly and unfriendly cars and trucks sold in the United States. The scatter diagram at the right shows the fuel usage, in miles per gallon of gasoline, both in the city and on the highway, for six of the 2003 model vehicles ranked worst for the environment. **a.** What was the fuel use on the highway, in miles per gallon, for the car that got 10 mpg in the city? **b.** What was the fuel use in the city, in miles per gallon, for the car that got 13 mpg on the highway?

37. 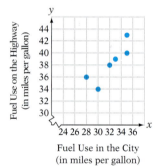 *Fuel Efficiency* The American Council for an Energy-Efficient Economy releases rankings of environmentally friendly and unfriendly cars and trucks sold in the United States. The scatter diagram at the right shows the fuel usage, in miles per gallon of gasoline, both in the city and on the highway, for six of the 2003 compact cars ranked best for the environment. **a.** What was the fuel use on the highway, in miles per gallon, for the car that got 30 mpg in the city? **b.** What was the fuel use in the city, in miles per gallon, for the car that got 36 mpg on the highway?

Critical Thinking

38. Decide on two quantities that may be related and collect at least 10 pairs of values. Here are some examples: height and weight, time studying for a test and grade on the test, age of a car and its cost. Draw a scatter diagram of the data. Is there any trend? That is, as the values on the horizontal axis increase, do the values on the vertical axis either increase or decrease?

39. There is a coordinate system on Earth that consists of *longitude* and *latitude*. Write a report on how location is determined on the surface of Earth. Include in your report the longitude and latitude coordinates of your school.

Copyright © Houghton Mifflin Company. All rights reserved.

Copyright © Houghton Mifflin Company. All rights reserved.

SECTION 6.6   **Graphs of Straight Lines**

OBJECTIVE **A**

Solutions of linear equations in two variables

Some equations express a relationship between *two* variables. For example, the relationship between the Fahrenheit temperature scale F and the Celsius temperature scale C is given by $F = \frac{9}{5}C + 32$. Using this equation, we can determine the Fahrenheit temperature for any Celsius temperature. For example, when the Celsius temperature is 30 degrees,

$$F = \frac{9}{5}(30) + 32$$
$$= 54 + 32$$
$$= 86$$

The Fahrenheit temperature is 86 degrees. The equation above is an example of a *linear equation in two variables.*

An equation of the form $y = mx + b$, where m and b are constants, is a **linear equation in two variables.** Examples of linear equations in two variables are shown at the right.

$$y = 2x + 1 \quad (m = 2, b = 1)$$
$$y = -2x - 5 \quad (m = -2, b = -5)$$
$$y = -\frac{3}{4}x \quad \left(m = -\frac{3}{4}, b = 0\right)$$

The equation $y = x^2 + 4x + 3$ is not a linear equation in two variables because there is a term with a variable squared. The equation $y = \frac{3}{x - 4}$ is not a linear equation because a variable occurs in the denominator of a fraction.

A **solution of an equation in two variables** is an ordered pair (x, y) whose coordinates make the equation a true statement.

Is $(-3, 7)$ a solution of $y = -2x + 1$?

Replace x by -3 and y by 7.

Compare the results. If the results are equal, the ordered pair is a solution of the equation. If the results are not equal, the ordered pair is not a solution of the equation.

$$\begin{array}{c|c} y = -2x + 1 \\ \hline 7 & -2(-3) + 1 \\ 7 & 6 + 1 \\ 7 = 7 \end{array}$$

Yes, the ordered pair $(-3, 7)$ is a solution of the equation.

> **Take Note**
>
> An ordered pair is of the form (x, y). For the ordered pair $(-3, 7)$, -3 is the x value and 7 is the y value. Substitute -3 for x and 7 for y.

Besides the ordered pair $(-3, 7)$, there are many other ordered-pair solutions of the equation $y = -2x + 1$. For example, $(-5, 11)$, $(0, 1)$, $\left(-\frac{3}{2}, 4\right)$, and $(4, -7)$ are also solutions of the equation.

In general, a linear equation in two variables has an infinite number of solutions. By choosing any value of x and substituting that value into the equation, we can calculate a corresponding value of y.

Find the ordered-pair solution of $y = \frac{2}{3}x - 3$ that corresponds to $x = 6$.

$$y = \frac{2}{3}x - 3$$

Replace x by 6. $y = \frac{2}{3}(6) - 3$

Solve for y. $y = 4 - 3$

$$y = 1$$

The ordered-pair solution is (6, 1).

---

1 Example

Is $(-3, 2)$ a solution of the equation $y = -2x - 5$?

Solution

$$y = -2x - 5$$

2	$-2(-3) - 5$	▶ Replace x by -3 and y by 2.
2	$6 - 5$	

$2 \neq 1$

No, $(-3, 2)$ is not a solution of the equation $y = -2x - 5$.

1 You Try It

Is $(-2, 4)$ a solution of the equation $y = -\frac{1}{2}x + 3$?

Your Solution

---

2 Example

Find the ordered-pair solution of the equation $y = -3x + 1$ corresponding to $x = 2$.

Solution

$$y = -3x + 1$$
$$y = -3(2) + 1 \quad \text{▶ Replace } x \text{ by 2.}$$
$$y = -6 + 1$$
$$y = -5$$

The ordered-pair solution is (2, −5).

2 You Try It

Find the ordered-pair solution of the equation $y = 2x - 3$ corresponding to $x = 0$.

Your Solution

Solutions on p. S17

VIDEO & DVD   CD TUTOR   WWW WEB   SSM

OBJECTIVE B

Equations of the form $y = mx + b$

The **graph of an equation in two variables** is a graph of the ordered-pair solutions of the equation.

x	$2x + 1$	y	(x, y)
-2	$2(-2) + 1$	-3	$(-2, -3)$
-1	$2(-1) + 1$	-1	$(-1, -1)$
0	$2(0) + 1$	1	$(0, 1)$
1	$2(1) + 1$	3	$(1, 3)$
2	$2(2) + 1$	5	$(2, 5)$

Consider $y = 2x + 1$. Choosing $x = -2, -1, 0, 1,$ and 2 and determining the corresponding values of y produces some of the ordered-pair solutions of the equation. These are recorded in the table at the left. See the graph of the ordered pairs in Figure 6.1.

Copyright © Houghton Mifflin Company. All rights reserved.

Choosing values of x that are not integers produces more ordered pairs to graph, such as $\left(-\frac{5}{2}, -4\right)$ and $\left(\frac{3}{2}, 4\right)$, as shown in Figure 6.2. Choosing still other values of x would result in more and more ordered pairs being graphed. The result would be so many dots that the graph would appear as the straight line shown in Figure 6.3, which is the graph of $y = 2x + 1$.

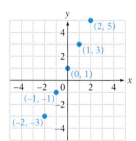

Figure 6.1

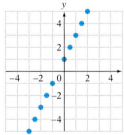

Figure 6.2

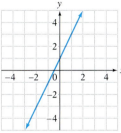

Figure 6.3

Equations in two variables have characteristic graphs. The equation $y = 2x + 1$ is an example of a *linear equation* because its graph is a straight line.

Linear Equation in Two Variables

Any equation of the form $y = mx + b$, where m and b are constants, is a **linear equation in two variables.** The graph of a linear equation in two variables is a straight line.

To graph a linear equation, choose some values of x and then find the corresponding values of y. Because a straight line is determined by two points, it is sufficient to find only two ordered-pair solutions. However, it is recommended that at least three ordered-pair solutions be used to ensure accuracy.

Graph $y = -\frac{3}{2}x + 2$.

This is a linear equation with $m = -\frac{3}{2}$ and $b = 2$. Find at least three solutions. Because m is a fraction, choose values of x that will simplify the calculations. We have chosen -2, 0, and 4 for x. (Any values of x could have been selected.)

Take Note

If the three points you graph do not lie on a straight line, you have made an arithmetic error in calculating a point or you have plotted a point incorrectly.

x	$y = -\frac{3}{2}x + 2$	y	(x, y)
-2	$-\frac{3}{2}(-2) + 2$	5	$(-2, 5)$
0	$-\frac{3}{2}(0) + 2$	2	$(0, 2)$
4	$-\frac{3}{2}(4) + 2$	-4	$(4, -4)$

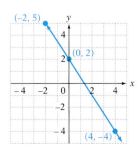

Copyright © Houghton Mifflin Company. All rights reserved.

Remember that a graph is a drawing of the ordered-pair solutions of the equation. Therefore, every point on the graph is a solution of the equation and every solution of the equation is a point on the graph.

The graph at the right is the graph of $y = x + 2$. Note that $(-4, -2)$ and $(1, 3)$ are points on the graph and that these points are solutions of $y = x + 2$. The point whose coordinates are $(4, 1)$ is not a point on the graph and is not a solution of the equation.

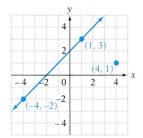

---

③ Example Graph $y = 3x - 2$.

Solution

x	y
0	-2
-1	-5
2	4

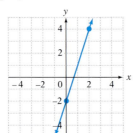

③ You Try It Graph $y = 3x + 1$.

Your Solution

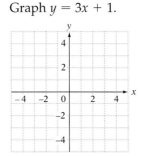

---

④ Example For the graph shown below, what is the y value when $x = 1$?

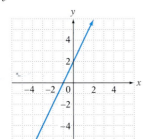

④ You Try It For the graph shown below, what is the x value when $y = 5$?

Your Solution

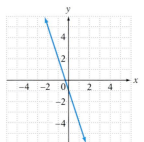

Solution Locate 1 on the x-axis. Follow the vertical line from 1 to a point on the graph. Follow a horizontal line from that point to the y-axis. Read the number where that line intersects the y-axis.

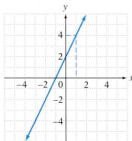

The y value is 4 when $x = 1$.

Solutions on p. S18

Copyright © Houghton Mifflin Company. All rights reserved.

6.6 EXERCISES

OBJECTIVE **A**

For Exercises 1–6, is the equation a linear equation in two variables?

1. $y = -x^2 - 3x + 4$

2. $y = \dfrac{1}{z + 5}$

3. $y = 6x - 3$

4. $y = -7x - 1$

5. $x = 2x - 8$

6. $y = -\dfrac{2}{3}y + 1$

7. Is $(3, 4)$ a solution of $y = -x + 7$?

8. Is $(2, -3)$ a solution of $y = x + 5$?

9. Is $(-1, 2)$ a solution of $y = \dfrac{1}{2}x - 1$?

10. Is $(1, -3)$ a solution of $y = -2x - 1$?

11. Is $(4, 1)$ a solution of $y = \dfrac{1}{4}x + 1$?

12. Is $(-5, 3)$ a solution of $y = -\dfrac{2}{5}x + 1$?

13. Is $(0, 4)$ a solution of $y = \dfrac{3}{4}x + 4$?

14. Is $(-2, 0)$ a solution of $y = -\dfrac{1}{2}x - 1$?

15. Is $(0, 0)$ a solution of $y = 3x + 2$?

16. Is $(0, 0)$ a solution of $y = -\dfrac{3}{4}x$?

17. Find the ordered-pair solution of $y = 3x - 2$ corresponding to $x = 3$.

18. Find the ordered-pair solution of $y = 4x + 1$ corresponding to $x = -1$.

19. Find the ordered-pair solution of $y = \dfrac{2}{3}x - 1$ corresponding to $x = 6$.

20. Find the ordered-pair solution of $y = \dfrac{3}{4}x - 2$ corresponding to $x = 4$.

21. Find the ordered-pair solution of $y = -3x + 1$ corresponding to $x = 0$.

22. Find the ordered-pair solution of $y = \dfrac{2}{5}x - 5$ corresponding to $x = 0$.

23. Find the ordered-pair solution of $y = \dfrac{2}{5}x + 2$ corresponding to $x = -5$.

24. Find the ordered-pair solution of $y = -\dfrac{1}{6}x - 2$ corresponding to $x = 12$.

Copyright © Houghton Mifflin Company. All rights reserved.

Copyright © Houghton Mifflin Company. All rights reserved.

OBJECTIVE **B**

For Exercises 25–28, is the graph of the equation a straight line?

25. $y = -\dfrac{1}{2}x + 5$

26. $y = \dfrac{1}{x} + 5$

27. $y = 2x^2 + 5$

28. $y = -2x - 5$

For Exercises 29–58, graph the equation.

29. $y = 2x - 4$

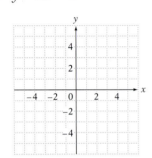

30. $y = x - 1$

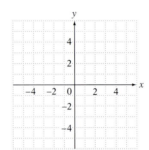

31. $y = -x + 2$

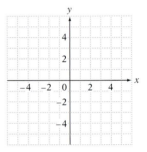

32. $y = x + 3$

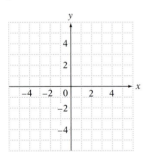

33. $y = x - 3$

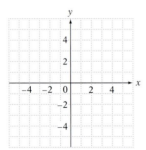

34. $y = -2x + 1$

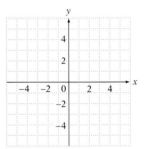

35. $y = -2x + 3$

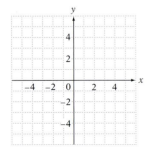

36. $y = -4x + 1$

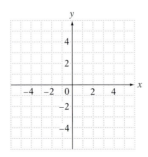

37. $y = -3x + 4$

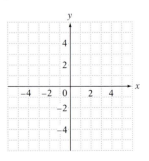

38. $y = 4x - 5$

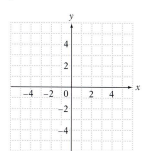

39. $y = 2x - 1$

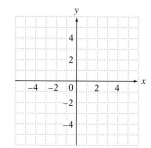

40. $y = 2x$

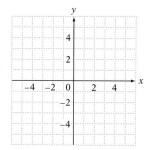

41. $y = 3x$

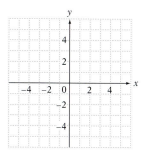

42. $y = \dfrac{3}{2}x$

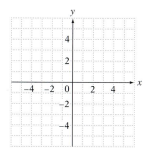

43. $y = \dfrac{1}{3}x$

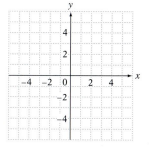

44. $y = -\dfrac{5}{2}x$

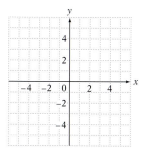

45. $y = -\dfrac{4}{3}x$

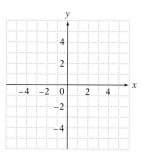

46. $y = \dfrac{2}{3}x + 1$

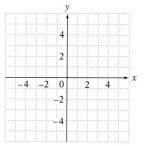

47. $y = \dfrac{3}{2}x - 1$

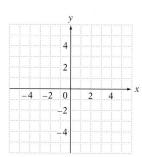

48. $y = \dfrac{1}{4}x + 2$

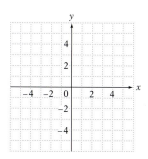

49. $y = \dfrac{2}{5}x - 1$

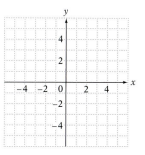

Copyright © Houghton Mifflin Company. All rights reserved.

50. $y = -\dfrac{1}{2}x + 3$

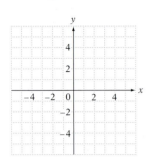

51. $y = -\dfrac{2}{3}x + 1$

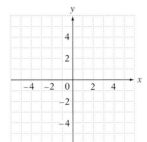

52. $y = -\dfrac{3}{4}x - 3$

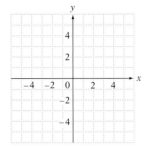

53. $y = -\dfrac{5}{3}x - 2$

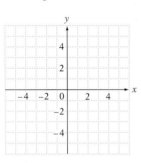

54. $y = \dfrac{1}{2}x - 1$

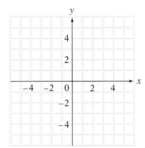

55. $y = \dfrac{5}{2}x - 1$

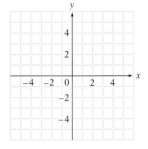

56. $y = -\dfrac{1}{4}x + 1$

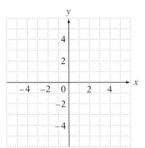

57. $y = x$

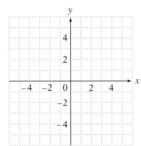

58. $y = -x$

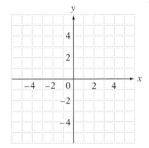

59. For the graph shown below, what is the y value when $x = 3$?

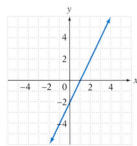

60. For the graph shown below, what is the y value when $x = -2$?

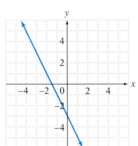

61. For the graph shown below, what is the y value when $x = 4$?

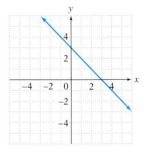

Copyright © Houghton Mifflin Company. All rights reserved.

62. For the graph shown below, what is the y value when $x = 2$?

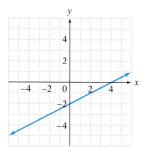

63. For the graph shown below, what is the y value when $x = 3$?

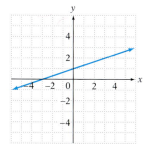

64. For the graph shown below, what is the y value when $x = -1$?

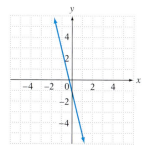

65. For the graph shown below, what is the x value when $y = -2$?

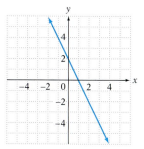

66. For the graph shown below, what is the x value when $y = -5$?

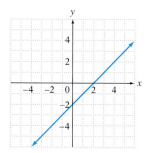

67. For the graph shown below, what is the x value when $y = -2$?

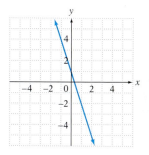

68. For the graph shown below, what is the x value when $y = 2$?

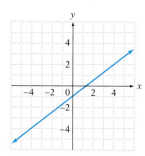

69. For the graph shown below, what is the x value when $y = -1$?

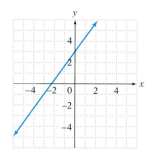

70. For the graph shown below, what is the x value when $y = 3$?

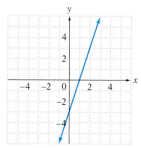

Critical Thinking

71. **a.** What are the coordinates of the point at which the graph of $y = 2x + 1$ crosses the y-axis?

b. What are the coordinates of the point at which the graph of $y = 3x - 6$ crosses the x-axis?

Copyright © Houghton Mifflin Company. All rights reserved.

72. Draw a line through all points with a first co-ordinate of 2.

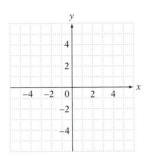

73. Draw a line through all points with a first co-ordinate of −4.

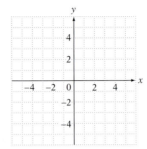

74. Draw a line through all points with a second coordinate of 3.

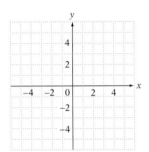

75. Draw a line through all points with a second cordinate of −5.

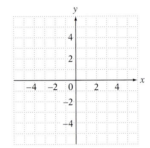

76. Select the correct word and fill in the blank.
  a. If $y = 3x - 4$ and the value of x changes from 3 to 4, then the value of y increases/decreases by __?__ .
  b. If $y = -2x + 1$ and the value of x changes from 3 to 4, then the value of y increases/decreases by __?__ .

77. Suppose you are checking whether an ordered pair is a solution of an equation and the result is $4 = -1$. What does this mean?

78. Graph $y = 2x + b$ for $b = -1, 0, 1,$ and 2. From the graphs, what observations can you make about the graphs as the value of b changes?

79. Graph $y = mx + 1$ for $m = \frac{1}{2}, 1, \frac{3}{2}, 2,$ and $\frac{5}{2}$. From the graphs, what observations can you make about the graphs as the value of m increases?

80. Graph $y = mx + 1$ for $m = -2, -1, 1,$ and 2. From the graphs, what observations can you make about the graphs of the lines when m is negative and when m is positive?

Copyright © Houghton Mifflin Company. All rights reserved.

Focus on **Problem Solving**

Making a Table

Sometimes a table can be used to organize information so that it is in a useful form. For example, in this chapter, we used tables to organize ordered-pair solutions of equations.

A basketball player scored 11 points in a game. The player can score 1 point for making a free throw, 2 points for making a field goal from within the three-point line, and 3 points for making a field goal from outside the three-point line. Find the number of possible combinations in which the player could have scored 11 points.

The following table lists the possible combinations for scoring 11 points.

	Points															
Free throws	0	2	1	3	5	0	2	4	6	8	1	3	5	7	9	11
2-point field goal	1	0	2	1	0	4	3	2	1	0	5	4	3	2	1	0
3-point field goal	3	3	2	2	2	1	1	1	1	1	0	0	0	0	0	0
Total points	11	11	11	11	11	11	11	11	11	11	11	11	11	11	11	11

There are 16 possible ways in which the basketball player could have scored 11 points.

1. A football team scores 17 points. A touchdown counts as 6 points, an extra point as 1 point, a field goal as 3 points, and a safety as 2 points. Find the number of possible combinations in which the team can score 17 points. Remember that the number of extra points cannot exceed the number of touchdowns scored.

2. Repeat Exercise 1. Assume that no safety was scored.

3. Repeat Exercise 1. Assume that no safety was scored and that the team scored two field goals.

4. Find the number of possible combinations of nickels, dimes, and quarters that one might get when receiving $.85 in change.

5. Repeat Exercise 4. Assume no combination contains coins that could be exchanged for a larger coin. That is, the combination of three quarters and two nickels would not be allowed because the two nickels could be exchanged for a dime.

6. Find the number of possible combinations of $1, $5, $10, and $20 bills that one might get when receiving $33.

Copyright © Houghton Mifflin Company. All rights reserved.

Projects & Group Activities

Modular Arithmetic

Sun	Mon	Tue	Wed	Thu	Fri	Sat
					1	2
3	4	5	6	7	8	9
10	11	12	13	14	15	16
17	18	19	20	21	22	23
24	25	26	27	28	29	30

For the calendar at the left, note that if each of the dates under Friday were divided by 7, the remainder would be the same. For example,

$$15 \div 7 = 2 \text{ with remainder } 1 \qquad 29 \div 7 = 4 \text{ with remainder } 1$$

Dividing each of the dates under Tuesday by 7 results in a remainder of 5.

$$19 \div 7 = 2 \text{ with remainder } 5$$

The same idea can be applied to each of the seven days of the week. Numbers that have the same remainder when divided by a given number n are said to be **congruent modulo n.** For example, 5, 12, 19, and 26 are congruent modulo 7.

The reason the remainders are the same is that there are 7 days in one week. For the given calendar, Tuesday is on the 5th, 12th (5 + 7), 19th (12 + 7), and 26th (19 + 7).

The notation $a \equiv b \pmod{n}$ is used to denote that a and b have the same remainder when divided by n. For example, $19 \equiv 26 \pmod 7$ because 19 and 26 have the same remainder when divided by 7. The remainder is 5.

For each of the problems below, mark the statement true or false.

1. $34 \equiv 9 \pmod 5$ **2.** $78 \equiv 23 \pmod 9$

3. $16 \equiv 52 \pmod{12}$ **4.** $20 \equiv 0 \pmod{10}$

There are many applications of congruence. The Universal Product Code (UPC) that is used by grocery stores is one application. The UPC identification number consists of 12 digits.

To be a valid UPC number, the following modular equation must be true:

$$13a_1 + a_2 + 13a_3 + a_4 + 13a_5 + a_6 + 13a_7 + a_8 + 13a_9 + a_{10} + 13a_{11} + a_{12} \equiv 0 \pmod{10}$$

Each a in this equation is one of the numbers of the UPC identification number. The first 11 numbers identify the country in which the product was made, the manufacturer, and type of product. The twelfth digit, a_{12}, is called the **check digit** and is chosen so that the equation is true.

ISBN 0−395−75524−7

0 4644277018 7

7 is the check digit

For example, the first 11 numbers of the UPC shown at the left identify an edition of *The Information Please Almanac* published by Houghton Mifflin Company. Substituting the first 11 numbers into the equation gives

$$13(0) + 4 + 13(6) + 4 + 13(4) + 2 + 13(7) + 7 + 13(0) + 1 + 13(8) + a_{12} \equiv 0 \pmod{10}$$
$$0 + 4 + 78 + 4 + 52 + 2 + 91 + 7 + 0 + 1 + 104 + a_{12} \equiv 0 \pmod{10}$$
$$343 + a_{12} \equiv 0 \pmod{10}$$

To have a valid UPC number, a_{12} is chosen so that the result is congruent to 0 (mod 10). For $343 + a_{12} \equiv 0 \pmod{10}$, $343 + a_{12}$ must be divisible by 10. The single digit that can be added to 343 so that the sum is divisible by 10 is 7. Therefore, $a_{12} = 7$, which is the check digit shown in the UPC number.

Copyright © Houghton Mifflin Company. All rights reserved.

If a bookstore ordered this book and incorrectly wrote the number 046443770187, a computer processing the order would be able to determine that there had been a mistake because the number is not congruent to 0 (mod 10) and therefore does not belong to any product.

Another number shown above the bar coding is 0-395-75524-7, which is the International Standard Book Number (ISBN). The first number identifies the book as being published in an English-speaking country. The next group of numbers is the specific publisher, the next group of five digits identifies the particular book, and the last digit is the check digit. In this case, a certain sum must be congruent to 0 (mod 11). The formula for an ISBN is

$$10a_1 + 9a_2 + 8a_3 + 7a_4 + 6a_5 + 5a_6 + 4a_7 + 3a_8 + 2a_9 + a_{10} \equiv 0 \ (\text{mod } 11)$$

5. Use this formula to verify the ISBN for *The Information Please Almanac*.

6. Verify the UPC identification number and the ISBN number on this textbook.

Chapter Summary

Key Words

An **equation** expresses the equality of two mathematical expressions. [6.1A, p. 377]	$5x + 6 = 7x - 3$ $y = 4x - 10$ $3a^2 - 6a + 4 = 0$
In a **first-degree equation in one variable,** the equation has only one variable, and each instance of the variable is the first power (the exponent on the variable is 1). [6.1A, p. 377]	$3x - 8 = 4$ $z = -11$ $6(x + 7) = 2 - (x - 9)$
A **solution** of an equation is a number that, when substituted for the variable, results in a true equation. [6.1A, p. 377]	6 is a solution of $x - 4 = 2$ because $6 - 4 = 2$ is a true equation.
To **solve** an equation means to find a solution of the equation. The goal is to rewrite the equation in the form **variable = constant.** [6.1A, p. 377]	$x = 5$ is in the form *variable = constant*. The solution of the equation $x = 5$ is the constant 5 because $5 = 5$ is a true equation.
Some of the words and phrases that translate to "equals" are **equals, is, is the same as, amounts to, totals,** and **represents.** [6.4A, p. 399]	"Eight plus a number is ten" translates to $8 + x = 10$.
A **rectangular coordinate system** is formed by two number lines, one horizontal and one vertical, that intersect at the zero point of each line. The point of intersection is called the **origin.** The number lines that make up a rectangular coordinate system are called **coordinate axes.** A rectangular coordinate system divides the plane into four regions called **quadrants.** [6.5A, p. 407]	

Examples

Copyright © Houghton Mifflin Company. All rights reserved.

An **ordered pair** (x, y) is used to locate a point in a rectangular coordinate system. The first number of the pair is the **abscissa**. The second number is the **ordinate**. The **coordinates** of the point are the numbers in the ordered pair associated with the point. To **graph**, or **plot**, a point in the plane, place a dot at the location given by the ordered pair. The **graph of an ordered pair** is the dot drawn at the coordinates of the point in the plane. [6.5A, p. 407]

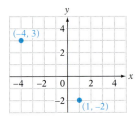

A **scatter diagram** is a graph of the ordered pairs of data. [6.5B, p. 409]

The distance, in miles, a house is from a fire station and the amount, in thousands of dollars, of fire damage that the house sustained in a fire are given in the scatter diagram.

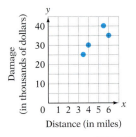

An equation of the form $y = mx + b$, where m and b are constants, is a **linear equation in two variables**. A solution of a linear equation in two variables is an ordered pair (x, y) whose coordinates make the equation a true statement. The **graph of an equation in two variables** is a graph of the ordered-pair solutions of the equation. The graph of a linear equation in two variables is a straight line. [6.6A, p. 415; 6.6B, p. 417]

$y = 3x + 2$ is a linear equation in two variables; $m = 3$ and $b = 2$. Ordered-pair solutions of $y = 3x + 2$ are shown below, along with the graph of the equation.

x	y
1	5
0	2
−1	−1

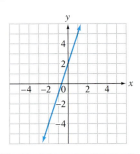

Essential Rules and Procedures

Addition Property of Equations [6.1A, p. 378]
The same number or variable expression can be added to each side of an equation without changing the solution of the equation.

$$x + 7 = 20$$
$$x + 7 + (-7) = 20 + (-7)$$
$$x = 13$$

Multiplication Property of Equations [6.1B, p. 381]
Each side of an equation can be multiplied by the same nonzero number without changing the solution of the equation.

$$\frac{3}{4}x = 24$$
$$\frac{4}{3} \cdot \frac{3}{4}x = \frac{4}{3} \cdot 24$$
$$x = 32$$

Steps in Solving General First-Degree Equations [6.3B, p. 393]
1. Use the Distributive Property to remove parentheses.
2. Combine like terms on each side of the equation.
3. Rewrite the equation with only one variable term.
4. Rewrite the equation with only one constant term.
5. Rewrite the equation so that the coefficient of the variable is 1.

$$8 - 4(2x + 3) = 2(1 - x)$$
$$8 - 8x - 12 = 2 - 2x$$
$$-8x - 4 = 2 - 2x$$
$$-6x - 4 = 2$$
$$-6x = 6$$
$$x = -1$$

Copyright © Houghton Mifflin Company. All rights reserved.

Chapter Review Exercises

1. Solve: $z + 5 = 2$

2. Solve: $-8x + 4x = -12$

3. Solve: $7 = 8a - 5$

4. Solve: $7 + a = 0$

5. Solve: $40 = -\dfrac{5}{3}y$

6. Solve: $-\dfrac{3}{8} = \dfrac{4}{5}z$

7. Solve: $9 - 5y = -1$

8. Solve: $-4(2 - x) = x + 9$

9. Solve: $3a + 8 = 12 - 5a$

10. Solve: $12p - 7 = 5p - 21$

11. Solve: $3(2n - 3) = 2n + 3$

12. Solve: $3m = -12$

13. Solve: $4 - 3(2p + 1) = 3p + 11$

14. Solve: $1 + 4(2c - 3) = 3(3c - 5)$

15. Solve: $\dfrac{3x}{4} + 10 = 7$

16. Is $(-10, 0)$ a solution of $y = \dfrac{1}{5}x + 2$?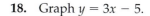

17. Graph the points whose coordinates are $(-2, 3)$, $(4, 5)$, $(0, -2)$, and $(-4, 0)$.

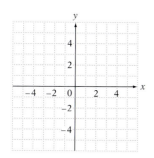

18. Graph $y = 3x - 5$.

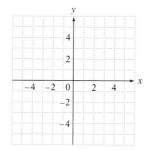

Copyright © Houghton Mifflin Company. All rights reserved.

19. Graph $y = -\frac{1}{2}x + 3$.

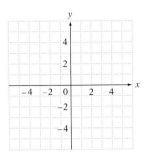

20. Find the ordered-pair solution of $y = 4x - 9$ that corresponds to $x = 2$.

21. Translate "the difference between seven and the product of five and a number is thirty-seven" into an equation and solve.

22. *Music*   A piano wire 24 in. long is cut into two pieces. Twice the length of the shorter piece is equal to the length of the longer piece. Find the length of the longer piece.

23. *Business*   The consulting fee for a security specialist was $1,300. This included $250 for supplies and $150 for each hour of consultation. Find the number of hours of consultation.

24. *Landmarks*   The height of the Eiffel Tower, including the television tower, is 322 m. This is 8 m less than six times the height of the leaning tower of Pisa. Find the height of the leaning tower of Pisa.

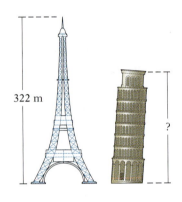

322 m

25. *Education*   The math midterm scores and the final exam scores for six students are given in the following table. Graph the scatter diagram for these data.

Midterm score, x	90	85	75	80	85	70
Final exam score, y	95	75	80	75	90	70

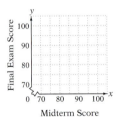

26. *Physics*   A lever is 18 ft long. A force of 25 lb is applied at a distance of 6 ft from the fulcrum. How large a force must be applied to the other end of the lever so that the system will balance? Use the lever system equation $F_1 x = F_2(d - x)$.

27. *Business*   A business analyst has determined that the cost per unit for an electric guitar amplifier is $127 and that the fixed costs per month are $20,000. Find the number of amplifiers produced during a month in which the total cost was $38,669. Use the equation $T = U \cdot N + F$, where T is the total cost, U is the cost per unit, N is the number of units produced, and F is the fixed cost.

Copyright © Houghton Mifflin Company. All rights reserved.

Chapter Test

1. Solve: $7 + x = 2$

2. Solve: $-\dfrac{3}{5}y = 6$

3. Solve: $2d - 7 = -13$

4. Solve: $4 - 5c = -11$

5. Solve: $3x + 4 = 24 - 2x$

6. Solve: $7 - 5y = 6y - 26$

7. Solve: $2t - 3(4 - t) = t - 8$

8. Solve: $12 - 3(n - 5) = 5n - 3$

9. Solve: $\dfrac{3}{8} - n = \dfrac{2}{3}$

10. Solve: $3p - 2 + 5p = 2p + 12$

11. What are the coordinates of point A?

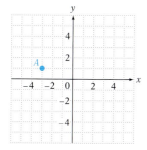

12. Graph the ordered pairs $(-1, 2)$, $(2, -4)$, and $(0, 1)$.

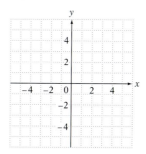

13. Graph $y = -x + 3$.

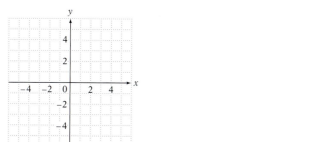

14. Graph $y = 2x - 3$.

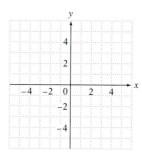

Copyright © Houghton Mifflin Company. All rights reserved.

15. Graph $y = -\dfrac{2}{3}x$.

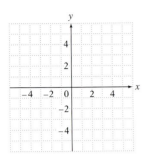

16. Graph $y = \dfrac{1}{4}x + 1$.

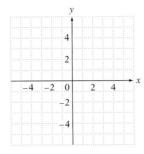

17. Solve: $2(4b - 14) = b - 7$

18. Solve: $\dfrac{5y}{3} + 12 = 2$

19. Find the ordered-pair solution of $y = \dfrac{1}{3}x - 4$ corresponding to $x = 6$.

20. Translate "four plus one third of a number is nine" into an equation and solve.

21. Translate "the sum of eight and the product of two and a number is negative four" into an equation and solve.

22. The sum of two numbers is 17. The total of four times the smaller number and two times the larger number is 44. Find the two numbers.

23. *Physics* A physics student recorded the speed of a steel ball as it rolled down a ramp. The results are recorded in the table below. Graph a scatter diagram for these data.

Time (in seconds), x	0	1	2	3
Speed (in feet per second), y	0	3	6	9

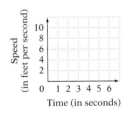

24. *Consumerism* An auto repair bill was $455. This included $165 for parts and $58 an hour for labor. How many hours of labor did the job require?

25. *Oceanography* The pressure P, in pounds per square inch, at a certain depth in the ocean can be approximated by the equation $P = 15 + \dfrac{1}{2}D$, where D is the depth in feet. Use this formula to find the depth when the pressure is 65 pounds per square inch.

Copyright © Houghton Mifflin Company. All rights reserved.

Cumulative Review Exercises

1. Evaluate $-3ab$ when $a = -2$ and $b = 3$.

2. Simplify: $-3(4p - 7)$

3. Simplify: $\left(\dfrac{2}{3}\right)\left(-\dfrac{9}{8}\right) + \dfrac{3}{4}$

4. Solve: $-\dfrac{2}{3}y = 12$

5. Evaluate $(-b)^3$ when $b = -2$.

6. Evaluate $4xy^2 - 2xy$ when $x = -2$ and $y = 3$.

7. Simplify: $\sqrt{121}$

8. Simplify: $\sqrt{48}$

9. Simplify: $4(3v - 2) - 5(2v - 3)$

10. Simplify: $-4(-3m)$

11. Is -9 a solution of the equation $-5d = -45$?

12. Solve: $5 - 7a = 3 - 5a$

13. Simplify: $6 - 2(7z - 3) + 4z$

14. Evaluate $\dfrac{a^2 + b^2}{2ab}$ when $a = -2$ and $b = -1$.

15. Solve: $8z - 9 = 3$

16. Simplify: $(2m^2n^5)^5$

17. Multiply: $-3a^3(2a^2 + 3ab - 4b^2)$

18. Multiply: $(2x - 3)(3x + 1)$

19. Simplify: 2^{-4}

20. Simplify: $\dfrac{x^8}{x^2}$

Copyright © Houghton Mifflin Company. All rights reserved.

21. Simplify: $(-5x^3y)(-3x^5y^2)$

22. Solve: $5 - 3(2x - 8) = -2(1 - x)$

23. Graph $y = \dfrac{5}{3}x + 1$.

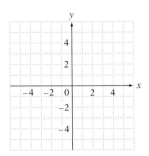

24. Graph $y = -\dfrac{2}{5}x$.

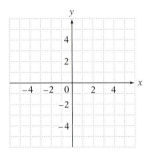

25. Write 3.5×10^{-8} in decimal notation.

26. Translate "the product of five and the sum of a number and two" into a variable expression. Then simplify the variable expression.

27. *Physics* Find the time it takes a falling object to increase its speed from 50 ft/s to 98 ft/s. Use the equation $v = v_0 + 32t$, where v is the final velocity, v_0 is the initial velocity, and t is the time it takes for the object to fall.

28. *Zoology* The number of dogs in the world is 1,000 times the number of wolves in the world. Express the number of dogs in the world in terms of the number of wolves in the world.

29. *The Film Industry* The figure at the right shows the top-grossing movies in the United States in the 1970s. Find the total box office gross for these four films.

30. *Finances* A homeowner's mortgage payment for one month for principal and interest was $949. The principal payment was $204 less than the interest payment. Find the amount of the interest payment.

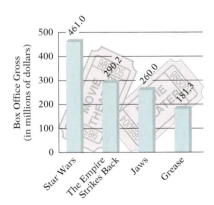

Top-Grossing Movies in the 1970s
Source: **www.worldwideboxoffice.com**

31. *Geography* The Aleutian Trench in the Pacific Ocean is 8,100 m deep. Each story of an average skyscraper is 4.2 m tall. How many stories, to the nearest whole number, would a skyscraper as tall as the Aleutian Trench have?

32. *Charities* A donation of $12,000 is given to two charities. One charity received twice the amount of the other charity. How much did each charity receive?

Copyright © Houghton Mifflin Company. All rights reserved.

Copyright © Houghton Mifflin Company. All rights reserved.

Measurement and Proportion

The New York Yankees Orlando Hernandez throws a pitch to one of the Arizona Diamondbacks during game four of the 2001 World Series. If the Diamondbacks score a run during this play, Hernandez' earned run average (ERA) will suffer. His ERA is the number of earned runs that have been scored for every 9 innings he has pitched. A pitcher's ERA can be calculated by setting up a proportion, as seen in the **Project on page 476.**

Need help? For online student resources, visit this website: **math.college.hmco.com**

Prep Test

1. Simplify: $\dfrac{8}{10}$

2. Write as a decimal: $\dfrac{372}{15}$

For Exercises 3–14, add, subtract, multiply, or divide.

3. $36 \times \dfrac{1}{9}$

4. $\dfrac{5}{3} \times 6$

5. $5\dfrac{3}{4} \times 8$

6. $3\overline{)714}$

7. $3.732 \times 10{,}000$

8. $41.07 \div 1{,}000$

9. $6 - 0.875$

10. $5 + 0.96$

11. 3.25×0.04

12. $35 \times \dfrac{1.61}{1}$

13. $1.67 \times \dfrac{1}{3.34}$

14. $315 \div 84$

Go Figure

Suppose you threw six darts and all six hit the target shown.
Which of the following could be your score?

4   15   58   28   29   31

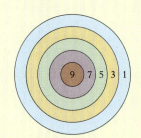

Copyright © Houghton Mifflin Company. All rights reserved.

OBJECTIVE   **A**

VIDEO & DVD   CD TUTOR   WEB   SSM

The metric system

International trade, or trade between nations, is a vital and growing segment of business in the world today. The opening of McDonald's restaurants around the globe is testimony to the expansion of international business.

The United States, as a nation, is dependent on world trade. And world trade is dependent on internationally standardized units of measurement: the metric system. In this section we will present the metric system of measurement and explain how to convert between different units.

The basic unit of *length,* or distance, in the metric system is the **meter** (m). One meter is approximately the distance from a doorknob to the floor. All units of length in the metric system are derived from the meter. Prefixes to the basic unit denote the length of each unit. For example, the prefix "centi-" means one-hundredth; therefore, one centimeter is 1 one-hundredth of a meter (0.01 m).

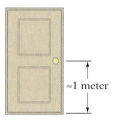
≈1 meter

kilo-	= 1 000	1 kilometer (km)	= 1 000 meters (m)
hecto-	= 100	1 hectometer (hm)	= 100 m
deca-	= 10	1 decameter (dam)	= 10 m
		1 meter (m)	= 1 m
deci-	= 0.1	1 decimeter (dm)	= 0.1 m
centi-	= 0.01	1 centimeter (cm)	= 0.01 m
milli-	= 0.001	1 millimeter (mm)	= 0.001 m

Note that in this list 1,000 is written as 1 000, with a space between the 1 and the zeros. When writing numbers using metric units, separate each group of three numbers by a space instead of a comma. A space is also used after each group of three numbers to the right of a decimal point. For example, 31,245.2976 is written 31 245.297 6 in metric notation.

> ▶ **Point of Interest**
>
> Originally the meter (spelled metre in some countries) was defined as $\frac{1}{10{,}000{,}000}$ of the distance from the equator to the north pole. Modern scientists have redefined the meter as 1,650,753.73 wavelengths of the orange-red light given off by the element krypton.

Mass and weight are closely related. *Weight* is a measure of how strongly gravity is pulling on an object. Therefore, an object's weight is less in space than on Earth's surface. However, the amount of material in the object, its *mass,* remains the same. On the surface of Earth, the terms *mass* and *weight* can be used interchangeably.

The basic unit of mass in the metric system is the **gram** (g). If a box that is 1 centimeter long on each side is filled with water, the mass of that water is 1 gram.

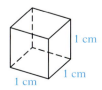
1 cm
1 cm
1 cm

1 gram = the mass of water in a box that is
1 centimeter long on each side

Copyright © Houghton Mifflin Company. All rights reserved.

The units of mass in the metric system have the same prefixes as the units of length.

$$
\begin{aligned}
1 \text{ kilogram (kg)} &= 1\,000 \text{ grams (g)}\\
1 \text{ hectogram (hg)} &= 100 \text{ g}\\
1 \text{ decagram (dag)} &= 10 \text{ g}\\
1 \text{ gram (g)} &= 1 \text{ g}\\
1 \text{ decigram (dg)} &= 0.1 \text{ g}\\
1 \text{ centigram (cg)} &= 0.01 \text{ g}\\
1 \text{ milligram (mg)} &= 0.001 \text{ g}
\end{aligned}
$$

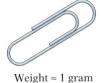

Weight ≈ 1 gram

The gram is a very small unit of mass. A paperclip weighs about one gram. In applications, the kilogram (1 000 grams) is a more useful unit of mass. This textbook weighs about 1 kilogram.

Liquid substances are measured in units of *capacity*.

The basic unit of capacity in the metric system is the **liter** (L). One liter is defined as the capacity of a box that is 10 centimeters long on each side.

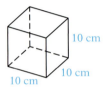

10 cm
10 cm
10 cm

1 liter = the capacity of a box that is 10 centimeters long on each side

The units of capacity in the metric system have the same prefixes as the units of length.

$$
\begin{aligned}
1 \text{ kiloliter (kl)} &= 1\,000 \text{ liters (L)}\\
1 \text{ hectoliter (hl)} &= 100 \text{ L}\\
1 \text{ decaliter (dal)} &= 10 \text{ L}\\
1 \text{ liter (L)} &= 1 \text{ L}\\
1 \text{ deciliter (dl)} &= 0.1 \text{ L}\\
1 \text{ centiliter (cl)} &= 0.01 \text{ L}\\
1 \text{ milliliter (ml)} &= 0.001 \text{ L}
\end{aligned}
$$

Converting between units in the metric system involves moving the decimal point to the right or to the left. Listing the units in order from largest to smallest will indicate how many places to move the decimal point and in which direction.

To convert 3 800 cm to meters, write the units of length in order from largest to smallest.

km   hm   dam   m   dm   cm   mm

2 positions

3 800 cm = 38.00 m

2 places

Converting from cm to m requires moving 2 places to the left.

Move the decimal point the same number of places and in the same direction.

Copyright © Houghton Mifflin Company. All rights reserved.

Convert 2.1 kg to grams.

kg hg dag g dg cg mg

3 positions

Write the units of mass in order from largest to smallest.

Converting from kg to g requires moving 3 positions to the right.

2.1 kg = 2 100 g

3 places

Move the decimal point the same number of places and in the same direction.

Take Note

In the metric system, all prefixes represent powers of 10. Therefore, when converting between units, we are multiplying or dividing by a power of 10.

---

1 Example

What unit in the metric system is used to measure the distance from San Francisco to Dallas?

Solution

The meter is the basic unit for measuring distance.

The distance from San Francisco to Dallas is measured in kilometers.

1 You Try It

What unit in the metric system is used to measure the amount of protein in a glass of milk?

Your Solution

---

2 Example

a. Convert 4.08 m to centimeters.
b. Convert 5.93 g to milligrams.
c. Convert 824 ml to liters.
d. Convert 9 kl to liters.

Solution

a. km hm dam (m) dm (cm) mm

Move the decimal point 2 places to the right.

4.08 m = 408 cm

b. kg hg dag (g) dg cg (mg)

Move the decimal point 3 places to the right.

5.93 g = 5 930 mg

c. kl hl dal (L) dl cl (ml)

Move the decimal point 3 places to the left.

824 ml = 0.824 L

d. (kl) hl dal (L) dl cl ml

Move the decimal point 3 places to the right.

9 kl = 9 000 L

2 You Try It

a. Convert 1 295 m to kilometers.
b. Convert 7 543 g to kilograms.
c. Convert 6.3 L to milliliters.
d. Convert 2 kl to liters.

Your Solution

Solutions on p. S18

Copyright © Houghton Mifflin Company. All rights reserved.

3 *Example*

The thickness of a single sheet of paper is 0.07 mm. Find the height in centimeters of a ream of paper. A ream is 500 sheets of paper.

Strategy

To find the height:

▶ Multiply the height of each sheet (0.07 mm) by the number of sheets in a ream (500). This will be the height in millimeters.
▶ Convert millimeters to centimeters.

Solution

$0.07(500) = 35$

$35 \text{ mm} = 3.5 \text{ cm}$

The height of a ream of paper is 3.5 cm.

3 *You Try It*

One egg contains 274 mg of cholesterol. How many grams of cholesterol are in one dozen eggs?

Your Strategy

Your Solution

Solution on p. S18

Other prefixes in the metric system are becoming more commonly used as a result of technological advances in the computer industry. For example,

tera-	= 1 000 000 000 000
giga-	= 1 000 000 000
mega-	= 1 000 000
micro-	= 0.000 001
nano-	= 0.000 000 001
pico-	= 0.000 000 000 001

A **bit** is the smallest unit of code that computers can read; it is a <u>b</u>inary dig<u>it</u>, either a 0 or a 1. Usually bits are grouped into bytes of 8 bits. Each byte stands for a letter, number, or any other symbol we might use in communicating information. For example, the letter W can be represented 01010111. The amount of memory in a computer hard drive is measured in terabytes, gigabytes, and megabytes. The speed of a computer used to be measured in microseconds and then nanoseconds, but now the speeds are measured in picoseconds.

Here are a few more examples of how these prefixes are used.

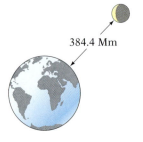

384.4 Mm

The mass of Earth gains 40 Gg (gigagrams) each year from captured meteorites and cosmic dust.

The average distance from Earth to the moon is 384.4 Mm (megameters), and the average distance from Earth to the sun is 149.5 Gm (gigameters).

The wavelength of yellow light is 590 nm (nanometers).

The diameter of a hydrogen atom is about 70 pm (picometers).

There are additional prefixes to the metric system, both larger and smaller. We may hear them more and more often as computer chips hold more and more information, as computers get faster and faster, and as we learn more and more about objects in our universe and beyond that are great distances away.

Copyright © Houghton Mifflin Company. All rights reserved.

7.1 EXERCISES

Copyright © Houghton Mifflin Company. All rights reserved.

OBJECTIVE   **A**

1. In the metric system, what is the basic unit of length? Of liquid measure? Of weight?

2.   **a.**   Explain how to convert meters to centimeters.
         b.   Explain how to convert milliliters to liters.

3. **a.** Complete the table.

Metric System Prefix	Symbol	Magnitude	Means Multiply the Basic Unit By:
tera-	T	10^{12}	1 000 000 000 000
giga-	G	_____	1 000 000 000
mega-	M	10^{6}	_____
kilo-	_____	_____	1 000
hecto-	h	_____	100
deca-	da	10^{1}	_____
deci-	d	$\frac{1}{10}$	_____
centi-	_____	$\frac{1}{10^{2}}$	_____
milli-	_____	_____	0.001
micro-	μ	$\frac{1}{10^{6}}$	_____
nano-	n	$\frac{1}{10^{9}}$	_____
pico-	p	_____	0.000 000 000 001

    b. How can the magnitude column in the table above be used to determine how many places to move the decimal point when converting to the basic unit in the metric system?

Name the unit in the metric system that would be used to measure each of the following.

4. the distance from New York to London

5. the weight of a truck

6. a person's waist

7. the amount of coffee in a mug

8. the weight of a thumbtack

9. the amount of water in a swimming pool

10. the distance a baseball player hits a baseball

11. a person's hat size

12. the amount of fat in a slice of cheddar cheese

13. a person's weight

14. the maple syrup served with pancakes

15. the amount of water in a watercooler

16. the amount of vitamin C in a vitamin tablet

17. a serving of cereal

18. the width of a hair

19. a person's height

20. the amount of medication in an aspirin

21. the weight of a lawn mower

22. the weight of a slice of bread

23. the contents of a bottle of salad dressing

24. the amount of water a family uses monthly

25. the newspapers collected at a recycling center

26. the amount of liquid in a bowl of soup

27. the distance to the bank

Convert.

28. 42 cm = _____ mm

29. 91 cm = _____ mm

30. 360 g = _____ kg

31. 1 856 g = _____ kg

32. 5 194 ml = _____ L

33. 7 285 ml = _____ L

34. 2 m = _____ mm

35. 8 m = _____ mm

36. 217 mg = _____ g

37. 34 mg = _____ g

38. 4.52 L = _____ ml

39. 0.029 7 L = _____ ml

40. 8 406 m = _____ km

41. 7 530 m = _____ km

42. 2.4 kg = _____ g

43. 9.2 kg = _____ g

44. 6.18 kl = _____ L

45. 0.036 kl = _____ L

46. 9.612 km = _____ m

47. 2.35 km = _____ m

48. 0.24 g = _____ mg

49. 0.083 g = _____ mg

50. 298 cm = _____ m

51. 71.6 cm = _____ m

Copyright © Houghton Mifflin Company. All rights reserved.

52. 2 431 L = _____ kl

53. 6 302 L = _____ kl

54. 0.66 m = _____ cm

55. 4.58 m = _____ cm

56. 243 mm = _____ cm

57. 92 mm = _____ cm

 Solve.

58. *The Olympics* **a.** One of the events in the summer Olympics is the 50 000-meter walk. How many kilometers do the entrants in this event walk? **b.** One of the events in the winter Olympic Games is the 10 000-meter speed skating. How many kilometers do the entrants in this event skate?

59. *Gemology* A carat is a unit of weight equal to 200 mg. Find the weight in grams of a 10-carat precious stone.

60. *Crafts* How many pieces of material, each 75 cm long, can be cut from a bolt of fabric that is 6 m long?

61. *Fundraising* A walkathon had two checkpoints. One checkpoint was 1 400 m from the starting point. The second checkpoint was 1 200 m from the first checkpoint. The second checkpoint was 1 800 m from the finish line. How long was the walk? Express the answer in kilometers.

62. *Consumerism* How many 240-milliliter servings are in a 2-liter bottle of cola? Round to the nearest whole number.

63. *Business* An athletic club uses 800 ml of chlorine each day for its swimming pool. How many liters of chlorine are used in a month of 30 days?

64. *Carpentry* Each of the four shelves in a bookcase measures 175 cm. Find the cost of the shelves when the price of lumber is $15.75 per meter.

65. *Consumerism* Find the cost of three packages of ground meat weighing 540 g, 670 g, and 890 g if the price per kilogram is $9.89. Round to the nearest cent.

66. *Agriculture* How many kilograms of fertilizer are necessary to fertilize 400 trees in an apple orchard if 300 g of fertilizer are used for each tree?

67. *Consumerism* The printed label from a container of milk is shown at the right. To the nearest whole number, how many 230-milliliter servings are in the container?

68. *Health* A patient is advised to supplement her diet with 2 g of calcium per day. The calcium tablets she purchases contain 500 mg of calcium per tablet. How many tablets per day should the patient take?

Apolo Anton Ohau and Fabio Carto

Dairy Hill

Skim Milk
Vitamin A & D Added
Pasteurized • Homogenized

INGREDIENTS: PASTEURIZED SKIM MILK, NONFAT MILK SOLIDS, VITAMIN A PALMITATE AND VITAMIN D3 ADDED.

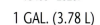

0 15400 20209 1

1 GAL. (3.78 L)

Copyright © Houghton Mifflin Company. All rights reserved.

69. *Consumerism* A 1.19-kilogram container of Quaker Oats contains 30 servings. Find the number of grams in one serving of the oatmeal. Round to the nearest gram.

70. *Mechanical Drawing* Find the missing dimension, in centimeters, in the diagram at the right.

71. *Chemistry* A laboratory assistant is in charge of ordering acid for three chemistry classes of 30 students each. Each student requires 80 ml of acid. How many liters of acid should be ordered? (The assistant must order by the whole liter.)

72. *Consumerism* A case of 12 one-liter bottles of apple juice costs $19.80. A case of 24 cans, each can containing 340 ml of apple juice, costs $14.50. Which case of apple juice costs less per milliliter?

73. *Consumerism* The nutrition label for a corn bread mix is shown at the right. **a.** How many kilograms of mix are in the package? **b.** How many grams of sodium are contained in two servings of the corn bread?

74. *Business* For $199, a cosmetician buys 5 L of moisturizer and repackages it in 125-milliliter jars. Each jar costs the cosmetician $.70. Each jar of moisturizer is sold for $11.90. Find the profit on the 5 L of moisturizer.

75. *Business* For $190, a pharmacist purchases 5 L of cough syrup and repackages it in 250-milliliter bottles. Each bottle costs the pharmacist $.50. Each bottle of cough syrup is sold for $15.78. Find the profit on the 5 L of cough syrup.

76. *Business* A service station operator bought 85 kl of gasoline for $28,875. The gasoline was sold for $.489 per liter. Find the profit on the 85 kl of gasoline.

77. *Business* A wholesale distributor purchased 32 kl of cooking oil for $44,480. The wholesaler repackaged the cooking oil in 1.25-liter bottles. The bottles cost $.21 each. Each bottle of cooking oil was sold for $2.97. Find the distributor's profit on the 32 kl of cooking oil.

78. *Construction* A column assembly is being constructed in a building. The components are shown in the diagram at the right. What length column must be cut?

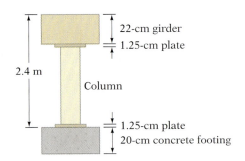

Critical Thinking

79. A 280-milliliter serving is taken from a 3-liter bottle of water. How much water remains in the container? Write the answer in two different ways.

80. ✏ Why is it necessary to have internationally standardized units of measurement?

Diagram top right: 274 mm, 4 cm, ?, 156 mm

Nutrition Facts
Serving Size ⅙ pkg. (31g mix)
Servings Per Container 6

Amount Per Serving	Mix	Prepared
Calories	110	160
Calories from Fat	10	50

	% Daily Value*	
Total Fat 1g	1%	9%
Saturated Fat 0g	0%	7%
Cholesterol 0mg	0%	12%
Sodium 210mg	9%	11%
Total Carbohydrate 24g	8%	8%
Sugars 6g		
Protein 2g		

Column diagram labels: 2.4 m, 22-cm girder, 1.25-cm plate, Column, 1.25-cm plate, 20-cm concrete footing

Copyright © Houghton Mifflin Company. All rights reserved.

SECTION 7.2 **Ratios and Rates**

OBJECTIVE **A**

Ratios and rates

In previous work, we have used quantities with units, such as 12 ft, 3 h, 2 ¢, and 15 acres. In these examples, the units are feet, hours, cents, and acres.

A **ratio** is the quotient or comparison of two quantities with the *same* unit. We can compare the measure of 3 ft to the measure of 8 ft by writing a quotient.

$$\frac{3 \text{ ft}}{8 \text{ ft}} = \frac{3}{8} \qquad 3 \text{ ft is } \frac{3}{8} \text{ of 8 ft.}$$

A ratio can be written in three ways:

1. As a fraction $\frac{3}{8}$

2. As two numbers separated by a colon   3:8

3. As two numbers separated by the word *to*   3 to 8

The ratio of 15 mi to 45 mi is written as

$$\frac{15 \text{ mi}}{45 \text{ mi}} = \frac{15}{45} = \frac{1}{3} \text{ or } 1:3 \text{ or } 1 \text{ to } 3$$

A ratio is in **simplest form** when the two numbers do not have a common factor. The units are not written in a ratio.

A **rate** is the comparison of two quantities with *different* units.

A catering company prepares 9 gal of coffee for every 50 people at a reception. This rate is written

$$\frac{9 \text{ gal}}{50 \text{ people}}$$

You traveled 200 mi in 6 h. The rate is written

$$\frac{200 \text{ mi}}{6 \text{ h}} = \frac{100 \text{ mi}}{3 \text{ h}}$$

A rate is in **simplest form** when the numbers have no common factors. The units are written as part of the rate.

Many rates are written as unit rates. A **unit rate** is a rate in which the number in the denominator is 1. The word *per* generally indicates a unit rate. It means "for each" or "for every." For example,

23 mi per gallon ▶ The unit rate is $\frac{23 \text{ mi}}{1 \text{ gal}}$.

65 mi per hour ▶ The unit rate is $\frac{65 \text{ mi}}{1 \text{ h}}$.

$4.78 per pound ▶ The unit rate is $\frac{\$4.78}{1 \text{ lb}}$.

> ▶ **Point of Interest**
>
> It is believed that billiards was invented in France during the reign of Louis XI (1423–1483). In the United States, the standard billiard table is 4 ft 6 in. by 9 ft. This is a ratio of 1:2. The same ratio holds for carom and snooker tables, which are 5 ft by 10 ft.

Copyright © Houghton Mifflin Company. All rights reserved.

Unit rates make comparisons easier. For example, if you travel 37 mph and I travel 43 mph, we know that I am traveling faster than you are. It is more difficult to compare speeds if we are told that you are traveling $\dfrac{111 \text{ mi}}{3 \text{ h}}$ and I am traveling $\dfrac{172 \text{ mi}}{4 \text{ h}}$

To find a unit rate, divide the number in the numerator of the rate by the number in the denominator of the rate. A unit rate is often written in decimal form.

A student received $57 for working 6 h at the bookstore. Find the wage per hour (the unit rate).

Write the rate as a fraction. $\dfrac{\$57}{6 \text{ h}}$

Divide the number in the numerator of the rate
(57) by the number in the denominator (6). $57 \div 6 = 9.5$

The unit rate is $\dfrac{\$9.50}{1 \text{ h}} = \$9.50/\text{h}$. This is read "$9.50 per hour."

---

① Example

Write the comparison of 12 to 8 as a ratio in simplest form using a fraction, using a colon, and using the word *to*.

Solution

$\dfrac{12}{8} = \dfrac{3}{2}$

$12{:}8 = 3{:}2$

$12 \text{ to } 8 = 3 \text{ to } 2$

① You Try It

Write the comparison of 12 to 20 as a ratio in simplest form using a fraction, using a colon, and using the word *to*.

Your Solution

---

② Example

Write "12 hits in 26 times at bat" as a rate in simplest form.

Solution

$\dfrac{12 \text{ hits}}{26 \text{ at-bats}} = \dfrac{6 \text{ hits}}{13 \text{ at-bats}}$

② You Try It

Write "20 bags of grass seed for 8 acres" as a rate in simplest form.

Your Solution

---

③ Example

Write "285 mi in 5 h" as a unit rate.

Solution

$\dfrac{285 \text{ mi}}{5 \text{ h}}$

$285 \div 5 = 57$

The unit rate is 57 mph.

③ You Try It

Write "$8.96 for 3.5 lb" as a unit rate.

Your Solution

Solutions on p. S18

Copyright © Houghton Mifflin Company. All rights reserved.

7.2 EXERCISES

OBJECTIVE   **A**

Write the comparison as a ratio in simplest form using a fraction, using a colon, and using the word *to*.

1. 16 in. to 24 in.

2. 8 lb to 60 lb

3. 9 h to 24 h

4. $55 to $150

5. 9 ft to 2 ft

6. 50 min to 6 min

7. 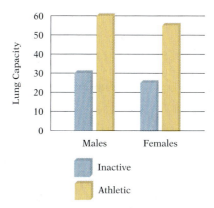 *Physical Fitness*   The figure at the right shows the lung capacity of inactive versus athletic 45-year-olds. Write the comparison of the lung capacity of an inactive male to that of an athletic male as a ratio in simplest form using a fraction, a colon, and the word *to*.

Lung Capacity (in milliliters of oxygen per kilogram of body weight per minute)

Write as a ratio in simplest form using a fraction.

8. *Construction*   The cost of building a patio cover was $1,200 for labor and $3,200 for materials. Find the ratio of the cost of materials to the cost of labor.

9. *Sports*   A baseball player had 3 errors in 42 fielding attempts. What is the ratio of the number of times the player did not make an error to the total number of attempts?

10. *Sports*   A basketball team won 18 games and lost 8 games during the season. What is the ratio of the number of games won to the total number of games?

11. *Mechanics*   Find the ratio of two meshed gears if one gear has 24 teeth and the other gear has 36 teeth.

Write as a rate in simplest form.

12. $85 for 3 shirts

13. 150 mi in 6 h

14. $76 for 8 h work

15. $6.56 for 6 candy bars

16. 252 avocado trees on 6 acres

17. 9 children in 4 families

Copyright © Houghton Mifflin Company. All rights reserved.

For Exercises 18–23, write as a unit rate.

18. $460 earned for 40 h of work

19. $38,700 earned in 12 months

20. 387.8 mi in 7 h

21. 364.8 mi on 9.5 gal of gas

22. $19.08 for 4.5 lb

23. $20.16 for 15 oz

24. *Sports* NCAA statistics show that for every 2,800 college seniors playing college basketball, only 50 will play as rookies in the National Basketball Association. Write the ratio of the number of National Basketball Association rookies to the number of college seniors playing basketball.

25. *Energy* A transformer has 40 turns in the primary coil and 480 turns in the secondary coil. State the ratio of the number of turns in the primary coil to the number of turns in the secondary coil.

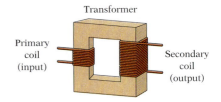

Transformer

Primary coil (input)

Secondary coil (output)

26. *Travel* An airplane flew 1,155 mi in 2.5 h. Find the rate of travel.

27. *Population Density* The table at the right shows the population and the area of three countries. Find the population density (people per square mile) for each country. Round to the nearest tenth.

Country	Population	Area (in square miles)
Australia	19,547,000	2,968,000
India	1,045,845,000	1,269,000
United States	281,422,000	3,718,000

28. *Investments* An investor purchased 100 shares of stock for $2,500. One year later the investor sold the stock for $3,200. What was the investor's profit per share?

Critical Thinking

29. *Publishing* In January 2003, *USA Today* reported that for every 10 copies of *Dr. Atkins' New Diet Revolution* sold, John Grisham's *The Summons* sold 7.3 copies. Write this as a ratio in fraction form without decimals.

30. *Social Security* According to the Social Security Administration, the number of workers per retiree is expected to be as given in the table below.

Year	2010	2020	2030	2040
Number of workers per retiree	3.1	2.5	2.1	2.0

Why is the shrinking number of workers per retiree important to the Social Security Administration?

Copyright © Houghton Mifflin Company. All rights reserved.

OBJECTIVE   **A**

The U.S. Customary System of Measurement

The standard U.S. Customary System units of length are **inch, foot, yard,** and **mile.** The abbreviations for these units of length are in., ft, yd, and mi. Equivalences between units of length in the U.S. Customary System are

$$1 \text{ ft} = 12 \text{ in.}$$
$$1 \text{ yd} = 3 \text{ ft}$$
$$1 \text{ yd} = 36 \text{ in.}$$
$$1 \text{ mi} = 5{,}280 \text{ ft}$$

Weight is a measure of how strongly Earth is pulling on an object. The U.S. Customary System units of weight are **ounce, pound,** and **ton.** The abbreviation for ounces is oz, and the abbreviation for pounds is lb. Equivalences between units of weight in the U.S. Customary System are

$$1 \text{ lb} = 16 \text{ oz}$$
$$1 \text{ ton} = 2{,}000 \text{ lb}$$

Liquids are measured in units of **capacity.** The standard U.S. Customary System units of capacity (and their abbreviations) are the **fluid ounce** (fl oz), **cup** (c), **pint** (pt), **quart** (qt), and **gallon** (gal). Equivalences between units of capacity in the U.S. Customary System are

$$1 \text{ c} = 8 \text{ fl oz}$$
$$1 \text{ pt} = 2 \text{ c}$$
$$1 \text{ qt} = 4 \text{ c}$$
$$1 \text{ gal} = 4 \text{ qt}$$

Area is a measure of the amount of surface in a region. The standard U.S. Customary System units of area are **square inch** (in^2), **square foot** (ft^2), **square yard** (yd^2), **square mile** (mi^2), and **acre.** Equivalences between units of area in the U.S. Customary System are

$$1 \text{ ft}^2 = 144 \text{ in}^2$$
$$1 \text{ yd}^2 = 9 \text{ ft}^2$$
$$1 \text{ acre} = 43{,}560 \text{ ft}^2$$
$$1 \text{ mi}^2 = 640 \text{ acres}$$

In solving application problems, scientists, engineers, and other professionals find it useful to include the units as they work through the solutions to problems so that the answers are in the proper units. Using units to organize and check the correctness of an application is called **dimensional analysis.** Applying dimensional analysis to application problems requires converting units as well as multiplying and dividing units.

▶ **Point of Interest**

The ancient Greeks devised the foot measurement which they usually divided into 16 fingers. It was the Romans who subdivided the foot into 12 units called inches. The word *inch* is derived from the Latin word *uncia* meaning a twelfth part.

The Romans also used a unit called "pace" which equaled two steps. One thousand paces equaled 1 mi. The word *mile* is derived from the Latin word *mille* which means 1,000.

The word *quart* has its root in the Latin word *quartus* which means one-fourth; a quart is one-fourth of a gallon. The same Latin word is responsible for other English words such as *quartet, quadrilateral,* and *quarter.*

Copyright © Houghton Mifflin Company. All rights reserved.

The equivalent measures listed above can be used to form conversion rates to change one unit of measurement to another. For example, the equivalent measures **1 mi** and **5,280 ft** are used to form the following conversion rates:

$$\frac{1 \text{ mi}}{5{,}280 \text{ ft}} \qquad \frac{5{,}280 \text{ ft}}{1 \text{ mi}}$$

Because **1 mi = 5,280 ft,** both of the conversion rates $\frac{1 \text{ mi}}{5{,}280 \text{ ft}}$ and $\frac{5{,}280 \text{ ft}}{1 \text{ mi}}$ are equal to 1.

To convert 3 mi to feet, multiply 3 mi by the conversion rate $\frac{5{,}280 \text{ ft}}{1 \text{ mi}}$.

$$3 \text{ mi} = 3 \text{ mi} \cdot \boxed{1} = \frac{3 \text{ mi}}{1} \cdot \boxed{\frac{5{,}280 \text{ ft}}{1 \text{ mi}}} = \frac{3 \text{ mi} \cdot 5{,}280 \text{ ft}}{1 \text{ mi}} = 3 \cdot 5{,}280 \text{ ft} = 15{,}840 \text{ ft}$$

There are three important points to notice in the above example. First, **you can think of dividing the numerator and denominator by the common unit "mile" just as you would divide the numerator and denominator of a fraction by a common factor.** Second, **the conversion rate** $\frac{5{,}280 \text{ ft}}{1 \text{ mi}}$ **is equal to 1, and multiplying an expression by 1 does not change the value of the expression.**

In the above example, we had the choice of two conversion rates, $\frac{1 \text{ mi}}{5{,}280 \text{ ft}}$ or $\frac{5{,}280 \text{ ft}}{1 \text{ mi}}$. In the conversion rate chosen, **the unit in the numerator is the same as the unit desired in the answer** (ft). **The unit in the denominator is the same as the unit in the given measurement** (mi).

Example

Convert 36 fl oz to cups.

Solution

The equivalence is 1 c = 8 fl oz.

The conversion rate must have c in the numerator and fl oz in the denominator: $\frac{1 \text{ c}}{8 \text{ fl oz}}$

$$36 \text{ fl oz} = 36 \text{ fl oz} \cdot 1 = \frac{36 \text{ fl oz}}{1} \cdot \frac{1 \text{ c}}{8 \text{ fl oz}}$$

$$= \frac{36 \text{ fl oz} \cdot 1 \text{ c}}{8 \text{ fl oz}}$$

$$= \frac{36 \text{ c}}{8} = 4\frac{1}{2} \text{ c}$$

You Try It

Convert 40 in. to feet.

Your Solution

Solution on p. S18

Copyright © Houghton Mifflin Company. All rights reserved.

2 Example

Convert $4\frac{1}{2}$ tons to pounds.

Solution

The equivalence is 1 ton = 2,000 lb.

The conversion rate must have lb in the numerator and ton in the denominator: $\frac{2{,}000\ \text{lb}}{1\ \text{ton}}$.

$$4\frac{1}{2}\ \text{tons} = 4\frac{1}{2}\ \text{tons} \cdot 1 = \frac{9}{2}\ \text{tons} \cdot \frac{2{,}000\ \text{lb}}{1\ \text{ton}}$$

$$= \frac{9\ \text{tons}}{2} \cdot \frac{2{,}000\ \text{lb}}{1\ \text{ton}}$$

$$= \frac{9\ \cancel{\text{tons}} \cdot 2{,}000\ \text{lb}}{2 \cdot 1\ \cancel{\text{ton}}}$$

$$= \frac{9 \cdot 2{,}000\ \text{lb}}{2}$$

$$= \frac{18{,}000\ \text{lb}}{2}$$

$$= 9{,}000\ \text{lb}$$

2 You Try It

Convert $2\frac{1}{2}$ yd to inches.

Your Solution

3 Example

How many seconds are in one hour?

Solution

We need to convert hours to minutes and minutes to seconds. The equivalences are
1 h = 60 min and 1 min = 60 s.

Choose the conversion rates so that we can divide by the unit "hours" and by the unit "minutes."

$$1\ \text{h} = 1\ \text{h} \cdot 1 \cdot 1 = \frac{1\ \text{h}}{1} \cdot \frac{60\ \text{min}}{1\ \text{h}} \cdot \frac{60\ \text{s}}{1\ \text{min}}$$

$$= \frac{1\ \cancel{\text{h}} \cdot 60\ \cancel{\text{min}} \cdot 60\ \text{s}}{1 \cdot 1\ \cancel{\text{h}} \cdot 1\ \cancel{\text{min}}}$$

$$= \frac{60 \cdot 60\ \text{s}}{1} = 3{,}600\ \text{s}$$

There are 3,600 s in 1 h.

3 You Try It

How many yards are in one mile?

Your Solution

Solutions on p. S18

OBJECTIVE B

Applications

In solving these application problems, we will keep the units throughout the solution as we work through the arithmetic. Note that, just as in the conversions in Objective A, conversion rates are used to set up the units before the arithmetic is performed.

Copyright © Houghton Mifflin Company. All rights reserved.

Empire Maker

In 2003, a horse named Empire Maker ran a 1.125-mile race in 1.8175 min. Find Empire Maker's average speed for that race in miles per hour. Round to the nearest tenth.

Strategy   To find the average speed in miles per hour:

▶ Write Empire Maker's speed as a rate in fraction form. The distance (1.125 mi) is in the numerator, and the time (1.8175 min) is in the denominator.

▶ Multiply the fraction by the conversion rate $\frac{60 \text{ min}}{1 \text{ h}}$.

Solution   Empire Maker's rate is $\frac{1.125 \text{ mi}}{1.8175 \text{ min}}$.

$$\frac{1.125 \text{ mi}}{1.8175 \text{ min}} = \frac{1.125 \text{ mi}}{1.8175 \text{ min}} \cdot \frac{60 \text{ min}}{1 \text{ h}}$$

$$= \frac{67.5 \text{ mi}}{1.8175 \text{ h}} \approx 37.1 \text{ mph}$$

Empire Maker's average speed was 37.1 mph.

---

④ Example

A carpet is to be placed in a room that is 20 ft wide and 30 ft long. At $28.50 per square yard, how much will it cost to carpet the area? Use the formula $A = LW$, where A is the area, L is the length, and W is the width, to find the area of the room.

Strategy

To find the cost of the carpet:

▶ Use the formula $A = LW$ to find the area.

▶ Use the conversion rate $\frac{1 \text{ yd}^2}{9 \text{ ft}^2}$ to find the area in square yards.

▶ Multiply by $\frac{\$28.50}{1 \text{ yd}^2}$ to find the cost.

Solution

$A = LW = 30 \text{ ft} \cdot 20 \text{ ft} = 600 \text{ ft}^2$ ▶ ft · ft = ft²

$$600 \text{ ft}^2 = \frac{600 \text{ ft}^2}{1} \cdot \frac{1 \text{ yd}^2}{9 \text{ ft}^2} = \frac{600 \text{ yd}^2}{9} = \frac{200 \text{ yd}^2}{3}$$

$$\text{Cost} = \frac{200 \text{ yd}^2}{3} \cdot \frac{\$28.50}{1 \text{ yd}^2} = \$1,900$$

The cost of the carpet is $1,900.

④ You Try It

Find the number of gallons of water in a fish tank that is 36 in. long and 24 in. wide and is filled to a depth of 16 in. Use the formula $V = LWH$, where V is the volume, L is the length, W is the width, and H is the depth of the water. (1 gal = 231 in³) Round to the nearest tenth.

Your Strategy

Your Solution

Solution on pp. S18–S19

Copyright © Houghton Mifflin Company. All rights reserved.

OBJECTIVE **C**

Conversion between the U.S. Customary System and the metric system

Because more than 90% of the world's population uses the metric system of measurement, converting U.S. Customary units to metric units is essential in trade and commerce, for example, in importing foreign goods and exporting domestic goods. Also, metric units are being used throughout the United States today. Cereal is packaged by the gram, 35-millimeter film is available, and soda is sold by the liter.

Approximate equivalences between the U.S. Customary System and the metric system are shown below.

Units of Length	*Units of Weight*	*Units of Capacity*
1 in. = 2.54 cm	1 oz ≈ 28.35 g	1 L ≈ 1.06 qt
1 m ≈ 3.28 ft	1 lb ≈ 454 g	1 gal ≈ 3.79 L
1 m ≈ 1.09 yd	1 kg ≈ 2.2 lb	
1 mi ≈ 1.61 km		

▶ **Point of Interest**

The definition of 1 in. has been changed as a consequence of the wide acceptance of the metric system. One inch is now exactly 25.4 mm.

These equivalences can be used to form conversion rates to change one measurement to another. For example, because 1 mi ≈ 1.61 km, the conversion rates $\dfrac{1 \text{ mi}}{1.61 \text{ km}}$ and $\dfrac{1.61 \text{ km}}{1 \text{ mi}}$ are each approximately equal to 1.

The procedure used to convert from one system to the other is identical to the conversions performed on the U.S. Customary System in Objective A.

Convert 55 mi to kilometers.

The equivalence is 1 mi ≈ 1.61 km. The conversion rate must have km in the numerator and mi in the denominator: $\dfrac{1.61 \text{ km}}{1 \text{ mi}}$.

$$55 \text{ mi} = \frac{55 \text{ mi}}{1}$$

Note that we are multiplying 55 mi by 1, so we are not changing its value.

$$\approx \frac{55 \text{ mi}}{1} \cdot \boxed{\frac{1.61 \text{ km}}{1 \text{ mi}}}$$

Divide the numerator and denominator by the common unit "mile."

$$\approx \frac{55 \cancel{\text{ mi}}}{1} \cdot \frac{1.61 \text{ km}}{1 \cancel{\text{ mi}}}$$

Multiply 55 times 1.61.

$$\approx \frac{88.55 \text{ km}}{1}$$

55 mi ≈ 88.55 km

Copyright © Houghton Mifflin Company. All rights reserved.

5 **Example**

Convert 200 m to feet.

Solution

$$200 \text{ m} = \frac{200 \text{ m}}{1} \approx \frac{200 \text{ m}}{1} \cdot \frac{3.28 \text{ ft}}{1 \text{ m}}$$
$$\approx 656 \text{ ft}$$

5 **You Try It**

Convert 45 cm to inches. Round to the nearest hundredth.

Your Solution

6 **Example**

Convert 45 mph to kilometers per hour.

Solution

$$45 \text{ mph} = \frac{45 \text{ mi}}{1 \text{ h}} \approx \frac{45 \text{ mi}}{1 \text{ h}} \cdot \frac{1.61 \text{ km}}{1 \text{ mi}}$$
$$\approx 72.45 \text{ km per hour}$$

6 **You Try It**

Convert 75 km per hour to miles per hour. Round to the nearest hundredth.

Your Solution

7 **Example**

The price of gasoline is $1.99 per gallon. Find the price per liter. Round to the nearest tenth of a cent.

Solution

$$\$1.99 \text{ per gallon} = \frac{\$1.99}{\text{gal}} \approx \frac{\$1.99}{\text{gal}} \cdot \frac{1 \text{ gal}}{3.79 \text{ L}}$$
$$\approx \$.525 \text{ per liter}$$

The price is approximately $.525 per liter.

7 **You Try It**

The price of milk is $3.59 per gallon. Find the price per liter. Round to the nearest cent.

Your Solution

8 **Example**

The price of gasoline is $.499 per liter. Find the price per gallon. Round to the nearest cent.

Solution

$$\$.499 \text{ per liter} = \frac{\$.499}{1 \text{ L}} \approx \frac{\$.499}{1 \text{ L}} \cdot \frac{3.79 \text{ L}}{1 \text{ gal}}$$
$$\approx \$1.89 \text{ per gallon}$$

The price is approximately $1.89 per gallon.

8 **You Try It**

The price of ice cream is $2.65 per liter. Find the price per gallon. Round to the nearest cent.

Your Solution

Solutions on p. S19

Copyright © Houghton Mifflin Company. All rights reserved.

7.3 EXERCISES

OBJECTIVE A

1. Convert 64 in. to feet.

2. Convert 14 ft to yards.

3. Convert 42 oz to pounds.

4. Convert 4,400 lb to tons.

5. Convert 7,920 ft to miles.

6. Convert 42 c to quarts.

7. Convert 500 lb to tons.

8. Convert 90 oz to pounds.

9. Convert 10 qt to gallons.

10. How many pounds are in $1\frac{1}{4}$ tons?

11. How many fluid ounces are in $2\frac{1}{2}$ c?

12. How many ounces are in $2\frac{5}{8}$ lb?

13. Convert $2\frac{1}{4}$ mi to feet.

14. Convert 17 c to quarts.

15. Convert $7\frac{1}{2}$ in. to feet.

16. Convert $2\frac{1}{4}$ gal to quarts.

17. Convert 60 fl oz to cups.

18. Convert $1\frac{1}{2}$ qt to cups.

19. Convert $7\frac{1}{2}$ pt to quarts.

20. Convert 20 fl oz to pints.

21. How many yards are in $1\frac{1}{2}$ mi?

22. How many seconds are in one day?

OBJECTIVE B

 Solve.

23. *Time*   When a person reaches the age of 35, for how many seconds has that person lived?

Copyright © Houghton Mifflin Company. All rights reserved.

24. *Interior Decorating* Fifty-eight feet of material are purchased for making pleated draperies. Find the total cost of the material if the price is $39 per yard.

25. *Catering* The Concord Theater serves punch during intermission. If each of 200 people drink 1 c of punch, how many gallons of punch should be prepared?

26. *Consumerism* A can of cranberry juice contains 25 fl oz. How many quarts of cranberry juice are in a case of 24 cans?

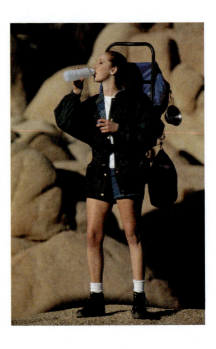

27. *Hiking* Five students are going backpacking in the desert. Each student requires 2 qt of water per day. How many gallons of water should they take for a 3-day trip?

28. *Hiking* A hiker is carrying 5 qt of water. Water weighs $8\frac{1}{3}$ lb per gallon. Find the weight of the water carried by the hiker.

29. *Business* A garage mechanic purchases oil in a 50-gallon container for changing oil in customers' cars. After 35 oil changes requiring 5 qt each, how much oil is left in the 50-gallon container?

30. *Sound* The speed of sound is about 1,100 ft per second. Find the speed of sound in miles per hour. $\left(Hint:\ 1{,}100\ \text{ft per second} = \dfrac{1{,}100\ \text{ft}}{1\ \text{s}}\right)$

31. *Interior Decorating* Wall-to-wall carpeting is to be laid in the living room of the home shown in the floor plan below. At $33 per square yard, how much will the carpeting cost? (The formula for the area of a rectangle is $A = LW$.)

Deck

Nook
9'8" x 9'

Dining
12' x 15'
cathedral

Kitchen
11'6" x 14'4"

Mstr Br
13'6" x 17'8"

Living
15' x 20'

open
to
above

UP   crawl access

W
D

FIRST FLOOR

Copyright © Houghton Mifflin Company. All rights reserved.

32. *Real Estate*   A building lot with the dimensions shown in the diagram at the right is priced at $20,000 per acre. Find the price of the building lot. (The formula for the area of a rectangle is $A = LW$.)

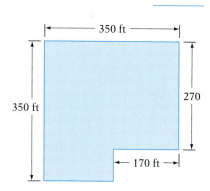

33. *Real Estate*   A one-half-acre commercial lot is selling for $3 per square foot. Find the price of the commercial lot.

OBJECTIVE   C

 Convert. Round to the nearest hundredth or to the nearest cent.

34. Convert the 100-yard dash to meters.

35. Find the weight in kilograms of a 145-pound person.

36. Find the height in meters of a person 5 ft 8 in. tall.

37. Find the number of cups in 2 L of soda.

38. How many kilograms does a 15-pound turkey weigh?

39. Find the number of liters in 14.3 gal of gasoline.

40. Find the distance in feet of the 1 500-meter race.

41. Find the weight in pounds of an 86-kilogram person.

42. Find the number of gallons in 6 L of antifreeze.

43. Find the width in inches of 35-millimeter film.

44. Find the weight in ounces of 327 g of cereal.

45. How many gallons of water does a 24-liter aquarium hold?

46. Express 65 mph in kilometers per hour.

47. Convert 60 ft/s to meters per second.

48. Fat free hot dogs cost $3.49 per pound. Find the price per kilogram.

49. Seedless watermelon costs $.99 per pound. Find the cost per kilogram.

50. Deck stain costs $24.99 per gallon. Find the cost per liter.

51. Express 80 km/h in miles per hour.

52. Express 30 m/s in feet per second.

53. Gasoline costs 48.5¢ per liter. Find the cost per gallon.

Copyright © Houghton Mifflin Company. All rights reserved.

54. *Sports* The largest trout ever caught in the state of Utah weighed 51 lb 8 oz. Find the trout's weight in kilograms.

55. *Earth Science* The distance around Earth at the equator is 24,887 mi. What is this distance around Earth in kilometers?

56. *Astronomy* The distance from Earth to the sun is 93,000,000 mi. Calculate the distance from Earth to the sun in kilometers.

Critical Thinking

57. a. Page 453 lists equivalences between the U.S. Customary System and the metric system of measurement. Explain why the word *rate* can be used in describing these equivalences.
  b. Express each of the following as a rate: 1 in. = 2.54 cm; 1 kg ≈ 2.2 lb; 1 L ≈ 1.06 qt

58. Determine whether the statement is true or false.
  a. A liter is more than a gallon.
  b. A meter is less than a yard.
  c. 30 mph is less than 60 km/h.
  d. A kilogram is greater than a pound.
  e. An ounce is less than a gram.

59. For the following U.S. Customary units, make an estimate of the metric equivalent. Then perform the conversion and see how close you came to the actual measurement.
  60 mph ≈   120 lb ≈
  6 ft ≈   1 mi ≈
  1 gal ≈   1 quarter-mile ≈

60. Find examples of U.S. Customary System units that were not included in this section. Provide equivalences for each unit. For example, 1 furlong = $\frac{1}{8}$ mi.

61. Explain why we can multiply an expression by the conversion rate $\frac{5{,}280 \text{ ft}}{1 \text{ mi}}$.

62. Write a paragraph describing the growing need for precision in our measurements as civilization progresses. Include a discussion of the need for precision in the space industry.

63. Should the United States keep the U.S. Customary System or convert to the metric system? Justify your position.

Copyright © Houghton Mifflin Company. All rights reserved.

Proportion

OBJECTIVE **A**

Proportion

A **proportion** is the equality of two ratios or rates.

The equality $\dfrac{250\ \text{mi}}{5\ \text{h}} = \dfrac{50\ \text{mi}}{1\ \text{h}}$ is a proportion.

> **Definition of Proportion**
>
> If $\dfrac{a}{b}$ and $\dfrac{c}{d}$ are equal ratios or rates, then $\dfrac{a}{b} = \dfrac{c}{d}$ is a proportion.

Each of the four numbers in a proportion is called a **term.** Each term is numbered according to the following diagram.

$$\begin{array}{ccc}\text{first term} \longleftarrow & a & c \longrightarrow \text{third term}\\ \text{second term} \longleftarrow & \dfrac{}{b} & \dfrac{}{d} \longrightarrow \text{fourth term}\end{array}$$

The first and fourth terms of the proportion are called the **extremes** and the second and third terms are called the **means.**

If we multiply the proportion by the least common multiple of the denominators, we obtain the following result:

$$\frac{a}{b} = \frac{c}{d}$$

$$bd\left(\frac{a}{b}\right) = bd\left(\frac{c}{d}\right)$$

$$ad = bc \qquad \textcolor{blue}{ad \text{ is the product of the extremes.}}$$
$$\textcolor{blue}{bc \text{ is the product of the means.}}$$

In any true proportion, **the product of the means equals the product of the extremes.** This is sometimes phrased as "the cross products are equal."

In the true proportion $\dfrac{3}{4} = \dfrac{9}{12}$, the cross products are equal.

$$\frac{3}{4} \diagdown\!\!\!\!\diagup \frac{9}{12} \quad \longrightarrow \quad 4 \cdot 9 = \textcolor{blue}{36} \quad \longleftarrow \text{ Product of the means}$$
$$\longrightarrow \quad 3 \cdot 12 = \textcolor{blue}{36} \quad \longleftarrow \text{ Product of the extremes}$$

Determine whether the proportion $\dfrac{47\ \text{mi}}{2\ \text{gal}} = \dfrac{304\ \text{mi}}{13\ \text{gal}}$ is a true proportion.

The product of the means: The product of the extremes:

$$2 \cdot 304 = \textcolor{blue}{608} \qquad\qquad 47 \cdot 13 = \textcolor{blue}{611}$$

The proportion is not true because $608 \neq 611$.

▶ Point of Interest

Proportions were studied by the earliest mathematicians. Clay tablets uncovered by archeologists show evidence of proportions in Egyptian and Babylonian cultures dating from 1800 B.C.

Copyright © Houghton Mifflin Company. All rights reserved.

When three terms of a proportion are given, the fourth term can be found. To solve a proportion for an unknown term, use the fact that the product of the means equals the product of the extremes.

■ **Calculator Note**

To use a calculator to solve the proportion at the right, multiply the second and third terms and divide by the fourth term. Enter

5 × 9 ÷ 16 =

The display reads 2.8125.

Solve: $\dfrac{n}{5} = \dfrac{9}{16}$

Find the number (n) that will make the proportion true.

The product of the means equals the product of the extremes. Solve for n.

$$\frac{n}{5} = \frac{9}{16}$$

$$5 \cdot 9 = n \cdot 16$$

$$45 = 16n$$

$$\frac{45}{16} = \frac{16n}{16}$$

$$2.8125 = n$$

---

1 *Example* Determine whether $\dfrac{15}{3} = \dfrac{90}{18}$ is a true proportion.

Solution

$$\frac{15}{3} \diagdown\diagup \frac{90}{18} \longrightarrow 3 \cdot 90 = 270$$
$$\longrightarrow 15 \cdot 18 = 270$$

The product of the means equals the product of the extremes.

The proportion is true.

1 *You Try It* Is $\dfrac{50 \text{ mi}}{3 \text{ gal}} = \dfrac{250 \text{ mi}}{12 \text{ gal}}$ a true proportion?

Your Solution

---

2 *Example* Solve: $\dfrac{5}{9} = \dfrac{x}{45}$

Solution

$$\frac{5}{9} = \frac{x}{45}$$
$$9 \cdot x = 5 \cdot 45$$
$$9x = 225$$
$$\frac{9x}{9} = \frac{225}{9}$$
$$x = 25$$

2 *You Try It* Solve: $\dfrac{7}{12} = \dfrac{42}{x}$

Your Solution

---

3 *Example* Solve: $\dfrac{6}{n} = \dfrac{45}{124}$. Round to the nearest tenth.

Solution

$$\frac{6}{n} = \frac{45}{124}$$
$$n \cdot 45 = 6 \cdot 124$$
$$45n = 744$$
$$\frac{45n}{45} = \frac{744}{45}$$
$$n \approx 16.5$$

3 *You Try It* Solve: $\dfrac{5}{n} = \dfrac{3}{322}$. Round to the nearest hundredth.

Your Solution

Solutions on p. S19

Copyright © Houghton Mifflin Company. All rights reserved.

4 **Example**   Solve: $\dfrac{x+2}{3} = \dfrac{7}{8}$

4 **You Try It**   Solve: $\dfrac{4}{5} = \dfrac{3}{x-3}$

Solution $\dfrac{x+2}{3} = \dfrac{7}{8}$

$3 \cdot 7 = (x+2)8$

$21 = 8x + 16$

$5 = 8x$

$0.625 = x$

Your Solution

Solution on p. S19

OBJECTIVE   **B**

VIDEO & DVD   CD TUTOR   WEB   SSM

Applications

Proportions are useful in many types of application problems. In recipes, proportions are used when a larger batch of ingredients is used than the recipe calls for. In mixing cement, the amounts of cement, sand, and rock are mixed in the same ratio. A map is drawn on a proportional basis, such as 1 in. representing 50 mi.

In setting up a proportion, keep the same units in the numerators and the same units in the denominators. For example, if *feet* is in the numerator on one side of the proportion, then *feet* must be in the numerator on the other side of the proportion.

A customer sees an ad in a newspaper advertising 2 tires for $162.50. The customer wants to buy 5 tires and use one for the spare. How much will the 5 tires cost?

Write a proportion.
Let c = the cost of the 5 tires.

$$\dfrac{2 \text{ tires}}{\$162.50} = \dfrac{5 \text{ tires}}{c}$$

$$162.50 \cdot 5 = 2 \cdot c$$

$$812.50 = 2c$$

$$\dfrac{812.50}{2} = \dfrac{2c}{2}$$

$$406.25 = c$$

The 5 tires will cost $406.25.

> **Take Note**
>
> It is also correct to write the proportion with the costs in the numerators and the number of tires in the denominators:
>
> $$\dfrac{\$162.50}{2 \text{ tires}} = \dfrac{c}{5 \text{ tires}}.$$ The solution will be the same.

Copyright © Houghton Mifflin Company. All rights reserved.

Example

During a Friday, the ratio of stocks declining in price to those advancing was 5 to 3. If 450,000 shares advanced, how many shares declined on that day?

Strategy

To find the number of shares declining in price, write and solve a proportion using n to represent the number of shares declining in price.

Solution

$$\frac{5 \text{ (declining)}}{3 \text{ (advancing)}} = \frac{n \text{ shares declining}}{450{,}000 \text{ shares advancing}}$$

$$3n = 5 \cdot 450{,}000$$

$$3n = 2{,}250{,}000$$

$$\frac{3n}{3} = \frac{2{,}250{,}000}{3}$$

$$n = 750{,}000$$

750,000 shares declined in price.

 You Try It

An automobile can travel 396 mi on 11 gal of gas. At the same rate, how many gallons of gas would be necessary to travel 832 mi? Round to the nearest tenth.

Your Strategy

Your Solution

 Example

From previous experience, a manufacturer knows that in an average production of 5,000 calculators, 40 will be defective. What number of defective calculators can be expected from a run of 45,000 calculators?

Strategy

To find the number of defective calculators, write and solve a proportion using n to represent the number of defective calculators.

Solution

$$\frac{40 \text{ defective calculators}}{5{,}000 \text{ calculators}} = \frac{n \text{ defective calculators}}{45{,}000 \text{ calculators}}$$

$$5{,}000 \cdot n = 40 \cdot 45{,}000$$

$$5{,}000n = 1{,}800{,}000$$

$$\frac{5{,}000n}{5{,}000} = \frac{1{,}800{,}000}{5{,}000}$$

$$n = 360$$

The manufacturer can expect 360 defective calculators.

 You Try It

An automobile recall was based on tests that showed 15 transmission defects in 1,200 cars. At this rate, how many defective transmissions will be found in 120,000 cars?

Your Strategy

Your Solution

Copyright © Houghton Mifflin Company. All rights reserved.

7.4 EXERCISES

Copyright © Houghton Mifflin Company. All rights reserved.

OBJECTIVE A

 Determine whether the proportion is true or not true.

1. $\dfrac{27}{8} = \dfrac{9}{4}$

2. $\dfrac{3}{18} = \dfrac{4}{19}$

3. $\dfrac{45}{135} = \dfrac{3}{9}$

4. $\dfrac{3}{4} = \dfrac{54}{72}$

5. $\dfrac{16}{3} = \dfrac{48}{9}$

6. $\dfrac{15}{5} = \dfrac{3}{1}$

7. $\dfrac{6 \text{ min}}{5 \text{ cents}} = \dfrac{30 \text{ min}}{25 \text{ cents}}$

8. $\dfrac{7 \text{ tiles}}{4 \text{ ft}} = \dfrac{42 \text{ tiles}}{20 \text{ ft}}$

9. $\dfrac{15 \text{ ft}}{3 \text{ yd}} = \dfrac{90 \text{ ft}}{18 \text{ yd}}$

10. $\dfrac{\$65}{5 \text{ days}} = \dfrac{\$26}{2 \text{ days}}$

11. $\dfrac{1 \text{ gal}}{4 \text{ qt}} = \dfrac{7 \text{ gal}}{28 \text{ qt}}$

12. $\dfrac{300 \text{ ft}}{4 \text{ rolls}} = \dfrac{450 \text{ ft}}{7 \text{ rolls}}$

 Solve. Round to the nearest hundredth.

13. $\dfrac{2}{3} = \dfrac{n}{15}$

14. $\dfrac{7}{15} = \dfrac{n}{15}$

15. $\dfrac{n}{5} = \dfrac{12}{25}$

16. $\dfrac{n}{8} = \dfrac{7}{8}$

17. $\dfrac{3}{8} = \dfrac{n}{12}$

18. $\dfrac{5}{8} = \dfrac{40}{n}$

19. $\dfrac{3}{n} = \dfrac{7}{40}$

20. $\dfrac{7}{12} = \dfrac{25}{n}$

21. $\dfrac{16}{n} = \dfrac{25}{40}$

22. $\dfrac{15}{45} = \dfrac{72}{n}$

23. $\dfrac{120}{n} = \dfrac{144}{25}$

24. $\dfrac{65}{20} = \dfrac{14}{n}$

25. $\dfrac{0.5}{2.3} = \dfrac{n}{20}$

26. $\dfrac{1.2}{2.8} = \dfrac{n}{32}$

27. $\dfrac{0.7}{1.2} = \dfrac{6.4}{n}$

28. $\dfrac{2.5}{0.6} = \dfrac{165}{n}$

29. $\dfrac{x}{6.25} = \dfrac{16}{87}$

30. $\dfrac{x}{2.54} = \dfrac{132}{640}$

31. $\dfrac{1.2}{0.44} = \dfrac{y}{14.2}$

32. $\dfrac{12.5}{y} = \dfrac{102}{55}$

33. $\dfrac{n+2}{5} = \dfrac{1}{2}$

34. $\dfrac{5+n}{8} = \dfrac{3}{4}$

35. $\dfrac{4}{3} = \dfrac{n-2}{6}$

36. $\dfrac{3}{5} = \dfrac{n-7}{8}$

37. $\dfrac{2}{n+3} = \dfrac{7}{12}$

38. $\dfrac{5}{n+1} = \dfrac{7}{3}$

39. $\dfrac{7}{10} = \dfrac{3+n}{2}$

40. $\dfrac{3}{2} = \dfrac{5+n}{4}$

41. $\dfrac{x-4}{3} = \dfrac{3}{4}$

42. $\dfrac{x-1}{8} = \dfrac{5}{2}$

43. $\dfrac{6}{1} = \dfrac{x-2}{5}$

44. $\dfrac{7}{3} = \dfrac{x-4}{8}$

45. $\dfrac{5}{8} = \dfrac{2}{x-3}$

46. $\dfrac{5}{2} = \dfrac{1}{x-6}$

47. $\dfrac{3}{x-4} = \dfrac{5}{3}$

48. $\dfrac{8}{x-6} = \dfrac{5}{4}$

 OBJECTIVE **B**

 Solve.

49. *Biology* In a drawing, the length of an amoeba is 2.6 in. The scale of the drawing is 1 in. on the drawing equals 0.002 in. on the amoeba. Find the actual length of the amoeba.

50. *Insurance* A life insurance policy costs $15.22 for every $1,000 of insurance. At this rate, what is the cost of $75,000 of insurance?

51. *Sewing* Six children's robes can be made from 6.5 yd of material. How many robes can be made from 26 yd of material?

52. *Computers* A computer manufacturer finds that an average of 3 defective hard drives are found in every 100 drives manufactured. How many defective drives are expected to be found in the production of 1,200 hard drives?

53. *Taxes* The property tax on a $180,000 home is $4,320. At this rate, what is the property tax on a home appraised at $280,000?

54. *Medicine* The dosage of a certain medication is 2 mg for every 80 lb of body weight. How many milligrams of this medication are required for a person who weighs 220 lb?

55. *Travel* An automobile was driven 84 mi and used 3 gal of gasoline. At the same rate of consumption, how far would the car travel on 14.5 gal of gasoline?

Copyright © Houghton Mifflin Company. All rights reserved.

56. *Nutrition*   If a 56-gram serving of pasta contains 7 g of protein, how many grams of protein are in a 454-gram box of the pasta?

57. *Consumerism*   If 4 grapefruit sell for $1.28, how much do 14 grapefruit cost?

58. *Sports*   A halfback on a college football team has rushed for 435 yd in 5 games. At this rate, how many rushing yards will the halfback have in 12 games?

59. *Construction*   A building contractor estimates that five overhead lights are needed for every 400 ft² of office space. Using this estimate, how many light fixtures are necessary for an office building of 35,000 ft²?

60. *Sports*   A softball player has hit 9 home runs in 32 games. At the same rate, how many home runs will the player hit in a 160-game schedule?

61. *Health*   A dieter has lost 3 lb in 5 weeks. At this rate, how long will it take the dieter to lose 36 lb?

62. *Consumerism*   Steak costs $25.20 for 3 lb. At this rate, how much does 8 lb of steak cost?

63. *Business*   An automobile recall was based on engineering tests that showed 22 defects in 1,000 cars. At this rate, how many defects would be found in 125,000 cars?

64. *Health*   Walking 5 mi in 2 h will use 650 calories. Walking at the same rate, how many miles would a person need to walk to lose 1 lb? (The burning of 3,500 calories is equivalent to the loss of 1 lb.) Round to the nearest hundredth.

65. *Travel*   An account executive bought a new car and drove 22,000 mi in the first 4 months. At the same rate, how many miles will the account executive drive in 3 years?

66. *Investments*   An investment of $1,500 earns $120 each year. At the same rate, how much additional money must be invested to earn $300 each year?

$1,500		$1,500 + x
earns		earns
$120		$300

67. *Investments*   A stock investment of $3,500 earns a dividend of $280. At the same rate, how much additional money would have to be invested so that the total dividend is $400?

Copyright © Houghton Mifflin Company. All rights reserved.

68. *Cartography* The scale on a map is $\frac{1}{2}$ in. equals 8 mi. What is the actual distance between two points that are $1\frac{1}{4}$ in. apart on the map?

69. *Energy* A slow-burning candle will burn 1.5 in. in 40 min. How many inches of the candle will burn in 4 h?

70. *Mixtures* A saltwater solution is made by dissolving $\frac{2}{3}$ lb of salt in 5 gal of water. At this rate, how many pounds of salt are required for 12 gal of water?

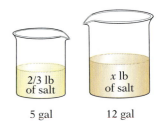

2/3 lb of salt   *x* lb of salt

5 gal   12 gal

71. *Business* A management consulting firm recommends that the ratio of midmanagement salaries to junior management salaries be 7:5. Using this recommendation, find the yearly midmanagement salary when the junior management salary is $90,000.

Critical Thinking

72. Determine whether the statement is true or false.

  a. A quotient ($a \div b$) is a ratio.

  b. If $\frac{a}{b} = \frac{c}{d}$, then $\frac{b}{a} = \frac{d}{c}$.

  c. If $\frac{a}{b} = \frac{c}{d}$, then $\frac{a}{c} = \frac{b}{d}$.

  d. If $\frac{a}{b} = \frac{c}{d}$, then $\frac{a}{d} = \frac{c}{b}$.

73. If $\frac{a}{b} = \frac{c}{d}$, does $\frac{a}{b} = \frac{a+c}{b+d}$? Explain your answer.

74. If $\frac{a}{b} = \frac{c}{d}$, show that $\frac{a+b}{b} = \frac{c+d}{d}$.

75. *Elections* A survey of voters in a city claimed that 2 people of every 5 who voted cast a ballot in favor of city amendment A, and 3 people of every 4 who voted cast a ballot against amendment A. Is this possible? Explain your answer.

76. *Compensation* In June 2002, *Time* magazine reported, "In 1980 the average CEO made 40 times the pay of the average factory worker; by 2000 the ratio had climbed to 531 to 1." What information would you need to know in order to determine the average pay of a CEO in 2000? With that information, how would you calculate the average pay of a CEO in 2000?

77. Write a paragraph describing how proportional representation is used to select the members in the U.S. House of Representatives.

Copyright © Houghton Mifflin Company. All rights reserved.

Direct and Inverse Variation

OBJECTIVE   **A**

Direct variation

An equation of the form $y = kx$ describes many important relationships in business, science, and engineering. The equation $y = kx$, where k is a constant, is an example of a **direct variation.** The constant k is called the **constant of variation** or the **constant of proportionality.** The equation $y = kx$ is read "y varies directly as x."

For example, the distance traveled by a car traveling at a constant rate of 55 mph is represented by $y = 55x$, where x is the number of hours and y is the total distance traveled. The number 55 is called the constant of proportionality.

The distance (d) sound travels varies directly as the time (t) it travels. If sound travels 8,920 ft in 8 s, find the distance that sound travels in 3 s.

This is a direct variation. k is the constant of proportionality.

$$d = kt$$

Substitute 8,920 for d and 8 for t.

$$8,920 = k \cdot 8$$

Solve for k.

$$\frac{8,920}{8} = \frac{k \cdot 8}{8}$$
$$1,115 = k$$

Write the direct variation equation for d, substituting 1,115 for k.
Find d when $t = 3$.

$$d = 1,115t$$

$$d = 1,115 \cdot 3$$
$$d = 3,345$$

Sound travels 3,345 ft in 3 s.

A direct variation equation can be written in the form $y = kx^n$, where n is a positive integer. For example, the equation $y = kx^2$ is read "y varies directly as the square of x."

The load (L) that a horizontal beam can safely support is directly proportional to the square of the depth (d) of the beam. A beam with a depth of 8 in. can support 800 lb. Find the load that a beam with a depth of 6 in. can support.

This is a direct variation. k is the constant of proportionality.

$$L = kd^2$$

Substitute 800 for L and 8 for d.

$$800 = k \cdot 8^2$$
$$800 = k \cdot 64$$

Solve for k.

$$\frac{800}{64} = \frac{k \cdot 64}{64}$$
$$12.5 = k$$

Write the direct variation equation for L, substituting 12.5 for k.
Find L when $d = 6$.

$$L = 12.5d^2$$

$$L = 12.5 \cdot 6^2$$
$$L = 12.5 \cdot 36$$
$$L = 450$$

The beam can support a load of 450 lb.

Copyright © Houghton Mifflin Company. All rights reserved.

1 Example

Find the constant of variation if y varies directly as x, and $y = 5$ when $x = 35$.

Strategy

To find the constant of variation, substitute 5 for y and 35 for x in the direct variation equation $y = kx$ and solve for k.

Solution

$$y = kx$$
$$5 = k \cdot 35$$
$$\frac{5}{35} = \frac{k \cdot 35}{35}$$
$$\frac{1}{7} = k$$

The constant of variation is $\frac{1}{7}$.

1 You Try It

Find the constant of variation if y varies directly as x, and $y = 120$ when $x = 8$.

Your Strategy

Your Solution

2 Example

Given that L varies directly as P, and $L = 24$ when $P = 16$, find P when $L = 80$. Round to the nearest tenth.

Strategy

To find P when $L = 80$:

▶ Write the basic direct variation equation, replace the variables by the given values, and solve for k.
▶ Write the direct variation equation, replacing k by its value. Substitute 80 for L and solve for P.

Solution

$$L = kP$$
$$24 = k \cdot 16$$
$$\frac{24}{16} = \frac{k \cdot 16}{16}$$
$$1.5 = k$$
$$L = 1.5P$$
$$80 = 1.5P$$
$$\frac{80}{1.5} = \frac{1.5P}{1.5}$$
$$53.3 \approx P$$

The value of P is approximately 53.3 when $L = 80$.

2 You Try It

Given that S varies directly as R, and $S = 8$ when $R = 30$, find S when $R = 200$. Round to the nearest tenth.

Your Strategy

Your Solution

Copyright © Houghton Mifflin Company. All rights reserved.

Example

The distance (d) required for a car to stop varies directly as the square of the velocity (v) of the car. If a car traveling 40 mph requires 130 ft to stop, find the stopping distance for a car traveling 60 mph.

Strategy

To find the stopping distance:

▶ Write the basic direct variation equation, replace the variables by the given values, and solve for k.
▶ Write the direct variation equation, replacing k by its value. Substitute 60 for v and solve for d.

Solution

$$d = kv^2$$
$$130 = k \cdot 40^2$$
$$130 = k \cdot 1{,}600$$
$$0.08125 = k$$
$$d = 0.08125 \cdot v^2 = 0.08125 \cdot 60^2$$
$$= 0.08125 \cdot 3{,}600 = 292.5$$

The stopping distance is 292.5 ft.

 You Try It

The distance (d) a body falls from rest varies directly as the square of the time (t) of the fall. An object falls 64 ft in 2 s. How far will the object fall in 9 s?

Your Strategy

Your Solution

Solution on p. S20

OBJECTIVE **B**

Inverse variation

The equation $y = \dfrac{k}{x}$, where k is a constant, is an example of an **inverse variation**. The equation $y = \dfrac{k}{x}$ is read "y varies inversely as x" or "y is inversely proportional to x." In general, an inverse variation equation can be written $y = \dfrac{k}{x^n}$, where n is a positive integer. For example, the equation $y = \dfrac{k}{x^2}$ is read "y varies inversely as the square of x."

The volume (V) of a gas at a fixed temperature varies inversely as the pressure (P). The inverse variation equation would be written as

$$V = \frac{k}{P}$$

The gravitational force (F) between two planets is inversely proportional to the square of the distance (d) between the planets. This inverse variation would be written as

$$F = \frac{k}{d^2}$$

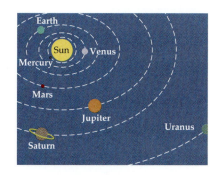

Copyright © Houghton Mifflin Company. All rights reserved.

Given that y varies inversely as the square of x, and $y = 5$ when $x = 2$, find y when $x = 40$.

Write the basic inverse variation equation where y varies inversely as the square of x.

$$y = \frac{k}{x^2}$$

Replace x and y by the given values.

$$5 = \frac{k}{2^2}$$

Solve for the constant of variation.

$$20 = k$$

Write the inverse variation equation by substituting the value of k into the basic inverse variation equation.

$$y = \frac{20}{x^2}$$

To find y when $x = 40$, substitute 40 for x in the equation and solve for y.

$$y = \frac{20}{40^2}$$

$$y = 0.0125$$

---

4 *Example*

A company that produces personal computers has determined that the number of computers it can sell (S) is inversely proportional to the price (P) of the computer. Two thousand computers can be sold when the price is $2,500. How many computers can be sold if the price of a computer is $2,000?

Strategy

To find the number of computers:

 Write the basic inverse variation equation, replace the variables by the given values, and solve for k.

 Write the inverse variation equation, replacing k by its value. Substitute 2,000 for the price and solve for the number sold.

Solution

$$S = \frac{k}{P}$$

$$2,000 = \frac{k}{2,500}$$

$$5,000,000 = k$$

$$S = \frac{5,000,000}{P} = \frac{5,000,000}{2,000} = 2,500$$

If the price is $2,000, then 2,500 computers can be sold.

4 *You Try It*

The resistance (R) to the flow of electric current in a wire of fixed length is inversely proportional to the square of the diameter (d) of the wire. If a wire of diameter 0.01 cm has a resistance of 0.5 ohm, what is the resistance in a wire that is 0.02 cm in diameter?

Your Strategy

Your Solution

Solution on p. S20

Copyright © Houghton Mifflin Company. All rights reserved.

7.5 EXERCISES

OBJECTIVE **A**

1. Which of the following are direct variations? Why?

   a. $y = kx$ **b.** $y = \dfrac{k}{x}$ **c.** $y = k + x$ **d.** $y = \dfrac{k}{x^2}$

2. Which of the following equations represents "d varies directly as t"? Explain your answer.

   a. $d = 400t$ **b.** $d = \dfrac{16}{t}$ **c.** $d = 25t$ **d.** $d = t - 50$

3. Find the constant of variation when y varies directly as x, and $y = 15$ when $x = 2$.

4. Find the constant of variation when t varies directly as s, and $t = 24$ when $s = 120$.

5. Find the constant of variation when n varies directly as the square of m, and $n = 64$ when $m = 2$.

6. Find the constant of variation when y varies directly as the square of x, and $y = 30$ when $x = 3$.

7. Given that P varies directly as R, and $P = 20$ when $R = 5$, find P when $R = 6$.

8. Given that T varies directly as S, and $T = 36$ when $S = 9$, find T when $S = 2$.

9. Given that M is directly proportional to P, and $M = 15$ when $P = 30$, find M when $P = 20$.

10. Given that A is directly proportional to B, and $A = 6$ when $B = 18$, find A when $B = 21$.

11. Given that y is directly proportional to the square of x, and $y = 10$ when $x = 2$, find y when $x = 0.5$.

12. Given that W is directly proportional to the square of V, and $W = 50$ when $V = 5$, find W when $V = 12$.

Copyright © Houghton Mifflin Company. All rights reserved.

Solve.

13. *Compensation* A worker's wage (w) is directly proportional to the number of hours (h) worked. If $82 is earned for working 8 h, how much is earned for working 30 h?

14. *Mechanics* The distance (d) a spring will stretch varies directly as the force (F) applied to the spring. If a force of 12 lb is required to stretch a spring 3 in., what force is required to stretch the spring 5 in.?

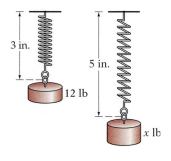

15. *Oceanography* The pressure (P) on a diver in the water varies directly as the depth (d). If the pressure is 2.25 lb/in^2 when the depth is 5 ft, what is the pressure when the depth is 12 ft?

16. *Computers* The number of words typed (w) is directly proportional to the time (t) spent typing. A typist can type 260 words in 4 min. Find the number of words typed in 15 min.

17. *Travel* The stopping distance (s) of a car varies directly as the square of its speed (v). If a car traveling 50 mph requires 170 ft to stop, find the stopping distance for a car traveling 65 mph.

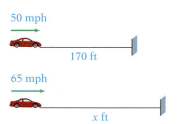

18. *Physics* The distance (d) an object falls is directly proportional to the square of the time (t) of the fall. If an object falls a distance of 8 ft in 0.5 s, how far will the object fall in 5 s?

19. *Energy* The current (I) varies directly as the voltage (V) in an electric circuit. If the current is 4 amps when the voltage is 100 volts, find the current when the voltage is 75 volts.

20. *Travel* The distance traveled (d) varies directly as the time (t) of travel, assuming that the speed is constant. If it takes 45 min to travel 50 mi, how many hours would it take to travel 180 mi?

OBJECTIVE B

21. Find the constant of variation when y varies inversely as x and $y = 10$ when $x = 5$.

22. Find the constant of proportionality when T varies inversely as S and $T = 0.2$ when $S = 8$.

Copyright © Houghton Mifflin Company. All rights reserved.

23. Find the constant of variation when p varies inversely as the square of q and $p = 4$ when $q = 5$.

24. Find the constant of variation when W varies inversely as the square of V and $W = 5$ when $V = 0.5$.

25. If y varies inversely as x and $y = 500$ when $x = 4$, find y when $x = 10$.

26. If W varies inversely as L and $W = 20$ when $L = 12$, find L when $W = 90$.

27. If y varies inversely as the square of x and $y = 40$ when $x = 4$, find y when $x = 10$.

28. If L varies inversely as the square of d and $L = 25$ when $d = 2$, find L when $d = 5$.

 Solve.

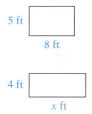

29. *Geometry*   The length (L) of a rectangle of fixed area varies inversely as the width (W). If the length of the rectangle is 8 ft when the width is 5 ft, find the length of the rectangle when the width is 4 ft.

30. *Travel*   The time (t) of travel of an automobile trip varies inversely as the speed (v). At an average speed of 65 mph, a trip took 4 h. The return trip took 5 h. Find the average speed of the return trip.

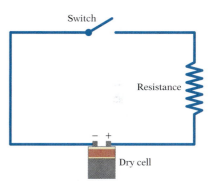

31. *Energy*   The current (I) in an electric circuit is inversely proportional to the resistance (R). If the current is 0.25 amp when the resistance is 8 ohms, find the resistance when the current is 1.2 amps.

32. *Physics*   The volume (V) of a gas varies inversely as the pressure (P) on the gas. If the volume of the gas is 12 ft^3 when the pressure is 15 lb/ft^2, find the volume of the gas when the pressure is 4 lb/ft^2.

33. *Business*   A computer company that produces personal computers has determined that the number of computers it can sell (S) is inversely proportional to the price (P) of the computer. Eighteen hundred computers can be sold if the price is $1,800. How many computers can be sold if the price is $1,500?

34. *Mechanics*   The speed (s) of a gear varies inversely as the number of teeth (t). If a gear that has 40 teeth makes 15 rpm (revolutions per minute), how many revolutions per minute will a gear that has 32 teeth make?

Copyright © Houghton Mifflin Company. All rights reserved.

35. *Energy* The intensity (I) of a light source is inversely proportional to the square of the distance (d) from the source. If the intensity is 20 lumens at a distance of 8 ft, what is the intensity when the distance is 5 ft?

36. *Magnetism* The repulsive force (f) between the north poles of two magnets is inversely proportional to the square of the distance (d) between them. If the repulsive force is 18 lb when the distance is 3 in., find the repulsive force when the distance is 1.2 in.

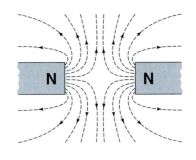

37. *Physics* For a constant temperature, the pressure (P) of a gas varies inversely as the volume (V). If the pressure is 25 lb/in² when the volume is 400 ft³, find the pressure when the volume is 150 ft³. Round to the nearest hundredth.

38. *Consumerism* The number of items (N) that can be purchased for a given amount of money is inversely proportional to the cost (C) of an item. If 390 items can be purchased when the cost per item is $6.40, how many items can be purchased when the cost per item is $6.00?

Critical Thinking

39. Determine whether the statement is true or false.
  a. In the direct variation equation $f = kx$, if x increases, then f increases.
  b. In the inverse variation equation $y = \dfrac{k}{x}$, if x increases, then y increases.
  c. In the direct variation equation $T = ks^2$, if s is doubled, then T doubles.

40. If y varies directly as x, and $x = 10$ when $y = 4$, find y when $x = 15$. Show that this variation can be solved by using a proportion.

41. **a.** The variable y varies directly as the cube of x. What is the effect on y when x is doubled?
  b. The variable y varies inversely as the cube of x. What is the effect on y when x is doubled?

42. Discuss how variation is used in prescribing medication for individuals.

43. Explain how proportions may be used in pricing large quantities of a purchase as compared to small quantities of a purchase. Is the unit price of a large purchase always smaller than the unit price of a small purchase?

44. Write a short history of pi, one of the universal constants of nature.

45. Explain the relationship between direct variation and proportion.

Copyright © Houghton Mifflin Company. All rights reserved.

Focus on Problem Solving

Relevant Information

Problems in mathematics or real life involve a question or a need and information or circumstances relating to that need. Solving problems in the sciences usually involves a question, an observation, and measurements of some kind.

One of the challenges of problem solving in the sciences is to separate the information that is relevant to the problem from other information. Following is an example from the physical sciences in which some relevant information was omitted.

Hooke's Law states that the distance that a weight will stretch a spring is directly proportional to the weight on the spring. That is, $d = kF$, where d is the distance the spring is stretched and F is the force. In an experiment to verify this law, some physics students were continually getting inconsistent results. Finally, the instructor discovered that the heat produced when the lights were turned on was affecting the experiment. In this case, relevant information was omitted—namely, that the temperature of the spring can affect the distance it will stretch.

A lawyer drove 8 miles to the train station. After a 35-minute ride of 18 miles, the lawyer walked 10 minutes to the office. Find the total time it took the lawyer to get to work.

From this situation, answer the following questions before reading on.

a. What is asked for?

b. Is there enough information to answer the question?

c. Is information given that is not needed?

Here are the answers.

a. We want the total time for the lawyer to get to work.

b. No. We do not know the time it takes the lawyer to get to the train station.

c. Yes. Neither the distance to the train station nor the distance of the train ride is necessary to answer the question.

For each of the following problems, answer these questions:

a. What is asked for?

b. Is there enough information to answer the question?

c. Is information given that is not needed?

1. A customer bought 6 boxes of strawberries and paid with a $20 bill. What was the change?

2. A board is cut into two pieces. One piece is 3 feet longer than the other piece. What is the length of the original board?

3. A family rented a car for their vacation and drove 680 miles. The cost of the rental car was $21 per day with 150 free miles per day and $.15 for each mile driven above the number of free miles allowed. How many miles did the family drive per day?

4. An investor bought 8 acres of land for $80,000. One and one-half acres were set aside for a park, and the remainder was developed into one-half-acre lots. How many lots were available for sale?

Copyright © Houghton Mifflin Company. All rights reserved.

5. You wrote checks of $43.67, $122.88, and $432.22 after making a deposit of $768.55. How much do you have left in your checking account?

Projects & Group Activities

Earned Run Average

One measure of a pitcher's success is earned run average. **Earned run average (ERA)** is the number of earned runs a pitcher gives up for every nine innings pitched. The definition of an earned run is somewhat complicated, but basically an earned run is a run that is scored as a result of hits and base running that involve no errors on the part of the pitcher's team. If the opposing team scores a run on an error (for example, a fly ball that should have been caught in the outfield was fumbled), then that is not an earned run.

A proportion is used to calculate a pitcher's ERA. Remember that the statistic involves the number of earned runs per *nine innings*. The answer is always rounded to the nearest hundredth. Here is an example:

Earned Run Average Leaders		
Year	Player, club	ERA
National League		
1990	Danny Darwin, Houston	2.21
1991	Dennis Martinez, Montreal	2.39
1992	Bill Swift, San Francisco	2.06
1993	Greg Maddux, Atlanta	2.36
1994	Greg Maddux, Atlanta	1.56
1995	Greg Maddux, Atlanta	1.63
1996	Kevin Brown, Florida	1.89
1997	Pedro Martinez, Montreal	1.90
1998	Greg Maddux, Atlanta	2.22
1999	Randy Johnson, Arizona	2.48
2000	Kevin K. Brown, Los Angeles	2.58
2001	Randy Johnson, Arizona	2.49
2002	Randy Johnson, Arizona	2.32
American League		
1990	Roger Clemens, Boston	1.93
1991	Roger Clemens, Boston	2.62
1992	Roger Clemens, Boston	2.41
1993	Kevin Appler, Kansas City	2.56
1994	Steve Ontiveros, Oakland	2.65
1995	Randy Johnson, Seattle	2.48
1996	Juan Guzman, Toronto	2.93
1997	Roger Clemens, Toronto	2.05
1998	Roger Clemens, Toronto	2.65
1999	Pedro Martinez, Boston	2.07
2000	Pedro Martinez, Boston	1.74
2001	Freddy Garcia, Seattle	3.05
2002	Pedro Martinez, Boston	2.26

During the 2001 regular baseball season, Chan Ho Park gave up 91 earned runs and pitched 234 innings for the Los Angeles Dodgers. Chan Ho Park's ERA is calculated as shown below.

Strategy

To find Chan Ho Park's ERA, let x = the number of earned runs for every nine innings pitched. Then set up a proportion. Solve the proportion for x.

Solution

$$\frac{91 \text{ earned runs}}{234 \text{ innings}} = \frac{x \text{ earned runs}}{9 \text{ innings}}$$
$$234x = 91(9)$$
$$234x = 819$$
$$\frac{234x}{234} = \frac{819}{234}$$
$$x = 3.5$$

Chan Ho Park's ERA for 2001 was 3.50.

1. In 1979, Jeff Reardon's rookie year, he pitched 21 innings for the Mets and gave up 4 earned runs. Calculate Reardon's ERA for 1979.

2. Roger Clemens's first year with the Boston Red Sox was 1984. During that season, he pitched 133.1 innings and gave up 64 earned runs. Calculate Clemens's ERA for 1984.

Copyright © Houghton Mifflin Company. All rights reserved.

3. During the 1998 baseball season, Pedro Martinez of the Boston Red Sox pitched 233.2 innings and gave up 75 earned runs. During the 1999 season, he gave up 49 earned runs and pitched 213.1 innings. During which season was his ERA lower? How much lower?

4. In 1987, Nolan Ryan had the lowest ERA of any pitcher in the major leagues. He gave up 65 earned runs and pitched 211.2 innings for the Astros. Calculate Ryan's ERA for 1987.

5. Find the necessary statistics for the pitcher of your "home team," and calculate that pitcher's ERA.

Joint Variation

A variation may involve more than two variables. If a quantity varies directly as the product of two or more variables, it is known as a **joint variation.**

The weight of a rectangular metal box is directly proportional to the volume of the box, given by length · width · height. Thus weight = $kLWH$.

The weight of a box with $L = 24$ in., $W = 12$ in., and $H = 12$ in. is 72 lb. Find the weight of another box with $L = 18$ in., $W = 9$ in., and $H = 18$ in.

$$\text{weight} = kLWH$$

$$72 = k(24)(12)(12)$$

$$\frac{72}{(24)(12)(12)} = k$$

$$\frac{1}{48} = k$$

$$\text{weight} = kLWH$$

$$\text{weight} = \frac{1}{48}(18)(9)(18)$$ ▶ Substitute $\frac{1}{48}$ for k. Substitute the dimensions of the other box into the equation.

$$\text{weight} = 60.75$$

The weight of the other box is 60.75 lb.

1. *Physics* The pressure (p) in a liquid varies directly as the product of the depth (d) and the density (D) of the liquid.
   a. Write the joint variation.
   b. If the pressure is 37.5 lb/in² when the depth is 100 in. and the density is 1.2 lb/in², find the pressure when the density remains the same and the depth is 60 in.

Copyright © Houghton Mifflin Company. All rights reserved.

2. *Electricity* The power (P) in an electrical circuit is directly proportional to the product of the current (I) and the square of the resistance (R).
   a. Write the joint variation.
   b. If the power is 100 watts when the current is 4 amps and the resistance is 5 ohms, find the power when the current is 2 amps and the resistance is 10 ohms.

3. *Sailing* The wind force (w) on a vertical surface varies directly as the product of the area (A) of the surface and the square of the wind velocity (v).
   a. Write the joint variation.
   b. When the wind is blowing at 30 mph, the force on an area of 10 ft² is 45 lb. Find the force on this area when the wind is blowing at 60 mph.
   c. What effect does doubling the area have on the force of the wind?
   d. What effect does doubling the speed of the wind have on the force of the wind?

Chapter Summary

Key Words

A **ratio** is the comparison of two quantities with the same unit. A ratio can be written in three ways: as a fraction, as two numbers separated by a colon, or as two numbers separated by the word *to*. A ratio is in simplest form when the two quantities do not have a common factor. [7.2A, p. 445]

Examples

The comparison 16 oz to 24 oz can be written as a ratio in simplest form as

$\frac{2}{3}$, 2 : 3, or 2 to 3.

A **rate** is the comparison of two quantities with different units. A rate is in simplest form when the two quantities do not have a common factor. [7.2A, p. 445]

You earned \$63 for working 6 h. The rate is written $\frac{\$21}{2\,\text{h}}$.

A **unit rate** is a rate in which the denominator is 1. [7.2A, p. 445]

You traveled 144 mi in 3 h. The unit rate is 48 mph.

The **metric system of measurement** is an internationally standardized system of measurement. It is based on the decimal system. The basic unit of length in the metric system is the **meter.** The basic unit of mass is the **gram.** The basic unit of capacity is the **liter.** In the metric system, prefixes to the basic unit denote the magnitude of each unit. [7.1A, pp. 437–438]

kilo-	= 1 000	deci-	= 0.1
hecto-	= 100	centi-	= 0.01
deca-	= 10	milli-	= 0.001

1 km = 1 000 m
1 kg = 1 000 g
1 kl = 1 000 L

1 m = 100 cm
1 m = 1 000 mm
1 g = 1 000 mg
1 L = 1 000 ml

Copyright © Houghton Mifflin Company. All rights reserved.

The standard **U.S. Customary System** units of length are inch, foot, yard, and mile. The standard units of weight are ounce, pound, and ton. The standard units of capacity are fluid ounce, cup, pint, quart, and gallon. The standard units of area are square inch, square foot, square yard, square mile, and acre. [7.3A, p. 449]

1 ft = 12 in.
1 yd = 3 ft
1 yd = 36 in.
1 mi = 5,280 ft

1 lb = 16 oz
1 ton = 2,000 lb

1 c = 8 fl oz
1 pt = 2 c
1 qt = 4 c
1 gal = 4 qt

$1 \text{ ft}^2 = 144 \text{ in}^2$
$1 \text{ yd}^2 = 9 \text{ ft}^2$
$1 \text{ acre} = 43,560 \text{ ft}^2$
$1 \text{ mi}^2 = 640 \text{ acres}$

Using units to organize and check the correctness of an application is called **dimensional analysis.** Equivalent measures are used to form conversion rates to change one unit in the U.S. Customary System of measurement to another. [7.3A, p. 449]

$$5 \text{ c} = 5 \text{ c} \cdot 1 = \frac{5 \text{ c}}{1} \cdot \frac{8 \text{ fl oz}}{1 \text{ c}}$$
$$= (5 \cdot 8) \text{ fl oz} = 40 \text{ fl oz}$$

Approximate equivalences between the U.S. Customary System and the metric system of measurement are used to form conversion rates to change one measurement to another. [7.3C, p. 453]

Units of Length

1 in. = 2.54 cm
1 m ≈ 3.28 ft
1 m ≈ 1.09 yd
1 mi ≈ 1.61 km

Units of Weight

1 oz ≈ 28.35 g
1 lb ≈ 454 g
1 kg ≈ 2.2 lb

Units of Capacity

1 L ≈ 1.06 qt
1 gal ≈ 3.79 L

A **proportion** is the equality of two ratios or rates. Each of the four members in a proportion is called a **term.**

first term ⟵ $\dfrac{a}{b}$ $=$ $\dfrac{c}{d}$ ⟶ third term
second term ⟵ $\phantom{\dfrac{a}{b}}$ ⟶ fourth term

The second and third terms of the proportion are called the **means,** and the first and fourth terms are called the **extremes.** [7.4A, p. 459]

In the proportion $\frac{3}{5} = \frac{12}{20}$, 5 and 12 are the means; 3 and 20 are the extremes.

Copyright © Houghton Mifflin Company. All rights reserved.

The equation $y = kx$, where k is a constant, is an example of a **direct variation.** The equation $y = kx$ is read "y varies directly as x" or "y is directly proportional to x." Some direct variation equations are written in the form $y = kx^2$, which is read "y varies directly as the square of x" or "y is directly proportional to the square of x." The constant k is called the **constant of variation** or the **constant of proportionality.** [7.5A, p. 467]

R varies directly as S, and $R = 9$ when $S = 6$.

$$R = kS$$
$$9 = k \cdot 6$$

An **inverse variation** is one that can be written in the form $y = \dfrac{k}{x}$, where k is a constant. The equation $y = \dfrac{k}{x}$ is read "y varies inversely as x" or "y is inversely proportional to x." Some inverse variation equations are written in the form $y = \dfrac{k}{x^2}$. This equation is read "y varies inversely as the square of x" or "y is inversely proportional to the square of x." [7.5B, p. 469]

T varies inversely as V, and $T = 12$ when $V = 8$.

$$T = \frac{k}{V}$$
$$12 = \frac{k}{8}$$

Essential Rules and Procedures

To find a unit rate, divide the number in the numerator of the rate by the number in the denominator of the rate. [7.2A, p. 446]

You earned $41 for working 4 hours.

$$41 \div 4 = 10.25$$

The unit rate is $10.25 per hour.

To set up a proportion, keep the same units in the numerator and the same units in the denominator. [7.4B, p. 461]

Three machines fill 5 cereal boxes per minute. How many boxes can 8 machines fill per minute?

$$\frac{3 \text{ machines}}{5 \text{ cereal boxes}} = \frac{8 \text{ machines}}{x \text{ cereal boxes}}$$

To solve a proportion, use the fact that the product of the means equals the product of the extremes. For the proportion $\dfrac{a}{b} = \dfrac{c}{d}$, $ad = bc$. [7.4A, p. 459]

$$\frac{6}{25} = \frac{9}{x}$$
$$25 \cdot 9 = 6 \cdot x$$
$$225 = 6x$$
$$\frac{225}{6} = \frac{6x}{6}$$
$$37.5 = x$$

Converting between units in the metric system involves moving the decimal point to the right or to the left. Listing the units in order from largest to smallest will indicate how many places to move the decimal point and in which direction. [7.1A, p. 438]

3.7 kg = 3 700 g
2 387 m = 2.387 km
9.5 L = 9 500 ml

Copyright © Houghton Mifflin Company. All rights reserved.

Chapter Review Exercises

1. Convert 1.25 km to meters.

2. Convert 0.450 g to milligrams.

3. Write the comparison 100 lb to 100 lb as a ratio in simplest form using a fraction, a colon, and the word *to.*

4. Write 18 roof supports for every 9 ft as a rate in simplest form.

5. Write $628 earned in 40 h as a unit rate.

6. Write 8 h to 15 h as a ratio in simplest form using a fraction.

7. Convert 96 in. to yards.

8. Convert 72 oz to pounds.

9. Convert 36 fl oz to cups.

10. Convert $1\frac{1}{4}$ mi to feet.

11. Solve: $\frac{n}{3} = \frac{8}{15}$

12. Write 15 lb of fertilizer for 12 trees as a rate in simplest form.

13. Write 171 mi driven in 3 h as a unit rate.

14. Solve $\frac{2}{3.5} = \frac{n}{12}$. Round to the nearest hundredth.

15. Find the number of milliliters in 1 qt. Round to the nearest hundredth.

16. The winning long jump at a track meet was 29 ft. Convert this distance to meters. Round to the nearest hundredth.

17. Convert the 100-meter dash to feet.

18. A backpack tent weighs 2.1 kg. Find its weight in pounds.

19. Express 30 mph in kilometers per hour.

20. Convert 75 km/h to miles per hour. Round to the nearest hundredth.

Copyright © Houghton Mifflin Company. All rights reserved.

21. Write 18 c of milk for 24 pints of ice cream as a rate in simplest form.

22. Find the constant of variation when y varies directly as x and $x = 30$ when $y = 10$.

23. Given that T varies directly as the square of S, and $T = 50$ when $S = 5$, find T when $S = 120$.

24. Given that y varies inversely as x, and $y = 0.2$ when $x = 5$, find y when $x = 25$.

25. *Mechanical Drawing* Find the total length, in centimeters, of the shaft in the diagram at the right.

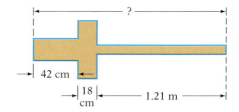

26. *Technology* In 3 years, the price of a graphing calculator went from $125 to $75. What is the ratio of the decrease in price to the original price?

27. *The Food Industry* In the butcher department of a supermarket, hamburger meat weighing 12 lb is equally divided and placed into 16 containers. How many ounces of hamburger meat is in each container?

28. *Investments* An investment of $8,000 earns $520 in dividends. At the same rate, how much money must be invested to earn $780 in dividends?

29. *Physics* Hooke's Law states that the distance (d) a spring will stretch is directly proportional to the weight (w) on the spring. A weight of 5 lb will stretch the spring 2 in. How far will a weight of 28 lb stretch a spring?

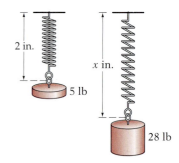

30. *Lawn Care* The directions on a bag of plant food recommend $\frac{1}{2}$ lb for every 50 ft² of lawn. How many pounds of plant food should be used on a lawn of 275 ft²?

31. *Sports* Find the speed in feet per second of a baseball pitched at 87 mph.

32. *Physics* Boyle's Law states that the volume (V) of a gas varies inversely as the pressure (P), assuming the temperature remains constant. The pressure of the gas in a balloon is 6 lb/in² when the volume is 2.5 ft³. Find the volume of the balloon if the pressure increases to 12 lb/in².

33. *Business* Two attorneys share the profits of their firm in the ratio 3:2. If the attorney receiving the larger amount of this year's profits receives $96,000, what amount does the other attorney receive?

Copyright © Houghton Mifflin Company. All rights reserved.

Chapter Test

1. Convert 4 650 cm to meters.

2. Convert 4.1 L to milliliters.

3. Write the comparison 3 yd to 24 yd as a ratio in simplest form using a fraction, using a colon, and using the word *to*.

4. Write 16 oz of sugar for 64 cookies as a rate in simplest form.

5. Write 120 mi driven in 200 min as a unit rate.

6. Write 200 ft to 100 ft as a ratio in simplest form using a fraction.

7. Convert $2\frac{3}{5}$ tons to pounds.

8. Convert $2\frac{1}{2}$ c to fluid ounces.

9. How many ounces are in $3\frac{1}{4}$ lb?

10. How many inches are in $8\frac{1}{2}$ ft?

11. Solve: $\dfrac{n}{5} = \dfrac{3}{20}$

12. Write 8 ft walked in 4 s as a unit rate.

13. Convert 4.3 c to ounces.

14. Convert 42 yd to feet.

15. Write 2,860 ft² mowed in 6 h as a unit rate. Round to the nearest hundredth.

16. Solve: $\dfrac{n}{4} = \dfrac{8}{9}$. Round to the nearest hundredth.

17. Convert 12 oz to grams.

18. A record ski jump was 547 ft. Convert this distance to meters. Round to the nearest hundredth.

19. Convert the 1 000-meter run to yards.

20. A backpack tent weighs 1.9 kg. Find its weight in pounds.

Copyright © Houghton Mifflin Company. All rights reserved.

21. Express 35 mph in kilometers per hour.

22. Convert 60 km/h to miles per hour. Round to the nearest hundredth.

23. Find the constant of proportionality when y varies inversely as x, and $y = 10$ when $x = 2$.

24. Given that R varies directly as P, and $P = 20$ when $R = 4$, find P when $R = 15$.

25. Given that U varies inversely as the square of V, and $V = 4$ when $U = 20$, find U when $V = 2$.

26. *Physical Fitness* A body builder who had been lifting weights for 2 years went from an original weight of 165 lb to 190 lb. What is the ratio of the original weight to the increased weight?

27. *Taxes* The sales tax on a $95 purchase is $7.60. Find the sales tax on a car costing $39,200.

28. *Elections* A preelection survey showed that 3 out of 4 registered voters would vote in a county election. At this rate, how many registered voters would vote in a county with 325,000 registered voters?

29. *The Food Industry* A gourmet food store buys 24 lb of cheese for $126. The cheese is cut and packaged and sold in 12-ounce packages for $7.50 each. Find the difference between the store's purchase price of the cheese and the store's income from selling the packages of cheese.

30. *Architecture* The scale on the architectural drawings for a new gymnasium is 1 in. equals 4 ft. How long is one of the rooms if it measures $12\frac{1}{2}$ in. on the drawing?

31. *Sports* Find the speed in feet per second of a soccer ball kicked at 52 mph. Round to the nearest hundredth.

32. *Travel* The stopping distance (d) of a car varies directly as the square of the speed (v) of the car. If a car traveling 40 mph requires 130 ft to stop, find the stopping distance for a car traveling 60 mph.

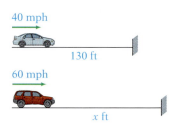

33. *Mechanics* The speed of a gear varies inversely as the number of teeth. If a gear that has 25 teeth makes 160 rpm, how many revolutions per minute will a gear that has 40 teeth make?

Copyright © Houghton Mifflin Company. All rights reserved.

Cumulative Review Exercises

1. Simplify: $18 \div \frac{6-3}{9} - (-3)$

2. Convert 1.2 gal to quarts.

3. Subtract: $7\frac{5}{12} - 3\frac{5}{9}$

4. Simplify: $\frac{4}{5} \div \frac{4}{5} + \frac{2}{3}$

5. Find the quotient of 342 and -3.

6. Evaluate $2a - 3ab$ when $a = 2$ and $b = -3$.

7. Solve: $5x - 20 = 0$

8. Solve: $3(x - 4) + 2x = 3$

9. Graph -3.5 on the number line.

10. Graph $x < -3$.

11. Simplify: $(-5)^2 - (-8) \div (7 - 5)^2 \cdot 2 - 8$

12. Simplify: $\left(-\frac{2}{3}\right)\left(-\frac{3}{4}\right)^2$

13. Simplify: $\sqrt{169}$

14. Simplify: $5 - 2(1 - 3a) + 2(a - 3)$

15. Multiply: $(4a^3b)(-5a^2b^3)$

16. Simplify: $-3y^2 + 3y - y^2 - 6y$

17. Find the ordered-pair solution of $y = 3x - 2$ that corresponds to $x = -1$.

18. Write 30 cents to 1 dollar as a ratio in simplest form.

19. Write $19,425 in 5 months as a unit rate.

20. The price of gasoline is $1.97 per gallon. Find the price per liter.

21. Solve: $\frac{2}{3} = \frac{n}{48}$

22. Simplify: $\dfrac{\frac{1}{2} + \frac{3}{4}}{2 - \frac{5}{8}}$

Copyright © Houghton Mifflin Company. All rights reserved.

23. Evaluate $-2\sqrt{x^2 - 3y}$ when $x = 4$ and $y = -3$.

24. Solve: $3x + 3(x + 4) = 4(x + 2)$

25. *Public Transportation* The figure at the right shows the average amount spent annually per household on public transportation, by region, in the United States. Find the difference between the average amount spent monthly per household in the northeast and in the south. Round to the nearest cent.

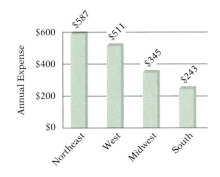

Average Annual Expense per Household on Public Transportation in the United States
Source: Bureau of Labor Statistics consumer expenditure survey

26. Five less than two-thirds of a number is three. Find the number.

27. Translate "the difference between four times a number and three times the sum of the number and two" into a variable expression. Then simplify.

28. *Travel* Your odometer reads 18,325 mi before you embark on a 125-mile trip. After you have driven $1\frac{1}{2}$ h, the odometer reads 18,386 mi. How many miles are left to drive?

29. *Banking* You had a balance of $422.89 in your checking account. You then made a deposit of $122.35 and wrote a check for $279.76. Find the new balance in your checking account.

PAYMENT/ DEBIT (−)		√ T	FEE (IF ANY) (−)	DEPOSIT/ CREDIT (+)	BALANCE $ 422	89
$			$	$ 122	35	
279	76					

30. *Computers* A data processor finished $\frac{2}{5}$ of a job on the first day and $\frac{1}{3}$ on the second day. What part of the job is to be finished on the third day?

31. *Elections* In a recent city election, $\frac{2}{3}$ of the registered voters voted. How many votes were cast if the city had 31,281 registered voters?

32. *Travel* A car is driven 402.5 mi on 11.5 gal of gas. Find the number of miles traveled per gallon of gas.

33. *Mechanics* At a certain speed, the engine revolutions per minute of a car in fourth gear is 2,500. This is two-thirds of the rpm of the engine in third gear. Find the rpm of the engine in third gear.

Copyright © Houghton Mifflin Company. All rights reserved.

Percent

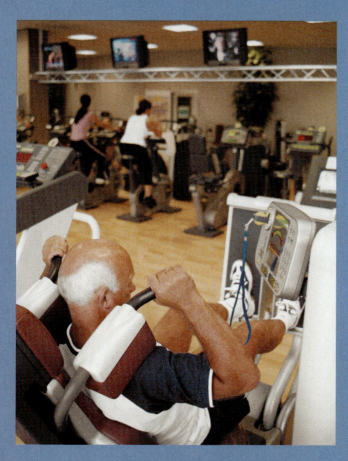

Everyone knows that exercise is very important for your health. This man works out on a machine in a gym, one of the many ways to get the exercise your body needs. Exercise is also related to diet. If you are active, your body needs more nutrients, such as protein, than if you are inactive. To determine how many grams of protein an active person needs versus a sedentary person, use the formula given in the **Project on page 520.**

Copyright © Houghton Mifflin Company. All rights reserved.

Need help? For online student resources, visit this website: **math.college.hmco.com**

Prep Test

For Exercises 1–6, multiply or divide.

1. $19 \times \dfrac{1}{100}$

2. 23×0.01

3. 0.47×100

4. $0.06 \times 47,500$

5. $60 \div 0.015$

6. $8 \div \dfrac{1}{4}$

7. Multiply $\dfrac{5}{8} \times 100$. Write the answer as a decimal.

8. Write $\dfrac{200}{3}$ as a mixed number.

9. Divide $28 \div 16$. Write the answer as a decimal.

Go Figure

I have 2 brothers and 1 sister. My father's parents have 10 grandchildren. My mother's parents have 11 grandchildren. If no divorces or remarriages occurred, how many first cousins do I have?

Copyright © Houghton Mifflin Company. All rights reserved.

Copyright © Houghton Mifflin Company. All rights reserved.

SECTION 8.1   Percent

OBJECTIVE **A**

Percents as decimals or fractions

Percent means "parts of 100." The figure at the right has 100 parts. Because 19 of the 100 parts are shaded, 19% of the figure is shaded.

19 parts to 100 parts can be expressed as the ratio $\frac{19}{100}$. One percent can be expressed as 1 part to 100, or $\frac{1}{100}$. Thus 1% is $\frac{1}{100}$ or 0.01.

"A population growth rate of 5%," "a manufacturer's discount of 40%," and "an 8% increase in pay" are typical examples of the many ways in which percent is used in applied problems. When solving problems involving a percent, it is usually necessary either to rewrite the percent as a fraction or a decimal, or to rewrite a fraction or a decimal as a percent.

To write a percent as a fraction, remove the percent sign and multiply by $\frac{1}{100}$.

Write 67% as a fraction.

Remove the percent sign and multiply by $\frac{1}{100}$. $67\% = 67\left(\frac{1}{100}\right) = \frac{67}{100}$

To write a percent as a decimal, remove the percent sign and multiply by 0.01.

Write 19% as a decimal.   19%   =   19(0.01)   =   0.19

Remove the percent sign and multiply by 0.01. This is the same as moving the decimal point two places to the left.

> Move the decimal point two places to the left. Then remove the percent sign.

---

1 *Example*   Write 150% as a fraction and as a decimal.

Solution $150\% = 150\left(\frac{1}{100}\right) = \frac{150}{100} = 1\frac{1}{2}$
$150\% = 150(0.01) = 1.50$

1 *You Try It*   Write 110% as a fraction and as a decimal.

Your Solution

---

2 *Example*   Write $66\frac{2}{3}\%$ as a fraction.

Solution $66\frac{2}{3}\% = 66\frac{2}{3}\left(\frac{1}{100}\right)$

$= \frac{200}{3}\left(\frac{1}{100}\right) = \frac{2}{3}$

2 *You Try It*   Write $16\frac{3}{8}\%$ as a fraction.

Your Solution

Solutions on p. S20

③ *Example* Write 0.35% as a decimal.

Solution $0.35\% = 0.35(0.01) = 0.0035$

③ *You Try It* Write 0.8% as a decimal.

Your Solution

Solution on p. S20

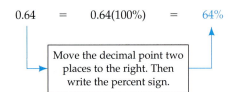

OBJECTIVE **B**

Fractions and decimals as percents

A fraction or decimal can be written as a percent by multiplying by 100%. Since 100% is $\frac{100}{100} = 1$, **multiplying by 100% is the same as multiplying by 1.**

Write $\frac{7}{8}$ as a percent.

Multiply $\frac{7}{8}$ by 100%. $\frac{7}{8} = \frac{7}{8}(100\%) = \frac{700}{8}\% = 87.5\%$

Write 0.64 as a percent. $0.64 \quad = \quad 0.64(100\%) \quad = \quad 64\%$

Multiply by 100%. This is the same as moving the decimal point two places to the right.

> Move the decimal point two places to the right. Then write the percent sign.

④ *Example* Write 1.78 as a percent.

Solution $1.78 = 1.78(100\%) = 178\%$

④ *You Try It* Write 0.038 as a percent.

Your Solution

⑤ *Example* Write $\frac{3}{11}$ as a percent. Write the remainder in fractional form.

Solution $\frac{3}{11} = \frac{3}{11}(100\%) = \frac{300}{11}\%$

$= 27\frac{3}{11}\%$

⑤ *You Try It* Write $\frac{9}{7}$ as a percent. Write the remainder in fractional form.

Your Solution

⑥ *Example* Write $1\frac{1}{7}$ as a percent. Round to the nearest tenth of a percent.

Solution $1\frac{1}{7} = \frac{8}{7} = \frac{8}{7}(100\%)$

$= \frac{800}{7}\% \approx 114.3\%$

⑥ *You Try It* Write $1\frac{5}{9}$ as a percent. Round to the nearest tenth of a percent.

Your Solution

Solutions on p. S20

Copyright © Houghton Mifflin Company. All rights reserved.

8.1 EXERCISES

OBJECTIVE **A**

1. a. Explain how to convert a percent to a fraction.
    b. Explain how to convert a percent to a decimal.

2. Explain why multiplying a number by 100% does not change the value of the number.

Write as a fraction and as a decimal.

3. 5%   4. 60%   5. 30%   6. 90%

7. 250%   8. 140%   9. 28%   10. 66%

11. 35%   12. 8%   13. 29%   14. 83%

Write as a fraction.

15. $11\frac{1}{9}\%$ 16. $12\frac{1}{2}\%$ 17. $37\frac{1}{2}\%$ 18. $31\frac{1}{4}\%$

19. $66\frac{2}{3}\%$ 20. $45\frac{5}{11}\%$ 21. $6\frac{2}{3}\%$ 22. $68\frac{3}{4}\%$

23. $\frac{1}{2}\%$ 24. $83\frac{1}{3}\%$ 25. $6\frac{1}{4}\%$ 26. $3\frac{1}{3}\%$

Write as a decimal.

27. 7.3%   28. 9.1%   29. 15.8%   30. 16.7%

31. 0.3%   32. 0.9%   33. 121.2%   34. 18.23%

35. 62.14%   36. 0.15%   37. 8.25%   38. 5.05%

39. *Pets* The figure at the right shows some ways in which owners pamper their dogs. What fraction of the owners surveyed would buy a house or a car with their dog in mind?

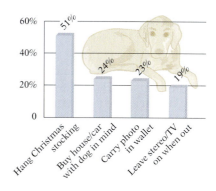

How Owners Pamper Their Dogs
Source: Purina Survey

Copyright © Houghton Mifflin Company. All rights reserved.

OBJECTIVE **B**

Write as a percent.

40. 0.15

41. 0.37

42. 0.05

43. 0.02

44. 0.175

45. 0.125

46. 1.15

47. 1.36

48. 0.62

49. 0.96

50. 2.09

51. 0.07

Write as a percent. Round to the nearest tenth of a percent.

52. $\dfrac{27}{50}$

53. $\dfrac{83}{100}$

54. $\dfrac{37}{200}$

55. $\dfrac{1}{3}$

56. $\dfrac{5}{11}$

57. $\dfrac{4}{9}$

58. $\dfrac{7}{8}$

59. $\dfrac{9}{20}$

60. $1\dfrac{2}{3}$

61. $2\dfrac{1}{2}$

62. $\dfrac{2}{5}$

63. $\dfrac{1}{6}$

Write as a percent. Write the remainder in fractional form.

64. $\dfrac{17}{50}$

65. $\dfrac{17}{25}$

66. $\dfrac{3}{8}$

67. $\dfrac{9}{16}$

68. $1\dfrac{1}{4}$

69. $2\dfrac{5}{8}$

70. $1\dfrac{5}{9}$

71. $2\dfrac{5}{6}$

72. $\dfrac{12}{25}$

73. $\dfrac{7}{30}$

74. $\dfrac{3}{7}$

75. $\dfrac{2}{9}$

Critical Thinking

76. Determine whether the statement is true or false. If the statement is false, give an example to show that it is false.
  a. Multiplying a number by a percent always decreases the number.
  b. Dividing by a percent always increases the number.
  c. The word *percent* means "per hundred."
  d. A percent is always less than one.

77. *Compensation* Employee A had an annual salary of $42,000, Employee B had an annual salary of $48,000, and Employee C had an annual salary of $46,000 before each employee was given a 5% raise. Which of the three employee's annual salary is now the highest? Explain how you arrived at your answer.

78. *Compensation* Each of three employees earned an annual salary of $45,000 before Employee A was given a 3% raise, Employee B was given a 6% raise, and Employee C was given a 4.5% raise. Which of the three employees now has the highest annual salary? Explain how you arrived at your answer.

Copyright © Houghton Mifflin Company. All rights reserved.

Copyright © Houghton Mifflin Company. All rights reserved.

 SECTION 8.2   ## The Basic Percent Equation

OBJECTIVE **A**

The basic percent equation

A real estate broker receives a payment that is 6% of a $275,000 sale. To find the amount the broker receives requires answering the question "6% of $275,000 is what?" This sentence can be written using mathematical symbols and then solved for the unknown number. Recall that **of** is written as · (times), **is** is written as = (equals), and **what** is written as n (the unknown number).

6%	of	$275,000	is	what?
↓	↓	↓	↓	↓
percent	·	base	=	amount
6%		$275,000		n

$$0.06 \cdot \$275{,}000 = n$$
$$\$16{,}500 = n$$

The broker receives a payment of $16,500.

The solution was found by solving the basic percent equation for amount.

The Basic Percent Equation

Percent · base = amount

Find 2.5% of 800.

Use the basic percent equation.
Percent = 2.5% = 0.025,
base = 800, amount = n

Percent · base = amount
$$0.025 \cdot 800 = n$$
$$20 = n$$

2.5% of 800 is 20.

A recent promotional game at a grocery store listed the probability of winning a prize as "1 chance in 2." A percent can be used to describe the chance of winning. This requires answering the question, "What percent of 2 is 1?"

The chance of winning can be found by solving the basic percent equation for percent.

What	percent	of	2	is	1?
	↓	↓	↓	↓	↓
	percent	·	base	=	amount
	n		2		1

$$n \cdot 2 = 1$$
$$n = \frac{1}{2}$$
$$n = \frac{1}{2}(100\%) = 50\%$$ ▶ Write the fraction as a percent.

There is a 50% chance of winning a prize.

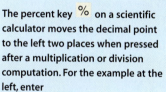

Calculator Note

The percent key % on a scientific calculator moves the decimal point to the left two places when pressed after a multiplication or division computation. For the example at the left, enter

800 ✕ 2 · 5 % =

The display reads 20.

Take Note

We have written $n \cdot 20 = 32$ because that is the form of the basic percent equation. We could have written $20n = 32$. The important point is that each side of the equation is divided by 20, the coefficient of n.

32 is what percent of 20?

Use the basic percent equation.
Percent = n, base = 20, amount = 32

Write 1.6 as a percent.

32 is 160% of 20.

$$\text{Percent} \cdot \text{base} = \text{amount}$$
$$n \cdot 20 = 32$$
$$\frac{20n}{20} = \frac{32}{20}$$
$$n = 1.6$$
$$n = 160\%$$

Each year an investor receives a payment that equals 8% of the value of an investment. This year that payment amounted to $640. To find the value of the investment this year, we must answer the question, "8% of what value is $640?"

The value of the investment can be found by solving the basic percent equation for the base.

$$\begin{array}{ccccc} 8\% & \text{of} & \text{what} & \text{is} & \$640? \\ \downarrow & \downarrow & \downarrow & \downarrow & \downarrow \\ \text{Percent} & \cdot & \text{base} & = & \text{amount} \\ 8\% & & n & & 640 \end{array}$$

$$0.08 \cdot n = 640$$
$$\frac{0.08n}{0.08} = \frac{640}{0.08}$$
$$n = 8{,}000$$

This year the investment is worth $8,000.

Take Note

The base in the basic percent equation will usually follow the phrase *percent of*. Some percent problems may use the word *find*. In this case, we can substitute *what is* for find. See Example 1 below.

62% of what is 800? Round to the nearest tenth.

Use the basic percent equation.
Percent = 62% = 0.62, base = n,
amount = 800

62% of 1,290.3 is approximately 800.

$$\text{Percent} \cdot \text{base} = \text{amount}$$
$$0.62 \cdot n = 800$$
$$\frac{0.62n}{0.62} = \frac{800}{0.62}$$
$$n \approx 1{,}290.3$$

Note from the previous three problems that if any two parts of the basic percent equation are given, the third part can be found.

① Example Find 9.4% of 240.

Strategy To find the amount, solve the basic percent equation.
Percent = 9.4% = 0.094,
base = 240, amount = n

Solution $\text{Percent} \cdot \text{base} = \text{amount}$
$0.094 \cdot 240 = n$
$22.56 = n$

22.56 is 9.4% of 240.

① You Try It Find $33\frac{1}{3}\%$ of 45.

Your Strategy

Your Solution

Solution on p. S20

Copyright © Houghton Mifflin Company. All rights reserved.

2 Example   What percent of 30 is 12?

Strategy   To find the percent, solve the basic percent equation. Percent = n, base = 30, amount = 12

Solution   Percent · base = amount
$$n \cdot 30 = 12$$
$$\frac{30n}{30} = \frac{12}{30}$$
$$n = 0.4$$
$$n = 40\%$$

12 is 40% of 30.

2 You Try It   25 is what percent of 40?

Your Strategy

Your Solution

---

 3 Example   60 is 2.5% of what?

Strategy   To find the base, solve the basic percent equation. Percent = 2.5% = 0.025, base = n, amount = 60

Solution   Percent · base = amount
$$0.025 \cdot n = 60$$
$$\frac{0.025n}{0.025} = \frac{60}{0.025}$$
$$n = 2,400$$

60 is 2.5% of 2,400.

3 You Try It $16\frac{2}{3}$% of what is 15?

Your Strategy

Your Solution

Solutions on p. S20

OBJECTIVE B

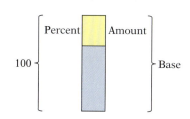

Percent problems using proportions

Percent problems can also be solved by using proportions. The proportion method is based on writing two ratios with quantities that can be found in the basic percent equation. One ratio is the percent ratio, written as $\frac{\text{percent}}{100}$. The second ratio is the amount-to-base ratio, written as $\frac{\text{amount}}{\text{base}}$. These two ratios form the proportion

$$\frac{\textbf{percent}}{\textbf{100}} = \frac{\textbf{amount}}{\textbf{base}}$$

The proportion method can be illustrated by a diagram. The rectangle at the right is divided into two parts. The whole rectangle is represented by 100 and the part by percent. On the other side, the whole rectangle is represented by the base and the part by amount. The ratio of the percent to 100 is equal to the ratio of the *amount* to the *base*.

Copyright © Houghton Mifflin Company. All rights reserved.

What is 32% of 85?

Sketch a diagram.

Percent = 32,
base = 85,
amount = n

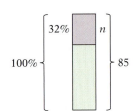

$$\frac{\text{percent}}{100} = \frac{\text{amount}}{\text{base}}$$

$$\frac{32}{100} = \frac{n}{85}$$

$$100 \cdot n = 32 \cdot 85$$
$$100n = 2{,}720$$
$$\frac{100n}{100} = \frac{2{,}720}{100}$$
$$n = 27.2$$

32% of 85 is 27.2.

---

4 **Example** 24% of what is 16? Round to the nearest hundredth.

Solution
$$\frac{\text{percent}}{100} = \frac{\text{amount}}{\text{base}}$$

$$\frac{24}{100} = \frac{16}{n}$$
$$24 \cdot n = 100 \cdot 16$$
$$24n = 1{,}600$$
$$n = \frac{1{,}600}{24} \approx 66.67$$

16 is approximately 24% of 66.67.

4 **You Try It** 8 is 25% of what?

Your Solution

---

5 **Example** Find 1.2% of 42.

Solution
$$\frac{\text{percent}}{100} = \frac{\text{amount}}{\text{base}}$$

$$\frac{1.2}{100} = \frac{n}{42}$$
$$1.2 \cdot 42 = 100 \cdot n$$
$$50.4 = 100n$$
$$\frac{50.4}{100} = \frac{100n}{100}$$
$$0.504 = n$$

1.2% of 42 is 0.504.

5 **You Try It** Find 0.74% of 1,200.

Your Solution

---

6 **Example** What percent of 52 is 13?

Solution
$$\frac{\text{percent}}{100} = \frac{\text{amount}}{\text{base}}$$

$$\frac{n}{100} = \frac{13}{52}$$
$$n \cdot 52 = 100 \cdot 13$$
$$52n = 1{,}300$$
$$\frac{52n}{52} = \frac{1{,}300}{52}$$
$$n = 25$$

25% of 52 is 13.

6 **You Try It** What percent of 180 is 54?

Your Solution

Solutions on pp. S20–S21

Copyright © Houghton Mifflin Company. All rights reserved.

OBJECTIVE **C**

VIDEO & DVD   CD TUTOR   WEB   SSM

Applications

The circle graph at the right shows the causes of death for all police officers who died while on duty during a recent year. What percent of the deaths were due to traffic accidents? Round to the nearest tenth of a percent.

Strategy To find the percent:

▶ Find the total number of officers who died in the line of duty.
▶ Use the basic percent equation.
Percent = n, base = total number killed, amount = number of deaths due to traffic accidents = 73.

Solution $58 + 73 + 6 + 19 = 156$

$$\text{Percent} \cdot \text{base} = \text{amount}$$
$$n \cdot 156 = 73$$
$$\frac{156n}{156} = \frac{73}{156}$$
$$n \approx 0.468$$
$$n \approx 46.8\%$$

46.8% of the deaths were due to traffic accidents.

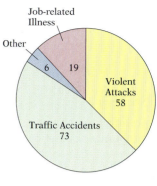

Figure 8.1 Causes of Death for Police Officers Killed in the Line of Duty
Source: International Union of Police Associations

---

7 Example

During a recent year, 276 billion product coupons were issued by manufacturers. Shoppers redeemed 4.8 billion of these coupons. (*Source:* NCH NuWorld Consumer Behavior Study, America Coupon Council) What percent of the coupons issued were redeemed by customers? Round to the nearest tenth of a percent.

Strategy

To find the percent, use the basic percent equation.
Percent = n, base = number of coupons issued = 276 billion, amount = number of coupons redeemed = 4.8 billion

Solution

$$\text{Percent} \cdot \text{base} = \text{amount}$$
$$n \cdot 276 = 4.8$$
$$\frac{276n}{276} = \frac{4.8}{276}$$
$$n \approx 0.017$$
$$n \approx 1.7\%$$

Of the product coupons issued, 1.7% were redeemed by customers.

7 You Try It

An instructor receives a monthly salary of $4,330, and $649.50 is deducted for income tax. Find the percent of the instructor's salary deducted for income tax.

Your Strategy

Your Solution

Solution on p. S21

Copyright © Houghton Mifflin Company. All rights reserved.

8 Example

A taxpayer pays a tax rate of 35% for state and federal taxes. The taxpayer has an income of $47,500. Find the amount of state and federal taxes paid by the taxpayer.

Strategy

To find the amount, solve the basic percent equation.
Percent = 35% = 0.35, base = 47,500, amount = n

Solution

Percent · base = amount
$$0.35 \cdot 47{,}500 = n$$
$$16{,}625 = n$$

The amount of taxes paid is $16,625.

8 You Try It

According to Board-Trac, approximately 19% of the country's 2.4 million surfers are women. Estimate the number of female surfers in this country. Write the number in standard form.

Your Strategy

Your Solution

9 Example

A department store has a blue blazer on sale for $114, which is 60% of the original price. What is the difference between the original price and the sale price?

Strategy

To find the difference between the original price and the sale price:

▶ Find the original price. Solve the basic percent equation.
Percent = 60% = 0.60, amount = 114, base = n
▶ Subtract the sale price from the original price.

Solution

Percent · base = amount
$$0.60 \cdot n = 114$$
$$\frac{0.60n}{0.60} = \frac{114}{0.60}$$
$$n = 190$$

$$190 - 114 = 76$$

The difference in price is $76.

9 You Try It

An electrician's wage this year is $30.13 per hour, which is 115% of last year's wage. What was the increase in the hourly wage over last year?

Your Strategy

Your Solution

Solutions on p. S21

Copyright © Houghton Mifflin Company. All rights reserved.

8.2 EXERCISES

OBJECTIVE   **A**

Solve. Use the basic percent equation.

1. 8% of 100 is what?

2. 16% of 50 is what?

3. 0.05% of 150 is what?

4. 0.075% of 625 is what?

5. 15 is what percent of 90?

6. 24 is what percent of 60?

7. What percent of 16 is 6?

8. What percent of 24 is 18?

9. 10 is 10% of what?

10. 37 is 37% of what?

11. 2.5% of what is 30?

12. 10.4% of what is 52?

13. Find 10.7% of 485.

14. Find 12.8% of 625.

15. 80% of 16.25 is what?

16. 26% of 19.5 is what?

17. 54 is what percent of 2,000?

18. 8 is what percent of 2,500?

19. 16.4 is what percent of 4.1?

20. 5.3 is what percent of 50?

21. 18 is 240% of what?

22. 24 is 320% of what?

OBJECTIVE   **B**

Solve. Use the proportion method.

23. 26% of 250 is what?

24. Find 18% of 150.

25. 37 is what percent of 148?

26. What percent of 150 is 33?

Copyright © Houghton Mifflin Company. All rights reserved.

27. 68% of what is 51?

28. 126 is 84% of what?

29. What percent of 344 is 43?

30. 750 is what percent of 50?

31. 82 is 20.5% of what?

32. 2.4% of what is 21?

33. What is 6.5% of 300?

34. Find 96% of 75.

35. 7.4 is what percent of 50?

36. What percent of 1,500 is 693?

37. Find 50.5% of 124.

38. What is 87.4% of 225?

39. 120% of what is 6?

40. 14 is 175% of what?

41. What is 250% of 18?

42. 325% of 4.4 is what?

43. 87 is what percent of 29?

44. What percent of 38 is 95?

OBJECTIVE **C**

Solve.

45. *Automotive Technology* A mechanic estimates that the brakes of an RV still have 6,000 mi of wear. This amount is 12% of the estimated safe-life use of the brakes. What is the estimated safe-life use of the brakes?

46. *Demographics* Of the 281,422,000 people in the United States, 28.6% are under the age of 20. (*Source:* U.S. Census 2000) How many people in the United States are under the age of 20?

47. *The Labor Force* In Arkansas, Wal-Mart's home state, 41,000 workers are Wal-Mart employees. This is 3.3% of the Arkansas labor force. (*Source:* Wal-Mart Company Reports; Bureau of Labor Statistics) Find the number of workers in the Arkansas labor force.

48. *Astronomy* The aphelion of Earth is its distance when it is farthest from the sun. The perihelion is its distance when it is nearest the sun, as shown in the figure at the right. What percent of the aphelion is the perihelion? Round to the nearest tenth of a percent.

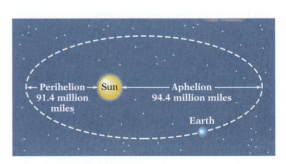

Copyright © Houghton Mifflin Company. All rights reserved.

49. *Fire Science*   A fire department received 24 false alarms out of a total of 200 alarms received. What percent of the alarms received were false alarms?

50. *Demographics*   The table at the right shows the projected increase in population from 2000 to 2040 for each of four counties in the Central Valley of California. What percent of the 2000 population of Sacramento County is the increase in population?

County	2000 Population	Projected Increase by 2040
Sacramento	1,200,000	900,000
Kern	651,700	948,300
Fresno	794,200	705,800
San Joaquin	562,000	737,400

Source: California Department of Finance

51. *Demographics*   The table at the right shows the projected increase in population from 2000 to 2040 for each of four counties in the Central Valley of California. What percent of the 2000 population of Kern County is the increase in population? Round to the nearest tenth of a percent.

52. *Business*   An antiques shop owner expects to receive $16\frac{2}{3}\%$ of the shop's sales as profit. What is the expected profit in a month when the total sales are $24,000?

53. *Poultry*   In a recent year, North Carolina produced 1,300,000 lb of turkey. This was 18.6% of the U.S. total in that year. (*Source:* U.S. Census Bureau) Calculate the U.S. total turkey production for that year. Round to the nearest million.

54. *Depreciation*   A used mobile home was purchased for $43,600. This amount was 64% of the cost of the mobile home when it was new. What was the new mobile home cost?

55. *Agriculture*   A farmer is given an income tax credit of 15% of the cost of some farm machinery. What tax credit would the farmer receive on farm equipment that cost $85,000?

56. *Financing*   A used car is sold for $18,900. The buyer of the car makes a down payment of $3,780. What percent of the selling price is the down payment?

57. *Medicine*   The active ingredient in a prescription skin cream is clobetasol propionate. It is 0.05% of the total ingredients. How many grams of clobetasol propionate are in a 30-gram tube of this cream?

58. *Charitable Giving*   In a recent year, Americans gave $212 billion to charities. Use the figure at the right to determine how much of that amount came from individuals.

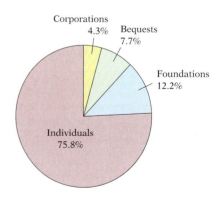

Charitable Giving
Sources: American Association of Fundraising Counsel; AP

59. *Astronomy*   The diameter of Earth is approximately 8,000 mi, and the diameter of the sun is approximately 870,000 mi. What percent of Earth's diameter is the sun's diameter?

Copyright © Houghton Mifflin Company. All rights reserved.

60. ● *Pets* The average costs associated with owning a dog over an average 11-year life span are shown in the graph at the right. These costs do not include the price of the puppy when purchased. The category labeled "Other" includes such expenses as fencing and repairing furniture damaged by the pet. What percent of the total cost is spent on food? Round to the nearest tenth of a percent.

61. *Manufacturing* During a quality control test, a manufacturer of computer boards found that 56 boards were defective. This was 0.7% of the total number of computer boards tested. How many of the tested computer boards were not defective?

62. ● *Agriculture* Of the 572 million pounds of cranberries grown in the United States in a recent year, Wisconsin growers produced 291.72 million pounds. What percent of the total cranberry crop was produced in Wisconsin?

63. ● *Politics* The results of a survey in which 32,840 full-time college and university faculty members were asked to describe their political views are shown at the right. How many more faculty members described their political views as liberal than described their views as far left?

64. ● *Politics* The results of a survey in which 32,840 full-time college and university faculty members were asked to describe their political views are shown at the right. How many fewer faculty members described their political views as conservative than described their views as middle of the road?

Cost of Owning a Dog
Source: American Kennel Club, *USA Today* research

Political View	Percent of Faculty Members Responding
Far left	5.3%
Liberal	42.3%
Middle of the road	34.3%
Conservative	17.7%
Far right	0.3%

Source: Higher Education Research Institute, UCLA

Critical Thinking

65. Find 10% of a number and subtract it from the original number. Now take 10% of the new number and subtract it from the new number. Is this the same as taking 20% of the original number? Explain.

66. Increase a number by 10%. Now decrease the new number by 10%. Is the result the original number? Explain.

67. *Compensation* Your employer agrees to give you a 5% raise after one year on the job, a 6% raise the next year, and a 7% raise the following year. Is your salary after the third year greater than, less than, or the same as it would be if you had received a 6% raise each year?

68. ✏ Visit a savings and loan institution or credit union to research and write a report on the meaning of *points* as it relates to a loan.

69. ✏ Find five different uses of percents and explain why percent was used in those instances.

Copyright © Houghton Mifflin Company. All rights reserved.

Copyright © Houghton Mifflin Company. All rights reserved.

SECTION 8.3   ## Percent Increase and Percent Decrease

OBJECTIVE   **A**

Percent increase

Percent increase is used to show how much a quantity has increased over its original value. The statements "sales volume increased by 11% over last year's sales volume" and "employees received an 8% pay increase" are illustrations of the use of percent increase.

▶ **Point of Interest**

The largest 1-day percent increase in the Dow Jones Industrial Average occurred on October 6, 1931. The Dow gained approximately 15% of its value.

 The population of the world is expected to increase from 6.1 billion people in 2000 to 9.3 billion people in 2050. (*Source:* U.N. Population Division, State of the World 2002, Worldwatch Institute) Find the percent increase in the world population from 2000 to 2050. Round to the nearest tenth of a percent.

Strategy   To find the percent increase:

▶ Find the increase in the population from 2000 to 2050.
▶ Use the basic percent equation.
  Percent = n, base = 6.1, amount = the amount of increase

Solution $9.3 - 6.1 = 3.2$

$$\text{Percent} \cdot \text{base} = \text{amount}$$
$$n \cdot 6.1 = 3.2$$
$$\frac{6.1n}{6.1} = \frac{3.2}{6.1}$$
$$n \approx 0.525 = 52.5\%$$

The population is expected to increase 52.5% from 2000 to 2050.

 Example

A sales associate was earning $11.60 per hour before an 8% increase in pay. What is the new hourly wage? Round to the nearest cent.

Strategy

To find the new hourly wage:

▶ Use the basic percent equation to find the increase in pay.
  Percent = 8% = 0.08, base = 11.60,
  amount = n
▶ Add the amount of increase to the original wage.

Solution

$$\text{Percent} \cdot \text{base} = \text{amount}$$
$$0.08 \cdot 11.60 = n$$
$$0.93 \approx n$$
$$\$11.60 + \$.93 = \$12.53$$

The new hourly wage is $12.53.

You Try It

An automobile manufacturer increased the average mileage on a car from 17.5 mi/gal to 18.2 mi/gal. Find the percent increase in mileage.

Your Strategy

Your Solution

Solution on p. S21

OBJECTIVE **B**

Percent decrease

Point of Interest

The largest 1-day percent decrease in the Dow Jones Industrial Average occurred on October 19, 1987. The Dow lost approximately 23% of its value.

Percent decrease is used to show how much a quantity has decreased from its original value. The statements "the president's approval rating has decreased 9% over last month" and "there has been a 15% decrease in the number of industrial accidents" are illustrations of the use of percent decrease.

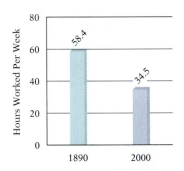

Figure 8.3 Average Number of Hours Worked per Week
Source: Bureau of Labor Statistics

Figure 8.3 shows the average number of hours worked per week by Americans in 1890 and in 2000. Find the percent decrease in the number of hours worked per week. Round to the nearest tenth of a percent.

Strategy   To find the percent decrease:
▶ Find the decrease in the number of hours worked per week.
▶ Use the basic percent equation.
Percent = n, base = 58.4, amount = the amount of decrease

Solution $58.4 - 34.5 = 23.9$

$$\text{Percent} \cdot \text{base} = \text{amount}$$
$$n \cdot 58.4 = 23.9$$
$$\frac{58.4n}{58.4} = \frac{23.9}{58.4}$$
$$n \approx 0.409 = 40.9\%$$

▶ The base is the *original value*. The amount is the *amount of decrease.*

The number of hours worked per week decreased 40.9% from 1890 to 2000.

---

2 *Example*

Violent crime in a small city decreased from 27 per 1,000 people to 24 per 1,000 people. Find the percent decrease in violent crime. Round to the nearest tenth of a percent.

Strategy

To find the percent decrease in crime:
▶ Find the decrease in the number of crimes.
▶ Use the basic percent equation to find the percent decrease in crime.

Solution

$27 - 24 = 3$
$\text{Percent} \cdot \text{base} = \text{amount}$
$n \cdot 27 = 3$
$n = \dfrac{3}{27} \approx 0.111$

▶ The amount is the amount of decrease.

Violent crime decreased by approximately 11.1% during the year.

2 *You Try It*

The market value of a luxury car decreased 24% during the year. Find the value of a luxury car that cost $47,000 last year. Round to the nearest dollar.

Your Strategy

Your Solution

Solution on p. S21

Copyright © Houghton Mifflin Company. All rights reserved.

8.3 EXERCISES

Copyright © Houghton Mifflin Company. All rights reserved.

OBJECTIVE   A

Solve. Round percents to the nearest tenth of a percent.

1. *Commuting*   From 1990 to 2000, the average commuting time for workers in Atlanta, Georgia, increased from 26 min to 31.2 min. (*Source:* U.S. Census and analysis by The Road Information Program) What was the percent increase in the average commuting time?

2. *The Olympics*   In 1924, the number of events in the Winter Olympics was 14. The 2002 Winter Olympics in Salt Lake City held 78 medal events. (*Source:* David Wallenchinsky's *The Complete Book of the Winter Olympics*) Find the percent increase in the number of events in the Winter Olympics from 1924 to 2002.

3. *Education*   The graph at the right shows the number of women attending four-year colleges. Find the percent increase in the number of women at four-year colleges from 1983 to 2003.

4. *Incarceration*   In 1990, 4.4 million adults were in the nation's correctional system. By 2001, the number had increased to 6.6 million. (*Source:* Associated Press) Find the percent increase in the number of people in the correctional system in the United States from 1990 to 2001.

5. *Wealth*   The table at the right shows the estimated number of millionaire households in the United States. Find the percent increase in the estimated number of households containing millionaires from 1975 to 2005.

6. *Nutrition*   According to the U.S. Department of Agriculture, Americans consumed an average of 3,100 calories per person per day in the 1960s and an average of 3,700 calories per person per day in the 1990s. From the 1960s to the 1990s, what was the percent increase in the average number of calories consumed per person per day?

7. *Telecommuting*   The graph at the right shows the projected growth in the number of telecommuters.
    a. During which 2-year period is the percent increase in the number of telecommuters the greatest?
    b. During which 2-year period is the percent increase in the number of telecommuters the least?
    c. Does the growth in telecommuting increase more slowly or more rapidly as we move from 1998 to 2006?

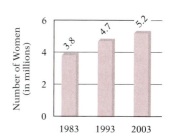

Women Attending Four-Year Colleges
Source: Copyright © 2000, *USA Today*. Reprinted with permission.

Year	Number of Households Containing Millionaires
1975	350,000
1997	3,500,000
2005	5,600,000

Source: Affluent Market Institute

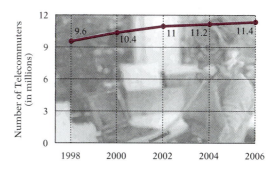

Projected Growth in Telecommuting
Source: Find/SVP

OBJECTIVE   B

Solve. Round percents to the nearest tenth of a percent.

8. *Consumerism*   A family reduced its normal monthly food bill of $320 by $50. What percent decrease does this represent?

9.  *The Atmosphere* The ozone hole over Antarctica shrank from 26.5 million square kilometers in 2001 to 15.6 million square kilometers in September 2002. (*Source:* Associated Press) Find the percent decrease in the size of the ozone hole. Round to the nearest tenth of a percent.

10. *Manufacturing* A new production method reduced the time needed to clean a piece of metal from 8 min to 5 min. What percent decrease does this represent?

11. *Marketing* As a result of an increased number of service lines at a grocery store, the average amount of time a customer waits in line has decreased from 3.8 min to 2.5 min. Find the percent decrease.

12. *The Environment* As a consequence of the shrinking ice fields on Mt. Kilimanjaro, the surface of Africa's Lake Chad has shrunk from 135,000 square miles to 6,500 square miles. (*Source:* Associated Press) Find the percent decrease in the size of the surface of Lake Chad. Round to the nearest tenth of a percent.

13. *Consumerism* It is estimated that the value of a new car is reduced 30% after 1 year of ownership. Find the value of a $21,900 new car after 1 year.

14. *Business* A sales manager's average monthly expense for gasoline was $92. After joining a car pool, the manager was able to decrease gasoline expenses by 22%. What is the average monthly gasoline bill now?

15. *The Military* The graph at the right shows the number of active-duty U.S. military personnel in 1990 and in 2000. Which branch of the military had the greatest percent decrease in personnel from 1990 to 2000? What was the percent decrease for this branch of the service?

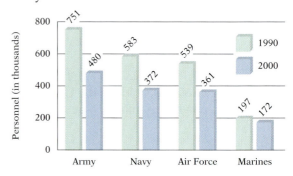

Number of Active-Duty U.S. Military Personnel
Source: Fiscal 2000 Annual Report to the President and Congress by the Secretary of Defense

Critical Thinking

16. *Consumerism* A department store gives you three discount coupons: a 10% discount, a 20% discount, and a 30% discount off any item. You have decided to purchase a CD player costing $225 and use all three coupons. Is there a particular order in which you should ask to have the discount coupons applied to your purchase so that your purchase price is as small as possible? Explain.

17. *Consumerism* A wide-screen TV costing $3,000 was on sale for 30% off. An additional 10% off the sale price was offered to customers who paid by check. Calculate the sales price after the two discounts. Is this the same as a discount of 40%? Find the equivalent discount of the successive discounts.

18. Define *per millage.* Explain its relation to percent.

Copyright © Houghton Mifflin Company. All rights reserved.

Copyright © Houghton Mifflin Company. All rights reserved.

SECTION 8.4 ▶ Markup and Discount

OBJECTIVE **A**

Markup

Cost is the price a merchandising business or retailer pays for a product. **Selling price,** or **retail price,** is the price for which a merchandising business or retailer sells a product to a customer.

The difference between selling price and cost is called **markup.** Markup is added to cost to cover the expenses of operating a business and provide a profit to the owners.

Markup can be expressed as a percent of the cost, or it can be expressed as a percent of the selling price. Here we present markup as a percent of the cost, which is the most common practice.

The percent markup is called the **markup rate,** and it is expressed as the markup based on the cost.

A diagram is useful when expressing the markup equations. In the diagram at the right, the total length is the selling price. One part of the diagram is the cost, and the other part is the markup.

> ### The Markup Equations
>
> $M = S - C$ where M = markup
> $S = C + M$ S = selling price
> $M = r \cdot C$ C = cost
> r = markup rate

Take Note

If C is added to both sides of the first equation, $M = S - C$, the result is the second equation listed, $S = C + M$.

The manager of a clothing store buys a sports jacket for $80 and sells the jacket for $116. Find the markup rate.

Strategy   To find the markup rate:

▶ Find the markup by solving the formula $M = S - C$ for M. $S = 116$, $C = 80$
▶ Solve the formula $M = r \cdot C$ for r.
   M = the markup, $C = 80$

Solution $M = S - C$
         $M = 116 - 80$
         $M = 36$

         $M = r \cdot C$
         $36 = r \cdot 80$
         $\dfrac{36}{80} = \dfrac{80r}{80}$
         $0.45 = r$

The markup rate is 45%.

1 *Example*

A soft top for a convertible that costs a dealer $250 has a markup rate of 35%. Find the markup.

Strategy

To find the markup, solve the formula $M = r \cdot C$ for M.
$r = 35\% = 0.35, C = 250$

Solution

$M = r \cdot C$

$M = 0.35 \cdot 250$

$M = 87.50$

The markup is $87.50.

1 *You Try It*

An outboard motor costing $650 has a markup rate of 45%. Find the markup.

Your Strategy

Your Solution

2 *Example*

A graphing calculator costing $45 is sold for $80. Find the markup rate. Round to the nearest tenth of a percent.

Strategy

To find the markup rate:

▶ Find the markup by solving the formula $M = S - C$ for M.
$S = 80, C = 45$
▶ Solve the formula $M = r \cdot C$ for r.
M = the markup, $C = 45$

Solution

$M = S - C$

$M = 80 - 45$

$M = 35$

$M = r \cdot C$

$35 = r \cdot 45$

$\dfrac{35}{45} = \dfrac{45r}{45}$

$0.778 \approx r$

The markup rate is 77.8%.

2 *You Try It*

A laser printer costing $950 is sold for $1,450. Find the markup rate. Round to the nearest tenth of a percent.

Your Strategy

Your Solution

Solutions on p. S21

Copyright © Houghton Mifflin Company. All rights reserved.

3 *Example*

A fishing reel with a cost of $50 has a markup rate of 22%. Find the selling price.

Strategy

To find the selling price:

▶ Find the markup by solving the equation $M = r \cdot C$ for M.
$r = 22\% = 0.22$, $C = 50$
▶ Solve the formula $S = C + M$ for S.
$C = 50$, $M =$ the markup

Solution

$M = r \cdot C$
$M = 0.22 \cdot 50$
$M = 11$

$S = C + M$
$S = 50 + 11$
$S = 61$

The selling price is $61.

3 *You Try It*

A basketball with a cost of $82.50 has a markup rate of 42%. Find the selling price.

Your Strategy

Your Solution

Solution on pp. S21–S22

OBJECTIVE **B**

Discount

A retailer may reduce the regular price of a product for a promotional sale because the goods are damaged, odd sizes or colors, or discontinued items. The **discount,** or **markdown,** is the amount by which a retailer reduces the regular price of a product.

The percent discount is called the **discount rate** and is usually expressed as a percent of the original price (the regular selling price).

The Discount Equations

$M = R - S$ where $M =$ discount or markdown
$M = r \cdot R$ $\quad\quad\quad S =$ sale price
$S = (1 - r)R$ $\quad\quad R =$ regular price
$\quad\quad\quad\quad\quad\quad\quad r =$ discount rate

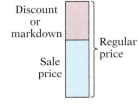

Copyright © Houghton Mifflin Company. All rights reserved.

④ Example

A laptop computer that regularly sells for $1,850 is on sale for $1,480. Find the discount rate.

Strategy

To find the discount rate:

▶ Find the discount by solving the formula $M = R - S$ for M. $R = 1,850$, $S = 1,480$
▶ Solve the formula $M = r \cdot R$ for r. M = the discount, $R = 1,850$

Solution

$$M = R - S \qquad\qquad M = r \cdot R$$
$$M = 1,850 - 1,480 \qquad 370 = r \cdot 1,850$$
$$M = 370 \qquad\qquad \frac{370}{1,850} = \frac{1,850r}{1,850}$$
$$0.2 = r$$

The discount rate is 20%.

④ You Try It

A camera has a regular price of $325. It is currently on sale for $253.50. Find the discount rate.

Your Strategy

Your Solution

---

⑤ Example

A necklace with a regular price of $450 is on sale for 32% off the regular price. Find the sale price.

Strategy

To find the sale price, solve the formula $S = (1 - r)R$ for S. $r = 32\% = 0.32$, $R = 450$

Solution

$$S = (1 - r)R$$
$$S = (1 - 0.32)450$$
$$S = (0.68)450$$
$$S = 306$$

The sale price is $306.

⑤ You Try It

A garage door opener regularly priced at $312 is on sale for 25% off the regular price. Find the sale price.

Your Strategy

Your Solution

---

⑥ Example

A video game is on sale for $104 after a markdown of 35%. Find the regular price.

Strategy

To find the regular price, solve the formula $S = (1 - r)R$ for R. $S = 104$, $r = 35\% = 0.35$

Solution

$$S = (1 - r)R$$
$$104 = (1 - 0.35)R$$
$$104 = 0.65R$$
$$\frac{104}{0.65} = \frac{0.65R}{0.65}$$
$$160 = R$$

The regular price is $160.

⑥ You Try It

A large-screen TV, marked down 35%, is on sale for $1,495. Find the regular price.

Your Strategy

Your Solution

Solutions on p. S22

Copyright © Houghton Mifflin Company. All rights reserved.

8.4 EXERCISES

OBJECTIVE A

Solve.

1. A bicycle costing $110 has a markup rate of 55%. Find the markup.

2. A television set costing $315 has a markup rate of 30%. Find the markup.

3. A watch which cost $98 is being sold for $156.80. Find the markup rate.

4. A set of golf clubs costing $360 is sold for $630. Find the markup rate.

5. A freezer selling for $520 cost the appliance dealer $360. Find the markup rate. Round to the nearest tenth of a percent.

6. A digital camera costing $320 is sold for $479. What is the markup rate on the digital camera? Round to the nearest tenth of a percent.

7. A markup rate of 25% is applied to a printer costing $1,750. Find the selling price.

8. A flat of strawberries costing $7.60 has a markup rate of 125%. Find the selling price.

9. A PC game costing $47 has a markup rate of 75%. Find the selling price.

10. A markup rate of 58% is applied to a leather jacket costing $225. Find the selling price.

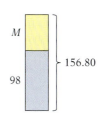

OBJECTIVE B

Solve.

11. An exercise bicycle that regularly sells for $460 is on sale for $350. Find the markdown.

12. A suit with a regular price of $179 is on sale for $119. Find the markdown.

13. Find the markdown rate on an oak bedroom set that has a regular price of $1,295 and is on sale for $995. Round to the nearest tenth of a percent.

14. A DVD player with a regular price of $495 is on sale for $380. Find the markdown rate. Round to the nearest tenth of a percent.

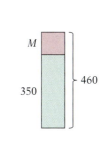

Copyright © Houghton Mifflin Company. All rights reserved.

15. A compact disc player with a regular price of $325 is on sale for $201.50. Find the discount rate.

16. Find the discount rate on a scooter that has a regular price of $178 and is on sale for $103.24.

17. A computer with a regular price of $1,995 is on sale for 30% off the regular price. Find the sale price.

18. What is the sale price on a painting that has a regular price of $1,600 and is on sale for 45% off the regular price?

19. A soccer ball that has a regular price of $42 is on sale for 40% off the regular price. What is the sale price of the soccer ball?

20. A gold ring with a regular price of $415 is on sale for 55% off the regular price. Find the sale price.

21. A mechanic's tool set is on sale for $180 after a markdown of 40% off the regular price. Find the regular price.

22. A battery with a discount price of $65 is on sale for 22% off the regular price. Find the regular price. Round to the nearest cent.

23. A cell phone that is on sale for $80 has a markdown of 35% off the regular price. What is the regular price of the cell phone? Round to the nearest cent.

Critical Thinking

24. During a recent promotion, Nestle offered a 25¢-off coupon on a bag of Toll House Morsels. If the grocery store offers double coupons for this item (the customer receives a discount that is double the value of a manufacturer's coupon), what is the percent discount in price on a 12-ounce bag costing $3.89? Round to the nearest tenth of a percent.

25. A computer is on sale for a discount of 20% off the regular price of $5,500. An additional 10% discount on the sale price was offered. Is the result a 30% discount? What is the single discount that would give the same sale price?

26. ✏️ A promotional sale at a department store offers 25% off the sale price of merchandise that has already been discounted 25% off the regular price. Is this the same as a sale that offers 50% off the regular price? If not, which sale price gives the lower price? Explain your answer.

Save 25¢
on one bag of Nestlé
Toll House Morsels
any variety
(except 6 oz. size)

Copyright © Houghton Mifflin Company. All rights reserved.

Simple Interest

VIDEO & DVD   CD TUTOR   WEB   SSM

OBJECTIVE   **A**

Simple interest

If you deposit money in a savings account at a bank, the bank will pay you for the privilege of using that money. The amount you deposit in the savings account is called the **principal.** The amount the bank pays you for the privilege of using the money is called **interest.**

If you borrow money from the bank in order to buy a car, the amount you borrow is called the **principal.** The additional amount of money you must pay the bank, above and beyond the amount borrowed, is called **interest.**

Whether you deposit money or borrow it, the amount of interest paid is usually computed as a percent of the principal. The percent used to determine the amount of interest to be paid is the **interest rate.** Interest rates are given for specific periods of time, such as months or years.

Interest computed on the original principal is called **simple interest.** Simple interest is the cost of a loan that is for a period of about 1 year or less.

The Simple Interest Formula

$I = Prt$,   where I = simple interest earned, P = principal,
r = annual simple interest rate, t = time (in years)

In the simple interest formula, t is the time in years. If a time period is given in days or months, it must be converted to years and then substituted in the formula for t. For example,

120 days is $\dfrac{120}{365}$ of a year.   6 months is $\dfrac{6}{12}$ of a year.

Shannon O'Hara borrowed $5,000 for 90 days at an annual simple interest rate of 7.5%. Find the simple interest due on the loan.

Strategy   To find the simple interest owed, use the simple interest formula.

$$P = 5,000, \ r = 7.5\% = 0.075, \ t = \dfrac{90}{365}$$

Solution $I = Prt$

$$I = 5,000(0.075)\left(\dfrac{90}{365}\right)$$

$$I \approx 92.47$$

The simple interest due on the loan is $92.47.

In the example above, we calculated that the simple interest due on Shannon O'Hara's 90-day, $5,000 loan was $92.47. This means that at the end of the 90 days, Shannon owes $5,000 + $92.47 = $5,092.47. The principal plus the interest owed on a loan is called the **maturity value.**

Copyright © Houghton Mifflin Company. All rights reserved.

Formula for the Maturity Value of a Simple Interest Loan

$M = P + I,$ where M = the maturity value, P = the principal, I = the simple interest

The example below illustrates solving the simple interest formula for the interest rate. The solution requires the Multiplication Property of Equations.

Ed Pabas took out a 45-day, $12,000 loan. The simple interest on the loan was $168. To the nearest tenth of a percent, what is the simple interest rate?

Strategy To find the simple interest rate, use the simple interest formula. $P = 12{,}000, t = \dfrac{45}{365}, I = 168$

Solution

$$I = Prt$$

$$168 = 12{,}000r\left(\frac{45}{365}\right)$$

$$168 = \frac{540{,}000}{365}r$$

$$\frac{365}{540{,}000}(168) = \frac{365}{540{,}000} \cdot \frac{540{,}000}{365}r$$

$$0.114 \approx r$$

The simple interest rate on the loan is 11.4%.

1 **Example**	**1** **You Try It**
You arrange for a 9-month bank loan of $9,000 at an annual simple interest rate of 8.5%. Find the total amount you must repay to the bank.	William Carey borrowed $12,500 for 8 months at an annual simple interest rate of 9.5%. Find the total amount due on the loan.
Strategy	**Your Strategy**

To calculate the maturity value:

► Find the simple interest due on the loan by solving the simple interest formula for I.

$t = \dfrac{9}{12}, P = 9{,}000, r = 8.5\% = 0.085$

► Use the formula for the maturity value of a simple interest loan, $M = P + I$.

Solution   **Your Solution**

$I = Prt$

$I = 9{,}000(0.085)\left(\dfrac{9}{12}\right)$

$I = 573.73$

$M = P + I$
$M = 9{,}000 + 573.75$
$M = 9{,}573.75$

The total amount owed to the bank is $9,573.75.

Solution on p. S22

Copyright © Houghton Mifflin Company. All rights reserved.

8.5 EXERCISES

Copyright © Houghton Mifflin Company. All rights reserved.

OBJECTIVE A

1. Explain what each variable in the simple interest formula represents.

2. Explain the difference between interest and interest rate.

Solve.

3. **a.** In the table below, the interest rate is an annual simple interest rate. Complete the table by calculating the simple interest due on the loan.

Loan Amount	Interest Rate	Period	Interest
$5,000	6%	1 month	_____
$5,000	6%	2 months	_____
$5,000	6%	3 months	_____
$5,000	6%	4 months	_____
$5,000	6%	5 months	_____

Use the pattern of your answers in the table to find the simple interest due on a $5,000 loan that has an annual simple interest of rate 6% for a period of:
b. 6 months   **c.** 7 months   **d.** 8 months   **e.** 9 months

4. Use your solutions to Exercise 3 to answer the following questions:
   a. If you know the simple interest due on a 1-month loan, explain how you can use that figure to calculate the simple interest due on a 7-month loan for the same principal and the same interest rate.
   b. If the time period of a loan is doubled but the principal and interest rate remain the same, how many times greater is the simple interest due on the loan?

5. Kristi Yang borrowed $15,000. The term of the loan was 90 days, and the annual simple interest rate was 7.4%. Find the simple interest due on the loan.

6. Hector Elizondo took out a 75-day loan of $7,500 at an annual interest rate of 9.5%. Find the simple interest due on the loan.

7. To finance the purchase of 15 new cars, the Lincoln Car Rental Agency borrowed $100,000 for 9 months at an annual interest rate of 9%. What is the simple interest due on the loan?

8. A home builder obtained a preconstruction loan of $50,000 for 8 months at an annual interest rate of 9.5%. What is the simple interest due on the loan?

9. Assume that Visa charges Francesca 1.6% per month on her unpaid balance. Find the interest owed to Visa when her unpaid balance for the month is $1,250.

10. The Mission Valley Credit Union charges its customers an interest rate of 2% per month on money that is transferred into an account that is over-drawn. Find the interest owed to the credit union for 1 month when $800 is transferred into an overdrawn account.

11. Find the simple interest Jacob Zucker owes on a 2-year loan of $8,000 at an annual interest rate of 9%.

12. Find the simple interest Kara Tanamachi owes on a $1\frac{1}{2}$-year loan of $1,500 at an annual interest rate of 7.5%.

13. An auto parts dealer borrowed $150,000 at a 9.5% annual simple interest rate for 1 year. Find the maturity value of the loan.

14. A corporate executive took out a $25,000 loan at an 8.2% annual simple interest rate for 1 year. Find the maturity value of the loan.

15. Capitol City Bank approves a home-improvement loan application for $14,000 at an annual simple interest rate of 10.25% for 270 days. What is the maturity value of the loan?

16. A credit union loans a member $5,000 for the purchase of a used car. The loan is made for 18 months at an annual simple interest rate of 6.9%. What is the maturity value of the car loan?

17. A $12,000 investment earned $462 in interest in 6 months. Find the annual simple interest rate on the loan.

18. Michele Gabrielle borrowed $3,000 for 9 months and paid $168.75 in simple interest on the loan. Find the annual simple interest rate that Michele paid on the loan.

19. An investor earned $937.50 on an investment of $50,000 in 75 days. Find the annual simple interest rate earned on the investment.

20. Don Glover borrowed $18,000 for 210 days and paid $604.80 in simple interest on the loan. What annual simple interest rate did Don pay on the loan?

Critical Thinking

21. Interest has been described as a rental fee for money. Explain this description of interest.

22. Visit a savings and loan officer to collect information about the different kinds of home loans. Write a short essay describing the different kinds of loans available.

Copyright © Houghton Mifflin Company. All rights reserved.

Focus on Problem Solving

Using a Calculator as a Problem-Solving Tool

A calculator is an important tool of problem solving. Here are a few problems to solve with a calculator. You may need to research some of the questions to find information you do not know.

1. Choose any single-digit positive number. Multiply the number by 1,507 and 7,373. What is the answer? Choose another positive single-digit number and again multiply by 1,507 and 7,373. What is the answer? What pattern do you see? Why does this work?

2. The gross domestic product in 2002 was $10,445,600,000,000. Is this more or less than the amount of money that would be placed on the last square of a standard checkerboard if 1 cent were placed on the first square, 2 cents were placed on the second square, 4 cents were placed on the third square, 8 cents were placed on the fourth square, and so on until the 64th square was reached?

3. Which of the reciprocals of the first 16 natural numbers have a terminating decimal representation and which have a repeating-decimal representation?

4. What is the largest natural number n for which
$$4^n > 1 \cdot 2 \cdot 3 \cdot 4 \cdot 5 \cdot \cdots \cdot n?$$

5. If $1,000 bills are stacked one on top of another, is the height of $1 billion less than or greater than the height of the Washington Monument?

6. What is the value of $1 + \cfrac{1}{1 + \cfrac{1}{1 + \cfrac{1}{1 + \cfrac{1}{1 + 1}}}}$?

7. Calculate 15^2, 35^2, 65^2, and 85^2. Study the results. Make a conjecture about a relationship between a number ending in 5 and its square. Use your conjecture to find 75^2 and 95^2. Does your conjecture work for 125^2?

8. Find the sum of the first 1,000 natural numbers. (*Hint:* You could just start adding $1 + 2 + 3 + \cdots$, but even if you performed one operation every 3 seconds, it would take you an hour to find the sum. Instead, try pairing the numbers and then adding the pairs. Pair 1 and 1,000, 2 and 999, 3 and 998, and so on. What is the sum of each pair? How many pairs are there? Use this information to answer the original question.)

9. For a borrower to qualify for a home loan, a bank requires that the monthly mortgage payment be less than 25% of a borrower's monthly take-home income. A laboratory technician has deductions for taxes, insurance, and retirement that amount to 25% of the technician's monthly gross income. What minimum monthly income must this technician earn to receive a bank loan that has a mortgage payment of $1,200 per month?

Copyright © Houghton Mifflin Company. All rights reserved.

Projects & Group Activities

Buying a Car

Suppose a student has an after-school job to earn money to buy and maintain a car. We will make assumptions about the monthly costs in several categories in order to determine how many hours per week the student must work to support the car. Assume that the student earns $8.50 per hour.

1. Monthly payment

   Assume that the car cost $8,500 with a down payment of $1,020. The remainder is financed for 3 years at an annual simple interest rate of 9%.

   Monthly payment = _____

2. Insurance

   Assume that insurance costs $1,500 per year.

   Monthly insurance payment = _____

3. Gasoline

   Assume that the student travels 750 miles per month, that the car travels 25 miles per gallon of gasoline, and that gasoline costs $1.50 per gallon.

   Number of gallons of gasoline purchased per month = _____

   Monthly cost for gasoline = _____

4. Miscellaneous

   Assume $.33 per mile for upkeep.

   Monthly expense for upkeep = _____

5. Total monthly expenses for the monthly payment, insurance, gasoline, and miscellaneous = _____

6. To find the number of hours per month that the student must work to finance the car, divide the total monthly expenses by the hourly rate.

   Number of hours per month = _____

7. To find the number of hours per week that the student must work, divide the number of hours per month by 4.

   Number of hours per week = _____

   The student has to work _____ hours per week to pay the monthly car expenses.

If you own a car, make out your own expense record. If you do not own a car, make assumptions about the kind of car that you would like to purchase, and calculate the total monthly expenses that you would have. An insurance company will give you rates on different kinds of insurance. An automobile club can give you approximations of miscellaneous expenses.

Copyright © Houghton Mifflin Company. All rights reserved.

Demography Demography is the statistical study of human populations. Many groups are interested in the size of certain segments of the population and projections of population growth. For example, public school administrators want estimates of the number of school-aged children who will be living in their districts 10 years from now.

The U.S. government provides estimates of what the population in the United States will be in the future. You can find these projections at the Census Bureau web site at **www.census.gov.** Provided there are three different projections, a lowest series, a middle series, and a highest series. These reflect different theories on how fast the population of this country will grow.

The table below contains data from that website. The figures are from the Census Bureau's middle series of projections.

	Age	Under 5	5 - 17	18 - 24	25 - 34	35 - 44	45 - 54	55 - 64	65 & older
2010	Male	10,272	26,639	15,388	19,286	19,449	21,706	16,973	16,966
2010	Female	9,827	25,363	14,774	19,565	19,993	22,455	18,457	22,749
2050	Male	13,748	35,227	18,734	25,034	24,550	22,340	21,127	36,289
2050	Female	13,165	33,608	18,069	25,425	25,039	23,106	22,517	45,710

Population Projections of the United States, by Age and Sex (in thousands)

For the exercises below, round all percents to the nearest hundredth of a percent.

1. Which of the age groups listed are of interest to public school officials?

2. Which age group is of interest to retirement home administrators?

3. Which age group is of concern to accountants determining future benefits to be paid out by the Social Security Administration?

4. The population of which age group is of interest to manufacturers of disposable diapers?

5. The population of which age group is of primary concern to college and university admissions officers?

6. In which age groups do the males outnumber the females? In which do the females outnumber the males?

7. What percent of the projected population aged 65 and older in the year 2010 is female? Does this percent decrease in the projected population in 2050? If so, by how much?

8. Find the difference between the percent of the population that will be under 18 in 2010 and the percent that will be under 18 in 2050.

9. Assume that the work force consists of people aged 25 to 64. Find the percent increase in that population from 2010 to 2050.

Copyright © Houghton Mifflin Company. All rights reserved.

10. What is the percent increase in the population aged 65 and over from 2010 to 2050?

11. Why are the sizes of the work force and the population aged 65 and over of concern to the Social Security Administration?

12. Describe any patterns you see in the table.

13. Calculate a statistic based on the data in the table and explain why it would be of interest to an institution (such as a school system) or a manufacturer of consumer goods (such as baby food).

Health and Nutrition   A formula used to determine recommended daily protein intake is

$$P = \frac{W}{2.2}$$

where P is the amount of protein in grams and W is the individual's weight in pounds. However, if you are sedentary, the result should be multiplied by 80%. If you exercise regularly (at least 30 min 5 times per week), multiply the result by 120%.

1. Explain why the result is multiplied by 80% for a sedentary person and by 120% for an active person.

2. Calculate the amount of protein you should consume each day.

The American College of Sports Medicine (ACSM) recommends that you know how to determine your target heart rate in order to get the full benefit of exercise. Your **target heart rate** is the rate at which your heart should beat during any aerobic exercise such as running, cycling, fast walking, or participating in an aerobics class. According to the ACSM, you should reach your target rate and then maintain it for 20 minutes or more to achieve cardiovascular fitness. The intensity level varies for different individuals. A sedentary person might begin at the 60% level and gradually work up to 70%, whereas athletes and very fit individuals might work at the 85% level. The ACSM suggests you calculate both 50% and 85% of your maximum heart rate. This will give you the low and high ends of the range within which your heart rate should stay.

To calculate your target heart rate:

	Example
Subtract your age from 220. This is your maximum heart rate.	$220 - 20 = 200$
Multiply your maximum heart rate by 50%. This is the low end of your range.	$200(0.50) = 100$
Divide the low end by 6. This is your low 10-second heart rate.	$100 \div 6 \approx 17$
Multiply your maximum heart rate by 85%. This is the high end of your range.	$200(0.85) = 170$
Divide the high end by 6. This is your high 10-second heart rate.	$170 \div 6 \approx 28$

Copyright © Houghton Mifflin Company. All rights reserved.

3. Why are the low end and high end divided by 6 in order to determine the 10-second heart rate?

4. Calculate your target heart rate, both the low and high ends of your range.

5. What is the percent increase in your 10-second heart rate from the low end to the high end?

Chapter Summary

Key Words

Key Words	Examples
Percent means "parts of 100." [8.1A, p. 489]	23% means 23 of 100 equal parts.
Percent increase is used to show how much a quantity has increased over its original value. [8.3A, p. 503]	The city's population increased 5%, from 10,000 people to 10,500 people.
Percent decrease is used to show how much a quantity has decreased from its original value. [8.3B, p. 504]	Sales decreased 10%, from 10,000 units in the third quarter to 9,000 units in the fourth quarter.
Cost is the price a business pays for a product. **Selling price** is the price for which a business sells a product to a customer. **Markup** is the difference between selling price and cost. The percent markup is called the **markup rate.** In this text, the markup rate is expressed as the markup rate based on cost. [8.4A, p. 507]	A business pays $90 for a pair of cross trainers; the cost is $90. The business sells the cross trainers for $135; the selling price is $135. The markup is $135 − $90 = $45. The markup rate is 45 ÷ 90 = 0.5 = 50%.
Discount or **markdown** is the difference between the regular price and the discount price. The discount is frequently stated as a percent, called the **discount rate.** [8.4B, p. 509]	A movie video that regularly sells for $50 is on sale for $40. The discount is $50 − $40 = $10. The discount rate is 10 ÷ 50 = 0.2 = 20%.
Principal is the amount of money originally deposited or borrowed. **Interest** is the amount paid for the privilege of using someone else's money. The percent used to determine the amount of interest is the **interest rate.** Interest computed on the original amount is called **simple interest.** The principal plus the interest owed on a loan is called the **maturity value.** [8.5A, p. 513]	Consider a 1-year loan of $5,000 at an annual simple interest rate of 8%. The principal is $5,000. The interest rate is 8%. The interest paid on the loan is $400. The maturity value is $5,000 + $400 = $5,400.

Essential Rules and Procedures

To write a percent as a fraction, drop the percent sign and multiply by $\frac{1}{100}$. [8.1A, p. 489]	$56\% = 56\left(\frac{1}{100}\right) = \frac{56}{100} = \frac{14}{25}$
To write a percent as a decimal, drop the percent sign and multiply by 0.01. [8.1A, p. 489]	$87\% = 87(0.01) = 0.87$

Copyright © Houghton Mifflin Company. All rights reserved.

To write a fraction as a percent, multiply by 100%. [8.1B, p. 490]

$$\frac{7}{20} = \frac{7}{20}(100\%) = \frac{700}{20} = 35\%$$

To write a decimal as a percent, multiply by 100%. [8.1B, p. 490]

$$0.325 = 0.325(100\%) = 32.5\%$$

The Basic Percent Equation [8.2A, p. 493]
Percent · base = amount

8% of 250 is what number?
Percent · base = amount
$$0.08 \cdot 250 = n$$
$$20 = n$$

Proportion Method of Solving a Percent Problem [8.2B, p. 495]
$$\frac{\text{percent}}{100} = \frac{\text{amount}}{\text{base}}$$

8% of 250 is what number?
$$\frac{\text{percent}}{100} = \frac{\text{amount}}{\text{base}}$$
$$\frac{8}{100} = \frac{n}{250}$$
$$100 \cdot n = 8 \cdot 250$$
$$100n = 2{,}000$$
$$n = 20$$

Markup Equations: [8.4A, p. 507]
M = markup, S = selling price, C = cost, r = markup rate:
$M = S - C$
$S = C + M$
$M = r \cdot C$

The manager of a sporting goods store buys a golf club for $275 and sells the golf club for $343.75. Find the markup rate.
$M = S - C$
$M = 343.75 - 275 = 68.75$
$M = r \cdot C$
$68.75 = r \cdot 275$
$0.25 = r$
The markup rate is 25%.

Discount Equations: [8.4B, p. 509]
M = discount or markdown, S = sale price, R = regular price, r = discount rate:
$M = R - S$
$M = r \cdot R$
$S = (1 - r)R$

A golf club that regularly sells for $343.75 is on sale for $275. Find the discount rate.
$M = S - C$
$M = 343.75 - 275 = 68.75$
$M = r \cdot R$
$68.75 = r \cdot 343.75$
$0.2 = r$
The markup rate is 20%.

Simple Interest Formula [8.5A, p. 513]
I = simple interest earned, P = principal, r = annual simple interest rate, t = time (in years):
$I = Prt$

You borrow $10,000 for 180 days at an annual interest rate of 8%. Find the simple interest due on the loan.
$I = Prt$
$$I = 10{,}000(0.08)\left(\frac{180}{365}\right)$$
$$I \approx 394.52$$

Formula for the Maturity Value of a Simple Interest Loan [8.5A, p. 514]
M = maturity value, P = principal, I = simple interest:
$M = P + I$

Suppose you paid $400 in interest on a 1-year loan of $5,000. The maturity value of the loan is $5,000 + $400 = $5,400.

Copyright © Houghton Mifflin Company. All rights reserved.

Chapter Review Exercises

1. Write 32% as a fraction.

2. Write 22% as a decimal.

3. Write 25% as a fraction and as a decimal.

4. Write $3\frac{2}{5}$% as a fraction.

5. Write $\frac{7}{40}$ as a percent.

6. Write $1\frac{2}{7}$ as a percent. Round to the nearest tenth of a percent.

7. Write 2.8 as a percent.

8. 42% of 50 is what?

9. What percent of 3 is 15?

10. 12 is what percent of 18? Round to the nearest tenth of a percent.

11. 150% of 20 is what number?

12. Find 18% of 85.

13. 32% of what number is 180?

14. 4.5 is what percent of 80?

15. Find 0.58% of 2.54.

16. 0.0048 is 0.05% of what number?

17. *Tourism* The table at the right shows the countries with the highest projected numbers of tourists visiting in 2020. What percent of the tourists visiting these countries will be visiting China? Round to the nearest tenth of a percent.

Country	Projected Number of Tourists in 2020
China	137 million
France	93 million
Spain	71 million
USA	102 million

Source: The State of the World Atlas by Dan Smith

18. *Business* A company spent 7% of its $120,000 budget for advertising. How much did the company spend for advertising?

19. *Manufacturing* A quality control inspector found that 1.2% of 4,000 cellular telephones were defective. How many of the phones were not defective?

20. *Television* According to the Cabletelevision Advertising Bureau, cable households watch an average of 61.35 h of television per week. On average, what percent of the week do cable households spend watching TV? Round to the nearest tenth of a percent.

Copyright © Houghton Mifflin Company. All rights reserved.

21. *Business* A resort lodge expects to make a profit of 22% of total income. What is the expected profit on $750,000 of income?

22. *Sports* A basketball auditorium increased its 9,000 seating capacity by 18%. How many seats were added to the auditorium?

23. *Travel* An airline knowingly overbooks flights by selling 12% more tickets than there are seats available. How many tickets would this airline sell for an airplane that has 175 seats?

24. *Elections* In a recent city election, 25,400 out of 112,000 registered voters voted. What percent of the registered voters voted in the election? Round to the nearest tenth of a percent.

25. *Compensation* A sales clerk was earning $10.50 an hour before an 8% increase in pay. What is the clerk's new hourly wage?

26. *Computers* A computer system that sold for $2,400 one year ago can now be bought for $1,800. What percent decrease does this represent?

27. *Business* A car dealer advertises a 6% markup rate on a car that cost the dealer $18,500. Find the selling price of the car.

28. *Business* A parka costing $110 is sold for $181.50. Find the markup rate.

29. *Business* A tennis racket that regularly sells for $80 is on sale for 30% off the regular price. Find the sale price.

30. *Travel* An airline is offering a 40% discount on round-trip air fares. Find the sale price of a round-trip ticket that normally sells for $650.

31. *Finance* Find the simple interest on a 45-day loan of $3,000 at an annual simple interest rate of 8.6%.

32. *Finance* A corporation borrowed $500,000 for 60 days and paid $7,397.26 in simple interest. What annual simple interest rate did the corporation pay on the loan? Round to the nearest hundredth of a percent.

33. *Finance* A realtor took out a $10,000 loan at an 8.4% annual simple interest rate for 9 months. Find the maturity value of the loan.

Copyright © Houghton Mifflin Company. All rights reserved.

Chapter Test

1. Write 86.4% as a decimal.

2. Write 0.4 as a percent.

3. Write $\frac{5}{4}$ as a percent.

4. Write $83\frac{1}{3}\%$ as a fraction.

5. Write 32% as a fraction.

6. Write 1.18 as a percent.

7. 18 is 20% of what number?

8. What is 68% of 73?

9. What percent of 320 is 180?

10. 28 is 14% of what number?

11. *Insurance*   An insurance company expects 2.2% of a company's employees will have an industrial accident. How many accidents are expected for a company that employs 1,500 people?

12. *Education*   A student missed 16 questions on a history exam of 90 questions. What percent of the questions did the student answer correctly? Round to the nearest tenth.

13. *Compensation*   An administrative assistant has a wage of $480 per week. This is 120% of last year's wage. What is the dollar increase in the assistant's weekly wage over last year?

14. *Education*   The table at the right shows the average cost of tuition, room, and board at both public and private colleges in the United States. What is the percent increase in cost for a student who goes from public college to private college? Round to the nearest tenth of a percent.

Average Tuition, Room, and Board	
Public college	$7,628
Private college	$19,143

Source: U.S. Department of Education

15. *Business*   The number of management trainees working for a company has increased from 36 to 42. What percent increase does this represent?

Copyright © Houghton Mifflin Company. All rights reserved.

16. 🥧 *Nutrition* The table at the right shows the fat, saturated fat, cholesterol, and calorie content in a 90-gram ground beef burger and in a 90-gram soy burger.
  a. As compared with the beef burger, by what percent is the fat content decreased in the soy burger?
  b. What is the percent decrease in cholesterol in the soy burger as compared with the beef burger?
  c. Calculate the percent decrease in calories in the soy burger as compared with the beef burger.

	Beef Burger	Soy Burger
Fat	24 g	4 g
Saturated Fat	10 g	1.5 g
Cholesterol	75 mg	0 mg
Calories	280	140

17. *Business* Last year a company's travel expenses totaled $25,000. This year the travel expenses totaled $23,000. What percent decrease does this represent?

18. *Art* A painting has a value of $1,500. This is 125% of the painting's value last year. What is the dollar increase in the value of the painting?

19. *Business* The manager of a stationery store uses a markup rate of 60%. What is the markup on a box of notepaper that costs the store $21?

20. *Business* An electric keyboard costing $225 is sold for $349. Find the markup rate on the keyboard. Round to the nearest tenth of a percent.

21. *Business* A telescope is on sale for $180 after a markdown of 40% off the regular price. Find the regular price.

22. *Business* The regular price of a 10-foot × 10-foot dome tent is $370. The tent is now marked down $51.80. Find the discount rate on the tent.

23. *Finance* Find the simple interest on a 9-month loan of $5,000 when the annual interest rate is 8.4%.

24. *Finance* Maribeth Bakke took out a 150-day, $40,000 business loan that had an annual simple interest rate of 9.25%. Find the maturity value of the loan.

25. *Finance* Gene Connery paid $672 in simple interest on an 8-month loan for $12,000. Find the simple interest rate on the loan.

Copyright © Houghton Mifflin Company. All rights reserved.

Cumulative Review Exercises

1. Evaluate $a - b$ when $a = 102.5$ and $b = 77.546$.

2. Evaluate 5^4.

3. Find the product of 4.67 and 3.007.

4. Multiply: $(2x - 3)(2x - 5)$

5. Divide: $3\frac{5}{8} \div 2\frac{7}{12}$

6. Multiply: $-2a^2b(-3ab^2 + 4a^2b^3 - ab^3)$

7. 120% of 35 is what?

8. Solve: $x - 2 = -5$

9. Find the product of 1.005 and 10^5.

10. Simplify: $-\frac{5}{8} - \left(-\frac{3}{4}\right) + \frac{5}{6}$

11. Simplify: $\dfrac{3 - \frac{7}{8}}{\frac{11}{12} + \frac{1}{4}}$

12. Multiply: $(-3a^2b)(4a^5b^4)$

13. Graph $y = -2x + 5$.

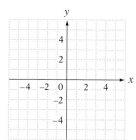

14. Graph $y = \frac{5}{3}x - 2$.

15. Find the quotient of $\frac{7}{8}$ and $\frac{5}{16}$.

16. Simplify: $4 - (-3) + 5 - 8$

17. Solve: $\frac{3}{4}x = -9$

18. Solve: $6x - 9 = -3x + 36$

Copyright © Houghton Mifflin Company. All rights reserved.

19. Write 322.4 mi in 5 h as a unit rate.

20. Solve: $\frac{32}{n} = \frac{5}{7}$

21. 2.5 is what percent of 30? Round to the nearest tenth of a percent.

22. Find 42% of 160.

23. Simplify: $44 - (-6)^2 \div (-3) + 2$

24. Solve: $3(x - 2) + 2 = 11$

Solve.

25. *Health* According to the table at the right, what fraction of the population aged 75–84 are affected by Alzheimer's disease?

Age Group	Percent Affected by Alzheimer's Disease
65 – 74	4%
75 – 84	10%
85 +	17%

Source: Mayo Clinic Family Health Book, Encyclopedia Americana, Associated Press

26. *Business* A suit that regularly sells for $202.50 is on sale for 36% off the regular price. Find the sale price.

27. *Business* A graphing calculator with a selling price of $67.20 has a markup of 60%. Find the cost of the graphing calculator.

28. *Sports* A baseball team has won 13 out of the first 18 games played. At this rate, how many games will the team win in a 162-game season?

29. *Sports* A wrestler needs to lose 8 lb in three days in order to make the proper weight class. The wrestler loses $3\frac{1}{2}$ lb the first day and $2\frac{1}{4}$ lb the second day. How many pounds must the wrestler lose the third day in order to make the weight class?

30. *Physics* The speed of a falling object is given by the formula $v = \sqrt{64d}$, where v is the speed of the falling object in feet per second and d is the distance in feet that the object has fallen. Find the speed of an object that has fallen 81 ft.

31. *Sports* In the 2000 Olympics, Michael Johnson ran the 400-meter dash in 43.84 s. Convert this speed to kilometers per hour. Round to the nearest hundredth.

32. *Contractor* A plumber charged $1,632 for work done on a medical building. This charge included $192 for materials and $40 per hour for labor. Find the number of hours the plumber worked on the medical building.

33. *Physics* The current (I) in an electric circuit is inversely proportional to the resistance (R). If the current is 2 amperes when the resistance is 20 ohms, find the resistance when the current is 8 amperes.

Copyright © Houghton Mifflin Company. All rights reserved.

Geometry

The best way to appreciate the different shapes and sizes of grassy areas in Buenos Aires' Rosedal Park is to view the garden from overhead. Each geometric shape, having its own set of dimensions, combines to form the entire park. The **Example 4 on page 556** illustrates how to use a geometric formula to determine how large an area is and how much grass seed is needed for an area that size.

Copyright © Houghton Mifflin Company. All rights reserved.

WEB *Need help? For online student resources, visit this website:* **math.college.hmco.com**

Prep Test

1. Simplify: $2(18) + 2(10)$

2. Evaluate abc when $a = 2$, $b = 3.14$, and $c = 9$.

3. Evaluate xyz^3 when $x = \dfrac{4}{3}$, $y = 3.14$, and $z = 3$.

4. Solve: $x + 47 = 90$

5. Solve: $32 + 97 + x = 180$

6. Solve: $\dfrac{5}{12} = \dfrac{6}{x}$

Go Figure

Draw the figure that would come next.

Copyright © Houghton Mifflin Company. All rights reserved.

SECTION 9.1   Introduction to Geometry

OBJECTIVE **A**

Problems involving lines and angles

The word *geometry* comes from the Greek words for *earth* and *measure*. In ancient Egypt, geometry was used by the Egyptians to measure land and to build structures such as the pyramids. Today geometry is used in many fields, such as physics, medicine, and geology. Geometry is also used in applied fields such as mechanical drawing and astronomy. Geometric forms are used in art and design.

Three basic concepts of geometry are point, line, and plane. A **point** is symbolized by drawing a dot. A **line** is determined by two distinct points and extends indefinitely in both directions, as the arrows on the line shown at the right indicate. This line contains points *A* and *B* and is represented by $\overleftrightarrow{AB}$. A line can also be represented by a single letter, such as ℓ.

A **ray** starts at a point and extends indefinitely in *one* direction. The point at which a ray starts is called the **endpoint** of the ray. The ray shown at the right is denoted by $\overrightarrow{AB}$. Point *A* is the endpoint of the ray.

A **line segment** is part of a line and has two endpoints. The line segment shown at the right is denoted by $\overline{AB}$.

The distance between the endpoints of $\overline{AC}$ is denoted by *AC*. If *B* is a point on $\overline{AC}$, then *AC* (the distance from *A* to *C*) is the sum of *AB* (the distance from *A* to *B*) and *BC* (the distance from *B* to *C*).

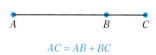

$AC = AB + BC$

Given the figure above and the fact that *AB* = 22 cm and *AC* = 31 cm, find *BC*.

Write an equation for the distances between points on the line segment. $AC = AB + BC$

Substitute the given distances for *AB* and *AC* into the equation. $31 = 22 + BC$

Solve for *BC*. $9 = BC$

$BC = 9$ cm

In this section we will be discussing figures that lie in a plane. A **plane** is a flat surface and can be pictured as a table top or blackboard that extends in all directions. Figures that lie in a plane are called **plane figures.**

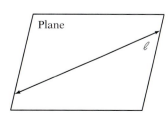

Point of Interest

Geometry is one of the oldest branches of mathematics. Around 350 B.C., Euclid of Alexandria wrote *Elements*, which contained all of the known concepts of geometry. Euclid's contribution was to unify various concepts into a single deductive system that was based on a set of axioms.

Copyright © Houghton Mifflin Company. All rights reserved.

Lines in a plane can be intersecting or parallel. **Intersecting lines** cross at a point in the plane. **Parallel lines** never intersect. The distance between them is always the same.

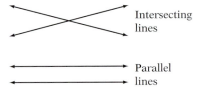

Intersecting lines

Parallel lines

The symbol $\parallel$ means "is parallel to." In the figure at the right, $j \parallel k$ and $\overline{AB} \parallel \overline{CD}$. Note that j contains $\overline{AB}$ and k contains $\overline{CD}$. Parallel lines contain parallel line segments.

An **angle** is formed by two rays with the same endpoint. The **vertex** of the angle is the point at which the two rays meet. The rays are called the **sides** of the angle.

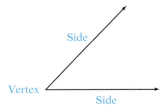

If A and C are points on rays r_1 and r_2, and B is the vertex, then the angle is called $\angle B$ or $\angle ABC$, where $\angle$ is the symbol for angle. Note that the angle is named by the vertex, or the vertex is the second point listed when the angle is named by giving three points. $\angle ABC$ could also be called $\angle CBA$.

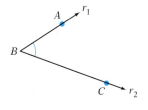

An angle can also be named by a variable written between the rays close to the vertex. In the figure at the right, $\angle x = \angle QRS$ and $\angle y = \angle SRT$. Note that in this figure, more than two rays meet at R. In this case, the vertex cannot be used to name an angle.

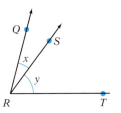

▶ **Point of Interest**

The Babylonians knew that Earth is in approximately the same position in the sky every 365 days. Historians suggest that one complete revolution of a circle is called 360° because 360 is the closest number to 365 that is divisible by many natural numbers.

An angle is measured in **degrees**. The symbol for degrees is a small raised circle, °. Probably because early Babylonians believed that Earth revolves around the sun in approximately 360 days, the angle formed by a circle has a measure of 360° (360 degrees).

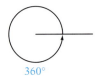

360°

Copyright © Houghton Mifflin Company. All rights reserved.

A **protractor** is used to measure an angle. Place the center of the protractor at the vertex of the angle with the edge of the protractor along a side of the angle. The angle shown in the figure below measures 58°.

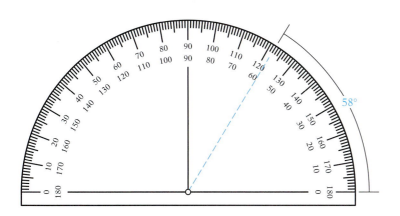

▶ **Point of Interest**

The Leaning Tower of Pisa is the bell tower of the Cathedral in Pisa, Italy. Its construction began on August 9, 1173, and continued for about 200 years. The tower was designed to be vertical, but it started to lean during its construction. By 1350 it was 2.5° off from the vertical; by 1817, it was 5.1° off; and by 1990, it was 5.5° off. In 2001, work on the structure that returned its list to 5° was completed. (*Source: Time* magazine, June 25, 2001, pp. 34–35)

A 90° angle is called a **right angle.** The symbol ∟ represents a right angle.

Perpendicular lines are intersecting lines that form right angles.

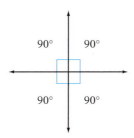

The symbol ⊥ means "is perpendicular to." In the figure at the right, $p \perp q$ and $\overline{AB} \perp \overline{CD}$. Note that line p contains $\overline{AB}$ and line q contains $\overline{CD}$. Perpendicular lines contain perpendicular line segments.

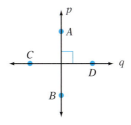

Complementary angles are two angles whose measures have the sum 90°.

$$\angle A + \angle B = 70° + 20° = 90°$$

∠A and ∠B are complementary angles.

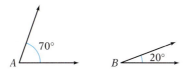

Copyright © Houghton Mifflin Company. All rights reserved.

A 180° angle is called a **straight angle.**

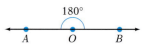
180°

$\angle AOB$ is a straight angle.

Supplementary angles are two angles whose measures have the sum 180°.

130°   50°

$$\angle A + \angle B = 130° + 50° = 180°$$

$\angle A$ and $\angle B$ are supplementary angles.

An **acute angle** is an angle whose measure is between 0° and 90°. $\angle B$ above is an acute angle. An **obtuse angle** is an angle whose measure is between 90° and 180°. $\angle A$ above is an obtuse angle.

Two angles that share a common side are **adjacent angles.** In the figure at the right, $\angle DAC$ and $\angle CAB$ are adjacent angles. $\angle DAC = 45°$ and $\angle CAB = 55°$.

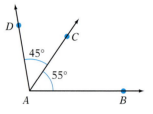
45°
55°

$$\angle DAB = \angle DAC + \angle CAB$$
$$= 45° + 55° = 100°$$

In the figure at the right, $\angle EDG = 80°$. $\angle FDG$ is three times the measure of $\angle EDF$. Find the measure of $\angle EDF$.

Let x = the measure of $\angle EDF$. Then $3x$ = the measure of $\angle FDG$. Write an equation and solve for x, the measure of $\angle EDF$.

$$\angle EDF + \angle FDG = \angle EDG$$
$$x + 3x = 80$$
$$4x = 80$$
$$x = 20$$

$\angle EDF = 20°$

---

1 *Example*

Given $MN = 15$ mm, $NO = 18$ mm, and $MP = 48$ mm, find OP.

M   N   O   P ℓ

Solution

$$MN + NO + OP = MP$$
$$15 + 18 + OP = 48$$
$$33 + OP = 48$$
$$OP = 15$$

$OP = 15$ mm

1 *You Try It*

Given $QR = 24$ cm, $ST = 17$ cm, and $QT = 62$ cm, find RS.

Q   R   S   T ℓ

Your Solution

Solution on p. S22

Copyright © Houghton Mifflin Company. All rights reserved.

2 *Example*

Given $XY = 9$ m and YZ is twice XY, find XZ.

$$X \qquad Y \qquad\qquad Z \qquad \ell$$

Solution

$XZ = XY + YZ$
$XZ = XY + 2(XY)$
$XZ = 9 + 2(9)$
$XZ = 9 + 18$
$XZ = 27$

$XZ = 27$ m

2 *You Try It*

Given $BC = 16$ ft and $AB = \frac{1}{4}(BC)$, find AC.

$$A \quad B \qquad\qquad C \qquad \ell$$

Your Solution

---

3 *Example*

Find the complement of a 38° angle.

Strategy

Complementary angles are two angles whose sum is 90°. To find the complement, let x represent the complement of a 38° angle. Write an equation and solve for x.

Solution

$x + 38° = 90°$
$\quad\;\; x = 52°$

The complement of a 38° angle is a 52° angle.

3 *You Try It*

Find the supplement of a 129° angle.

Your Strategy

Your Solution

---

4 *Example*

Find the measure of $\angle x$.

Strategy

To find the measure of $\angle x$, write an equation using the fact that the sum of the measure of $\angle x$ and 47° is 90°. Solve for $\angle x$.

Solution

$\angle x + 47° = 90°$
$\quad\;\; \angle x = 43°$

The measure of $\angle x$ is 43°.

4 *You Try It*

Find the measure of $\angle a$.

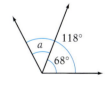

Your Strategy

Your Solution

Solutions on p. S22

Copyright © Houghton Mifflin Company. All rights reserved.

VIDEO & DVD CD TUTOR WEB SSM

OBJECTIVE **B**

Problems involving angles formed by intersecting lines

▶ Point of Interest

Many cities in the New World, unlike those in Europe, were designed using rectangular street grids. Washington, D.C. was planned that way except that diagonal avenues were added, primarily for the purpose of enabling quick troop movement in the event that the city required defense. As an added precaution, monuments were constructed at major intersections so that attackers would not have a straight shot down a boulevard.

Four angles are formed by the intersection of two lines. If the two lines are perpendicular, each of the four angles is a right angle. If the two lines are not perpendicular, then two of the angles formed are acute angles and two of the angles are obtuse angles. The two acute angles are always opposite each other, and the two obtuse angles are always opposite each other.

In the figure at the right, $\angle w$ and $\angle y$ are acute angles. $\angle x$ and $\angle z$ are obtuse angles.

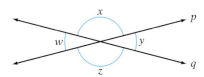

Two angles that are on opposite sides of the intersection of two lines are called **vertical angles.** Vertical angles have the same measure. $\angle w$ and $\angle y$ are vertical angles. $\angle x$ and $\angle z$ are vertical angles.

Vertical angles have the same measure.

$$\angle w = \angle y$$
$$\angle x = \angle z$$

Recall that two angles that share a common side are called **adjacent angles.** For the figure shown above, $\angle x$ and $\angle y$ are adjacent angles, as are $\angle y$ and $\angle z$, $\angle z$ and $\angle w$, and $\angle w$ and $\angle x$. Adjacent angles of intersecting lines are supplementary angles.

Adjacent angles of intersecting lines are supplementary angles.

$$\angle x + \angle y = 180°$$
$$\angle y + \angle z = 180°$$
$$\angle z + \angle w = 180°$$
$$\angle w + \angle x = 180°$$

Given that $\angle c = 65°$, find the measures of angles a, b, and d.

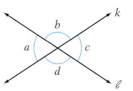

$\angle a = \angle c$ because $\angle a$ and $\angle c$ are vertical angles.

$\angle a = 65°$

$\angle b$ is supplementary to $\angle c$ because $\angle b$ and $\angle c$ are adjacent angles of intersecting lines.

$\angle b + \angle c = 180°$
$\angle b + 65° = 180°$
$\angle b = 115°$

$\angle d = \angle b$ because $\angle d$ and $\angle b$ are vertical angles.

$\angle d = 115°$

Copyright © Houghton Mifflin Company. All rights reserved.

A line that intersects two other lines at different points is called a **transversal**.

If the lines cut by a transversal *t* are parallel lines and the transversal is perpendicular to the parallel lines, all eight angles formed are right angles.

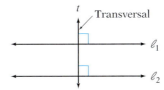

If the lines cut by a transversal *t* are parallel lines and the transversal is not perpendicular to the parallel lines, all four acute angles have the same measure and all four obtuse angles have the same measure. For the figure at the right,

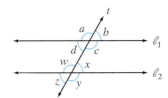

$$\angle b = \angle d = \angle x = \angle z$$

$$\angle a = \angle c = \angle w = \angle y$$

Alternate interior angles are two nonadjacent angles that are on opposite sides of the transversal and between the parallel lines. In the figure above, $\angle c$ and $\angle w$ are alternate interior angles; $\angle d$ and $\angle x$ are alternate interior angles. Alternate interior angles have the same measure.

Alternate interior angles have the same measure.

$$\angle c = \angle w$$
$$\angle d = \angle x$$

Alternate exterior angles are two nonadjacent angles that are on opposite sides of the transversal and outside the parallel lines. In the figure above, $\angle a$ and $\angle y$ are alternate exterior angles; $\angle b$ and $\angle z$ are alternate exterior angles. Alternate exterior angles have the same measure.

Alternate exterior angles have the same measure.

$$\angle a = \angle y$$
$$\angle b = \angle z$$

Corresponding angles are two angles that are on the same side of the transversal and are both acute angles or are both obtuse angles. For the figure above, the following pairs of angles are corresponding angles: $\angle a$ and $\angle w$, $\angle d$ and $\angle z$, $\angle b$ and $\angle x$, $\angle c$ and $\angle y$. Corresponding angles have the same measure.

Corresponding angles have the same measure.

$$\angle a = \angle w$$
$$\angle d = \angle z$$
$$\angle b = \angle x$$
$$\angle c = \angle y$$

Copyright © Houghton Mifflin Company. All rights reserved.

Given that $\ell_1 \| \ell_2$ and $\angle c = 58°$, find the measures of $\angle f$, $\angle h$, and $\angle g$.

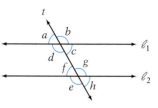

$\angle c$ and $\angle f$ are alternate interior angles.

$\angle f = \angle c = 58°$

$\angle c$ and $\angle h$ are corresponding angles.

$\angle h = \angle c = 58°$

$\angle g$ is supplementary to $\angle h$.

$\angle g + \angle h = 180°$
$\angle g + 58° = 180°$
$\angle g = 122°$

---

5 *Example*

Find x.

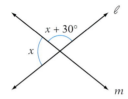

Strategy

The angles labeled are adjacent angles of intersecting lines and are therefore supplementary angles. To find x, write an equation and solve for x.

Solution

$x + (x + 30°) = 180°$
$2x + 30° = 180°$
$2x = 150°$
$x = 75°$

5 *You Try It*

Find x.

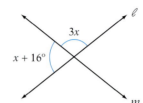

Your Strategy

Your Solution

---

6 *Example*

Given $\ell_1 \| \ell_2$, find x.

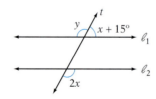

Strategy

$2x = y$ because alternate exterior angles have the same measure. $(x + 15°) + y = 180°$ because adjacent angles of intersecting lines are supplementary angles. Substitute $2x$ for y and solve for x.

Solution

$(x + 15°) + 2x = 180°$
$3x + 15° = 180°$
$3x = 165°$
$x = 55°$

6 *You Try It*

Given $\ell_1 \| \ell_2$, find x.

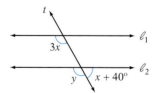

Your Strategy

Your Solution

Solutions on pp. S22–S23

Copyright © Houghton Mifflin Company. All rights reserved.

VIDEO & DVD   CD TUTOR   WWW WEB   SSM

OBJECTIVE **C**

Problems involving the angles of a triangle

If the lines cut by a transversal are not parallel lines, the three lines will intersect at three points. In the figure at the right, the transversal *t* intersects lines *p* and *q*. The three lines intersect at points *A*, *B*, and *C*. These three points define three line segments, $\overline{AB}$, $\overline{BC}$, and $\overline{AC}$. The plane figure formed by these three line segments is called a **triangle.**

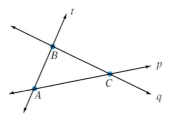

Each of the three points of intersection is the vertex of four angles. The angles within the region enclosed by the triangle are called **interior angles.** In the figure at the right, angles *a*, *b*, and *c* are interior angles. The sum of the measures of the interior angles of a triangle is 180°.

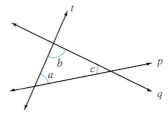

$\angle a + \angle b + \angle c = 180°$

The Sum of the Measures of the Interior Angles of a Triangle

The sum of the measures of the interior angles of a triangle is 180°.

An angle adjacent to an interior angle is an **exterior angle.** In the figure at the right, angles *m* and *n* are exterior angles for angle *a*. The sum of the measures of an interior and an exterior angle is 180°.

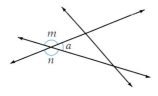

$\angle a + \angle m = 180°$
$\angle a + \angle n = 180°$

Given that $\angle c = 40°$ and $\angle d = 100°$, find the measure of $\angle e$.

$\angle d$ and $\angle b$ are supplementary angles.

$\angle d + \angle b = 180°$
$100° + \angle b = 180°$
$\angle b = 80°$

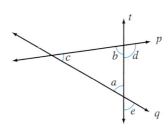

The sum of the interior angles is 180°.

$\angle c + \angle b + \angle a = 180°$
$40° + 80° + \angle a = 180°$
$120° + \angle a = 180°$
$\angle a = 60°$

$\angle a$ and $\angle e$ are vertical angles.

$\angle e = \angle a = 60°$

Copyright © Houghton Mifflin Company. All rights reserved.

7 *Example 7*

Given that $\angle y = 55°$, find the measures of angles a, b, and d.

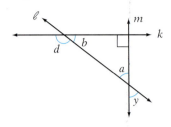

Strategy

▶ To find the measure of angle a, use the fact that $\angle a$ and $\angle y$ are vertical angles.
▶ To find the measure of angle b, use the fact that the sum of the measures of the interior angles of a triangle is 180°.
▶ To find the measure of angle d, use the fact that the sum of an interior and an exterior angle is 180°.

Solution

$\angle a = \angle y = 55°$

$\angle a + \angle b + 90° = 180°$
$55° + \angle b + 90° = 180°$
$\angle b + 145° = 180°$
$\angle b = 35°$

$\angle d + \angle b = 180°$
$\angle d + 35° = 180°$
$\angle d = 145°$

7 *You Try It 7*

Given that $\angle a = 45°$ and $\angle x = 100°$, find the measures of angles b, c, and y.

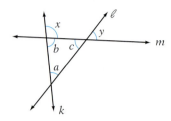

Your Strategy

Your Solution

8 *Example 8*

Two angles of a triangle measure 53° and 78°. Find the measure of the third angle.

Strategy

To find the measure of the third angle, use the fact that the sum of the measures of the interior angles of a triangle is 180°. Write an equation using x to represent the measure of the third angle. Solve the equation for x.

Solution

$x + 53° + 78° = 180°$
$x + 131° = 180°$
$x = 49°$

The measure of the third angle is 49°.

8 *You Try It 8*

One angle in a triangle is a right angle, and one angle measures 34°. Find the measure of the third angle.

Your Strategy

Your Solution

Solutions on p. S23

Copyright © Houghton Mifflin Company. All rights reserved.

9.1 EXERCISES

OBJECTIVE **A**

In Exercises 1–6, use a protractor to measure the angle. State whether the angle is acute, obtuse, or right.

1.
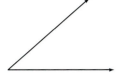

2.

3.
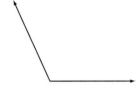

4.
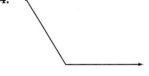

5.
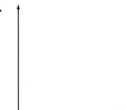

6.

7. Find the complement of a 62° angle.

8. Find the complement of a 31° angle.

9. Find the supplement of a 162° angle.

10. Find the supplement of a 72° angle.

11. Given $AB = 12$ cm, $CD = 9$ cm, and $AD = 35$ cm, find the length of BC.

12. Given $AB = 21$ mm, $BC = 14$ mm, and $AD = 54$ mm, find the length of CD.
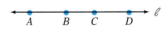

13. Given $QR = 7$ ft and RS is three times the length of QR, find the length of QS.

14. Given $QR = 15$ in. and RS is twice the length of QR, find the length of QS.

15. Given $EF = 20$ m and FG is $\frac{1}{2}$ the length of EF, find the length of EG.

Copyright © Houghton Mifflin Company. All rights reserved.

16. Given $EF = 18$ cm and FG is $\frac{1}{3}$ the length of EF, find the length of EG.

17. Given $\angle LOM = 53°$ and $\angle LON = 139°$, find the measure of $\angle MON$.

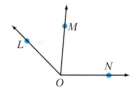

18. Given $\angle MON = 38°$, and $\angle LON = 85°$, find the measure of $\angle LOM$.

Find the measure of $\angle x$.

19.

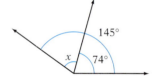

20.

Given that $\angle LON$ is a right angle, find the measure of $\angle x$.

21.

22.

23.

24.

Find the measure of $\angle a$.

25.

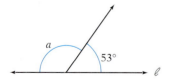

26.

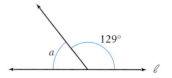

27.

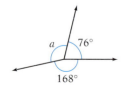

28.

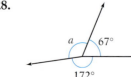

Copyright © Houghton Mifflin Company. All rights reserved.

In Exercises 29–34, find x.

29.
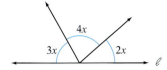

30.
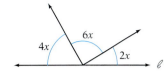

31.
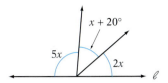

32.
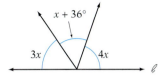

33.

34.

35. Given $\angle a = 51°$, find the measure of $\angle b$.

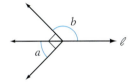

36. Given $\angle a = 38°$, find the measure of $\angle b$.

Copyright © Houghton Mifflin Company. All rights reserved.

OBJECTIVE   B

Find the measure of $\angle x$.

37.
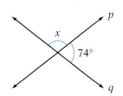

38.
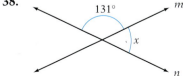

Find x.

39.

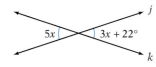

$5x$ $3x + 22°$

40.

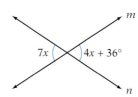

$7x$ $4x + 36°$

Given that $\ell_1 \| \ell_2$, find the measures of angles a and b.

41.

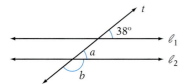

$38°$
a
b

42.

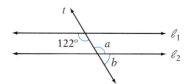

$122°$ a
b

43.

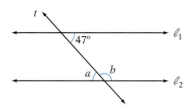

$47°$
a b

44.

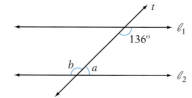

$136°$
b a

Given that $\ell_1 \| \ell_2$, find x.

45.

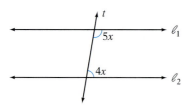

$5x$
$4x$

46.

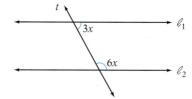

$3x$
$6x$

47.

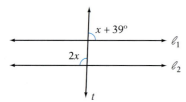

$x + 39°$
$2x$

48.

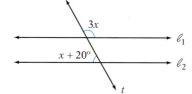

$3x$
$x + 20°$

OBJECTIVE **C**

49. Given that $\angle a = 95°$ and $\angle b = 70°$, find the measures of angles x and y.

a
b
x
y

Copyright © Houghton Mifflin Company. All rights reserved.

50. Given that $\angle a = 35°$ and $\angle b = 55°$, find the measures of angles x and y.

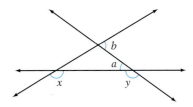

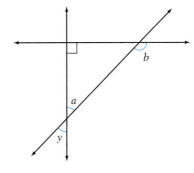

51. Given that $\angle y = 45°$, find the measures of angles a and b.

52. Given that $\angle y = 130°$, find the measures of angles a and b.

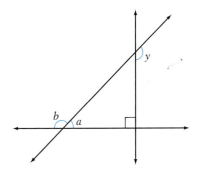

53. Given that $\overline{AO} \perp \overline{OB}$, express in terms of x the number of degrees in $\angle BOC$.

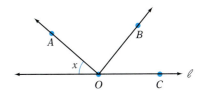

54. Given that $\overline{AO} \perp \overline{OB}$, express in terms of x the number of degrees in $\angle AOC$.

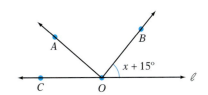

55. One angle in a triangle is a right angle, and one angle is equal to 30°. What is the measure of the third angle?

Copyright © Houghton Mifflin Company. All rights reserved.

56. A triangle has a 45° angle and a right angle. Find the measure of the third angle.

57. Two angles of a triangle measure 42° and 103°. Find the measure of the third angle.

58. A triangle has a 13° angle and a 65° angle. What is the measure of the third angle?

Critical Thinking

59. (a) What is the smallest possible whole number of degrees in an angle of a triangle? (b) What is the largest possible whole number of degrees in an angle of a triangle?

60. Cut out a triangle and then tear off two of the angles, as shown at the right. Position the pieces you tore off so that angle a is adjacent to angle b and angle c is adjacent to angle b. Describe what you observe. What does this demonstrate?

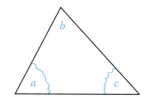

61. Construct a triangle with the given angle measures.
   a. 45°, 45°, and 90°   **b.** 30°, 60°, and 90°   **c.** 40°, 40°, and 100°

62. Determine whether the statement is always true, sometimes true, or never true.
   a. Two lines that are parallel to a third line are parallel to each other.
   b. A triangle contains at least two acute angles.
   c. Vertical angles are complementary angles.

63. For the figure at the right, find the sum of the measures of angles x, y, and z.

64. 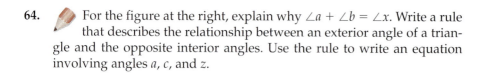 For the figure at the right, explain why $\angle a + \angle b = \angle x$. Write a rule that describes the relationship between an exterior angle of a triangle and the opposite interior angles. Use the rule to write an equation involving angles a, c, and z.

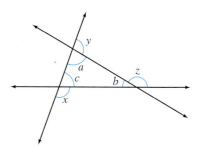

65. If $\overline{AB}$ and $\overline{CD}$ intersect at point O, and $\angle AOC = \angle BOC$, explain why $\overline{AB} \perp \overline{CD}$.

66. Do some research on the principle of reflection. Explain how this principle applies to the operation of a periscope and to the game of billiards.

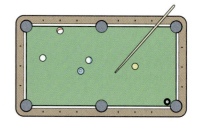

Copyright © Houghton Mifflin Company. All rights reserved.

Copyright © Houghton Mifflin Company. All rights reserved.

SECTION 9.2   Plane Geometric Figures

OBJECTIVE   **A**

Perimeter of a plane geometric figure

A **polygon** is a closed figure determined by three or more line segments that lie in a plane. The line segments that form the polygon are called its **sides.** The figures below are examples of polygons.

 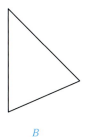 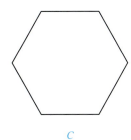 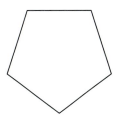 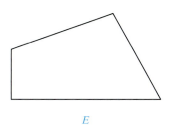

A *A*   B *B*   C *C*   D *D*   E *E*

A **regular polygon** is one in which all sides have the same length and all angles have the same measure. The polygons in Figures *A, C,* and *D* above are regular polygons.

The name of a polygon is based on the number of its sides. The table below lists the names of polygons that have from 3 to 10 sides.

Number of Sides	Name of the Polygon
3	Triangle
4	Quadrilateral
5	Pentagon
6	Hexagon
7	Heptagon
8	Octagon
9	Nonagon
10	Decagon

▶ **Point of Interest**

Although a polygon is defined in terms of its sides, the word actually comes from the Latin word *polygonum,* meaning "many angles."

The Pentagon in Arlington, Virginia

Triangles and quadrilaterals are two of the most common types of polygons. Triangles are distinguished by the number of equal sides and also by the measures of their angles.

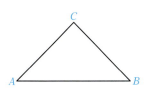   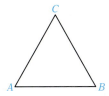   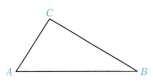

An **isosceles triangle** has two sides of equal length. The angles opposite the equal sides are of equal measure.

$AC = BC$

$\angle A = \angle B$

The three sides of an **equilateral triangle** are of equal length. The three angles are of equal measure.

$AB = BC = AC$

$\angle A = \angle B = \angle C$

A **scalene triangle** has no two sides of equal length. No two angles are of equal measure.

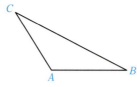

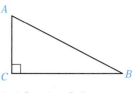

An **acute triangle** has three acute angles.

An **obtuse triangle** has an obtuse angle.

A **right triangle** has a right angle.

Quadrilaterals are also distinguished by their sides and angles, as shown below. Note that a rectangle, a square, and a rhombus are different forms of a parallelogram.

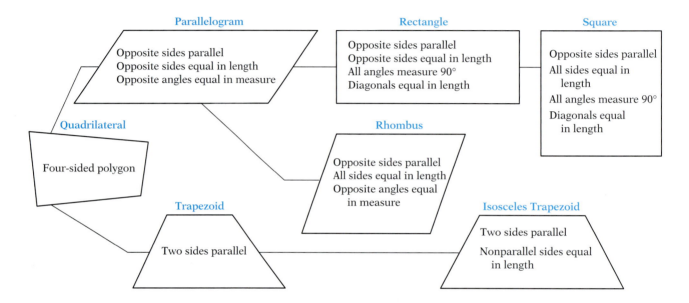

Parallelogram
Opposite sides parallel
Opposite sides equal in length
Opposite angles equal in measure

Rectangle
Opposite sides parallel
Opposite sides equal in length
All angles measure 90°
Diagonals equal in length

Square
Opposite sides parallel
All sides equal in length
All angles measure 90°
Diagonals equal in length

Quadrilateral
Four-sided polygon

Rhombus
Opposite sides parallel
All sides equal in length
Opposite angles equal in measure

Trapezoid
Two sides parallel

Isosceles Trapezoid
Two sides parallel
Nonparallel sides equal in length

The **perimeter** of a plane geometric figure is a measure of the distance around the figure. Perimeter is used in buying fencing for a lawn or determining how much baseboard is needed for a room.

The perimeter of a triangle is the sum of the lengths of the three sides.

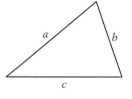

Perimeter of a Triangle

Let a, b, and c be the lengths of the sides of a triangle. The perimeter, P, of the triangle is given by
$P = a + b + c$.

$P = a + b + c$

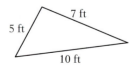

Find the perimeter of the triangle shown at the right.

$P = 5 + 7 + 10 = 22$

The perimeter is 22 ft.

Copyright © Houghton Mifflin Company. All rights reserved.

The perimeter of a quadrilateral is the sum of the lengths of its four sides.

A rectangle has four right angles and opposite sides of equal length. Usually the length, L, of a rectangle refers to the length of one of the longer sides of the rectangle, and the width, W, refers to the length of one of the shorter sides. The perimeter can then be represented $P = L + W + L + W$.

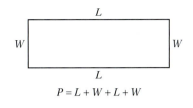

$$P = L + W + L + W$$

The formula for the perimeter of a rectangle is derived by combining like terms.

$$P = 2L + 2W$$

Copyright © Houghton Mifflin Company. All rights reserved.

> ### Perimeter of a Rectangle
>
> Let L represent the length and W the width of a rectangle. The perimeter, P, of the rectangle is given by $P = 2L + 2W$.

Find the perimeter of the rectangle shown at the right.

The length is 5 m. Substitute 5 for L.
The width is 2 m. Substitute 2 for W.
Solve for P.

$P = 2L + 2W$
$P = 2(5) + 2(2)$
$P = 10 + 4$
$P = 14$

5 m

2 m

The perimeter is 14 m.

A square is a rectangle in which each side has the same length. If we let s represent the length of each side of a square, the perimeter of a square can be represented $P = s + s + s + s$.

$$P = s + s + s + s$$

The formula for the perimeter of a square is derived by combining like terms.

$$P = 4s$$

> ### Perimeter of a Square
>
> Let s represent the length of a side of a square. The perimeter, P, of the square is given by $P = 4s$.

Find the perimeter of the square shown at the right.

$P = 4s = 4(8) = 32$

8 in.

The perimeter is 32 in.

A **circle** is a plane figure in which all points are the same distance from point O, called the **center** of the circle.

The **diameter** of a circle is a line segment across the circle through point O. AB is a diameter of the circle at the right. The variable d is used to designate the diameter of a circle.

The **radius** of a circle is a line segment from the center of the circle to a point on the circle. OC is a radius of the circle at the right. The variable r is used to designate a radius of a circle.

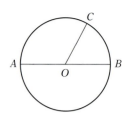

The length of the diameter is twice the length of the radius.

$$d = 2r \text{ or } r = \frac{1}{2}d$$

The distance around a circle is called the **circumference**. The circumference, C, of a circle is equal to the product of π (pi) and the diameter.

$$C = \pi d$$

Because $d = 2r$, the formula for the circumference can be written in terms of r.

$$C = 2\pi r$$

▶ **Point of Interest**

Archimedes (c. 287–212 B.C.) is the person who calculated that $\pi \approx 3\frac{1}{7}$. He actually showed that $3\frac{10}{71} < \pi < 3\frac{1}{7}$. The approximation $3\frac{10}{71}$ is a more accurate approximation of π than $3\frac{1}{7}$, but it is more difficult to use.

> ### The Circumference of a Circle
>
> The circumference, C, of a circle with diameter d and radius r is given by $C = \pi d$ or $C = 2\pi r$.

The formula for circumference uses the number π, which is an irrational number. The value of π can be approximated by a fraction or by a decimal.

$$\pi \approx \frac{22}{7} \text{ or } \pi \approx 3.14$$

The π key on a scientific calculator gives a closer approximation of π than 3.14. Use a scientific calculator to find approximate values in calculations involving π.

Calculator Note

The π key on your calculator can be used to find decimal approximations to formulas that contain π. To perform the calculation at the right, enter

6 × π = .

Find the circumference of a circle with a diameter of 6 in.

The diameter of the circle is given. Use the circumference formula that involves the diameter. $d = 6$.

$$C = \pi d$$
$$C = \pi(6)$$

The exact circumference of the circle is 6π in.

$$C = 6\pi$$

An approximate measure is found by using the π key on a calculator.

$$C \approx 18.85$$

The approximate circumference is 18.85 in.

Copyright © Houghton Mifflin Company. All rights reserved.

1 Example

A carpenter is designing a square patio with a perimeter of 44 ft. What is the length of each side?

Strategy

To find the length of each side, use the formula for the perimeter of a square. Substitute 44 for P and solve for s.

Solution

$P = 4s$
$44 = 4s$
$11 = s$

The length of each side of the patio is 11 ft.

1 You Try It

The infield for a softball field is a square with each side of length 60 ft. Find the perimeter of the infield.

Your Strategy

Your Solution

---

2 Example

The dimensions of a triangular sail are 18 ft, 11 ft, and 15 ft. What is the perimeter of the sail?

Strategy

To find the perimeter, use the formula for the perimeter of a triangle. Substitute 18 for a, 11 for b, and 15 for c. Solve for P.

Solution

$P = a + b + c$
$P = 18 + 11 + 15$
$P = 44$

The perimeter of the sail is 44 ft.

2 You Try It

What is the perimeter of a standard piece of typing paper that measures $8\frac{1}{2}$ in. by 11 in.?

Your Strategy

Your Solution

---

3 Example

Find the circumference of a circle with a radius of 15 cm. Round to the nearest hundredth.

Strategy

To find the circumference, use the circumference formula that involves the radius. An approximation is asked for; use the π key on a calculator. $r = 15$.

Solution

$C = 2\pi r = 2\pi(15) = 30\pi \approx 94.25$

The circumference is 94.25 cm.

3 You Try It

Find the circumference of a circle with a diameter of 9 in. Give the exact measure.

Your Strategy

Your Solution

Solutions on p. S23

Copyright © Houghton Mifflin Company. All rights reserved.

OBJECTIVE **B**

VIDEO & DVD
CD TUTOR
WEB
SSM

Area of a plane geometric figure

Area is the amount of surface in a region. Area can be used to describe the size of a rug, a parking lot, a farm, or a national park. Area is measured in square units.

A square that measures 1 in. on each side has an area of 1 square inch, written 1 in^2.

A square that measures 1 cm on each side has an area of 1 square centimeter, written 1 cm^2.

1 in^2
1 cm^2

Larger areas can be measured in square feet (ft^2), square meters (m^2), acres (43,560 ft^2), square miles (mi^2), or any other square unit.

The area of a geometric figure is the number of squares that are necessary to cover the figure. In the figures below, two rectangles have been drawn and covered with squares. In the figure on the left, 12 squares, each of area 1 cm^2, were used to cover the rectangle. The area of the rectangle is 12 cm^2. In the figure on the right, 6 squares, each of area 1 in^2, were used to cover the rectangle. The area of the rectangle is 6 in^2.

Copyright © Houghton Mifflin Company. All rights reserved.

► **Point of Interest**

Figurate numbers are whole numbers that can be represented as regular geometric figures. For example, a square number is one that can be represented as a square array.

o   oo   ooo   oooo
     oo   ooo   oooo
         ooo   oooo
             oooo

1   4   9   16

The square numbers are 1, 4, 9, 16, 25, … They can be represented as $1^2, 2^2, 3^2, 4^2, 5^2, \ldots$

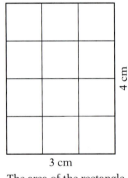
3 cm

The area of the rectangle is 12 cm^2.

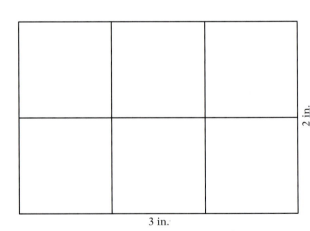
4 cm
2 in.
3 in.

The area of the rectangle is 6 in^2.

Note from the above figures that the area of a rectangle can be found by multiplying the length of the rectangle by its width.

> ### Area of a Rectangle
>
> Let L represent the length and W the width of a rectangle. The area, A, of the rectangle is given by $A = LW$.

Find the area of the rectangle shown at the right.

$A = LW = 11(7) = 77$

The area is 77 m^2.

7 m
11 m

A square is a rectangle in which all sides are the same length. Therefore, both the length and the width of a square can be represented by s, and $A = LW = s \cdot s = s^2$.

Area of a Square

Let s represent the length of a side of a square. The area, A, of the square is given by $A = s^2$.

$A = s \cdot s = s^2$

Find the area of the square shown at the right.

$A = s^2 = 9^2 = 81$

The area is 81 mi².

9 mi

Figure $ABCD$ is a parallelogram. BC is the **base**, b, of the parallelogram. AE, perpendicular to the base, is the **height**, h, of the parallelogram.

Any side of a parallelogram can be designated as the base. The corresponding height is found by drawing a line segment perpendicular to the base from the opposite side.

A rectangle can be formed from a parallelogram by cutting a right triangle from one end of the parallelogram and attaching it to the other end. The area of the resulting rectangle will equal the area of the original parallelogram.

Area of a Parallelogram

Let b represent the length of the base and h the height of a parallelogram. The area, A, of the parallelogram is given by $A = bh$.

Find the area of the parallelogram shown at the right.

$A = bh = 12 \cdot 6 = 72$

The area is 72 m².

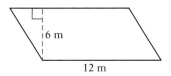

6 m

12 m

Copyright © Houghton Mifflin Company. All rights reserved.

Figure *ABC* is a triangle. *AB* is the **base,** *b*, of the triangle. *CD*, perpendicular to the base, is the **height,** *h*, of the triangle.

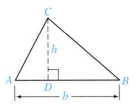

Any side of a triangle can be designated as the base. The corresponding height is found by drawing a line segment perpendicular to the base from the vertex opposite the base.

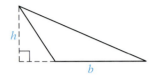

Consider the triangle with base *b* and height *h* shown at the right. By extending a line from *C* parallel to the base *AB* and equal in length to the base, a parallelogram is formed. The area of the parallelogram is *bh* and is twice the area of the triangle. Therefore, the area of the triangle is one-half the area of the parallelogram, or $\frac{1}{2}bh$.

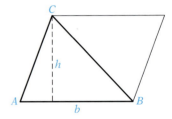

Area of a Triangle

Let *b* represent the length of the base and *h* the height of a triangle. The area, *A*, of the triangle is given by $A = \frac{1}{2}bh$.

Find the area of a triangle with a base of 18 cm and a height of 6 cm.

$$A = \frac{1}{2}bh = \frac{1}{2} \cdot 18 \cdot 6 = 54$$

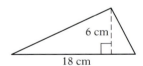

The area is 54 cm².

Figure *ABCD* is a trapezoid. *AB* is one **base,** b_1, of the trapezoid, and *CD* is the other base, b_2. *AE*, perpendicular to the two bases, is the **height,** *h*.

In the trapezoid at the right, the line segment *BD* divides the trapezoid into two triangles, *ABD* and *BCD*. In triangle *ABD*, b_1 is the base and *h* is the height. In triangle *BCD*, b_2 is the base and *h* is the height. The area of the trapezoid is the sum of the areas of the two triangles.

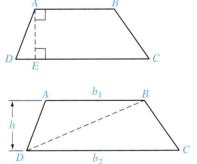

Area of trapezoid *ABCD* = area of triangle *ABD* + area of triangle *BCD*

$$= \frac{1}{2}b_1 h + \frac{1}{2}b_2 h = \frac{1}{2}h(b_1 + b_2)$$

Copyright © Houghton Mifflin Company. All rights reserved.

> ### *Area of a Trapezoid*
>
> Let b_1 and b_2 represent the lengths of the bases and h the height of a trapezoid. The area, A, of the trapezoid is given by
>
> $$A = \frac{1}{2}h(b_1 + b_2).$$

Find the area of a trapezoid that has bases measuring 15 in. and 5 in. and a height of 8 in.

$$A = \frac{1}{2}h(b_1 + b_2)$$

$$= \frac{1}{2} \cdot 8(15 + 5) = 4(20) = 80$$

The area is 80 in².

The area of a circle is equal to the product of π and the square of the radius.

$A = \pi r^2$

> ### *The Area of a Circle*
>
> The area, A, of a circle with radius r is given by $A = \pi r^2$.

Find the area of a circle that has a radius of 6 cm.

Use the formula for the area of a circle. $A = \pi r^2$
$r = 6$. $A = \pi(6)^2$
                                                 $A = \pi(36)$

The exact area of the circle is 36π cm². $A = 36\pi$

An approximate measure is found by
using the π key on a calculator. $A \approx 113.10$

The approximate area of the circle is
113.10 cm².

Calculator Note

To approximate 36π on your

calculator, enter 36 ⊠ π = .

For your reference, all of the formulas for the perimeter and the area of the geometric figures presented in this section are listed in the Chapter Summary which begins on page 598.

Copyright © Houghton Mifflin Company. All rights reserved.

4 *Example*

The Parks and Recreation Department of a city plans to plant grass seed in a playground that has the shape of a trapezoid, as shown below. Each bag of grass seed will seed 1,500 ft². How many bags of grass seed should the department purchase?

80 ft

64 ft

115 ft

Strategy

To find the number of bags to be purchased:

▶ Use the formula for the area of a trapezoid to find the area of the playground.
▶ Divide the area of the playground by the area one bag will seed (1,500).

Solution

$$A = \frac{1}{2}h(b_1 + b_2)$$

$$A = \frac{1}{2} \cdot 64(80 + 115)$$

$A = 6{,}240$ ▶ The area of the playground is 6,240 ft².

$6{,}240 \div 1{,}500 = 4.16$

Because a portion of a fifth bag is needed, 5 bags of grass seed should be purchased.

4 *You Try It*

An interior designer decides to wallpaper two walls of a room. Each roll of wallpaper will cover 30 ft². Each wall measures 8 ft by 12 ft. How many rolls of wallpaper should be purchased?

Your Strategy

Your Solution

5 *Example*

Find the area of a circle with a diameter of 5 ft. Give the exact measure.

Strategy

To find the area:

▶ Find the radius of the circle.
▶ Use the formula for the area of a circle. Leave the answer in terms of π.

Solution

$$r = \frac{1}{2}d = \frac{1}{2}(5) = 2.5$$

$$A = \pi r^2 = \pi(2.5)^2 = \pi(6.25) = 6.25\pi$$

The area of the circle is 6.25π ft².

5 *You Try It*

Find the area of a circle with a radius of 11 cm. Round to the nearest hundredth.

Your Strategy

Your Solution

Solutions on p. S23

Copyright © Houghton Mifflin Company. All rights reserved.

9.2 EXERCISES

OBJECTIVE   A

Name each polygon.

1.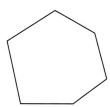

2.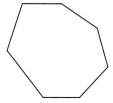

3.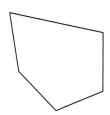

4.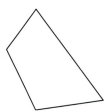

Classify the triangle as isosceles, equilateral, or scalene.

5.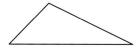

6.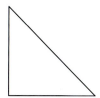

7.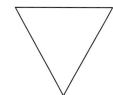

8.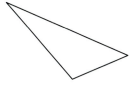

Classify the triangle as acute, obtuse, or right.

9.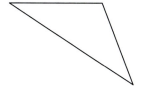

10.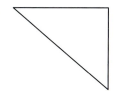

11.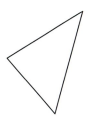

12.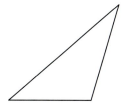

Find the perimeter of the figure.

13.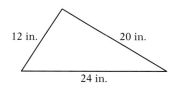

12 in.   20 in.

24 in.

14.

7 cm

11 cm

15.

3.5 ft

3.5 ft

16.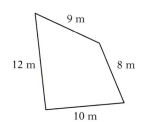

9 m

12 m   8 m

10 m

17.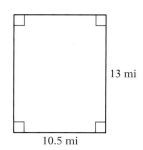

13 mi

10.5 mi

18.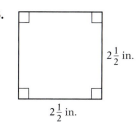

$2\frac{1}{2}$ in.

$2\frac{1}{2}$ in.

Copyright © Houghton Mifflin Company. All rights reserved.

In Exercises 19–24, find the circumference of the figure.
Give both the exact value and an approximation to the nearest hundredth.

19.

4 cm

20.

12 m

21.

5.5 mi

22.

18 in.

23.

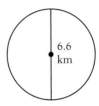

17 ft

24.

6.6 km

25. The lengths of the three sides of a triangle are 3.8 cm, 5.2 cm, and 8.4 cm. Find the perimeter of the triangle.

26. The lengths of the three sides of a triangle are 7.5 m, 6.1 m, and 4.9 m. Find the perimeter of the triangle.

27. The length of each of two sides of an isosceles triangle is $2\frac{1}{2}$ cm. The third side measures 3 cm. Find the perimeter of the triangle.

28. The length of each side of an equilateral triangle is $4\frac{1}{2}$ in. Find the perimeter of the triangle.

29. A rectangle has a length of 8.5 m and a width of 3.5 m. Find the perimeter of the rectangle.

30. Find the perimeter of a rectangle that has a length of $5\frac{1}{2}$ ft and a width of 4 ft.

31. The length of each side of a square is 12.2 cm. Find the perimeter of the square.

32. Find the perimeter of a square that is 0.5 m on each side.

33. Find the perimeter of a regular pentagon that measures 3.5 in. on each side.

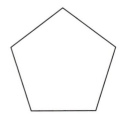

Copyright © Houghton Mifflin Company. All rights reserved.

34. What is the perimeter of a regular hexagon that measures 8.5 cm on each side?

35. Find the circumference of a circle that has a diameter of 1.5 in. Give the exact value.

36. 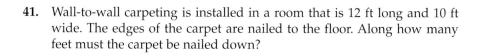 The diameter of a circle is 4.2 ft. Find the circumference of the circle. Round to the nearest hundredth.

37. 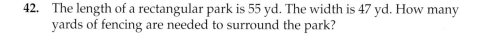 The radius of a circle is 36 cm. Find the circumference of the circle. Round to the nearest hundredth.

38. Find the circumference of a circle that has a radius of 2.5 m. Give the exact value.

39. How many feet of fencing should be purchased for a rectangular garden that is 18 ft long and 12 ft wide?

40. How many meters of binding are required to bind the edge of a rectangular quilt that measures 3.5 m by 8.5 m?

41. Wall-to-wall carpeting is installed in a room that is 12 ft long and 10 ft wide. The edges of the carpet are nailed to the floor. Along how many feet must the carpet be nailed down?

42. The length of a rectangular park is 55 yd. The width is 47 yd. How many yards of fencing are needed to surround the park?

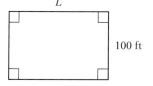

43. The perimeter of a rectangular playground is 440 ft. If the width is 100 ft, what is the length of the playground?

44. A rectangular vegetable garden has a perimeter of 64 ft. The length of the garden is 20 ft. What is the width of the garden?

45. Each of two sides of a triangular banner measures 18 in. If the perimeter of the banner is 46 in., what is the length of the third side of the banner?

46. The perimeter of an equilateral triangle is 13.2 cm. What is the length of each side of the triangle?

47. The perimeter of a square picture frame is 48 in. Find the length of each side of the frame.

Copyright © Houghton Mifflin Company. All rights reserved.

48. A square rug has a perimeter of 32 ft. Find the length of each edge of the rug.

 Solve. For Exercises 49 to 55, round to the nearest hundredth.

49. The circumference of a circle is 8 cm. Find the length of a diameter of the circle.

50. The circumference of a circle is 15 in. Find the length of a radius of the circle.

51. Find the length of molding needed to put around a circular table that is 4.2 ft in diameter.

52. How much binding is needed to bind the edge of a circular rug that is 3 m in diameter?

53. A bicycle tire has a diameter of 24 in. How many feet does the bicycle travel when the wheel makes eight revolutions?

54. A tricycle tire has a diameter of 12 in. How many feet does the tricycle travel when the wheel makes 12 revolutions?

55. The distance from the surface of Earth to its center is 6,356 km. What is the circumference of Earth?

56. Bias binding is to be sewed around the edge of a rectangular tablecloth measuring 72 in. by 45 in. If the bias binding comes in packages containing 15 ft of binding, how many packages of bias binding are needed for the tablecloth?

OBJECTIVE **B**

 Find the area of the figure.

57.

5 ft

12 ft

58.

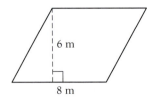

6 m

8 m

59.

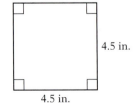

4.5 in.

4.5 in.

Copyright © Houghton Mifflin Company. All rights reserved.

60.

12 in.

20 in.

61.

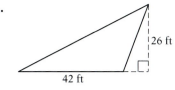

26 ft

42 ft

62.

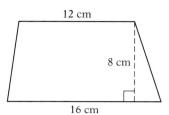

12 cm

8 cm

16 cm

 In Exercises 63–68, find the area of the figure.
Give both the exact value and an approximation to the nearest hundredth.

63.

4 cm

64.

12 m

65.

5.5 mi

66.

18 in.

67.

17 ft

68.

6.6 km

Solve.

69. The length of a side of a square is 12.5 cm. Find the area of the square.

70. Each side of a square measures $3\frac{1}{2}$ in. Find the area of the square.

71. The length of a rectangle is 38 in., and the width is 15 in. Find the area of the rectangle.

72. Find the area of a rectangle that has a length of 6.5 m and a width of 3.8 m.

73. The length of the base of a parallelogram is 16 in., and the height is 12 in. Find the area of the parallelogram.

74. The height of a parallelogram is 3.4 m, and the length of the base is 5.2 m. Find the area of the parallelogram.

Copyright © Houghton Mifflin Company. All rights reserved.

75. The length of the base of a triangle is 6 ft. The height is 4.5 ft. Find the area of the triangle.

76. The height of a triangle is 4.2 cm. The length of the base is 5 cm. Find the area of the triangle.

77. The length of one base of a trapezoid is 35 cm, and the length of the other base is 20 cm. If the height is 12 cm, what is the area of the trapezoid?

78. The height of a trapezoid is 5 in. The bases measure 16 in. and 18 in. Find the area of the trapezoid.

79. The radius of a circle is 5 in. Find the area of the circle. Give the exact value.

80. The diameter of a circle is 6.5 m. Find the area of the circle. Give the exact value.

81. The lens on the Hale telescope at Mount Palomar, California, has a diameter of 200 in. Find its area. Give the exact value.

82. An irrigation system waters a circular field that has a 50-foot radius. Find the area watered by the irrigation system. Give the exact value.

83. Find the area of a rectangular flower garden that measures 14 ft by 9 ft.

84. What is the area of a square patio that measures 8.5 m on each side?

85. Artificial turf is being used to cover a playing field. If the field is rectangular with a length of 100 yd and a width of 75 yd, how much artificial turf must be purchased to cover the field?

86. A fabric wall hanging is to fill a space that measures 5 m by 3.5 m. Allowing for 0.1 m of the fabric to be folded back along each edge, how much fabric must be purchased for the wall hanging?

87. The area of a rectangle is 300 in². If the length of the rectangle is 30 in., what is the width?

88. The width of a rectangle is 12 ft. If the area is 312 ft², what is the length of the rectangle?

Copyright © Houghton Mifflin Company. All rights reserved.

89. The height of a triangle is 5 m. The area of the triangle is 50 m². Find the length of the base of the triangle.

90. The area of a parallelogram is 42 m². If the height of the parallelogram is 7 m, what is the length of the base?

91. You plan to stain the wooden deck attached to your house. The deck measures 10 ft by 8 ft. If a quart of stain will cover 50 ft², how many quarts of stain should you buy?

92. You want to tile your kitchen floor. The floor measures 12 ft by 9 ft. How many tiles, each a square with side $1\frac{1}{2}$ ft, should you purchase for the job?

93. You are wallpapering two walls of a child's room, one measuring 9 ft by 8 ft and the other measuring 11 ft by 8 ft. The wallpaper costs $18.50 per roll, and each roll of the wallpaper will cover 40 ft². What is the cost to wallpaper the two walls?

94. An urban renewal project involves reseeding a park that is in the shape of a square, 60 ft on each side. Each bag of grass seed costs $5.75 and will seed 1,200 ft². How much money should be budgeted for buying grass seed for the park?

95. A circle has a radius of 8 in. Find the increase in area when the radius is increased by 2 in. Round to the nearest hundredth.

96. A circle has a radius of 6 cm. Find the increase in area when the radius is doubled. Round to the nearest hundredth.

97. You want to install wall-to-wall carpeting in your living room, which measures 15 ft by 24 ft. If the cost of the carpet you would like to purchase is $15.95 per square yard, what is the cost of the carpeting for your living room? (*Hint:* 9 ft² = 1 yd²)

98. You want to paint the walls of your bedroom. Two walls measure 15 ft by 9 ft, and the other two walls measure 12 ft by 9 ft. The paint you wish to purchase costs $19.98 per gallon, and each gallon will cover 400 ft² of wall. Find the total amount you will spend on paint.

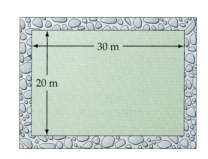

99. A walkway 2 m wide surrounds a rectangular plot of grass. The plot is 30 m long and 20 m wide. What is the area of the walkway?

Copyright © Houghton Mifflin Company. All rights reserved.

100. Pleated draperies for a window must be twice as wide as the width of the window. Draperies are being made for four windows, each 2 ft wide and 4 ft high. Since the drapes will fall slightly below the window sill and extra fabric will be needed for hemming the drapes, 1 ft must be added to the height of the window. How much material must be purchased to make the drapes?

Critical Thinking

101. Find the ratio of the areas of two squares if the ratio of the lengths of their sides is 2:3.

102. If both the length and the width of a rectangle are doubled, how many times larger is the area of the resulting rectangle?

103. If the formula $C = \pi d$ is solved for π, the resulting equation is $\pi = \dfrac{C}{d}$.

Therefore, π is the ratio of the circumference of a circle to the length of its diameter. Use several circular objects, such as coins, plates, tin cans, and wheels, to show that the ratio of the circumference of each object to its diameter is approximately equal to 3.14.

104. Derive a formula for the area of a circle in terms of the diameter of the circle.

105. Determine whether the statement is always true, sometimes true, or never true.
  a. Two triangles that have the same perimeter have the same area.
  b. Two rectangles that have the same area have the same perimeter.
  c. If two squares have the same area, then the sides of the squares have the same length.
  d. An equilateral triangle is also an isosceles triangle.
  e. All the radii (plural of radius) of a circle are equal.
  f. All the diameters of a circle are equal.

106. Suppose a circle is cut into 16 equal pieces, which are then arranged as shown at the right. The figure formed resembles a parallelogram. What variable expression could describe the base of the parallelogram? What variable could describe its height? Explain how the formula for the area of a circle is derived from this approach.

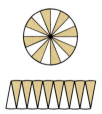

107. Prepare a report on the history of quilts in the United States. Find examples of quilt patterns that incorporate regular polygons. Use pieces of cardboard to create the shapes needed for one block of one of the quilt patterns you learned about.

108. The **apothem** of a regular polygon is the distance from the center of the polygon to a side. Explain how to derive a formula for the area of a regular polygon using the apothem. (*Hint:* Use the formula for the area of a triangle.)

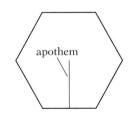

apothem

Copyright © Houghton Mifflin Company. All rights reserved.

Copyright © Houghton Mifflin Company. All rights reserved.

SECTION 9.3 Triangles

OBJECTIVE **A**

The Pythagorean Theorem

A **right triangle** contains one right angle. The side opposite the right angle is called the **hypotenuse.** The other two sides are called **legs.**

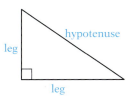

The angles in a right triangle are usually labeled with the capital letters A, B, and C, with C reserved for the right angle. The side opposite angle A is side a, the side opposite angle B is side b, and c is the hypotenuse.

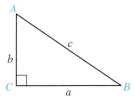

The Greek mathematician Pythagoras is generally credited with the discovery that the square of the hypotenuse of a right triangle is equal to the sum of the squares of the two legs. This is called the **Pythagorean Theorem.**

The figure at the right is a right triangle with legs measuring 3 units and 4 units and a hypotenuse measuring 5 units. Each side of the triangle is also the side of a square. The number of square units in the area of the largest square is equal to the sum of the numbers of square units in the areas of the smaller squares.

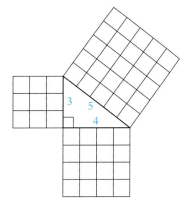

Square of the hypotenuse = sum of the squares of the two legs

$$5^2 = 3^2 + 4^2$$
$$25 = 9 + 16$$
$$25 = 25$$

▶ **Point of Interest**

The first known proof of the Pythagorean Theorem is in a Chinese textbook that dates from 150 B.C. The book is called *Nine Chapters on the Mathematical Art.* The diagram below is from that book and was used in the proof of the theorem.

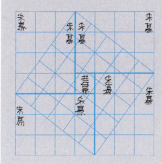

> **Pythagorean Theorem**
>
> If a and b are the lengths of the legs of a right triangle and c is the length of the hypotenuse, then $c^2 = a^2 + b^2$.

If the lengths of two sides of a right triangle are known, the Pythagorean Theorem can be used to find the length of the third side.

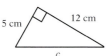

5 cm · 12 cm · c

Consider a right triangle with legs that measure 5 cm and 12 cm. Use the Pythagorean Theorem, with $a = 5$ and $b = 12$, to find the length of the hypotenuse. (If you let $a = 12$ and $b = 5$, the result is the same.)

$$c^2 = a^2 + b^2$$
$$c^2 = 5^2 + 12^2$$
$$c^2 = 25 + 144$$
$$c^2 = 169$$

This equation states that the square of c is 169. Since $13^2 = 169$, $c = 13$, and the length of the hypotenuse is 13 cm. We can find c by taking the square root of 169: $\sqrt{169} = 13$. This suggests the following property.

Calculator Note

The way in which you evaluate the square root of a number depends on the type of calculator you have. Here are two possible keystrokes to find $\sqrt{35}$:

35 √ =

or

√ 35 ENTER

The first method is used on many scientific calculators. The second method is used on many graphing calculators.

The Principal Square Root Property

If $r^2 = s$, then $r = \sqrt{s}$, and r is called the square root of s.

The Principal Square Root Property and its application can be illustrated as follows: Because $5^2 = 25$, $5 = \sqrt{25}$. Therefore, if $c^2 = 25$, $c = \sqrt{25} = 5$.

Recall that numbers whose square roots are integers, such as 25, are perfect squares. If a number is not a perfect square, a calculator can be used to find an approximate square root when a decimal approximation is required.

The length of one leg of a right triangle is 8 in. The hypotenuse is 12 in. Find the length of the other leg. Round to the nearest hundredth.

12 in. · 8 in.

Use the Pythagorean Theorem.
$a = 8, c = 12$
Solve for b^2.
(If you let $b = 8$ and solve for a^2, the result is the same.)

$$a^2 + b^2 = c^2$$
$$8^2 + b^2 = 12^2$$
$$64 + b^2 = 144$$
$$b^2 = 80$$

Use the Principal Square Root Property.
Since $b^2 = 80$, b is the square root of 80.

$$b = \sqrt{80}$$

Use a calculator to approximate $\sqrt{80}$.

$$b \approx 8.94$$

The length of the other leg is approximately 8.94 in.

---

1 Example

The two legs of a right triangle measure 12 ft and 9 ft. Find the hypotenuse of the right triangle.

Strategy

To find the hypotenuse, use the Pythagorean Theorem. $a = 12$, $b = 9$

Solution

$$c^2 = a^2 + b^2$$
$$c^2 = 12^2 + 9^2$$
$$c^2 = 144 + 81$$
$$c^2 = 225$$
$$c = \sqrt{225}$$
$$c = 15$$

The length of the hypotenuse is 15 ft.

1 You Try It

The hypotenuse of a right triangle measures 6 m, and one leg measures 2 m. Find the measure of the other leg. Round to the nearest hundredth.

Your Strategy

Your Solution

Solution on pp. S23–S24

Copyright © Houghton Mifflin Company. All rights reserved.

OBJECTIVE **B**

Similar triangles

Similar objects have the same shape but not necessarily the same size. A tennis ball is similar to a basketball. A model ship is similar to an actual ship.

Similar objects have corresponding parts; for example, the rudder on the model ship corresponds to the rudder on the actual ship. The relationship between the sizes of each of the corresponding parts can be written as a ratio, and each ratio will be the same. If the rudder on the model ship is $\frac{1}{100}$ the size of the rudder on the actual ship, then the model wheelhouse is $\frac{1}{100}$ the size of the actual wheelhouse, the width of the model is $\frac{1}{100}$ the width of the actual ship, and so on.

The two triangles ABC and DEF shown at the right are similar. Side $\overline{AB}$ corresponds to side $\overline{DE}$, side $\overline{BC}$ corresponds to side $\overline{EF}$, and side $\overline{AC}$ corresponds to side $\overline{DF}$. The ratios of corresponding sides are equal.

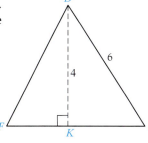

$$\frac{AB}{DE} = \frac{2}{6} = \frac{1}{3}, \frac{BC}{EF} = \frac{3}{9} = \frac{1}{3}, \text{ and } \frac{AC}{DF} = \frac{4}{12} = \frac{1}{3}.$$

Since the ratios of corresponding sides are equal, three proportions can be formed.

$$\frac{AB}{DE} = \frac{BC}{EF}, \frac{AB}{DE} = \frac{AC}{DF}, \text{ and } \frac{BC}{EF} = \frac{AC}{DF}.$$

The corresponding angles in similar triangles are equal. Therefore,

$$\angle A = \angle D, \angle B = \angle E, \text{ and } \angle C = \angle F.$$

Triangles ABC and DEF at the right are similar triangles. AH and DK are the heights of the triangles. The ratio of heights of similar triangles equals the ratio of corresponding sides.

Ratio of corresponding sides $= \dfrac{1.5}{6} = \dfrac{1}{4}$

Ratio of heights $= \dfrac{1}{4}$

> ### *Properties of Similar Triangles*
>
> For similar triangles, the ratios of corresponding sides are equal. The ratio of corresponding heights is equal to the ratio of corresponding sides.

Copyright © Houghton Mifflin Company. All rights reserved.

► **Point of Interest**

Many mathematicians have studied similar objects. Thales of Miletus (c. 624 B.C.–547 B.C.) discovered that he could determine the heights of pyramids and other objects by measuring a small object and the length of its shadow and then making use of similar triangles.

The two triangles at the right are similar triangles. Find the length of side $\overline{EF}$. Round to the nearest tenth.

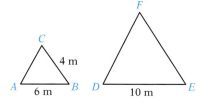

The triangles are similar, so the ratios of corresponding sides are equal.

$$\frac{EF}{BC} = \frac{DE}{AB}$$

$$\frac{EF}{4} = \frac{10}{6}$$

$$6(EF) = 4(10)$$
$$6(EF) = 40$$
$$EF \approx 6.7$$

The length of side EF is approximately 6.7 m.

---

2 *Example*

Triangles ABC and DEF are similar. Find FG, the height of triangle DEF.

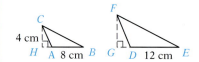

Strategy

To find FG, write a proportion using the fact that, in similar triangles, the ratio of corresponding sides equals the ratio of corresponding heights. Solve the proportion for FG.

Solution

$$\frac{AB}{DE} = \frac{CH}{FG}$$

$$\frac{8}{12} = \frac{4}{FG}$$

$$8(FG) = 12(4)$$
$$8(FG) = 48$$
$$FG = 6$$

The height FG of triangle DEF is 6 cm.

2 *You Try It*

Triangles ABC and DEF are similar. Find FG, the height of triangle DEF.

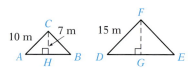

Your Strategy

Your Solution

Solution on p. S24

Copyright © Houghton Mifflin Company. All rights reserved.

Copyright © Houghton Mifflin Company. All rights reserved.

OBJECTIVE   C

Congruent triangles

Congruent objects have the same shape *and* the same size.

The two triangles at the right are congruent. They have the same size.

 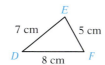

Congruent and similar triangles differ in that the corresponding sides and angles of congruent triangles must be equal, whereas for similar triangles, corresponding angles are equal, but corresponding sides are not necessarily the same length.

The three major rules used to determine whether two triangles are congruent are given below.

> ### Side-Side-Side Rule (SSS)
>
> Two triangles are congruent if the three sides of one triangle equal the corresponding three sides of a second triangle.

In the triangles at the right, $AC = DE$, $AB = EF$, and $BC = DF$. The corresponding sides of triangles ABC and DEF are equal. The triangles are congruent by the SSS Rule.

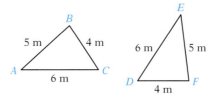

> ### Side-Angle-Side Rule (SAS)
>
> If two sides and the included angle of one triangle equal two sides and the included angle of a second triangle, the two triangles are congruent.

In the two triangles at the right, $AB = EF$, $AC = DE$, and $\angle BAC = \angle DEF$. The triangles are congruent by the SAS Rule.

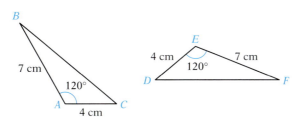

> ### Angle-Side-Angle Rule (ASA)
>
> If two angles and the included side of one triangle equal two angles and the included side of a second triangle, the two triangles are congruent.

For triangles *ABC* and *DEF* at the right, ∠*A* = ∠*F*, ∠*C* = ∠*E*, and *AC* = *EF*. The triangles are congruent by the ASA Rule.

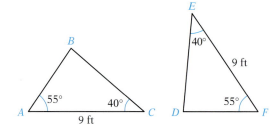

Given triangle *PQR* and triangle *MNO*, do the conditions ∠*P* = ∠*O*, ∠*Q* = ∠*M* and *PQ* = *MO* guarantee that triangle *PQR* is congruent to triangle *MNO*?

Draw a sketch of the two triangles and determine whether one of the rules for congruence is satisfied.

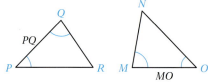

Because two angles and the included side of one triangle equal two angles and the included side of the second triangle, the triangles are congruent by the ASA Rule.

---

3 ***Example***

In the figure below, is triangle *ABC* congruent to triangle *DEF*?

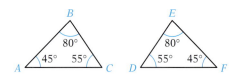

Strategy

To determine whether the triangles are congruent, determine whether one of the rules for congruence is satisfied.

Solution

The triangles do not satisfy the SSS Rule, the SAS Rule, or the ASA Rule. The triangles are not necessarily congruent.

3 ***You Try It***

In the figure below, is triangle *PQR* congruent to triangle *MNO*?

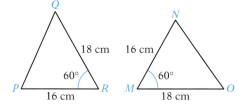

Your Strategy

Your Solution

Solution on p. S24

Copyright © Houghton Mifflin Company. All rights reserved.

9.3 EXERCISES

OBJECTIVE   **A**

Find the unknown side of the triangle. Round to the nearest tenth.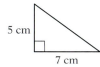

1.

3 in.
4 in.

2.
 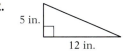
5 in.
12 in.

3.
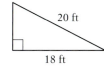
5 cm
7 cm

4.

7 cm
9 cm

5.

15 ft
10 ft

6.

20 ft
18 ft

7.

4 cm   6 cm

8.

9 m   12 m

9.
9 yd
9 yd

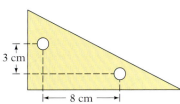

Solve. Round to the nearest tenth.

10. A ladder 8 m long is leaning against a building. How high on the building will the ladder reach when the bottom of the ladder is 3 m from the building?

8 m

3 m

11. Find the distance between the centers of the holes in the metal plate.

3 cm

8 cm

12. If you travel 18 mi east and then 12 mi north, how far are you from your starting point?

13. Find the perimeter of a right triangle with legs that measure 5 cm and 9 cm.

14. Find the perimeter of a right triangle with legs that measure 6 in. and 8 in.

Copyright © Houghton Mifflin Company. All rights reserved.

OBJECTIVE **B**

Find the ratio of corresponding sides for the similar triangles.

15.

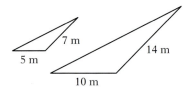

16.

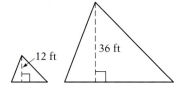

17.

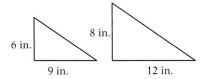

18.

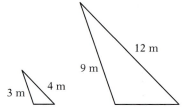

In Exercises 19–28, triangles *ABC* and *DEF* are similar triangles. Solve and round to the nearest tenth.

19. Find side *DE*.

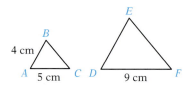

20. Find side *DE*.

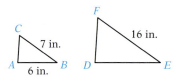

21. Find the height of triangle *DEF*.

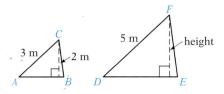

22. Find the height of triangle *ABC*.

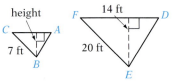

23. Find the perimeter of triangle *ABC*.

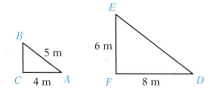

24. Find the perimeter of triangle *DEF*.

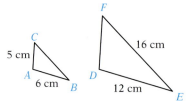

25. Find the perimeter of triangle *ABC*.

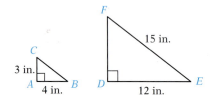

26. Find the area of triangle *DEF*.

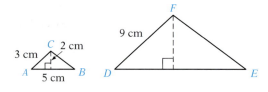

Copyright © Houghton Mifflin Company. All rights reserved.

27. Find the area of triangle *ABC*.

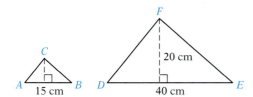

28. Find the area of triangle *DEF*.

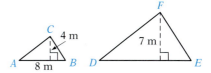

The sun's rays, objects on Earth, and the shadows cast by them form similar triangles. Use this fact to solve Exercises 29–32.

29. Find the height of the flagpole.

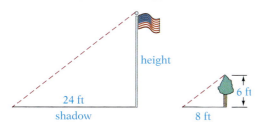

30. Find the height of the flagpole.

31. Find the height of the building.

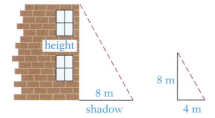

32. Find the height of the building.

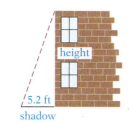

OBJECTIVE **C**

In Exercises 33–38, determine whether the two triangles are congruent. If they are congruent, state by what rule they are congruent.

33.

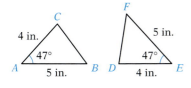

34.

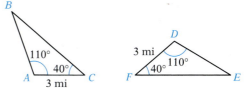

35.

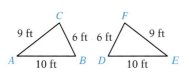

36.

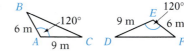

Copyright © Houghton Mifflin Company. All rights reserved.

37.

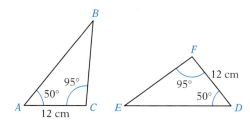

38.

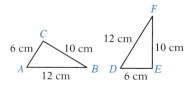

39. Given triangle *ABC* and triangle *DEF*, do the conditions ∠*C* = ∠*E*, *AC* = *EF*, and *BC* = *DE* guarantee that triangle *ABC* is congruent to triangle *DEF*? If they are congruent, by what rule are they congruent?

40. Given triangle *PQR* and triangle *MNO*, do the conditions *PR* = *NO*, *PQ* = *MO*, and *QR* = *MN* guarantee that triangle *PQR* is congruent to triangle *MNO*? If they are congruent, by what rule are they congruent?

41. Given triangle *LMN* and triangle *QRS*, do the conditions ∠*M* = ∠*S*, ∠*N* = ∠*Q*, and ∠*L* = ∠*R* guarantee that triangle *LMN* is congruent to triangle *QRS*? If they are congruent, by what rule are they congruent?

42. Given triangle *DEF* and triangle *JKL*, do the conditions ∠*D* = ∠*K*, ∠*E* = ∠*L*, and *DE* = *KL* guarantee that triangle *DEF* is congruent to triangle *JKL*? If they are congruent, by what rule are they congruent?

43. Given triangle *ABC* and triangle *PQR*, do the conditions ∠*B* = ∠*P*, *BC* = *PQ*, and *AC* = *QR* guarantee that triangle *ABC* is congruent to triangle *PQR*? If they are congruent, by what rule are they congruent?

Critical Thinking

44. Determine whether the statement is always true, sometimes true, or never true.
   a. If two angles of one triangle are equal to two angles of a second triangle, then the triangles are similar triangles.
   b. Two isosceles triangles are similar triangles.
   c. Two equilateral triangles are similar triangles.

45. *Home Maintenance* You need to clean the gutters of your home. The gutters are 24 ft above the ground. For safety, the distance a ladder reaches up a wall should be four times the distance from the bottom of the ladder to the base of the side of the house. Therefore, the ladder must be 6 ft from the base of the house. Will a 25-foot ladder be long enough to reach the gutters? Explain how you determined your answer.

46. What is a Pythagorean triple? Provide at least three examples of Pythagorean triples.

Copyright © Houghton Mifflin Company. All rights reserved.

SECTION 9.4   **Solids**

OBJECTIVE **A**

Volume of a solid

Geometric solids are figures in space. Five common geometric solids are the rectangular solid, the sphere, the cylinder, the cone, and the pyramid.

A **rectangular solid** is one in which all six sides, called **faces,** are rectangles. The variable L is used to represent the length of a rectangular solid, W its width, and H its height.

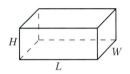

A **sphere** is a solid in which all points are the same distance from point O, called the **center** of the sphere. The **diameter,** d, of a sphere is a line across the sphere going through point O. The **radius,** r, is a line from the center to a point on the sphere. AB is a diameter and OC is a radius of the sphere shown at the right.

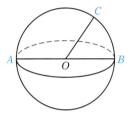

$$d = 2r \quad \text{or} \quad r = \frac{1}{2}d$$

The most common cylinder, called a **right circular cylinder,** is one in which the bases are circles and are perpendicular to the height of the cylinder. The variable r is used to represent the radius of a base of a cylinder, and h represents the height. In this text, only right circular cylinders are discussed.

A **right circular cone** is obtained when one base of a right circular cylinder is shrunk to a point, called the **vertex,** V. The variable r is used to represent the radius of the base of the cone, and h represents the height. The variable l is used to represent the **slant height,** which is the distance from a point on the circumference of the base to the vertex. In this text, only right circular cones are discussed.

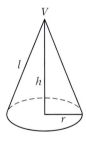

The base of a **regular pyramid** is a regular polygon, and the sides are isosceles triangles. The height, h, is the distance from the vertex, V, to the base and is perpendicular to the base. The variable l is used to represent the **slant height,** which is the height of one of the isosceles triangles on the face of the pyramid. The regular square pyramid at the right has a square base. This is the only type of pyramid discussed in this text.

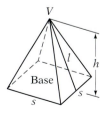

Copyright © Houghton Mifflin Company. All rights reserved.

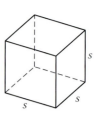

A **cube** is a special type of rectangular solid. Each of the six faces of a cube is a square. The variable s is used to represent the length of one side of a cube.

Volume is a measure of the amount of space occupied by a geometric solid. Volume can be used to describe the amount of trash in a landfill, the amount of concrete poured for the foundation of a house, or the amount of water in a town's reservoir.

► **Point of Interest**

Originally, the human body was used as the standard of measure. A mouthful was used as a unit of measure in ancient Egypt; it was later referred to as a *half jigger*. In French, the word for *inch* is *pouce*, which means thumb. A *span* was the distance from the tip of the outstretched thumb to the tip of the little finger. The *cubit* referred to the distance from the elbow to the end of the fingers. A *fathom* was the distance from the tip of the fingers on one hand to the tip of the fingers on the other hand when standing with arms fully extended out from the sides. The *hand,* where 1 hand = 4 inches, is still used today to measure horses.

A cube that is 1 ft on each side has a volume of 1 cubic foot, which is written 1 ft^3. A cube that measures 1 cm on each side has a volume of 1 cubic centimeter, written 1 cm^3.

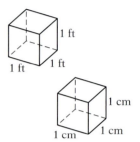

The volume of a solid is the number of cubes that are necessary to exactly fill the solid. The volume of the rectangular solid at the right is 24 cm^3 because it will hold exactly 24 cubes, each 1 cm on a side. Note that the volume can be found by multiplying the length times the width times the height.

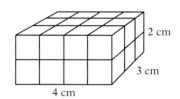

The formulas for the volumes of the geometric solids described above are given below.

Volumes of Geometric Solids

The volume, V, of a **rectangular solid** with length L, width W, and height H is given by $V = LWH$.

The volume, V, of a **cube** with side s is given by $V = s^3$.

The volume, V, of a **sphere** with radius r is given by $V = \frac{4}{3}\pi r^3$.

The volume, V, of a **right circular cylinder** is given by $V = \pi r^2 h$, where r is the radius of the base and h is the height.

The volume, V, of a **right circular cone** is given by $V = \frac{1}{3}\pi r^2 h$, where r is the radius of the circular base and h is the height.

The volume, V, of a **regular square pyramid** is given by $V = \frac{1}{3}s^2 h$, where s is the length of a side of the base and h is the height.

Copyright © Houghton Mifflin Company. All rights reserved.

Find the volume of a sphere with a diameter of 6 in.

First find the radius of the sphere.

$$r = \frac{1}{2}d = \frac{1}{2}(6) = 3$$

Use the formula for the volume of a sphere.

$$V = \frac{4}{3}\pi r^3$$

$$V = \frac{4}{3}\pi(3)^3$$

$$V = \frac{4}{3}\pi(27)$$

The exact volume of the sphere is 36π in³.

$$V = 36\pi$$

An approximate measure can be found by using the π key on a calculator.

$$V \approx 113.10$$

The approximate volume is 113.10 in³.

■ **Calculator Note**

To approximate 36π on your calculator, enter 36 ☒ π ☐ .

---

1 *Example*

The length of a rectangular solid is 5 m, the width is 3.2 m, and the height is 4 m. Find the volume of the solid.

Strategy

To find the volume, use the formula for the volume of a rectangular solid. $L = 5$, $W = 3.2$, $H = 4$

Solution

$V = LWH = 5(3.2)(4) = 64$

The volume of the rectangular solid is 64 m³.

1 *You Try It*

Find the volume of a cube that measures 2.5 m on a side.

Your Strategy

Your Solution

---

2 *Example*

The radius of the base of a cone is 8 cm. The height is 12 cm. Find the volume of the cone. Round to the nearest hundredth.

Strategy

To find the volume, use the formula for the volume of a cone. An approximation is asked for; use the π key on a calculator. $r = 8$, $h = 12$

Solution

$$V = \frac{1}{3}\pi r^2 h$$

$$V = \frac{1}{3}\pi(8)^2(12) = \frac{1}{3}\pi(64)(12) = 256\pi \approx 804.25$$

The volume is approximately 804.25 cm³.

2 *You Try It*

The diameter of the base of a cylinder is 8 ft. The height of the cylinder is 22 ft. Find the exact volume of the cylinder.

Your Strategy

Your Solution

Solutions on p. S24

Copyright © Houghton Mifflin Company. All rights reserved.

OBJECTIVE **B**

VIDEO & DVD · CD TUTOR · WEB · SSM

Surface area of a solid

The **surface area** of a solid is the total area on the surface of the solid. Suppose you want to cover a geometric solid with wallpaper. The amount of wallpaper needed is equal to the surface area of the figure.

When a rectangular solid is cut open and flattened out, each face is a rectangle. The surface area, SA, of the rectangular solid is the sum of the areas of the six rectangles:

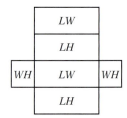

$$SA = LW + LH + WH + LW + WH + LH$$

which simplifies to

$$SA = 2LW + 2LH + 2WH$$

The surface area of a cube is the sum of the areas of the six faces of the cube. The area of each face is s^2. Therefore, the surface area, SA, of a cube is given by the formula $SA = 6s^2$.

When a cylinder is cut open and flattened out, the top and bottom of the cylinder are circles. The side of the cylinder flattens out to a rectangle. The length of the rectangle is the circumference of the base, which is $2\pi r$; the width is h, the height of the cylinder. Therefore, the area of the rectangle is $2\pi rh$. The area of each circle is πr^2. The surface area, SA, of the cylinder is

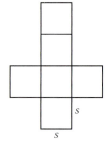 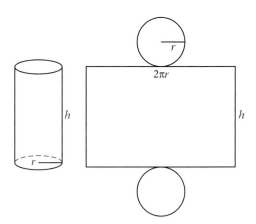

$$SA = \pi r^2 + 2\pi rh + \pi r^2$$

which simplifies to

$$SA = 2\pi r^2 + 2\pi rh$$

Copyright © Houghton Mifflin Company. All rights reserved.

The surface area of a pyramid is the area of the base plus the area of the four isosceles triangles. A side of the square base is s; therefore, the area of the base is s^2. The slant height, l, is the height of each triangle, and s is the base of each triangle. The surface area, SA, of a pyramid is

$$SA = s^2 + 4\left(\frac{1}{2}sl\right)$$

which simplifies to

$$SA = s^2 + 2sl$$

Formulas for the surface areas of geometric solids are given below.

Surface Areas of Geometric Solids

The surface area, SA, of a **rectangular solid** with length L, width W, and height H is given by $SA = 2LW + 2LH + 2WH$.

The surface area, SA, of a **cube** with side s is given by $SA = 6s^2$.

The surface area, SA, of a **sphere** with radius r is given by $SA = 4\pi r^2$.

The surface area, SA, of a **right circular cylinder** is given by $SA = 2\pi r^2 + 2\pi rh$, where r is the radius of the base and h is the height.

The surface area, SA, of a **right circular cone** is given by $SA = \pi r^2 + \pi rl$, where r is the radius of the circular base and l is the slant height.

The surface area, SA, of a **regular pyramid** is given by $SA = s^2 + 2sl$, where s is the length of a side of the base and l is the slant height.

Find the surface area of a sphere with a diameter of 18 cm.

First find the radius of the sphere. $r = \dfrac{1}{2}d = \dfrac{1}{2}(18) = 9$

Use the formula for the surface area of a sphere.

$$SA = 4\pi r^2$$
$$SA = 4\pi(9)^2$$
$$SA = 4\pi(81)$$
$$SA = 324\pi$$

The exact surface area of the sphere is 324π cm².

An approximate measure can be found by using the π key on a calculator. $SA \approx 1{,}017.88$

The approximate surface area is 1,017.88 cm².

Calculator Note

To approximate 324π on your calculator, enter 324 ✕ 𝜋 = .

Copyright © Houghton Mifflin Company. All rights reserved.

③ Example

The diameter of the base of a cone is 5 m, and the slant height is 4 m. Find the surface area of the cone. Give the exact measure.

Strategy

To find the surface area of the cone:

▶ Find the radius of the base of the cone.
▶ Use the formula for the surface area of a cone. Leave the answer in terms of π.

Solution

$$r = \frac{1}{2}d = \frac{1}{2}(5) = 2.5$$

$SA = \pi r^2 + \pi r l$
$SA = \pi(2.5)^2 + \pi(2.5)(4)$
$SA = \pi(6.25) + \pi(2.5)(4)$
$SA = 6.25\pi + 10\pi$
$SA = 16.25\pi$

The surface area of the cone is 16.25π m².

③ You Try It

The diameter of the base of a cylinder is 6 ft, and the height is 8 ft. Find the surface area of the cylinder. Round to the nearest hundredth.

Your Strategy

Your Solution

④ Example

Find the area of a label used to cover a soup can that has a radius of 4 cm and a height of 12 cm. Round to the nearest hundredth.

Strategy

To find the area of the label, use the fact that the surface area of the sides of a cylinder is given by $2\pi r h$. An approximation is asked for; use the π key on a calculator. $r = 4$, $h = 12$

Solution

Area of the label $= 2\pi r h$
Area of the label $= 2\pi(4)(12) = 96\pi \approx 301.59$

The area is approximately 301.59 cm².

④ You Try It

Which has a larger surface area, a cube with a side measuring 10 cm or a sphere with a diameter measuring 8 cm?

Your Strategy

Your Solution

Solutions on p. S24

Copyright © Houghton Mifflin Company. All rights reserved.

9.4 EXERCISES

OBJECTIVE   **A**

 In Exercises 1–6, find the volume of the figure. For calculations involving π, give both the exact value and an approximation to the nearest hundredth.

1.
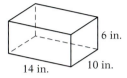
6 in.
14 in.   10 in.

2.
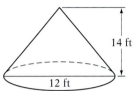
14 ft
12 ft

3.
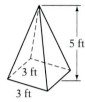
5 ft
3 ft
3 ft

4.
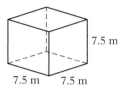
7.5 m
7.5 m   7.5 m

5.

3 cm

6.
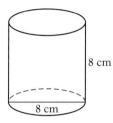
8 cm
8 cm

 Solve.

7. A rectangular solid has a length of 6.8 m, a width of 2.5 m, and a height of 2 m. Find the volume of the solid.

8. Find the volume of a rectangular solid that has a length of 4.5 ft, a width of 3 ft, and a height of 1.5 ft.

9. Find the volume of a cube whose side measures 2.5 in.

10. The length of a side of a cube is 7 cm. Find the volume of the cube.

11. The diameter of a sphere is 6 ft. Find the volume of the sphere. Give the exact measure.

12. Find the volume of a sphere that has a radius of 1.2 m. Round to the nearest tenth.

13. The diameter of the base of a cylinder is 24 cm. The height of the cylinder is 18 cm. Find the volume of the cylinder. Round to the nearest hundredth.

14. The radius of the base of a cone is 5 in. The height of the cone is 9 in. Find the volume of the cone. Give the exact measure.

15. The height of a cone is 15 cm. The diameter of the cone is 10 cm. Find the volume of the cone. Round to the nearest hundredth.

Copyright © Houghton Mifflin Company. All rights reserved.

16. The length of a side of the base of a pyramid is 6 in., and the height is 10 in. Find the volume of the pyramid.

17. The height of a pyramid is 8 m, and the length of a side of the base is 9 m. What is the volume of the pyramid?

18. The index finger on the Statue of Liberty is 8 ft long. The circumference at the second joint is 3.5 ft. Use the formula for the volume of a cylinder to approximate the volume of the index finger on the Statue of Liberty. Round to the nearest hundredth.

19. The volume of a freezer with a length of 7 ft and a height of 3 ft is 52.5 ft³. Find the width of the freezer.

20. The length of an aquarium is 18 in., and the width is 12 in. If the volume of the aquarium is 1,836 in³, what is the height of the aquarium?

21. The volume of a cylinder with a height of 10 in. is 502.4 in³. Find the radius of the base of the cylinder. Round to the nearest hundredth.

22. The diameter of the base of a cylinder is 14 cm. If the volume of the cylinder is 2,310 cm³, find the height of the cylinder. Round to the nearest hundredth.

23. A rectangular solid has a square base and a height of 5 in. If the volume of the solid is 125 in³, find the length and the width.

24. The volume of a rectangular solid is 864 m³. The rectangular solid has a square base and a height of 6 m. Find the dimensions of the solid.

25. An oil storage tank, which is in the shape of a cylinder, is 4 m high and has a diameter of 6 m. The oil tank is two-thirds full. Find the number of cubic meters of oil in the tank. Round to the nearest hundredth.

26. A silo, which is in the shape of a cylinder, is 16 ft in diameter and has a height of 30 ft. The silo is three-fourths full. Find the volume of the portion of the silo that is not being used for storage. Round to the nearest hundredth.

 OBJECTIVE ▮ B

 Find the surface area of the figure.

27.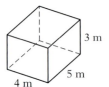
3 m
5 m
4 m

28.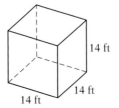
14 ft
14 ft
14 ft

29.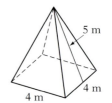
5 m
4 m
4 m

Copyright © Houghton Mifflin Company. All rights reserved.

In Exercises 30–32, find the surface area of the figure. Give both the exact value and an approximation to the nearest hundredth.

30.

2 cm

31.

2 in.
6 in.

32.
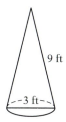
9 ft
3 ft

Solve.

33. The height of a rectangular solid is 5 ft. The length is 8 ft, and the width is 4 ft. Find the surface area of the solid.

34. The width of a rectangular solid is 32 cm. The length is 60 cm, and the height is 14 cm. What is the surface area of the solid?

35. The side of a cube measures 3.4 m. Find the surface area of the cube.

36. Find the surface area of a cube that has a side measuring 1.5 in.

37. Find the surface area of a sphere with a diameter of 15 cm. Give the exact value.

38. The radius of a sphere is 2 in. Find the surface area of the sphere. Round to the nearest hundredth.

39. The radius of the base of a cylinder is 4 in. The height of the cylinder is 12 in. Find the surface area of the cylinder. Round to the nearest hundredth.

40. The diameter of the base of a cylinder is 1.8 m. The height of the cylinder is 0.7 m. Find the surface area of the cylinder. Give the exact value.

41. The slant height of a cone is 2.5 ft. The radius of the base is 1.5 ft. Find the surface area of the cone. Give the exact value.

42. The diameter of the base of a cone is 21 in. The slant height is 16 in. What is the surface area of the cone? Round to the nearest hundredth.

43. The length of a side of the base of a pyramid is 9 in., and the slant height is 12 in. Find the surface area of the pyramid.

44. The slant height of a pyramid is 18 m, and the length of a side of the base is 16 m. What is the surface area of the pyramid?

Copyright © Houghton Mifflin Company. All rights reserved.

45. The surface area of a rectangular solid is 108 cm². The height of the solid is 4 cm, and the length is 6 cm. Find the width of the rectangular solid.

46. The length of a rectangular solid is 12 ft. The width is 3 ft. If the surface area is 162 ft², find the height of the rectangular solid.

47. A can of paint will cover 300 ft². How many cans of paint should be purchased in order to paint a cylinder that has a height of 30 ft and a radius of 12 ft?

48. A hot air balloon is in the shape of a sphere. Approximately how much fabric was used to construct the balloon if its diameter is 32 ft? Round to the nearest whole number.

49. How much glass is needed to make a fish tank that is 12 in. long, 8 in. wide, and 9 in. high? The fish tank is open at the top.

50. Find the area of a label used to cover a can of juice that has a diameter of 16.5 cm and a height of 17 cm. Round to the nearest hundredth.

51. The length of a side of the base of a pyramid is 5 cm, and the slant height is 8 cm. How much larger is the surface area of this pyramid than the surface area of a cone with a diameter of 5 cm and a slant height of 8 cm? Round to the nearest hundredth.

Critical Thinking

52. Half of a sphere is called a **hemisphere.** Derive formulas for the volume and surface area of a hemisphere.

53. Determine whether the statement is always true, sometimes true, or never true.
   a. The slant height of a regular pyramid is longer than the height.
   b. The slant height of a cone is shorter than the height.
   c. The four triangular faces of a regular pyramid are equilateral triangles.

54. **a.** What is the effect on the surface area of a rectangular solid if the width and height are doubled?
   b. What is the effect on the volume of a rectangular solid if both the length and the width are doubled?
   c. What is the effect on the volume of a cube if the length of each side of the cube is doubled?
   d. What is the effect on the surface area of a cylinder if the radius and height are doubled?

55. Explain how you could cut through a cube so that the face of the resulting solid is (a) a square, (b) an equilateral triangle, (c) a trapezoid, (d) a hexagon.

Copyright © Houghton Mifflin Company. All rights reserved.

SECTION 9.5 **Composite Figures**

OBJECTIVE   **A**

Perimeter of a composite plane figure

Composite geometric figures are made from two or more geometric figures. The composite figure below is made from parts of a rectangle and a circle.

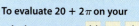

Composite figure = 3 sides of a rectangle + $\frac{1}{2}$ the circumference of a circle

Perimeter   = $2L + W$ + $\frac{1}{2}\pi d$

> Find the perimeter of the composite figure shown above if the width of the rectangle is 4 m and the length of the rectangle is 8 m. Round to the nearest hundredth.
>
> Use the equation given above. $L = 8$, $W = 4$. $P = 2L + W + \frac{1}{2}\pi d$
> The diameter of the circle equals the width
> of the rectangle, 4. $P = 2(8) + 4 + \frac{1}{2}\pi(4)$
>
> Use the π key on a calculator to $P = 20 + 2\pi$
> approximate the perimeter. $P \approx 26.28$
>
> To the nearest hundredth, the perimeter of the figure is 26.28 m.

Calculator Note

To evaluate $20 + 2\pi$ on your calculator, enter 2 ✕ π + 20 = . Round the number in the display to two decimal places.

---

1 *Example*

Find the perimeter of the figure. Round to the nearest hundredth.

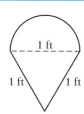

Strategy

The perimeter is equal to 2 sides of a triangle plus $\frac{1}{2}$ the circumference of a circle. An approximation is asked for; use the π key on a calculator.

Solution

$P = a + b + \frac{1}{2}\pi d$

$P = 1 + 1 + \frac{1}{2}\pi(1) = 2 + 0.5\pi \approx 3.57$

The perimeter is approximately 3.57 ft.

1 *You Try It*

The circumference of the circle in the figure is 6π cm. Find the perimeter of square *ABCD*.

Your Strategy

Your Solution

Solution on p. S24

Copyright © Houghton Mifflin Company. All rights reserved.

VIDEO & DVD · CD TUTOR · WWW WEB · SSM

OBJECTIVE **B**

Area of a composite plane figure

The area of the composite figure shown below is found by calculating the area of the rectangle and then subtracting the area of the triangle.

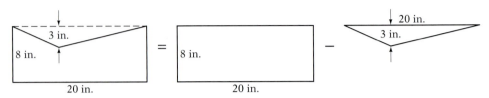

Area of the composite figure = area of the rectangle − area of the triangle

$$= \quad LW \quad - \quad \frac{1}{2}bh$$

$$= \quad 20(8) \quad - \quad \frac{1}{2}(20)(3)$$

$$= \quad 160 \quad - \quad 30$$

$$= \quad 130$$

The area of the composite figure is 130 in².

---

2 *Example*

Find the area of the shaded portion of the figure. Round to the nearest hundredth.

Strategy

The area is equal to the area of the square minus the area of the circle. The radius of the circle is one-half the length of a side of the square (8). An approximation is asked for; use the π key on a calculator.

Solution

$$r = \frac{1}{2}s = \frac{1}{2}(8) = 4$$

$$A = s^2 - \pi r^2$$
$$A = (8)^2 - \pi(4)^2 = 64 - 16\pi \approx 13.73$$

The area is approximately 13.73 m².

2 *You Try It*

Find the area of the composite figure.

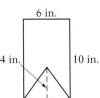

Your Strategy

Your Solution

Solution on p. S25

Copyright © Houghton Mifflin Company. All rights reserved.

OBJECTIVE **C**

Volume of a composite solid

Composite geometric solids are solids made from two or more geometric solids. The following solid is made from a cylinder and one-half of a sphere.

Composite solid   = a cylinder + one-half of a sphere

Volume of the composite solid = $\pi r^2 h$ + $\dfrac{1}{2} \cdot \dfrac{4}{3} \pi r^3$

> Find the volume of the solid shown above if the radius of the base of the cylinder is 3 in. and the height of the cylinder is 10 in. Give the exact measure.
>
> Use the equation given above. $r = 3$, $h = 10$. The radius of the sphere equals the radius of the base of the cylinder, 3.
>
> $$V = \pi r^2 h + \frac{1}{2} \cdot \frac{4}{3} \pi r^3$$
>
> $$V = \pi(3)^2(10) + \frac{1}{2} \cdot \frac{4}{3} \pi(3)^3$$
>
> $$V = \pi(9)(10) + \frac{2}{3}\pi(27)$$
>
> $$V = 90\pi + 18\pi = 108\pi$$
>
> The volume of the solid is 108π in³.

---

3 *Example*

Find the volume of the solid. Round to the nearest hundredth.

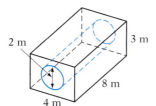

Strategy

The volume is equal to the volume of the rectangular solid minus the volume of the cylinder. The radius of the circle is one-half the diameter of the circle. An approximation is asked for; use the π key on a calculator.

Solution

$$r = \frac{1}{2}d = \frac{1}{2}(2) = 1$$

$$V = LWH - \pi r^2 h$$
$$V = 8(4)(3) - \pi(1)^2(8) = 96 - 8\pi \approx 70.87$$

The volume is approximately 70.87 m³.

3 *You Try It*

Find the volume of the solid. Give the exact measure.

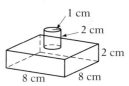

Your Strategy

Your Solution

Solution on p. S25

Copyright © Houghton Mifflin Company. All rights reserved.

OBJECTIVE **D**

Surface area of a composite solid

The composite solid shown below is made from a cone, a cylinder, and one-half of a sphere.

Surface area of the solid = the surface area of a cone minus the base +
the surface area of the sides of a cylinder +
one-half of the surface area of a sphere

$$= \pi r l \quad + \quad 2\pi r h \quad + \quad \frac{1}{2}(4\pi r^2)$$

Find the surface area of the solid shown above. The radius of the base of the cylinder is 4 m and the height is 5 m. The slant height of the cone is 6 m. Give the exact measure.

Use the equation given above. $r = 4$, $h = 5$, $l = 6$. The radius of the base of the cone and the radius of the sphere equal the radius of the base of the cylinder, 4.

$$SA = \pi r l + 2\pi r h + \frac{1}{2}(4\pi r^2)$$

$$SA = \pi(4)(6) + 2\pi(4)(5) + \frac{1}{2}[4\pi(4)^2]$$

$$SA = \pi(24) + 2\pi(20) + 2\pi(16)$$

$$SA = 24\pi + 40\pi + 32\pi = 96\pi$$

The surface area of the solid is 96π m².

---

4 *Example*

Find the surface area of the solid. Round to the nearest hundredth.

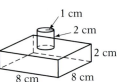

1 cm
2 cm
2 cm
8 cm   8 cm

Strategy

The total surface area equals the surface area of the rectangular solid, minus the bottom of the cylinder, plus the surface area of the cylinder, minus the bottom of the cylinder.

Solution

$$SA = 2LW + 2LH + 2HW - \pi r^2 + 2\pi r^2$$
$$\quad + 2\pi r h - \pi r^2$$
$$SA = 2LW + 2LH + 2HW + 2\pi r h$$
$$SA = 2(8)(8) + 2(8)(2) + 2(2)(8) + 2\pi(1)(2)$$
$$SA = 128 + 32 + 32 + 4\pi$$
$$SA = 192 + 4\pi \approx 204.57$$

The surface area is approximately 204.57 cm².

4 *You Try It*

Find the surface area of the solid. Round to the nearest hundredth.

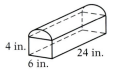

4 in.
24 in.
6 in.

Your Strategy

Your Solution

Solution on p. S25

Copyright © Houghton Mifflin Company. All rights reserved.

9.5 EXERCISES

Copyright © Houghton Mifflin Company. All rights reserved.

OBJECTIVE   **A**

In Exercises 1–12, find the perimeter of the composite figure. For calculations involving π, give both the exact value and an approximation to the nearest hundredth.

1.
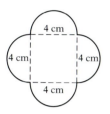
5 cm
8 cm
20 cm
27 cm
19 cm
42 cm

2.
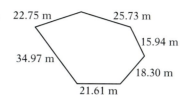
22.75 m   25.73 m
15.94 m
34.97 m
18.30 m
21.61 m

3.
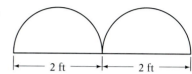
2 ft   2 ft

4.
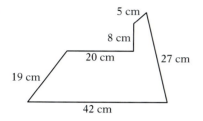
4 cm
4 cm   4 cm
4 cm

5.
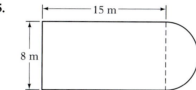
15 m
8 m

6.
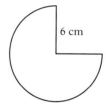
6 cm

7.
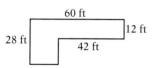
60 ft
28 ft   12 ft
42 ft

8.
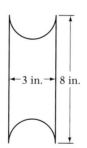
3 in.   8 in.

9.
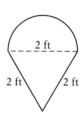
2 ft
2 ft   2 ft

10.
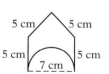
5 cm   5 cm
5 cm   5 cm
7 cm

11.
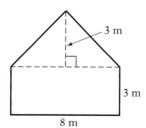
3 m
3 m
8 m

12.
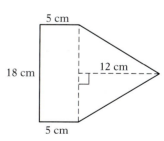
5 cm
18 cm   12 cm
5 cm

 Solve.

13. Find the length of weather stripping installed around the arched door shown in the figure at the right. Round to the nearest hundredth.

6.5 ft
3 ft

14. Find the perimeter of the roller rink shown in the figure at the right. Round to the nearest hundredth.

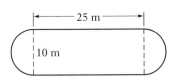

25 m

10 m

15. The rectangular lot shown in the figure at the right is being fenced. The fencing along the road will cost $6.70 per foot. The rest of the fencing will cost $5.10 per foot. Find the total cost to fence the lot.

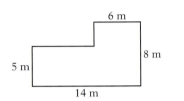

800 ft

1,250 ft

16. A rain gutter is being installed on a home that has the dimensions shown in the figure at the right. At a cost of $22.60 per meter, how much will it cost to install the rain gutter?

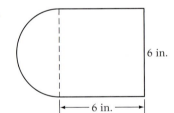

6 m

8 m

5 m

14 m

▌ OBJECTIVE **B**

In Exercises 17–28, find the area of the composite figure. For calculations involving π, give both the exact value and an approximation to the nearest hundredth.

17.

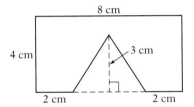

8 cm

4 cm

3 cm

2 cm   2 cm

18.

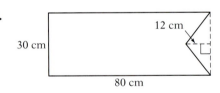

30 cm

12 cm

80 cm

19.

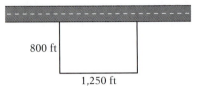

6 in.

6 in.

20.

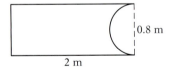

0.8 m

2 m

21.

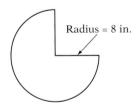

Radius = 8 in.

22.

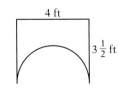

4 ft

$3\frac{1}{2}$ ft

23.

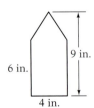

9 in.

6 in.

4 in.

24.

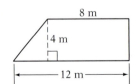

8 m

4 m

12 m

25.

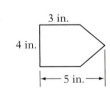

3 in.

4 in.

5 in.

Copyright © Houghton Mifflin Company. All rights reserved.

26.

22 cm
22 cm

27.
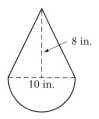
8 in.
10 in.

28.
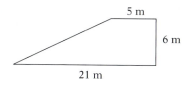
5 m
6 m
21 m

Solve.

29. A carpet is to be installed in one room and a hallway, as shown in the diagram at the right. At a cost of $28.50 per square meter, how much will it cost to carpet the area?

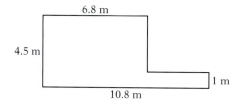
6.8 m
4.5 m
10.8 m
1 m

30. Find the area of the 2-meter boundary around the swimming pool shown in the figure at the right.

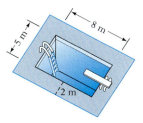
5 m
8 m
2 m

31. How much hardwood floor is needed to cover the roller rink shown in the figure at the right? Round to the nearest hundredth.

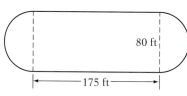
80 ft
175 ft

32. Find the total area of a national park with the dimensions shown in the figure at the right. Round to the nearest hundredth.

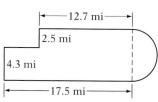
12.7 mi
2.5 mi
4.3 mi
17.5 mi

OBJECTIVE **C**

In Exercises 33–44, find the volume of the composite figure. For calculations involving π, give both the exact value and an approximation to the nearest hundredth.

33.
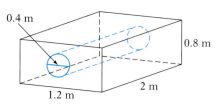
0.4 m
0.8 m
1.2 m
2 m

34.
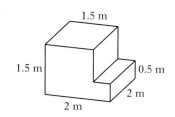
1.5 m
1.5 m
0.5 m
2 m
2 m

35.
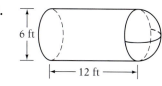
6 ft
12 ft

Copyright © Houghton Mifflin Company. All rights reserved.

36.

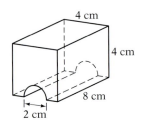

37.

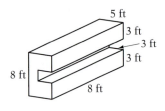

38.

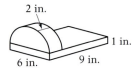

39.

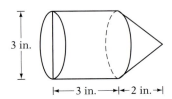

40.

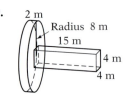

41.

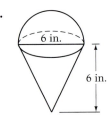

42.

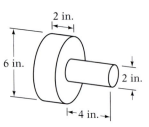

43.

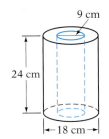

44.

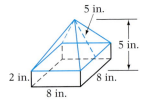

Solve.

45. Find the volume of the bushing shown in the figure at the right. Round to the nearest hundredth.

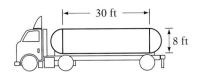

46. A truck is carrying an oil tank, as shown in the figure at the right. If the tank is half full, how many cubic feet of oil is the truck carrying? Round to the nearest hundredth.

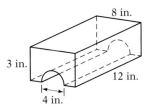

47. The concrete floor of a building is shown in the figure at the right. At a cost of $6.15 per cubic foot, find the cost of having the floor poured. Round to the nearest cent.

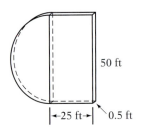

Copyright © Houghton Mifflin Company. All rights reserved.

48. How many liters of water are needed to fill the swimming pool shown at the right? (1 m³ contains 1 000 L.)

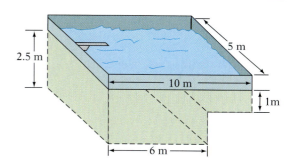

OBJECTIVE   **D**

 In Exercises 49–60, find the surface area of the composite figure. For calculations involving π, give both the exact value and an approximation to the nearest hundredth.

49.

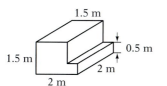

50.

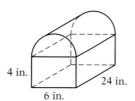

51.

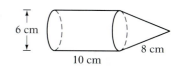

52.

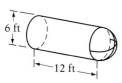

53.

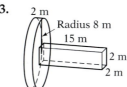

54.

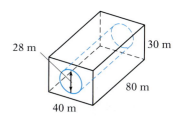

55.

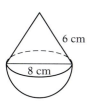

56.

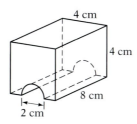

57.

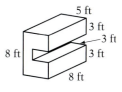

58.

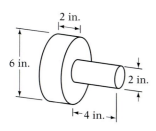

59.

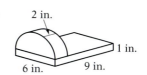

60.

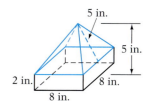

Copyright © Houghton Mifflin Company. All rights reserved.

Solve.

61. A can of paint will cover 250 ft². Find the number of cans that should be purchased in order to paint the exterior of the auditorium shown in the figure at the right.

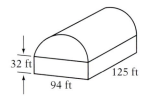

62. A piece of sheet metal is cut and formed into the shape shown at the right. Given that there are 0.24 g in 1 cm² of the metal, find the total number of grams of metal used. Round to the nearest hundredth.

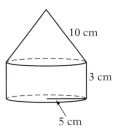

63. The walls of a room that is 25.5 ft long, 22 ft wide, and 8 ft high are being plastered. There are two doors in the room, each 2.5 ft by 7 ft. Each of the six windows in the room measures 2.5 ft by 4 ft. At a cost of $1.50 per square foot, find the cost of plastering the walls of the room.

Critical Thinking

64. *Painting* You plan on painting the bookcase shown at the right. The bookcase is 10 in. deep, 36 in. high, and 36 in. long. The wood is 1 in. thick. You do not plan to paint the back side or the bottom. Find the surface area of the wood that needs to be painted.

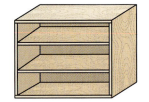

65. *Shipping* Bottles of apple juice are being packaged six to a carton for shipping. The diameter of the base of the bottles is 4 in. The height of the bottles is 8 in. The cartons are made of corrugated cardboard that is $\frac{1}{8}$ in. thick. Pieces of cardboard, each $\frac{1}{16}$ in. thick, are placed between bottles. Find the dimensions of the shipping carton.

66. A sphere fits inside a cylinder as shown at the right. The height of the cylinder equals the diameter of the sphere. Show that the surface area of the sphere equals the surface area of the sides of the cylinder.

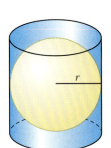

67. Explain the meaning of the "vanishing point" in a drawing. Find examples of its use.

68. Prepare a report on the use of geometric forms in architecture. Include examples of both plane geometric figures and geometric solids.

69. Write a paper on the artist M. C. Escher. Explain how he used mathematics and geometry in his works.

Copyright © Houghton Mifflin Company. All rights reserved.

Focus on Problem Solving

Trial and Error

Some problems in mathematics are solved by using **trial and error**. The trial-and-error method of arriving at a solution to a problem involves repeated tests or experiments until a satisfactory conclusion is reached.

Many of the Critical Thinking exercises in this text require a trial-and-error method of solution. For example, an exercise on page 584 reads as follows:

Explain how you could cut through a cube so that the face of the resulting solid is (a) a square, (b) an equilateral triangle, (c) a trapezoid, (d) a hexagon.

There is no formula to apply to this problem; there is no computation to perform. This problem requires picturing a cube and the results after it is cut through at different places on its surface and at different angles. For part (a), cutting perpendicular to the top and bottom of the cube and parallel to two of its sides will result in a square. The other shapes may prove more difficult.

When solving problems of this type, keep an open mind. Sometimes when using the trial-and-error method, we are hampered by narrowness of vision; we cannot expand our thinking to include other possibilities. Then when we see someone else's solution, it appears so obvious to us! For example, for the question above, it is necessary to conceive of cutting through the cube at places other than the top surface; we need to be open to the idea of beginning the cut at one of the corner points of the cube.

A topic of the Projects and Group Activities in this chapter is symmetry. Here again, the trial-and-error method is used to determine the lines of symmetry inherent in an object. For example, in determining lines of symmetry for a square, begin by drawing a square. The horizontal line of symmetry and the vertical line of symmetry may be immediately obvious to you.

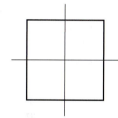

But there are two others. Do you see that a line drawn through opposite corners of the square is also a line of symmetry?

Many of the questions in this text that require an answer of "always true," "sometimes true," or "never true" are best solved by the trial-and-error method. For example, consider the statement presented in Section 2 of this chapter:

Two rectangles that have the same area have the same perimeter.

Try some numbers. Each of two rectangles, one measuring 6 units by 2 units and another measuring 4 units by 3 units, has an area of 12 square units, but the perimeter of the first is 16 units and the perimeter of the second is 14 units, so the answer "always true" has been eliminated. We still need to determine whether there is a case for which it *is* true. After experimenting with a lot of numbers, you may come to realize that we are trying to determine whether it is possible for two different pairs of factors of a number to have the same sum. Is it?

Don't be afraid to make many experiments, and remember that *errors*, or tests that "don't work," are a part of the trial-and-*error* process.

Copyright © Houghton Mifflin Company. All rights reserved.

Projects & Group Activities

Lines of Symmetry

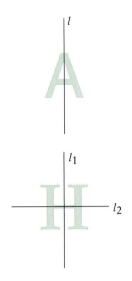

Look at the letter A printed at the left. If the letter were folded along line *l*, the two sides of the letter would match exactly. This letter has **symmetry** with respect to line *l*. Line *l* is called the **axis of symmetry.**

Now consider the letter H printed below at the left. Both lines l_1 and l_2 are axes of symmetry for this letter; the letter could be folded along either line and the two sides would match exactly.

1. Does the letter A have more than one axis of symmetry?

2. Find axes of symmetry for other capital letters of the alphabet.

3. Which lower-case letters have one axis of symmetry?

4. Do any of the lower-case letters have more than one axis of symmetry?

5. Find the numbers of axes of symmetry for the plane geometric figures presented in this chapter.

6. There are other types of symmetry. Look up the meaning of *point symmetry* and *rotational symmetry.* Which plane geometric figures provide examples of these types of symmetry?

7. Find examples of symmetry in nature, art, and architecture.

Preparing a Circle Graph

In Section 1 of this chapter, a protractor was used to measure angles. Preparing a circle graph requires the ability to use a protractor to draw angles.

To draw an angle of 142°, first draw a ray. Place a dot at the endpoint of the ray. This dot will be the vertex of the angle.

Place the straight bottom edge of the protractor on the ray as shown in the figure at the right. Make sure the center of the bottom edge of the protractor is located directly over the vertex point. Locate the position of the 142° mark. Place a dot next to the mark.

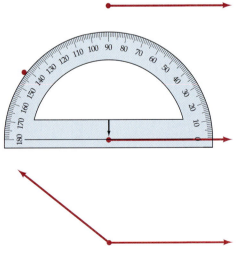

Remove the protractor and draw a ray from the vertex to the dot at the 142° mark.

An example of preparing a circle graph is given on the next page.

Copyright © Houghton Mifflin Company. All rights reserved.

The revenues (in thousands of dollars) from four segments of a car dealership for the first quarter of a recent year were

| New Car Sales: | $2,100 | Used Car/Truck Sales: | $1,500 |
| New Truck Sales: | $1,200 | Parts/Service: | $700 |

To draw a circle graph to represent the percent that each segment contributed to the total revenue from all four segments, proceed as follows:

Find the total revenue from all four segments.

$2,100 + 1,200 + 1,500 + 700 = 5,500$

Find what percent each segment is of the total revenue of $5,500.

New car sales: $\dfrac{2,100}{5,500} \approx 38.2\%$

New truck sales: $\dfrac{1,200}{5,500} \approx 21.8\%$

Used car/truck sales: $\dfrac{1,500}{5,500} \approx 27.3\%$

Parts/service: $\dfrac{700}{5,500} \approx 12.7\%$

Each percent represents the part of the circle for that sector. Because the circle contains 360°, multiply each percent by 360° to find the measure of the angle for each sector. Round to the nearest whole number.

New car sales:

$0.382 \times 360° \approx 138°$

New truck sales:

$0.218 \times 360° \approx 78°$

Used car/truck sales:

$0.273 \times 360° \approx 98°$

Parts/service:

$0.127 \times 360° \approx 46°$

Draw a circle and use a protractor to draw the sectors representing the percents that each segment contributed to the total revenue from all four segments.

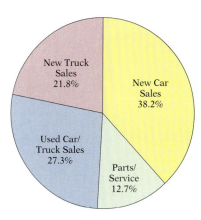

Collect data appropriate for display in a circle graph. [Some possibilities are last year's sales for the top three car manufacturers in the United States, votes cast in the last election for your state governor, the majors of the students in your math class, and the number of students enrolled in each class (senior, junior, etc.) at your college.] Then prepare the circle graph.

Copyright © Houghton Mifflin Company. All rights reserved.

Chapter Summary

Key Words	Examples

A **line** is determined by two distinct points and extends indefinitely in both directions. A **line segment** is part of a line and has two endpoints. **Parallel lines** never meet; the distance between them is always the same. **Perpendicular lines** are intersecting lines that form right angles. [9.1A, pp. 531–533]

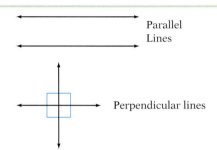

Parallel Lines

Perpendicular lines

A **ray** starts at a point and extends indefinitely in one direction. The point at which a ray starts is the **endpoint** of the ray. An **angle** is formed by two rays with the same endpoint. The **vertex** of an angle is the point at which the two rays meet. An angle is measured in **degrees**. A 90° angle is a **right angle**. A 180° angle is a **straight angle**. An **acute angle** is an angle whose measure is between 0° and 90°. An **obtuse angle** is an angle whose measure is between 90° and 180°. **Complementary angles** are two angles whose measures have the sum 90°. **Supplementary angles** are two angles whose measures have the sum 180°. [9.1A, pp. 531–534]

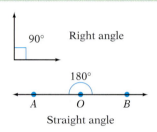

90° Right angle

180°

A   O   B

Straight angle

Two angles that are on opposite sides of the intersection of two lines are **vertical angles;** vertical angles have the same measure. Two angles that share a common side are **adjacent angles;** adjacent angles of intersecting lines are supplementary angles. [9.1A, p. 534; 9.1B, p. 536]

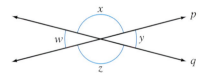

Angles w and y are vertical angles.
Angles x and y are adjacent angles.

A line that intersects two other lines at two different points is a **transversal.** If the lines cut by a transversal are parallel lines, equal angles are formed: **alternate interior angles, alternate exterior angles,** and **corresponding angles.** [9.1B, p. 537]

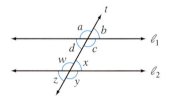

Parallel lines ℓ_1 and ℓ_2 are cut by transversal t. All four acute angles have the same measure. All four obtuse angles have the same measure.

A **polygon** is a closed figure determined by three or more line segments. The line segments that form the polygon are its **sides.** A **regular polygon** is one in which all sides have the same length and all angles have the same measure. Polygons are classified by the number of sides. A **quadrilateral** is a four-sided polygon. A parallelogram, a rectangle, a square, a rhombus, and a trapezoid are all quadrilaterals. [9.2A, pp. 547–548]

Number of Sides	Name of the Polygon
3	Triangle
4	Quadrilateral
5	Pentagon
6	Hexagon
7	Heptagon
8	Octagon
9	Nonagon
10	Decagon

Copyright © Houghton Mifflin Company. All rights reserved.

A **triangle** is a plane figure formed by three line segments. An **isosceles triangle** has two sides of equal length. The three sides of an **equilateral triangle** are of equal length. A **scalene triangle** has no two sides of equal length. An **acute triangle** has three acute angles. An **obtuse triangle** has one obtuse angle. A **right triangle** has a right angle. [9.1C, p. 539; 9.2A, pp. 547–548]

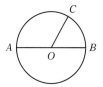

A right triangle

A **circle** is a plane figure in which all points are the same distance from the center of the circle. A **diameter** of a circle is a line segment across the circle through the center. A **radius** of a circle is a line segment from the center of the circle to a point on the circle. [9.2A, p. 550]

AB is a diameter of the circle.
OC is a radius.

Similar triangles have the same shape but not necessarily the same size. The ratios of corresponding sides are equal. The ratio of corresponding heights is equal to the ratio of corresponding sides. **Congruent triangles** have the same shape and the same size. [9.3B, p. 567; 9.3C, p. 569]

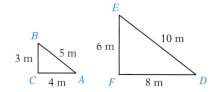

Triangles *ABC* and *DEF* are similar triangles. The ratio of corresponding sides is $\frac{1}{2}$.

Essential Rules and Procedures

Triangles [9.1C, p. 539, 9.3C, pp. 569–570]
Sum of the measures of the interior angles = 180°
Sum of an interior and the corresponding exterior angle = 180°
Rules to determine congruence: SSS Rule, SAS Rule, ASA Rule

In a right triangle, the measure of one acute angle is 12°. Find the measure of the other acute angle.

$$x + 12° + 90° = 180°$$
$$x + 102° = 180°$$
$$x = 78°$$

Formulas for Perimeter (the distance around a figure) [9.2A, pp. 548–550]
Triangle: $P = a + b + c$
Rectangle: $P = 2L + 2W$
Square: $P = 4s$
Circumference of a circle: $C = \pi d$ or $C = 2\pi r$

The length of a rectangle is 8 m. The width is 5.5 m. Find the perimeter of the rectangle.

$$P = 2L + 2W$$
$$P = 2(8) + 2(5.5)$$
$$P = 16 + 11$$
$$P = 27$$

The perimeter is 27 m.

Copyright © Houghton Mifflin Company. All rights reserved.

Formulas for Area (the amount of surface in a region) [9.2B, pp. 552–555]

Triangle: $A = \dfrac{1}{2}bh$

Rectangle: $A = LW$
Square: $A = s^2$
Circle: $A = \pi r^2$
Parallelogram: $A = bh$

Trapezoid: $A = \dfrac{1}{2}h(b_1 + b_2)$

The length of the base of a parallelogram is 12 cm, and the height is 4 cm. Find the area of the parallelogram.

$A = bh$

$A = 12(4)$

$A = 48$

The area is 48 cm².

Formulas for Volume (the amount of space inside a figure in space) [9.4A, p. 576]
Rectangular solid: $V = LWH$
Cube: $V = s^3$

Sphere: $V = \dfrac{4}{3}\pi r^3$

Right circular cylinder: $V = \pi r^2 h$

Right circular cone: $V = \dfrac{1}{3}\pi r^2 h$

Regular pyramid: $V = \dfrac{1}{3}s^2 h$

Find the volume of a cube that measures 3 in. on a side.

$V = s^3$

$V = 3^3$

$V = 27$

The volume is 27 in³.

Formulas for Surface Area (the total area on the surface of the solid) [9.4B, p. 579]
Rectangular solid: $SA = 2LW + 2LH + 2WH$
Cube: $SA = 6s^2$
Sphere: $SA = 4\pi r^2$
Right circular cylinder: $SA = 2\pi r^2 + 2\pi rh$
Right circular cone: $SA = \pi r^2 + \pi rl$
Regular pyramid: $SA = s^2 + 2sl$

Find the surface area of a sphere with a diameter of 10 cm. Give the exact value.

$r = \dfrac{1}{2}d = \dfrac{1}{2}(10) = 5$

$SA = 4\pi r^2$

$SA = 4\pi(5^2)$

$SA = 4\pi(25)$

$SA = 100\pi$

The surface area is 100π cm².

Pythagorean Theorem [9.3A, p. 565]
If a and b are the legs of a right triangle and c is the length of the hypotenuse, then $c^2 = a^2 + b^2$.

Two legs of a right triangle measure 6 ft and 8 ft. Find the hypotenuse of the right triangle.

$c^2 = a^2 + b^2$

$c^2 = 6^2 + 8^2$

$c^2 = 36 + 64$

$c^2 = 100$

$c = \sqrt{100}$

$c = 10$

The length of the hypotenuse is 10 ft.

Principal Square Root Property [9.3A, p. 566]
If $r^2 = s$, then $r = \sqrt{s}$, and r is called the square root of s.

If $c^2 = 16$, then $c = \sqrt{16} = 4$.

Copyright © Houghton Mifflin Company. All rights reserved.

Chapter Review Exercises

1. Given that ∠a = 74° and ∠b = 52°, find the measures of angles x and y.

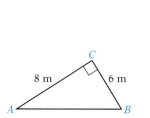

2. Triangles ABC and DEF are similar. Find the perimeter of triangle ABC.

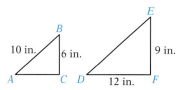

3. Find the volume of the composite figure.

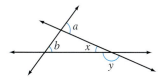

4. Find the measure of ∠x.

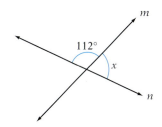

5. Determine whether the two triangles are congruent. If they are congruent, state by what rule they are congruent.

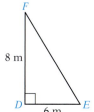

6. Find the surface area of the composite figure. Round to the nearest hundredth.

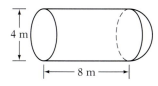

7. Given that BC = 11 cm and AB is three times the length of BC, find the length of AC.

8. Find x.

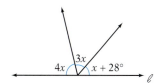

9. Find the area of the composite figure. Round to the nearest hundredth.

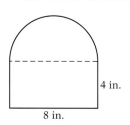

10. Find the volume of the figure.

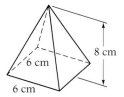

Copyright © Houghton Mifflin Company. All rights reserved.

11. Find the perimeter of the composite figure. Round to the nearest hundredth.

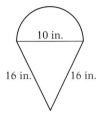

10 in.

16 in.   16 in.

12. Given that $\ell_1 \| \ell_2$, find the measures of angles a and b.

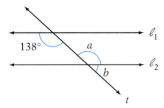

138° a ℓ_1

b ℓ_2

t

13. Find the surface area of the figure.

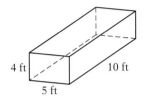

4 ft   10 ft

5 ft

14. Find the unknown side of the triangle. Round to the nearest hundredth.

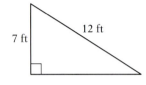

12 ft

7 ft

15. Find the volume of a cube whose side measures 3.5 in.

16. Find the supplement of a 32° angle.

17. Find the volume of a rectangular solid with a length of 6.5 ft, a width of 2 ft, and a height of 3 ft.

18. Two angles of a triangle measure 37° and 48°. Find the measure of the third angle.

19. The height of a triangle is 7 cm. The area of the triangle is 28 cm². Find the length of the base of the triangle.

20. Find the volume of a sphere that has a diameter of 12 mm. Give the exact value.

21. The perimeter of a square picture frame is 86 cm. Find the length of each side of the frame.

22. A can of paint will cover 200 ft². How many cans of paint should be purchased in order to paint a cylinder that has a height of 15 ft and a radius of 6 ft?

23. The length of a rectangular park is 56 yd. The width is 48 yd. How many yards of fencing are needed to surround the park?

24. What is the area of a square patio that measures 9.5 m on each side?

25. A walkway 2 m wide surrounds a rectangular plot of grass. The plot is 40 m long and 25 m wide. What is the area of the walkway?

Copyright © Houghton Mifflin Company. All rights reserved.

Chapter Test

1. For the right triangle shown below, determine the length of side BC. Round to the nearest hundredth.

2. Determine whether the two triangles are congruent. If they are congruent, state by what rule they are congruent.

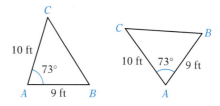

3. Determine the area of a rectangle with a length of 15 m and a width of 7.4 m.

4. Determine the area of a triangle whose base is 7 ft and whose height is 12 ft.

5. Determine the exact volume of a right circular cone whose radius is 7 cm and whose height is 16 cm.

6. Determine the exact surface area of a pyramid whose square base is 3 m on each side and whose slant height is 11 m.

7. Determine the volume of the composite solid shown below. Round to the nearest hundredth.

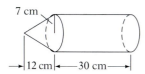

8. Determine the area of the trapezoid shown below.

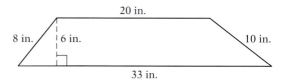

9. Determine the perimeter of the composite figure shown below. Round to the nearest tenth.

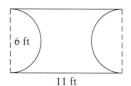

10. Determine the exact surface area of the composite figure shown below.

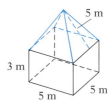

11. Find x.

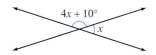

12. Name the figure shown below.

Copyright © Houghton Mifflin Company. All rights reserved.

13. Determine whether the two triangles are congruent. If they are congruent, state by what rule they are congruent.

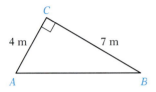

14. Determine the volume of the rectangular solid shown below.

15. Figure *ABC* is a right triangle. Determine the length of side *AB*. Round to the nearest hundredth.

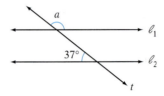

16. Given that l_1 and l_2 are parallel lines, determine the measure of angle *a*.

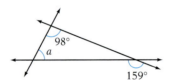

17. Determine the exact surface area of the right circular cylinder shown below.

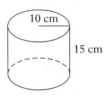

18. Determine the measure of angle *a*.

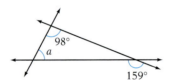

19. Triangles *ABC* and *DEF* are similar triangles. Determine the length of line segment *FG*. Round to the nearest hundredth.

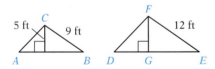

20. Triangles *ABC* and *DEF* are similar triangles. Determine the length of side *BC*. Round to the nearest hundredth.

21. Determine the perimeter of a square whose side is 5 m.

22. Determine the perimeter of a rectangle whose length is 8 cm and whose width is 5 cm.

23. Find the perimeter of a right triangle with legs that measure 12 ft and 18 ft. Round to the nearest tenth.

24. Two angles of a triangle measure 41° and 37°. Find the measure of the third angle.

25. Determine the area of the composite figure shown at the right. Round to the nearest tenth.

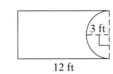

Copyright © Houghton Mifflin Company. All rights reserved.

Cumulative Review Exercises

1. Find 8.5% of 2,400.

2. Find all the factors of 78.

3. Divide: $4\frac{2}{3} \div 5\frac{3}{5}$

4. Add: $(3x^2 + 5x - 2) + (4x^2 - x + 7)$

5. Divide and round to the nearest tenth: $82.93 \div 6.5$

6. Write 0.000029 in scientific notation.

7. Find the measure of $\angle x$.

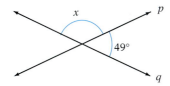

8. Find the unknown side of the triangle.

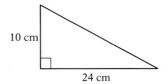

9. Find the area of the composite figure.

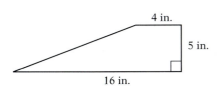

10. Find the volume of the composite figure. Round to the nearest hundredth.

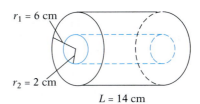

11. Multiply: $(4x^2y^2)(-3x^3y)$

12. Solve: $3(2x + 5) = 18$

13. Find the perimeter of the figure. Round to the nearest hundredth.

14. Graph $x > -3$.

-6 -5 -4 -3 -2 -1 0 1 2 3 4 5 6

15. Simplify: $5(2x + 4) - (3x + 2)$

16. Evaluate $2x + 3y^2z$ when $x = 5$, $y = -1$, and $z = -4$.

17. Evaluate $x^2y - 2z$ when $x = \frac{1}{2}$, $y = \frac{4}{5}$, and $z = -\frac{3}{10}$.

18. Convert 60 mph to kilometers per hour. Round to the nearest tenth. (1 mi $\approx$ 1.61 km)

Copyright © Houghton Mifflin Company. All rights reserved.

19. Solve: $4x + 2 = 6x - 8$

20. Graph $y = -\dfrac{3}{2}x + 3$.

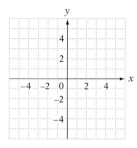

21. Convert 3 482 m to kilometers.

22. Write $\dfrac{3}{8}$ as a percent.

23. Find the simple interest on a 270-day loan of $20,000 at an annual interest rate of 8.875%.

24. *Catering* Two hundred fifty people are expected to attend a reception. Assuming that each person drinks 12 oz of coffee, how many gallons of coffee should be prepared? Round to the nearest whole number.

25. *Business* The charge for cellular phone service for a business executive is $22 per month plus $.25 per minute of phone use. In a month when the executive's phone bill was $43.75, how many minutes did the executive use the cellular phone?

26. *Taxes* If the sales tax on a $12.50 purchase is $.75, what is the sales tax on a $75 purchase?

27. *Foreign Trade* The figure at the right shows the value of the imports and exports during the first and second quarters of a recent year. Find the percent increase in the value of the imports from the first quarter to the second quarter. Round to the nearest tenth of a percent.

28. *Geometry* The volume of a box is 144 ft³. The length of the box is 12 ft, and the width is 4 ft. Find the height of the box.

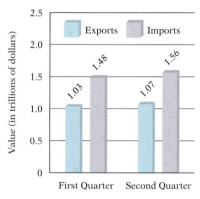

Value of Imports and Exports
Source: Bureau of Economic Analysis

29. *Oceanography* The pressure, *P*, in pounds per square inch, at a certain depth in the ocean can be approximated by the equation $P = 15 + \dfrac{1}{2}D$, where *D* is the depth in feet. Use this equation to find the depth when the pressure is 35 lb/in².

30. *Astronautics* The weight of an object is related to the distance the object is above the surface of Earth. A formula for this relationship is $d = 4{,}000\sqrt{\dfrac{E}{S}} - 4{,}000$, where *E* is the object's weight on the surface of Earth and *S* is the object's weight at a distance of *d* miles above Earth's surface. A space explorer who weighs 196 lb on the surface of Earth weighs 49 lb in space. How far above Earth's surface is the space explorer?

Copyright © Houghton Mifflin Company. All rights reserved.

Statistics and Probability

In the 1982 Kentucky Derby, Gato del Sol was considered a "longshot," meaning the odds were against him winning the race. Air Forbes Won was considered the "favorite," meaning the odds were in his favor. *Odds in favor* and *odds against* are closely linked to probability. **Example 3 on page 632** shows how to calculate the probability of a horse winning a race given the odds against that horse winning.

Copyright © Houghton Mifflin Company. All rights reserved.

Need help? For online student resources, visit this website: **math.college.hmco.com**

Prep Test

1. Simplify: $\dfrac{3}{2+7}$

2. Approximate $\sqrt{13}$ to the nearest thousandth.

3. Bill-related mail accounted for 49 billion of the 102 billion pieces of first-class mail handled by the U.S. Postal Service during a recent year. (*Source:* USPS) What percent of the pieces of first-class mail handled by the U.S. Postal Service was bill-related mail? Round to the nearest tenth of a percent.

4. The table at the right shows the estimated costs of funding an education at a public college. Between which two enrollment years is the increase in cost greatest? What is the increase between these two years?

Enrollment Year	Cost of Public College
2005	$70,206
2006	$74,418
2007	$78,883
2008	$83,616
2009	$88,633
2010	$93,951

Source: The College Board's Annual Survey of Colleges

5. During the 1924 Summer Olympics in Paris, France, the United States won 45 gold medals, 27 silver medals, and 27 bronze medals. (*Source: The Ultimate Book of Sports List*)
   a. Find the ratio of gold medals won by the United States to silver medals won by the United States during the 1924 Summer Olympics. Write the ratio as a fraction in simplest form.

   b. Find the ratio of silver medals won by the United States to bronze medals won by the United States during the 1924 Summer Olympics. Write the ratio using a colon.

6. The table (below right) shows the number of television viewers, in millions, who watch pay-cable channels, such as HBO and Showtime, each night of the week. (*Source:* Neilsen Media Research analyzed by Initiative Media North America)
   a. Arrange the numbers in the table from smallest to largest.

Mon	Tue	Wed	Thu	Fri	Sat	Sun
3.9	4.5	4.2	3.9	5.2	7.1	5.5

   b. Find the average number of viewers per night.

7. Approximately 15% of the nation's 1.4 million service members on active military duty are women. (*Source:* Women in Military Service for America Memorial Foundation)
   a. Approximately how many women are on active military duty?

   b. What fraction of the service members on active military duty are women?

Go Figure

In the addition at the right, each letter stands for a different digit. If N = 1, I = 5, U = 7, F = 2, and T = 3, what is the value of S?

```
  FUN
   IN
+ THE
-----
  SUN
```

Copyright © Houghton Mifflin Company. All rights reserved.

OBJECTIVE   **A**

Frequency distributions

Statistics is the study of collecting, organizing, and interpreting data. Data are collected from a **population,** which is the set of all observations of interest. Here are some examples of populations.

> An auto insurance company wants to determine information about the size of claims for auto accidents. The population for the insurance company is the dollar amount of each claim.

> A medical researcher wants to determine the effectiveness of a new drug to control blood pressure. The population for the researcher is the amount of change in blood pressure for each patient receiving the medication.

> The quality control inspector of a precision instrument company wants to determine the diameters of ball bearings. The population for the inspector is the measure of the diameter of each ball bearing.

A **frequency distribution** is one method of organizing the data collected from a population. A frequency distribution is constructed by dividing the data gathered from the population into **classes.** Here is an example:

A ski association surveys 40 of its members, asking them to report the percent of their ski terrain that is rated expert. The results of the survey follow.

Percent of Expert Terrain at 40 Ski Resorts

14	24	8	31	27	9	12	32	24	27
12	21	24	23	12	31	30	31	26	34
13	18	29	33	34	21	28	23	11	10
25	20	14	18	15	11	17	29	21	25

To organize these data into a frequency distribution:

1. Find the smallest number (8) and the largest number (34) in the table. The difference between these two numbers is the **range** of the data.

    Range = 34 − 8 = 26

2. Decide how many classes the frequency distribution will contain. Usually frequency distributions have from 6 to 12 classes. The frequency distribution for this example will contain 6 classes.

3. Divide the range by the number of classes. If necessary, round the quotient to a whole number. This number is called the **class width.**

    $\frac{26}{6} \approx 4$. The class width is 4.

4. Form the classes of the frequency distribution.

Classes	
8–12	Add 4 to the smallest number.
13–17	Add 4 again.
18–22	
23–27	Continue until a class contains
28–32	the largest number in the set
33–37	of data.

Calculator Note

Recall that $\frac{26}{6}$ can be read 26 ÷ 6. To calculate the class width, press

26 ÷ 6 = .

The display reads 4.3333333.

Copyright © Houghton Mifflin Company. All rights reserved.

Take Note

For your convenience, the data presented on page 609 are repeated below.

14	12	13	25
24	21	18	20
8	24	29	14
31	23	33	18
27	12	34	15
9	31	21	11
12	30	28	17
32	31	23	29
24	26	11	21
27	34	10	25

5. Complete the table by tabulating the data for each class. For each number from the data, place a slash next to the class that contains the number. Count the number of tallies in each class. This is the **class frequency.**

Frequency Distribution for Ski Resort Data

Classes	Tally	Frequency
8–12	////////	8
13–17	/////	5
18–22	//////	6
23–27	//////////	10
28–32	////////	8
33–37	///	3

Organizing data into a frequency distribution enables us to make statements about the data. For example, twenty-seven (6 + 10 + 8 + 3) of the ski resorts reported that 18% or more of their terrain was rated expert.

An insurance adjuster had tabulated the dollar amount of 50 auto accident claims. The results are given in the following table. Use these data for Example 1 and You Try It 1.

Dollar Amount of 50 Auto Insurance Claims

475	224	722	721	815	351	596	625	981	748
993	881	361	560	574	742	703	998	435	873
882	278	455	803	985	305	522	900	638	810
677	688	410	505	890	186	829	631	882	991
484	339	950	579	539	422	326	793	453	118

① Example

For the table of Auto Insurance Claims, make a frequency distribution that has 6 classes.

Strategy

To make the frequency distribution:

▶ Find the range.
▶ Divide the range by 6, the number of classes. Round the quotient to the nearest whole number. This is the class width.
▶ Tabulate the data for each class.

Solution

Range $= 998 - 118 = 880$

Class width $= \dfrac{880}{6} \approx 147$

Dollar Amount of Insurance Claims

Classes	Tally	Frequency
118–265	///	3
266–413	////////	7
414–561	//////////	10
562–709	/////////	9
710–857	/////////	9
858–1,005	////////////	12

① You Try It

For the table of Auto Insurance Claims, make a frequency distribution that has 8 classes.

Your Strategy

Your Solution

Solution on p. S25

Copyright © Houghton Mifflin Company. All rights reserved.

OBJECTIVE ▪ B

Histograms

A **histogram** is a bar graph that represents the data in a frequency distribution. The width of a bar represents each class, and the height of the bar corresponds to the frequency of the class.

A survey of 105 households is conducted, and the number of kilowatt-hours (kWh) of electricity that are used by each in a 1-month period is recorded in the frequency distribution, shown at the left below. The histogram for the frequency distribution is shown in Figure 10.1.

Classes (kWh)	Frequency
850–900	9
900–950	14
950–1,000	17
1,000–1,050	25
1,050–1,100	16
1,100–1,150	14
1,150–1,200	10

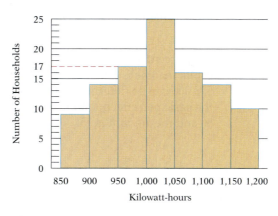

Figure 10.1

From the frequency distribution or the histogram, we can see that 17 households used between 950 kWh and 1,000 kWh during the 1-month period.

---

2 *Example*

Use the histogram in Figure 10.1 to find the number of households that used 950 kWh of electricity or less during the month.

Strategy

To find the number:

▶ Read the histogram to find the number of households whose use was between 850 and 900 kWh and the number whose use was between 900 and 950 kWh.

▶ Add the two numbers.

Solution

Number between 850 and 900 kWh: 9
Number between 900 and 950 kWh: 14

9 + 14 = 23

23 households used 950 kWh of electricity or less during the month.

2 *You Try It*

Use the histogram in Figure 10.1 to find the number of households that used 1,100 kWh of electricity or more during the month.

Your Strategy

Your Solution

Solution on p. S25

Copyright © Houghton Mifflin Company. All rights reserved.

OBJECTIVE **C**

Frequency polygons

A **frequency polygon** is a graph that displays information in a manner similar to a histogram. A dot is placed above the center of each class interval at a height corresponding to that class's frequency. The dots are then connected to form a broken-line graph. The center of a class interval is called the **class midpoint.**

The per capita incomes in a recent year for the 50 states are recorded in the frequency polygon in Figure 10.2. The number of states with a per capita income between $24,000 and $28,000 is 15.

The percent of states for which the per capita income is between $24,000 and $28,000 can be determined by solving the basic percent equation. The base is 50 and the amount is 15.

$$pB = A$$
$$p(50) = 15$$
$$\frac{50p}{50} = \frac{15}{50}$$
$$p = 0.3$$

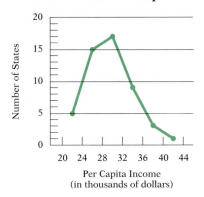

Figure 10.2
Source: Bureau of Economic Analysis

30% of the states had a per capita income between $24,000 and $28,000.

 Example

Use Figure 10.2 to find the number of states for which the per capita income was $36,000 or above.

Strategy

To find the number of states:

▶ Read the frequency polygon to find the number of states with a per capita income between $36,000 and $40,000 and the number of states with a per capita income between $40,000 and $44,000.

▶ Add the numbers.

Solution

Number with per capita income between $36,000 and $40,000: 3

Number with per capita income between $40,000 and $44,000: 1

$3 + 1 = 4$

The per capita income was $36,000 or above in 4 states.

 You Try It

Use Figure 10.2 to find the ratio of the number of states with a per capita income between $20,000 and $24,000 to the number with a per capita income between $24,000 and $28,000.

Your Strategy

Your Solution

Solution on p. S25

Copyright © Houghton Mifflin Company. All rights reserved.

10.1 EXERCISES

1. In your own words, describe a frequency distribution.

2. Explain what *class* and *frequency* indicate in a frequency distribution.

Education   Use the table below for Exercises 3–12.

Annual Tuition at 40 Universities (hundreds of dollars)

85	87	95	48	41	91	88	92
71	74	63	51	70	87	84	95
72	94	61	52	88	49	55	60
77	53	89	91	45	96	49	58
83	36	39	32	36	59	95	67

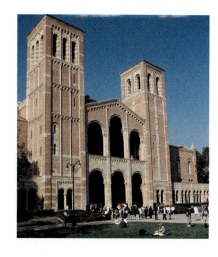

3. What is the range of data in the Annual Tuition table?

4. Make a frequency distribution for the Annual Tuition table. Use 8 classes.

5. Which class has the greatest frequency?

6. How many universities charge a tuition that is between $7,700 and $8,500?

7. How many universities charge a tuition that is between $5,000 and $5,800?

8. How many universities charge a tuition that is less than or equal to $6,700?

9. What percent of the universities charge a tuition that is between $9,500 and $10,300?

10. What percent of the universities charge a tuition that is between $3,200 and $4,000?

11. What percent of the universities charge a tuition that is greater than or equal to $6,800?

12. What percent of the universities charge a tuition that is less than or equal to $7,600?

Copyright © Houghton Mifflin Company. All rights reserved.

The Hotel Industry   Use the table below for Exercises 13–22.

Corporate Room Rate for 50 Hotels

60	87	77	117	114	82	91	65	69	63
106	71	74	86	106	78	101	100	107	109
57	106	103	100	95	68	99	112	107	77
64	68	99	112	107	76	116	100	82	86
81	98	92	78	95	89	91	102	115	127

13. Make a frequency distribution for the hotel room rates. Use 7 classes.

14. How many hotels charge a corporate room rate that is between $79 and $89 per night?

15. How many hotels charge a corporate room rate that is between $57 and $67 per night?

16. How many hotels charge a corporate room rate that is between $112 and $133 per night?

17. How many hotels charge a corporate room rate that is less than or equal to $100?

18. What percent of the hotels charge a corporate room rate that is between $101 and $111 per night?

19. What percent of the hotels charge a corporate room rate that is between $90 and $100 per night?

20. What percent of the hotels charge a corporate room rate that is greater than or equal to $101 per night?

21. What percent of the hotels charge a corporate room rate that is less than or equal to $78 per night?

22. What is the ratio of the number of hotels whose room rates are between $79 and $89 to those whose room rates are between $90 and $100?

Copyright © Houghton Mifflin Company. All rights reserved.

OBJECTIVE **B**

Consumer Credit   A total of 50 monthly credit account balances are recorded in the figure at the right.

23. How many account balances were between $1,500 and $2,000?

24. How many account balances were less than $2,000?

25. What percent of the account balances were between $2,000 and $2,500?

26. What percent of the account balances were greater than $1,500?

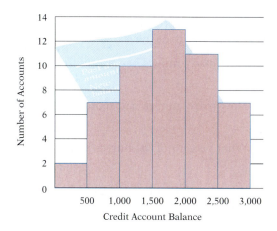

Marathons   The times (in minutes) for 100 runners in a marathon are recorded in the figure at the right.

27. What is the ratio of the number of runners with times that were between 150 min and 155 min to those with times between 175 min and 180 min?

28. What is the ratio of the number of runners with times that were between 165 min and 170 min to those with times between 155 min and 160 min?

29. What percent of the runners had times greater than 165 min?

30. What percent of the runners had times less than 170 min?

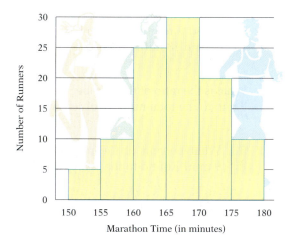

Cost of Living   A total of 40 apartment complexes were surveyed to find the monthly rent for a one-bedroom apartment. The results are recorded in the figure at the right.

31. What percent of the apartments had rents between $1,250 and $1,500?

32. What percent of the apartments had rents between $500 and $750?

33. What percent of the apartments had rents greater than $1,000?

34. What percent of the apartments had rents less than $1,250?

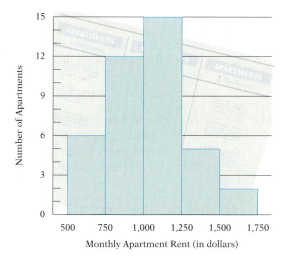

Copyright © Houghton Mifflin Company. All rights reserved.

Copyright © Houghton Mifflin Company. All rights reserved.

OBJECTIVE **C**

Education The scores of 50 nurses taking a state board exam are given in the figure at the right.

35. How many nurses had scores that were greater than 80?

36. How many nurses had scores that were less than 70?

37. What percent of the nurses had scores between 70 and 90?

38. What percent of the nurses had scores greater than 70?

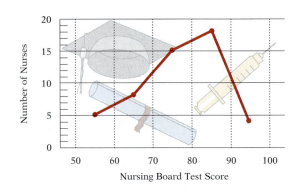

Emergency Calls The response times for 75 emergency 911 calls for a city are recorded in the figure at the right.

39. What is the ratio of the number of response times between 6 min and 9 min to the number of response times between 15 min and 18 min?

40. What is the ratio of the number of response times of 0 min to 3 min to the total number of recorded response times?

41. What percent of the response times are greater than 9 min? Round to the nearest tenth of a percent.

42. What percent of the response times are less than 12 min? Round to the nearest tenth of a percent.

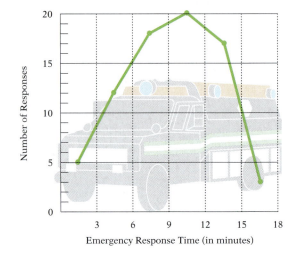

Critical Thinking

43. Toss two dice 100 times. Record the sum of the dots on the upward faces. Make a histogram showing the number of times the sum was 2, 3, . . . , 12.

44. If each frequency in a frequency table is divided by the total number of observations, a relative frequency table is formed. A relative frequency histogram is a histogram of the relative frequencies. Draw a relative frequency histogram for the data in Exercise 43.

45. How are a frequency table and a histogram alike? How are they different?

46. The frequency table at the right contains data from a survey of the type of vehicle a prospective buyer would consider. Explain why these data could be shown in a bar graph but not in a histogram.

Body Style	Frequency
Sedan	25
Convertible	16
SUV	9
Sports car	20
Truck	23

OBJECTIVE   **A**

The mean, median, and mode of a distribution

The average annual rainfall in Mobile, Alabama, is 67 in. The average annual snowfall in Syracuse, New York, is 111 in. The average daily low temperature in January in Bismarck, North Dakota, is −4°F. Each of these statements uses one number to describe an entire collection of numbers. Such a number is called an *average*. In statistics, there are various ways to calculate an average. Three of the most common—*mean, median,* and *mode*—are discussed here.

An automotive engineer tests the miles-per-gallon ratings of 15 cars and records the results as follows:

Miles-per-Gallon Ratings of 15 Cars

25   22   21   27   25   35   29   31   25   26   21   39   34   32   28

The **mean** of the data is the sum of the measurements divided by the number of measurements. The symbol for the mean is $\bar{x}$.

> ### Formula for the Mean
>
> $$\bar{x} = \frac{\text{sum of all data values}}{\text{number of data values}}$$

To find the mean for the data above, add the numbers and then divide by 15.

$$\bar{x} = \frac{25 + 22 + 21 + 27 + 25 + 35 + 29 + 31 + 25 + 26 + 21 + 39 + 34 + 32 + 28}{15}$$

$$= \frac{420}{15} = 28$$

The mean number of miles per gallon for the 15 cars tested was 28 mi/gal.

The mean is one of the most frequently computed averages. It is the one that is commonly used to calculate a student's performance in a class.

> The scores for a history student on 5 tests were 78, 82, 91, 87, and 93. What was the mean score for this student?
>
> To find the mean, add the numbers. $\bar{x} = \dfrac{78 + 82 + 91 + 87 + 93}{5}$
>
> Then divide by 5.
>
> $$= \frac{431}{5} = 86.2$$
>
> The mean score for the history student was 86.2.

The **median** of data is the number that separates the data into two equal parts when the numbers are arranged from smallest to largest (or largest to smallest). There are always an equal number of values above the median and below the median.

Copyright © Houghton Mifflin Company. All rights reserved.

To find the median of a set of numbers, first arrange the numbers from smallest to largest. The median is the number in the middle. The result of arranging the data, given on the previous page, for the miles-per-gallon ratings from smallest to largest is given below.

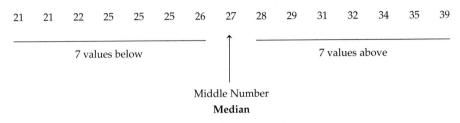

Take Note

Half the data values are less than 27, and half the data values are greater than 27. The median indicates the center, or middle, of the set of data.

The median is 27.

If the data contain an *even* number of values, the median is the sum of the two middle numbers divided by 2.

Take Note

If the data contain an even number of values, the median is the mean of the two middle values.

The selling prices of the last six homes sold by a real estate agent were $175,000, $150,000, $250,000, $130,000, $245,000, and $190,000. Find the median selling price of these homes.

Arrange the numbers from smallest to largest. Because there are an even number of values, the median is the sum of the two middle numbers divided by two.

130,000   150,000   <u>175,000   190,000</u>   245,000   250,000

middle 2 numbers

$$\text{Median} = \frac{175,000 + 190,000}{2} = 182,500$$

The median selling price of a home was $182,500.

▶ **Point of Interest**

A set of data can have no mode, one mode, or several modes. The data

3, 7, 7, 12, 14, 15, 15, 19

has two modes, 7 and 15. A set of data with two modes is called bimodal.

The **mode** of a set of numbers is the value that occurs most frequently. If a set of numbers has no number that occurs more than once, then the data have no mode.

Here again are the data for the gasoline mileage ratings of cars.

Miles-per-Gallon Ratings of 15 Cars

25   22   21   27   25   35   29   31   25   26   21   39   34   32   28

↑   ↑   ↑

25 is the number that occurs most frequently. The mode is 25.

The gasoline mileage rating data show that the mean, median, and mode of a set of numbers do not have to be the same value. For the data on rating the miles per gallon for 15 cars,

Mean = 28   Median = 27   Mode = 25

Although any of the averages can be used when the data collected consist of numbers, the mean and median are not appropriate for *qualitative* data. Examples of qualitative data are recording a person's favorite color or recording a person's preference from among classical, hard rock, jazz, rap, and country western music. It does not make sense to say that the *average* favorite color is red or the *average* musical choice is jazz. The mode is used to indicate the most frequently chosen color or musical category. The **modal response** is the category that receives the greatest number of responses.

Copyright © Houghton Mifflin Company. All rights reserved.

A survey asked people to state whether they strongly disagree, disagree, have no opinion, agree, or strongly agree with the position a state's governor has taken on increasing taxes for health care. What was the modal response for these data?

Strongly disagree   57
Disagree   68
No opinion   12
Agree   45
Strongly agree   58

Because a response of "disagree" was recorded most frequently, the modal response was disagree.

---

1 *Example*

Twenty students were asked the number of units in which they were currently enrolled. The responses were

| 15 | 12 | 13 | 15 | 17 | 18 | 13 | 20 | 9 | 16 |

| 14 | 10 | 15 | 12 | 17 | 16 | 6 | 14 | 15 | 12 |

Find the mean and median number of units taken by these students.

Strategy

To find the mean number of units taken by the 20 students:
▶ Determine the sum of the numbers.
▶ Divide the sum by 20.

To find the median number of units taken by the 20 students:
▶ Arrange the numbers from smallest to largest.
▶ Because there is an even number of values, the median is the sum of the two middle numbers divided by 2.

Solution

The sum of the numbers is 279.

$$\bar{x} = \frac{279}{20} = 13.95$$

The mean is 13.95 units.

| 6 | 9 | 10 | 12 | 12 | 12 | 13 | 13 | 14 | 14 |

| 15 | 15 | 15 | 15 | 16 | 16 | 17 | 17 | 18 | 20 |

$$\text{Median} = \frac{14 + 15}{2} = 14.5$$

The median is 14.5 units.

1 *You Try It*

The amounts spent by the last 10 customers at a fast-food restaurant were

| 4.32 | 6.21 | 5.45 | 5.90 | 5.58 | 4.45 | 5.05 | 6.00 | 3.59 | 4.75 |

Find the mean and median amount spent on lunch by these customers.

Your Strategy

Your Solution

Solution on pp. S25–S26

Copyright © Houghton Mifflin Company. All rights reserved.

Example

A bowler has scores of 165, 172, 168, and 185 for four games. What score must the bowler have on the next game so that the mean for the five games is 174?

Strategy

To find the score, use the formula for the mean, letting n be the score on the fifth game.

Solution

$$174 = \frac{165 + 172 + 168 + 185 + n}{5}$$

$$174 = \frac{690 + n}{5}$$

$870 = 690 + n$ ▶ Multiply each side by 5.

$180 = n$

The score on the fifth game must be 180.

 You Try It

You have scores of 82, 91, 79, and 83 on four exams. What score must you receive on the fifth exam to have a mean of 84 for the five exams?

Your Strategy

Your Solution

Solution on p. S26

OBJECTIVE **B**

Box-and-whiskers plots

The purpose of calculating a mean or median is to obtain one number that describes some measurements. That one number alone, however, may not adequately represent the data. A **box-and-whiskers plot** is a graph that gives a more comprehensive picture of the data. A box-and-whiskers plot shows five numbers: the smallest value, the *first quartile*, the median, the *third quartile*, and the largest value. The **first quartile**, symbolized by Q_1, is the number below which one-quarter of the data lie. The **third quartile**, symbolized by Q_3, is the number above which one-quarter of the data lie.

Find the first quartile Q_1 and the third quartile Q_3 for the prices of 15 half-gallon cartons of ice cream.

| 3.26 | 4.71 | 4.18 | 4.45 | 5.49 | 3.18 | 3.86 | 3.58 | 4.29 | 5.44 | 4.83 | 4.56 | 4.36 | 2.39 | 2.66 |

To find the quartiles, first arrange the data from the smallest value to the largest value. Then find the median.

| 2.39 | 2.66 | 3.18 | 3.26 | 3.58 | 3.86 | 4.18 | 4.29 | 4.36 | 4.45 | 4.56 | 4.71 | 4.83 | 5.44 | 5.49 |

The median is 4.29.

Now separate the data into two groups: those values below the median and those values above the median.

Values Less Than the Median	*Values Greater Than the Median*
2.39 2.66 3.18 3.26 3.58 3.86 4.18	4.36 4.45 4.56 4.71 4.83 5.44 5.49
↑	↑
Q_1	Q_3

The first quartile Q_1 is the median of the lower half of the data: $Q_1 = 3.26$.
The third quartile Q_3 is the median of the upper half of the data: $Q_3 = 4.71$.

Copyright © Houghton Mifflin Company. All rights reserved.

The **interquartile range** is the difference between Q_3 and Q_1.

$$\text{Interquartile range} = Q_3 - Q_1 = 4.71 - 3.26 = 1.45$$

A box-and-whiskers plot shows the data in the interquartile range as a box. The box-and-whiskers plot for the data on the cost of ice cream is shown below.

Note that the box-and-whiskers plot labels five values: the smallest, 2.39; the first quartile Q_1, 3.26; the median, 4.29; the third quartile Q_3, 4.71; and the largest value, 5.49.

Take Note

To draw this box-and-whiskers plot, think of a number line that includes the five values listed. With this in mind, mark off the five values. Draw a box that spans the distance from Q_1 to Q_3. Draw a vertical line the height of the box at the median, Q_2.

For Example 3 and You Try It 3, use the data in the table at the right, which gives the number of people registered for a software training program.

Participants in Software Training

30	45	54	24	48	38	43
38	46	53	62	64	40	35

3 *Example*

Find Q_1 and Q_3 for the data in the software training table.

Strategy

To find Q_1 and Q_3, arrange the data from smallest to largest. Find the median. Then find Q_1, the median of the lower half of the data, and Q_3, the median of the upper half of the data.

Solution

24	30	35	**38**	38	40	43
45	46	48	**53**	54	62	64

$$\text{Median} = \frac{43 + 45}{2} = 44$$

$Q_1 = 38$ The median of the top row of data.
$Q_3 = 53$ The median of the bottom row of data.

3 *You Try It*

Draw the box-and-whiskers plot for the data in the software training table.

Your Strategy

Your Solution

Solution on p. S26

OBJECTIVE **C**

VIDEO & DVD   CD TUTOR   WEB   SSM

The standard deviation of a distribution

Consider two students, each of whom has taken five exams.

Student A

84	86	83	85	87

$$\bar{x} = \frac{84 + 86 + 83 + 85 + 87}{5} = \frac{425}{5} = 85$$

The mean for Student A is 85.

Student B

90	75	94	68	98

$$\bar{x} = \frac{90 + 75 + 94 + 68 + 98}{5} = \frac{425}{5} = 85$$

The mean for Student B is 85.

Copyright © Houghton Mifflin Company. All rights reserved.

For each of these students, the mean (average) for the 5 tests is 85. However, Student A has a more consistent record of scores than Student B. One way to measure the consistency, or "clustering," of data near the mean is the **standard deviation**.

To calculate the standard deviation:

1. Sum the squares of the differences between each value of data and the mean.

2. Divide the result in Step 1 by the number of items in the set of data.

3. Take the square root of the result in Step 2.

Here is the calculation for Student A. The symbol for standard deviation is the Greek letter *sigma*, denoted by σ.

Step 1	x	$(x - \bar{x})$	$(x - \bar{x})^2$
	84	$(84 - 85)$	$(-1)^2 = 1$
	86	$(86 - 85)$	$1^2 = 1$
	83	$(83 - 85)$	$(-2)^2 = 4$
	85	$(85 - 85)$	$0^2 = 0$
	87	$(87 - 85)$	$2^2 = 4$
			Total $= 10$

Step 2 $\dfrac{10}{5} = 2$

Step 3 $\sigma = \sqrt{2} \approx 1.414$

The standard deviation for Student A's scores is approximately 1.414.

> **Take Note**
>
> The standard deviation of Student B's scores are greater than the standard deviation of Student A's scores, and the range of Student B's scores $(98 - 68 = 30)$ is greater than the range of Student A's scores $(87 - 83 = 4)$.

Following a similar procedure for Student B, we find that the standard deviation for Student B's scores is approximately 11.524. Because the standard deviation of Student B's scores is greater than that of Student A's $(11.524 > 1.414)$, Student B's scores are not as consistent as those of Student A.

In this text, standard deviations are rounded to the nearest thousandth.

---

4 *Example*

The weights in pounds of the five-man front line of a college football team are 210, 245, 220, 230, and 225. Find the standard deviation of the weights.

Strategy

To calculate the standard deviation:

▶ Find the mean of the weights.
▶ Use the procedure for calculating standard deviation.

Solution

$$\bar{x} = \frac{210 + 245 + 220 + 230 + 225}{5} = 226$$

Step 1

x	$(x - \bar{x})^2$
210	$(210 - 226)^2 = 256$
245	$(245 - 226)^2 = 361$
220	$(220 - 226)^2 = 36$
230	$(230 - 226)^2 = 16$
225	$(225 - 226)^2 = 1$
	Total $= 670$

Step 2

$$\frac{670}{5} = 134$$

Step 3

$$\sigma = \sqrt{134} \approx 11.576$$

The standard deviation of the weights is approximately 11.576 lb.

4 *You Try It*

The number of miles a runner recorded for the last six days of running were 5, 7, 3, 6, 9, and 6. Find the standard deviation of the miles run.

Your Strategy

Your Solution

Solution on p. S26

Copyright © Houghton Mifflin Company. All rights reserved.

10.2 EXERCISES

OBJECTIVE **A**

1. ✎ Explain how to determine the **a.** mean, **b.** median, and **c.** mode of a set of data.

2. ✎ A set of data has a mean of 16, a median of 15, and a mode of 14. Which of these numbers must be a value in the data? Explain your answer.

Solve.

3. *Business*   The number of big-screen televisions sold each month for one year was recorded by an electronics store. The results were 15, 12, 20, 20, 19, 17, 22, 24, 17, 20, 15, and 27. Calculate the mean and the median number of televisions sold per month.

4. *The Airline Industry*   The number of seats occupied on a jet for 16 transatlantic flights was recorded. The numbers were 309, 422, 389, 412, 401, 352, 367, 319, 410, 391, 330, 408, 399, 387, 411, and 398. Calculate the mean and the median number of occupied seats.

5. *Sports*   The times, in seconds, for a 100-meter dash at a college track meet were 10.45, 10.23, 10.57, 11.01, 10.26, 10.90, 10.74, 10.64, 10.52, and 10.78. Calculate the mean and median times for the 100-meter dash.

6. *Consumerism*   A consumer research group purchased identical items in 8 grocery stores. The costs for the purchased items were $45.89, $52.12, $41.43, $40.67, $48.73, $42.45, $47.81, and $45.82. Calculate the mean and the median cost of the purchased items.

7. *Education*   Your scores on six history tests were 78, 92, 95, 77, 94, and 88. If an "average score" of 90 receives an A for the course, which average, the mean or the median, would you prefer the instructor to use?

8. *Health Insurance*   Eight health maintenance organizations (HMOs) presented group health insurance plans to a company. The monthly rates per employee were $423, $390, $405, $396, $426, $355, $404, and $430. Calculate the mean and the median monthly rates for these eight companies.

9. *Sports*   The number of yards gained by a college running back was recorded for 6 games. The numbers were 98, 105, 120, 90, 111, and 104. How many yards must this running back gain in the next game so that the average for the seven games is 100 yd?

10. *Sports*   The number of unforced errors a tennis player made in four sets of tennis was recorded. The numbers were 15, 22, 24, and 18. How many unforced errors did this player make in the fifth set so that the mean number of unforced errors for the five sets was 20?

Copyright © Houghton Mifflin Company. All rights reserved.

11. *Sports*   The last five golf scores for a player were 78, 82, 75, 77, and 79. What score on the next round of golf will give the player a mean score of 78 for all six rounds?

12. *Business*   A survey by an ice cream store asked people to name their favorite ice cream from five flavors. The responses were mint chocolate chip, 34; pralines and cream, 27; German chocolate cake, 44; chocolate raspberry swirl, 34; rocky road, 42. What was the modal response?

13. *Physical Characteristics*   The eye colors of 100 students were recorded. The results were blue, 35; brown, 38; hazel, 14; green, 3; grey, 9. What was the modal eye color?

14. *Politics*   A newspaper survey asked people to rate the performance of the city's mayor. The responses were very unsatisfactory, 230; unsatisfactory, 403; satisfactory, 1,237; very satisfactory, 403. What was the modal response for this survey?

15. *Business*   The patrons of a restaurant were asked to rate the quality of food. The responses were bad, 8; good, 21; very good, 43; excellent, 21. What was the modal response for this survey?

OBJECTIVE   **B**

16. *U.S. Presidents*   The box-and-whiskers plot at the right shows the distribution of the ages of presidents of the United States at the time of their inaugurations. What is the youngest age in the set of data? The oldest age? The median? Find the range.

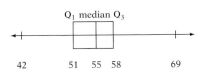

17. *Compensation*   The hourly wage for an entry-level position at various firms was recorded by a labor research firm. The results were as follows.

Starting Hourly Wages for 16 Companies

8.09	11.50	7.46	7.70	9.85	9.03	11.40	9.31
10.35	7.45	7.35	8.64	9.02	8.12	10.05	8.94

Find the first quartile and the third quartile, and draw a box-and-whiskers plot of the data.

18. *Health*   The cholesterol levels for 14 adults are recorded in the table below. Find the first quartile and the third quartile, and draw a box-and-whiskers plot of the data.

Cholesterol Levels for 14 Adults

375	185	254	221	183	251	258
292	214	172	233	208	198	211

Copyright © Houghton Mifflin Company. All rights reserved.

19. *Fuel Efficiency* The gasoline consumption of 19 cars was tested and the results recorded in the following table. Find the first quartile and the third quartile, and draw a box-and-whiskers plot of the data.

Miles per Gallon for 19 Cars

33	21	30	32	20	31	25	20	16	24
22	31	30	28	26	19	21	17	26	

20. *Education* The ages of the accountants who passed the certified public accountant (CPA) exam at one test center are recorded in the table below. Find the first quartile and the third quartile, and draw a box-and-whiskers plot of the data.

Ages of Accountants Passing the CPA Exam

24	42	35	26	24	37	27	26	28
34	43	46	29	34	25	30	28	

21. *Manufacturing* The times for new employees to learn how to assemble a toy are recorded in the table below. Find the first quartile and the third quartile, and draw a box-and-whiskers plot of the data.

Time to Train Employees (in hours)

4.3	3.1	5.3	8.0	2.6	3.5	4.9	4.3
6.2	6.8	5.4	6.0	5.1	4.8	5.3	6.7

22. *Manufacturing* A manufacturer of light bulbs tested the life of 20 light bulbs. The results are recorded in the table below. Find the first quartile and the third quartile, and draw a box-and-whiskers plot of the data.

Life of 20 Light Bulbs (in hours)

1,010	1,235	1,200	998	1,400	789	986	905	1,050	1,100
1,180	1,020	1,381	992	1,106	1,298	1,268	1,309	1,390	890

OBJECTIVE C

 Solve.

23. *The Airline Industry* An airline recorded the times for a ground crew to unload the baggage from an airplane. The recorded times, in minutes, were 12, 18, 20, 14, and 16. Find the standard deviation of these times.

Copyright © Houghton Mifflin Company. All rights reserved.

24. *Health* The weight in ounces of newborn infants was recorded by a hospital. The weights were 96, 105, 84, 90, 102, and 99. Find the standard deviation of the weights.

25. *Business* The numbers of rooms occupied in a hotel on six consecutive days were 234, 321, 222, 246, 312, and 396. Find the standard deviation for the number of rooms occupied.

26. *Coin Tosses* Seven coins were tossed 100 times. The numbers of heads recorded were 56, 63, 49, 50, 48, 53, and 52. Find the standard deviation of the number of heads.

27. *Meteorology* The high temperatures for eleven consecutive days at a desert resort were 95°, 98°, 98°, 104°, 97°, 100°, 96°, 97°, 108°, 93°, and 104°. For the same days, the high temperatures in Antarctica were 27°, 28°, 28°, 30°, 28°, 27°, 30°, 25°, 24°, 26°, and 21°. Which location has the greater standard deviation of high temperatures?

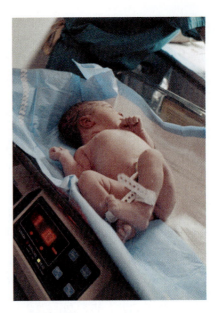

28. *Sports* The scores for five college basketball games were 56, 68, 60, 72, and 64. The scores for five professional basketball games were 106, 118, 110, 122, and 114. Which scores have the greater standard deviation?

Critical Thinking

29. *Education* One student received scores of 85, 92, 86, and 89. A second student received scores of 90, 97, 91, and 94 (exactly 5 points more on each test). Are the means of the two students the same? If not, what is the relationship between the means of the two students? Are the standard deviations of the scores of the two students the same? If not, what is the relationship between the standard deviations of the two students?

30. Determine whether the statement is always true, sometimes true, or never true.
    a. If there is an odd number of values in a set of data, the median is one of the numbers in the set.
    b. If there is an even number of values in a set of data, the median is not one of the numbers in the set.
    c. The median of a set of data is less than the highest number in the set.

31. A set of data has a mean of 16, a median of 15, and a mode of 14. Create a set of data that meets these criteria.

32. A company is negotiating with its employees for a raise in salary. One proposal would add $1,500 a year to each employee's salary. The second proposal would give each employee a 4% raise. Explain how these proposals would affect the current mean and standard deviation of salaries for the company.

Copyright © Houghton Mifflin Company. All rights reserved.

OBJECTIVE   **A**

VIDEO & DVD   CD TUTOR   WEB   SSM

The probability of simple events

A weather forecaster estimates that there is a 75% chance of rain. A state lottery director claims that there is a $\frac{1}{9}$ chance of winning a prize offered by the lottery. Each of these statements involves uncertainty to some extent. The degree of uncertainty is called **probability**. For the statements above, the probability of rain is 75% and the probability of winning a prize in the lottery is $\frac{1}{9}$.

A probability is determined from an **experiment**, which is any activity that has an observable outcome. Examples of experiments are

> Tossing a coin and observing whether it lands heads or tails
> Interviewing voters to determine their preference for a political candidate
> Recording the percent change in the price of a stock

All the possible outcomes of an experiment are called the **sample space** of the experiment. The outcomes of an experiment are listed between braces and frequently designated by S.

For each experiment, list all the possible outcomes.

a. A number cube, which has the numbers from 1 to 6 written on its sides, is rolled once.
b. A fair coin is tossed once.
c. The spinner at the right is spun once.

a. Any of the numbers from 1 to 6 could show on the top of the cube. $S = \{1, 2, 3, 4, 5, 6\}$.
b. A fair coin is one for which heads and tails have an equal chance of being tossed. $S = \{H, T\}$, where H represents heads and T represents tails.
c. Assuming that the spinner does not come to rest on a line, the arrow could come to rest in any one of the four sectors. $S = \{1, 2, 3, 4\}$.

An **event** is one or more outcomes of an experiment. Events are denoted by capital letters. Consider the experiment of rolling the number cube given above. Some possible events are:

> The number is even. $E = \{2, 4, 6\}$
> The number is a prime number. $P = \{2, 3, 5\}$
> The number is less than 10. $T = \{1, 2, 3, 4, 5, 6\}$. Note that in this case, the event is the entire sample space.
> The number is greater than 20. This event is impossible for the given sample space. The impossible event is symbolized by $\varnothing$.

When discussing experiments and events, it is convenient to refer to the *favorable outcomes* of an experiment. These are the outcomes of an experiment that satisfy the requirements of the particular event. For instance, consider the experiment of rolling a fair die once. The sample space is $\{1, 2, 3, 4, 5, 6\}$, and one possible event E would be rolling a number that is divisible by 3. The outcomes of the experiment that are favorable to E are 3 and 6, and $E = \{3, 6\}$.

Copyright © Houghton Mifflin Company. All rights reserved.

► **Point of Interest**

It was dice playing that led Antoine Gombaud, Chevalier de Mere, to ask Blaise Pascal, a French mathematician, to figure out the probability of throwing two sixes. Pascal and Pierre Fermat solved the problem, and their explorations led to the birth of probability theory.

Probability Formula

The probability of an event E, written $P(E)$, is the ratio of the number of favorable outcomes of an experiment to the total number of possible outcomes of the experiment.

$$P(E) = \frac{\text{number of favorable outcomes}}{\text{number of possible outcomes}}$$

The outcomes of the experiment of tossing a fair coin are *equally likely*. Any one of the outcomes is just as likely as another. If a fair coin is tossed once, the probability of a head or a tail is $\frac{1}{2}$. Each event, heads or tails, is equally likely.

The probability formula applies to experiments for which the outcomes are equally likely.

Not all experiments have equally likely outcomes. Consider an exhibition baseball game between a professional team and a college team. Although either team *could* win the game, the probability that the professional team will win is greater than that of the college team. The outcomes are not equally likely. For the experiments in this section, assume that the outcomes of an experiment are equally likely.

There are five choices, *a* through *e*, for each question on a multiple-choice test. By just guessing, what is the probability of choosing the correct answer for a certain question?

It is possible to select any of the letters *a*, *b*, *c*, *d*, or *e*.	There are 5 possible outcomes of the experiment.
The event *E* is the correct answer.	There is 1 favorable outcome, guessing the correct answer.
Use the probability formula.	$P(E) = \dfrac{\text{number of favorable outcomes}}{\text{number of possible outcomes}} = \dfrac{1}{5}$

The probability of guessing the correct answer is $\frac{1}{5}$.

Each of the letters of the word *Tennessee* is written on a card, and the cards are placed in a hat. If one card is drawn at random from the hat, what is the probability that the card has the letter *e* on it?

The phrase "at random" means that each card has an equal chance of being drawn.	There are 9 letters in *Tennessee*. Therefore, there are 9 possible outcomes of the experiment.
There are 4 cards with an *e* on them.	There are 4 favorable outcomes of the experiment, the 4 *e*'s.
Use the probability formula.	$P(E) = \dfrac{\text{number of favorable outcomes}}{\text{number of possible outcomes}} = \dfrac{4}{9}$

The probability is $\frac{4}{9}$.

Copyright © Houghton Mifflin Company. All rights reserved.

Calculating the probability of an event requires counting the number of possible outcomes of an experiment and the number of outcomes that are favorable to the event. One way to do this is to list the outcomes of the experiment in some systematic way. Using a table is often very helpful.

A professor writes three true/false questions for a test. If the professor randomly chooses which questions will have a true answer and which will have a false answer, what is the probability that the test will have 2 true questions and 1 false question?

Q_1	Q_2	Q_3
T	T	T
T	T	F
T	F	T
T	F	F
F	T	T
F	T	F
F	F	T
F	F	F

The experiment S consists of choosing T or F for each of the 3 questions. The possible outcomes of the experiment are shown in the table at the right.

$S = \{\text{TTT, TTF, TFT, TFF,}$
$\qquad \text{FTT, FTF, FFT, FFF}\}$

There are 8 outcomes for S.

The event E consists of 2 true questions and 1 false question.

$E = \{\text{TTF, TFT, FTT}\}$

There are 3 outcomes for E.

Use the probability formula.

$P(E) = \dfrac{3}{8}$

The probability that there are 2 true questions and 1 false question is $\dfrac{3}{8}$.

The probabilities that we have calculated so far are referred to as *mathematical* or *theoretical probabilities*. The calculations are based on theory—for example, that either side of a coin is equally likely to land face up or that each of the six sides of a fair die is equally likely to be face up. Not all probabilities arise from such assumptions.

Empirical probabilities are based on observations of certain events. For instance, a weather forecast of a 75% chance of rain is an empirical probability. From historical records kept by the weather bureau, when a similar weather pattern existed, rain occurred 75% of the time. It is theoretically impossible to predict the weather, and only observations of past weather patterns can be used to predict future weather conditions.

Empirical Probability Formula

The empirical probability of an event E is the ratio of the number of observations of E to the total number of observations.

$$P(E) = \frac{\text{number of observations of } E}{\text{total number of observations}}$$

Records of an insurance company show that of 2,549 claims for theft filed by policyholders, 927 were claims for more than $5,000. What is the empirical probability that the next claim for theft this company receives will be a claim for more than $5,000?

The empirical probability of E is the ratio of the number of claims for over $5,000 to the total number of claims.

$$P(E) = \frac{927}{2,549} \approx 0.36$$

The probability is approximately 0.36.

Copyright © Houghton Mifflin Company. All rights reserved.

Two dice are rolled, one after the other. The sample space is shown below. There are 36 possible outcomes.

► **Point of Interest**

Romans called a die that was marked on four faces a *talus*, which meant "anklebone." The anklebone was considered an ideal die because it is roughly a rectangular solid and it has no marrow, so loose ones from sheep were more likely to be lying around after the wolves had left their prey.

Possible Outcomes from Rolling Two Dice

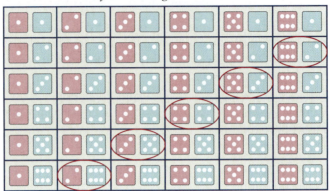

The outcomes that the sum of the numbers on the two dice is 8 are circled above. There are 5 possible outcomes that have a sum of 8. The probability that the sum of the number on the two dice is 8 is $P(E) = \dfrac{5}{36}$.

---

1 *Example*

Two dice are rolled once. Calculate the probability that the sum of the numbers on the two dice is 7.

Strategy

To find the probability:

► Count the number of possible outcomes of the experiment.
► Count the outcomes of the experiment that are favorable to the event the sum is 7.
► Use the probability formula.

Solution

There are 36 possible outcomes.

There are 6 outcomes favorable for E: (1, 6), (2, 5), (3, 4), (4, 3), (5, 2), and (6, 1).

$$P(E) = \dfrac{6}{36} = \dfrac{1}{6}$$

The probability that the sum is 7 is $\dfrac{1}{6}$.

1 *You Try It*

Two dice are rolled once. Calculate the probability that the two numbers on the dice are equal.

Your Strategy

Your Solution

Solution on p. S26

Copyright © Houghton Mifflin Company. All rights reserved.

2 Example

A large box contains 25 red, 35 blue, and 40 white balls. If one ball is randomly selected from the box, what is the probability that it is blue? Write the answer as a percent.

2 You Try It

There are 8 covered circles on a "scratcher card" that is given to each customer at a fast-food restaurant. Under one of the circles is a symbol for a free soft drink. If the customer scratches off one circle, what is the probability that the soft drink symbol will be uncovered?

Strategy

To find the probability:
▶ Count the number of outcomes of the experiment.
▶ Count the number of outcomes of the experiment favorable to the event E that the ball is blue.
▶ Use the probability formula.

Your Strategy

Solution

There are 100 (25 + 35 + 40) balls in the box.

35 balls of the 100 are blue.

$$P(E) = \frac{35}{100} = 0.35 = 35\%$$

There is a 35% chance of selecting a blue ball.

Your Solution

Solution on p. S26

OBJECTIVE **B**

The odds of an event

Sometimes the chances of an event occurring are given in terms of *odds*. This concept is closely related to probability.

Odds in Favor of an Event

The **odds in favor** of an event is the ratio of the number of favorable outcomes of an experiment to the number of unfavorable outcomes.

$$\text{Odds in favor} = \frac{\text{number of favorable outcomes}}{\text{number of unfavorable outcomes}}$$

Odds Against an Event

The **odds against** an event is the ratio of the number of unfavorable outcomes of an experiment to the number of favorable outcomes.

$$\text{Odds against} = \frac{\text{number of unfavorable outcomes}}{\text{number of favorable outcomes}}$$

Copyright © Houghton Mifflin Company. All rights reserved.

To find the odds in favor of a 4 when a die is rolled once, list the favorable outcomes and the unfavorable outcomes.

favorable outcomes: 4   unfavorable outcomes: 1, 2, 3, 5, 6

$$\text{Odds in favor of a 4} = \frac{\text{number of favorable outcomes}}{\text{number of unfavorable outcomes}} = \frac{1}{5}$$

Frequently the odds of an event are expressed as a ratio using the word *to*. For the last problem, the odds in favor of a 4 are 1 to 5.

It is possible to compute the probability of an event from the odds in favor fraction. The probability of an event is the ratio of the numerator to the sum of the numerator and denominator.

The odds in favor of winning a prize in a charity drawing are 1 to 19. What is the probability of winning a prize?

Write the ratio 1 to 19 as a fraction. $1 \text{ to } 19 = \frac{1}{19}$

The probability of winning a prize is the ratio of the numerator to the sum of the numerator and denominator. $\text{Probability} = \frac{1}{1 + 19} = \frac{1}{20}$

The probability of winning a prize is $\frac{1}{20}$.

 Example

In a horse race, the odds against a horse winning the race are posted as 9 to 2. What is the probability of the horse's winning the race?

Strategy

To calculate the probability of winning:

▶ Restate the odds against as odds in favor.
▶ Using the odds-in-favor fraction, the probability of winning is the ratio of the numerator to the sum of the numerator and denominator.

Solution

The odds against winning are 9 to 2. Therefore, the odds in favor of winning are 2 to 9.

$$\text{Probability of winning} = \frac{2}{2 + 9} = \frac{2}{11}$$

The probability of the horse's winning the race is $\frac{2}{11}$.

You Try It

The odds in favor of contracting the flu during a flu epidemic are 2 to 13. Calculate the probability of getting the flu.

Your Strategy

Your Solution

Solution on p. S27

Copyright © Houghton Mifflin Company. All rights reserved.

10.3 EXERCISES

OBJECTIVE **A**

1. 🖊 Describe two situations in which probabilities are cited.

2. 🖊 Why can the probability of an event not be $\frac{5}{3}$?

3. A coin is tossed 4 times. List all the possible outcomes of the experiment as a sample space. The table on page 629 shows an example of a systematic way of recording results for a similar problem.

4. Three cards—one red, one green, and one blue—are to be arranged in a stack. Using R for red, G for green, and B for blue, make a list of all the different stacks that can be formed. (Some computer monitors are called RGB monitors for the colors red, green, and blue.)

5. A tetrahedral die is one with four triangular sides. If two tetrahedral dice are rolled, list all the possible outcomes of the experiment as a sample space. (See the table for Two Dice, page 630, for assistance in listing the outcomes.)

6. A coin is tossed and then a die is rolled. List all the possible outcomes of the experiment as a sample space. (To get you started, (H, 1) is one of the possible outcomes.)

7. Some people who cheat at gambling use dice that are loaded so that 7 occurs more frequently than expected. If these dice are used, are the probabilities of the outcomes equal? Why or why not?

8. If the spinner at the right is spun once, is each of the numbers 1 through 5 equally likely? Why or why not?

9. A coin is tossed four times. What is the probability that the outcomes of the tosses are exactly in the order HHTT? (See Exercise 3.)

10. A coin is tossed four times. What is the probability that the outcomes of the tosses are exactly in the order HTTH? (See Exercise 3.)

Copyright © Houghton Mifflin Company. All rights reserved.

11. A coin is tossed four times. What is the probability that the outcomes of the tosses consist of two heads and two tails? (See Exercise 3.)

12. A coin is tossed four times. What is the probability that the outcomes of the tosses consist of one head and three tails? (See Exercise 3.)

13. If two dice are rolled, what is the probability that the sum of the dots on the upward faces is 5?

14. If two dice are rolled, what is the probability that the sum of the dots on the upward faces is 9?

15. If two dice are rolled, what is the probability that the sum of the dots on the upward faces is 15?

16. If two dice are rolled, what is the probability that the sum of the dots on the upward faces is less than 15?

17. If two dice are rolled, what is the probability that the sum of the dots on the upward faces is 2?

18. If two dice are rolled, what is the probability that the sum of the dots on the upward faces is 12?

19. A dodecahedral die has 12 sides. If the die is rolled once, what is the probability that the upward face shows an 11?

20. A dodecahedral die has 12 sides. If the die is rolled once, what is the probability that the upward face shows 5?

21. If two tetrahedral dice are rolled (see Exercise 5), what is the probability that the sum on the upward faces is 4?

22. If two tetrahedral dice are rolled (see Exercise 5), what is the probability that the sum on the upward faces is 6?

23. A dodecahedral die has 12 sides. If the die is rolled once, what is the probability that the upward face shows a number divisible by 4?

Copyright © Houghton Mifflin Company. All rights reserved.

24. A dodecahedral die has 12 sides. If the die is rolled once, what is the probability that the upward face shows a number that is a multiple of 3?

25. A survey of 95 people showed that 37 preferred a cash discount of 2% if an item was purchased using cash or a check. On the basis of this survey, what is the empirical probability that a person prefers a cash discount?

26. A survey of 725 people showed that 587 had a group health insurance plan where they worked. Based on this survey, what is the empirical probability that an employee has a group health insurance plan?

27. A signal light is green for 3 min, yellow for 15 s, and red for 2 min. If you drive up to this light, what is the probability that it will be green when you reach the intersection?

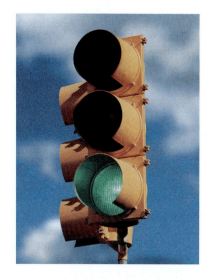

28. In a history class, a professor gave 4 A's, 8 B's, 22 C's, 10 D's, and 3 F's. If a single student's paper is chosen from this class, what is the probability that it received a B?

29. A television cable company surveyed some of its customers and asked them to rate the cable service as excellent, satisfactory, average, unsatisfactory, or poor. The results are recorded in the table at the right. What is the probability that a customer who was surveyed rated the service as satisfactory or excellent?

Quality of Service	Number Who Voted
Excellent	98
Satisfactory	87
Average	129
Unsatisfactory	42
Poor	21

30. Using the television cable survey in Exercise 29, what is the probability that a customer who was surveyed rated the service as unsatisfactory or poor?

OBJECTIVE **B**

31. A fair coin is tossed once. What are the odds of its showing heads?

32. A fair coin is tossed twice. What are the odds of its showing tails both times?

33. The odds in favor of a candidate winning an election are 3 to 2. What is the probability of the candidate winning the election?

34. On a board game, the odds in favor of winning $5,000 are 3 to 7. What is the probability of winning the $5,000?

Copyright © Houghton Mifflin Company. All rights reserved.

35. Two fair dice are rolled. What are the odds in favor of rolling a 7?

36. Two fair dice are rolled. What are the odds in favor of rolling a 12?

37. A single card is selected from a regular deck of playing cards. What are the odds against its being an ace?

38. A single card is selected from a regular deck of playing cards. What are the odds against its being a heart?

39. At the beginning of the professional football season, one team was given 40 to 1 odds against its winning the Super Bowl. What is the probability of this team winning the Super Bowl?

40. At the beginning of the professional baseball season, one team was given 25 to 1 odds against its winning the World Series. What is the probability of this team winning the World Series?

41. A stock market analyst estimates that the odds in favor of a stock going up in value are 2 to 1. What is the probability of the stock's not going up in value?

42. The odds in favor of the occurrence of an event A are given as 5 to 2, and the odds against a second event B are given as 1 to 7. Which event, A or B, has the greater probability of occurring?

Critical Thinking

43. A box contains only white, blue, and red balls. You are told that the probability of choosing a white ball is $\frac{1}{2}$, that of choosing a blue ball is $\frac{1}{3}$, and that of choosing a red ball is $\frac{1}{9}$. What is wrong with that statement?

44. Three line segments are randomly chosen from line segments whose lengths are 1 cm, 2 cm, 3 cm, 4 cm, and 5 cm. What is the probability that a triangle can be formed from the line segments?

45. The probability of tossing a fair coin and having it land heads is $\frac{1}{2}$.

    Does this mean that if that coin is tossed 100 times, it will land heads 50 times? Explain your answer.

46. Suppose a surgeon tells you that an operation you need has a 90% success rate. Explain what this means.

Copyright © Houghton Mifflin Company. All rights reserved.

Focus on Problem Solving

Applying Solutions to Other Problems

Problem solving in the previous chapters concentrated on solving specific problems. After a problem is solved, however, there is an important question to be asked: "Does the solution to this problem apply to other types of problems?"

To illustrate this extension to problem solving, we will consider *triangular numbers*, which were studied by ancient Greek mathematicians. The numbers 1, 3, 6, 10, 15, and 21 are the first six triangular numbers. What is the next triangular number?

To answer this question, note in the diagram below that a triangle can be formed using the number of dots that correspond to a triangular number.

1	3	6	10	15	21

Observe that the number of dots in each row is one more than the number of dots in the row above. The total number of dots can be found by addition.

$$1 = 1 \qquad 1 + 2 = 3 \qquad 1 + 2 + 3 = 6 \qquad 1 + 2 + 3 + 4 = 10$$
$$1 + 2 + 3 + 4 + 5 = 15 \qquad 1 + 2 + 3 + 4 + 5 + 6 = 21$$

The pattern suggests that the next triangular number (the 7th one) is the sum of the first 7 natural numbers.

$$1 + 2 + 3 + 4 + 5 + 6 + 7 = 28$$

The 7th triangular number is 28. The diagram at the left shows the 7th triangular number.

Using the pattern for triangular numbers, the 10th triangular number is

$$1 + 2 + 3 + 4 + 5 + 6 + 7 + 8 + 9 + 10 = 55$$

Now consider a situation that may seem to be totally unrelated to triangular numbers. Suppose you are in charge of scheduling softball games for a league. There are seven teams in the league, and each team must play every other team once. How many games must be scheduled?

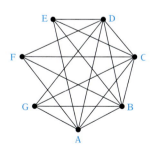

We label the teams A, B, C, D, E, F, and G. (See the figure at the left). A line between two teams indicates that the two teams play each other. Beginning with A, there are 6 lines for the 6 teams that A must play.

Now consider B. There are 6 teams that B must play, but the line between A and B has already been drawn, so there are only 5 remaining games to schedule for B. Now move on to C. The lines between C and A and between C and B have already been drawn, so there are 4 additional lines to be drawn to represent the teams that C will play. Moving on to D, the lines between D and A, D and B, and D and C have already been drawn, so there are 3 more lines to be drawn to represent the teams that D will play.

Note that each time we move from team to team, one fewer line needs to be drawn. When we reach F, there is only one line to be drawn, the one between F and G. The total number of lines drawn is $6 + 5 + 4 + 3 + 2 + 1 = 21$, the sixth triangular number. For a league with 7 teams, the number of games that must be scheduled so that each team plays every other team once is the 6th

Copyright © Houghton Mifflin Company. All rights reserved.

triangular number. If there were 10 teams in the league, the number of games that must be scheduled would be the 9th triangular number, which is 45.

A college chess team wants to schedule a match so that each of its 15 members plays each other member of the team once. How many matches must be scheduled?

Projects & Group Activities

Random Samples

▶ **Point of Interest**

In the 1936 presidential election, a telephone survey of more than 2 million people indicated that Landon would defeat Roosevelt by 57% to 43%. However, when the votes were counted, Roosevelt won by a margin of 62% to 38%.

The error was a consequence of the fact that those surveyed were drawn from a population of telephone owners and, in 1936, that population was largely people with high incomes. It was not representative of the population of all voters in the United States.

When a survey is taken to determine, for example, who is the most popular choice for a political office such as a governor or president, it would not be appropriate to survey only one ethnic group, or survey only one religious group, or survey only one political group. A survey done in that way would reflect only the views of that particular group of people. Instead, a *random sample* of people must be chosen. This sample would include people from different ethnic, religious, political, and income groups. The purpose of the random sample is to identify the popular choice of all the people by interviewing only a few people, the people in the random sample.

One way of choosing a random sample is to use a table of random numbers. An example of a portion of such a table is shown below.

Random Digits

40784	38916	12949
29798	57707	57392
42228	94940	10668
02218	89355	76117
15736	08506	29759
42658	32502	99698
98670	57794	64795
38266	30138	61250
68249	32459	41627
36910	85225	78541

In a random number table, each digit should occur with approximately the same frequency as any other digit.

1. Make a frequency table for the table of random digits above. Do the digits occur with approximately the same frequency?

As an example of how to use a random number table, suppose there are 33 students in your class and you want to randomly select 6 students. Using the numbers 01, 02, 03, . . . , 31, 32, 33, assign each student a two-digit number. Starting with the first column of random numbers, move down the column looking at only the first two digits. If the two digits are one of the numbers 01 through 33, write them down. If not, move to the next number. Continue in this way until six numbers have been selected.

For instance, the first two digits of 40784 are 40, which is not between 01 and 33. Therefore, move to the next number. The first two digits of 29798 are 29. Since 29 is between 01 and 33, write it down. Continuing in this way reveals that the random sample would be the students with numbers 29, 02, 15, 08, 32, and 30.

Copyright © Houghton Mifflin Company. All rights reserved.

2. Once a random sample has been selected, it is possible to use this sample to estimate characteristics of the entire class. For example, find the mean height of the students in the random sample and compare it to the mean height of the entire class.

Random numbers also are used to *simulate* random events. In a simulation, it is not the actual activity that is performed, but instead one that is very similar.

For instance, suppose a coin is tossed repeatedly, and the result, head or tail, is recorded. The actual activity is tossing the coin. A simulation of tossing a coin uses a random number table. Starting with the first column of random numbers, move down the column looking at the first digit. Associate an even digit with heads and an odd digit with tails. Using the first column of the table on the previous page, a simulation of the first ten tosses of the coin would be H, H, H, H, T, H, T, T, H, and T.

Probabilities can be approximated by simulating an event. Consider the event of tossing two heads when a fair coin is tossed twice. We can simulate the event by associating a two-digit number with two even digits with two heads, one with two odd digits with two tails, and any other pair of digits with a head and tail or tail and head.

3. Using the 30 numbers in the table on the previous page, show that the probability of tossing two heads would be $\frac{7}{30} \approx 0.233$. Explain why the actual probability is 0.25.

4. Go to the library and find a table of random numbers. Use this table to simulate rolling a pair of dice. If you use the first two digits of a column, a valid roll consists of two digits that are both between 1 and 6, inclusive. Simulate the empirical probability that the sum of the dice is 7 by using 100 rolls of the dice.

Working Backwards

Sometimes the solution to a problem can be found by *working backwards*. This technique can be used to find a winning strategy for a game called Nim.

There are many variations of this game. For our game, there are two players, Player A and Player B, who alternately place 1, 2, or 3 matchsticks in a pile. The object of the game is to place the 32nd matchstick in the pile. Is there a strategy that Player A can use to guarantee winning the game?

Working backwards, if there are 29, 30, or 31 matchsticks in the pile when it is A's turn to play, A can win by placing 3 (29 + 3 = 32), 2 (30 + 2 = 32), or 1 (31 + 1 = 32) matchsticks on the pile. If there are to be 29, 30, or 31 match sticks in the pile when it is A's turn, there must be 28 matchsticks in the pile when it is B's turn.

Working backwards from 28, if there are to be 28 matches in the pile at B's turn, there must be 25, 26, or 27 at A's turn. Player A can then add 3, 2, or 1 matchsticks to the pile to bring the number to 28. For there to be 25, 26, or 27 matchsticks in the pile at A's turn, there must be 24 matchsticks in the pile at B's turn.

Copyright © Houghton Mifflin Company. All rights reserved.

Now working backwards from 24, if there are to be 24 matches in the pile at B's turn, there must be 21, 22, or 23 at A's turn. Player A can then add 3, 2, or 1 matchsticks to the pile to bring the number to 24. For there to be 21, 22, or 23 matchsticks in the pile at A's turn, there must be 20 matchsticks in the pile at B's turn.

So far, we have found that for Player A to win, there must be 28, 24, or 20 matchsticks in the pile when it is B's turn to play. Note that each time, the number is decreasing by 4. Continuing this pattern, Player A will win if there are 16, 12, 8, or 4 matchsticks in the pile when it is B's turn.

Player A can guarantee winning by making sure that the number of matchsticks in the pile is a multiple of 4. To ensure this, Player A allows Player B to go first and then adds exactly enough matchsticks to the pile to bring the total to a multiple of 4.

For example, suppose B places 3 matchsticks in the pile; then A places 1 matchstick $(3 + 1 = 4)$ in the pile. Now B places 2 matchsticks in the pile. The total is now 6 matchsticks. Player A then places 2 matchsticks in the pile to bring the total to 8, a multiple of 4. If play continues in this way, Player A will win.

Here are some variations of Nim. See whether you can develop a winning strategy for Player A.

1. Suppose the goal is to place the last matchstick in a pile of 30 matches.

2. Suppose the players make two piles of matchsticks, with the final number of matchsticks in each pile to be 20.

3. In this variation of Nim, there are 40 matchsticks in a pile. Each player alternately selects 1, 2, or 3 matches from the pile. The player who selects the last match wins.

Chapter Summary

Key Words

Statistics is the collecting, organizing, and interpreting of data. Data are collected from a **population,** which is the set of all observations of interest. A **frequency distribution** is one method of organizing data. A frequency distribution is constructed by dividing the data gathered from a population into **classes.** The **range** of a set of numerical data is the difference between the largest and smallest values. [10.1A, p. 609]

Examples

An Internet service provider (ISP) surveyed 1,000 of its subscribers to determine the time required for each subscriber to download a particular file. The results are summarized in the frequency distribution at the top of the next page. The distribution has 12 classes.

Copyright © Houghton Mifflin Company. All rights reserved.

Download Time (in seconds)	Number of Subscribers
0–5	6
5–10	17
10–15	43
15–20	92
20–25	151
25–30	192
30–35	190
35–40	149
40–45	90
45–50	45
50–55	15
55–60	10

A **histogram** is a bar graph that represents the data in a frequency distribution. [10.1B, p. 611]

Below is a histogram for the frequency distribution given above.

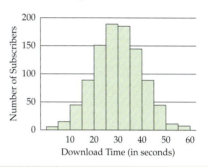

A **frequency polygon** is a graph that displays information in a manner similar to a histogram. A dot is placed above the center of each class interval at a height corresponding to that class's frequency. The dots are connected to form a broken-line graph. The center of a class interval is called the **class midpoint.** [10.1C, p. 612]

Below is a frequency polygon for the frequency distribution given above.

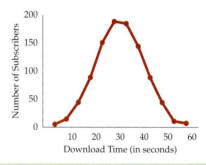

The **mean, median,** and **mode** are three types of averages used in statistics. The **median** of data is the number that separates the data into two equal parts when the data have been arranged from smallest to largest (or largest to smallest). The **mode** is the most frequently occurring data value. [10.2A, pp. 617–618]

Consider the following set of data:

24, 28, 33, 45, 45

The median is 33.
The mode is 45.

A **box-and-whiskers plot** is a graph that shows five numbers: the smallest value, the first quartile, the median, the third quartile, and the largest value. The **first quartile** Q_1 is the number below which one-fourth of the data lie. The **third quartile** Q_3 is a number above which one-fourth of the data lie. The box is placed around the values between the first quartile and the third quartile. The difference between Q_3 and Q_1 is the **interquartile range.** [10.2B, pp. 620–621]

The box-and-whiskers plot for a set of test scores is shown below.

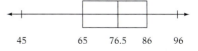

The interquartile range = $Q_3 - Q_1$
$$= 86 - 65 = 21$$

Copyright © Houghton Mifflin Company. All rights reserved.

An **experiment** is an activity with an observable outcome. All the possible outcomes of an experiment are called the **sample space** of the experiment. An **event** is one or more outcomes of an experiment. [10.3A, p. 627]

Tossing a single die is an example of an experiment. The sample space for this experiment is the set of possible outcomes: $S = \{1, 2, 3, 4, 5, 6\}$. The event that the number landing face up is an odd number is represented by $E = \{1, 3, 5\}$.

Essential Rules and Procedures

Mean of a set of data [10.2A, p. 617]

$$\bar{x} = \frac{\text{sum of all data values}}{\text{number of data values}}$$

Consider the following set of data:

24, 28, 33, 45, 45

The mean is $\frac{24 + 28 + 33 + 45 + 45}{5} = 35$.

Standard deviation [10.2C, p. 622]
To determine standard deviation, which is a measure of the clustering of data near the mean:
1. Sum the squares of the differences between each value of data and the mean.
2. Divide the result in Step 1 by the number of items in the set of data.
3. Take the square root of the result in Step 2.

Consider the following sets of data:

24, 28, 33, 45, 45

The mean is 35.

1. x	$(x - \bar{x})$	$(x - \bar{x})^2$
24	$(24 - 35)$	$(-11)^2 = 121$
28	$(28 - 35)$	$(-7)^2 = 49$
33	$(33 - 35)$	$(-2)^2 = 4$
45	$(45 - 35)$	$(10)^2 = 100$
45	$(45 - 35)$	$(10)^2 = \underline{100}$
		Total $= 374$

2. $\frac{374}{5} = 74.8$

3. $\sigma = \sqrt{74.8} \approx 8.649$

Probability Formula [10.3A, p. 628]

$$P(E) = \frac{\text{number of favorable outcomes}}{\text{number of possible outcomes}}$$

A die is rolled. The probability of rolling a 2 or a 4 is

$$P(E) = \frac{2}{6} = \frac{1}{3}$$

Empirical Probability Formula [10.3A, p. 629]

$$P(E) = \frac{\text{number of observations of } E}{\text{total number of observations}}$$

A thumbtack is tossed 100 times. It lands point up 15 times and lands on its side 85 times. From this experiment, the empirical probability of "point up" is

$$P(\text{point up}) = \frac{15}{100} = \frac{3}{20}$$

Odds in Favor of an Event [10.3B, p. 631]

$$\text{Odds in favor} = \frac{\text{number of favorable outcomes}}{\text{number of unfavorable outcomes}}$$

A die is rolled.
The odds in favor of rolling a 2 or a 4:

$$\text{Odds in favor} = \frac{2}{4} = \frac{1}{2}$$

Odds Against an Event [10.3B, p. 631]

$$\text{Odds against} = \frac{\text{number of unfavorable outcomes}}{\text{number of favorable outcomes}}$$

The odds against rolling a 2 or a 4:

$$\text{Odds against} = \frac{4}{2} = \frac{2}{1}$$

Copyright © Houghton Mifflin Company. All rights reserved.

Chapter Review Exercises

Education   Use the data in the table below for Exercises 1–5.

Number of Students in 40 Mathematics Classes

30	45	54	24	48	12	38	31
15	36	37	27	40	35	55	32
42	14	21	18	29	25	16	42
44	41	28	32	27	24	30	24
21	35	27	32	39	41	35	48

1. Make a frequency distribution for these data using 6 classes.

2. Which class has the greatest frequency?

3. How many math classes have 35 or fewer students?

4. What percent of the math classes have 44 or more students?

5. What percent of the math classes have 27 or fewer students?

Meteorology   The high temperatures at a ski resort during a 125-day ski season are recorded in the figure at the right. Use this histogram for Exercises 6 and 7.

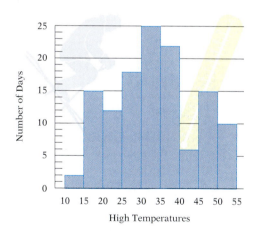

6. Find the number of days the high temperature was 45° or above.

7. How many days had a high temperature below 25°?

8. *Health*   A health clinic administered a test for cholesterol to 11 people. The results were 180, 220, 160, 230, 280, 200, 210, 250, 190, 230, and 210. Find the mean and median of these data.

9. *Health*   The weights, in pounds, of 10 babies born at a hospital were recorded as 6.3, 5.9, 8.1, 6.5, 7.2, 5.6, 8.9, 9.1, 6.9, and 7.2. Find the mean and median of these data.

10. *The Arts*   People leaving a new movie were asked to rate the movie as bad, good, very good, or excellent. The responses were bad, 28; good, 65; very good, 49; excellent, 28. What was the modal response for this survey?

Copyright © Houghton Mifflin Company. All rights reserved.

Investments The frequency polygon in the figure at the right shows the number of shares of stock that were sold on a stock exchange. Use the frequency polygon for Exercises 11–13.

11. How many shares of stock were sold between 7 A.M. and 10 A.M.?

12. Between which hours were less than 15 million shares sold?

13. What is the ratio of the number of shares of stock sold between 10 A.M. and 11 A.M. to the number that were sold between 11 A.M. and 12 P.M.?

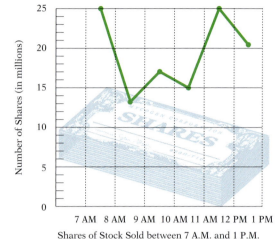

Shares of Stock Sold between 7 A.M. and 1 P.M.

14. *Sports* The numbers of points scored by a basketball team for fifteen games were 89, 102, 134, 110, 121, 124, 111, 116, 99, 120, 105, 109, 110, 124, and 131. Find the first quartile, median, and third quartile. Draw a box-and-whiskers plot.

15. *Fuel Efficiency* A consumer research group tested the average miles per gallon for six cars. The results were 24, 28, 22, 35, 41, and 27. Find the standard deviation of the gasoline mileage ratings.

16. A charity raffle sells 2,500 raffle tickets for a big-screen television set. If you purchase 5 tickets, what is the probability that you will win the television?

17. A box contains 50 balls, of which 15 are red. If one ball is randomly selected from the box, what are the odds in favor of the ball's being red?

18. In professional jai alai, a gambler can wager on who will win the event. The odds against one of the players winning are given as 5 to 2. What is the probability of that player's winning?

19. A dodecahedral die has 12 sides numbered from 1 to 12. If this die is rolled once, what is the probability that a number divisible by 6 will be on the upward face?

20. One student is randomly selected from 3 first-year students, 4 sophomores, 5 juniors, and 2 seniors. What is the probability that the student is a junior?

Copyright © Houghton Mifflin Company. All rights reserved.

Chapter Test

1. *Communications*   The histogram below shows the cost for telephone service for 100 residences. For how many residences was the cost for telephone service $60 or more?

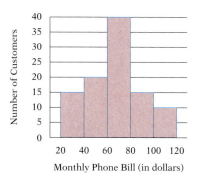

Number of Customers
Monthly Phone Bill (in dollars)

2. *Business*   The frequency polygon below shows the gross sales at a record store for 6 months. Find the total gross sales for January and February.

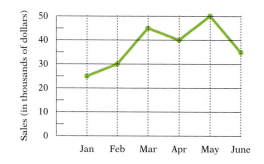

Sales (in thousands of dollars)

Jan   Feb   Mar   Apr   May   June

3. *Business*   The annual sales for the 50 restaurants in a restaurant chain are given in the frequency table at the right. What percent of the restaurants had annual sales between $750,000 and $1,000,000?

Annual Sales (in dollars)	Frequency
0–250,000	4
250,000–500,000	10
500,000–750,000	17
750,000–1,000,000	8
1,000,000–1,250,000	8
1,250,000–1,500,000	3

4. *Sports*   The bowling scores for eight people were 138, 125, 162, 144, 129, 168, 184, and 173. What was the mean score for these eight people?

5. *Emergency Calls*   The response times by an ambulance service to emergency calls were recorded by a public safety commission. The times (in minutes) were 17, 21, 11, 8, 22, 15, 11, 14, and 8. Determine the median response time for these calls.

6. *Education*   Recent college graduates were asked to rate the quality of their education. The responses were 47, excellent; 86, very good; 32, good; 20, poor. What was the modal response?

7. *Business*   The numbers of digital assistants sold by a store for the first 5 months of the year were 34, 28, 31, 36, and 38. How many digital assistants must be sold in the sixth month so that the mean number sold per month for the 6 months is 35?

8. *Manufacturing*   The average time, in minutes, it takes for a factory worker to assemble 14 different toys is given in the table. Determine the first quartile of the data.

10.5	21.0	17.3	11.2	9.3	6.5	8.6
9.8	20.3	19.6	9.8	10.5	11.9	18.5

Copyright © Houghton Mifflin Company. All rights reserved.

9. *Business* The number of vacation days taken last year by each of the employees of a firm was recorded. The box-and-whiskers plot at the right represents the data. **a.** Determine the range of the data. **b.** What was the median number of vacation days taken?

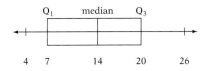

10. *Sports* The scores of the 14 leaders in a college golf tournament are given in the table. Draw a box-and-whiskers plot of the data.

| 80 | 76 | 70 | 71 | 74 | 68 | 72 |
| 74 | 70 | 70 | 73 | 75 | 69 | 73 |

11. *Testing* An employee at a department of motor vehicles analyzed the written tests of the last 10 applicants for a driver's license. The number of incorrect answers for each of these applicants were 2, 0, 3, 1, 0, 4, 5, 1, 3, and 1. What is the standard deviation of the number of incorrect answers? Round to the nearest hundredth.

12. A coin is tossed and then a regular die is rolled. How many elements are in the sample space?

13. Three coins—a nickel, a dime, and a quarter—are stacked. List the elements in the sample space.

14. A cross-country flight has 14 passengers in first class, 32 passengers in business class, and 202 passengers in coach. If one passenger is selected at random, what is the probability that the person is in business class?

15. Three playing cards—an ace, a king, and a queen—are randomly arranged and stacked. What is the probability that the ace is on top of the stack?

16. A quiz contains three true/false questions. If a student attempts to answer the questions by just guessing, what is the probability that the student will answer all three questions correctly?

17. A package of flower seeds contains 15 seeds for red flowers, 20 seeds for white flowers, and 10 seeds for pink flowers. If one seed is selected at random, what is the probability that it is not a seed for a red flower?

18. The odds of winning a prize in a lottery are given as 1 to 12. What is the probability of winning a prize?

19. The probability of rolling a sum of nine on two standard dice is $\frac{1}{9}$. What are the odds in favor of rolling a nine on these dice?

20. A dodecahedral die has 12 sides. If this die is tossed once, what is the probability that the number on the upward face is less than six?

Copyright © Houghton Mifflin Company. All rights reserved.

Cumulative Review Exercises

1. Simplify: $\sqrt{200}$

2. Solve: $7p - 2(3p - 1) = 5p + 6$

3. Evaluate $3a^2b - 4ab^2$ when $a = -1$ and $b = 2$.

4. Simplify: $-2[2 - 4(3x - 1) + 2(3x - 1)]$

5. Solve: $-\dfrac{2}{3}y - 5 = 7$

6. Simplify: $-\dfrac{4}{5}\left[\dfrac{3}{4} - \dfrac{7}{8} - \left(\dfrac{2}{3}\right)^2\right]$

7. Graph $y = \dfrac{4}{3}x - 3$.

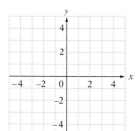

8. Graph $y = \dfrac{1}{3}x$.

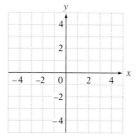

9. Subtract: $(7y^2 + 5y - 8) - (4y^2 - 3y + 1)$

10. Simplify: $(4a^2b)^3$

11. $16\dfrac{2}{3}\%$ of what number is 24?

12. Solve: $\dfrac{9}{8} = \dfrac{3}{n}$

13. Write $87,600,000,000$ in scientific notation.

14. A landscape architect designed the cement patio shown below. Determine the area of the patio.

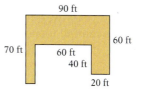

Copyright © Houghton Mifflin Company. All rights reserved.

15. Multiply: $(5c^2d^4)(-3cd^6)$

16. Convert 40 km to meters.

17. What is the measure of $\angle n$ in the figure below?

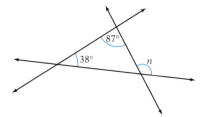

18. Find the area of the parallelogram shown below.

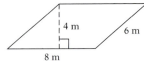

Solve.

19. Simple Interest Find the simple interest on a 3-month loan of $25,000 at an annual interest rate of 7.5%.

20. Probability A box contains 12 white, 15 blue, and 9 red balls. If one ball is randomly chosen from the box, what is the probability that the ball is not white?

21. Education The scores for six students on an achievement test were 24, 38, 22, 34, 37, and 31. What were the mean and median scores for these students?

22. Elections The results of a recent city election showed that 55,000 people voted, out of a possible 230,000 registered voters. What percent of the registered voters did not vote in the election? Round to the nearest tenth of a percent.

23. Measurement Eratosthenes (circa 300 B.C.), a Greek mathematician, calculated that an angle of 7.5° at Earth's center cuts an arc of 1 600 km on Earth's surface. Using this information, what is the approximate circumference of Earth?

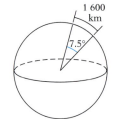

24. Meteorology The annual rainfall totals, in inches, for a certain region for the last five years were 12, 16, 20, 18, and 14. Find the standard deviation of these rainfall totals.

25. Compensation A chef's helper received a 10% hourly wage increase to $19.80 per hour. What was the chef's helper's hourly wage before the increase?

Copyright © Houghton Mifflin Company. All rights reserved.

Final Examination

1. Estimate the sum of 672, 843, 509, and 417.

2. Simplify: $18 + 3(6 - 4)^2 \div 2$

3. Simplify: $-8 - (-13) - 10 + 7$

4. Evaluate $|a - b| - 3bc^3$ when $a = -2$, $b = 4$, and $c = -1$.

5. What is $5\frac{3}{8}$ minus $2\frac{11}{16}$?

6. Find the quotient of $\frac{7}{9}$ and $\frac{5}{6}$.

7. Simplify: $\dfrac{\frac{3}{4} - \frac{1}{2}}{\frac{5}{8} + \frac{1}{2}}$

8. Place the correct symbol, $<$ or $>$, between the two numbers.

   $\dfrac{5}{16}$ 0.313

9. Evaluate $-10qr$ when $q = -8.1$ and $r = -9.5$.

10. Divide and round to the nearest hundredth:
    $-15.32 \div 4.67$

11. Is -0.5 a solution of the equation $-90y = 45$?

12. Simplify: $\sqrt{162}$

13. Graph $x \geq -4$.

14. Simplify: $-\dfrac{5}{6}(-12t)$

15. Simplify: $2(x - 3y) - 4(x + 2y)$

16. Subtract: $(5z^3 + 2z^2 - 1) - (4z^3 + 6z - 8)$

Copyright © Houghton Mifflin Company. All rights reserved.

17. Multiply: $(4x^2)(2x^5y)$

18. Multiply: $2a^2b^2(5a^2 - 3ab + 4b^2)$

19. Multiply: $(3x - 2)(5x + 3)$

20. Simplify: $(3x^2y)^4$

21. Evaluate: 4^{-3}

22. Simplify: $\dfrac{m^5n^8}{m^3n^4}$

23. Solve: $2 - \dfrac{4}{3}y = 10$

24. Solve: $6z + 8 = 5 - 3z$

25. Solve: $8 + 2(6c - 7) = 4$

26. Convert 2.48 m to centimeters.

27. Convert 2.6 mi to feet.

28. Solve: $\dfrac{n + 2}{8} = \dfrac{5}{12}$

29. Given that $\ell_1 \| \ell_2$, find the measures of angles a and b.

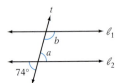

30. Find the unknown side of the triangle. Round to the nearest tenth.

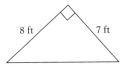

31. Find the perimeter of the composite figure. Round to the nearest hundredth.

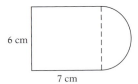

32. Find the volume of the composite figure. Round to the nearest hundredth.

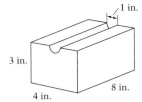

Copyright © Houghton Mifflin Company. All rights reserved.

33. Graph $y = -2x + 3$.

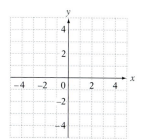

34. Graph $y = \dfrac{3}{5}x - 4$.

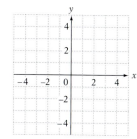

35. *Physics*   Find the ground speed of an airplane traveling into a 22-mph wind with an air speed of 386 mph. Use the formula $g = a - h$, where g is the ground speed, a is the air speed, and h is the speed of the head wind.

36. *Manufacturing*   A factory worker can inspect a product in $1\dfrac{1}{2}$ min. How many products can the worker inspect during an 8-hour day?

37. *Chemistry*   The boiling point of bromine is 58.78°C. The melting point of bromine is -7.2°C. Find the difference between the boiling point and the melting point of bromine.

38. *Physics*   One light-year, which is the distance that light travels through empty space in one year, is approximately 5,880,000,000,000 mi. Write this number in scientific notation.

39. *Physics*   Two children are sitting on a seesaw that is 10 ft long. One child weighs 50 lb, and the second child weighs 75 lb. How far from the 50-pound child should the fulcrum be placed so that the seesaw balances? Use the formula $F_1 x = F_2(d - x)$.

40. *Consumerism*   The fee charged by a ticketing agency for a concert is $10.50 plus $52.50 for each ticket purchased. If your total charge for tickets is $325.50, how many tickets are you purchasing?

41. *Taxes*   The property tax on a $250,000 house is $3,750. At this rate, what is the property tax on a home appraised at $314,000?

Copyright © Houghton Mifflin Company. All rights reserved.

42. 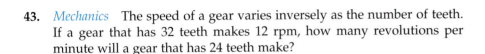 *Geography* The figure at the right represents the land area of the states in the United States. What percent of the states have a land area of 75,000 mi² or more?

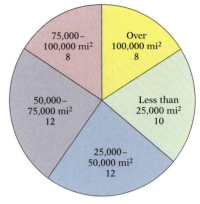

43. *Mechanics* The speed of a gear varies inversely as the number of teeth. If a gear that has 32 teeth makes 12 rpm, how many revolutions per minute will a gear that has 24 teeth make?

Land Area of the States in the United States

44. *Consumerism* A customer purchased a car for $32,500 and paid a sales tax of 5.5% of the cost. Find the total cost of the car including sales tax.

45. *Economics* Due to a recession, the number of housing starts in a community decreased from 124 to 96. What percent decrease does this represent? Round to the nearest tenth of a percent.

46. *Business* A necklace with a regular price of $245 is on sale for 35% off the regular price. Find the sale price.

47. *Finances* Find the simple interest on a 9-month loan of $25,000 at an annual interest rate of 8.6%.

48. *Labor Force* The numbers of hours per week that 80 twelfth grade students spend at paid jobs are given in the figure at the right. What percent of the students work more than 15 h per week?

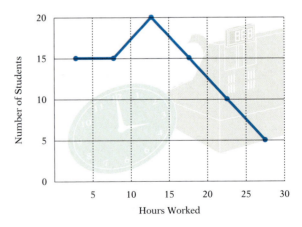

49. *Insurance* You requested rates for term life insurance from five different insurance companies. The annual premiums were $297, $425, $362, $281, and $309. Calculate the mean and median annual premiums for these five insurance companies.

50. *Probability* If two dice are tossed, what is the probability that the sum of the dots on the upward faces is divisible by 3?

Copyright © Houghton Mifflin Company. All rights reserved.

Solutions to Chapter 1 *You Try Its*

Section 1.1 *(pages 3–12)*

You Try It 1

You Try It 2

7 is 4 units to the left of 11.

You Try It 3   **a.** $47 > 19$ **b.** $26 > 0$

You Try It 4   0, 3, 17, 52, 68, 94

You Try It 5   forty-six million thirty-two thousand seven hundred fifteen

You Try It 6   920,008

You Try It 7 $70,000 + 6,000 + 200 + 40 + 5$

You Try It 8

> ⌐――――――― Given place value
> 529,374
> └――――――― $9 > 5$

529,374 rounded to the nearest ten-thousand is 530,000.

You Try It 9

> ⌐――――――― Given place value
> 7,985
> └――――――― $8 > 5$

7,985 rounded to the nearest hundred is 8,000.

You Try It 10

Strategy   To find the sport named by the greatest number of people, find the largest number given in the circle graph.

Solution   The largest number given in the graph is 80.

The sport named by the greatest number of people was football.

You Try It 11

Strategy   To find the shorter distance, compare the numbers 347 and 387.

Solution $347 < 387$

The shorter distance is between Los Angeles and San Jose.

You Try It 12

Strategy   To determine which state has fewer sanctioned league bowlers, compare the numbers 239,951 and 239,010.

Solution $239,010 < 239,951$

Ohio has fewer sanctioned league bowlers.

You Try It 13

Strategy   To find the land area to the nearest thousand square miles, round 3,851,809 to the nearest thousand.

Solution   3,851,809 rounded to the nearest thousand is 3,852,000.

To the nearest thousand, the land area of Canada is 3,852,000 mi^2.

Section 1.2 *(pages 19–32)*

You Try It 1

$$
\begin{array}{rcr}
6,285 & \longrightarrow & 6,000 \\
3,972 & \longrightarrow & 4,000 \\
5,140 & \longrightarrow & +\ 5,000 \\
\hline
 & & 15,000
\end{array}
$$

You Try It 2   The Addition Property of Zero

You Try It 3

$$
\begin{array}{r}
111,100,000 \\
61,600,000 \\
24,100,000 \\
+\ \ \ 1,600,000 \\
\hline
198,400,000
\end{array}
$$

A total of 198,400,000 cases of eggs were produced during the year.

You Try It 4

$x + y + z$
$1,692 + 4,783 + 5,046$

$$
\begin{array}{r}
\overset{1\ \ 21}{1,692} \\
4,783 \\
+\ 5,046 \\
\hline
11,521
\end{array}
$$

You Try It 5

$13 = b + 6$

$13 \mid 7 + 6$

$13 = 13$

Yes, 7 is a solution of the equation.

You Try It 6

$$
\begin{array}{r}
\overset{8\ \ 9\,9\,12}{49,\cancel{0}\cancel{0}2} \\
-31,865 \\
\hline
17,137
\end{array}
\qquad
\text{Check:}
\qquad
\begin{array}{r}
31,865 \\
+17,137 \\
\hline
49,002
\end{array}
$$

You Try It 7

$$
\begin{array}{rcrr}
8,544 & \longrightarrow & 9,000 & 8,544 \\
3,621 & \longrightarrow & -4,000 & -3,621 \\
\hline
 & & 5,000 & 4,923
\end{array}
$$

You Try It 8

2020: 612 quadrillion Btu
1990: 346 quadrillion Btu

$$
\begin{array}{r}
612 \\
-346 \\
\hline
266
\end{array}
$$

The difference is 266 quadrillion Btu.

You Try It 9

$x - y$
$7,061 - 3,229$

$$
\begin{array}{r}
\overset{6\ 10\ 5\ 11}{7,\cancel{0}\cancel{6}\cancel{1}} \\
-3,229 \\
\hline
3,832
\end{array}
$$

Copyright © Houghton Mifflin Company. All rights reserved.

You Try It 10

$$46 = 58 - p$$

$$46 \mid 58 - 11$$

$$46 \neq 47$$

No, 11 is not a solution of the equation.

You Try It 11

Strategy   To find the total number of fatal accidents during 1991 through 1999:

▶ Find the number of fatal accidents each year.

▶ Add the 9 numbers.

Solution

1991: 3	1994: 2	1997: 4
1992: 2	1995: 3	1998: 5
1993: 4	1996: 3	1999: 6

$$3 + 2 + 4 + 2 + 3 + 3 + 4 + 5 + 6 = 32$$

During 1991 through 1999, there were 32 fatal accidents on amusement rides.

You Try It 12

Strategy   To find the price, replace C by 148 and M by 74 in the given formula and solve for P.

Solution

$$P = C + M$$

$$P = 148 + 74$$

$$P = 222$$

The price of the leather jacket is $222.

You Try It 13

Strategy   Draw a diagram.

60 ft

60 ft

To find the length of fencing needed, use the formula for the perimeter of a rectangle, $P = L + W + L + W$. $L = 60$ and $W = 60$.

Solution

$$P = L + W + L + W$$

$$P = 60 + 60 + 60 + 60$$

$$P = 240$$

240 ft of fencing are needed.

Section 1.3 *(pages 41–58)*

You Try It 1   Average monthly savings in France are $175.

$$\begin{array}{r} 175 \\ \times\ 12 \\ \hline 350 \\ 175 \\ \hline 2{,}100 \end{array}$$

The average annual savings of individuals in France is $2,100.

You Try It 2

$$8{,}704 \longrightarrow 9{,}000$$

$$93 \longrightarrow 90$$

$$9{,}000 \cdot 90 = 810{,}000$$

You Try It 3 $5xy$

$$5(20)(60) = 100(60)$$

$$= 6{,}000$$

You Try It 4 $90(7{,}000) = 630{,}000$

You Try It 5 $0 \cdot 10 = 0$

You Try It 6

$$7a = 77$$

$$7 \cdot 11 \mid 77$$

$$77 = 77$$

Yes, 11 is a solution of the equation.

You Try It 7 $2 \cdot 2 \cdot 2 \cdot 3 \cdot 3 \cdot 3 \cdot 3 = 2^3 \cdot 3^4$

You Try It 8 $6^4 = 6 \cdot 6 \cdot 6 \cdot 6 = 36 \cdot 6 \cdot 6$

$$= 216 \cdot 6 = 1{,}296$$

You Try It 9 $10^8 = 100{,}000{,}000$

You Try It 10 $2^4 \cdot 3^2 = (2 \cdot 2 \cdot 2 \cdot 2) \cdot (3 \cdot 3)$

$$= 16 \cdot 9 = 144$$

You Try It 11 $x^4 y^2$

$$1^4 \cdot 3^2 = (1 \cdot 1 \cdot 1 \cdot 1) \cdot (3 \cdot 3)$$

$$= 1 \cdot 9$$

$$= 9$$

You Try It 12

$$\begin{array}{r} 320\ \text{r}14 \\ 24\overline{)7{,}694} \\ -7\,2 \\ \hline 49 \\ -48 \\ \hline 14 \\ -\ 0 \\ \hline 14 \end{array}$$

Check: $(320 \cdot 24) + 14 = 7{,}680 + 14$

$$= 7{,}694$$

You Try It 13   The annual expense for food is $7,200.

$$7{,}200 \div 12 = 600$$

The monthly expense for food is $600.

You Try It 14

$$216{,}936 \longrightarrow 200{,}000$$

$$207 \longrightarrow 200$$

$$200{,}000 \div 200 = 1{,}000$$

You Try It 15

$$\frac{x}{y}$$

$$\frac{672}{8} = 84$$

You Try It 16

$$\frac{60}{y} = 2$$

$$\frac{60}{12} \mid 2$$

$$5 \neq 2$$

No, 12 is not a solution of the equation.

Copyright © Houghton Mifflin Company. All rights reserved.

You Try It 17

$30 \div 1 = 30$
$30 \div 2 = 15$
$30 \div 3 = 10$
$30 \div 4$ Does not divide evenly.
$30 \div 5 = 6$
$30 \div 6 = 5$ The factors are repeating.

The factors of 30 are 1, 2, 3, 5, 6, 10, 15, and 30.

You Try It 18

$$\begin{array}{r} 11 \\ 2\overline{)22} \\ 2\overline{)44} \\ 2\overline{)88} \end{array}$$

$88 = 2 \cdot 2 \cdot 2 \cdot 11 = 2^3 \cdot 11$

You Try It 19

$$\begin{array}{r} 59 \\ 5\overline{)295} \end{array}$$

$295 = 5 \cdot 59$

You Try It 20

Strategy   To find how many times more expensive a stamp was, divide the cost in 1997 (32) by the cost in 1960 (4).

Solution $32 \div 4 = 8$

A stamp was 8 times more expensive in 1997.

You Try It 21

Strategy   Draw a diagram.

 6 m

To find the amount of carpet that should be purchased, use the formula for the area of a square, $A = s^2$, with $s = 6$.

Solution $A = s^2$
$A = 6^2$
$A = 36$

36 m² of carpet should be purchased.

You Try It 22

Strategy   To find the speed, replace d by 486 and t by 9 in the given formula and solve for r.

Solution $r = \dfrac{d}{t}$

$r = \dfrac{486}{9}$

$r = 54$

You would need to travel at a speed of 54 mph.

Section 1.4 *(pages 67–70)*

You Try It 1

$37 = a + 12$
$37 - 12 = a + 12 - 12$
$25 = a + 0$
$25 = a$

Check: $\dfrac{37 = a + 12}{37 \mid 25 + 12}$
$37 = 37$

The solution is 25.

You Try It 2

$3z = 36$
$\dfrac{3z}{3} = \dfrac{36}{3}$
$1z = 12$
$z = 12$

Check: $\dfrac{3z = 36}{3(12) \mid 36}$
$36 = 36$

The solution is 12.

You Try It 3

The unknown number: n

A number increased by four	is	seventeen

$n + 4 = 17$
$n + 4 - 4 = 17 - 4$
$n = 13$

The number is 13.

You Try It 4

Strategy   To find the number of face lifts performed, write and solve an equation using x to represent the number of face lifts performed.

Solution

The number of liposuctions performed	was	220,159 more than the number of face lifts performed

$354{,}015 = x + 220{,}159$
$354{,}015 - 220{,}159 = x + 220{,}159 - 220{,}159$
$133{,}856 = x$

There were 133,856 face lifts performed during the year.

You Try It 5

Strategy   To find the interest earned, replace A by 21,060 and P by 18,000 in the given formula and solve for I.

Solution $A = P + I$
$21{,}060 = 18{,}000 + I$
$21{,}060 - 18{,}000 = 18{,}000 - 18{,}000 + I$
$3{,}060 = I$

The interest earned on the investment is $3,060.

Section 1.5 *(pages 73–74)*

You Try It 1

$4 \cdot (8 - 3) \div 5 - 2 = 4 \cdot 5 \div 5 - 2$
$= 20 \div 5 - 2$
$= 4 - 2$
$= 2$

Copyright © Houghton Mifflin Company. All rights reserved.

You Try It 2

$$16 + 3(6 - 1)^2 \div 5 = 16 + 3(5)^2 \div 5$$
$$= 16 + 3(25) \div 5$$
$$= 16 + 75 \div 5$$
$$= 16 + 15$$
$$= 31$$

You Try It 3

$$(a - b)^2 + 5c$$
$$(7 - 2)^2 + 5(4) = 5^2 + 5(4)$$
$$= 25 + 5(4)$$
$$= 25 + 20$$
$$= 45$$

Solutions to Chapter 2 *You Try Its*

Section 2.1 *(pages 89–94)*

You Try It 1

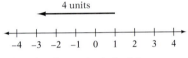

4 units

-3 is 4 units to the left of 1.

You Try It 2

A is -5, and C is -3.

You Try It 3

a. 2 is to the right of -5 on the number line.

   $2 > -5$

b. -4 is to the left of 3 on the number line.

   $-4 < 3$

You Try It 4

$-7, -1, 0, 4, 8$

You Try It 5

a. -24 b. 13   c. b

You Try It 6

a. negative three minus twelve

b. eight plus negative five

You Try It 7

a. $-(-59) = 59$ b. $-(y) = -y$

You Try It 8

a. $|-8| = 8$ b. $|12| = 12$

You Try It 9

a. $|0| = 0$ b. $-|35| = -35$

You Try It 10

$|-y| = |-2| = 2$

You Try It 11

$|6| = 6, |-2| = 2, -(-1) = 1, -|-8| = -8$
$-8, -4, 1, 2, 6$
$-|-8|, -4, -(-1), |-2|, |6|$

You Try It 12

Strategy   To find the player that came in third, find the player with the third lowest number for a score.

Solution $-14 < -12 < -9 < -7 < -6$

The third lowest number among the scores is -9.

Gilder came in third in the tournament.

You Try It 13

Strategy   To determine which is closer to blastoff, find the absolute value of each number. The number with the smaller absolute value is closer to zero and, therefore, closer to blastoff.

Solution $|-9| = 9, |-7| = 7$

$7 < 9$

-7 s and counting is closer to blastoff than -9 s and counting.

Section 2.2 *(pages 101–110)*

You Try It 1 $-38 + (-62) = -100$

You Try It 2 $47 + (-53) = -6$

You Try It 3 $-36 + 17 + (-21) = -19 + (-21)$
$$= -40$$

You Try It 4 $-154 + (-37) = -191$

You Try It 5
$$-x + y$$
$$-(-3) + (-10) = 3 + (-10)$$
$$= -7$$

You Try It 6
$$2 = 11 + a$$

2	$11 + (-9)$

$$2 = 2$$

Yes, -9 is a solution of the equation.

You Try It 7 $-35 - (-34) = -35 + 34$
$$= -1$$

You Try It 8 $83 - (-29) = 83 + 29$
$$= 112$$

You Try It 9   The boiling point of xenon is -108. The melting point of xenon is -112.

$$-108 - (-112) = -108 + 112$$
$$= 4$$

The difference is 4°C.

You Try It 10 $-8 - 14 = -8 + (-14)$
$$= -22$$

You Try It 11 $25 - 68 = 25 + (-68)$
$$= -43$$

You Try It 12
$$-4 - (-3) + 12 - (-7) - 20$$
$$= -4 + 3 + 12 + 7 + (-20)$$
$$= -1 + 12 + 7 + (-20)$$
$$= 11 + 7 + (-20)$$
$$= 18 + (-20)$$
$$= -2$$

You Try It 13
$$x - y$$
$$-9 - 7 = -9 + (-7)$$
$$= -16$$

Copyright © Houghton Mifflin Company. All rights reserved.

You Try It 14

$$a - 5 = -8$$

$$
\begin{array}{c|c}
-3 - 5 & -8 \\
\hline
-3 + (-5) & -8 \\
-8 = -8
\end{array}
$$

Yes, -3 is a solution of the equation.

You Try It 15

Strategy To find the difference, subtract the lowest melting point shown (-259) from the highest melting point shown (181).

Solution
$$181 - (-259) = 181 + 259$$
$$= 440$$

The difference is 440°C.

You Try It 16

Strategy To find the temperature, add the increase (10) to the previous temperature (-3).

Solution $-3 + 10 = 7$

The temperature is 7°C.

You Try It 17

Strategy To find the difference, subtract the lower temperature (-70) from the higher temperature (57).

Solution
$$57 - (-70) = 57 + 70$$
$$= 127$$

The difference between the average temperatures is 127°F.

You Try It 18

Strategy To find d, replace a by -6 and b by 5 in the given formula and solve for d.

Solution
$$d = |a - b|$$
$$d = |-6 - 5|$$
$$d = |-11|$$
$$d = 11$$

The distance between the two points is 11 units.

Section 2.3 *(pages 117–122)*

You Try It 1 $-38(51) = -1,938$

You Try It 2
$$-7(-8)(9)(-2) = 56(9)(-2)$$
$$= 504(-2)$$
$$= -1,008$$

You Try It 3
$-9y$
$$-9(20) = -180$$

You Try It 4

$$12 = -4a$$

$$
\begin{array}{c|c}
12 & -4(-3) \\
\hline
12 = 12
\end{array}
$$

Yes, -3 is a solution of the equation.

You Try It 5 $0 \div (-17) = 0$

You Try It 6 $\dfrac{84}{-6} = -14$

You Try It 7 Any number divided by one is the number.
$$x \div 1 = x$$

You Try It 8 $\dfrac{a}{-b}$

$$\frac{-14}{-(-7)} = \frac{-14}{7} = -2$$

You Try It 9 $\dfrac{-6}{y} = -2$

$$
\begin{array}{c|c}
\dfrac{-6}{-3} & -2 \\
\hline
2 \neq -2
\end{array}
$$

No, -3 is not a solution of the equation.

You Try It 10

Strategy To find the average daily high temperature:
   ▶ Add the seven temperature readings.
   ▶ Divide by 7.

Solution
$$-7 + (-8) + 0 + (-1) + (-6) + (-11) + (-2) = -35$$
$$-35 \div 7 = -5$$

The average daily high temperature was $-5°$.

Section 2.4 *(pages 129–132)*

You Try It 1
$$-12 = x + 12$$
$$-12 - 12 = x + 12 - 12$$
$$-24 = x$$

The solution is -24.

You Try It 2
$$14a = -28$$
$$\frac{14a}{14} = \frac{-28}{14}$$
$$a = -2$$

The solution is -2.

You Try It 3

Strategy To find the number of mothers who gave birth to twins, write and solve an equation using M to represent the unknown number of mothers.

Solution

The number of mothers who gave birth to triplets	was	112,906 less than the number of mothers who gave birth to twins

$$6,742 = M - 112,906$$
$$6,742 + 112,906 = M - 112,906 + 112,906$$
$$119,648 = M$$

119,648 mothers gave birth to twins.

Copyright © Houghton Mifflin Company. All rights reserved.

You Try It 4

Strategy To find the air speed, replace g by 250 and h by 50 in the given formula and solve for a.

Solution
$$g = a - h$$
$$250 = a - 50$$
$$250 + 50 = a - 50 + 50$$
$$300 = a$$

The air speed of the plane is 300 mph.

Section 2.5 *(pages 135–136)*

You Try It 1
$$(-5)^2 = (-5)(-5) = 25$$
$$-5^2 = -(5 \cdot 5) = -25$$

You Try It 2
$$8 \div 4 \cdot 4 - (-2)^2 = 8 \div 4 \cdot 4 - 4$$
$$= 2 \cdot 4 - 4$$
$$= 8 - 4$$
$$= 4$$

You Try It 3
$$(-2)^2(3 - 7)^2 - (-16) \div (-4)$$
$$= (-2)^2(-4)^2 - (-16) \div (-4)$$
$$= (4)(16) - (-16) \div (-4)$$
$$= 64 - (-16) \div (-4)$$
$$= 64 - 4$$
$$= 60$$

You Try It 4
$$3a - 4b$$
$$3(-2) - 4(5) = -6 - 4(5)$$
$$= -6 - 20$$
$$= -6 + (-20)$$
$$= -26$$

Solutions to Chapter 3 *You Try Its*

Section 3.1 *(pages 153–156)*

You Try It 1
$$16 = \boxed{2^4}$$
$$24 = 2^3 \cdot \boxed{3}$$
$$28 = 2^2 \cdot \boxed{7}$$

LCM $= 2^4 \cdot 3 \cdot 7 = 16 \cdot 3 \cdot 7 = 336.$

You Try It 2
$$25 = 5^2$$
$$52 = 2^2 \cdot 13$$

No prime factor occurs in both factorizations.
GCF $= 1$.

You Try It 3
$$32 = 2^5$$
$$40 = \boxed{2^3} \cdot 5$$
$$56 = 2^3 \cdot 7$$

GCF $= 2^3 = 8.$

You Try It 4

Strategy To find the number of CDs to be packaged together, find the GCF of 20, 50, and 100.

Solution
$$20 = 2^2 \cdot \boxed{5}$$
$$50 = \boxed{2} \cdot 5^2$$
$$100 = 2^2 \cdot 5^2$$

GCF $= 2 \cdot 5 = 10.$

Each package should contain 10 CDs.

You Try It 5

Strategy To find how long it will be before both of you are at the starting point again, find the LCM of 3 and 4.

Solution
$$3 = \boxed{3}$$
$$4 = \boxed{2^2}$$

LCM $= 3 \cdot 2^2 = 12.$

In 12 min both of you will be at the starting point again.

You will not have passed each other at some other point on the track prior to that time. It is as if it takes the faster runner 4 laps to "catch up to" the slower runner.

Section 3.2 *(pages 159–166)*

You Try It 1
$$\frac{19}{6} ; 3\frac{1}{6}$$

You Try It 2
$$\begin{array}{r} 8 \\ 3\overline{)26} \\ -24 \\ \hline 2 \end{array} \qquad \frac{26}{3} = 8\frac{2}{3}$$

You Try It 3
$$\begin{array}{r} 9 \\ 4\overline{)36} \\ -36 \\ \hline 0 \end{array} \qquad \frac{36}{4} = 9$$

You Try It 4
$$9\frac{4}{7} = \frac{(7 \cdot 9) + 4}{7} = \frac{63 + 4}{7} = \frac{67}{7}$$

You Try It 5
$$3 = \frac{3}{1}$$

You Try It 6
$$48 \div 8 = 6$$
$$\frac{5}{8} = \frac{5 \cdot 6}{8 \cdot 6} = \frac{30}{48}$$
$$\frac{30}{48} \text{ is equivalent to } \frac{5}{8}.$$

You Try It 7
$$8 = \frac{8}{1} \qquad 12 \div 1 = 12$$
$$8 = \frac{8}{1} = \frac{8 \cdot 12}{1 \cdot 12} = \frac{96}{12}$$
$$\frac{96}{12} \text{ is equivalent to } 8.$$

You Try It 8
$$\frac{21}{84} = \frac{3 \cdot 7}{2 \cdot 2 \cdot 3 \cdot 7} = \frac{1}{4}$$

Copyright © Houghton Mifflin Company. All rights reserved.

You Try It 9 $\dfrac{32}{12} = \dfrac{2 \cdot 2 \cdot 2 \cdot 2 \cdot 2}{2 \cdot 2 \cdot 3} = \dfrac{8}{3}$

You Try It 10 $\dfrac{11t}{11} = \dfrac{11 \cdot t}{11} = t$

You Try It 11   The LCM of 9 and 21 is 63.

$$\dfrac{4}{9} = \dfrac{28}{63} \qquad \dfrac{8}{21} = \dfrac{24}{63}$$

$$\dfrac{28}{63} > \dfrac{24}{63}$$

$$\dfrac{4}{9} > \dfrac{8}{21}$$

You Try It 12   The LCM of 24 and 9 is 72.

$$\dfrac{17}{24} = \dfrac{51}{72} \qquad \dfrac{7}{9} = \dfrac{56}{72}$$

$$\dfrac{51}{72} < \dfrac{56}{72}$$

$$\dfrac{17}{24} < \dfrac{7}{9}$$

You Try It 13

Strategy   To find the fraction:

▶ Add the populations of all the segments to find the total U.S. population.

▶ Write a fraction with the population 65 and older in the numerator and the total population in the denominator. Write the fraction in simplest form.

Solution $19 + 61 + 104 + 62 + 34 = 280$

$$\dfrac{34}{280} = \dfrac{17}{140}$$

$\dfrac{17}{140}$ of the U.S. population 65 and older.

You Try It 14

Strategy   To find the fraction, write a fraction with the amount spent for physicians in the numerator and the number of cents in one dollar (100) in the denominator.

Solution $\dfrac{32}{100} = \dfrac{8}{25}$

$\dfrac{8}{25}$ of every dollar spent for health care is for physicians.

Section 3.3 (pages 173–182)

You Try It 1 $\dfrac{7}{12} + \dfrac{3}{8} = \dfrac{14}{24} + \dfrac{9}{24} = \dfrac{23}{24}$

You Try It 2 $\dfrac{3}{5} + \dfrac{2}{3} + \dfrac{5}{6} = \dfrac{18}{30} + \dfrac{20}{30} + \dfrac{25}{30} = \dfrac{63}{30}$

$$= 2\dfrac{3}{30} = 2\dfrac{1}{10}$$

You Try It 3 $16 + 8\dfrac{5}{9} = 24\dfrac{5}{9}$

You Try It 4 $-\dfrac{5}{12} + \dfrac{5}{8} + \left(-\dfrac{1}{6}\right) = \dfrac{-5}{12} + \dfrac{5}{8} + \dfrac{-1}{6}$

$$= \dfrac{-10}{24} + \dfrac{15}{24} + \dfrac{-4}{24}$$

$$= \dfrac{-10 + 15 + (-4)}{24}$$

$$= \dfrac{1}{24}$$

You Try It 5 $x + y + z$

$$3\dfrac{5}{6} + 2\dfrac{1}{9} + 5\dfrac{5}{12} = 3\dfrac{30}{36} + 2\dfrac{4}{36} + 5\dfrac{15}{36}$$

$$= 10\dfrac{49}{36}$$

$$= 11\dfrac{13}{36}$$

You Try It 6 $-\dfrac{5}{6} - \dfrac{7}{9} = \dfrac{-5}{6} - \dfrac{7}{9}$

$$= \dfrac{-15}{18} - \dfrac{14}{18}$$

$$= \dfrac{-15 - 14}{18}$$

$$= \dfrac{-29}{18}$$

$$= -\dfrac{29}{18} = -1\dfrac{11}{18}$$

You Try It 7 $9\dfrac{7}{8} - 5\dfrac{2}{3} = 9\dfrac{21}{24} - 5\dfrac{16}{24} = 4\dfrac{5}{24}$

You Try It 8 $6 - 4\dfrac{2}{11} = 5\dfrac{11}{11} - 4\dfrac{2}{11} = 1\dfrac{9}{11}$

You Try It 9 $\dfrac{2}{3} - v = \dfrac{11}{12}$

$\dfrac{2}{3} - \left(-\dfrac{1}{4}\right)$	$\dfrac{11}{12}$
$\dfrac{2}{3} + \dfrac{1}{4}$	$\dfrac{11}{12}$
$\dfrac{8}{12} + \dfrac{3}{12}$	$\dfrac{11}{12}$
$\dfrac{11}{12}$ =	$\dfrac{11}{12}$

Yes, $-\dfrac{1}{4}$ is a solution of the equation.

You Try It 10

Strategy   To find the fraction of the respondents that did not name glazed, filled, or frosted:

▶ Add the three fractions to find the fraction that named glazed, filled, or frosted.

Copyright © Houghton Mifflin Company. All rights reserved.

▶ Subtract the fraction that named glazed, filled, or frosted from 1, the entire group surveyed.

Solution

$$\frac{2}{5} + \frac{8}{25} + \frac{3}{20} = \frac{40}{100} + \frac{32}{100} + \frac{15}{100}$$

$$= \frac{87}{100}$$

$$1 - \frac{87}{100} = \frac{100}{100} - \frac{87}{100} = \frac{13}{100}$$

$\frac{13}{100}$ of the respondents did not name glazed, filled, or frosted as their favorite type of doughnut.

You Try It 11

Strategy To find the size of the penny nail needed:

▶ Find the thickness of one board by subtracting $\frac{1}{4}$ in. from the given thickness (1 in.).

▶ Find the thickness of 3 boards.

▶ To the thickness of 3 boards, add $\frac{1}{2}$ in., as we want the nail to extend $\frac{1}{2}$ in. into the fourth board. This is the length of the penny nail needed.

▶ To calculate the size penny nail needed, use the facts that the length of a nail increases by $\frac{1}{4}$ in. for each 1 penny increase in size and that a 4-penny nail is $1\frac{1}{2}$ in. long.

Solution Given thickness of one board minus $\frac{1}{4}$ in. $= 1 - \frac{1}{4} = \frac{3}{4}$

Thickness of 3 boards $= \frac{3}{4} + \frac{3}{4} + \frac{3}{4}$

$$= \frac{9}{4} = 2\frac{1}{4}$$

Length of penny nail needed $= 2\frac{1}{4} + \frac{1}{2}$

$$= 2\frac{1}{4} + \frac{2}{4} = 2\frac{3}{4}$$

A 4-penny nail is $1\frac{1}{2}$ in. long.

A 5-penny nail is $1\frac{1}{2} + \frac{1}{4} = 1\frac{3}{4}$ in. long.

A 6-penny nail is $1\frac{3}{4} + \frac{1}{4} = 2$ in. long.

A 7-penny nail is $2 + \frac{1}{4} = 2\frac{1}{4}$ in. long.

An 8-penny nail is $2\frac{1}{4} + \frac{1}{4} = 2\frac{1}{2}$ in. long.

A 9-penny nail is $2\frac{1}{2} + \frac{1}{4} = 2\frac{3}{4}$ in. long.

A 9-penny nail is needed.

Section 3.4 *(pages 191–200)*

You Try It 1

$$\frac{5}{12} \cdot \frac{9}{35} \cdot \frac{7}{8} = \frac{5 \cdot 9 \cdot 7}{12 \cdot 35 \cdot 8}$$

$$= \frac{5 \cdot 3 \cdot 3 \cdot 7}{2 \cdot 2 \cdot 3 \cdot 5 \cdot 7 \cdot 2 \cdot 2 \cdot 2}$$

$$= \frac{3}{32}$$

You Try It 2

$$\frac{y}{10} \cdot \frac{z}{7} = \frac{y \cdot z}{10 \cdot 7} = \frac{yz}{70}$$

You Try It 3

$$-\frac{1}{3}\left(-\frac{5}{12}\right)\left(\frac{8}{15}\right) = \frac{1}{3} \cdot \frac{5}{12} \cdot \frac{8}{15}$$

$$= \frac{1 \cdot 5 \cdot 8}{3 \cdot 12 \cdot 15}$$

$$= \frac{1 \cdot 5 \cdot 2 \cdot 2 \cdot 2}{3 \cdot 2 \cdot 2 \cdot 3 \cdot 3 \cdot 5}$$

$$= \frac{2}{27}$$

You Try It 4

$$\frac{8}{9} \cdot 6 = \frac{8}{9} \cdot \frac{6}{1} = \frac{8 \cdot 6}{9 \cdot 1}$$

$$= \frac{2 \cdot 2 \cdot 2 \cdot 2 \cdot 3}{3 \cdot 3 \cdot 1} = \frac{16}{3} = 5\frac{1}{3}$$

You Try It 5

$$3\frac{6}{7} \cdot 2\frac{4}{9} = \frac{27}{7} \cdot \frac{22}{9} = \frac{27 \cdot 22}{7 \cdot 9}$$

$$= \frac{3 \cdot 3 \cdot 3 \cdot 2 \cdot 11}{7 \cdot 3 \cdot 3} = \frac{66}{7} = 9\frac{3}{7}$$

You Try It 6 xy

$$5\frac{1}{8} \cdot \frac{2}{3} = \frac{41}{8} \cdot \frac{2}{3}$$

$$= \frac{41 \cdot 2}{8 \cdot 3}$$

$$= \frac{41 \cdot 2}{2 \cdot 2 \cdot 2 \cdot 3}$$

$$= \frac{41}{12} = 3\frac{5}{12}$$

You Try It 7

$$\frac{5}{6} \div \frac{10}{27} = \frac{5}{6} \cdot \frac{27}{10} = \frac{5 \cdot 27}{6 \cdot 10}$$

$$= \frac{5 \cdot 3 \cdot 3 \cdot 3}{2 \cdot 3 \cdot 2 \cdot 5} = \frac{9}{4} = 2\frac{1}{4}$$

You Try It 8

$$\frac{x}{8} \div \frac{y}{6} = \frac{x}{8} \cdot \frac{6}{y}$$

$$= \frac{x \cdot 6}{8 \cdot y} = \frac{x \cdot 2 \cdot 3}{2 \cdot 2 \cdot 2 \cdot y} = \frac{3x}{4y}$$

Copyright © Houghton Mifflin Company. All rights reserved.

You Try It 9

$$4 \div \left(-\frac{6}{7}\right) = -\left(\frac{4}{1} \div \frac{6}{7}\right)$$

$$= -\left(\frac{4}{1} \cdot \frac{7}{6}\right)$$

$$= -\frac{4 \cdot 7}{1 \cdot 6}$$

$$= -\frac{2 \cdot 2 \cdot 7}{1 \cdot 2 \cdot 3} = -\frac{14}{3} = -4\frac{2}{3}$$

You Try It 10

$$4\frac{3}{8} \div 3\frac{1}{2} = \frac{35}{8} \div \frac{7}{2} = \frac{35}{8} \cdot \frac{2}{7} = \frac{35 \cdot 2}{8 \cdot 7}$$

$$= \frac{5 \cdot 7 \cdot 2}{2 \cdot 2 \cdot 2 \cdot 7} = \frac{5}{4} = 1\frac{1}{4}$$

You Try It 11 $x \div y$

$$2\frac{1}{4} \div 9 = \frac{9}{4} \div \frac{9}{1} = \frac{9}{4} \cdot \frac{1}{9} = \frac{9 \cdot 1}{4 \cdot 9}$$

$$= \frac{3 \cdot 3 \cdot 1}{2 \cdot 2 \cdot 3 \cdot 3} = \frac{1}{4}$$

You Try It 12

Strategy   To find the area, use the formula for the area of a triangle, $A = \frac{1}{2}bh$. $b = 18$ and $h = 9$.

Solution $A = \frac{1}{2}bh$

$A = \frac{1}{2}(18)(9)$

$A = 81$

81 in² of felt are needed.

You Try It 13

Strategy   To find the total cost:

▶ Multiply the amount of material per sash $\left(1\frac{3}{8}\right)$ by the number of sashes (22) to find the total number of yards of material needed.

▶ Multiply the total number of yards of material needed by the cost per yard (12).

Solution

$$1\frac{3}{8} \cdot 22 = \frac{11}{8} \cdot \frac{22}{1} = \frac{11 \cdot 22}{8 \cdot 1} = \frac{11 \cdot 2 \cdot 11}{2 \cdot 2 \cdot 2 \cdot 1}$$

$$= \frac{121}{4} = 30\frac{1}{4}$$

$$30\frac{1}{4} \cdot 12 = \frac{121}{4} \cdot \frac{12}{1} = \frac{121 \cdot 12}{4 \cdot 1}$$

$$= \frac{11 \cdot 11 \cdot 2 \cdot 2 \cdot 3}{2 \cdot 2 \cdot 1} = 363$$

The total cost of the material is \$363.

Section 3.5 *(pages 209–212)*

You Try It 1

$$-\frac{1}{5} = z - \frac{5}{6}$$

$$-\frac{1}{5} + \frac{5}{6} = z - \frac{5}{6} + \frac{5}{6}$$

$$-\frac{6}{30} + \frac{25}{30} = z$$

$$\frac{-6 + 25}{30} = z$$

$$\frac{19}{30} = z$$

The solution is $\frac{19}{30}$.

You Try It 2

$$26 = 4x$$

$$\frac{26}{4} = \frac{4x}{4}$$

$$\frac{13}{2} = x$$

$$6\frac{1}{2} = x$$

The solution is $6\frac{1}{2}$.

You Try It 3   The unknown number: x

Negative five-sixths	is equal to	ten-thirds of a number

$$-\frac{5}{6} = \frac{10}{3}x$$

$$\frac{3}{10}\left(-\frac{5}{6}\right) = \frac{3}{10} \cdot \frac{10}{3}x$$

$$-\frac{15}{60} = x$$

$$-\frac{1}{4} = x$$

The number is $-\frac{1}{4}$.

You Try It 4

Strategy   To find the total number of software products sold in January, write and solve an equation using s to represent the number of software products sold in January.

Solution

The number of computer software games sold in January	was	three-fifths of all the software products sold

$$450 = \frac{3}{5}s$$

$$\frac{5}{3} \cdot 450 = \frac{5}{3} \cdot \frac{3}{5}s$$

$$750 = s$$

Copyright © Houghton Mifflin Company. All rights reserved.

BAL Software sold a total of 750 software products in January.

You Try It 5

Strategy To find the total number of points scored, replace A by 73 and N by 5 in the given formula and solve for T.

Solution
$$A = \frac{T}{N}$$

$$73 = \frac{T}{5}$$

$$5 \cdot 73 = 5 \cdot \frac{T}{5}$$

$$365 = T$$

The total number of points scored was 365.

Section 3.6 *(pages 215–220)*

You Try It 1
$$\left(\frac{2}{9}\right)^2 \cdot (-3)^4 = \frac{2}{9} \cdot \frac{2}{9} \cdot (-3)(-3)(-3)(-3)$$

$$= \frac{2}{9} \cdot \frac{2}{9} \cdot 3 \cdot 3 \cdot 3 \cdot 3$$

$$= \frac{2}{9} \cdot \frac{2}{9} \cdot \frac{3}{1} \cdot \frac{3}{1} \cdot \frac{3}{1} \cdot \frac{3}{1}$$

$$= \frac{2 \cdot 2 \cdot 3 \cdot 3 \cdot 3 \cdot 3}{9 \cdot 9 \cdot 1 \cdot 1 \cdot 1 \cdot 1} = 4$$

You Try It 2 $x^4 y^3$
$$\left(2\frac{1}{3}\right)^4 \cdot \left(\frac{3}{7}\right)^3 = \left(\frac{7}{3}\right)^4 \cdot \left(\frac{3}{7}\right)^3$$

$$= \frac{7}{3} \cdot \frac{7}{3} \cdot \frac{7}{3} \cdot \frac{7}{3} \cdot \frac{3}{7} \cdot \frac{3}{7} \cdot \frac{3}{7}$$

$$= \frac{7 \cdot 7 \cdot 7 \cdot 7 \cdot 3 \cdot 3 \cdot 3}{3 \cdot 3 \cdot 3 \cdot 3 \cdot 7 \cdot 7 \cdot 7} = \frac{7}{3} = 2\frac{1}{3}$$

You Try It 3
$$\frac{2y - 3}{y} = -2$$

$\dfrac{2\left(-\dfrac{1}{2}\right) - 3}{-\dfrac{1}{2}}$	-2
$\dfrac{-1 - 3}{-\dfrac{1}{2}}$	-2
$\dfrac{-4}{-\dfrac{1}{2}}$	-2
$-4(-2)$	-2
$8 \neq -2$	

No, $-\dfrac{1}{2}$ is not a solution of the equation.

You Try It 4 $\dfrac{x}{y - z}$

$$\frac{2\dfrac{4}{9}}{3 - 1\dfrac{1}{3}} = \frac{\dfrac{22}{9}}{\dfrac{5}{3}} = \frac{22}{9} \div \frac{5}{3} = \frac{22}{9} \cdot \frac{3}{5}$$

$$= \frac{22}{15} = 1\frac{7}{15}$$

You Try It 5
$$\left(-\frac{1}{2}\right)^3 \cdot \frac{7 - 3}{4 - 9} + \frac{4}{5}$$

$$= \left(-\frac{1}{2}\right)^3 \cdot \frac{4}{-5} + \frac{4}{5}$$

$$= -\frac{1}{8} \cdot \frac{4}{-5} + \frac{4}{5}$$

$$= \frac{1}{10} + \frac{4}{5} = \frac{1}{10} + \frac{8}{10} = \frac{9}{10}$$

Solutions to Chapter 4   *You Try Its*

Section 4.1 *(pages 239–244)*

You Try It 1 The digit 4 is in the thousandths place.

You Try It 2 $\dfrac{501}{1,000} = 0.501$
(five hundred one thousandths)

You Try It 3 $0.67 = \dfrac{67}{100}$ (sixty-seven hundredths)

You Try It 4 fifty-five and six thousand eighty-three ten-thousandths

You Try It 5 806.00491

You Try It 6
$0.065 = 0.0650$
$0.0650 < 0.0802$
$0.065 < 0.0802$

You Try It 7
3.03, 0.33, 0.30, 3.30, 0.03
0.03, 0.30, 0.33, 3.03, 3.30
0.03, 0.3, 0.33, 3.03, 3.3

You Try It 8
Given place value
3.675849
$4 < 5$

3.675849 rounded to the nearest ten-thousandth is 3.6758.

You Try It 9
Given place value
48.907
$0 < 5$

48.907 rounded to the nearest tenth is 48.9.

Copyright © Houghton Mifflin Company. All rights reserved.

You Try It 10

$$\text{Given place value}$$
31.8652

$$8 > 5$$

31.8652 rounded to the nearest whole number is 32.

You Try It 11

Strategy   To determine who had more home runs for every 100 times at bat, compare the numbers 7.03 and 7.09.

Solution   7.09 > 7.03

Ralph Kiner had more home runs for every 100 times at bat.

You Try It 12

Strategy   To determine the average annual precipitation to the nearest inch, round the number 2.65 to the nearest whole number.

Solution   2.65 rounded to the nearest whole number is 3.

To the nearest inch, the average annual precipitation in Yuma is 3 in.

Section 4.2 *(pages 249–264)*

You Try It 1

$$\begin{array}{r} {}^{1\ \ 1}\\ 8.64\\ 52.7\\ +\ 0.39105\\ \hline 61.73105 \end{array}$$

You Try It 2

$$4.002 - 9.378 = 4.002 + (-9.378)$$
$$= -5.376$$

You Try It 3

$$\begin{array}{r} {}^{4\ \ 9\ 10}\\ 2\cancel{5}.\cancel{0}\,\cancel{0}\\ -\ 4.91\\ \hline 20.09 \end{array} \qquad \text{Check:} \qquad \begin{array}{r} 4.91\\ +20.09\\ \hline 25.00 \end{array}$$

You Try It 4

$$\begin{array}{r} 6.514 \longrightarrow\ 7\\ 8.903 \longrightarrow\ 9\\ 2.275 \longrightarrow\ +\ 2\\ \hline 18 \end{array}$$

You Try It 5

$$x + y + z$$
$$-7.84 + (-3.05) + 2.19$$
$$= -10.89 + 2.19$$
$$= -8.7$$

You Try It 6

$$\begin{array}{c|c} -m + 16.9 = 40.7 & \\ \hline -(-23.8) + 16.9 & 40.7\\ 23.8 + 16.9 & 40.7\\ 40.7 = 40.7 \end{array}$$

Yes, −23.8 is a solution of the equation.

You Try It 7

$$\begin{array}{r} 0.000081\\ \times\ \ \ \ 0.025\\ \hline 405\\ 162\\ \hline 0.000002025 \end{array}$$

You Try It 8

$$\begin{array}{r} 6.407 \longrightarrow\ \ \ 6\\ 0.959 \longrightarrow\ \times 1\\ \hline 6 \end{array}$$

You Try It 9

$$1.756 \cdot 10^4 = 17{,}560$$

You Try It 10

$$(-0.7)(-5.8) = 4.06$$

You Try It 11

$$25xy$$
$$25(-0.8)(0.6) = -20(0.6) = -12$$

You Try It 12

$$\begin{array}{r} 48.2\\ 6.53.\overline{)314.74.6}\\ -\ 261\ 2\\ \hline 53\ 54\\ -\ 52\ 24\\ \hline 1\ 30\ 6\\ -\ 1\ 30\ 6\\ \hline 0 \end{array}$$

You Try It 13

$$62.7 \longrightarrow\ 60$$
$$3.45 \longrightarrow\ 3$$
$$60 \div 3 = 20$$

You Try It 14

$$\begin{array}{r} 6.0391 \approx 6.039\\ 86\overline{)519.3700}\\ -\ 516\\ \hline 3\ 3\\ -\ \ 0\\ \hline 3\ 37\\ -\ 2\ 58\\ \hline 790\\ -\ 774\\ \hline 160\\ -\ 86\\ \hline 74 \end{array}$$

You Try It 15

$$63.7 \div 100 = 0.637$$

You Try It 16

The quotient is negative.
$$-25.7 \div 0.31 \approx -82.9$$

You Try It 17

$$\dfrac{x}{y}$$
$$\dfrac{-40.6}{-0.7} = 58$$

You Try It 18

$$-2 = \dfrac{d}{-0.6}$$

$$\begin{array}{c|c} & \dfrac{-1.2}{-0.6}\\ -2 & \end{array}$$

$$-2 \neq 2$$

No, −1.2 is not a solution of the equation.

You Try It 19

$$\begin{array}{r} 0.8\\ 5\overline{)4.0} \end{array} \qquad \dfrac{4}{5} = 0.8$$

You Try It 20

$$\begin{array}{r} 0.8333\\ 6\overline{)5.0000} \end{array} \qquad 1\dfrac{5}{6} = 1.8\overline{3}$$

Copyright © Houghton Mifflin Company. All rights reserved.

You Try It 21 $6.2 = 6\dfrac{2}{10} = 6\dfrac{1}{5}$

You Try It 22 $\dfrac{7}{12} \approx 0.5833$

$0.5880 > 0.5833$

$0.588 > \dfrac{7}{12}$

You Try It 23

Strategy   To find the change you receive:
- ▶ Multiply the number of stamps (12) by the cost of each stamp (37¢) to find the total cost of the stamps.
- ▶ Convert the total cost of the stamps to dollars and cents.
- ▶ Subtract the total cost of the stamps from $10.

Solution $12(37) = 444$ The stamps cost 444¢.
444¢ = \$4.44   The stamps cost \$4.44.
$10.00 - 4.44 = \$5.56$
You receive \$5.56 in change.

You Try It 24

Strategy   To make the comparison:
- ▶ Divide the number of hearing impaired who are aged 65–74 (5.41 million) by the number who are aged 0–17 (1.37 million).
- ▶ Compare the quotient to the number 4.

Solution $5.41 \div 1.37 \approx 3.9$
$3.9 < 4$

The number of hearing-impaired individuals who are aged 65–74 is less than 4 times the number of hearing impaired who are aged 0–17.

You Try It 25

Strategy   To find the profit:
- ▶ Divide the number of pounds per 100-pound container (100) by the number of pounds packaged in each bag (2) to find the number of bags sold.
- ▶ Multiply the number of bags sold by the selling price per bag (12.50) to find the income from selling the nuts.
- ▶ Multiply the number of bags sold by the cost for each bag (.06) to find the total cost of the bags.
- ▶ Subtract the cost of the bags and the cost of the nuts (475) from the income.

Solution $100 \div 2 = 50$ Each container makes 50 bags of nuts.

$50(12.50) = 625$ The income from the 50 bags is \$625.

$50(.06) = 3$ The total cost of the bags is \$3.

$625 - 3 - 475 = 147$
The profit is \$147.

You Try It 26

Strategy   To find the insurance premium due, replace B by 276.25 and F by 1.8 in the given formula and solve for P.

Solution $P = BF$
$P = 276.25(1.8)$
$P = 497.25$

The insurance premium due is \$497.25.

Section 4.3 *(pages 277–278)*

You Try It 1 $a - 1.23 = -6$
$a - 1.23 + 1.23 = -6 + 1.23$
$a = -4.77$

The solution is -4.77.

You Try It 2 $-2.13 = -0.71c$
$\dfrac{-2.13}{-0.71} = \dfrac{-0.71c}{-0.71}$
$3 = c$

The solution is 3.

You Try It 3

Strategy   To find the assets, replace N by 24.3 and L by 17.9 in the given formula and solve for A.

Solution $N = A - L$
$24.3 = A - 17.9$
$24.3 + 17.9 = A - 17.9 + 17.9$
$42.2 = A$

The assets of the business are \$42.2 billion.

You Try It 4

Strategy   To find the markup, write and solve an equation using M to represent the amount of the markup.

Solution

The selling price	is	the sum of the amount paid by the store and the amount of the markup

$295.50 = 223.75 + M$
$295.50 - 223.75 = 223.75 - 223.75 + M$
$71.75 = M$

The markup is \$71.75.

Section 4.4 *(pages 281–286)*

You Try It 1   Since $12^2 = 144$, $-\sqrt{144} = -12$.

You Try It 2   Since $\left(\dfrac{9}{10}\right)^2 = \dfrac{81}{100}$, $\sqrt{\dfrac{81}{100}} = \dfrac{9}{10}$.

You Try It 3 $4\sqrt{16} - \sqrt{9} = 4 \cdot 4 - 3$
$= 16 - 3 = 13$

You Try It 4 $5\sqrt{a + b}$
$5\sqrt{17 + 19} = 5\sqrt{36}$
$= 5 \cdot 6$
$= 30$

Copyright © Houghton Mifflin Company. All rights reserved.

You Try It 5 $5\sqrt{23} \approx 23.9792$

You Try It 6   57 is between the perfect squares 49 and 64.
$$\sqrt{49} = 7 \quad \text{and} \quad \sqrt{64} = 8$$
$$7 < \sqrt{57} < 8$$

You Try It 7 $9^2 = 81$; 81 is too big.
$8^2 = 64$; 64 is not a factor of 80.
$7^2 = 49$; 49 is not a factor of 80.
$6^2 = 36$; 36 is not a factor of 80.
$5^2 = 25$; 25 is not a factor of 80.
$4^2 = 16$; 16 is a factor of 80. $(80 = 16 \cdot 5)$
$$\sqrt{80} = \sqrt{16 \cdot 5} = \sqrt{16} \cdot \sqrt{5}$$
$$= 4 \cdot \sqrt{5} = 4\sqrt{5}$$

You Try It 8

Strategy   To find the range, replace h by 6 in the given formula and solve for R.

Solution $R = 1.4\sqrt{h}$
$R = 1.4\sqrt{6}$
$R \approx 3.43$

The range of the periscope is 3.43 mi.

Section 4.5 *(pages 291–296)*

You Try It 1

You Try It 2

You Try It 3

You Try It 4   **a.** $x \geq 4$
$-1 \geq 4$ False
b. $x \geq 4$
$0 \geq 4$ False
c. $x \geq 4$
$4 \geq 4$ True
d. $x \geq 4$
$\sqrt{26} \geq 4$ True

The numbers 4 and $\sqrt{26}$ make the inequality true.

You Try It 5   All real numbers greater than -7 make the inequality $x > -7$ true.

You Try It 6

You Try It 7

Strategy   ▶ To write the inequality, let s represent the speeds at which a motorist is ticketed. Motorists are ticketed at speeds greater than 55.

▶ To determine whether a motorist traveling at 58 mph will be ticketed, replace s in the inequality by 58. If the inequality is true, the motorist will be ticketed. If the inequality is false, the motorist will not be ticketed.

Solution $s > 55$
$58 > 55$ True

Yes, a motorist traveling at 58 mph will be ticketed.

Solutions to Chapter 5   *You Try Its*

Section 5.1 *(pages 315–320)*

You Try It 1 $-6(-3p) = [-6(-3)]p = 18p$

You Try It 2 $(-2m)(-8n) = [(-2)(-8)](m \cdot n)$
$$= 16mn$$

You Try It 3 $(-12)(-d) = (-12)(-1d)$
$$= [(-12)(-1)]d$$
$$= 12d$$

You Try It 4 $6n + 9 + (-6n) = 6n + (-6n) + 9$
$$= [6n + (-6n)] + 9$$
$$= 0 + 9$$
$$= 9$$

You Try It 5 $-7(2k - 5) = -7(2k) - (-7)(5)$
$$= -14k + 35$$

You Try It 6 $-4(x - 2y) = -4(x) - (-4)(2y)$
$$= -4x + 8y$$

You Try It 7 $3(-2v + 3w - 7) = 3(-2v) + 3(3w) - 3(7)$
$$= -6v + 9w - 21$$

You Try It 8 $-4(2x - 7y - z)$
$$= -4(2x) - (-4)(7y) - (-4)(z)$$
$$= -8x + 28y + 4z$$

You Try It 9 $-(c - 9d + 1) = -c + 9d - 1$

Section 5.2 *(pages 325–328)*

You Try It 1 $\dfrac{x}{5} + \dfrac{2x}{5} = \dfrac{x + 2x}{5} = \dfrac{1x + 2x}{5} = \dfrac{(1 + 2)x}{5} = \dfrac{3x}{5}$

You Try It 2 $12a^2 - 8a + 3 - 16a^2 + 8a$
$$= 12a^2 - 16a^2 - 8a + 8a + 3$$
$$= -4a^2 + 0a + 3$$
$$= -4a^2 + 3$$

You Try It 3 $-7x^2 + 4xy + 8x^2 - 12xy$
$$= -7x^2 + 8x^2 + 4xy - 12xy$$
$$= x^2 - 8xy$$

You Try It 4 $-2r + 7s - 12 - 8r + s + 8$
$$= -2r - 8r + 7s + s - 12 + 8$$
$$= -10r + 8s - 4$$

You Try It 5 $8x^2y - 15xy^2 + 12xy^2 - 7x^2y$
$$= 8x^2y - 7x^2y - 15xy^2 + 12xy^2$$
$$= x^2y - 3xy^2$$

You Try It 6 $6 - 4(2x - y) + 3(x - 4y)$
$$= 6 - 8x + 4y + 3x - 12y$$
$$= -5x - 8y + 6$$

Copyright © Houghton Mifflin Company. All rights reserved.

You Try It 7 $8c - 4(3c - 8) - 5(c + 4)$
$= 8c - 12c + 32 - 5c - 20$
$= -9c + 12$

You Try It 8 $6p + 5[3(2 - 3p) - 2(5 - 4p)]$
$= 6p + 5[6 - 9p - 10 + 8p]$
$= 6p + 5[-p - 4]$
$= 6p - 5p - 20$
$= p - 20$

Section 5.3 *(pages 333–336)*

You Try It 1
$(-4x^3 + 2x^2 - 8) + (4x^3 + 6x^2 - 7x + 5)$
$= (-4x^3 + 4x^3) + (2x^2 + 6x^2) - 7x + (-8 + 5)$
$= 8x^2 - 7x - 3$

You Try It 2
$$\begin{array}{r} 6x^3 \quad\quad + 2x + 8 \\ -9x^3 + 2x^2 - 12x - 8 \\ \hline -3x^3 + 2x^2 - 10x \end{array}$$

You Try It 3 $(6a^4 - 5a^2 + 7) + (8a^4 + 3a^2 - 1)$
$= (6a^4 + 8a^4) + (-5a^2 + 3a^2) + (7 - 1)$
$= 14a^4 - 2a^2 + 6$

You Try It 4
The opposite of $3w^3 - 4w^2 + 2w - 1$ is
$-3w^3 + 4w^2 - 2w + 1$.

Add the opposite of $3w^3 - 4w^2 + 2w - 1$ to the first polynomial.

$(-4w^3 + 8w - 8) - (3w^3 - 4w^2 + 2w - 1)$
$= (-4w^3 + 8w - 8) + (-3w^3 + 4w^2 - 2w + 1)$
$= (-4w^3 - 3w^3) + 4w^2 + (8w - 2w) + (-8 + 1)$
$= -7w^3 + 4w^2 + 6w - 7$

You Try It 5
$$\begin{array}{r} 13y^3 \quad\quad - 6y - 7 \\ - 4y^2 + 6y + 9 \\ \hline 13y^3 - 4y^2 \quad\quad + 2 \end{array}$$

You Try It 6 $(-6n^4 + 5n^2 - 10) - (4n^2 + 2)$
$= (-6n^4 + 5n^2 - 10) + (-4n^2 - 2)$
$= -6n^4 + n^2 - 12$

You Try It 7 $(5y^2 - y) + (7y^2 + 4) = (5y^2 + 7y^2) - y + 4$
$= 12y^2 - y + 4$

The distance from Dover to Farley is
$(12y^2 - y + 4)$ miles.

Section 5.4 *(pages 341–344)*

You Try It 1 $(-7a^4)(4a^2) = [-7(4)](a^4 \cdot a^2)$
$= -28a^{4+2}$
$= -28a^6$

You Try It 2 $(8m^3n)(-3n^5) = [8(-3)](m^3)(n \cdot n^5)$
$= -24m^3n^{1+5}$
$= -24m^3n^6$

You Try It 3 $(12p^4q^3)(-3p^5q^2) = [12(-3)](p^4 \cdot p^5)(q^3 \cdot q^2)$
$= -36p^{4+5}q^{3+2}$
$= -36p^9q^5$

You Try It 4 $(-y^4)^5 = [(-1)y^4]^5$
$= (-1)^{1\cdot5}y^{4\cdot5}$
$= (-1)^5y^{20}$
$= -1y^{20}$
$= -y^{20}$

You Try It 5 $(-3a^4bc^2)^3 = (-3)^{1\cdot3}a^{4\cdot3}b^{1\cdot3}c^{2\cdot3}$
$= (-3)^3a^{12}b^3c^6 = -27a^{12}b^3c^6$

Section 5.5 *(pages 347–348)*

You Try It 1 $-3a(-6a + 5b) = (-3a)(-6a) + (-3a)(5b)$
$= 18a^2 - 15ab$

You Try It 2 $3mn^2(2m^2 - 3mn - 1)$
$= (3mn^2)(2m^2) - (3mn^2)(3mn) - (3mn^2)1$
$= 6m^3n^2 - 9m^2n^3 - 3mn^2$

You Try It 3 $(3c + 7)(3c - 7)$
$= (3c)(3c) + (3c)(-7) + 7(3c) + (7)(-7)$
$= 9c^2 - 21c + 21c - 49$
$= 9c^2 - 49$

Section 5.6 *(pages 351–354)*

You Try It 1 $\dfrac{1}{d^{-6}} = d^6$

You Try It 2 $4^{-2} = \dfrac{1}{4^2} = \dfrac{1}{16}$

You Try It 3 $\dfrac{n^6}{n^{11}} = n^{6-11} = n^{-5} = \dfrac{1}{n^5}$

You Try It 4 $0.000000961 = 9.61 \times 10^{-7}$

You Try It 5 $7.329 \times 10^6 = 7,329,000$

Section 5.7 *(pages 357–360)*

You Try It 1   twice x divided by the difference between x and 7

$\dfrac{2x}{x - 7}$

You Try It 2   the product of negative three and the square of d

$-3d^2$

You Try It 3   The smaller number is x.

The larger number is $16 - x$.

the difference between the larger number and twice the smaller number

$(16 - x) - 2x$
$-3x + 16$

Copyright © Houghton Mifflin Company. All rights reserved.

You Try It 4   the <u>difference</u> between fourteen and the <u>sum</u> of a number and seven

Let the unknown number be x.

$14 - (x + 7)$

$14 - x - 7$

$-x + 7$

You Try It 5   the pounds of caramel: c

the pounds of milk chocolate: $c + 3$

Solutions to Chapter 6   *You Try Its*

Section 6.1 *(pages 377–382)*

You Try It 1
$$7 + y = 12$$
$$7 - 7 + y = 12 - 7$$
$$y = 5$$

The solution is 5.

You Try It 2
$$b - \frac{3}{8} = \frac{1}{2}$$
$$b - \frac{3}{8} + \frac{3}{8} = \frac{1}{2} + \frac{3}{8}$$
$$b = \frac{4}{8} + \frac{3}{8}$$
$$b = \frac{7}{8}$$

The solution is $\frac{7}{8}$.

You Try It 3
$$-5r + 3 + 6r = 1$$
$$r + 3 = 1 \quad \blacktriangleright \text{ Combine like terms.}$$
$$r + 3 - 3 = 1 - 3$$
$$r = -2$$

The solution is -2.

You Try It 4
$$-60 = 5d$$
$$\frac{-60}{5} = \frac{5d}{5}$$
$$-12 = d$$

The solution is -12.

You Try It 5
$$10 = \frac{-2x}{5}$$
$$\left(-\frac{5}{2}\right)10 = \left(-\frac{5}{2}\right)\left(-\frac{2}{5}x\right) \quad \blacktriangleright \; -\frac{2x}{5} = -\frac{2}{5}x$$
$$-25 = x$$

The solution is -25.

You Try It 6
$$\frac{1}{3}x - \frac{5}{6}x = 4$$
$$\frac{2}{6}x - \frac{5}{6}x = 4$$
$$-\frac{1}{2}x = 4 \quad \blacktriangleright \; -\frac{3}{6} = -\frac{1}{2}$$
$$-2\left(-\frac{1}{2}x\right) = -2(4)$$
$$x = -8$$

Check:
$$\frac{1}{3}x - \frac{5}{6}x = 4$$

$$\begin{array}{c|c} \frac{1}{3}(-8) - \frac{5}{6}(-8) & 4 \\[6pt] -\frac{8}{3} - \left(-\frac{20}{3}\right) & 4 \\[6pt] -\frac{8}{3} + \frac{20}{3} & 4 \\[6pt] 4 = 4 \end{array}$$

-8 checks as the solution.

The solution is -8.

Section 6.2 *(pages 385–386)*

You Try It 1
$$5v + 3 - 9v = 9$$
$$-4v + 3 = 9 \quad \blacktriangleright \text{ Combine like terms.}$$
$$-4v + 3 - 3 = 9 - 3$$
$$-4v = 6$$
$$\frac{-4v}{-4} = \frac{6}{-4}$$
$$v = -\frac{3}{2}$$

The solution is $-\frac{3}{2}$.

You Try It 2

Strategy   To find the pressure, replace P by its value and solve for D. $P = 45$.

Solution
$$P = 15 + \frac{1}{2}D$$
$$45 = 15 + \frac{1}{2}D$$
$$45 - 15 = 15 - 15 + \frac{1}{2}D$$
$$30 = \frac{1}{2}D$$
$$2(30) = 2\left(\frac{1}{2}D\right)$$
$$60 = D$$

When the pressure is 45 pounds per square inch, the depth is 60 ft.

Section 6.3 *(pages 391–394)*

You Try It 1
$$r - 7 = 5 - 3r$$
$$r + 3r - 7 = 5 - 3r + 3r$$
$$4r - 7 = 5$$
$$4r - 7 + 7 = 5 + 7$$
$$4r = 12$$
$$\frac{4r}{4} = \frac{12}{4}$$
$$r = 3$$

The solution is 3.

Copyright © Houghton Mifflin Company. All rights reserved.

You Try It 2

$$4a - 2 + 5a = 2a - 2 + 3a$$
$$9a - 2 = 5a - 2$$
$$9a - 5a - 2 = 5a - 5a - 2$$
$$4a - 2 = -2$$
$$4a - 2 + 2 = -2 + 2$$
$$4a = 0$$
$$\frac{4a}{4} = \frac{0}{4}$$
$$a = 0$$

The solution is 0.

You Try It 3

$$6 - 5(3y + 2) = 26$$
$$6 - 15y - 10 = 26$$
$$-15y - 4 = 26$$
$$-15y - 4 + 4 = 26 + 4$$
$$-15y = 30$$
$$\frac{-15y}{-15} = \frac{30}{-15}$$
$$y = -2$$

The solution is −2.

You Try It 4

$$2w - 7(3w + 1) = 5(5 - 3w)$$
$$2w - 21w - 7 = 25 - 15w$$
$$-19w - 7 = 25 - 15w$$
$$-19w + 15w - 7 = 25 - 15w + 15w$$
$$-4w - 7 = 25$$
$$-4w - 7 + 7 = 25 + 7$$
$$-4w = 32$$
$$\frac{-4w}{-4} = \frac{32}{-4}$$
$$w = -8$$

The solution is −8.

You Try It 5

Strategy To find the location of the fulcrum when the system balances, replace the variables, F_1, F_2, and d in the lever system equation by the given values and solve for x. $F_1 = 45$, $F_2 = 80$, $d = 25$.

Solution
$$F_1x = F_2(d - x)$$
$$45x = 80(25 - x)$$
$$45x = 2000 - 80x$$
$$45x + 80x = 2000 - 80x + 80x$$
$$125x = 2000$$
$$\frac{125x}{125} = \frac{2000}{125}$$
$$x = 16$$

The fulcrum is 16 ft from the 45-pound force.

Section 6.4 (pages 399–402)

You Try It 1

The unknown number: x

| Six more than one-half a number | is | the total of the number and nine |

$$\frac{1}{2}x + 6 = x + 9$$
$$\frac{1}{2}x - x + 6 = x - x + 9$$

$$-\frac{1}{2}x + 6 = 9$$
$$-\frac{1}{2}x + 6 - 6 = 9 - 6$$
$$-\frac{1}{2}x = 3$$
$$(-2)\left(-\frac{1}{2}x\right) = (-2)3$$
$$x = -6$$

−6 checks as the solution. The solution is −6.

You Try It 2

The unknown number: x

| Seven less than a number | is equal to | five more than three times the number |

$$x - 7 = 3x + 5$$
$$x - 3x - 7 = 3x - 3x + 5$$
$$-2x - 7 = 5$$
$$-2x - 7 + 7 = 5 + 7$$
$$-2x = 12$$
$$\frac{-2x}{-2} = \frac{12}{-2}$$
$$x = -6$$

−6 checks as the solution.

The solution is −6.

You Try It 3

The smaller number: n
The larger number: $14 - n$

| One more than three times the smaller number | equals | the sum of the larger number and three |

$$3n + 1 = (14 - n) + 3$$
$$3n + 1 = 17 - n$$
$$3n + n + 1 = 17 - n + n$$
$$4n + 1 = 17$$
$$4n + 1 - 1 = 17 - 1$$
$$4n = 16$$
$$\frac{4n}{4} = \frac{16}{4}$$
$$n = 4$$

$$14 - n = 14 - 4 = 10$$

These numbers check as solutions.

The smaller number is 4.
The larger number is 10.

You Try It 4

Strategy To find how many miles Americans put on their cars in 1980, write and solve an equation using m to represent the unknown number of miles.

Solution

| 11,988 | was | 3,175 miles more than they put on their cars in 1980 |

Copyright © Houghton Mifflin Company. All rights reserved.

$$11{,}988 = m + 3{,}175$$
$$11{,}988 - 3{,}175 = m + 3{,}175 - 3{,}175$$
$$8{,}813 = m$$

On average, Americans put 8,813 miles on their cars in 1980.

You Try It 5

Strategy To find the number of tickets that you are purchasing, write and solve an equation using x to represent the number of tickets purchased.

Solution

| $9.50 plus $57.50 for each ticket | equals | $527 |

$$9.50 + 57.50x = 527$$
$$9.50 - 9.50 + 57.50x = 527 - 9.50$$
$$57.50x = 517.50$$
$$\frac{57.50x}{57.50} = \frac{517.50}{57.50}$$
$$x = 9$$

You are purchasing 9 tickets.

You Try It 6

Strategy To find the number of days, write and solve an equation using d to represent the number of days.

Solution

| 10 + 5.50 per day | equals | 43 |

$$10 + 5.50d = 43$$
$$10 - 10 + 5.50d = 43 - 10$$
$$5.50d = 33$$
$$\frac{5.50d}{5.50} = \frac{33}{5.50}$$
$$d = 6$$

The customer posted the ad for 6 days.

You Try It 7

Strategy To find the length of each piece, write and solve an equation using x to represent the length of the shorter piece and $18 - x$ to represent the length of the longer piece.

Solution

| 1 ft more than twice the shorter piece | is | 2 ft less than the longer piece |

$$2x + 1 = (18 - x) - 2$$
$$2x + 1 = 16 - x$$
$$2x + x + 1 = 16 - x + x$$
$$3x + 1 = 16$$
$$3x + 1 - 1 = 16 - 1$$
$$3x = 15$$
$$\frac{3x}{3} = \frac{15}{3}$$
$$x = 5$$
$$18 - x = 18 - 5 = 13$$

The length of the shorter piece is 5 ft, and the length of the longer piece is 13 ft.

Section 6.5 *(pages 407–410)*

You Try It 1

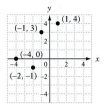

You Try It 2 $A(4, 2)$
$B(-3, 4)$
$C(-3, 0)$
$D(0, 0)$

You Try It 3

Strategy Graph the ordered pairs on a rectangular coordinate system where the horizontal axis represents the number of yards gained and the vertical axis represents the number of points scored.

Solution

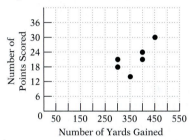

You Try It 4 Locate 20 in. on the x-axis. Follow the vertical line from 20 to a point plotted in the diagram. Follow a horizontal line from that point to the y-axis. Read the number where that line intersects the y-axis.

The ordered pair is $(20, 37)$, which indicates that when the space between seats was 20 in., the evacuation time was 37 s.

Section 6.6 *(pages 415–418)*

You Try It 1 Replace x by -2 and y by 4.

$$y = -\frac{1}{2}x + 3$$

4	$\left(-\frac{1}{2}\right)(-2) + 3$
4	$1 + 3$
$4 = 4$	

Yes, $(-2, 4)$ is a solution of the equation $y = -\frac{1}{2}x + 3$.

You Try It 2 $y = 2x - 3$
$y = 2(0) - 3$ ▶ Replace x by 0.
$y = 0 - 3$
$y = -3$

The ordered-pair solution is $(0, -3)$.

Copyright © Houghton Mifflin Company. All rights reserved.

You Try It 3 $y = 3x + 1$

x	y
0	1
-1	-2
1	4

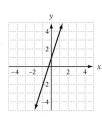

You Try It 4 Locate 5 on the y-axis. Follow the horizontal line from 5 to a point plotted on the graph. Follow a vertical line from that point to the x-axis. Read the number where that line intersects the x-axis.

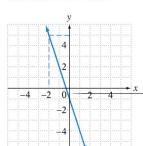

The x value is -2.

Solutions to Chapter 7 *You Try Its*

Section 7.1 *(pages 437–440)*

You Try It 1 The gram is the basic unit for measuring mass.

The amount of protein in a glass of milk is measured in grams.

You Try It 2
a. $1\,295\text{ m} = 1.295\text{ km}$
b. $7\,543\text{ g} = 7.543\text{ kg}$
c. $6.3\text{ L} = 6\,300\text{ ml}$
d. $2\text{ kl} = 2\,000\text{ L}$

You Try It 3

Strategy To find the number of grams of cholesterol in one dozen eggs:

▶ Multiply the amount of cholesterol in one egg (274 mg) by number of eggs (12). This will be the amount of cholesterol in milligrams.

▶ Convert milligrams to grams.

Solution $274(12) = 3{,}288$

$3\,288\text{ mg} = 3.288\text{ g}$

One dozen eggs contains 3.288 g of cholesterol.

Section 7.2 *(pages 445–446)*

You Try It 1 $\dfrac{12}{20} = \dfrac{3}{5}$

$12:20 = 3:5$

12 to 20 = 3 to 5

You Try It 2 $\dfrac{20\text{ bags}}{8\text{ acres}} = \dfrac{5\text{ bags}}{2\text{ acres}}$

You Try It 3 $\dfrac{\$8.96}{3.5\text{ lb}}$

$8.96 \div 3.5 = 2.56$

The unit rate is $2.56/lb.

Section 7.3 *(pages 449–454)*

You Try It 1 The equivalence is 1 ft = 12 in.

The conversion rate must have ft in the numerator and in. in the denominator: $\dfrac{1\text{ ft}}{12\text{ in.}}$

$$40\text{ in.} = 40\text{ in.} \cdot 1 = \frac{40\text{ in.}}{1} \cdot \frac{1\text{ ft}}{12\text{ in.}}$$
$$= \frac{40\text{ in.} \cdot 1\text{ ft}}{12\text{ in.}}$$
$$= \frac{40\text{ ft}}{12} = 3\frac{1}{3}\text{ ft}$$

You Try It 2 The equivalence is 1 yd = 36 in.

The conversion rate must have in. in the numerator and yd in the denominator: $\dfrac{36\text{ in.}}{1\text{ yd}}$

$$2\frac{1}{2}\text{ yd} = 2\frac{1}{2}\text{ yd} \cdot 1 = \frac{5}{2}\text{ yd} \cdot \frac{36\text{ in.}}{1\text{ yd}}$$
$$= \frac{5\text{ yd}}{2} \cdot \frac{36\text{ in.}}{1\text{ yd}}$$
$$= \frac{5\text{ yd} \cdot 36\text{ in.}}{2 \cdot 1\text{ yd}}$$
$$= \frac{5 \cdot 36\text{ in.}}{2}$$
$$= 90\text{ in.}$$

You Try It 3 We need to convert a mile to feet and feet to yards. The equivalences are 1 mi = 5,280 ft and 1 yd = 3 ft.

Choose the conversion rates so that we can divide by the unit "miles" and by the unit "feet."

$$1\text{ mi} = 1\text{ mi} \cdot 1 \cdot 1$$
$$= \frac{1\text{ mi}}{1} \cdot \frac{5{,}280\text{ ft}}{1\text{ mi}} \cdot \frac{1\text{ yd}}{3\text{ ft}}$$
$$= \frac{1\text{ mi} \cdot 5{,}280\text{ ft} \cdot 1\text{ yd}}{1 \cdot 1\text{ mi} \cdot 3\text{ ft}}$$
$$= \frac{5{,}280\text{ yd}}{3} = 1{,}760\text{ yd}$$

You Try It 4

Strategy To find the number of gallons:

▶ Use the formula $V = LWH$ to find the volume in cubic inches.

▶ Use the conversion factor $\dfrac{1\text{ gal}}{231\text{ in}^3}$ to convert cubic inches to gallons.

Copyright © Houghton Mifflin Company. All rights reserved.

Solution

$$V = LWH = 36 \text{ in.} \cdot 24 \text{ in.} \cdot 16 \text{ in.} = 13{,}824 \text{ in}^3$$

$$13{,}824 \text{ in}^3 = \frac{13{,}824 \text{ in}^3}{1} \cdot \frac{1 \text{ gal}}{231 \text{ in}^3} \approx 59.8 \text{ gal}$$

The fishtank holds 59.8 gal of water.

You Try It 5

$$45 \text{ cm} = \frac{45 \text{ cm}}{1}$$

$$\approx \frac{45 \text{ cm}}{1} \cdot \frac{1 \text{ in.}}{2.54 \text{ cm}}$$

$$\approx 17.72 \text{ in.}$$

You Try It 6

$$75 \text{ km per hour} = \frac{75 \text{ km}}{1 \text{ h}}$$

$$\approx \frac{75 \text{ km}}{1 \text{ h}} \cdot \frac{1 \text{ mi}}{1.61 \text{ km}}$$

$$\approx \frac{46.58 \text{ mi}}{\text{h}}$$

75 km per hour is approximately 46.58 mph.

You Try It 7

$$\$3.59 \text{ per gallon} = \frac{\$3.59}{\text{gal}}$$

$$\approx \frac{\$3.59}{\text{gal}} \cdot \frac{1 \text{ gal}}{3.79 \text{ L}}$$

$$\approx \$.95 \text{ per liter}$$

The price is approximately $.95 per liter.

You Try It 8

$$\$2.65 \text{ per liter} = \frac{\$2.65}{1 \text{ L}}$$

$$\approx \frac{\$2.65}{1 \text{ L}} \cdot \frac{3.79 \text{ L}}{1 \text{ gal}}$$

$$\approx \$10.04 \text{ per gal}$$

The price is approximately $10.04 per gallon.

Section 7.4 *(pages 459–462)*

You Try It 1

$$\frac{50}{3} \diagdown \frac{250}{12} \longrightarrow 3 \cdot 250 = 750$$
$$\longrightarrow 50 \cdot 12 = 600$$

$$750 \neq 600$$

The proportion is not true.

You Try It 2

$$\frac{7}{12} = \frac{42}{x}$$

$$12 \cdot 42 = 7 \cdot x$$

$$504 = 7x$$

$$72 = x$$

You Try It 3

$$\frac{5}{n} = \frac{3}{322}$$

$$n \cdot 3 = 5 \cdot 322$$

$$3n = 1{,}610$$

$$\frac{3n}{3} = \frac{1{,}610}{3}$$

$$n \approx 536.67$$

You Try It 4

$$\frac{4}{5} = \frac{3}{x - 3}$$

$$5 \cdot 3 = 4(x - 3)$$

$$15 = 4x - 12$$

$$27 = 4x$$

$$6.75 = x$$

You Try It 5

Strategy

To find the number of gallons, write and solve a proportion using n to represent the number of gallons needed to travel 832 mi.

Solution

$$\frac{396 \text{ mi}}{11 \text{ gal}} = \frac{832 \text{ mi}}{n \text{ gal}}$$

$$11 \cdot 832 = 396 \cdot n$$

$$9{,}152 = 396n$$

$$23.1 \approx n$$

To travel 832 mi, approximately 23.1 gal of gas are needed.

You Try It 6

Strategy

To find the number of defective transmissions, write and solve a proportion using n to represent the number of defective transmissions in 120,000 cars.

Solution

$$\frac{15 \text{ defective transmissions}}{1{,}200 \text{ cars}} = \frac{n \text{ defective transmissions}}{120{,}000 \text{ cars}}$$

$$1{,}200 \cdot n = 15 \cdot 120{,}000$$

$$1{,}200n = 1{,}800{,}000$$

$$n = 1{,}500$$

1,500 defective transmissions would be found in 120,000 cars.

Section 7.5 *(pages 467–470)*

You Try It 1

Strategy

To find the constant of variation, substitute 120 for y and 8 for x in the direct variation equation $y = kx$ and solve for k.

Solution

$$y = kx$$

$$120 = k \cdot 8$$

$$15 = k$$

The constant of variation is 15.

You Try It 2

Strategy

To find S when $R = 200$:

▶ Write the basic direct variation equation, replace the variables by the given values, and solve for k.

▶ Write the direct variation equation, replacing k by its value. Substitute 200 for R and solve for S.

Solution

$$S = kR$$

$$8 = k \cdot 30$$

$$\frac{8}{30} = k$$

$$\frac{4}{15} = k$$

$$S = \frac{4}{15}R = \frac{4}{15}(200) = \frac{160}{3} \approx 53.3$$

The value of S is approximately 53.3 when $R = 200$.

Copyright © Houghton Mifflin Company. All rights reserved.

You Try It 3

Strategy To find the distance:

▶ Write the basic direct variation equation, replace the variables by the given values, and solve for k.

▶ Write the direct variation equation, replacing k by its value. Substitute 9 for the time and solve for the distance.

Solution
$$d = kt^2$$
$$64 = k \cdot 2^2$$
$$64 = k \cdot 4$$
$$16 = k$$
$$d = 16t^2 = 16 \cdot 9^2 = 16 \cdot 81 = 1,296$$

The object will fall 1,296 ft.

You Try It 4

Strategy To find the resistance:

▶ Write the basic inverse variation equation, replace the variables by the given values, and solve for k.

▶ Write the inverse variation equation, replacing k by its value. Substitute 0.02 for the diameter and solve for the resistance.

Solution
$$R = \frac{k}{d^2}$$
$$0.5 = \frac{k}{(0.01)^2}$$
$$0.5 = \frac{k}{0.0001}$$
$$0.00005 = k$$
$$R = \frac{0.00005}{d^2} = \frac{0.00005}{(0.02)^2} = \frac{0.00005}{0.0004} = 0.125$$

The resistance is 0.125 ohm.

Solutions to Chapter 8 *You Try Its*

Section 8.1 *(pages 489–490)*

You Try It 1 $110\% = 110\left(\dfrac{1}{100}\right) = \left(\dfrac{110}{100}\right) = 1\dfrac{1}{10}$
$110\% = 110(0.01) = 1.10$

You Try It 2 $16\dfrac{3}{8}\% = 16\dfrac{3}{8}\left(\dfrac{1}{100}\right) = \dfrac{131}{8}\left(\dfrac{1}{100}\right)$
$$= \dfrac{131}{800}$$

You Try It 3 $0.8\% = 0.8(0.01) = 0.008$

You Try It 4 $0.038 = 0.038(100\%) = 3.8\%$

You Try It 5 $\dfrac{9}{7} = \dfrac{9}{7}(100\%) = \dfrac{900}{7}\% = 128\dfrac{4}{7}\%$

You Try It 6 $1\dfrac{5}{9} = \dfrac{14}{9} = \dfrac{14}{9}(100\%) = \dfrac{1,400}{9}\% \approx 155.6\%$

Section 8.2 *(pages 493–498)*

You Try It 1

Strategy To find the amount, solve the basic percent equation. Percent $= 33\dfrac{1}{3}\% = \dfrac{1}{3}$, base $= 45$, amount $= n$

Solution Percent $\cdot$ base $=$ amount
$$\dfrac{1}{3}(45) = n$$
$$15 = n$$

15 is $33\dfrac{1}{3}\%$ of 45.

You Try It 2

Strategy To find the percent, solve the basic percent equation. Percent $= n$, base $= 40$, amount $= 25$

Solution Percent $\cdot$ base $=$ amount
$$n \cdot 40 = 25$$
$$\dfrac{40n}{40} = \dfrac{25}{40} = 0.625$$
$$n = 62.5\%$$

25 is 62.5% of 40.

You Try It 3

Strategy To find the base, solve the basic percent equation. Percent $= 16\dfrac{2}{3}\% = \dfrac{1}{6}$, base $= n$, amount $= 15$

Solution Percent $\cdot$ base $=$ amount
$$\dfrac{1}{6} \cdot n = 15$$
$$6 \cdot \dfrac{1}{6}n = 15 \cdot 6$$
$$n = 90$$

$16\dfrac{2}{3}\%$ of 90 is 15.

You Try It 4

Percent $= 25$, base $= n$, amount $= 8$
$$\dfrac{25}{100} = \dfrac{8}{n}$$
$$25 \cdot n = 100 \cdot 8$$
$$25n = 800$$
$$\dfrac{25n}{25} = \dfrac{800}{25}$$
$$n = 32$$

8 is 25% of 32.

You Try It 5

Percent $= 0.74$, base $= 1,200$, amount $= n$
$$\dfrac{0.74}{100} = \dfrac{n}{1,200}$$
$$100 \cdot n = 0.74 \cdot 1,200$$
$$100n = 888$$
$$\dfrac{100n}{100} = \dfrac{888}{100}$$
$$n = 8.88$$

0.74% of 1,200 is 8.88.

Copyright © Houghton Mifflin Company. All rights reserved.

You Try It 6 Percent = n, base = 180, amount = 54

$$\frac{n}{100} = \frac{54}{180}$$
$$n \cdot 180 = 100 \cdot 54$$
$$180n = 5{,}400$$
$$\frac{180n}{180} = \frac{5{,}400}{180}$$
$$n = 30$$

30% of 180 is 54.

You Try It 7

Strategy To find the percent, use the basic percent equation. Percent = n, base = 4,330, amount = 649.50

Solution Percent · base = amount
$$n \cdot 4{,}330 = 649.50$$
$$\frac{4{,}330n}{4{,}330} = \frac{649.50}{4{,}330}$$
$$n = 0.15$$

15% of the instructor's salary is deducted for income tax.

You Try It 8

Strategy To find the number, solve the basic percent equation.

Percent = 19% = 0.19, base = 2.4 million, amount = n

Solution Percent · base = amount
$$0.19 \cdot 2.4 = n$$
$$0.456 = n$$

0.456 million = 456,000

There are approximately 456,000 female surfers in this country.

You Try It 9

Strategy To find the increase in the hourly wage:

► Find last year's wage. Solve the basic percent equation.

Percent = 115% = 1.15, base = n, amount = 30.13

► Subtract last year's wage from this year's wage.

Solution Percent · base = amount
$$1.15 \cdot n = 30.13$$
$$\frac{1.15n}{1.15} = \frac{30.13}{1.15}$$
$$n = 26.20$$
$$30.13 - 26.20 = 3.93$$

The increase in the hourly wage was $3.93.

Section 8.3 *(pages 503–504)*

You Try It 1

Strategy To find the percent increase in mileage:

► Find the amount of increase in mileage.

► Solve the basic percent equation.
Percent = n, base = 17.5,
amount = amount of increase

Solution $18.2 - 17.5 = 0.7$
Percent · base = amount
$$n \cdot 17.5 = 0.7$$
$$n = \frac{0.7}{17.5}$$
$$n = 0.04$$

The percent increase in mileage is 4%.

You Try It 2

Strategy To find the value of the car:

► Solve the basic percent equation to find the amount of decrease in value.
Percent = 24% = 0.24, base = 47,000,
amount = n

► Subtract the amount of decrease from the cost.

Solution Percent · base = amount
$$0.24 \cdot 47{,}000 = n$$
$$11{,}280 = n$$
$$47{,}000 - 11{,}280 = 35{,}720$$

The value of the car is $35,720.

Section 8.4 *(pages 507–510)*

You Try It 1

Strategy To find the markup, solve the formula $M = r \cdot C$ for M.
$r = 45\% = 0.45$, $C = 650$

Solution $M = r \cdot C$
$M = 0.45 \cdot 650$
$M = 292.50$

The markup is $292.50.

You Try It 2

Strategy To find the markup rate:

► Solve the formula $M = S - C$ for M.
$S = 1{,}450$, $C = 950$

► Solve the formula $M = r \cdot C$ for r.

Solution $M = S - C$
$M = 1{,}450 - 950$
$M = 500$

$M = r \cdot C$
$500 = r \cdot 950$
$$\frac{500}{950} = r$$
$0.526 \approx r$

The markup rate is 52.6%.

You Try It 3

Strategy To find the selling price:

► Find the markup by solving the equation $M = r \cdot C$ for M. $r = 42\% = 0.42$, $C = 82.50$

► Solve the formula $S = C + M$ for S.
$C = 82.50$, $M = $ the markup.

Copyright © Houghton Mifflin Company. All rights reserved.

Solution

$M = r \cdot C$
$M = 0.42 \cdot 82.50$
$M = 34.65$

$S = C + M$
$S = 82.50 + 34.65$
$S = 117.15$

The selling price is $117.15.

You Try It 4

Strategy

To find the discount rate:

▶ Find the discount by solving the formula $M = R - S$ for M. $R = 325$, $S = 253.50$

▶ Solve the formula $M = r \cdot R$ for r. M = the discount, $R = 325$

Solution

$M = R - S$
$M = 325 - 253.50$
$M = 71.50$

$M = r \cdot R$
$71.50 = r \cdot 325$
$\dfrac{71.50}{325} = \dfrac{325r}{325}$
$0.22 = r$

The discount rate is 22%.

You Try It 5

Strategy

To find the sale price, solve the formula $S = (1 - r)R$ for S. $r = 25\% = 0.25$, $R = 312$

Solution

$S = (1 - r)R$
$S = (1 - 0.25) \cdot 312$
$S = 0.75 \cdot 312$
$S = 234$

The sale price is $234.

You Try It 6

Strategy

To find the regular price, solve the formula $S = (1 - r)R$ for R. $S = 1{,}495$, $r = 35\% = 0.35$

Solution

$S = (1 - r)R$
$1{,}495 = (1 - 0.35)R$
$1{,}495 = 0.65R$
$\dfrac{1{,}495}{0.65} = \dfrac{0.65R}{0.65}$
$2{,}300 = R$

The regular price is $2,300.

Section 8.5 *(pages 513–514)*

You Try It 1

Strategy

To calculate the maturity value:

▶ Find the simple interest due on the loan by solving the simple interest formula for I.

$t = \dfrac{8}{12}$, $P = 12{,}500$, $r = 9.5\% = 0.095$

▶ Use the formula for the maturity value of a simple interest loan, $M = P + I$.

Solution

$I = Prt$
$I = 12{,}500(0.095)\left(\dfrac{8}{12}\right)$
$I \approx 791.67$

$M = P + I$
$M = 12{,}500 + 791.67$
$M = 13{,}291.67$

The total amount due on the loan is $13,291.67.

Solutions to Chapter 9 *You Try Its*

Section 9.1 *(pages 531–540)*

You Try It 1

$QR + RS + ST = QT$
$24 + RS + 17 = 62$
$41 + RS = 62$
$RS = 21$

$RS = 21$ cm

You Try It 2

$AC = AB + BC$
$AC = \dfrac{1}{4}(BC) + BC$
$AC = \dfrac{1}{4}(16) + 16$
$AC = 4 + 16$
$AC = 20$
$AC = 20$ ft

You Try It 3

Strategy

Supplementary angles are two angles whose sum is 180°. To find the supplement, let x represent the supplement of a 129° angle. Write an equation and solve for x.

Solution

$x + 129° = 180°$
$x = 51°$

The supplement of a 129° angle is a 51° angle.

You Try It 4

Strategy

To find the measure of $\angle a$, write an equation using the fact that the sum of the measure of $\angle a$ and 68° is 118°. Solve for $\angle a$.

Solution

$\angle a + 68° = 118°$
$\angle a = 50°$

The measure of $\angle a$ is 50°.

You Try It 5

Strategy

The angles labeled are adjacent angles of intersecting lines and are therefore supplementary angles. To find x, write an equation and solve for x.

Solution

$(x + 16°) + 3x = 180°$
$4x + 16° = 180°$
$4x = 164°$
$x = 41°$

Copyright © Houghton Mifflin Company. All rights reserved.

You Try It 6

Strategy $3x = y$ because corresponding angles have the same measure. $y + (x + 40°) = 180°$ because adjacent angles of intersecting lines are supplementary angles. Substitute $3x$ for y and solve for x.

Solution
$$3x + (x + 40°) = 180°$$
$$4x + 40° = 180°$$
$$4x = 140°$$
$$x = 35°$$

You Try It 7

Strategy
- To find the measure of angle b, use the fact that $\angle b$ and $\angle x$ are supplementary angles.
- To find the measure of angle c, use the fact that the sum of the interior angles of a triangle is 180°.
- To find the measure of angle y, use the fact that $\angle c$ and $\angle y$ are vertical angles.

Solution
$$\angle b + \angle x = 180°$$
$$\angle b + 100° = 180°$$
$$\angle b = 80°$$

$$\angle a + \angle b + \angle c = 180°$$
$$45° + 80° + \angle c = 180°$$
$$125° + \angle c = 180°$$
$$\angle c = 55°$$

$$\angle y = \angle c = 55°$$

You Try It 8

Strategy To find the measure of the third angle, use the fact that the measure of a right angle is 90° and the fact that the sum of the measures of the interior angles of a triangle is 180°. Write an equation using x to represent the measure of the third angle. Solve the equation for x.

Solution
$$x + 90° + 34° = 180°$$
$$x + 124° = 180°$$
$$x = 56°$$

The measure of the third angle is 56°.

Section 9.2 *(pages 547–556)*

You Try It 1

Strategy To find the perimeter, use the formula for the perimeter of a square. Substitute 60 for s and solve for P.

Solution
$$P = 4s$$
$$P = 4(60)$$
$$P = 240$$

The perimeter of the infield is 240 ft.

You Try It 2

Strategy To find the perimeter, use the formula for the perimeter of a rectangle. Substitute 11 for L and $8\frac{1}{2}$ for W and solve for P.

Solution
$$P = 2L + 2W$$
$$P = 2(11) + 2\left(8\frac{1}{2}\right)$$
$$P = 2(11) + 2\left(\frac{17}{2}\right)$$
$$P = 22 + 17$$
$$P = 39$$

The perimeter of a standard piece of typing paper is 39 in.

You Try It 3

Strategy To find the circumference, use the circumference formula that involves the diameter. Leave the answer in terms of π.

Solution
$$C = \pi d$$
$$C = \pi(9)$$
$$C = 9\pi$$

The circumference is 9π in.

You Try It 4

Strategy

To find the number of rolls of wallpaper to be purchased:
- Use the formula for the area of a rectangle to find the area of one wall.
- Multiply the area of one wall by the number of walls to be covered (2).
- Divide the area of wall to be covered by the area one roll of wallpaper will cover (30).

Solution

$A = LW$
$A = 12 \cdot 8 = 96$ The area of one wall is 96 ft².
$2(96) = 192$ The area of the two walls is 192 ft².
$192 \div 30 = 6.4$

Because a portion of a seventh roll is needed, 7 rolls of wallpaper should be purchased.

You Try It 5

Strategy To find the area, use the formula for the area of a circle. An approximation is asked for; use the π key on a calculator. $r = 11$.

Solution
$$A = \pi r^2$$
$$A = \pi(11)^2$$
$$A = 121\pi$$
$$A \approx 380.13$$

The area is approximately 380.13 cm².

Section 9.3 *(pages 565–570)*

You Try It 1

Strategy To find the measure of the other leg, use the Pythagorean Theorem. $a = 2, c = 6$

Solution
$$a^2 + b^2 = c^2$$
$$2^2 + b^2 = 6^2$$
$$4 + b^2 = 36$$
$$b^2 = 32$$

Copyright © Houghton Mifflin Company. All rights reserved.

$b = \sqrt{32}$
$b \approx 5.66$

The measure of the other leg is approximately 5.66 m.

You Try It 2

Strategy To find FG, write a proportion using the fact that, in similar triangles, the ratio of corresponding sides equals the ratio of corresponding heights. Solve the proportion for FG.

Solution
$$\frac{AC}{DF} = \frac{CH}{FG}$$
$$\frac{10}{15} = \frac{7}{FG}$$
$$10(FG) = 15(7)$$
$$10(FG) = 105$$
$$FG = 10.5$$

The height FG of triangle DEF is 10.5 m.

You Try It 3

Strategy To determine whether the triangles are congruent, determine whether one of the rules for congruence is satisfied.

Solution $PR = MN$, $QR = MO$, and $\angle QRP = \angle OMN$. Two sides and the included angle of one triangle equal two sides and the included angle of the other triangle.

The triangles are congruent by the SAS Rule.

Section 9.4 *(pages 575–580)*

You Try It 1

Strategy To find the volume, use the formula for the volume of a cube. $s = 2.5$.

Solution $V = s^2$
$V = (2.5)^3 = 15.625$

The volume of the cube is 15.625 m³.

You Try It 2

Strategy To find the volume:
▶ Find the radius of the base of the cylinder. $d = 8$.
▶ Use the formula for the volume of a cylinder. Leave the answer in terms of π.

Solution $r = \frac{1}{2}d = \frac{1}{2}(8) = 4$

$V = \pi r^2 h = \pi(4)^2(22) = \pi(16)(22) = 352\pi$

The volume of the cylinder is 352π ft³.

You Try It 3

Strategy To find the surface area of the cylinder.
▶ Find the radius of the base of the cylinder. $d = 6$.

▶ Use the formula for the surface area of a cylinder. An approximation is asked for; use the π key on a calculator.

Solution $r = \frac{1}{2}d = \frac{1}{2}(6) = 3$

$SA = 2\pi r^2 + 2\pi r h$
$SA = 2\pi(3)^2 + 2\pi(3)(8)$
$= 2\pi(9) + 2\pi(3)(8)$
$= 18\pi + 48\pi$
$= 66\pi$
≈ 207.35

The surface area of the cylinder is approximately 207.35 ft².

You Try It 4

Strategy To find which solid has the larger surface area:
▶ Use the formula for the surface area of a cube to find the surface area of the cube. $s = 10$.
▶ Find the radius of the sphere. $d = 8$.
▶ Use the formula for the surface area of a sphere to find the surface area of the sphere. Because this number is to be compared to another number, use the π key on a calculator to approximate the surface area.
▶ Compare the two numbers.

Solution $SA = 6s^2$
$SA = 6(10)^2 = 6(100) = 600$

The surface area of the cube is 600 cm².

$r = \frac{1}{2}d = \frac{1}{2}(8) = 4$

$SA = 4\pi r^2$
$SA = 4\pi(4)^2 = 4\pi(16) = 64\pi \approx 201.06$

The surface area of the sphere is 201.06 cm².

$600 > 201.06$

The cube has a larger surface area than the sphere.

Section 9.5 *(pages 585–588)*

You Try It 1

Strategy To find the perimeter of square $ABCD$:
▶ Use the circumference formula that involves the diameter to find a diameter of the circle. A diameter of the circle is equal to the length of a side of the square.
▶ Use the formula for the perimeter of a square.

Solution $C = \pi d$
$6\pi = \pi d$
$6 = d$ The diameter of the circle is 6 cm.

$P = 4s$
$P = 4(6) = 24$

The perimeter of square $ABCD$ is 24 cm.

Copyright © Houghton Mifflin Company. All rights reserved.

You Try It 2

Strategy   The area is equal to the area of the rectangle minus the area of the triangle. The base of the triangle is equal to the width of the rectangle.

Solution $A = LW - \dfrac{1}{2}bh$

$A = 10(6) - \dfrac{1}{2}(6)(4) = 60 - 12 = 48$

The area of the composite figure is 48 in^2.

You Try It 3

Strategy   The volume is equal to the volume of the rectangular solid plus the volume of the cylinder. Leave the answer in terms of π.

Solution $V = LWH + \pi r^2 h$
$V = (8)(8)(2) + \pi(1)^2(2)$
$= (8)(8)(2) + \pi(1)(2) = 128 + 2\pi$

The volume of the solid is $(128 + 2\pi)$ cm^3.

You Try It 4

Strategy

The total surface area equals the surface area of the rectangular solid, minus the top of the rectangular solid, plus one-half the surface area of the cylinder. The radius of the base of the cylinder is one-half the width of the rectangular solid. The height of the cylinder is equal to the length of the rectangular solid. An approximation is asked for; use the π key on a calculator.

Solution

$r = \dfrac{1}{2}(W) = \dfrac{1}{2}(6) = 3$

$SA = LW + 2LH + 2WH + \dfrac{1}{2}(2\pi r^2 + 2\pi rh)$

$SA = (24)(6) + 2(24)(4) + 2(6)(4) + \dfrac{1}{2}[2\pi(3)^2 + 2\pi(3)(24)]$

$= 144 + 192 + 48 + \dfrac{1}{2}(18\pi + 144\pi)$

$= 144 + 192 + 48 + \dfrac{1}{2}(162\pi)$

$= 384 + 81\pi$
≈ 638.47

The surface area of the solid is approximately 638.47 in^2.

Solutions to Chapter 10   *You Try Its*

Section 10.1   *(pages 609–612)*

You Try It 1

Strategy   To make the frequency distribution:
▶ Find the range.
▶ Divide the range by 8, the number of classes. The quotient is the class width.
▶ Tabulate the data for each class.

Solution   Range $= 998 - 118 = 880$

Class width $= \dfrac{880}{8} = 110$

Dollar Amount of Insurance Claims

Classes	Tally	Frequency
118–228	///	3
229–339	////	4
340–450	/////	5
451–561	////////	8
562–672	//////	6
673–783	///////	7
784–894	//////////	10
895–1,005	///////	7

You Try It 2

Strategy   To find the number:
▶ Read the histogram to find the number of households using between 1,100 and 1,150 kWh and the number using between 1,150 and 1,200 kWh.
▶ Add the two numbers.

Solution   Number between 1,100 and 1,150 kWh: 14
Number between 1,150 and 1,200 kWh: 10

$14 + 10 = 24$

24 households used 1,100 kWh of electricity or more during the month.

You Try It 3

Strategy   To find the ratio:
▶ Read the frequency polygon to find the number of states with a per capita income between $20,000 and $24,000 and between $24,000 and $28,000.
▶ Write the ratio of the number of states with a per capita income between $20,000 and $24,000 to the number of states with a per capita income between $24,000 and $28,000.

Solution   Number of states with a per capita income between $20,000 and $24,000: 5

Number of states with a per capita income between $24,000 and $28,000: 15

$\dfrac{\text{Income between \$20,000 and \$24,000}}{\text{Income between \$24,000 and \$28,000}}$

$= \dfrac{5}{15} = \dfrac{1}{3}$

The ratio is $\dfrac{1}{3}$ or 1 to 3.

Section 10.2   *(pages 617–622)*

You Try It 1

Strategy   To calculate the mean amount spent:
▶ Calculate the sum of the amounts spent by the customers.
▶ Divide the sum by the number of customers.

Copyright © Houghton Mifflin Company. All rights reserved.

To calculate the median amount spent by the customers:

▶ Arrange the numbers from smallest to largest.

▶ Because there is an even number of values, the median is the sum of the two middle numbers divided by 2.

Solution The sum of the numbers is 51.30.

$$\bar{x} = \frac{51.30}{10} = 5.13$$

The mean amount spent by the customers was $5.13.

Arrange the numbers from smallest to largest.

3.59 4.32 4.45 4.75 5.05
5.45 5.58 5.90 6.00 6.21

$$\text{Median} = \frac{5.05 + 5.45}{2} = 5.25$$

The median is $5.25.

You Try It 2

Strategy To find the score, use the formula for the mean, letting n be the score on the fifth exam.

Solution
$$84 = \frac{82 + 91 + 79 + 83 + n}{5}$$

$$84 = \frac{335 + n}{5}$$

$$5 \cdot 84 = 5\left(\frac{335 + n}{5}\right)$$

$$420 = 335 + n$$

$$85 = n$$

The score on the fifth test must be 85.

You Try It 3

Strategy To draw the box-and-whiskers plot:

▶ Use the value of the first quartile, the median, and the third quartile from Example 3.

▶ Determine the smallest and largest data value.

▶ Draw the box-and-whiskers plot.

Solution $Q_1 = 38$, median $= 44$, $Q_3 = 53$

Smallest value: 24
Largest value: 64

You Try It 4

Strategy To calculate the standard deviation:

▶ Find the mean of the number of miles run.

▶ Use the procedure for calculating the standard deviation.

Solution
$$\bar{x} = \frac{5 + 7 + 3 + 6 + 9 + 6}{6} = 6$$

Step 1

x	$(x - \bar{x})^2$	
5	$(5 - 6)^2 =$	1
7	$(7 - 6)^2 =$	1
3	$(3 - 6)^2 =$	9
6	$(6 - 6)^2 =$	0
9	$(9 - 6)^2 =$	9
6	$(6 - 6)^2 =$	0
	Total $=$	20

Step 2 $\frac{20}{6} = \frac{10}{3}$

Step 3 $\sigma = \sqrt{\frac{10}{3}} \approx 1.826$

The standard deviation is 1.826 mi.

Section 10.3 (pages 627–632)

You Try It 1

Strategy To calculate the probability:

▶ Count the number of possible outcomes of the experiment.

▶ Count the outcomes of the experiment that are favorable to the event the two numbers are the same.

▶ Use the probability formula.

Solution There are 36 possible outcomes.
There are 6 favorable outcomes: (1, 1), (2, 2), (3, 3), (4, 4), (5, 5), and (6, 6).

$$P(E) = \frac{6}{36} = \frac{1}{6}$$

The probability that the two numbers are equal is $\frac{1}{6}$.

You Try It 2

Strategy To calculate the probability:

▶ Count the number of possible outcomes of the experiment.

▶ Count the number of outcomes of the experiment that are favorable to the event E that the soft drink symbol is uncovered.

▶ Use the probability formula.

Solution There are 8 possible outcomes of the experiment.
There is 1 favorable outcome for E, uncovering the soft drink symbol.

$$P(E) = \frac{1}{8}$$

The probability is $\frac{1}{8}$ that the soft drink symbol will be uncovered.

Copyright © Houghton Mifflin Company. All rights reserved.

You Try It 3

Strategy

To calculate the probability:

▶ Write the odds in favor of contracting the flu as a fraction.

▶ The probability of contracting the flu is the numerator of the odds in favor fraction over the sum of the numerator and the denominator.

Solution

Odds in favor of contracting the flu $= \dfrac{2}{13}$

Probability of contracting the flu $= \dfrac{2}{2 + 13} = \dfrac{2}{15}$

The probability of contracting the flu is $\dfrac{2}{15}$.

Copyright © Houghton Mifflin Company. All rights reserved.

Answers to Chapter 1 *Exercises*

Prep Test *(page 2)*

1. 8   **2.** 1 2 3 4 5 6 7 8 9 10   **3.** a and D; b and E; c and A; d and B; e and F; f and C   **4.** 0   **5.** fifty

1.1 Exercises *(pages 13–18)*

3. (number line, point at 2; 0 1 2 3 4 5 6 7 8 9 10 11 12)   **5.** (number line, point at 9; 0 1 2 3 4 5 6 7 8 9 10 11 12)   **7.** (number line, point at 5; 0 1 2 3 4 5 6 7 8 9 10 11 12)
9. 5   **11.** 5   **13.** 0   **15.** $27 < 39$ **17.** $0 < 52$ **19.** $273 > 194$ **21.** $2{,}761 < 3{,}857$ **23.** $4{,}610 > 4{,}061$
25. $8{,}005 < 8{,}050$ **27.** 11, 14, 16, 21, 32   **29.** 13, 48, 72, 84, 93   **31.** 26, 49, 77, 90, 106   **33.** 204, 399, 662, 736, 981
35. 307, 370, 377, 3,077, 3,700   **37.** five hundred eight   **39.** six hundred thirty-five   **41.** four thousand seven hundred
ninety   **43.** fifty-three thousand six hundred fourteen   **45.** two hundred forty-six thousand fifty-three   **47.** three million
eight hundred forty-two thousand nine hundred five   **49.** 496   **51.** 53,340   **53.** 502,140   **55.** 9,706   **57.** 5,012,907
59. 8,005,010   **61.** $7{,}000 + 200 + 40 + 5$ **63.** $500{,}000 + 30{,}000 + 2{,}000 + 700 + 90 + 1$ **65.** $5{,}000 + 60 + 4$
67. $20{,}000 + 300 + 90 + 7$ **69.** $400{,}000 + 2{,}000 + 700 + 8$ **71.** $8{,}000{,}000 + 300 + 10 + 6$ **73.** 7,110   **75.** 5,000
77. 28,600   **79.** 7,000   **81.** 94,000   **83.** 630,000   **85.** 350,000   **87.** 72,000,000   **89.** Billy Hamilton   **91.** *Fiddler on the Roof*   **93.** two tablespoons of peanut butter   **95.** St. Louis to San Diego   **97.** Neptune   **99a.** 67 min   **b.** 2,000
101. 160,000 acres   **103a.** 1985   **b.** decrease   **105.** 300,000 km/s   **107.** 999; 10,000

1.2 Exercises *(pages 33–40)*

3. 1,383,659   **5.** 6,043   **7.** 112,152   **9.** 12,548   **11.** 199,556   **13.** 327,473   **15.** 168,574   **17.** 7,947   **19.** 99,637
21. 1,872 students   **23.** 15,000; 15,040   **25.** 1,400,000; 1,388,917   **27.** 2,000; 1,998   **29.** 307,000; 329,801   **31.** 1,272
33. 12,150   **35.** 89,900   **37.** 1,572   **39.** 14,591   **41.** 56,010   **43.** The Commutative Property of Addition
45. The Associative Property of Addition   **47.** The Addition Property of Zero   **49.** 28   **51.** 4   **53.** 15   **55.** Yes
57. No   **59.** Yes   **63.** 416   **65.** 188   **67.** 464   **69.** 208   **71.** 3,557   **73.** 2,836   **75.** 1,437   **77.** 20,148
79. 1,618   **81.** 7,378   **83.** 17,548   **85.** 15 ft   **87.** 2,000; 2,136   **89.** 40,000; 38,283   **91.** 35,000; 31,195
93. 100,000; 125,665   **95.** 13   **97.** 643   **99.** 355   **101.** 5,211   **103.** 766   **105.** 18,231   **107.** Yes   **109.** No
111. Yes   **113.** 210   **115.** 901   **117.** 370 calories   **119.** 78 m   **121.** 43 in.   **123.** 560 ft   **125.** 43 orbits
127. 13,729 seats   **129.** $1,645   **131.** 20,000 mi   **133.** January to February; 24 cars   **135.** $13,275   **137.** $261,000
139. 350 mph   **141a.** 9,571 drivers   **b.** 4,211 drivers   **143.** No   **145.** 11   **147a.** always true   **b.** always true

1.3 Exercises *(pages 59–66)*

3. 1,143   **5.** 46,963   **7.** 470,152   **9.** 48,493   **11.** 324,438   **13.** 3,206,160   **15.** 1,500   **17.** 2,000   **19.** 0
21. qrs **23.** 1,200,000; 1,244,653   **25.** 1,200,000; 1,138,134   **27.** 42,000; 46,935   **29.** 6,300,000; 6,491,166   **31.** 14,880
33. 3,255   **35.** 1,800   **37.** 3,082   **39.** The Multiplication Property of One   **41.** The Commutative Property of
Multiplication   **43.** 30   **45.** 0   **47.** Yes   **49.** No   **51.** Yes   **53.** $2^3 \cdot 7^5$ **55.** $2^2 \cdot 3^3 \cdot 5^4$ **57.** c^2
59. $x^3 y^3$ **61.** 32   **63.** 1,000,000   **65.** 200   **67.** 9,000   **69.** 0   **71.** 540   **73.** 144   **75.** 512   **77.** a^4 **79.** 24
81. 320   **83.** 225   **87.** 307   **89.** 309 r4   **91.** 2,550   **93.** 21 r9   **95.** 147 r38   **97.** 200 r8   **99.** 404 r34
101. 16 r97   **103.** 907   **105.** 881 r1   **107.** $\dfrac{c}{d}$ **109.** 800; 776   **111.** 5,000; 5,129   **113.** 500; 493 r37   **115.** 1,500;
1,516   **117.** 48   **119.** undefined   **121.** 9,800   **123.** Yes   **125.** No   **127.** 1, 2, 5, 10   **129.** 1, 2, 3, 4, 6, 12
131. 1, 2, 4, 8   **133.** 1, 13   **135.** 1, 2, 3, 6, 9, 18   **137.** 1, 5, 25   **139.** 1, 2, 4, 7, 8, 14, 28, 56   **141.** 1, 2, 4, 7, 14, 28
143. 1, 2, 3, 4, 6, 8, 12, 16, 24, 48   **145.** 1, 2, 3, 6, 9, 18, 27, 54   **147.** 2^4 **149.** $2^2 \cdot 3$ **151.** $3 \cdot 5$ **153.** $2^3 \cdot 5$
155. prime   **157.** $5 \cdot 13$ **159.** $2^2 \cdot 7$ **161.** $2 \cdot 3 \cdot 7$ **163.** $3 \cdot 17$ **165.** $2 \cdot 23$ **167.** 460 calories   **169.** 4,325 gal
171a. 78 m   **b.** 360 m^2 **173.** 96 ft   **175.** 576 ft^2 **177.** 59,136 cm^2 **179.** $16,000   **181.** $6,840   **183.** 9 h
185. $21   **187.** an approximation   **189.** 222

1.4 Exercises *(pages 71–72)*

1. 14   **3.** 25   **5.** 5   **7.** 13   **9.** 7   **11.** 9   **13.** 8   **15.** 1   **17.** 0   **19.** 76   **21.** 21   **23.** 33   **25.** 24
27. 6   **29.** 12   **31.** 58   **33.** 12 in.   **35.** 190 mi   **37.** 24 payments   **39.** 8 h   **41.** Answers may vary. For
example: **a.** $5x = 0$ **b.** $8x = 8$

1.5 Exercises *(pages 75–76)*

3. 4   **5.** 29   **7.** 13   **9.** 19   **11.** 11   **13.** 6   **15.** 61   **17.** 54   **19.** 19   **21.** 24   **23.** 186   **25.** 39   **27.** 18
29. 14   **31.** 14   **33.** 2   **35.** 57   **37.** 8   **39.** 68   **41.** 16   **43.** $12 + (9 - 5) \cdot 3 > 11 + (8 + 4) \div 6$
45. $4 + 3 \cdot 12 > 81 - 8^2 > 27 \div 9 + 8 > 5(10 - 2) \div 4 > 2(1 + 4)^2 \div 10 > 50 - 6(8)$

Copyright © Houghton Mifflin Company. All rights reserved.

Chapter Review Exercises *(pages 83–84)*

Note: The numbers in brackets following the answers in the Chapter Review are a reference to the objective that corresponds with that problem. For example, the reference [1.2A] stands for Section 1.2, Objective A. This notation will be used for all Chapter Reviews, Chapter Tests, and Cumulative Reviews throughout the text.

1. [1.1A]   **2.** 10,000 [1.3B]   **3.** 2,583 [1.2B]   **4.** $3^2 \cdot 5^4$ [1.3B]   **5.** 1,389 [1.2A]
6. 38,700 [1.1C]   **7.** $247 > 163$ [1.1A]   **8.** 32,509 [1.1B]   **9.** 700 [1.3A]   **10.** 2,607 [1.3C]   **11.** 4,048 [1.2B]
12. 1,500 [1.2A]   **13.** 1, 2, 5, 10, 25, 50 [1.3D]   **14.** Yes [1.2B]   **15.** 18 [1.5A]   **16.** The Commutative Property of
Addition [1.2A]   **17.** four million nine hundred twenty-seven thousand thirty-six [1.1B]   **18.** 675 [1.3B]
19a. 16 times more   **b.** 61 times more [1.3E]   **20.** 67 r70 [1.3C]   **21.** 2,636 [1.3A]   **22.** 137 [1.2B]
23. $2 \cdot 3^2 \cdot 5$ [1.3D]   **24.** 80 [1.3C]   **25.** 1 [1.3A]   **26.** 9 [1.4A]   **27.** 932 [1.2A]   **28.** 432 [1.3A]
29. 56 [1.5A]   **30.** Kareem Abdul-Jabbar [1.2C]   **31.** $182,000 [1.3E]   **32a.** 74 m   **b.** 300 m² [1.2C, 1.3E]
33a. 1960s   **b.** 4,792,000 million students [1.2C]   **34.** 42 [1.3E]   **35.** $449 [1.2C]

Chapter Test *(pages 85–86)*

1. 329,700 [1.3A]   **2.** 16,000 [1.3B]   **3.** 4,029 [1.2B]   **4.** x^4y^3 [1.3B]   **5.** Yes [1.2A]   **6.** 3,000 [1.1C]
7. $7,177 < 7,717$ [1.1A]   **8.** 8,490 [1.1B]   **9.** three hundred eighty-two thousand nine hundred four [1.1B]
10. 2,000 [1.2A]   **11.** 11,008 [1.3A]   **12.** 2,400,000 [1.3A]   **13.** 1, 2, 4, 23, 46, 92 [1.3D]   **14.** $2^4 \cdot 3 \cdot 5$ [1.3D]
15. 30,866 [1.2B]   **16.** The Commutative Property of Addition [1.2A]   **17.** 897 [1.3C]   **18.** 26 [1.5A]   **19.** $13,900 [1.2B]
20. 51 [1.4A]   **21.** 44 [1.4A]   **22.** 56 [1.5A]   **23.** 7 [1.2A]   **24.** 78 [1.4B]   **25.** 720 [1.3E]   **26.** $556 [1.2C]
27a. 96 cm   **b.** 576 cm² [1.3E]   **28.** $4,456 [1.2C]   **29a.** 2001–2002   **b.** 125,000 vehicles [1.2C]   **30.** $960 [1.3E]
31. $11 [1.3E]

Answers to Chapter 2   *Exercises*

Prep Test *(page 88)*

1. $54 > 45$ [1.1A]   **2.** 4 units [1.1A]   **3.** 15,847 [1.2A]   **4.** 3779 [1.2B]   **5.** 26,432 [1.3A]   **6.** 6 [1.3B]   **7.** 13 [1.4A]
8. 5 [1.4A]   **9.** $172 [1.2C]   **10.** 31 [1.5A]

2.1 Exercises *(pages 95–100)*

1.   **3.**
5.   **7.**   **9.** 1   **11.** −1   **13.** 3
15. A is −4. C is −2.   **17.** A is −7. D is −4.   **19.** $-2 > -5$ **21.** $3 > -7$ **23.** $-42 < 27$ **25.** $53 > -46$
27. $-51 < -20$ **29.** $-131 < 101$ **31.** −7, −2, 0, 3   **33.** −5, −3, 1, 4   **35.** −4, 0, 5, 9   **37.** −10, −7, −5, 4, 12
39. −11, −7, −2, 5, 10   **41.** −45   **43.** 88   **45.** −n   **47.** d   **49.** the opposite of negative thirteen
51. the opposite of negative p   **53.** five plus negative ten   **55.** negative fourteen minus negative three   **57.** negative
thirteen minus eight   **59.** m plus negative n   **61.** 7   **63.** 61   **65.** −46   **67.** 73   **69.** z   **71.** −p   **73.**
4   **75.** 9   **77.** 11   **79.** 12   **81.** 23   **83.** −27   **85.** 25   **87.** −41   **89.** −93   **91.** 10   **93.** 8
95. 6   **97.** $|-12| > |8|$ **99.** $|6| < |13|$ **101.** $|-1| < |-17|$ **103.** $|x| = |-x|$ **105.** $-|6|, -(4), |-7|, -(-9)$
107. $-9, -|-7|, -(5), |4|$ **109.** $-|10|, -|-8|, -(-2), -(-3), |5|$ **111.** 11, −11   **113.** −6, −5, −4, −3, −2, −1, 0, 1, 2, 3,
4, 5, 6   **115.** −9°F   **117.** −35°F   **119.** −30°F with a 5 mph wind   **121a.** −27¢   **b.** −40¢   **123.** Yes; 2003
125. Stock B   **127.** third quarter   **129a.** −2 and 6   **b.** −2 and 8   **133.** −9, −8, −7, 7, 8, 9

2.2 Exercises *(pages 111–116)*

3. −11   **5.** −5   **7.** 8   **9.** −4   **11.** −2   **13.** −9   **15.** 1   **17.** −15   **19.** 0   **21.** −21   **23.** −14
25. 19   **27.** −5   **29.** −30   **31.** 9   **33.** −12   **35.** −28   **37.** −13   **39.** −18   **41.** 11   **43.** 1
45. $x + (-7)$ **47a.** −$85,509,000,000   **b.** −$76,987,000,000   **c.** −$68,071,000,000   **49.** 5   **51.** −2
53. −11   **55.** −17   **57.** The Addition Property of Zero   **59.** The Associative Property of Addition   **61.** 0
63. 18   **65.** No   **67.** Yes   **69.** No   **73.** −3   **75.** −13   **77.** 7   **79.** 0   **81.** −17   **83.** −3
85. 12   **87.** 27   **89.** −106   **91.** −67   **93.** −6   **95.** −15   **97.** $-t - r$ **99.** 82°C   **101.** −9
103. 11   **105.** 0   **107.** −138   **109.** 26   **111.** 13   **113.** −8   **115.** 5   **117.** 2   **119.** −6   **121.** 12
123. −3   **125.** 18   **127.** Yes   **129.** No   **131.** Yes   **133a.** 7,046 m   **b.** 6,051 m   **135.** Europe
137. 86°   **139.** 36°   **141.** −3   **143.** 19   **145a.** sometimes true   **b.** always true

Copyright © Houghton Mifflin Company. All rights reserved.

2.3 Exercises (pages 123–128)

3. -24 **5.** 6 **7.** 18 **9.** -20 **11.** -16 **13.** 25 **15.** 0 **17.** 42 **19.** -128 **21.** 208
23. -243 **25.** -115 **27.** 238 **29.** -96 **31.** -210 **33.** -224 **35.** -40 **37.** 180 **39.** $-qr$
41a. $-\$98,664,000$ **b.** $-\$26,772,000$ **c.** $-\$13,048,000$ **43.** The Multiplication Property of One **45.** The
Associative Property of Multiplication **47.** -6 **49.** 1 **51.** -24 **53.** -60 **55.** 357 **57.** -56
59. $-1,600$ **61.** No **63.** No **65.** Yes **67.** -6 **69.** 8 **71.** -49 **73.** 8 **75.** -11 **77.** 14

79. 13 **81.** 1 **83.** 26 **85.** 23 **87.** -110 **89.** 111 **91.** $\dfrac{-9}{x}$ **93.** $-\$236,000$ **95.** -9 **97.** 9

99. -6 **101.** 6 **103.** Yes **105.** No **107.** Yes **109.** -3 **111.** $-62°F$ **113.** $-4°$ **115.** $-45°F$
117. $-16, 32, -64$ **119.** $-125, -625, -3,125$ **121a.** 81 **b.** -17 **123.** $-3, -2, -1$

2.4 Exercises (pages 133–134)

1. 15 **3.** 11 **5.** -7 **7.** -16 **9.** -8 **11.** 0 **13.** -2 **15.** 20 **17.** -5 **19.** -2
21. 5 **23.** 10 **25.** -20 **27.** 0 **29.** -8 **31.** -4 **33.** 25 **35.** -15 **37.** -8 **39.** 0
41. $-\$24,565$ million **43.** $3°C$ **45.** $\$13,525$ **47.** $\$15$ million **49a.** False; for example, 0 is the solution of $3x = 0$.
b. False; for example, -2 is the solution of $-3x = 6$. **c.** False; for example, 5 is the solution of $-2x = -10$.

2.5 Exercises (pages 137–138)

1. -3 **3.** -6 **5.** -5 **7.** -12 **9.** -3 **11.** 19 **13.** 2 **15.** 1 **17.** 14 **19.** 42 **21.** -13
23. -12 **25.** 32 **27.** 30 **29.** -27 **31.** 27 **33.** 2 **35.** 8 **37.** 1 **39.** 15 **41.** 32 **43.** 1
45. 1 **47.** 5 **49.** 28 **51.** -4 **53a.** No **b.** Yes

Chapter Review Exercises (pages 145–146)

1. eight minus negative one [2.1B] **2.** -36 [2.1C] **3.** 200 [2.3A] **4.** -9 [2.3B] **5.** -14 [2.2A] **6.** 13 [2.1B]
7. [2.1A] **8.** 4 [2.4A] **9.** 17 [2.3B] **10.** -210 [2.3B] **11.** -2 [2.2B]
12. -18 [2.3A] **13.** -1 [2.2A] **14.** -72 [2.3A] **15.** -4 [2.5A] **16.** -2 [2.2B] **17.** 11 strokes [2.2B]
18. 13 [2.2B] **19.** The Commutative Property of Multiplication [2.3A] **20.** Yes [2.2B] **21.** 14 [2.2B]
22. 0 [2.3B] **23.** -60 [2.3A] **24.** -12 [2.2A] **25.** 5 [2.5A] **26.** $-8 > -10$ [2.1A] **27.** 21 [2.2A]
28. 27 [2.1C] **29.** -8 [2.4B] **30.** $-12°C$ [2.1D] **31.** $-238°C$ [2.3C] **32.** $-3°C$ [2.2C] **33.** 12 [2.2C]

Chapter Test (pages 147–148)

1. negative three plus negative five [2.1B] **2.** -34 [2.1C] **3.** 18 [2.2B] **4.** -20 [2.2A] **5.** 24 [2.3A] **6.** The
Commutative Property of Addition [2.2A] **7.** 12 [2.3B] **8.** 2 [2.2A] **9.** $16 > -19$ [2.1A] **10.** -2 [2.2B]
11. -3 [2.2B] **12.** 49 [2.1B] **13.** -250 [2.3A] **14.** $-|5|, -(3), |-9|, -(-11)$ [2.1C] **15.** No [2.2B] **16.** -3
[2.1A] **17.** 16 strokes [2.2B] **18.** 0 [2.3B] **19.** 19 [2.5A] **20.** -25 [2.1B] **21.** 16 [2.4A] **22.** -11 [2.2B]
23. 24 [2.3B] **24.** 10 [2.5A] **25.** -7 [2.3B] **26.** 60 [2.3A] **27.** -11 [2.4A] **28.** -10 [2.2B]
29. $5°C$ [2.2C] **30.** $-64°F$ [2.3C] **31.** $-5°C$ [2.2C] **32.** 16 units [2.2C] **33.** $\$24$ million [2.4B]

Cumulative Review Exercises (pages 149–150)

1. 5 [2.2B] **2.** 12,000 [1.3A] **3.** 3,209 [1.3C] **4.** 2 [1.5A] **5.** -82 [2.1C] **6.** 309,480 [1.1B]
7. 2,400 [1.3A] **8.** 21 [2.3B] **9.** -11 [2.2B] **10.** -40 [2.2A] **11.** 1, 2, 4, 11, 22, 44 [1.3D] **12.** 1,936 [1.3B]
13. 630,000 [1.1C] **14.** 1,300 [1.2A] **15.** 9 [2.2B] **16.** $-2,500$ [2.3A] **17.** $3 \cdot 23$ [1.3D] **18.** -16 [2.4A]
19. -32 [2.5A] **20.** -4 [2.3B] **21.** -3 [2.3B] **22.** $-62 < 26$ [2.1A] **23.** 126 [2.3A] **24.** -9 [2.4A]
25. $2^5 \cdot 7^2$ [1.3B] **26.** 47 [1.5A] **27.** 10,062 [1.2A] **28.** -26 [2.2B] **29.** 5,000 [1.2B] **30.** 2,025 [1.3B]
31. $1,722,685$ mi^2 [1.2C] **32.** 76 years old [1.2C] **33.** $\$14,200$ [1.2C] **34.** $\$92,250$ [1.3E] **35.** $-5°C$ [2.2C]
36a. $168°F$ **b.** Alaska [2.2C] **37.** $\$24,900$ [1.2C] **38.** -8 [2.2C]

Answers to Chapter 3 *Exercises*

Prep Test (page 152)

1. 20 [1.3A] **2.** 120 [1.3B] **3.** 9 [1.3A] **4.** -2 [2.2A] **5.** -13 [2.2B] **6.** 2 r 3 [1.3C] **7.** 24 [1.3C]
8. 4 [1.3C] **9.** 59 [1.5A] **10.** 7 [1.2A] **11.** $44 < 48$ [1.1A]

Copyright © Houghton Mifflin Company. All rights reserved.

3.1 Exercises (pages 157–158)

1. 8 3. 14 5. 30 7. 45 9. 48 11. 20 13. 42 15. 72 17. 120 19. 30 21. 24 23. 180
25. 90 27. 78 29. 3 31. 6 33. 14 35. 16 37. 1 39. 4 41. 4 43. 6 45. 12 47. 15
49. 2 51. 3 53. 21 55. 12 57. every 6 min 59. 25 copies 61. 12:20 P.M.; 12:20 P.M. 63. $2x$; x

3.2 Exercises (pages 167–172)

1. $\frac{4}{5}$ 3. $\frac{1}{4}$ 5. $\frac{4}{3}$; $1\frac{1}{3}$ 7. $\frac{13}{5}$; $2\frac{3}{5}$ 9. $3\frac{1}{4}$ 11. 4 13. $2\frac{7}{10}$ 15. 7 17. $1\frac{8}{9}$ 19. $2\frac{2}{5}$ 21. 18

23. $2\frac{2}{15}$ 25. 1 27. $9\frac{1}{3}$ 29. $\frac{9}{4}$ 31. $\frac{11}{2}$ 33. $\frac{14}{5}$ 35. $\frac{47}{6}$ 37. $\frac{7}{1}$ 39. $\frac{33}{4}$ 41. $\frac{31}{3}$ 43. $\frac{55}{12}$

45. $\frac{8}{1}$ 47. $\frac{64}{5}$ 49. $\frac{6}{12}$ 51. $\frac{9}{24}$ 53. $\frac{6}{51}$ 55. $\frac{24}{32}$ 57. $\frac{108}{18}$ 59. $\frac{30}{90}$ 61. $\frac{14}{21}$ 63. $\frac{42}{49}$ 65. $\frac{8}{18}$

67. $\frac{28}{4}$ 69. $\frac{1}{4}$ 71. $\frac{3}{4}$ 73. $\frac{1}{6}$ 75. $\frac{8}{33}$ 77. 0 79. $\frac{7}{6}$ 81. 1 83. $\frac{3}{5}$ 85. $\frac{4}{15}$ 87. $\frac{3}{5}$

89. $\frac{2m}{3}$ 91. $\frac{y}{2}$ 93. $\frac{2a}{3}$ 95. c 97. $6k$ 99. $\frac{3}{8} < \frac{2}{5}$ 101. $\frac{3}{4} < \frac{7}{9}$ 103. $\frac{2}{3} > \frac{7}{11}$ 105. $\frac{17}{24} > \frac{11}{16}$

107. $\frac{7}{15} > \frac{5}{12}$ 109. $\frac{5}{9} > \frac{11}{21}$ 111. $\frac{7}{12} < \frac{13}{18}$ 113. $\frac{4}{5} > \frac{7}{9}$ 115. $\frac{9}{16} > \frac{5}{9}$ 117. $\frac{5}{8} < \frac{13}{20}$ 119. $\frac{1}{8}$

121. $\frac{5}{6}$ 123. $\frac{3}{4}$ 125. location 127. more 129. job market 131. enough money to pay bills 133. $\frac{1}{5}$

135a. $\frac{3}{40}$ b. $\frac{7}{80}$ 137. $m + (n - 1)$ 139a. 2005 b. 2006 141. 2^{1003}

3.3 Exercises (pages 183–190)

1. $\frac{9}{11}$ 3. 1 5. $1\frac{2}{3}$ 7. $1\frac{1}{6}$ 9. $\frac{16}{b}$ 11. $\frac{9}{c}$ 13. $\frac{11}{x}$ 15. $\frac{11}{12}$ 17. $\frac{11}{12}$ 19. $1\frac{7}{12}$ 21. $2\frac{2}{15}$

23. $-\frac{1}{12}$ 25. $-\frac{1}{3}$ 27. $\frac{11}{24}$ 29. $\frac{1}{12}$ 31. $15\frac{2}{3}$ 33. $5\frac{2}{3}$ 35. $15\frac{1}{20}$ 37. $10\frac{7}{36}$ 39. $7\frac{5}{12}$ 41. $-\frac{7}{18}$

43. $\frac{3}{4}$ 45. $-1\frac{1}{2}$ 47. $6\frac{5}{24}$ 49. $2\frac{5}{24}$ 51. $1\frac{2}{5}$ 53. $-\frac{1}{12}$ 55. $1\frac{13}{18}$ 57. $-\frac{19}{24}$ 59. $1\frac{5}{24}$ 61. $11\frac{2}{3}$

63. $14\frac{3}{4}$ 65. Yes 67. Yes 69. $\frac{31}{50}$ 71. $\frac{1}{6}$ 73. $\frac{1}{6}$ 75. $\frac{5}{d}$ 77. $-\frac{5}{n}$ 79. $\frac{1}{14}$ 81. $\frac{1}{2}$ 83. $\frac{1}{4}$

85. $-\frac{7}{8}$ 87. $-1\frac{1}{10}$ 89. $\frac{1}{4}$ 91. $\frac{13}{36}$ 93. $2\frac{1}{3}$ 95. $6\frac{3}{4}$ 97. $1\frac{1}{12}$ 99. $3\frac{3}{8}$ 101. $5\frac{1}{9}$ 103. $2\frac{3}{4}$

105. $1\frac{17}{24}$ 107. $4\frac{19}{24}$ 109. $1\frac{7}{10}$ 111. $-1\frac{13}{36}$ 113. $-\frac{5}{24}$ 115. $6\frac{5}{12}$ 117. $\frac{1}{3}$ 119. $-1\frac{1}{3}$ 121. $\frac{1}{12}$

123. $\frac{1}{6}$ 125. $1\frac{1}{9}$ 127. $4\frac{2}{5}$ 129. $2\frac{2}{9}$ 131. $4\frac{11}{12}$ 133. No 135. Yes 137. $\frac{28}{75}$ 139. $1\frac{3}{4}$ acres

141. $7\frac{3}{4}$ h 143. $6\frac{3}{4}$ lb 145a. 5 meals b. $\frac{23}{100}$ c. $\frac{49}{100}$; less than $\frac{1}{2}$ 147. $\frac{3}{32}$ in. 149. $29\frac{1}{2}$ ft

151. $55\frac{3}{4}$ ft 153. $1\frac{1}{3}$ ft 155. No, because the parts are not equal in size.

3.4 Exercises (pages 201–208)

3. $\frac{3}{5}$ 5. $-\frac{11}{14}$ 7. $\frac{4}{5}$ 9. 0 11. $\frac{1}{10}$ 13. $-\frac{3}{8}$ 15. $\frac{63}{xy}$ 17. $-\frac{yz}{30}$ 19. $\frac{1}{9}$ 21. $-\frac{7}{30}$ 23. $\frac{3}{16}$

25. 1 27. 6 29. $-7\frac{1}{2}$ 31. $-3\frac{11}{15}$ 33. 0 35. $\frac{1}{2}$ 37. 19 39. $-2\frac{1}{3}$ 41. 1 43. $7\frac{7}{9}$

45. -30 47. 42 49. $5\frac{1}{2}$ 51. $\frac{7}{10}$ 53. $-\frac{1}{12}$ 55. $-\frac{1}{21}$ 57. $1\frac{4}{5}$ 59. $4\frac{1}{2}$ 61. $13,000

63. $-\frac{7}{48}$ 65. $3\frac{1}{2}$ 67. $-17\frac{1}{2}$ 69. -8 71. $\frac{1}{5}$ 73. $\frac{1}{6}$ 75. $-3\frac{2}{3}$ 77. Yes 79. No 81. No

83. $1\frac{11}{14}$ 85. -1 87. 0 89. $-\frac{2}{3}$ 91. $\frac{5}{6}$ 93. 0 95. 8 97. $-\frac{1}{8}$ 99. undefined 101. $-\frac{8}{9}$

Copyright © Houghton Mifflin Company. All rights reserved.

103. $\dfrac{1}{6}$ **105.** $-\dfrac{32}{xy}$ **107.** $\dfrac{bd}{30}$ **109.** $5\dfrac{1}{3}$ **111.** -8 **113.** $-\dfrac{6}{7}$ **115.** $\dfrac{1}{2}$ **117.** $5\dfrac{2}{7}$ **119.** -12

121. $1\dfrac{29}{31}$ **123.** $1\dfrac{1}{5}$ **125.** $-1\dfrac{1}{24}$ **127.** $\dfrac{7}{26}$ **129.** $-\dfrac{10}{11}$ **131.** $\dfrac{1}{12}$ **133.** undefined **135.** -48

137. $\dfrac{4}{29}$ **139.** $-1\dfrac{3}{5}$ **141.** 2 **143.** 32 servings **145.** 30 min **147.** $16\dfrac{1}{2}$ ft; 198 in. **149.** 234 h

151. 30 houses **153.** 14 in. by 7 in. by $1\dfrac{3}{4}$ in. **155a.** $23\dfrac{1}{3}$ cans **b.** 3,500 calories **c.** 7 cans **157.** $42\dfrac{1}{2}$ yd^2

159. 96 m^2 **161.** 2 bags **163.** $3\dfrac{1}{2}$ mph **165.** 1,250 mi **167.** Examples will vary.

3.5 Exercises *(pages 213–214)*

1. 36 **3.** -12 **5.** 25 **7.** -12 **9.** $\dfrac{1}{2}$ **11.** $\dfrac{7}{12}$ **13.** $\dfrac{3}{4}$ **15.** $-\dfrac{4}{5}$ **17.** $\dfrac{2}{3}$ **19.** $1\dfrac{1}{2}$ **21.** $-1\dfrac{1}{3}$

23. $-\dfrac{4}{9}$ **25.** $\dfrac{5}{6}$ **27.** $1\dfrac{1}{2}$ **29.** -3 **31.** $-\dfrac{2}{9}$ **33.** \$80,000 **35.** 25 qt **37.** \$1,500 **39.** 532 mi

41. a

3.6 Exercises *(pages 221–224)*

1. $\dfrac{9}{16}$ **3.** $-\dfrac{1}{216}$ **5.** $5\dfrac{1}{16}$ **7.** $\dfrac{5}{128}$ **9.** $\dfrac{4}{45}$ **11.** $-\dfrac{1}{10}$ **13.** $1\dfrac{1}{7}$ **15.** $-\dfrac{27}{49}$ **17.** $\dfrac{16}{81}$ **19.** $\dfrac{25}{144}$

21. $\dfrac{2}{3}$ **23.** $\dfrac{3}{4}$ **25.** $-\dfrac{8}{9}$ **27.** $\dfrac{1}{6}$ **29.** 6 **31.** $\dfrac{18}{35}$ **33.** $-\dfrac{1}{2}$ **35.** $1\dfrac{7}{25}$ **37.** $3\dfrac{3}{11}$ **39.** 17 **41.** $-\dfrac{4}{5}$

43. 1 **45.** No **47.** $1\dfrac{1}{5}$ **49.** $\dfrac{5}{36}$ **51.** $\dfrac{11}{32}$ **53.** 1 **55.** 4 **57.** 0 **59.** $1\dfrac{3}{10}$ **61.** $1\dfrac{1}{9}$ **63.** $1\dfrac{15}{16}$

65. $\dfrac{1}{2}$ **67.** 1 **69.** No **71.** $\left(\dfrac{9}{10}\right)^3 < 1^5$ **73.** $\left(-1\dfrac{1}{10}\right)^2 > (0.9)^2$ **75.** $0; \dfrac{9}{16}$

Chapter Review Exercises *(pages 231–232)*

1. $9\dfrac{1}{2}$ [3.2A] **2.** $2\dfrac{5}{6}$ [3.3B] **3.** $1\dfrac{1}{2}$ [3.4B] **4.** -1 [3.4A] **5.** 2 [3.4B] **6.** $2\dfrac{2}{3}$ [3.4A] **7.** $2\dfrac{11}{12}$ [3.6B]

8. $\dfrac{3}{5} > \dfrac{7}{15}$ [3.2C] **9.** 150 [3.1A] **10.** $11\dfrac{13}{30}$ [3.3A] **11.** $3\dfrac{1}{3}$ [3.4A] **12.** $\dfrac{10}{7}; 1\dfrac{3}{7}$ [3.2A] **13.** $\dfrac{7}{8} > \dfrac{17}{20}$ [3.2C]

14. $\dfrac{3}{5}$ [3.6B] **15.** $\dfrac{32}{72}$ [3.2B] **16.** $-\dfrac{1}{3}$ [3.6A] **17.** $\dfrac{2}{7}$ [3.6C] **18.** 21 [3.1B] **19.** $\dfrac{33}{14}$ [3.2A] **20.** $\dfrac{3}{8}$ [3.3A]

21. $-\dfrac{5}{6}$ [3.4B] **22.** $1\dfrac{3}{40}$ [3.6C] **23.** -14 [3.4A] **24.** $\dfrac{1}{18}$ [3.3B] **25.** $1\dfrac{17}{24}$ [3.3B] **26.** $2\dfrac{1}{4}$ [3.6A] **27.** $9\dfrac{1}{12}$ [3.3A]

28. $\dfrac{2}{7}$ [3.2B] **29.** $4\dfrac{7}{10}$ [3.3B] **30.** $-\dfrac{13}{18}$ [3.5A] **31.** $\dfrac{2}{3}$ [3.2D] **32.** $68\dfrac{1}{6}$ yd [3.3C] **33.** $6\dfrac{1}{4}$ lb [3.3C]

34. 192 units [3.4C] **35.** \$150 [3.4C] **36.** 496 ft/s [3.4C]

Chapter Test *(pages 233–234)*

1. $2\dfrac{4}{7}$ [3.2A] **2.** $3\dfrac{11}{12}$ [3.3B] **3.** $22\dfrac{1}{2}$ [3.4A] **4.** $\dfrac{7}{12}$ [3.4A] **5.** 90 [3.1A] **6.** $\dfrac{13}{24}$ [3.3A] **7.** $2\dfrac{11}{32}$ [3.6A]

8. $\dfrac{19}{5}$ [3.2A] **9.** $\dfrac{7}{9}$ [3.4B] **10.** 2 [3.6C] **11.** 7 [3.6B] **12.** 18 [3.1B] **13.** $\dfrac{1}{6}$ [3.3B] **14.** $\dfrac{4}{5}$ [3.2B]

15. $2\dfrac{17}{24}$ [3.3A] **16.** $\dfrac{5}{6} > \dfrac{11}{15}$ [3.2C] **17.** $3\dfrac{16}{25}$ [3.6C] **18.** $\dfrac{5}{6}$ [3.6B] **19.** $\dfrac{1}{4}$ [3.6B] **20.** $-1\dfrac{1}{2}$ [3.4B]

21. $-\dfrac{1}{2}$ [3.5A] **22.** No [3.3A] **23.** $2\dfrac{1}{11}$ [3.4A] **24.** $\dfrac{1}{2}$ [3.5A] **25.** $\dfrac{12}{28}$ [3.2B] **26.** $\dfrac{5}{6}$ [3.5B]

27. $\dfrac{5}{12}$ [3.2D] **28.** $10\dfrac{5}{24}$ lb [3.3C] **29.** $17\dfrac{1}{2}$ lb [3.4C] **30.** 120 in^2 [3.4C] **31.** 10 h [3.3C] **32.** 80 units [3.4C]

33. \$5,100 [3.4C]

Copyright © Houghton Mifflin Company. All rights reserved.

Cumulative Review Exercises *(pages 235–236)*

1. 39 [1.5A] **2.** $3\frac{1}{2}$ [3.4A] **3.** $8\frac{11}{18}$ [3.3A] **4.** −15 [2.2B] **5.** 36 [3.1B] **6.** 16 [3.4A] **7.** $-1\frac{1}{9}$ [3.4B]

8. $-\frac{4}{15}$ [3.3B] **9.** 9 [3.6B] **10.** $\frac{7}{11} < \frac{4}{5}$ [3.2C] **11.** $-1\frac{22}{27}$ [3.4B] **12.** $\frac{1}{15}$ [3.4A] **13.** 2 [3.4A]

14. $7\frac{1}{28}$ [3.3B] **15.** $\frac{23}{24}$ [3.3B] **16.** $1\frac{7}{12}$ [3.6C] **17.** $1\frac{5}{8}$ [3.3B] **18.** $6\frac{3}{16}$ [3.3A] **19.** −4 [2.4A] **20.** $4\frac{5}{9}$ [3.2A]

21. $\frac{1}{7}$ [3.3B] **22.** $\frac{3}{28}$ [3.6A] **23.** −21 [2.5A] **24.** 11,272 [1.2A] **25.** 48 [1.5A] **26.** $-\frac{11}{20}$ [3.5A]

27. 20,000 [1.2B] **28.** −13 [2.2B] **29.** $\frac{31}{4}$ [3.2A] **30.** $2^2 \cdot 5 \cdot 7$ [1.3D] **31.** 40 calories [1.3E]

32. 1,740,000 people [1.2C] **33.** 9 years [3.5B] **34.** 66 ft [3.4C] **35.** $4\frac{1}{8}$ mi [3.4C] **36.** $22\frac{3}{8}$ lb/in² [3.4C]

Answers to Chapter 4 *Exercises*

Prep Test *(page 238)*

1. $\frac{3}{10}$ [3.2A] **2.** 36,900 [1.1C] **3.** four thousand seven hundred ninety-one [1.1B] **4.** 6842 [1.1B]

5. ⟨—+—+—◆—+—+—+—+—+—+—⟩
−5−4−3−2−1 0 1 2 3 4 5 **6.** 9320 [2.2A] **7.** 3168 [2.2B] **8.** 76,804 [2.3A] **9.** 278 r 18 [1.3C] **10.** 64 [1.3B]

4.1 Exercises *(pages 245–248)*

1. thousandths **3.** ten-thousandths **5.** hundredths **7.** 0.3 **9.** 0.21 **11.** 0.461 **13.** 0.093
15. $\frac{1}{10}$ **17.** $\frac{47}{100}$ **19.** $\frac{289}{1,000}$ **21.** $\frac{9}{100}$ **23.** thirty-seven hundredths **25.** nine and four tenths
27. fifty-three ten-thousandths **29.** forty-five thousandths **31.** twenty-six and four hundredths **33.** 3.0806
35. 407.03 **37.** 246.024 **39.** 73.02684 **41.** 0.7 > 0.56 **43.** 3.605 > 3.065 **45.** 9.004 < 9.04
47. 9.31 > 9.031 **49.** 4.6 < 40.6 **51.** 0.07046 > 0.07036 **53.** 0.609, 0.66, 0.696, 0.699 **55.** 1.237, 1.327,
1.372, 1.732 **57.** 21.78, 21.805, 21.87, 21.875 **59.** 5.4 **61.** 30.0 **63.** 413.60 **65.** 6.062 **67.** 97
69. 5,440 **71.** 0.0236 **73.** 0.18 oz **75.** 26.2 mi **77.** Barry Sanders **79.** 42.2 km **81a.** $2.40
b. $3.60 **c.** $6.00 **d.** $7.00 **e.** $4.70 **f.** $2.40 **g.** $2.40 **83.** For example: **a.** 0.15 **b.** 1.05
c. 0.001

4.2 Exercises *(pages 265–276)*

1. 65.9421 **3.** 190.857 **5.** 21.26 **7.** 21.26 **9.** 2.768 **11.** −50.7 **13.** −3.312 **15.** −5.905
17. −16.35 **19.** −9.55 **21.** −19.189 **23.** 56.361 **25.** 53.67 **27.** −98.38 **29.** −649.36 **31.** 31.09
33. 12; 12.325 **35.** 40; 33.63 **37.** 0.3; 0.303 **39.** 40; 38.618 **41a.** 53.446 million children
b. 40.49 million children **43.** −1.159 **45.** −25.665 **47.** 13.535 **49.** 28.3925 **51.** 10.737 **53.** −27.553
55. −1.412 **57.** Yes **59.** Yes **61.** 1.70 **63.** 0.03316 **65.** 15.12 **67.** −5.46 **69.** −0.00786
71. −473 **73.** 4,250 **75.** 67,100 **77.** 0.036 **79.** 8.0; 7.5537 **81.** 70; 68.5936 **83.** 30; 32.1485
85. 12,672 pounds **87.** 50.16 **89.** −48 **91.** −0.08338 **93.** 23.0867 **95.** Yes **97.** No **99.** 32.3
101. −67.7 **103.** 4.14 **105.** −6.1 **107.** 6.3 **109.** 5.8 **111.** 0.81 **113.** −0.08 **115.** 5.278
117. 0.4805 **119.** −25.4 **121.** −0.5 **123.** 10; 11.17 **125.** 1; 1.16 **127.** 50; 58.90 **129.** 6; 7.20
131. 2.5 times greater **133.** 5.06 **135.** −0.24 **137.** 2.06 **139.** −6.1 **141.** Yes **143.** No **145.** 0.375
147. $0.\overline{72}$ **149.** $0.58\overline{3}$ **151.** 1.75 **153.** 1.5 **155.** $4.1\overline{6}$ **157.** 2.25 **159.** $3.\overline{8}$ **161.** $\frac{1}{5}$ **163.** $\frac{3}{4}$
165. $\frac{1}{8}$ **167.** $2\frac{1}{2}$ **169.** $4\frac{11}{20}$ **171.** $1\frac{18}{25}$ **173.** $\frac{9}{200}$ **175.** $\frac{9}{10} > 0.89$ **177.** $\frac{4}{5} < 0.803$ **179.** $0.444 < \frac{4}{9}$
181. $0.13 > \frac{3}{25}$ **183.** $\frac{5}{16} > 0.312$ **185.** $\frac{10}{11} > 0.909$ **187.** $3,968.25 **189.** 32.22°C **191.** $.37 **193.** $3.42
195. $4,322.33 **197a.** 1970 to 1980 **b.** 1990 to 2000 **199.** $505.17 **201.** $808.50 **203.** $69 **205a.** Yes

Copyright © Houghton Mifflin Company. All rights reserved.

b. Females; 5.8 years   **c.** 1970   **207a.** $52.90   **b.** $79.60   **c.** $61.05   **209.** 15.5 in.   **211.** 14.625 in²
213. 13.95 m   **215.** 14 ft   **217.** $562.20   **219.** −41.65 newtons   **221.** $57,146.75   **223.** 31¢   **225.** 0.098

4.3 Exercises *(pages 279–280)*

1. 4.49   **3.** 5.7   **5.** −8.03   **7.** −1.2   **9.** 0.144   **11.** −0.3   **13.** −0.01   **15.** −1.86   **17.** −0.21
19. −2.5   **21.** −2.005   **23.** 16.38   **25.** $23.73   **27.** 100.8 ft/s   **29.** $.0025   **31.** $256.45   **33.** 12.5 ft
35. 7.5 ft   **37.** Answers will vary. For example: **a.** $x - 0.04 = -1$ **b.** $-3x = -6.3$

4.4 Exercises *(pages 287–290)*

3. 6   **5.** −3   **7.** 13   **9.** 15   **11.** −5   **13.** −10   **15.** 5   **17.** 10   **19.** 9   **21.** 27   **23.** −14
25. 16   **27.** 13   **29.** −6   **31.** 32   **33.** $\frac{1}{10}$ **35.** $\frac{3}{4}$ **37.** $\frac{5}{8}$ **39.** −24   **41.** 40   **43.** 23   **45.** 5
47. 5   **49.** 8   **51.** 1   **53.** −36   **55.** 1.7321   **57.** 3.1623   **59.** 4.8990   **61.** 11.2250   **63.** −5.6569
65. −43.8178   **67.** 4 and 5   **69.** 5 and 6   **71.** 7 and 8   **73.** 11 and 12   **75.** $2\sqrt{2}$ **77.** $3\sqrt{5}$ **79.** $2\sqrt{5}$
81. $3\sqrt{3}$ **83.** $4\sqrt{3}$ **85.** $5\sqrt{3}$ **87.** $3\sqrt{7}$ **89.** $7\sqrt{2}$ **91.** $4\sqrt{7}$ **93.** $5\sqrt{7}$ **95.** 30 ft/s   **97.** 3 s
99. 4,000 mi   **101.** 3, 4, 5, 6, 7, 8, 9   **103.** $\sqrt{\frac{1}{5}+\frac{1}{6}}, \sqrt{\frac{1}{4}+\frac{1}{8}}, \sqrt{\frac{1}{3}+\frac{1}{9}}$ **105a.** 36   **b.** 4 and 14

4.5 Exercises *(pages 297–300)*

1.   **3.**   **5.**
7.   **9.**   **11.**   **13.**
15.   **17.**   **19.**   **21.**
23.   **25.** $\sqrt{101}$ **27.** $-2, 0.4, \sqrt{17}$ **29.** all real numbers less than 3   **31.** all real
numbers greater than or equal to −1   **33.**   **35.**
37.   **39.**   **41.** $s \geq 50,000$; no   **43.** $h \leq 9$; yes   **45.** $b \leq 2,400$; yes
47. $T > 85$; no   **49a.** integer, negative integer, rational number, real number   **b.** whole number, integer, positive integer,
rational number, real number   **c.** rational number, real number   **d.** rational number, real number
e. rational number, real number   **f.** irrational number, real number   **51a.** −2.5, 0   **b.** −6.3, −3, 0, 6.7   **c.** 4, 13.6
d. −4.9, 0, 2.1, 5   **53a.** always true   **b.** always true   **c.** sometimes true

Chapter Review Exercises *(pages 307–308)*

1. 20.5670 [4.4B]   **2.** 91,800 [4.2B]   **3.** −11 [4.4A]   **4.** −8.301 [4.2A]   **5.** 89.243 [4.2A]   **6.** 5.034 [4.1A]
7. −4 [4.4A]   **8.** 0.0142 [4.2C]   **9.** −0.34 [4.3A]   **10.** 8.039 < 8.31 [4.1B]   **11.** 0.11 [4.2C]   **12.** 2.4622 [4.2B]
13. −1, −0.5, $\sqrt{10}$ [4.5B]   **14.** $\frac{3}{7} < 0.429$ [4.2D]   **15.** $\frac{7}{25}$ [4.2D]   **16.** −0.1 [4.2C]   **17.** 1.2 million workers [4.2A]
18. [4.5A]   **19.** [4.5B]   **20.** −441.2 [4.2A]   **21.** 6.143 [4.2C]
22. 50.743 [4.2A]   **23.** $3\sqrt{10}$ [4.4B]   **24.** −1,110 [4.2B]   **25.** 440 [4.2A]   **26.** $G \geq 3.5$; No [4.5C]
27. 66°C [4.2E]   **28a.** $2.72 trillion [4.2E]   **b.** 1.5 times [4.2E]   **29.** $1.68 [4.2E]   **30.** $415.74 [4.3B]
31. $499.49 [4.2E]   **32.** 40 ft/s [4.4C]

Chapter Test *(pages 309–310)*

1. 9.033 [4.1A]   **2.** 4.003 < 4.009 [4.1B]   **3.** 6.051 [4.1C]   **4.** −22.753 [4.2A]   **5.** 14.659 [4.2A]   **6.** 70 [4.2A]
7. 6.697 [4.2A]   **8.** −18.4 [4.2B]   **9.** 64 [4.2B]   **10.** −1.4 [4.3A]   **11.** −6 [4.4A]   **12.** 0.8496 [4.2C]
13. −8.5 [4.2C]   **14.** $0.22 < \frac{2}{9}$ [4.2D]   **15.** 13.5647 [4.4B]   **16.** $2\sqrt{17}$ [4.4B]   **17.** $40.8 million [4.2E]
18. No [4.2A]   **19.** 89,730 [4.2B]   **20.** [4.5A]   **21.** [4.5B]
22. −162.51 [4.2A]   **23.** −3.25 [4.3A]   **24.** 31.48°C [4.2E]   **25.** 32 ft/s [4.4C]   **26.** $20.6 million [4.3B]
27. 18.5 m [4.2E]   **28.** $x \geq 65,000$; no [4.5C]   **29.** −56.35 newtons [4.2E]   **30.** 23.34°C [4.2E]

Copyright © Houghton Mifflin Company. All rights reserved.

Cumulative Review Exercises *(pages 311–312)*

1. 0.03879 [4.2C]   **2.** 11 [2.5A]   **3.** 20 [4.3A]   **4.** 8,072,092 [1.1B]   **5.** [4.5A]

6. [4.5B]   **7.** -4 [2.2B]   **8.** 1,900 [1.2A]   **9.** $8\sqrt{3}$ [4.4B]   **10.** $1\frac{1}{2}$ [3.4B]

11. -18.42 [4.2A]   **12.** $\frac{1}{7}$ [3.4A]   **13.** 1,600 [1.3B]   **14.** $2^2 \cdot 5 \cdot 13$ [1.3D]   **15.** 0.76 [4.2D]   **16.** 95.3939 [4.4B]

17a. Sweden   **b.** 1.5 times [1.1D/4.2C]   **18.** undefined [2.3B]   **19.** $-\dfrac{11}{21}$ [3.3A]   **20.** 11 [4.4A]   **21.** 30 [4.2B]

22. 17 [2.5A]   **23.** $\dfrac{3}{10}$ [3.6B]   **24.** $\dfrac{1}{24}$ [3.3B]   **25.** 2.8 [4.2C]   **26.** \$67.74 [4.2E]   **27.** 46.62°C [4.2E]

28. \$3.12 [4.5C]   **29a.** 46.5 h   **b.** face-to-face selling [4.2E]   **30.** 30 mph [4.4C]

Answers to Chapter 5   *Exercises*

Prep Test *(page 314)*

1. $54 > 45$ [1.1A]   **2.** -11 [2.2A]   **3.** -12 [2.2B]   **4.** -88 [2.3A]   **5.** 6 [3.4A]   **6.** 39,700 [4.2B]   **7.** 9 [2.5A]
8. 31 [2.5A]

5.1 Exercises *(pages 321–324)*

1. The Associative Property of Multiplication   **3.** The Commutative Property of Addition   **5.** The Inverse Property
of Addition   **7.** The Inverse Property of Multiplication   **9.** The Commutative Property of Multiplication
11a. The Associative Property of Multiplication   **b.** The Inverse Property of Multiplication   **c.** The Multiplication
Property of One   **13.** $(x + 4) + y$ **15.** $\dfrac{1}{5}$ **17.** 0   **19.** $7y$ **21.** $-\dfrac{3}{2}$ **23.** $12x$ **25.** $-15x$ **27.** $21t$
29. $-21p$ **31.** $12q$ **33.** $2x$ **35.** $-15w$ **37.** x **39.** $6x^2$ **41.** $-27x^2$ **43.** x^2 **45.** x **47.** c
49. a **51.** $12w$ **53.** $16vw$ **55.** $-28bc$ **57.** 0   **59.** 0   **61.** 9   **63.** 7   **65.** -15 **67.** $-5y$
69. $13b$ **71.** $10z + 4$ **73.** $12y + 30z$ **75.** $21x - 27$ **77.** $-2x + 7$ **79.** $4x + 9$ **81.** $-5y - 15$
83. $-12x + 18$ **85.** $-20n + 40$ **87.** $48z - 24$ **89.** $24p + 42$ **91.** $10a + 15b + 5$ **93.** $12x - 4y - 4$
95. $36m - 9n + 18$ **97.** $12v - 18w - 42$ **99.** $20x + 4$ **101.** $20a - 25b + 5c$ **103.** $-18p + 12r + 54$
105. $-5a + 9b - 7$ **107.** $-11p + 2q + r$ **109.** No. Zero   **111.** No. Zero

5.2 Exercises *(pages 329–332)*

1. $3x^2, 4x, \underline{-9}$ **3.** $b, \underline{5}$ **5.** $9\underline{a}^2, -12\underline{a}, 4\underline{b}^2$ **7.** $3x^2$ **9.** $1, -6$ **11.** $12, 4$ **13.** $16a$ **15.** $27x$ **17.** $3z$
19. $8x$ **21.** $-7z$ **23.** $-6w$ **25.** 0   **27.** s **29.** $\dfrac{4n}{5}$ **31.** $\dfrac{x}{2}$ **33.** $\dfrac{4y}{7}$ **35.** $\dfrac{2c}{3}$ **37.** $6x - 3y$
39. $2r + 13p$ **41.** $-3w + 2v$ **43.** $-9p + 11$ **45.** $2p$ **47.** 6   **49.** $13y^2 + 1$ **51.** $12w^2 - 16$
53. $-14w$ **55.** $5a^2b + 8ab^2$ **57.** 5   **59.** $11x^2 - 2x$ **61.** $8b^2 - 2b$ **63.** $7x + 2$ **65.** $3n + 3$
67. $4a + 4$ **69.** $4a + 1$ **71.** $8x + 42$ **73.** $-12x + 28$ **75.** $-18m - 52$ **77.** $20c + 23$ **79.** $8a + 5b$
81. $15z - 12$ **83.** -19 **85.** $-13x - 2y$ **87.** $-2v + 13$ **89.** $-5c - 6$ **91.** $2a + 21$ **93.** $11n - 26$
95. $-9x + 6$ **97.** $111v - 246$ **99.** $-3r - 24$ **101.** $27z^2 - 24z - 90$
103. $6 \cdot 527 = 6(500 + 20 + 7) = 6 \cdot 500 + 6 \cdot 20 + 6 \cdot 7 = 3,000 + 120 + 42 = 3,162$

5.3 Exercises *(pages 337–340)*

1. Yes   **3.** No   **5.** Yes   **7.** No   **9.** Yes   **11.** No   **13.** 3   **15.** 1   **17.** binomial   **19.** monomial
21. $3x^3 + 8x^2 - 2x - 6$ **23.** $5a^3 - 3a^2 + 2a + 1$ **25.** $-b^2 + 4$ **27.** $11y^2 - 4y + 2$ **29.** $3b^2 - 3b - 2$
31. $3w^3 + 13w^2 - 8w - 5$ **33.** $-2a^3 - 9a^2 - 8a + 1$ **35.** $7t^3 + 7t^2 - 35$ **37.** $13t^2 - 35$ **39.** $11k^2 + 2k - 18$
41. $17x^3 - 5$ **43.** $16b^3 + 14b^2 + 1$ **45.** $8p^3 + 9p^2 - 6p - 7$ **47.** $-6a^3 - 17$ **49.** $3d^4 - 10d^2 - 3$
51. $-8x^3 - 5x^2 + 3x + 6$ **53.** $9a^3 - a^2 + 2a - 9$ **55.** $4y^2 - 10y - 1$ **57.** $6w^3 + 3w^2 + 9w - 19$ **59.** $-2t^3 + 16t + 5$
61. $8p^3 - 9p^2 + 14p + 12$ **63.** $-6v^3 + 9v^2 + v - 3$ **65.** $5m^2 - 2m - 11$ **67.** $3b^2 - 15b - 18$
69. $10y^3 - 6y^2 - 10y - 20$ **71.** $-3a^3 + a + 24$ **73.** $-2m^3 + m^2 + 7m - 6$ **75.** $-10q^3 + 16q$ **77.** $12x^4 + 11x^2 - 17$
79. $(7y^2 - 1)$ kilometers   **81.** $(3n^2 + 6)$ meters   **83.** $(-0.6n^2 + 200n - 4000)$ dollars   **85.** $(-n^2 + 700n - 1500)$ dollars
87. $-3x^2 - 4x - 3$

Copyright © Houghton Mifflin Company. All rights reserved.

5.4 Exercises (pages 345–346)

1. a^9　**3.** x^{16}　**5.** n^6　**7.** z^8　**9.** a^8b^3　**11.** $-m^9n^3$　**13.** $10x^7$　**15.** $8x^3y^6$　**17.** $-12m^7$
19. $-14v^3w$　**21.** $-2ab^5c^5$　**23.** $24a^3b^5c^2$　**25.** $40r^3t^7v$　**27.** $-27m^2n^4p^3$　**29.** $24x^7$　**31.** $6a^6b^5$
33. $-15x^3y^8$　**35.** $48r^4t^8v^4$　**37.** $-60a^3b^5c^4$　**39.** $-8a^6b^{10}$　**41.** b^8　**43.** p^{28}　**45.** c^{28}　**47.** $9x^2$
49. $x^{12}y^{18}$　**51.** $r^{12}t^4$　**53.** y^4　**55.** $8x^{12}$　**57.** $-8a^6$　**59.** $9x^4y^2$　**61.** $8a^9b^3c^6$　**63.** $m^4n^{20}p^{12}$
65. $(3a^6b^6)$ ft^2　**67.** $(2^3)^2 = 2^6 = 64,\ 2^{(3^2)} = 2^9 = 512,\ 2^{(3^2)} > (2^3)^2$

5.5 Exercises (pages 349–350)

1. $x^3 - 3x^2 - 4x$　**3.** $8a^3 + 12a^2 - 24a$　**5.** $-6a^3 - 18a^2 + 14a$　**7.** $4m^4 - 9m^3$　**9.** $10x^5 - 12x^4y + 4x^3y^2$
11. $-6r^7 + 12r^6 + 36r^5$　**13.** $12a^4 + 24a^3 - 28a^2$　**15.** $-6n^2 + 8n^5 + 10n^7$　**17.** $3a^3b^2 - 4a^2b^3 + ab^4$
19. $-4x^7y^5 + 5x^5y^4 + 7x^3y^3$　**21.** $6r^2t^3 - 6r^3t^4 - 6r^5t^6$　**23.** $36q^2 - 28q$　**25.** $y^2 + 12y + 27$　**27.** $x^2 + 11x + 30$
29. $a^2 - 11a + 24$　**31.** $10z^2 + 9z + 2$　**33.** $40c^2 - 11c - 21$　**35.** $10v^2 - 11v + 3$　**37.** $35t^2 + 18t - 8$
39. $24x^2 - x - 10$　**41.** $25r^2 - 4$　**43.** $21y^2 - 41y - 40$　**45.** $(2x^2 - 9x - 18)$ mi^2
47a. False. $(5 + x)^2 = 25 + 10x + x^2$　**b.** True　**c.** False. $(a - 4)^2 = a^2 - 8a + 16$

5.6 Exercises (pages 355–356)

1. 1　**3.** -1　**5.** $\dfrac{1}{9}$　**7.** $\dfrac{1}{8}$　**9.** $\dfrac{1}{x^5}$　**11.** $\dfrac{1}{w^8}$　**13.** $\dfrac{1}{y}$　**15.** a^5　**17.** b^3　**19.** a^6　**21.** q^4

23. mn^2　**25.** t^2u^3　**27.** $\dfrac{1}{x^5}$　**29.** $\dfrac{1}{b^4}$　**31.** 2.37×10^6　**33.** 4.5×10^{-4}　**35.** 3.09×10^5　**37.** 6.01×10^{-7}

39. 5.7×10^{10}　**41.** 1.7×10^{-8}　**43.** $710,000$　**45.** 0.000043　**47.** $671,000,000$　**49.** 0.00000713
51. $5,000,000,000,000$　**53.** 0.00801　**55.** 1.6×10^{10}　**57.** 1.6×10^{-19}　**59.** 1×10^{-12}　**61a.** $>$　**b.** $>$
c. $>$　**d.** $<$

5.7 Exercises (pages 361–364)

1. $t + 3$　**3.** $6m - 5$　**5.** $3b - 7$　**7.** $7n$　**9.** $2(3 + w)$　**11.** $4(2r - 5)$　**13.** $\dfrac{v}{v - 4}$　**15.** $4t^2$

17. $m^2 + m^3$　**19.** $(31 - s) + 5$　**21.** -12　**23.** $\dfrac{7}{24}x$　**25.** $14x + 12$　**27.** $14x$　**29.** $9x + 63$　**31.** $x + 12$

33. $7x - 28$　**35.** $7x$　**37.** $3x - 8$　**39.** $7x - 98$　**41.** $8x + 80$　**43.** $8x - 8$　**45.** $2x + 35$　**47.** $-x + 1$
49. $16x$　**51.** $5x + 6$　**53.** $-5y + 45$　**55.** $-3m + 42$　**57.** Let d be the distance from Earth to the moon; $390d$
59. Let G be the number of genes in the roundworm genome; $G + 11,000$　**61.** Let A be the amount of cashews in the
mixture; $3A$　**63.** Let p be the original price; $\dfrac{3}{4}p$　**65.** Let L be the length of the longer piece; $3 - L$
67. Let L be the length of the shorter piece; $12 - L$　**69.** $2x$　**71.** Let n be the number of turns made by the smaller
wheel; $\dfrac{4}{7}n$

Chapter Review Exercises (pages 369–370)

1. $6z^2 - 6z$ [5.2A]　**2.** $-18z - 2$ [5.1B]　**3.** $10z^2 - z - 15$ [5.3A]　**4.** $-8m^5n^2$ [5.4A]　**5.** $\dfrac{1}{243}$ [5.6A]

6. $-\dfrac{3}{7}$ [5.1A]　**7.** x [5.1A]　**8.** $-4s + 43t$ [5.2B]　**9.** $15x^3y^7$ [5.4A]　**10.** $21a^2 - 10a - 24$ [5.5B]

11. $-3b^3 + 4b - 18$ [5.3B]　**12.** $32z^{20}$ [5.4B]　**13.** $6w$ [5.1A]　**14.** $-15x^3yz^3 + 30xy^2z^4 - 5x^4y^5z^2$ [5.5A]　**15.** $-\dfrac{4}{9}$ [5.1A]

16. $-12c + 32$ [5.1B]　**17.** $-2m + 16$ [5.2A]　**18.** $-12a^5b^{15}$ [5.4A]　**19.** The Distributive Property [5.1B]

20. p^6q^9 [5.4B]　**21.** $\dfrac{1}{a^7}$ [5.6A]　**22.** 3.97×10^{-5} [5.6B]　**23.** The Commutative Property of Addition [5.1A]

24. $3y^3 + 8y^2 + 8y - 19$ [5.3A]　**25.** $12c - 4d$ [5.2B]　**26.** $14m - 42$ [5.1B]　**27.** x^2y^4 [5.6A]
28. $-5a^2 + 3a + 9$ [5.2A]　**29.** $12p^2 - 15p - 63$ [5.5B]　**30.** $-8a^5b + 10a^3b^3 - 6a^2b^5$ [5.5A]　**31.** $3x - 4y$ [5.2A]
32. $-a + 13b$ [5.2B]　**33.** $\dfrac{1}{c^5}$ [5.6A]　**34.** $6x^3 - 10x$ [5.3B]　**35.** $240,000$ [5.6B]　**36.** $(3b^2 + 3)$ ft [5.3C]

37. $\dfrac{4x}{7} - 9$ [5.7A]　**38.** $5x - 14$ [5.7B]　**39.** 6.023×10^{23} [5.6B]　**40.** Let p be the number of pounds of mocha java
beans; $30 - p$ [5.7C]

Copyright © Houghton Mifflin Company. All rights reserved.

Chapter Test (pages 371–372)

1. $-r$ [5.1A] **2.** $-15y + 21$ [5.1B] **3.** $3y + 3$ [5.2A] **4.** $-4x^2 + 5z$ [5.2A] **5.** $-3a - 6b + 18$ [5.2A]

6. $\dfrac{4}{5}$ [5.1A] **7.** $4x + 13y$ [5.2B] **8.** $-10a + b + 9$ [5.2B] **9.** 7.9×10^{-7} [5.6B] **10.** 4,900,000 [5.6B]

11. $6x^2 - 5x + 5$ [5.3A] **12.** v^8w^{20} [5.4B] **13.** $27m^6n^9$ [5.4B] **14.** $-10v^5z^3$ [5.4A] **15.** $6p^2 - p - 40$ [5.5B]

16. $-8m^3n^5 + 4m^5n^2 - 6m^2n^6$ [5.5A] **17.** $4w$ [5.1A] **18.** xy^3 [5.6A] **19.** $\dfrac{1}{a^5}$ [5.6A] **20.** The Associative Property

of Multiplication [5.1A] **21.** $-3a^3 + a^2 - 10$ [5.3B] **22.** c^6 [5.6A] **23.** The Distributive Property [5.1B]

24. 0 [5.1A] **25.** $9x^2 - 49y^2$ [5.5B] **26.** $\dfrac{4}{7}$ [5.1A] **27.** 7.2×10^8 [5.6B] **28.** $12a^2 - 18a - 12$ [5.5B]

29. $23a - 12b$ [5.2B] **30.** $\dfrac{m^2}{n^3}$ [5.6A] **31.** $3x + 5$ [5.7A] **32.** $5x - 28$ [5.7B]

33. Let s be the number of cups of sugar in the batter; $s + 3$. [5.7C]

Cumulative Review Exercises (pages 373–374)

1. -12.4 [4.2C] **2.** $14v - 2$ [5.2A] **3.** $6x^2 + 2x - 20$ [5.5B] **4.** $\dfrac{3}{8}$ [3.3B] **5.** 24 [4.4A]

6. ⟵——[——|——⟶ [4.5A] **7.** x^7 [5.6A] **8.** -9 [2.4A] **9.** 8.4×10^{-7} [5.6B] **10.** $9x^2 - 2x - 4$ [5.3A]
　　　　　　　　-3　0

11. -35 [4.4A] **12.** $\dfrac{11}{20}$ [3.6B] **13.** $-12a^7b^9$ [5.4A] **14.** $\dfrac{1}{x^2}$ [5.6A] **15.** $\dfrac{2}{5}$ [3.6A] **16.** $-48p$ [5.1A]

17. 200 [4.2A] **18.** $-12a^3b^3 - 15a^2b^3 + 6a^2b^4$ [5.5A] **19.** $18x$ [5.2B] **20.** -7 [2.3B] **21.** $\dfrac{9}{16}$ [4.2D] **22.** 4 [2.5A]

23. $10\sqrt{3}$ [4.4B] **24.** $5y^2 - 2y - 5$ [5.3B] **25.** 1 [3.4A] **26.** -1 [5.6A] **27.** $32a^{20}b^{15}$ [5.4B] **28.** 90 [2.5A]

29. $2\dfrac{2}{5}$ [3.4A] **30.** 0.0000623 [5.6B] **31.** $\dfrac{10}{x - 9}$ [5.7A] **32.** $2x + 6$ [5.7B] **33.** 30.78 in. [4.2E]

34. 657 lb [4.2E] **35.** Let d be the distance from Earth to the sun; $30d$ [5.7C] **36.** $3,075 [3.4C]

Answers to Chapter 6 Exercises

Prep Test (page 376)

1. -4 [2.2B] **2.** 1 [3.4A] **3.** -10 [3.4A] **4.** 1 [2.3B] **5.** $7y$ [5.2A] **6.** -9 [5.2A] **7.** -5 [2.5A] **8.** 13 [2.5A]

6.1 Exercises (pages 383–384)

3. 6 **5.** 9 **7.** 17 **9.** 1 **11.** -7 **13.** -9 **15.** 5 **17.** 9 **19.** 0 **21.** -8 **23.** -6 **25.** 12

27. 6 **29.** -12 **31.** $\dfrac{2}{7}$ **33.** $\dfrac{1}{2}$ **35.** $\dfrac{1}{12}$ **37.** $\dfrac{5}{8}$ **39.** 3 **41.** -3 **43.** -8 **45.** 7 **47.** 0

49. 6 **51.** -4 **53.** 4 **55.** $\dfrac{5}{2}$ **57.** $-\dfrac{7}{2}$ **59.** $-\dfrac{7}{3}$ **61.** $\dfrac{26}{9}$ **63.** 6 **65.** -36 **67.** -28

69. 12 **71.** 10 **73.** $\dfrac{7}{10}$ **75.** -5 **77.** -7 **79a.** $x = b - a$; yes **b.** $x = \dfrac{b}{a}$; no, $a \neq 0$

6.2 Exercises (pages 387–390)

3. 2 **5.** 10 **7.** 2 **9.** -1 **11.** -2 **13.** -4 **15.** 3 **17.** 9 **19.** 4 **21.** $\dfrac{3}{5}$ **23.** $\dfrac{7}{2}$ **25.** $\dfrac{3}{4}$

27. $\dfrac{3}{2}$ **29.** $\dfrac{7}{4}$ **31.** $-\dfrac{5}{2}$ **33.** $\dfrac{9}{2}$ **35.** 0 **37.** $-\dfrac{7}{6}$ **39.** $\dfrac{1}{3}$ **41.** 1 **43.** 10 **45.** 28 **47.** -45

49. -12 **51.** 10 **53.** $\dfrac{24}{7}$ **55.** $\dfrac{5}{6}$ **57.** $\dfrac{7}{5}$ **59.** $\dfrac{9}{4}$ **61.** 1.1 **63.** 5.8 **65.** 4 **67.** 1 **69.** $\dfrac{11}{3}$

71. 5 **73.** 2.8 years **75.** $15,655.11 **77.** 1987 **79.** 136 ft **81.** For example, $2x + 5 = -1$ **83.** No; it is not an equation.

Copyright © Houghton Mifflin Company. All rights reserved.

6.3 Exercises *(pages 395–398)*

1. 3 **3.** 3 **5.** −1 **7.** −4 **9.** 5 **11.** 2 **13.** 1 **15.** $\frac{7}{6}$ **17.** 3 **19.** $\frac{8}{5}$ **21.** $\frac{7}{2}$ **23.** $\frac{3}{2}$

25. $\frac{7}{2}$ **27.** 3 **29.** −1 **31.** 0 **33.** $\frac{5}{6}$ **35.** −3 **37.** 3 **39.** 2 **41.** 2 **43.** 2 **45.** 3

47. $\frac{1}{2}$ **49.** 8 **51.** −2 **53.** $-\frac{6}{5}$ **55.** $-\frac{1}{3}$ **57.** 6 ft **59.** 34.6 lb **61.** 325 units **63.** 400 units

65. 108

6.4 Exercises *(pages 403–406)*

1. $x + 12 = 20$; 8 **3.** $\frac{3}{5}x = -30$; −50 **5.** $3x + 4 = 13$; 3 **7.** $9x - 6 = 12$; 2 **9.** $x + 2x = 9$; 3

11. $5x - 17 = 3$; 4 **13.** $6x + 7 = 3x - 8$; −5 **15.** $40 = 7x - 9$; 7 **17.** $2(x - 25) = 3x$; −50

19. $4(x - 3) = 6x - 8$; −2 **21.** $2(3x + 8) + 6 = -2$; −4 **23.** $3x = 2(20 - x)$; 8 and 12

25. $2x = (21 - x) + 3$; 8 and 13 **27.** $23 - x = 2x + 5$; 6 and 17 **29.** $24,875 **31.** 2 million people **33.** 6 cents

35. 4 million tons **37.** 37 h **39.** 3 ft and 9 ft **41.** $5,000 **43.** 2 lb of Colombian; 3 lb of French Roast; 5 lb of

Java **45.** identity **47.** conditional **49.** contradiction

6.5 Exercises *(pages 411–414)*

3. I **5.** II **7a.** The abscissa is positive, and the ordinate is positive. **b.** The abscissa is negative, and the ordinate is

negative. **9.** 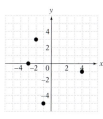 **11.** 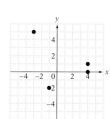 **13.**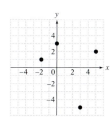

15. 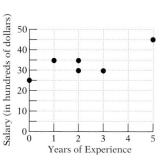 **17.** 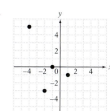 **19.** $A(0, 2)$, $B(-4, -1)$, $C(2, 0)$, $D(1, -3)$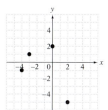

21. $A(0, 4)$, $B(-4, 3)$, $C(-2, 0)$, $D(2, -3)$ **23.** $A(3, 5)$, $B(1, -4)$, $C(-3, -5)$, $D(-5, 0)$

25. $A(1, -4)$, $B(-3, -6)$, $C(-2, 0)$, $D(3, 5)$ **27a.** 2; −4 **b.** 1; −3 **29a.** 4; −3 **b.** −2; 2

31. 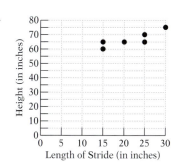 **33.** **35.** 200 s **37a.** 34 mpg **b.** 28 mpg

6.6 Exercises *(pages 419–424)*

1. No **3.** Yes **5.** No **7.** Yes **9.** No **11.** No **13.** Yes **15.** No **17.** (3, 7) **19.** (6, 3) **21.** (0, 1)

23. (−5, 0) **25.** Yes **27.** No

Copyright © Houghton Mifflin Company. All rights reserved.

29.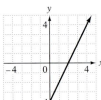

31.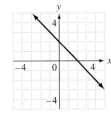

33.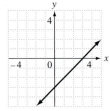

35.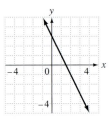

37.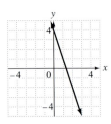

39.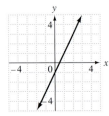

41.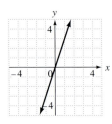

43.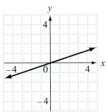

45.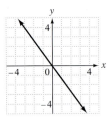

47.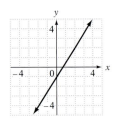

49.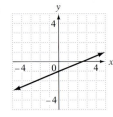

51.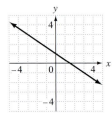

53.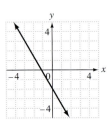

55.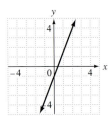

57.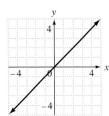

59. 4   **61.** −1   **63.** 2

65. 2   **67.** 1   **69.** −3   **71a.** (0, 1)   **b.** (2, 0)   **73.** 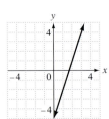   **75.**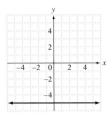

Chapter Review Exercises *(pages 429–430)*

1. −3 [6.1A]   **2.** 3 [6.1B]   **3.** $\frac{3}{2}$ [6.2A]   **4.** −7 [6.1A]   **5.** −24 [6.1B]   **6.** $-\frac{15}{32}$ [6.1B]   **7.** 2 [6.2A]

8. $\frac{17}{3}$ [6.3B]   **9.** $\frac{1}{2}$ [6.3A]   **10.** −2 [6.3A]   **11.** 3 [6.3B]   **12.** −4 [6.1B]   **13.** $-\frac{10}{9}$ [6.3B]   **14.** 4 [6.3B]

15. −4 [6.2A]   **16.** Yes [6.6A]   **17.** 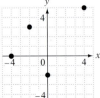 [6.5A]   **18.** [6.6B]

Copyright © Houghton Mifflin Company. All rights reserved.

19. 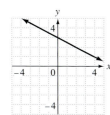 [6.6B]   **20.** $(2, -1)$ [6.6A]   **21.** $7 - 5x = 37$; -6 [6.4A]   **22.** 16 in. [6.4B]   **23.** 7 h [6.4B]

24. 55 m [6.4B]   **25.** 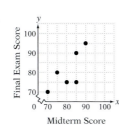 [6.5B]   **26.** 12.5 lb [6.3C]   **27.** 147 amplifiers [6.2B]

Chapter Test *(pages 431–432)*

1. -5 [6.1A]   **2.** -10 [6.1B]   **3.** -3 [6.2A]   **4.** 3 [6.2A]   **5.** 4 [6.3A]   **6.** 3 [6.3A]   **7.** 1 [6.3B]   **8.** $\dfrac{15}{4}$ [6.3B]   **9.** $-\dfrac{7}{24}$ [6.2A]   **10.** $\dfrac{7}{3}$ [6.3A]   **11.** $(-3, 1)$ [6.5A]   **12.** 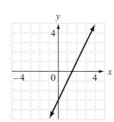 [6.5A]

13. 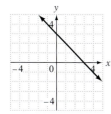 [6.6B]   **14.** 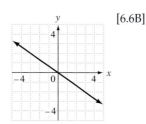 [6.6B]   **15.** [6.6B]

16. 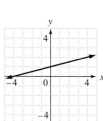 [6.6B]   **17.** 3 [6.3B]   **18.** -6 [6.2A]   **19.** $(6, -2)$ [6.6A]   **20.** $4 + \dfrac{1}{3}n = 9$; 15 [6.4A]

21. $8 + 2x = -4$; -6 [6.4A]   **22.** 5 and 12 [6.4A]   **23.** 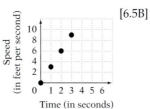 [6.5B]   **24.** 5 h [6.4B]

25. 100 ft [6.2B]

Cumulative Review Exercises *(pages 433–434)*

1. 18 [2.3A]   **2.** $-12p + 21$ [5.1B]   **3.** 0 [3.6C]   **4.** -18 [6.1B]   **5.** 8 [2.5A]   **6.** -60 [2.5A]   **7.** 11 [4.4A]   **8.** $4\sqrt{3}$ [4.4B]   **9.** $2v + 7$ [5.2B]   **10.** $12m$ [5.1A]   **11.** No [2.3A]   **12.** 1 [6.3A]   **13.** $-10z + 12$ [5.2B]   **14.** $\dfrac{5}{4}$ [3.6C]   **15.** $\dfrac{3}{2}$ [6.2A]   **16.** $32m^{10}n^{25}$ [5.4B]   **17.** $-6a^5 - 9a^4b + 12a^3b^2$ [5.5A]   **18.** $6x^2 - 7x - 3$ [5.5B]

Copyright © Houghton Mifflin Company. All rights reserved.

19. $\frac{1}{16}$ [5.6A] **20.** x^6 [5.6A] **21.** $15x^8y^3$ [5.4A] **22.** $\frac{31}{8}$ [6.3B] **23.** [6.6B]

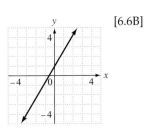

24. [6.6B] **25.** 0.000000035 [5.6B] **26.** $5(n + 2)$; $5n + 10$ [5.7B] **27.** 1.5 s [6.2B]

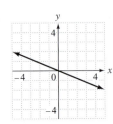

28. Let w be the number of wolves in the world; $1,000w$ [5.7C] **29.** \$1,192.5 million [4.2A] **30.** \$576.50 [6.4B]
31. 1,929 stories [4.2E] **32.** \$4,000 and \$8,000 [6.4B]

Answers to Chapter 7 *Exercises*

Prep Test *(page 436)*

1. $\frac{4}{5}$ [3.2B] **2.** 24.8 [4.2D] **3.** 4 [3.4A] **4.** 10 [3.4A] **5.** 46 [3.4A] **6.** 238 [1.3C] **7.** 37,320 [4.2B]
8. 0.04107 [4.2C] **9.** 5.125 [4.2A] **10.** 5.96 [4.2A] **11.** 0.13 [4.2B] **12.** 56.35 [3.4A, 4.2B] **13.** 0.5 [3.4A, 4.2B]
14. 3.75 [4.2C]

7.1 Exercises *(pages 441–444)*

3a. row 2: 10^9; row 3: 1 000 000; row 4: k, 10^3; row 5: 10^2; row 6: 10; row 7: 0.1; row 8: c, 0.01; row 9; m, $\frac{1}{10^3}$; row 10: 0.000 001;

row 11: 0.000 000 001; row 12: $\frac{1}{10^{12}}$ **b.** The exponent on 10 indicates the number of places to move the decimal point. For
prefixes tera-, giga-, mega-, kilo-, hecto-, and deca-, move the decimal point to the right. For the other prefixes shown, move the
decimal point to the left. **5.** kilogram **7.** milliliter **9.** kiloliter **11.** centimeter **13.** kilogram **15.** liter
17. gram **19.** meter or centimeter **21.** kilogram **23.** milliliter **25.** kilogram **27.** kilometer
29. 910 mm **31.** 1.856 kg **33.** 7.285 L **35.** 8 000 mm **37.** 0.034 g **39.** 29.7 ml **41.** 7.530 km
43. 9 200 g **45.** 36 L **47.** 2 350 m **49.** 83 mg **51.** 0.716 m **53.** 6.302 kl **55.** 458 cm
57. 9.2 cm **59.** 2 g **61.** 4.4 km **63.** 24 L **65.** \$20.77 **67.** 16 servings **69.** 40 g **71.** 8 L
73a. 0.186 kg **b.** 0.42 g **75.** \$115.60 **77.** \$26,176 **79.** 2 720 ml or 2.720 L

7.2 Exercises *(pages 447–448)*

1. $\frac{2}{3}$, 2:3, 2 to 3 **3.** $\frac{3}{8}$, 3:8, 3 to 8 **5.** $\frac{9}{2}$, 9:2, 9 to 2 **7.** $\frac{1}{2}$, 1:2, 1 to 2 **9.** $\frac{13}{14}$ **11.** $\frac{2}{3}$ **13.** $\frac{25 \text{ mi}}{1 \text{ h}}$

15. $\frac{\$3.28}{3 \text{ bars}}$ **17.** $\frac{9 \text{ children}}{4 \text{ families}}$ **19.** \$3,225/month **21.** 38.4 mi/gal **23.** \$1.344/oz **25.** $\frac{1}{12}$

27. Australia: 6.6 people/mi²; India: 824.1 people/mi²; U.S.: 75.7 people/mi² **29.** $\frac{100}{73}$

7.3 Exercises *(pages 455–458)*

1. $5\frac{1}{3}$ ft **3.** $2\frac{5}{8}$ lb **5.** $1\frac{1}{2}$ mi **7.** $\frac{1}{4}$ ton **9.** $2\frac{1}{2}$ gal **11.** 20 fl oz **13.** 11,880 ft **15.** $\frac{5}{8}$ ft

17. $7\frac{1}{2}$ c **19.** $3\frac{3}{4}$ qt **21.** 2,640 yd **23.** 1,103,760,000 s **25.** $12\frac{1}{2}$ gal **27.** $7\frac{1}{2}$ gal **29.** 25 qt

Copyright © Houghton Mifflin Company. All rights reserved.

31. $1,100   **33.** $65,340   **35.** 65.91 kg   **37.** 8.48 c   **39.** 54.20 L   **41.** 189.2 lb   **43.** 1.38 in.   **45.** 6.33 gal
47. 18.29 m/s   **49.** $2.18   **51.** 49.69 mph   **53.** $1.84   **55.** 40 068.07 km   **57a.** Answers will vary.
b. 2.54 cm/in.; 2.2 lb/kg; 1.06 qt/L   **59.** Answers will vary.

7.4 Exercises (pages 463–466)

1. Not true   **3.** True   **5.** True   **7.** True   **9.** True   **11.** True   **13.** 10   **15.** 2.4   **17.** 4.5   **19.** 17.14
21. 25.6   **23.** 20.83   **25.** 4.35   **27.** 10.97   **29.** 1.15   **31.** 38.73   **33.** 0.5   **35.** 10   **37.** 0.43
39. -1.6 **41.** 6.25   **43.** 32   **45.** 6.2   **47.** 5.8   **49.** 0.0052 in.   **51.** 24 robes   **53.** $6,720   **55.** 406 mi
57. $4.48   **59.** 438 lights   **61.** 60 weeks   **63.** 2,750 defects   **65.** 198,000 mi   **67.** $1,500   **69.** 9 in.
71. $126,000   **73.** Yes   **75.** No

7.5 Exercises (pages 471–474)

3. $\dfrac{15}{2}$ **5.** 16   **7.** 24   **9.** 10   **11.** 0.625   **13.** $307.50   **15.** 5.4 lb/in^2 **17.** 287.3 ft   **19.** 3 amps

21. 50   **23.** 100   **25.** 200   **27.** 6.4   **29.** 10 ft   **31.** $1.\overline{6}$ ohms   **33.** 2,160 computers   **35.** 51.2 lumens

37. 66.67 lb/in^2 **39a.** True   **b.** False   **c.** False   **41a.** y is 8 times larger.   **b.** y is $\dfrac{1}{8}$ as large.

Chapter Review Exercises (pages 481–482)

1. 1 250 m [7.1A]   **2.** 450 mg [7.1A]   **3.** $\dfrac{1}{1}$, 1:1, 1 to 1 [7.2A]   **4.** $\dfrac{2 \text{ roof supports}}{1 \text{ ft}}$ [7.2A]   **5.** $15.70/h [7.2A]

6. $\dfrac{8}{15}$ [7.2A]   **7.** $2\dfrac{2}{3}$ yd [7.3A]   **8.** $4\dfrac{1}{2}$ lb [7.3A]   **9.** $4\dfrac{1}{2}$ c [7.3A]   **10.** 6,600 ft [7.3A]   **11.** 1.6 [7.4A]

12. $\dfrac{5 \text{ lb}}{4 \text{ trees}}$ [7.2A]   **13.** 57 mph [7.2A]   **14.** 6.86 [7.4A]   **15.** 943.40 ml [7.3C]   **16.** 8.84 m [7.3C]   **17.** 328 ft
[7.3C]   **18.** 4.62 lb [7.3C]   **19.** 48.3 km/h [7.3C]   **20.** 46.58 mph [7.3C]   **21.** $\dfrac{3 \text{ c}}{4 \text{ pt}}$ [7.2A]   **22.** $\dfrac{1}{3}$ [7.5A]

23. 28,800 [7.5A]   **24.** 0.04 [7.5B]   **25.** 181 cm [7.1A]   **26.** $\dfrac{2}{5}$ [7.2A]   **27.** 12 oz [7.3B]   **28.** $12,000 [7.4B]

29. 11.2 in. [7.5A]   **30.** 2.75 lb [7.4B]   **31.** 127.6 ft/s [7.3B]   **32.** 1.25 ft^3 [7.5B]   **33.** $64,000 [7.4B]

Chapter Test (pages 483–484)

1. 46.5 m [7.1A]   **2.** 4 100 ml [7.1A]   **3.** $\dfrac{1}{8}$, 1:8, 1 to 8 [7.2A]   **4.** $\dfrac{1 \text{ oz}}{4 \text{ cookies}}$ [7.2A]   **5.** 0.6 mi/min [7.2A]

6. $\dfrac{2}{1}$ [7.2A]   **7.** 5,200 lb [7.3A]   **8.** 20 fl oz [7.3A]   **9.** 52 oz [7.3A]   **10.** 102 in. [7.3A]   **11.** 0.75 [7.4A]

12. 2 ft/s [7.2A]   **13.** 34.4 oz [7.3A]   **14.** 126 ft [7.3A]   **15.** 476.67 ft^2/h [7.2A]   **16.** 3.56 [7.4A]   **17.** 340.2 g
[7.3C]   **18.** 166.77 m [7.3C]   **19.** 1,090 yd [7.3C]   **20.** 4.18 lb [7.3C]   **21.** 56.35 km/h [7.3C]   **22.** 37.27 mph [7.3C]

23. 20 [7.5B]   **24.** 75 [7.5A]   **25.** 80 [7.5B]   **26.** $\dfrac{33}{38}$ [7.2A]   **27.** $3,136 [7.4B]   **28.** 243,750 voters [7.4B]

29. $114 [7.3B]   **30.** 50 ft [7.4B]   **31.** 76.27 ft/s [7.3B]   **32.** 292.5 ft [7.5A]   **33.** 100 rpm [7.5B]

Cumulative Review Exercises (pages 485–486)

1. 57 [3.6C]   **2.** 4.8 qt [7.3A]   **3.** $3\dfrac{31}{36}$ [3.3B]   **4.** $1\dfrac{2}{3}$ [3.6C]   **5.** -114 [2.3B]   **6.** 22 [2.5A]   **7.** 4 [6.2A]

8. 3 [6.3B]   **9.** [4.5A]   **10.** [4.5B]   **11.** 21 [2.5A]   **12.** $-\dfrac{3}{8}$ [3.6A]

13. 13 [4.4A]   **14.** $8a - 3$ [5.2B]   **15.** $-20a^5b^4$ [5.4A]   **16.** $-4y^2 - 3y$ [5.2A]   **17.** $(-1, -5)$ [6.6A]

18. $\dfrac{3}{10}$ [7.2A]   **19.** $3,885/month [7.2A]   **20.** $.52 [7.3C]   **21.** 32 [7.4A]   **22.** $\dfrac{10}{11}$ [3.6B]   **23.** -10 [4.4A]

24. -2 [6.3B]   **25.** $28.67 [4.2E]   **26.** 12 [6.4A]   **27.** $4x - 3(x + 2); x - 6$ [5.7B]   **28.** 64 mi [1.2C]

29. $265.48 [4.2E]   **30.** $\dfrac{4}{15}$ [3.3C]   **31.** 20,854 votes [7.4B]   **32.** 35 mi [7.2A]   **33.** 3,750 rpm [6.4B]

Copyright © Houghton Mifflin Company. All rights reserved.

Answers to Chapter 8 *Exercises*

Prep Test *(page 488)*

1. $\dfrac{19}{100}$ [3.4A]   **2.** 0.23 [4.2B]   **3.** 47 [4.2B]   **4.** 2850 [4.2B]   **5.** 4000 [4.2C]   **6.** 32 [3.4B]   **7.** 62.5 [3.4A, 4.2D]

8. $66\dfrac{2}{3}$ [3.2A]   **9.** 1.75 [4.2C]

8.1 Exercises *(pages 491–492)*

3. $\dfrac{1}{20}$, 0.05   **5.** $\dfrac{3}{10}$, 0.30   **7.** $\dfrac{5}{2}$, 2.50   **9.** $\dfrac{7}{25}$, 0.28   **11.** $\dfrac{7}{20}$, 0.35   **13.** $\dfrac{29}{100}$, 0.29   **15.** $\dfrac{1}{9}$ **17.** $\dfrac{3}{8}$

19. $\dfrac{2}{3}$ **21.** $\dfrac{1}{15}$ **23.** $\dfrac{1}{200}$ **25.** $\dfrac{1}{16}$ **27.** 0.073   **29.** 0.158   **31.** 0.003   **33.** 1.212   **35.** 0.6214

37. 0.0825   **39.** $\dfrac{6}{25}$ **41.** 37%   **43.** 2%   **45.** 12.5%   **47.** 136%   **49.** 96%   **51.** 7%   **53.** 83%

55. 33.3%   **57.** 44.4%   **59.** 45%   **61.** 250%   **63.** 16.7%   **65.** 68%   **67.** $56\dfrac{1}{4}$%   **69.** $262\dfrac{1}{2}$%

71. $283\dfrac{1}{3}$%   **73.** $23\dfrac{1}{3}$%   **75.** $22\dfrac{2}{9}$%

8.2 Exercises *(pages 499–502)*

1. 8   **3.** 0.075   **5.** $16\dfrac{2}{3}$%   **7.** 37.5%   **9.** 100   **11.** 1,200   **13.** 51.895   **15.** 13   **17.** 2.7%
19. 400%   **21.** 7.5   **23.** 65   **25.** 25%   **27.** 75   **29.** 12.5%   **31.** 400   **33.** 19.5   **35.** 14.8%
37. 62.62   **39.** 5   **41.** 45   **43.** 300%   **45.** 50,000 mi   **47.** 1,242,424 workers   **49.** 12%   **51.** 145.5%
53. 7 million pounds   **55.** $12,750   **57.** 0.015 g   **59.** 10,875%   **61.** 7,944 computer boards
63. 12,151 more faculty members   **65.** No   **67.** less than

8.3 Exercises *(pages 505–506)*

1. 20%   **3.** 36.8%   **5.** 1,500%   **7a.** 1998–2000   **b.** 2004–2006   **c.** more slowly   **9.** 41.1%
11. 34.2%   **13.** $15,330   **15.** Navy; 36.2%   **17.** $1,890; no; 37%

8.4 Exercises *(pages 511–512)*

1. $60.50   **3.** 60%   **5.** 44.4%   **7.** $2,187.50   **9.** $82.25   **11.** $110   **13.** 23.2%   **15.** 38%
17. $1,396.50   **19.** $25.20   **21.** $300   **23.** $123.08   **25.** No; 28%

8.5 Exercises *(pages 515–516)*

3a. $25, $50, $75, $100, $125   **b.** $150   **c.** $175   **d.** $200   **e.** $225   **5.** $273.70   **7.** $6,750   **9.** $20
11. $1,440   **13.** $164,250   **15.** $15,061.51   **17.** 7.7%   **19.** 9.125%

Chapter Review Exercises *(pages 523–524)*

1. $\dfrac{8}{25}$ [8.1A]   **2.** 0.22 [8.1A]   **3.** $\dfrac{1}{4}$, 0.25 [8.1A]   **4.** $\dfrac{17}{500}$ [8.1A]   **5.** 17.5% [8.1B]   **6.** 128.6% [8.1B]   **7.** 280%

[8.1B]   **8.** 21 [8.2A/8.2B]   **9.** 500% [8.2A/8.2B]   **10.** 66.7% [8.2A/8.2B]   **11.** 30 [8.2A/8.2B]   **12.** 15.3
[8.2A/8.2B]   **13.** 562.5 [8.2A/8.2B]   **14.** 5.625% [8.2A/8.2B]   **15.** 0.014732 [8.2A/8.2B]   **16.** 9.6 [8.2A/8.2B]
17. 34.0% [8.2C]   **18.** $8,400 [8.2C]   **19.** 3,952 telephones [8.2C]   **20.** 36.5% [8.2C]   **21.** $165,000 [8.2C]
22. 1,620 seats [8.2C]   **23.** 196 tickets [8.2C]   **24.** 22.7% [8.2C]   **25.** $11.34 [8.3A]   **26.** 25% [8.3B]
27. $19,610 [8.4A]   **28.** 65% [8.4A]   **29.** $56 [8.4B]   **30.** $390 [8.4B]   **31.** $31.81 [8.5A]   **32.** 9.00% [8.5A]
33. $10,630 [8.5A]

Copyright © Houghton Mifflin Company. All rights reserved.

Chapter Test *(pages 525–526)*

1. 0.864 [8.1A]   **2.** 40% [8.1B]   **3.** 125% [8.1B]   **4.** $\frac{5}{6}$ [8.1A]   **5.** $\frac{8}{25}$ [8.1A]   **6.** 118% [8.1B]   **7.** 90

[8.2A/8.2B]   **8.** 49.64 [8.2A/8.2B]   **9.** 56.25% [8.2A/8.2B]   **10.** 200 [8.2A/8.2B]   **11.** 33 accidents [8.2C]

12. 82.2% [8.2C]   **13.** $80 [8.2C]   **14.** 151.0% [8.3A]   **15.** $16\frac{2}{3}$% [8.3A]   **16a.** $83\frac{1}{3}$%   **b.** 100%

c. 50% [8.3B]   **17.** 8% [8.3B]   **18.** $300 [8.2C]   **19.** $12.60 [8.4A]   **20.** 55.1% [8.4A]   **21.** $300 [8.4B]
22. 14% [8.4B]   **23.** $315 [8.5A]   **24.** $41,520.55 [8.5A]   **25.** 8.4% [8.5A]

Cumulative Review Exercises *(pages 527–528)*

1. 24.954 [4.2A]   **2.** 625 [1.3B]   **3.** 14.04269 [4.2B]   **4.** $4x^2 - 16x + 15$ [5.5B]   **5.** $1\frac{25}{62}$ [3.4B]

6. $6a^3b^3 - 8a^4b^4 + 2a^3b^4$ [5.5A]   **7.** 42 [8.2A/8.2B]   **8.** -3 [6.1A]   **9.** 100,500 [4.2B]   **10.** $\frac{23}{24}$ [3.3B]

11. $1\frac{23}{28}$ [3.6B]   **12.** $-12a^7b^5$ [5.4A]   **13.**   [6.6B]   **14.**   [6.6B]

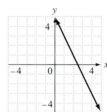   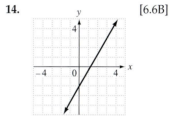

15. $2\frac{4}{5}$ [3.4B]   **16.** 4 [2.2B]   **17.** -12 [6.1B]   **18.** 5 [6.3A]   **19.** 64.48 mph [7.2A]   **20.** 44.8 [7.4A]

21. 8.3% [8.2A/8.2B]   **22.** 67.2 [8.2A/8.2B]   **23.** 58 [2.5A]   **24.** 5 [6.3B]   **25.** $\frac{1}{10}$ [8.1A]   **26.** $129.60 [8.4B]

27. $42 [8.4A]   **28.** 117 games [7.4B]   **29.** $2\frac{1}{4}$ lb [3.3C]   **30.** 72 ft/s [4.4C]   **31.** 32.85 km/h [7.1A]   **32.** 36 h

[6.4B]   **33.** 5 ohms [7.5B]

Answers to Chapter 9   *Exercises*

Prep Test *(page 530)*

1. 56 [1.5A]   **2.** 56.52 [4.2B]   **3.** 113.04 [4.2B]   **4.** 43 [6.1A]   **5.** 51 [6.1A]   **6.** 14.4 [7.4A]

9.1 Exercises *(pages 541–546)*

1. 40°; acute   **3.** 115°; obtuse   **5.** 90°; right   **7.** 28°   **9.** 18°   **11.** 14 cm   **13.** 28 ft   **15.** 30 m
17. 86°   **19.** 71°   **21.** 30°   **23.** 36°   **25.** 127°   **27.** 116°   **29.** 20°   **31.** 20°   **33.** 20°   **35.** 141°
37. 106°   **39.** 11°   **41.** $\angle a = 38°$, $\angle b = 142°$ **43.** $\angle a = 47°$, $\angle b = 133°$ **45.** 20°   **47.** 47°   **49.** $\angle x = 155°$,
$\angle y = 70°$ **51.** $\angle a = 45°$, $\angle b = 135°$ **53.** $90° - x$ **55.** 60°   **57.** 35°   **59a.** 1°   **b.** 179°   **63.** 360°

9.2 Exercises *(pages 557–564)*

1. hexagon   **3.** pentagon   **5.** scalene   **7.** equilateral   **9.** obtuse   **11.** acute   **13.** 56 in.   **15.** 14 ft
17. 47 mi   **19.** 8π cm; 25.13 cm   **21.** 11π mi; 34.56 mi   **23.** 17π ft; 53.41 ft   **25.** 17.4 cm   **27.** 8 cm
29. 24 m   **31.** 48.8 cm   **33.** 17.5 in.   **35.** 1.5π in.   **37.** 226.19 cm   **39.** 60 ft   **41.** 44 ft   **43.** 120 ft
45. 10 in.   **47.** 12 in.   **49.** 2.55 cm   **51.** 13.19 ft   **53.** 50.27 ft   **55.** 39,935.93 km   **57.** 60 ft²   **59.** 20.25 in²
61. 546 ft²   **63.** 16π cm²; 50.27 cm²   **65.** 30.25π mi²; 95.03 mi²   **67.** 72.25π ft²; 226.98 ft²   **69.** 156.25 cm²
71. 570 in²   **73.** 192 in²   **75.** 13.5 ft²   **77.** 330 cm²   **79.** 25π in²   **81.** $10,000\pi$ in²   **83.** 126 ft²
85. 7,500 yd²   **87.** 10 in.   **89.** 20 m   **91.** 2 qt   **93.** $74   **95.** 113.10 in²   **97.** $638   **99.** 216 m²
101. 4:9   **105a.** sometimes true   **b.** sometimes true   **c.** always true   **d.** always true   **e.** always true
f. always true

Copyright © Houghton Mifflin Company. All rights reserved.

9.3 Exercises *(pages 571–574)*

1. 5 in. **3.** 8.6 cm **5.** 11.2 ft **7.** 4.5 cm **9.** 12.7 yd **11.** 8.5 cm **13.** 24.3 cm **15.** $\frac{1}{2}$ **17.** $\frac{3}{4}$

19. 7.2 cm **21.** 3.3 m **23.** 12 m **25.** 12 in. **27.** 56.3 cm² **29.** 18 ft **31.** 16 m **33.** Yes, SAS Rule
35. Yes, SSS Rule **37.** Yes, ASA Rule **39.** Yes, SAS Rule **41.** No **43.** No

9.4 Exercises *(pages 581–584)*

1. 840 in³ **3.** 15 ft³ **5.** 4.5π cm³; 14.14 cm³ **7.** 34 m³ **9.** 15.625 in³ **11.** 36π ft³ **13.** 8,143.01 cm³
15. 392.70 cm³ **17.** 216 m³ **19.** 2.5 ft **21.** 4.00 in. **23.** length: 5 in.; width: 5 in. **25.** 75.40 m³ **27.** 94 m²
29. 56 m² **31.** 96π in²; 301.59 in² **33.** 184 ft² **35.** 69.36 m² **37.** 225π cm² **39.** 402.12 in² **41.** 6π ft²
43. 297 in² **45.** 3 cm **47.** 11 cans **49.** 456 in² **51.** 22.53 cm² **53a.** always true **b.** never true
c. sometimes true

9.5 Exercises *(pages 589–594)*

1. 121 cm **3.** (4 + 2π) ft; 10.28 ft **5.** (38 + 4π) m; 50.57 m **7.** 176 ft **9.** (4 + π) ft; 7.14 ft **11.** 24 m
13. 20.71 ft **15.** $22,910 **17.** 26 cm² **19.** (36 + 4.5π) in²; 50.14 in² **21.** 48π in²; 150.80 in² **23.** 30 in²
25. 16 in² **27.** (40 + 12.5π) in²; 79.27 in² **29.** $986.10 **31.** 19,026.55 ft² **33.** (1.92 − 0.08π) m³; 1.67 m³
35. 126π ft³; 395.84 ft³ **37.** 272 ft³ **39.** 8.25π in³; 25.92 in³ **41.** 36π in³; 113.10 in³ **43.** 1,458π cm³; 4,580.44 cm³
45. 212.60 in³ **47.** $6,862.62 **49.** 19 m² **51.** 93π cm²; 292.17 cm² **53.** (120 + 160π) m²; 622.65 m²
55. 56π cm²; 175.93 cm² **57.** 324 ft² **59.** (126 + 15π) in²; 173.12 in² **61.** 158 cans **63.** $997.50

65. $12\frac{3}{8}$ in. $\times$ $8\frac{5}{16}$ in. $\times$ $8\frac{1}{4}$ in.

Chapter Review Exercises *(pages 601–602)*

1. ∠x = 22°, ∠y = 158° [9.1C] **2.** 24 in. [9.3B] **3.** 240 in³ [9.5C] **4.** 68° [9.1B] **5.** Yes, by the SAS Rule [9.3C]
6. 138.23 m² [9.5D] **7.** 44 cm [9.1A] **8.** 19° [9.1A] **9.** 57.13 in² [9.5B] **10.** 96 cm³ [9.4A] **11.** 47.71 in. [9.5A]
12. ∠a = 138°, ∠b = 42° [9.1B] **13.** 220 ft² [9.4B] **14.** 9.75 ft [9.3A] **15.** 42.875 in³ [9.4A] **16.** 148° [9.1A]
17. 39 ft³ [9.4A] **18.** 95° [9.1C] **19.** 8 cm [9.2B] **20.** 288π mm³ [9.4A] **21.** 21.5 cm [9.2A] **22.** 4 cans [9.4B]
23. 208 yd [9.2A] **24.** 90.25 m² [9.2B] **25.** 276 m² [9.2B]

Chapter Test *(pages 603–604)*

1. 7.55 cm [9.3A] **2.** congruent, SAS [9.3C] **3.** 111 m² [9.2B] **4.** 42 ft² [9.2B] **5.** $\frac{784\pi}{3}$ cm³ [9.4A] **6.** 75 m²
[9.4B] **7.** 5,233.89 cm³ [9.5C] **8.** 159 in² [9.2B] **9.** 40.8 ft [9.5A] **10.** 135 m² [9.5D] **11.** 34° [9.1B]
12. octagon [9.2A] **13.** not necessarily congruent [9.3C] **14.** 168 ft³ [9.4A] **15.** 8.06 m [9.3A] **16.** 143° [9.1B]
17. 500π cm² [9.4B] **18.** 61° [9.1C] **19.** 6.67 ft [9.3B] **20.** 4.27 ft [9.3B] **21.** 20 m [9.2A] **22.** 26 cm [9.2A]
23. 51.6 ft [9.3A] **24.** 102° [9.1C] **25.** 57.9 ft² [9.5B]

Cumulative Review Exercises *(pages 605–606)*

1. 204 [8.2A/8.2B] **2.** 1, 2, 3, 6, 13, 26, 39, 78 [1.3D] **3.** $\frac{5}{6}$ [3.4B] **4.** 7x² + 4x + 5 [5.3A] **5.** 12.8 [4.2C]

6. 2.9 × 10⁻⁵ [5.6B] **7.** 131° [9.1B] **8.** 26 cm [9.3A] **9.** 50 in² [9.5B] **10.** 1,407.43 cm³ [9.5C]

11. −12x⁵y³ [5.4A] **12.** $\frac{1}{2}$ [6.3B] **13.** 11.14 cm [9.5A] **14.** [4.5B] **15.** 7x + 18 [5.2B]

16. −2 [2.5A] **17.** $\frac{4}{5}$ [3.6C] **18.** 96.6 km/h [7.3C] **19.** 5 [6.3A] **20.** [6.6B]

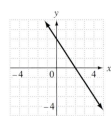

Copyright © Houghton Mifflin Company. All rights reserved.

21. 3.482 km [7.1A] **22.** 37.5% [8.1B] **23.** 1,313.01 [8.5A] **24.** 23 gal [7.3B] **25.** 87 min [6.4B]
26. $4.50 [7.4B] **27.** 5.4% [8.3A] **28.** 3 ft [9.4A] **29.** 40 ft [6.2B] **30.** 4,000 mi [4.4C]

Answers to Chapter 10 *Exercises*

Prep Test *(page 608)*

1. $\frac{1}{3}$ [3.2B] **2.** 3.606 [4.4B] **3.** 48.0% [8.2C] **4.** Between 2009 and 2010; $5318 [1.2C] **5a.** The ratio is $\frac{5}{3}$. [7.2A]

b. The ratio is 1:1. [7.2A] **6a.** 3.9, 3.9, 4.2, 4.5, 5.2, 5.5, 7.1 [4.1B] **b.** 4.9 million [4.2E] **7a.** 210,000 women [8.2C]

b. $\frac{3}{20}$ [8.1A]

10.1 Exercises *(pages 613–616)*

3. 64 **5.** 86–94 **7.** 5 universities **9.** 10% **11.** 52.5% **13.**

Classes	Tally	Frequency
57–67	/////	5
68–78	//////////	10
79–89	///////	7
90–100	///////////	11
101–111	//////////	10
112–122	//////	6
123–133	/	1

15. 5 hotels **17.** 33 hotels **19.** 22% **21.** 30% **23.** 13 account balances **25.** 22% **27.** $\frac{1}{2}$ **29.** 60%

31. 12.5% **33.** 55% **35.** 22 nurses **37.** 66% **39.** $\frac{6}{1}$ **41.** 53.3% **43.** Answers will vary.

10.2 Exercises *(pages 623–626)*

3. mean: 19 televisions; median: 19.5 televisions **5.** mean: 10.61 s; median: 10.605 s **7.** median **9.** 72 yd
11. 77 **13.** brown **15.** very good **17.** $Q_1 = 7.895, Q_3 = 9.95$

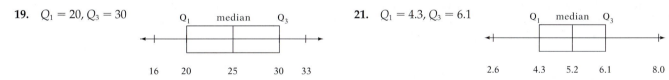

19. $Q_1 = 20, Q_3 = 30$ **21.** $Q_1 = 4.3, Q_3 = 6.1$

23. 2.828 min **25.** 61.051 rooms **27.** desert resort **29.** The mean score of the second student is 5 points higher.
The standard deviations are the same. **31.** Answers will vary; for example, 10, 14, 14, 14, 16, 18, 19, 23

10.3 Exercises *(pages 633–636)*

3. HHHH, HHHT, HHTH, HTHH, THHH, HHTT, HTTH, TTHH, HTHT, THTH, THHT, HTTT, TTTH, TTHT, THTT, TTTT
5. (1, 1), (1, 2), (1, 3), (1, 4), (2, 1), (2, 2), (2, 3), (2, 4), (3, 1), (3, 2), (3, 3), (3, 4), (4, 1), (4, 2), (4, 3), (4, 4) **7.** No. Because the

dice are weighted so that some numbers occur more often than other numbers. **9.** $\frac{1}{16}$ **11.** $\frac{3}{8}$ **13.** $\frac{1}{9}$ **15.** 0

17. $\frac{1}{36}$ **19.** $\frac{1}{12}$ **21.** $\frac{3}{16}$ **23.** $\frac{1}{4}$ **25.** $\frac{37}{95}$ **27.** $\frac{4}{7}$ **29.** $\frac{185}{377}$ **31.** 1 to 1 **33.** $\frac{3}{5}$ **35.** $\frac{1}{5}$ **37.** $\frac{12}{1}$

39. $\frac{1}{41}$ **41.** $\frac{1}{3}$ **43.** The sum of the probabilities is not 1.

Copyright © Houghton Mifflin Company. All rights reserved.

Chapter Review Exercises *(pages 643–644)*

1.

Classes	Tally	Frequency	[10.1A]
12–19	/////	5	
20–27	/////////	9	
28–35	///////////	11	
36–43	/////////	9	
44–51	////	4	
52–59	//	2	

2. 28–35 [10.1A] **3.** 25 classes [10.1A] **4.** 15% [10.1A]

5. 35% [10.1A] **6.** 25 days [10.1B] **7.** 29 days [10.1B] **8.** mean: $214.\overline{54}$; median: 210 [10.2A] **9.** mean: 7.17 lb; median: 7.05 lb [10.2A] **10.** good [10.2A] **11.** 55 million shares [10.1C] **12.** 8 A.M.–9 A.M. [10.1C] **13.** $\frac{3}{5}$ [10.1C]

14. $Q_1 = 105$, median $= 111$, $Q_3 = 124$

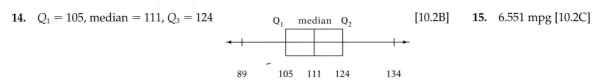

[10.2B] **15.** 6.551 mpg [10.2C]

16. $\frac{1}{500}$ [10.3A] **17.** $\frac{3}{7}$ [10.3B] **18.** $\frac{2}{7}$ [10.3B] **19.** $\frac{1}{6}$ [10.3A] **20.** $\frac{5}{14}$ [10.3A]

Chapter Test *(pages 645–646)*

1. 65 residences [10.1B] **2.** $55,000 [10.1C] **3.** 16% [10.1A] **4.** 152.875 [10.2A] **5.** 14 min [10.2A] **6.** very good [10.2A] **7.** 43 digital assistants [10.2A] **8.** 9.8 [10.2B] **9a.** 22 days **b.** 14 vacation days [10.2B]
10.

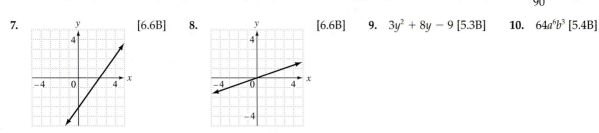

[10.2B] **11.** 1.612 incorrect answers [10.2C] **12.** 12 elements [10.3A]

13. (N, D, Q), (N, Q, D), (D, N, Q), (D, Q, N), (Q, N, D), (Q, D, N) [10.3A] **14.** $\frac{4}{31}$ [10.3A] **15.** $\frac{1}{3}$ [10.3A]

16. $\frac{1}{8}$ [10.3A] **17.** $\frac{2}{3}$ [10.3A] **18.** $\frac{1}{13}$ [10.3B] **19.** 1 to 8 [10.3B] **20.** $\frac{5}{12}$ [10.3A]

Cumulative Review Exercises *(pages 647–648)*

1. $10\sqrt{2}$ [4.4B] **2.** -1 [6.3B] **3.** 22 [2.5A] **4.** $12x - 8$ [5.2B] **5.** -18 [6.2A] **6.** $\frac{41}{90}$ [3.6C]

7. [6.6B] **8.** [6.6B] **9.** $3y^2 + 8y - 9$ [5.3B] **10.** $64a^6b^3$ [5.4B]

11. 144 [8.2A/B] **12.** $\frac{8}{3}$ [7.4A] **13.** 8.76×10^{10} [5.6B] **14.** 3,100 ft^2 [9.5B] **15.** $-15c^3d^{10}$ [5.4A]

16. 40,000 m [7.1A] **17.** 125° [9.1C] **18.** 32 m^2 [9.2B] **19.** $468.75 [8.5A] **20.** $\frac{2}{3}$ [10.3A]

21. mean: 31; median: 32.5 [10.2A] **22.** 76.1% [8.2C] **23.** 76 800 km [7.4B] **24.** 2.828 in. [10.2C]
25. $18.00 [8.3A]

Copyright © Houghton Mifflin Company. All rights reserved.

Answers to Final Examination *(pages 649–652)*

1. 2,400 [1.2A]   **2.** 24 [1.5A]   **3.** 2 [2.2B]   **4.** 18 [2.5A]   **5.** $2\frac{11}{16}$ [3.3B]   **6.** $\frac{14}{15}$ [3.4B]   **7.** $\frac{2}{9}$ [3.6B]

8. $\frac{5}{16} < 0.313$ [4.2D]   **9.** -769.5 [4.2B]   **10.** -3.28 [4.2C]   **11.** Yes [4.2B]   **12.** $9\sqrt{2}$ [4.4B]

13. 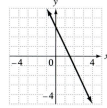 [4.5B]   **14.** $10t$ [5.1A]   **15.** $-2x - 14y$ [5.2B]   **16.** $z^3 + 2z^2 - 6z + 7$ [5.3B]   **17.** $8x^7y$

[5.4A]   **18.** $10a^4b^2 - 6a^3b^3 + 8a^2b^4$ [5.5A]   **19.** $15x^2 - x - 6$ [5.5B]   **20.** $81x^8y^4$ [5.4B]   **21.** $\frac{1}{64}$ [5.6A]

22. m^2n^4 [5.6A]   **23.** -6 [6.2A]   **24.** $-\frac{1}{3}$ [6.3A]   **25.** $\frac{5}{6}$ [6.3B]   **26.** 248 cm [7.1A]   **27.** 13,728 ft [7.3A]

28. $\frac{4}{3}$ [7.4A]   **29.** $\angle a = 74°;\ \angle b = 106°$ [9.1B]   **30.** 10.6 ft [9.3A]   **31.** 29.42 cm [9.5A]   **32.** 92.86 in³ [9.5C]

33. 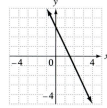 [6.6B]   **34.** 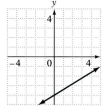 [6.6B]   **35.** 364 mph [1.2C]   **36.** 320 products [3.4C]

37. 65.98°C [4.2E]   **38.** 5.88×10^{12} [5.6B]   **39.** 6 ft [6.3C]   **40.** 6 tickets [6.4B]   **41.** $4,710 [7.4B]   **42.** 32%
[8.2C]   **43.** 16 rpm [7.5B]   **44.** $34,287.50 [8.2C]   **45.** 22.6% [8.3B]   **46.** $159.25 [8.4B]   **47.** $1,612.50 [8.5A]

48. 37.5% [10.1C]   **49.** mean: $334.80; median: $309 [10.2A]   **50.** $\frac{1}{3}$ [10.3A]

Copyright © Houghton Mifflin Company. All rights reserved.

Index

Abscissa, 407
Absolute value, 92–93
Abundant number, 53
Acre, 449
Acute angle, 534
Acute triangle, 548
Addends, 19
Addition, 19
  applications of, 20, 24, 29–32, 103, 110, 174, 182, 263–264
  Associative Property of, 22, 103, 315
  carrying in, 20
  Commutative Property of, 22, 103, 315
  of decimals, 249, 251, 252
  of fractions, 173–177
  of integers, 101–105
  Inverse Property of, 103, 316
  of mixed numbers, 175–176, 177
  Order of Operations Agreement and, 73, 135, 219
  of polynomials, 333–334
  verbal phrases for, 19, 31, 357
  of whole numbers, 19–24
Addition Property of Equations, 378
Addition Property of Zero, 22, 103, 316
Additive inverse, 103
Adjacent angles, 534, 536
Alternate exterior angles, 537
Alternate interior angles, 537
Amount, in percent calculation, 493, 495
Analytic geometry, 407
Angle(s), 29, 532
  acute, 534
  adjacent, 534, 536
  alternate exterior, 537
  alternate interior, 537
  complementary, 533
  corresponding, 537
  exterior, 539
  formed by intersecting lines, 536–538
  interior, 539
  measure of, 29, 532–533
  naming of, 532
  obtuse, 534
  right, 29, 533
  sides of, 532
  straight, 534
  supplementary, 534, 536
  of triangle, 539–540
  vertex of, 532
  vertical, 536

Angle-Side-Angle Rule (ASA), 569
Apothem, 564
Approximately equal to ($\approx$), 256
Approximation
  by rounding decimals, 242–243, 256
  by rounding whole numbers, 6–7
  of square roots, 284
  symbol for, 256
  See also Estimate
Area, 56, 552
  of circle, 555
  of composite figure, 586
  of parallelogram, 553
  of rectangle, 56, 552
  of square, 56–57, 552, 553
  surface area, 578–580, 588
  of trapezoid, 554–555
  of triangle, 199–200, 554
  units of, 449
ASA (Angle-Side-Angle Rule), 569
Associative Property
  of Addition, 22, 103, 315
  of Multiplication, 44, 118, 315
Averages, 617–620
Axes, 407
Axis of symmetry, 596

Bar graph, 9
  histogram, 611
  negative numbers on, 93–94
Base
  of cone, 575
  of cylinder, 575
  of exponential expression, 46
    fractional, 215
    variable, 341
  of parallelogram, 553
  in percent calculation, 493, 494, 495
  of pyramid, 575
  of trapezoid, 554
  of triangle, 199, 554
Base ten, 142
Base two, 142
Basic percent equation, 493–495
  applications of, 497–498
Binary digit, 440
Binary number system, 142
Binomials, 333
  multiplication of, 348
Bit, 440

Borrowing in subtraction
  of mixed numbers, 179
  of whole numbers, 25–26
Box-and-whiskers plot, 620–621
Bracket
  as grouping symbol, 317, 328
  on real number line, 295
Broken-line graph, 10
Byte, 440

Calculator
  arithmetic operations, 20
  checking solution of equation, 392
  decimal point, 253
  division by zero and, 49
  estimating correct answer, 21
  exponential expressions, 47
  fraction key, 174, 176
  negative numbers, 102, 107, 117, 120
  opposite of a number, 91
  Order of Operations Agreement and, 73
  parentheses keys, 73
  percent key, 493
  pi (π) key, 550, 555
  as problem-solving tool, 517
  radical expressions, 285
  rounding of decimals, 261
  solving equation, 386
  solving proportion, 460
  square of a number, 57, 135
  square root, 284, 566
  subtraction, 91, 107
  truncation of decimals, 261
Capacity, 438, 449, 453
Carrying
  in addition, 20
  in multiplication, 41
Center
  of circle, 550
  of sphere, 575
Chapter Review Exercises, 83, 145, 231, 307, 369, 429, 481, 523, 601, 643
Chapter Summary, 79, 143, 228, 304, 366, 427, 478, 521, 598, 640
Chapter Test, 85, 147, 233, 309, 371, 431, 483, 525, 603, 645
Circle, 550
  area of, 555
  circumference of, 550

Copyright © Houghton Mifflin Company. All rights reserved.

I1

Copyright © Houghton Mifflin Company. All rights reserved.

Copyright © Houghton Mifflin Company. All rights reserved.

Copyright © Houghton Mifflin Company. All rights reserved.

Copyright © Houghton Mifflin Company. All rights reserved.

Copyright © Houghton Mifflin Company. All rights reserved.

Copyright © Houghton Mifflin Company. All rights reserved.